Bacilli in Climate Resilient Agriculture and Bioprospecting

Microorganisms play an important role due to their presence all over the universe, and regulating biogeochemical systems in virtually all of our planet's environments. Among all microbes especially in prokaryotic genera, *Bacillus* spp. represents one of the most important unmapped pools of biodiversity with immense potential of applications in agriculture, industry and medicine. These Gram-positive spore forming bacteria are some of the most dominant groups that exist in various ecological niches on the earth due to their survivability in adverse environmental conditions by producing endospores. Recognizing the enormous potential of the bacteria in this genus, scientists all over the world have directed significant research towards selection and commercialization of the best organisms that may provide protection of plants from harmful microbes and/or enhance plant growth, produce industrially important enzymes, antibiotics, probiotics and other biochemicals. Innovative approaches are also being explored utilizing *Bacillus* mediated bioremediation of environmental pollutants such as pesticides, explosive wastes, dyes and polycyclic aromatic hydrocarbons. Discovery of insecticidal toxin from *Bacillus thuringiensis* (Bt) revolutionized insect pest management in many economically important crops by developing resistant crop varieties. Agricultural biotechnology that involves a wide range of insecticidal toxin producing genes from Bt has dominated the pest management research for last few decades. The emergence of resistant insect pests to Bt-based bioinsecticides and Bt-crops has created new challenges invoking more research on stacking resistant genes and further modifications of Bt toxin chemistries. While Bt transgenic crops are being released, questions on ecological aspects are getting louder provoking more research in this area. Genetics and genomics research on allied species of *Bacillus* has grown at a bewildering pace in the last decade.

More information about this series at https://link.springer.com/bookseries/16030

M. Tofazzal Islam • Mahfuz Rahman
Piyush Pandey
Editors

Bacilli in Agrobiotechnology

Plant Stress Tolerance, Bioremediation, and Bioprospecting

Springer

Editors
M. Tofazzal Islam
Institute of Biotechnology and Genetic Engineering (IBGE)
Bangabandhu Sheikh Mujibur Rahman Agricultural University
Gazipur, Bangladesh

Mahfuz Rahman
WVU Extension Service
West Virginia University
Morgantown, WV, USA

Piyush Pandey
Department of Microbiology
Assam University
Silchar, Assam, India

ISSN 2524-5120 ISSN 2524-5139 (electronic)
Bacilli in Climate Resilient Agriculture and Bioprospecting
ISBN 978-3-030-85464-5 ISBN 978-3-030-85465-2 (eBook)
https://doi.org/10.1007/978-3-030-85465-2

This Springer imprint is published by the registered company Springer Nature Switzerland AG
The registered company address is: Gewerbestrasse 11, 6330 Cham, Switzerland

Contents

Chapter 1
Heavy Metal Removal by *Bacillus* for Sustainable Agriculture

Sougata Ghosh, Joorie Bhattacharya, Rahul Nitnavare, and Thomas J. Webster

Abstract Microbial biosorbents are widely used for the removal of various toxic metals which pose a significant threat to agriculture. Metals like cadmium, chromium, cobalt, copper, iron, lead, manganese, mercury, nickel, palladium, platinum, and zinc are the predominant metal contaminants in our soils and water which call for instantaneous action to design microbiological techniques for effective bioremediation. The associated anthropogenic activities lead to a significant release of toxic metals into the environment purposely. Various industries related to mining, surface finishing, energy and fuel producing, fertilizer, pesticide, metallurgy, iron and steel, electroplating, electrolysis, paints and ceramic discharge metal laden effluents result in severe environmental pollution and health hazards. An indefinite persistence of heavy metals in the environment is a potential health hazard as it leads to bioaccumulation of toxic metals in the crops that eventually leads to biomagnification upon entering the food chain. This chapter highlights the promises of *Bacillus* as a potential biosorbent for the effective removal of toxic heavy metals from the environment. Numerous members of the genus *Bacillus*, like *B. subtilis, B. thuringiensis, B. sterothermophilus, B. megaterium, B. cereus, B. pumilus, B. licheniformis*, and *B. jeotgali* have been reported to remove heavy metals most effectively. Diverse functional groups like carboxyl, amino, amide, phosphate, and hydroxyl groups

S. Ghosh (✉)
Department of Microbiology, School of Science, RK University, Rajkot, Gujarat, India

Department of Chemical Engineering, Northeastern University, Boston, USA

J. Bhattacharya
Genetic Gains, International Crops Research Institute for the Semi-Arid Tropics, Hyderabad, India

R. Nitnavare
Division of Plant and Crop Sciences, School of Biosciences, University of Nottingham, Nottingham, UK

Department of Plant Sciences, Rothamsted Research, Harpenden, UK

T. J. Webster
Department of Chemical Engineering, Northeastern University, Boston, USA

M. T. Islam et al. (eds.), *Bacilli in Agrobiotechnology*, Bacilli in Climate Resilient Agriculture and Bioprospecting,
https://doi.org/10.1007/978-3-030-85465-2_1

associated to bacterial cell walls which attribute to biosorption capacity have been described herein. Numerous contributing factors like time, temperature, pH, cell density, and agitation are also discussed. *Bacillus*-mediated biosorption and bioaccumulation is a powerful strategy for the removal of toxic heavy metal stress in order to ensure sustainable agriculture.

Keywords *Bacillus* · Heavy metal toxicity · Bioremediation · Biosorption · Bioaccumulation · Detoxification

1.1 Introduction

Heavy metals are nondegradable, environmentally hazardous, and toxic. Various adverse effects of heavy metals include immune suppression, cancer, and neurotoxicity. Critical enzymes involved in vital metabolic processes may be inhibited by heavy metals. Numerous industrial activities such as mining minerals, smelting, and tannery result in the release of toxic metals such as cadmium, chromium, cobalt, copper, iron, lead, manganese, mercury, nickel, palladium, platinum, and zinc. These metals can accumulate within plants from contaminated soils or water leading to biomagnification. Consumption of metal-contaminated crops can cause damage to the lung, kidney, liver, and pancreas as well as lead to nervous disorders (Arivalagan et al. 2014). The indiscriminate use of excessive pesticides, weedicides, herbicides, and chemical fertilizers has contaminated soil and irrigational water with toxic elements metals (Ramirez et al. 2019).

Although heavy metals are found in trace amounts in the environment, upon bioaccumulation and biomagnification, their levels in the organism may be at different levels that can cause severe toxicity and impairment of regular metabolic activities. Conventional chemical and physical methods of metal removal like ion exchange, photo catalytic reduction, precipitation, ultrafiltration, and reverse osmosis are costly, time-consuming, inadequate, and energy-consuming. Thus, economical, environmentally benign, and efficient biological techniques are being designed with high heavy metal removing efficiency. Bacteria, yeast, fungi, and algae are capable of binding to and adsorbing metal ions. Many organisms dwelling in metal-contaminated ecological niches have developed chromosomally or extrachromosomally controlled detoxification mechanisms to overcome the detrimental effects of toxic heavy metals. Metal tolerance can be attributed to mechanisms such as extracellular precipitation and exclusion, binding to the cell surface, and intracellular sequestration. The most significant mode of bioremediation is binding of metal cations onto the negatively charged outer surface of bacterial cells, which can be later recovered as well (El-Helow et al. 2000). Thus, recently, bacterial biomass is considered as the most efficient biomaterial for removing metals. Specific metabolic processes and diverse active functional groups and proteins on bacterial cell walls, such as sulfate, amine, hydroxyl, and phosphates, enable microbes to interact and accumulate heavy metal ions on their surface (Dadrasnia et al. 2015).

In this chapter we have highlighted the potential of *Bacillus* as one of the most efficient heavy metal removing agents that can be beneficial for sustainable agriculture.

1.2 Heavy Metal Toxicity

Heavy metals, like cadmium, are extremely toxic pollutants which are widely used in rechargeable nickel–cadmium batteries, fertilizers, pesticides, pigments, metal refineries, electroplating, mining, etc. (Arivalagan et al. 2014). Its harmful effects include "itai-itai" disease, renal damage, emphysema, hypertension, testicular atrophy, atherosclerosis, and cancer (Ganguly et al. 2011). Chromium (Cr) is considered as a highly hazardous heavy metal as it does not easily degrade and is mostly released from textile industries, petroleum refining, and chemical and electronic manufacturing, as well as agriculture and mining activities. High levels of chromium consumption cause headache, nausea, vomiting, as well as skin, throat, and lung cancers. In nature, chromium exists as Cr (VI) or Cr(III), which differ in biochemical reactions and physicochemical properties. Cr (VI) is found as chromate (CrO_4^{2-}) which is soluble, mobile, toxic, and carcinogenic with a potential health hazard while Cr(III) is immobile, less toxic, and readily precipitated as $Cr(OH)_3$. Cr (III) is also an essential element in the metabolism of lipids and carbohydrates (Dadrasnia et al. 2015). Cobalt (Co) is radioactive cobalt ($^{58}Co/^{60}Co$) and has diverse applications in alloy production for orthopedic use, as a cold pasteurization agent in the food industry and radiotherapy (Paraneeiswaran et al. 2015). Iron (Fe) comprises 5% of the earth's outer crust and is also present in trace amounts in the human body. Anthropogenic activities greatly alter the natural biogeochemical composition of the soil and ecosystem. Iron in the form Fe(III)-O-arsenite is the most toxic (Huang et al. 1994). Lead (Pb) is highly toxic to the environment as well as humans. It is also known to cause neuronal disorders with long-term effects (Varghese et al. 2012). Manganese (Mn) is a trace element and is required by phytoplanktons and other higher plants for photosynthesis (Dismukes 1986). Higher concentrations Mn can be toxic to microorganisms when it is released in the environment from the effluent of industries associated with steel and alloys, paint, batteries, and metal coatings (Pakarinen and Paatero 2011). High doses of Mn have been recorded to be neurotoxic and are known to cause neurodegenerative diseases such as Parkinson's (Olanow 2004; Hussain et al. 2006). Bioaccumulation of Mn in the food chain causes grave consequences to human and animal health as well as the ecosystem. Mercury (Hg) in the form of mercuric chloride is toxic both to animals and the environment. They can penetrate the cell membrane causing an imbalance in cellular enzymatic regulation (Glendinning et al. 2005). Nickel (Ni) is one of the major constituents of the earth's crust and is used in several industries such as mineral processing, electroplating, power plants, paint, and porcelain manufacturing factories (Patterson 1985). Palladium (Pd) is used in industries having a greater operation in crystal engineering, photo-electrochemistry and catalysis which intend to use Pd in larger quantities. The automobile industry is the highest (50%) among the industries that use Pd (Matthey 2002). Extraction of Pd from waste and contaminated areas have previously been done through chemical methods and pyrometallurgical processes; however, these methods have proven to be highly labor- and time-intensive, as well as expensive (Jacobsen 2005). Platinum (Pt) is a noble element that forms an important chunk of jewellery and is present in the chemical, electrical, glass, and petroleum industries. A huge proportion of Pt is involved in the automobile industry as a vehicle exhaust catalyst (VEC) with emissions as high as 0.5–1.4 tons/year (Ravindra et al. 2004). Zinc (Zn) is a natural element and is found in the earth's crust imbibed in rocks and minerals. Years of weathering leads to demineralization of zinc and its subsequent release in water bodies and groundwater. Zinc is also an important industrial raw material used in batteries, electrical equipment, paint, rubber as well as in pharmaceuticals. In the human body, zinc exists as an inorganic metal and is vital for certain metabolic and redox reactions (Igiri et al. 2018). In dietary terms, the excessive intake of zinc has proven to be toxic leading to several gastrointestinal complications. The presence of high concentrations of zinc in aquatic ecosystems can result in subsequent accumulation in the food chain giving rise to the urge for bioremediation techniques for its efficient removal.

1.3 Heavy Metal Stress in Agriculture

Geologic and anthropogenic activities increase heavy metal concentration in soil. The rapid accumulation of heavy metals in agricultural soil, irrigational water, and living systems causes excessive harm in amounts that are harmful to crops. High concentrations of heavy metals in the soil result in their absorption and accumulation by plants, which ultimately pass into humans via the food chain. The absorption of toxic metals by aerial and underground plant organs directly or indirectly affect plant health and subsequently the yield of crops. Direct consequences include inhibition of cytoplasmic enzymes and damage to cell structures due to oxidative stress. The generation of reactive oxygen species (ROS) and cytotoxic compounds like methylglyoxal (MG) impair the equilibrium of ionic homeostasis within plant cells. Indirect oxidative stress includes multiple mechanisms such as glutathione depletion, binding to sulfhydryl groups of proteins, inhibiting antioxidative enzymes, or inducing ROS producing enzymes like NADPH oxidases. The cumulative effect of direct and indirect heavy metal toxicity leads to a reduction or even complete cessation of all metabolic activities of the plants. Beyond a certain limit of heavy metal concentration, defense strategies from plants to avoid or tolerate heavy metal intoxication fail protect it and the survival of the plant is jeopardized. Hence removal of heavy metals in order to ensure plant health and superior crop yield is most important, and therefore, the microbial approaches, which are cost effective, eco-friendly, and rapid, have been preferred over conventional strategies (Mishra et al. 2017).

1.4 *Bacillus* and Bioremediation

Several microorganisms can adapt to metal-polluted environments owing to their ability to protect themselves from various mechanisms that include adsorption, uptake, methylation, oxidation, and reduction. Bacteria like *Acinetobacter* sp., *Arthrobacter* sp., *Pseudomonas* sp., *Serratia marcescens*, *Ochrobactrum* sp., *Desulfovibrio vulgaris*, *Cellulomonas* sp., and *Bacillus* sp. are known for their significant role in the bioremediation of heavy metal-contaminated soils (Ramirez et al. 2019). Several species of *Bacillus* can detoxify and/or degrade the pollutants in the soil. These microbes play a crucial role in nutrient recycling in ecosystems, and especially can survive at toxic concentrations of metals as they develop a variety of resistance mechanisms to neutralize heavy metal stresses (Arivalagan et al. 2014). The *B. thuringiensis* strain DM55 expresses a variety of extracellular degradative enzymes such as amylase, mannanase, lichenase, and caseinase, and it is reported that their resistance to toxic metals like cadmium is not a plasmid-controlled character. Similarly, as the most characterized Gram-positive bacterium, *B. subtilis* 168 is a widely used model to investigate the influence of a histidine kinase, KinA, on the negative charge of the outer cell surface in *Bacillus* cells. Surface properties of *Bacillus* are very significant determinants of their metal binding capacity (El-Helow et al. 2000). *Bacillus* can adopt two ways for bioremediation; bioaccumulation, where the metal is taken up inside the cell, and biosorption, where the metals remain attached to the cell surface. Both dead and live cells can potentially exhibit biosorption; however, the percentage removal may vary. Dead bacterial cells as bioremediators are more beneficial over live cells as any external nutrient medium is not required for maintenance, and is also stored comparatively easily (Velásquez and Dussan 2009). The microbial systems according to their adaptive and physiological mechanisms can detoxify the metal by: (i) effluxing it out, (ii) accumulating in the cytoplasm, or by (iii) converting it into a less-toxic form.

1.5 Heavy Metal Removal by *Bacillus*

1.5.1 Cadmium

B. cereus (strain KTSMBNL 43) isolated from the electroplating industry-contaminated soil at Coimbatore, Tamilnadu, India exhibited Cd^{2+} tolerance. Isolation of cadmium-resistant bacteria was carried out on LB medium amended with 10, 50, 100, 200, and 300 mg L^{-1} of Cd^{2+}. The effect of different parameters like pH, temperature, initial metal concentration, and contact time on Cd^{2+} biosorption was evaluated using atomic absorption spectroscopy (AAS) and the biomass was determined using a UV-spectrophotometer at 595 nm. The maximum biosorption capacity (82%) was observed at pH 6.0 and at 35 °C with an initial metal concentration of 200 mg L^{-1} Cd^{2+} (Fig. 1.1). FTIR spectra of before and after Cd^{2+}

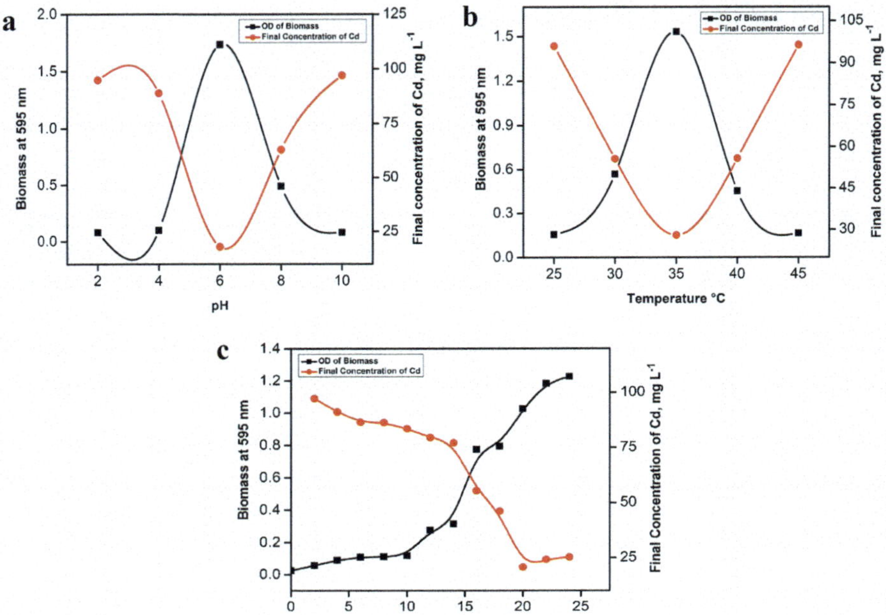

Fig. 1.1 Effect of various parameters on Cd^{2+} biosorption: (**a**) Effect of pH on Cd^{2+} biosorption by *B. cereus*. (Initial Cd^{2+}concentration—200 mg L^{-1}; temperature—35 °C; contact time—24 h; agitation speed—120 rpm); (**b**) Effect of temperature on Cd^{2+} biosorption by *B. cereus*. (Initial Cd^{2+} concentration—200 mg L^{-1}; pH —6.0; contact time—24 h; agitations peed—120 rpm); (**c**) Effect of contact time on Cd^{2+} biosorption by *B. cereus*. (Initial Cd^{2+} concentration—200 mg L^{-1}; pH —6.0; temperature—35 °C; agitation speed—120 rpm). (Reprinted with permission from Arivalagan et al. 2014. Copyright © 2014 Elsevier B.V)

biosorption revealed the cell surface-associated functional groups responsible for biosorption and thereby metal removal. A broad spectra of bands at 3364 and 3422 cm^{-1} corresponding to the presence of a hydrogen bond OH stretch was visible apart from the C-H and the C=O stretching at 3005 and 1729 cm^{-1}, respectively, in the Cd^{2+}-loaded biomass (Table 1.1). The alteration in peaks observed between 1244 and 1283 cm^{-1} might be attributed to the stretching vibration of SO_3 groups in the adsorption of Cd^{2+}. Hence, *B. cereus* may serve as an interesting low-cost, eco-friendly alternative for Cd^{2+} removal from polluted environments (Arivalagan et al. 2014). In another study, about 75% and 88% of the cadmium was removed by growing and nongrowing cells of *B. cereus* M^1_{16} from its aqueous solution at pH 6.8 ± 0.2, temperature 30 ± 1 °C and 120 rpm shaking speed. For growing cells, an inoculum size (2%) and medium volume (50 mL) was optimum. It is important to note that the effective desorption of cadmium was achieved using mineral acids such as HNO_3, H_2SO_4, HCl, and EDTA, which resulted in 98%, 93.5%, 90%, and 88.9% Cd(II) recovery from the loaded biomass, respectively. A greater affinity of the functional groups towards the H^+ ions might have attributed to this efficient metal recovery (Ganguly et al. 2011).

Table 1.1 Main functional groups on the surface of *B. cereus* as observed from FTIR spectroscopy

FTIR spectra range (cm^{-1})	Wavenumber range (cm^{-1}) Before Cd biosorption	After Cd biosorption	Functional group assignment
3600–3100	3364	3422	Bonded -OH stretch
3000–2800	Not observed	3005	C-H stretch
1730–1650	Not observed	1729	C=O stretch (it may represent carboxylic acids, ketones, and aldehydes)
1670–1640	1653	1652	C=O stretching (carboxylic groups)
1600–1500	1545	1545	N=O stretch (nitro groups)
1500–1440	1451	1448	H-C-H bend (alkanes)
1600–1500	1396	Not observed	COO- stretch (fatty acids)
1600–1500	1308	Not observed	N=O bend (nitro groups)
1300–1000	1244	1283	-SO_3 stretch
1300–1000	1082	1058	C-O stretch (alcohol groups)
600–500	539	559	C-I stretch (aliphatic iodo compounds)

Cadmium uptake by the metal-resistant bacterium *B. circulans* strain EB1 was checked by enriching 50 mL of Mueller–Hinton broth with cadmium (28.1 mg/L). Before inoculation with 1 mL of mid-logarithmic culture, the bacterial cells were preconditioned with 11.2 mg/L cadmium, *B. circulans* exhibited a specific biosorption capacity of 5.8 mg Cd/g dry wt biomass in the first 8 h. When preconditioning the cells with low concentrations of cadmium, the uptake was enhanced to 6.7 mg Cd/g dry weight biomass. The maximum uptake of cadmium was observed during the mid-logarithmic phase of growth. The specific biosorption capacities of wet and dry biomasses were 9.8 and 26.5 mg Cd/g dry weight biomass, respectively, at pH 7.0. Resting cells remarkably removed more cadmium compared to the growing cells (Yilmaz and Ensari 2005).

The biosorption of Cd(II) from aqueous solutions was studied in a batch method by using dead bacteria *B. licheniformis* sp. extracted from the soil in the area of the Tigris River. Initially, the bacterium was inoculated in liquid nutrient broth and the recovered biomass was washed, dried, and sieved to select particles with 180-μm size. The *B. licheniformis* sp. biomass was added to 50 mL of the Cd(II) solution. The monolayer adsorption capacity of *B. licheniformis* was 24.51 mg/g while the activation energy was determined as 23.24 kJ/mol for Cd(II) biosorption, indicating that the biosorption of metal ions onto the *B. licheniformis* biomass might be attributed to physical adsorption (Baran and Duz 2019).

Vijayakumar et al. (2011) reported the removal of Cd^{2+} ions from an aqueous solution using live and dead *B. subtilis*. An increase in initial concentration of Cd^{2+} leads to an increase in uptake capacity of the *B. subtilis* owing to the interactions between metal ions and different functional groups of the biomass. The percentage removal of Cd^{2+} increased from 33 to 60% with an increase in pH from 2.0 to 6.0

and a decrease to 29.78% at pH 10 for the sorption of live *B. subtilis* and dead *B. subtilis*, respectively. It was speculated that the adsorption of Cd $^{2+}$ ions on the adsorbent was dependent on the nature of the microbial surface and the species distribution of the metal cations.

A marine bacterial strain, *B. thuringiensis* strain DM55, with multiple heavy metal resistance and biosorption phenotypes, exhibited considerably high cadmium resistance. The successive transfers of DM55 cells into media containing up to 0.25 mM $CdCl_2$ helped to induce cadmium resistance. Cd^{2+}-amended culture exhibited a lag phase of 3.5 h with about a 1.85-fold increase in doubling time and a two-fold decrease in viable count when compared to the untreated control. Potassium phosphates and peptone were the most significant variables in cadmium removal. DM55 cells grew in the presence of 0.25 mM $CdCl_2$ to remove about 79% of the metal ions within 24 h with a specific biosorption capacity of 21.57 mg g^{-1} of biomass. Both fresh and dry cells of DM55 were able to remove Cd^{2+} under the influence of phosphate in the medium. Moreover KinA, a major phosphate provider in the phosphorelay of *Bacillus* cells, also regulated the magnitude of cell surface affinity for cadmium ions (El-Helow et al. 2000).

1.5.2 Chromium

Live and dead cells of *B. salmalaya* isolated from agricultural soil in Malaysia exhibited excellent chromium biosorption capacity. Initially, the *B. salmalaya* strain 139SI was inoculated into BHI medium and the cells in late exponential phase and live cells were harvested. The dead cell biomass was prepared by autoclaving. Live and dead pellets were mixed with an initial Cr(VI) concentration of 50 ppm (prepared from potassium dichromate). The maximum biosorption capacity was 20.35 mg/g at 25 °C, with pH 3 and a contact time of 50 min with a pseudo second-order mechanism. Alteration of the cellular morphology was observed. Cells of dead biomass that were long, thin, and rod-shaped before chromium adsorption became fat, spongy, and plumped after adsorption of chromium ions onto the cell surface (Fig. 1.2). Chromium accumulation occurred in the cell wall of *B. salmalaya* 139SI rather than intracellularly. High desorption efficiency up to 92% and 70% using dead and live cells, respectively, indicated the reusability and recyclability of *B. salmalaya* 139SI (Dadrasnia et al. 2015).

Bacillus sp. MH778713, close to the *B. cereus* group, was isolated from the nodules of mesquite trees (*Prosopis laevigata*) growing in aluminum, titanium, chromium, and zirconium-polluted soils of a semi-arid region in Nexapa River Chietla, Puebla, Mexico. This bacteria was the most resistant strain, which could tolerate up to 15,000 mg/L Cr (VI) and 10,000 mg/L of Al. *Bacillus* sp. MH778713 accumulated up to 100 mg Cr(VI)/g of cells when it was exposed to 1474 mg/L of Cr (VI). *Bacillus* sp. MH778713 further assisted *P. laevigata*-mediated phytoremediation when plants were inoculated with *Bacillus* sp. MH778713 and grown in a nitrogen-free Jensen's medium added with 0, 10, and 25 mg/L of Cr(VI). The growth of

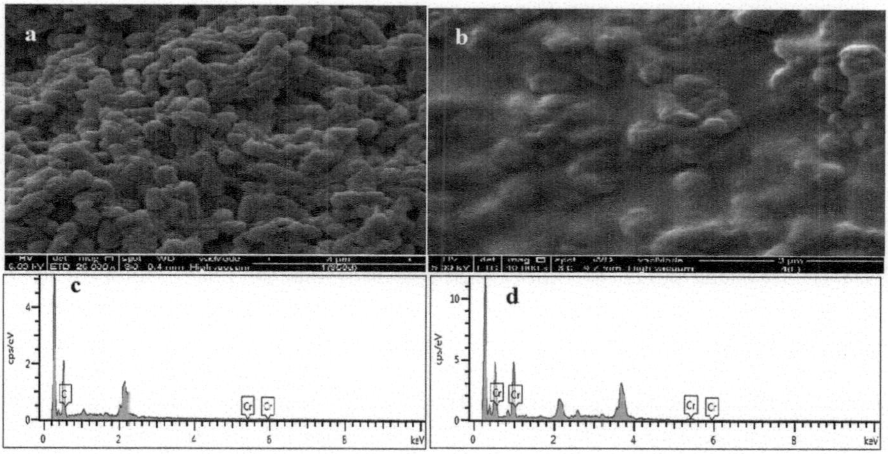

Fig. 1.2 SEM analysis of dead cells (**a**) before adsorption; (**b**) after adsorption; (**c**) EDX analysis of **a**; and (**d**) EDX analysis of **b**. (Reprinted from Dadrasnia et al. 2015. (Open access))

plants inoculated with *Bacillus* sp. in the presence of chromium indicated combinations of *P. laevigata* and *Bacillus* spp. as ideal candidates for soil restoration of chromium-contaminated arid and semiarid sites (Ramirez et al. 2019).

Chromium(VI)-resistant *Bacillus* sp. isolated from Cr-electroplating sludge from Daxing Electroplate Factory in Changsha, China, were grown with Cr(VI) (60 mg/L). The biosorbent was prepared by harvesting cells followed by washing. The live cells were freeze-dried for 12 h and used for biosorption. Living and freeze-dried biomasses are capable of absorbing up to 34.5 mg/g and 17.8 mg/g of chromium (VI) using an initial concentration of 60 mg/L, respectively, that followed pseudo-second order kinetics. The maximum metal uptake of 25.5 mg/g (living cells) and 18.3 mg/g (freeze-dried cells) were observed at pH 2.0. As the initial Cr (VI) concentration increased from 40 mg/L to 400 mg/L, the sorption capacity increased from 22.3 mg/g to 79.5 mg/g for a living biomass and 14.9 mg/g to 47.5 mg/g for freeze-dried biomass. A high efficiency for Cr(VI) removal was evident as a sorption capacity of 45 mg/g and 28.21 mg/g was achieved when the mass of cells used was just 0.01 g (dry mass) for living and freeze-dried cells, respectively (Yun-guo et al. 2008).

Sukumar et al. (2017) reported a fixed-bed column for using co-immobilized (activated carbon and *B. subtilis*) beads to remove Cr(VI) from aqueous solution. Various parameters like initial Cr(VI) concentration, flow rate, and bed depth were optimized for maximum Cr(VI) adsorption onto co-immobilized beads. The duration to reach the breakthrough (t_b) decreased from 150 to 25 min upon increasing the flow rate from 3 to 15 mL/min. Further, the Cr(VI) removal decreased from 57.8 to 10.8% as the flow rate increased to 15 mL/min. However, superior adsorption (57.8%) was observed at a lower flow rate (3 mL/min) which might be attributed to the availability of more residential time for the effective binding of Cr(VI) to co-immobilized beads. Significant elution (90%) was achieved using 0.1 M NaOH in all the six cycles emphasizing the potential of co-immobilized beads as effective agents for the removal of Cr(VI) from an aqueous solution.

1.5.3 Cobalt

Paraneeiswaran et al. (2015) reported the reduction of [Co(III)–EDTA] – to $[Co(II)–EDTA]^{2-}$ by bacterium *B. licheniformis* SPB-2 from solar salt pan locations of solar salt pan at Marakkanam, the Southern coast of the Bay of Bengal, Pondicherry, India. The bacterial colonies were anaerobically transferred to Tris–G broth supplemented with 1, 3 and 5 mM [Co(III)–EDTA] – and were incubated at 30 °C. *B. licheniformis* SPB-2 might have gained [Co(III)–EDTA] – resistance either by reducing the oxidation state of Co(III) to Co(II) or by physical adsorption. Moreover, *B. licheniformis* SPB-2 reduced 1 mM [Co(III)–EDTA] – in 14 days when grown under batch mode with a significant increase in the reduction activity up to four cycles (Fig. 1.3). It is interesting to note that [Co(III)–EDTA] – induced spore germination and the resulting vegetative cells could successfully reduce [Co(III)–EDTA]–. After reduction, the resulting $[Co(II)–EDTA]^{2-}$ complex could be effectively removed by *B. licheniformis* SPB-2-mediated biosorption. *B. licheniformis* SPB-2 showed a D_{10} value (radiation dose required to kill 90% cells) of~ 250 Gray (Gy) indicating its potential for bioremediation of moderately active nuclear waste.

Lee and Tebo (1994) reported cobalt (II) oxidation by a marine manganese (II)-oxidizing *Bacillus* sp. strain SG-1. This bacteria could oxidize Mn(II) by its mature spores which was attributed to a spore coat-associated protein. Similarly, SG-1 spores also oxidized Co(II) optimally around pH 8 and between 55 and 65 °C. SG-1 spores oxidized Co(II) at all Co(II) concentrations tested from the trace levels found in seawater to 100 mM. A kinetics study using an Eadie-Hofstee plot indicated that SG-1 spores exhibited two oxidation systems, a high-affinity-low-rate system (K_m, 3.3×10^{-8} M; V_{max}, 1.7×10^{-15} M .spore^{-1}. h^{-1}) and a low-affinity-high-rate system (K_m 5.2×10^{-6} M; V_{max}, 8.9×10^{-15} M .spore^{-1}. h^{-1}). SG-1 spores failed to oxidize Co(II) in the absence of oxygen, which proved the fact that oxidation was a biological associated with the preformed Mn (III,IV) oxide surface. This process may have potential applications in metal removal, recovery, and immobilization processes.

1.5.4 Copper

Copper was reported from an aqueous solution employing growing cells and washed cells of *B. cereus* M^1_{16}. Initially, the bacteria was grown in 50 mL medium (beef extract: 1.0; yeast extract: 2.0; peptone: 5.0; NaCl: 5.0 g/L; and pH: 6.0) containing 50 mg/L Cu (II) ions. Under optimum conditions, growing cells could remove 73.3% of Cu (II) ions from a 50 mL aqueous solution at 30 °C, pH 6.0, and after

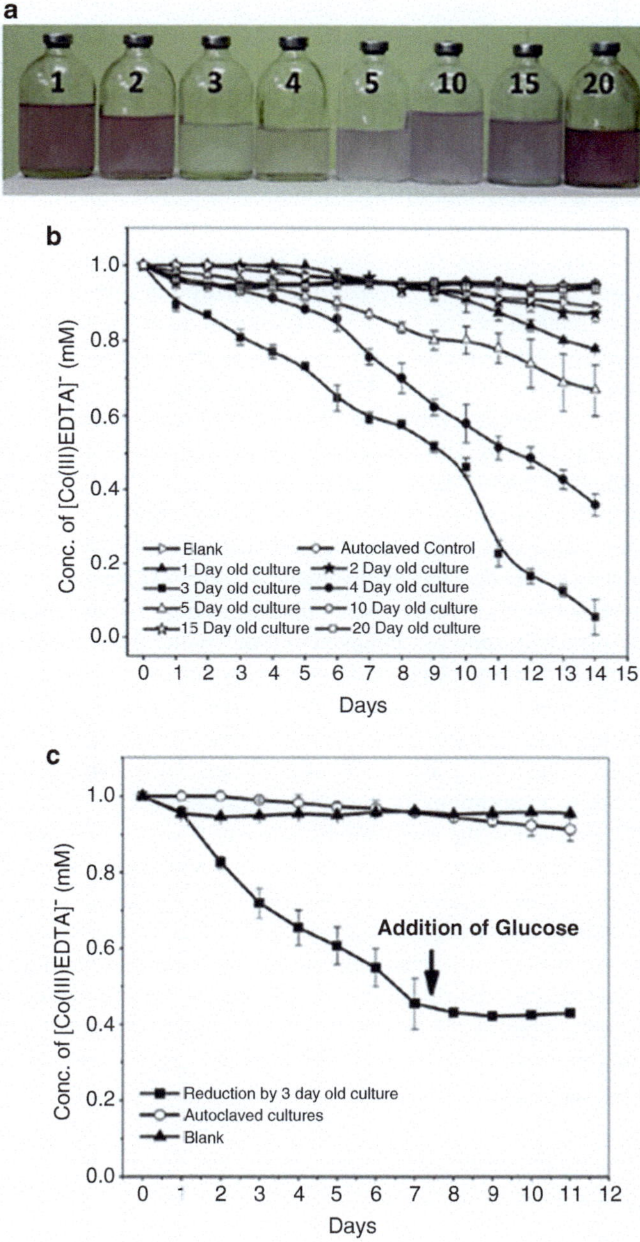

Fig. 1.3 (**a**) Image showing [Co(III)–EDTA] – reduction (decolouration) in serum bottles starting with different days old *B. licheniformis* SPB-2 cultures after one month of incubation. (**b**) Day-wise reduction of [Co(III)–EDTA] – by different days old *B. licheniformis* SPB-2 culture in Tris–G. (**c**) [Co(III)–EDTA] – reduction by *Bacillus licheniformis* SPB-2 during the addition of 0.2% glucose in the middle of the experiment. Each data point indicates the mean ± SD of triplicates. (Reprinted with permission from Paraneeiswaran et al. 2015. Copyright © 2014 Elsevier B.V)

24 h with an inoculum size (24 h cell growth) of 3% and initial metal ion concentration of 75 mg/L. Biosorption employing washed bacterial cells attained equilibrium within 180 min at 30 °C, pH 3.5 and with an initial metal ion concentration of 100 mg/L using 8.18 g/L biomass (dry weight) in a 50 mL reaction volume (Bairagi et al. 2010).

Copper(II)-resistant *B. cereus* KTSMBNL 81 was isolated from soil samples collected around an electroplating industry in Coimbatore, Tamil Nadu, India. Initial Cu(II) ion concentrations, pH of the solution, temperatures, and contact times played an important role in the biosorption of Cu(II) using *B. cereus*. Initial Cu(II) concentration was inversely proportional to the biosorption capacity. Saturation of binding sites due to the more number of Cu(II) ions might be the reason for reduced biosorption capacity. Although the removal capacity of *B. cereus* was low (3%) at pH 2.0, it increased up to 9% at pH 4.0. Maximum removal of Cu(II) (89%) was seen at pH 6.0. The highest removal of Cu(II) (86%) was observed at 35 °C. An enhancement in biosorption with a rise in temperature from 25 to 35 °C might be due to enlargement of pores that increased the surface area facilitating high sorption, diffusion, and penetration of Cu(II) into the biomass. Cu(II) sorption increased and a maximum (88%) was achieved with a contact time of 24 h. The FTIR spectrum of the biomass revealed the involvement of carboxyl, hydroxyl, and amino groups for the biosorption of Cu(II). As evident from Fig. 1.4, after biosorption of Cu(II), the cell surface morphology changed to rough with severe damage of the cell wall due to the toxicity of the heavy metals (Pugazhendhi et al. 2018).

Immobilized *B. subtilis* cells in chitosan beads (BICB) showed efficient removal of Cu(II) from aqueous solutions. The stable bacteria immobilized beads were filtered and washed with sterile water. A pH of 6, biosorbent dosage of 3% (w/v) and 37 °C temperature were optimum for maximum biosorption. Low Cu(II) concentration of 10 mg/L resulted in the removal efficiency of 98.4%. A desorption ratio of Cu(II) with EDTA and HCl was much lower than that with NaOH with a desorption rate of 76.02% even after 5 cycles (Liu et al. 2013).

A metal biosorbing *B. thuringiensis* strain OSM29 was isolated from rhizosphere of cauliflower grown in soil irrigated consistently with metal contaminating industrial effluents near Aligarh, North India. A dry biomass of *B. thuringiensis* OSM29 showed 91.8% of copper at pH 6 and 30 ± 2 °C within 30 min. FTIR analysis revealed that amino, carboxyl, hydroxyl, and carbonyl groups were involved in the biosorption (Oves et al. 2013).

1.5.5 Iron

Bioremediation offers a viable option for the safe removal of iron from the environment in an effective and efficient manner. A strain of *B. subtilis* D_{215} was tested for its potential in iron removal from effluents. Both live and dead cells were taken for a study at different pH, temperatures, and contact time. It was found that the above factors played an essential role in governing bioremoval capacity. With 600 μL of an

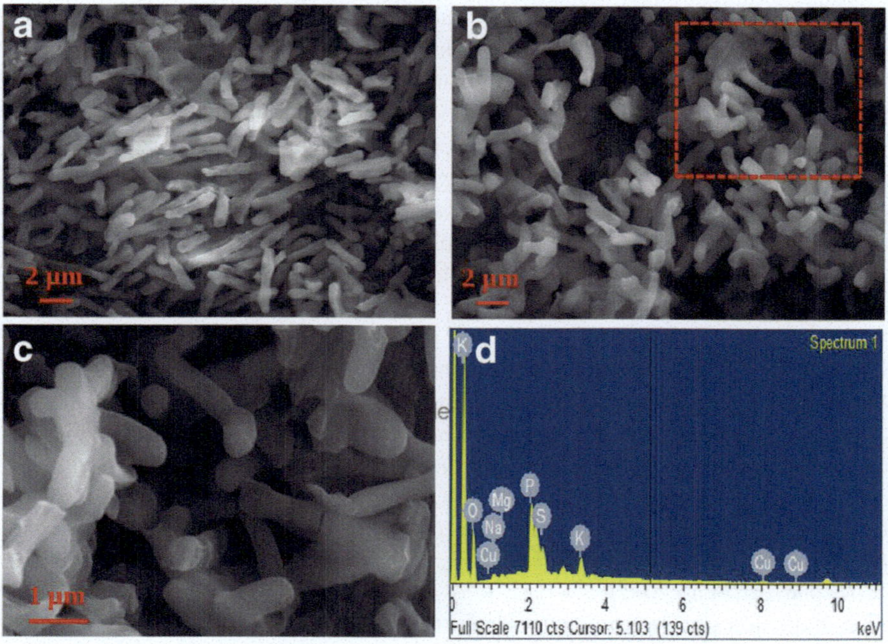

Fig. 1.4 Scanning electron micrographs and EDXA spectra of *B. cereus* KTSMBNL 81. (**a**) SEM image of Cu(II) unloaded; (**b**) SEM image of Cu(II) loaded; (**c**) SEM image of Cu(II) loaded (enlarged view); and (**d**) EDXA spectra of Cu(II) loaded. (Reprinted with permission from Pugazhendhi et al. 2018. Copyright © 2018 Springer International Publishing AG, part of Springer Nature 2018)

inoculum, an average of 70.35% of Fe^{2+} was removed at a pH 7, temperature of 30 °C, and contact time of 48 h. On the other hand, dead *B. subtilis* cells removed 21.30% of the heavy metal in the exact same biophysical conditions. Further, genetic studies of the bacterial genome has led to the discovery of several metalloregulatory proteins with Fur and DTxR homologues which a function of both metal and DNA binding sites, thereby facilitating metal uptake (Sabae et al. 2006). Further exploration of the genetic composition of *Bacillus* would help in highlighting genes responsible for Fe^{2+}tolerance and resistance in *Bacillus.*

The biosorption capacity of nine different *B. sphaericus* strains, isolated from beetle larvae and oaks forest (Lozano 1998), was studied in live conditions. These were: 2362, OT4b25, OT4b31, OT4b48, OT4b51, OT4b56, III(2)3, IV(1)8, and IV(4)10. The cells were grown at 30 °C for 24 h. Out of these, OT4b31 and IV(1)8 showed the highest biosorption capacity for Fe^{2+} at 2.1 mg/L and 2.5 mg/L, respectively. The presence of an S-layer was also studied by running it over sodium dodecyl sulfate-polyacrylamide gel electrophoresis (SDS-PAGE) and matrix-assisted laser desorption/ionization (MALDI) (Velásquez and Dussan 2009). The S-layer of the bacterial cell wall is present on the cell surface as a paracrystalline layer with a high porosity between 30–70% (Sleytr and Beveridge 1999) which aids in metal binding. The studies strongly rationalize the introduction of S-layer proteins into high metal-contaminated environments for effective bioremediation.

Hyper-alkaline and hypersaline aqueous environments are often developed due to contamination with heavy metals. One such region with high iron contamination was studied to understand the abundance of tolerant species surviving in such extreme conditions. 16S rRNA sequencing and terminal restriction fragment length polymorphism (T-RFLP) indicated *B. alkalinitrilicus* as the most copiously living microorganism in those conditions. At a pH above 9 and salinity of over 100 PSU, *Bacillus* sp. was previously reported to thrive in haloalkaline conditions (Sorokin et al. 2008).

Living cells of *B. licheniformis* were isolated for experiments regarding iron removal from sewage water. A high percentage removal of 95% (w/v) was observed for iron at pH 3.5, 28 °C and for 48 h with a maximum tolerable iron concentration (MTC) of 1500 mg/L. The *B. licheniformis* cell wall is composed of teichoic and teichouronic acid which are responsible for effective metal ion binding (Beveridge et al. 1982). Additionally, the biosorption rate was determined with dead *Bacillus* cells as well. However, in the case of bioremoval of iron with dead cells of *B. licheniformis*, the bioremoval percentage of only 60% implied greater efficiency of living cells (Samarth et al. 2012).

A reduction of ferric ions is essential to remove the toxic ferric oxyhydroxide forms from iron-contaminated soils. Several bacterial species have been described which possess the ability to reduce ferric ions (Lovely 1987). The *Bacillus* sp. SO-10 was isolated due to its metal-degrading ability and resistance. The *Bacillus* sp. was studied for various concentrations of iron from 100–1000 mg/L at 30 °C for 72 h. Maximum solubilization of iron was observed at pH 7; however, the pH was not a determining factor of solubilization rate. The study also concluded that an additional carbon source of 1% glucose enhanced the solubilization rate of iron by SO-10 (Sivakami et al. 2012).

In a rather elaborate study, 183 bacterial isolates growing in iron rich soil were isolated and studied. 12 bacterial strains were found to have notable growth in the presence of 1 g/L of iron. *B. subtilis* UFLA SCF590 showed the maximum bioremoval efficiency of 100% at 0.77 g/L at an acidic pH of 3.5 (Kelly et al. 2014).

Agricultural biowaste in the form of soil contaminants due to a high level of heavy metals present in fertilizers is a matter of great concern. One such fertilizer is phosphogypsum (PG) which contains metals such as cadmium, cobalt, chromium, copper, zinc, manganese, and iron. The concentration of iron in PG is the highest with 1199 mg/kg. The gradual bioaccumulation of PG in soil leads to an increase in soil metal concentration which poses an environmental threat as well as animal/human health hazard (Blaudez et al. 2000). The *B. megaterium* BM_{30} strain is a known remover of metal waste and is also a part of the biogeochemical cycle (Kumar and Achyuthan 2007; Shazia et al. 2002). This bacteria immobilizes metals through sorption and converts them into less toxic forms such as oxalates and phosphates. The bioaccumulation rate is directly related to the biomass. An increase in sporulation of *Bacillus megaterium* leads to an enhancement of biomass accumulation and subsequently increasing the bioremediation of metals. In this study, the pH was maintained at slightly alkaline conditions between 7.4–7.6 and a temperature of 30 °C. Optimal iron bioaccumulation was observed at 18 h when the bacteria were in the stationary phase. At varying concentrations of PG, 2 g/L, 6 g/L and 10 g/L, the maximum iron bioaccumulation was observed as 1 mg/L, 7 mg/L, and 9 mg/L, respectively. This study shows promising results for the bioremoval of iron using a new species of *Bacillus* (Stefanescu 2015).

Another agro-based industry includes the distilleries, which produce waste in the form of aromatic amines, organic and inorganic compounds as well as metal ions such as iron. Microbes isolated from a nearby soil source of such distilleries was identified as *B. subtilis.* The treatment of distillery effluents with *B. subtilis* saw a gradual decrease in iron concentration. In an experiment performed at an interval of 5 days, the level of iron was observed as 3.25 mg/L at the fifth day, 2.18 mg/L at the tenth day and 1.16 mg/L at the 15th day indicating a decrease with time. An overall percentage reduction of 80% was observed at a pH range of 7.4–7.8. This can be applied for water treatment due to severe metal contamination. Further, time based studies allow for the opening of newer avenues of research to comprehend and determine the minimum time interval required for the maximum reduction of heavy metals in optimized biophysical environmental conditions (Deborah and Sebastin Raj 2016).

1.5.6 Lead

Extracellular polysaccharides (EPS) from bacteria can be used for biosorption. The EPS produced by *B. firmus* MS-102 was studied to analyze the potential of the biosorption of lead. The biosorption capacity was found to be dependent on a wide array of factors such as pH, initial polysaccharide concentration, and the initial metal concentration. At neutral pH, the absorption was observed to be 55.6 mg/g. However, at higher pH levels the uptake increased. The optimal pH was found to be 4.5, while at neutral pH, 43.6% biosorption was achieved. This may be due to the interaction between acidic groups of polysaccharides with cations. Further studies with metal concentrations demonstrated that above 1000 mg/L, metal saturation occurred and there was a subsequent decrease in metal uptake. Additionally, the optimal temperature was found to be 25 °C in this study (Salehizadeh and Shojaosadati 2003).

In another study, *Bacillus* species isolated from animal excreta were exposed to increasing concentrations of industrial lead. It was found that *Bacillus* was able to survive and resist even high concentrations of lead up to 500 μg/mL (Akujobi et al. 2012). However, it was evident from the growth curve of the organism that with increasing dosages of lead, the rate of survival decreased (Akujobi et al. 2012). The inherent cellular resistance of bacteria to toxic heavy metals is attributed to several factors such as enzymatic conversion of the metal ions, decrease in sensitivity, and removal of ions by a cellular permeable barrier (Brewer and Taylor 1997).

Studies have demonstrated the biosorption capacity of Gram-positive bacteria, *B. subtilis* for lead (IV). Conditions like pH, temperature, and initial biomass play an important role in efficient bioremoval of toxic metals. In the case of *B. subtilis*, the maximum biosorption was observed to be 97.68% (w/w) at pH 4.5 and temperature 40 °C within 48 h at an initial load of 700 ppm (Hossain and Anantharaman 2006).

The presence of lead is diverse and can be seen in soil, water as well as sediment. A study performed by Varghese et al. (2012) confirmed *Bacillus* sp. as an excellent agent for the bioaccumulation of lead ions. A total of 164 isolates were collected from varying regions of wetlands, canals, and river water. 45% of the isolates were found to have very high tolerance to lead up to 6000 µg/mL, exhibiting a 50% lead reduction in in vitro conditions. Similar to previous results, an average of pH 5 was found to be the optimal condition. The increase in biomass of the bacteria corresponded with an increase in adsorption. This may be due to the associated increase in surface area of the microorganism (Bai et al. 2002). The cell wall is composed of different functional groups, such as amine, phosphates, and carboxyl groups. These may serve as scavengers for cations and hence increase the reduction of metal ions (Varghese et al. 2012).

Mines are an important source of lead contamination of nearby agricultural fields as well as water bodies through leaching. Biomagnification of the heavy metals is observed and different methods such as reduction, precipitation, and membrane processes have been adopted to manage the damage cause by lead (Wang and Chen 2009). *Bacillus* sp. KK1 exhibited maximum tolerance to lead (300 mg/L). Reduction capacity of the microorganism into nontoxic forms is an important attribute of microbes used for bioremediation. *Bacillus* sp. KK1 reduced lead nitrates into lead sulfides and lead silicone oxides (Govarthanan et al. 2013). Isolation of *B. licheniformis* from mining sites demonstrated 86 ± 5% removal of lead at pH 6 after 192 h at a flow rate of 10 mL/min. This study validated the role of biomass in the effective adsorption of heavy metals from contaminated soil (Akshatha Jain et al. 2017).

A lead-tolerant bacterium, *B. thuringiensis* strain OSM29, was isolated from the rhizosphere of cauliflower growing in an agriculture field irrigated with factory waste. It was observed that after a certain time period, the absorption rate decreases with an increase in metal ions which might be due to saturation of the adsorption sites (Oves et al. 2013). Bai et al. (2002) showed an increase in metal adsorption with an increase in biomass. Maximum biosorption was achieved at an optimal pH of 6 at a temperature of 30 ± 2 °C within 30 min of contact time. Fourier transform infrared spectroscopy (FTIR) revealed that the main functional groups like amino, hydroxyl, carboxyl, and carbonyl were responsible for heavy metal biosorption. An average of 25 ± 2 mg/g bioremoval of lead was observed at optimal conditions. Conclusive results deduced the efficient use of this strain for reliable bioremoval of Pb in aqueous conditions (Oves et al. 2013).

Govarthanan et al. (2015) reported *Bacillus* sp. SKK11 from a brackish environment that was tolerant to about 750 mg/L of lead and could convert lead nitrates to lead sulfides either intracellularly or extracellularly. Subsequently, enzymes like urease, amylase, cellulose, dehydrogenase, phosphatase, and invertase resulted in enhanced bioremediation of lead-contaminated soil. The conversion of lead into carbonate-bound lead is a phenomenon observed previously in another bacterium. This reduces the prevalence of free lead in the soil therefore reducing soil stress. Govarthanan et al. (2013) observed a 38% increase in carbonate-bound lead in soil treated with KK1. On similar lines, SKK1-treated soil also demonstrated an increase in 15.3% carbonate fraction (Govarthanan et al. 2015).

Bioremediation using *Bacillus* as a probiotic was also explored. Belapurkar et al. 2016, demonstrated *B. coagulans*-mediated reduction in the concentration of lead. *Bacillus* sp. is known to have a greater adsorptive nature due to the presence of teichoic acid and peptidoglycan in its cell wall (Gavrilescu 2004). *B. coagulans* reduced lead (II) up to 86% within 72 h in the gut. These studies can be further explored for other species of *Bacillus* in the development of similar probiotics and effective bioremediation.

B. badius AK isolated from water hyacinth green compost showed maximum removal of lead(II) with a pH of 4–4.5 with inoculums at a density of 1.1×10^{11} CFU/mL at 30 °C. Results showed that at optimal environmental conditions, *B. badius* was able to remove 100 mg/L of lead within 24 h with an initial bacterial biomass volume of 20 mL having a 1.7×10^{16} CFU/mL value (Vishan et al. 2017).

Studies based on the transformation of lead by rhizospheric bacteria showed that *Bacillus* inoculated in fulvic and humic soils can be a significant strategy. Increased mobilization of lead was observed due to a decrease in humic acids and an increase in fulvic acids. Technogenic soil facilitated the immobilization of Fe-hydroxide-bound, chelate-bound lead in the rhizosphere and in roots resulting in decreased lead uptake by the upper parts of plants (Belogolova et al. 2020).

1.5.7 Manganese

An isolate of *Bacillus* was found to have the capacity to generate manganese oxide nanoparticles. Nanoparticles of this nature are known to have tremendous potential in the pharmaceutical industry for drug delivery, as biosensors and also as magnetic storage media. The heavy metal-resistant strain of *Bacillus* was identified as MTCC10650 which reacted with a filter sterilized $MnCl_2{\cdot}4H_2O$ to form manganese oxide nanoparticles. Very few, small-sized, and randomly dispersed nanoparticles were found on the cell wall at 24 h (Fig. 1.5a). However, enhanced accumulation of manganese in the cytoplasm was observed after 48 h (Fig. 1.5b) and 72 h (Fig. 1.5c) of growing the cells. The accumulated particles were mostly spherical, lesser than 6 nm in diameter and uniformly dispersed into the cytoplasm. The average size of the bacteriogenic nanoparticles was 4.6 ± 0.14 nm, with some particles having 6–8 nm size and a very small percentage having a diameter greater than 10 nm (Sinha et al. 2011). This study implied two important conclusions: the production of nanoparticles for industrial purposes as well as the potential of *Bacillus* sp. in manganese bioremediation of industrial discharge.

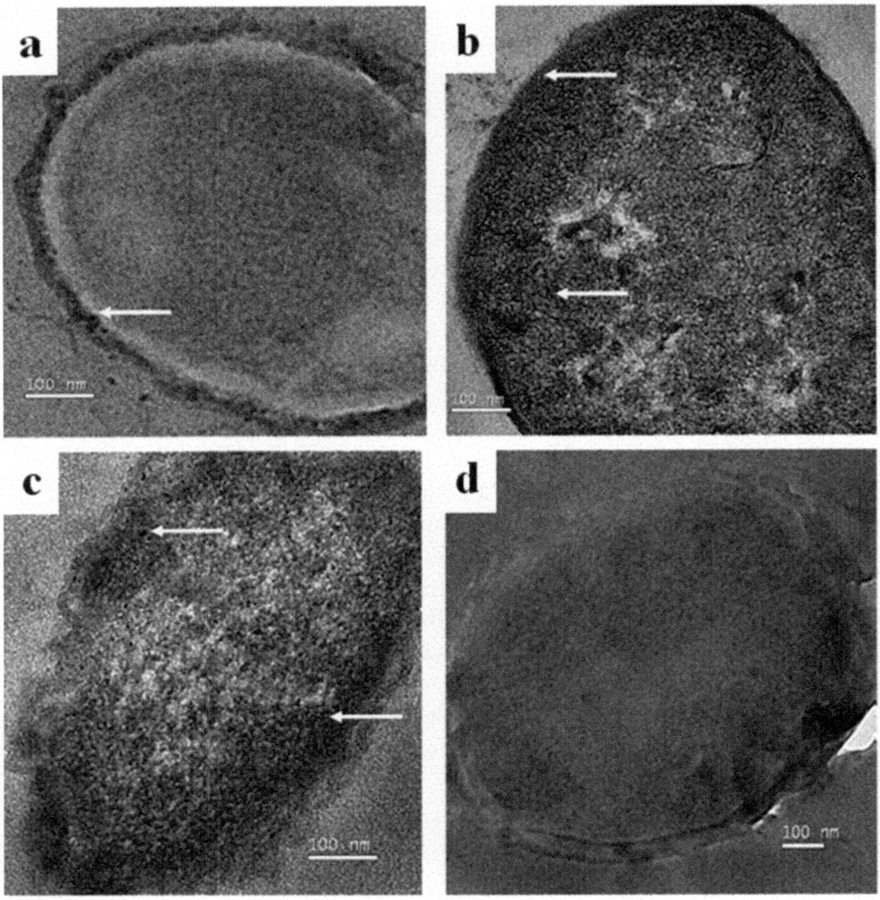

Fig. 1.5 Transmission electron micrograph (TEM) of *Bacillus* sp. cells taken at different times of growth. Cells grown in manganese for (**a**) 24 h; (**b**) 48 h; (**c**) 72 h; and (**d**) control, cells grown in the absence of manganese for 72 h. Arrow heads show manganese oxide accumulation. Bar scale 100 nm. (Reprinted with permission from Sinha et al. 2011. Copyright © 2011 Elsevier B.V)

Agro-based industries are also known to generate a copious amount of organic waste which contributes to contamination of water bodies (Govarthanan et al. 2013). In this regard, a consortium of immobilized bacteria with *B. subtilis* sp. and two other genus were used for the bioremediation of sugar mill effluents. Along with having high biochemical oxygen demand (BOD), chemical oxygen demand (COD), and dissolved solids, the discharge consists of a number of heavy metals, with toxic manganese being one of them. Treatment with a consortium of bacteria leads to a reduction in concentration to manganese below detection levels (Saranraj and Stella 2012). Agro-based products such as fertilizers like phosphogypsum (PG) also contribute to soil pollution to a certain extent due to the presence of heavy metals. PG contains manganese which leads to toxicity as a result of biomagnification.

B. megaterium, BM_{30}, is known for its ability to bioaccumulate heavy metals through sorption and conversion into compounds such as sulfites and phosphates. An increasing trend of bioaccumulation of manganese was seen with increasing levels of PG in contaminated soil in the presence of Bm_{30}. Therefore, bioremediation using BM_{30} bioaccumulates manganese as monitored in cell biomass which was about 80% (Stefanescu 2015).

Although several bacterial species can biosorb manganese, *Bacillus* sp. is the most efficient. *Bacillus* sp. HAH1 was able to biosorp 13.2 mg/g MnII at a pH range of 5–8. With an increase in pH, there was also an increase in biosorption (Hasan et al. 2010). In an experiment carried out in contaminated water at the laboratory scale, a *Bacillus* sp. was able to biosorp 43.5 mg/g Mn(II). The functional groups, such as carboxyl, carbonyl, and amide, were found to be responsible for effective biosorption (Hasan et al. 2012). This study formed the basis for the biosorption for drinking water treatment.

Regulators inherently present in bacteria enable them to modulate the tolerance levels to manganese toxicity at varying levels. This is achieved through metabolic mechanisms of the bacteria such as extracellular polysaccharides and cell walls as barriers, reduced intake or increased efflux and enzymatic regulation. Manganese in the II oxidation state is the most abundantly found compound which contributes to brackish water of water bodies. It is necessary to convert it into its oxidized form, MnO_2, which can be easily removed by filtration, absorption, or clarification processes (Barboza et al. 2016). Enzymatic reactions in various regions of the cell contribute to this oxidation process. This matrix is found in the exosporium in *Bacillus*. Maintenance of homeostasis of $Mn^{2+}$is achieved through regulation of the intake and efflux of the ion. Pumps present in the cytoplasm facilitate this in *B. subtilis*. Natural resistance-associated macrophage protein (Nramp) acts on various transporter uptake systems existing in the bacterial cell such as *mntH, mntA,* and *mntABCD* which are metal ion transporters (Que and Helmann 2000). During high levels of Mn^{2+}, a protein repressor *MntR* acts on *mntH* such that the Nramp is repressed; whereas during low Mn^{2+}, it activates the mntABCD transporter which in turn activates the ABC transporters (Que and Helmann 2000).

In seawater, bacteria are able to convert manganese into its oxide form with the help of multicopper oxidases (MCOs). Gene encoding for these enzymes have been identified as *mnxG*. Spores produced by the *Bacillus* sp. strain SG-1 formed amorphous MnO_2. Few Mn II- oxidizing proteins required during processes, like post-translational modifications, were also found in this species (Francist and Tebo 2002; Dick et al. 2008). This ability of the *Bacillus* sp. strain SG-1 could be used for effective bioremediation of manganese in toxic metal-contaminated saltwater.

The *B. cereus* strain HM-5 obtained from manganese ore was studied for biosorption capacity. Similar to most species of *Bacillus*, biosorption was dependent on factors such as pH, temperature, contact time, initial microbial biomass as well as the metal ion concentration. The most optimal pH and temperature for this strain was found to be 6 and 35 °C with a metal ion concentration of 600 mg/L. *B. cereus* reached a capacity of 98.9% with a contact time of 5 days. Fourier transform infrared (FTIR) analysis revealed the essential functional groups of this bacteria to be hydroxyl, amide, and phosphoryl groups for biosorption. Morphological analysis showed that without manganese bioabsorption, the cells were smooth, while post biosorption, the cells appeared wrinkled and damaged. This suggests that metal ions were trapped in the polymer matrix of the cell causing the observed deformation (Zhenggang et al. 2019). These observations indicated that *B. cereus* can be a viable option for the removal and recovery of toxic manganese (II) from polluted environments.

1.5.8 Mercury

The first incidence of mercury (Hg) resistance observed in *Bacillus* sp. was in the strain RC607 (Wang et al. 1989). No such occurrence of *Bacillus* sp. resistance was seen in extreme conditions until thermophilic bacteria, *Bacillus* sp. RC607, at temperatures as high as 62 °C was reported. *Bacillus* sp. RC607 had a minimum inhibitory concentration (MIC) of 60 µg/mL for $HgCl_2$ (Glendinning et al. 2005). The MIC values for the isolated *B. pallidus* and *Ureibacillus thermosphaericus* were 80 µg/mL and 60 µg/mL at 62 °C. Mercury resistance is an induced event due to genetic modification while on the other hand, tolerance is a detoxification process due to regular metabolism (Baldi 1997). In usual circumstances, the reduction of mercury results in the formation of sulfides like H_2S. However, neither of the two isolates demonstrated reductase activity, in spite of showing removal of mercury. This opens newer aspects of mercury removal without the production of toxic sulfides and possibly volatile thiol species (Glendinning et al. 2005).

The removal of mercuric ions from sludge wastewater of a chlor-alkali industry was studied by the introduction of a new bacterial strain, *B. cereus* JUBT1. The successful removal of Hg^{2+} was observed in both free and attached forms. Mercury removal using immobilized bacteria was studied with *B. megaterium* MB1. This species is nonpathogenic and possesses high mercury tolerance. Bacterium immobilized on alginate beads converted mercury into its volatile form. Also, it was able to trap a substantial amount of mercury in the beads which could be later treated for recovery. Alginate gel immobilized *B. megaterium* MB1 cells efficiently removed 80% of mercury from the solution containing 10 mg/L mercuric chloride within 24 h. The efficiency of the beads was not compromised by repeated exposure of the same beads to mercuric ions. 73.3% was seen to be converted into metallic mercury and out of which 44% was entrapped in the alginate beads (Chien et al. 2012).

On similar lines, immobilized *B. cereus* was used for the bioremediation of mercury with synthetic effluents. The experiment saw a biosorption concentration of 104.1 mg/g at biophysical conditions such as pH 7, temperature of 30 °C and contact time of 72 h. The initial biomass was 0.02 g/L. The immobilized beads had a removal capacity of 10 g/L of Hg^{2+} at a continuous pace of 11 h (Sinha et al. 2012).

Most of the mercury reducing bacteria consist of a *mer* operon which helps in the conversion of mercury into a nontoxic form and often encodes for reductase enzymes. The *mer* operon consists of specific functional genes along with a promoter, regulator, and operator. The most common functional genes are *mer*A and *mer*B, which code for mercuric ion reductase and organomercurial lyase, respectively. The lyase is responsible for reducing highly toxic organomercurial compounds, such as methylmercury and phenyl mercuric acetate, into almost nontoxic volatile elemental mercury with the help of the enzyme reductase (Kannan et al. 2006; Dash and Das 2012).

The occurrence of toxic mercury in marine environments is a common occurrence due to various anthropogenic activities. Several marine bacteria have therefore been studied for the bioremoval of toxic metals from seawater (Naik et al. 2012). A strain of *B. thuringiensis*, PW-05, was able to remove about 50 ppm of $HgCl_2$ from marine water. The isolate was found to possess the *merA* operon. The functional groups present in the cell wall of the bacteria, namely sulfhydryl and carboxyl, are responsible for the binding of mercury. Due to their high tolerance, *Bacillus* found in extreme conditions are highly beneficial for the bioremediation in saltwater. *B. thuringiensis* also formed biofilms that enhanced the bioremediation potential. This can be attributed to the secretion of biosurfactants and polysaccharides. The strain showed >90% volatilization of mercury and this ability can be further explored for bioremediation of inorganic mercury in extreme geophysical conditions (Dash et al. 2014).

Electroplating industries generate a mercury-laden effluent from which *B. cereus* MRS-1 was isolated that could convert mercury chloride into mercury sulfide nanoparticles (HgSNPs) as confirmed by transmission electron microscopy (TEM). The resulting HgSNPs were 10–100 nm in size with spherical morphology. Adsorbed nanoparticles on the bacterial cell wall indicated that reduction was the underlying mechanism of synthesis (Sathyavathi et al. 2013).

Another metal tolerating bacteria, *B. subtilis,* was able to remove 29.9% of Hg^{2+} in 5 days with an initial concentration of 500 ppm (Imam et al. 2016). Even though the bacteria were not able to bioaccumulate as much Hg^{2+} as its other counterparts, it still forms a basis for further studies of this species under differing environmental conditions to generate optimal results. As emphasized previously, environmental conditions play a significant role in the level of bacteria-mediated bioremediation and hence an optimum value is necessary for effective bioaccumulation. Along these lines, the ability of *B. licheniformis* for Hg^{2+} biosorption with respect to biophysical factors such as pH, temperature, contact time, and initial concentration of metal was studied. At pH 7, with initial bacterial biomass of 25 mg for 1 h, 70% bioremoval of mercury was observed. An increase in biosorbent concentration resulted in an enhancement of mercury biosorption (Upadhyay et al. 2017). The availability of metal binding sites and concentration of metal are inversely proportional to each other and therefore may limit the effectiveness of uptake of metal ions (Al-Homaidan et al. 2014). The role of extracellular polymeric substance (EPS) producing ability in *B. licheniformis* might have attributed to its efficient biosorption of mercury (200 μg/mg). Due to its extremely high efficacy, *B. licheniformis* is a prospective candidate for the bioremediation of Hg from contaminated water.

1.5.9 Nickel

Bacillus sp. KL1 isolated from municipal waste-contaminated sites in the Kermanshah province of Iran showed efficient nickel (Ni) removal, which was dependent on pH, temperature, and contact time of bacteria. It was concluded that the initial concentration of bacteria and biomass was the main contributing factor to the percentage of bioremoval. The highest bioremediation of nickel (II) was observed as 55.06% after 24 h at 30 °C, pH 7, and 100 ppm concentration. Among various interacting pairs of factors, time and concentration were found to be the most efficient (Taran et al. 2015).

The biosorption capacity of *B. thuringiensis* was studied, wherein, both its vegetative and sporulating stages were considered. Due to its unique ability to form a crystal protein during sporulation, which serves as a biopesticide to a wide range of insects, both the stages were taken into account. Both these forms were found to be efficient adsorbents for nickel(II)ion. The optimal temperature for maximum biosorption was 35 °C and pH was 6 for both stages, in which spore stages had a bioremoval capacity of 45.87 mg nickel (II)/g dry biomass while the vegetative cell was 35.46 mg nickel (II)/g dry biomass. At a 250 mg/L nickel (II) concentration, the adsorption percentage was 15.7% for spore-crystal cells and 10% for vegetative cells (Öztürk 2007). Another strain of *B. thuringiensis* KUNi1, isolated from industrial waste-contaminated soil, exhibited extremely high potential for nickel removal. It was able to remove 82% nickel from the culture tolerating a maximum of 7.5-10 mM nickel concentration within a pH range of 5–9. Further, dead cell biomass had no effect on the bioremoval percentage by KUNi1 (Das et al. 2014).

Thermophilic *Bacillus* sp. resistant to heavy metals have been used as living, immobilized, and dead cells to study their effects on the bioremoval of nickel. Four species were taken up for studies, namely, *B. sphaericus, B. pumilus, Panibacillus alvae*, and *Geobacillus sterothermophilus.* All of these species exhibited tolerance to nickel (II) at a concentration of 100 mg/L. Bioremoval in all the species appeared to be dependent on temperature and optimal results were obtained in the range of 37–70 °C. Dead cells had a higher adsorption capacity as compared to live cells due to the presence of functional groups in the cell wall and a lack of protons generated during metabolism (Pardo et al. 2003). The mean sorption efficiency percentage was observed as follows amongst the four species: immobilized cells 59–73%, dead cells 55–71%, and living cells 39–55% (Al-Daghistani 2012). An increase in adsorption capacity was observed at higher temperatures across all four species. This was attributed to an increase in diffusion rate through the adsorbate pore causing a reduction in viscosity of the liquid. However, at higher temperatures above 40 °C, the efficiency was also seen to decrease in *B. sphaericus*. This may be attributed to a decrease in surface activity of the microorganism (Sari et al. 2008). Further, the role of functional groups in the cell wall of the microorganisms for removal has been attributed to carboxyl and amino groups. Esterification of the carboxyl and methylation of amino groups occurred due to the treatment with sodium azide and formaldehyde resulting in a reduction in nickel removal in *B. subtilis* (Abdel-Monem et al. 2010).

Heavy metal-resistant *Bacillus* sp. EB1, isolated from contaminated soil, showed a considerable amount of nickel bioremoval percentage in aerobic conditions. EB1 was capable of removing 45% nickel during the active log phase with a biosorption capacity of 13 mg/L (Yilmaz 2003).

A probiotic species of *Bacillus* was used effectively for the removal of nickel as well. The activity of *B. clausii* and *B. coagulans* was demonstrated in which it exhibited both the properties of probiotics and also bioremediation potential (Belapurkar et al. 2016). Correspondingly, dead cells of *B. coagulans* were found to be able to reduce nickel at 68.4 mg/g of biomass (Lei et al. 2014). This property of certain bacterial species has been ascribed to the biochemistry of the cell wall associated with EPS (extracellular polysaccharides), techoic acid and lipotechoic acid.

Flagellin protein was produced more under metal stress. This protein facilitated the movement of the bacterium from a region of high metal concentration to that of a lower one. This study was carried out in *B. cereus* CMG2K4 demonstrating the role of genetics in bioremediation (Ottemann and Miller 1997).

Most of the studies conducted till date on the bioremoval of nickel with bacteria showed optimal activity at high temperatures. However, it is essential to note that bacteria which are functionally active in colder temperatures can remove nickel as well. Psychrotolerant bacteria are able to survive at temperatures at a maximum of 20 °C. *B. cereus* D2 was isolated from a nickel mining area in China that grew well in harsh environmental conditions at a temperature of 10 °C, and at a nickel (II) concentration up to 400 mg/L. Under a low temperature (10 °C), *B. cereus* D2 induced carbonate precipitation ($Ni_2CO_3(OH)_2{\cdot}H_2O$) through biomineralization for removing the high efficiency of nickel ions (73.47%) from the culture solution. *B. cereus* D2 was found to be able to remove nickel at 10 °C. It demonstrated urease enzyme activity wherein, during urea hydrolysis, it produced ammonia, carbonate, and hydroxyl ions which consequently altered the pH of the solution. The optimum pH for the growth of the bacterium was well within a range of 6–9 at 10 °C implying its peak activity at mostly alkaline conditions. However, at alkaline conditions above 9, there was a reduction in bacterial activity and growth (Do et al. 2020).

1.5.10 Other Metals

The reduction of palladium (II) into zero valent, nontoxic metallic forms requires microbial reductases (De Corte et al. 2012). *B. sphaericus* JG-A12, a Gram-positive bacteria, accumulated high concentrations of palladium on the surface. This is attributed to the S-layer on the bacterial cell wall which has a unique paracrystalline proteinaceous nature. In this study, a sol-gel technique was adopted that rendered high mechanical and photochemical stability along with biological inactivity and stable porosity. The *B. sphaericus* JG-A12 cells were encapsulated on a silica gel forming a biological matrix called a biocer or bioceramic. Effective and high metal binding on this surface was observed which was also found to be reversible. Due to its flexibility, stability, specificity, and ability of easy recovery of immobilized metal on a surface, biocers are a suitable option for bioremediation (Pollmann et al. 2006). Understanding the biochemical composition of the S-layer could open newer avenues into the synthesis of stable bionanoparticles. The S-layer of bacteria is composed of functional groups such as sulfhydryl, carboxyl, phosphates, amino and hydroxyl which serve as a template for nanoparticle generation (Dieluweit et al. 1998). Palladium nanoparticles synthesized by irradiation of the S-layer of *B. sphaericus* NCTC 9602 was achieved. The accumulation of palladium is a two-step process consisting of biosorption of the metal onto the S-layer surface and then the reduction of the metal by the addition of an electron donor, such as H_2 (Pollmann et al. 2005; Pollmann et al. 2006). X-ray absorption fine structure (EXAFS) spectroscopy and attenuated total reflectance Fourier transform-infrared-spectroscopy (ATR-FT-IR-spectroscopy) studies demonstrated the role of carboxyl functional groups in the formation of palladium nanostructures which in turn keeps the S-layer stable even in extremely acidic conditions (pH 0.8) (Pollmann et al. 2005). Thus, the *Bacillus*-mediated recovery of metals from waste water and the synthesis of nanoparticles can be further used as catalysts.

Platinum is an essential metal emerging in the catalysis and electronics industry. Therefore, their recovery from industrial wastes is of immense significance. Similar to palladium, the recovery of platinum also has a bioremediation perspective through the synthesis of nanoparticles. The significance of the S-layer is emphasized in the synthesis of platinum nanoparticles by irradiating *B. sphaericus* NCTC 9602 in the presence of K_2PtCl_4. The S-layer allows for the formation of an organized template which facilitates the synthesis of inorganic nanoparticles (Wahl et al. 2001).

Bacillus is a potent biosorbent for many metals including zinc. The *B. jeotgali* strain U3 was found to actively remove zinc. The bacteria was isolated from the sediment collected from the Urıas coastal lagoon, Sinaloa, Mexico. Maximum zinc removal was observed at pH 7, salinity of 0 (freshwater), and at 30 °C (Green-Ruiz et al. 2008).

Conventional methods such as coagulation, ion exchange, and filtration have many limitations which are now being replaced by more efficient and safe alternatives employing the use of extracellular polysaccharides (EPS) extracted from bacteria, yeast, and fungi (Bender et al. 1994; Volesky 1994; Salehizadeh and Shojaosadati 2001). An acidic EPS produced by the strain, *B. firmus*, could remove zinc metal from solutions. The maximum biosorption took place at a pH of 6 and a temperature of 25 °C. The sorption rate was found to increase as the solution moved towards more acidic. This may be due to the presence of negative polysaccharides and its interaction with the cations in the acidic solution (Salehizadeh and Shojaosadati 2003).

Living cells of the *B. subtilis* D_{215} strain was also used where bioremoval percentage up to 63.73 was achieved when incubating at 30 °C at a pH of 7. The incubation period also is an important factor governing the bioremediation process. In this case, the incubation of the bacterial strain with the metal solution was kept for 48 h.

On the other hand, zinc bioremoval with dead *B. subtilis* was 26.83% (Sabae et al. 2006).

Other strains such as *B. licheniformis* also demonstrated a significant reduction in zinc concentration from 21.32 to 5.31 mg/L. The mechanism of bioremoval was an interaction between the metal ions and cell wall through weak interactions such as Van der Waals forces and covalent bonding (Blencowe and Morby 2003). In the above study, the bacterial strain was isolated from textile dye effluents which contributes to a vast chunk of the water toxicity due to metal pollution (Basha and Rajaganesh 2014).

A *B. thuringiensis* strain, ISI, was studied in agricultural effluents contaminated with heavy metals and substantial removal of zinc was observed (54%). However, after the fourth day, degradation efficiency reduced to 31%. Even though the results do not indicate a robust system of bioremediation using this strain, nevertheless, it indicates its potential use from an agricultural perspective (Kumar et al. 2015).

1.6 Conclusions and Future Perspectives

Several strains of *Bacillus* are routinely isolated and tested for their potential to remove heavy metals from soils and effluents from large-scale industries. It is important to note that various isolates of *Bacillus* vary in their extent of metal removal, detoxification and recovery. Chemical modification methods can increase/activate the binding sites on the surface of the biomass that can be achieved by pretreatment, binding site enhancement, binding site modification, and polymerization. Functionalization of long polymer chains onto the bacterial surface either by direct grafting or polymerization of a monomer can introduce functional groups. Modification by graft polymerization of acrylic acid (AAc) on the surface of ozone-pretreated biomass can be a powerful strategy for enhancing metal binding. Adsorption of metals like copper and cadmium can be significantly enhanced when the carboxylic acid group on the biomass surfaces is converted to carboxylate ions using NaOH treatment.

Further, genetic manipulation can also be employed to develop novel recombinant *Bacillus* based biosorbents. This strategy can improve or empower microorganisms with enhanced selectivity as well as the bioaccumulating properties of the cells. Simultaneously, elucidation of the underlying molecular mechanism using biotechnology will enable one to design an optimized process for bioreactors where genetically engineered *Bacillus* can be used for effective metal removal. Understanding of metal uptake mechanisms, engineered technologies, including the cell surface display technology, and a further improvement in the metal removal performance of biomass from an aqueous solution can be achieved.

Metalloregulatory protein, MerR, exhibits a high affinity and selectivity toward mercury, and can be exploited for the construction of microbial biosorbents specific for mercury removal. Whole-cell biosorbents can be constructed with MerR genetically engineered onto the surface of *Bacillus* cells by using an ice nucleation protein anchor. This strategy for construction of surface-exposed MerR on the engineered strains can lead to a many-fold higher biosorption capacity compared to wild types of *Bacillus*. A variety of investigations have demonstrated that heavy metal removal using *Bacillus* is a useful alternative to conventional systems that can be significant for sustainable agriculture. Further investigation is required towards modelling, immobilization, and treatment of irrigational water in order to develop microbial processes for heavy metal bioremediation in agricultural fields.

Acknowledgements Dr. Sougata Ghosh acknowledges the Department of Science and Technology (DST), Ministry of Science and Technology, Government of India and Jawaharlal Nehru Centre for an Advanced Scientific Research, India for funding under the Post-doctoral Overseas Fellowship in Nano Science and Technology (Ref. JNC/AO/A.0610.1(4) 2019-2260 dated August19, 2019).

References

Abdel-Monem MO, Al-Zubeiry AHS, Al-Gheethi AAS (2010) Biosorption of nickel by *Pseudomonas cepacia* 120S and *Bacillus subtilis* 117S. Water Sci Technol 61:2994–3007

Akshatha Jain N, Udayashankara TH, Lokesh KS, Sudarshan BL (2017) Bioremediation of lead, nickel and copper by metal resistant *Bacillus licheniformis* isolated from mining site : optimization of operating parameters under laboratory conditions. Int J Res Eng Technol 5:13–32

Akujobi CO, Odu NN, Okorondu SI (2012) Bioaccumulation of lead by *Bacillus* species isolated from pig waste. J Res Biol 2:83–89

Al-Daghistani H (2012) Bio-remediation of Cu, Ni and Cr from rotogravure wastewater using immobilized, dead, and live biomass of indigenous thermophilic *Bacillus* species. Internet J Microbiol 10:1–10

Al-Homaidan AA, Al-Houri HJ, Al-Hazzani AA, Elgaaly G, Moubayed NMS (2014) Biosorption of copper ions from aqueous solutions by *Spirulina platensis* biomass. Arab J Chem 7:57–62

Arivalagan P, Singaraj D, Haridass V, Kaliannan T (2014) Removal of cadmium from aqueous solution by batch studies using *Bacillus cereus*. Ecol Eng 71:728–735

Bai B, Li Q, Holdsworth CH, Asma E, Tai YC, Chatziioannou A, Leahy RM (2002) Model-based normalization for iterative 3D PET image reconstruction. Phys Med Biol 47:2773–2784

Bairagi H, Ghati A, Ray L (2010) Biosorption of copper ions by *Bacillus cereus* M^{1}_{16} from aqueous solution. Indian Chem Eng 51(3):203–214

Baldi F (1997) Microbial transformation of mercury species and their importance in the biogeochemical cycle of mercury. Met Ions Biol Syst 34:213–257

Baran MF, Duz MZ (2019) Removal of cadmium (II) in the aqueous solutions by biosorption of Bacillus licheniformis isolated from soil in the area of Tigris River. Int J Environ Anal Chem. https://doi.org/10.1080/03067319.2019.1669583

Barboza NR, Guerra-Sá R, Leão VA (2016) Mechanisms of manganese bioremediation by microbes: an overview. J Chem Technol Biotechnol 91:2733–2739

Basha SA, Rajaganesh K (2014) Original research article microbial bioremediation of heavy metals from textile industry dye effluents using isolated bacterial strains. Int J Curr Microbiol App Sci 3:785–794

Belapurkar P, Goyal P, Kar A (2016) In vitro evaluation of bioremediation capacity of a commercial probiotic, *Bacillus coagulans*, for chromium (VI) and lead (II) toxicity. J Pharm Bioallied Sci 8:272–276

Belogolova G, Gordeeva O, Sokolova M, Pastukhov M, Vaishlya O, Poletaeva V, Belozerova O (2020) Transformation of lead compounds in the soil-plant system under the influence of *Bacillus* and *Azotobacter* rhizobacteria. Chem Ecol 36:220–235

Bender J, Rodriguez-Eaton S, Ekanemesang UM, Phillips P (1994) Characterization of metal-binding bioflocculants produced by the cyanobacterial component of mixed microbial mats. Appl Environ Microbiol 60:2311–2315

Beveridge TJ, Forsberg CW, Doyle RJ (1982) Major sites of metal binding in *Bacillus licheniformis* walls. J Bacteriol 150:1438–1448

Blaudez D, Botton B, Chalot M (2000) Effects of heavy metals on nitrogen uptake by *Paxillus involutus* and mycorrhizal birch seedlings. FEMS Microbiol Ecol 33:61–67

Blencowe DK, Morby AP (2003) Zn (II) metabolism in prokaryotes. FEMS Microbiol Rev 27:291–311

Brewer PA, Taylor MP (1997) The spatial distribution of heavy metal contaminated sediment across terraced floodplains. Catena 30(2–3):229–249

Chien M, Nakahata R, Ono T, Miyauchi K, Endo G (2012) Mercury removal and recovery by immobilized *Bacillus megaterium* MB1. Front Chem Sci Eng 6:192–197

Dadrasnia A, Wei KSC, Shahsavari N, Azirun MS, Ismail S (2015) Biosorption potential of *Bacillus salmalaya* strain 139SI for removal of Cr(VI) from aqueous solution. Int J Environ Res Public Health 12:15321–15338

Das P, Sinha S, Mukherjee SK (2014) Nickel bioremediation potential of *Bacillus thuringiensis* KUNi1 and some environmental factors in nickel removal. Biorem J 18:169–177

Dash HR, Das S (2012) Bioremediation of mercury and the importance of bacterial mer genes. Int Biodeterior Biodegrad 75:207–213

Dash HR, Mangwani N, Das S (2014) Characterization and potential application in mercury bioremediation of highly mercury-resistant marine bacterium *Bacillus thuringiensis* PW-05. Environ Sci Pollut Res 21:2642–2653

De Corte S, Hennebel T, De Gusseme B, Verstraete W, Boon N (2012) Bio-palladium: from metal recovery to catalytic applications. Microb Biotechnol 5:5–17

Deborah S, Sebastin Raj J (2016) Bioremediation of heavy metals from distilleries effluent using microbes. J Appl Adv Res 1(2):23–28

Dick GJ, Podell S, Johnson HA, Rivera-Espinoza Y, Bernier-Latmani R, McCarthy JK, Torpey JW, Clement BG, Gaasterland T, Tebo BM (2008) Genomic insights into Mn(II) oxidation by the marine alphaproteobacterium *Aurantimonas* sp. strain SI85-9A1. Appl. Environ Microbiol 74:2646–2658

Dieluweit S, Pum D, Sleytr UB (1998) Formation of a gold superlattice on an S-layer with square lattice symmetry. Supramol Science 5(1–2):15–19

Dismukes GC (1986) The metal centers of the photosynthetic oxygen-evolving complex. Photochem Photobiol 43:99–115

Do H, Wang Y, Long Z, Ketehouli T, Li X, Zhao Z, Li M (2020) A psychrotolerant Ni-resistant *Bacillus cereus* D2 induces carbonate precipitation of nickel at low temperature. Ecotoxicol Environ Saf 198:110672

El-Helow ER, Sabry SA, Amer RM (2000) Cadmium biosorption by a cadmium resistant strain of *Bacillus thuringiensis*: regulation and optimization of cell surface affinity for metal cations. Biometals 13:273–280

Francist CA, Tebo BM (2002) Enzymatic manganese(II) oxidation by metabolically dormant spores of diverse *Bacillus* species. Appl Environ Microbiol 68:874–880

Ganguly A, Guha AK, Ray L (2011) Adsorption behaviour of cadmium by *Bacillus cereus* M^1_{16}: some physical and biochemical studies. Chem Spec Bioavailab 23(3):175–182

Gavrilescu M (2004) Removal of heavy metals from the environment by biosorption. Eng Life Sci 4:219–232

Glendinning KJ, Macaskie LE, Brown NL (2005) Mercury tolerance of thermophilic *Bacillus* sp. and *Ureibacillus* sp. Biotechnol Lett 27:1657–1662

Govarthanan M, Lee K-J, Cho M, Kim JS, Kamala-Kannan S, Oh B-T (2013) Significance of autochthonous *Bacillus* sp. KK1 on biomineralization of lead in mine tailings. Nature 90:2267–2272

Govarthanan M, Park S-H, Park Y-J, Myung H, Krishnamurthy RR, Lee SH, Lovanh N, Kamala-Kannan S, Oh B-T (2015) Lead biotransformation potential of allochthonous *Bacillus* sp. SKK11 with sesame oil cake extract in mine soil. RSC Adv 5:54564–54570

Green-Ruiz C, Rodriguez-Tirado V, Gomez-Gil B (2008) Cadmium and zinc removal from aqueous solutions by *Bacillus jeotgali*: pH, salinity and temperature effects. Bioresour Technol 99:3864–3870

Hasan HA, Abdullah SRS, Kofli NT, Kamaruddin SK (2010) Biosorption of manganese in drinking water by isolated bacteria. J Appl Sci 10(21):2653–2657

Hasan HA, Abdullah SRS, Kofli NT, Kamarudin SK (2012) Isotherm equilibria of Mn^{2+} biosorption in drinking water treatment by locally isolated *Bacillus* species and sewage activated sludge. J Environ Manag 111:34–43

Hossain SM, Anantharaman N (2006) Studies on bacterial growth and lead(IV) biosorption using *Bacillus subtilis*. Indian J Chem Technol 13:591–596

Huang JP, Huang CP, Morehart AL (1994) Removal of heavy metals by fungal (Aspergillus oryzae) adsorption In Vernet JP (ed) Trace metals in the environmentl, 2nd edn. Science Publishers/ Elsevier, Amsterdam

Hussain SM, Javorina AK, Schrand AM, Duhart HM, Ali SF, Schlager JJ (2006) The interaction of manganese nanoparticles with PC-12 cells induces dopamine depletion. Toxicol Sci 92:456–463

Igiri BE, Okoduwa SIR, Idoko GO, Akabuogu EP, Adeyi AO, Ejiogu IK (2018) Toxicity and bioremediation of heavy metals contaminated ecosystem from tannery wastewater: a review. J Toxicol:Article ID 2568038

Imam SSA, Rajpoot IK, Gajjar B, Sachdeva A (2016) Comparative study of heavy metal bioremediation in soil by *Bacillus subtilis* and *Saccharomyces cerevisiae*. Indian J Sci Technol 9(47). https://doi.org/10.17485/ijst/2016/v9i47/106911

Jacobsen RT (2005) Catalyst recovery – part 3: removing contaminants from spent catalysts. Chem Eng Prog 101:41–43

Kannan SK, Mahadevan S, Krishnamoorthy R (2006) Characterization of a mercury-reducing *Bacillus cereus* strain isolated from the Pulicat Lake sediments, south east coast of India. Arch Microbiol 185:202–211

Kelly CDR, Cristina FS, Whasley FD, Rosane FS (2014) Bioaccumulation of Fe^{3+} by bacteria isolated from soil and fermented foods for use in bioremediation processes. African J Microbiol Res 8:2513–2521

Kumar KA, Achyuthan H (2007) Heavy metal accumulation in certain marine animals along the East Coast of Chennai, Tamil Nadu, India. J Environ Biol 28:637–643

Kumar V, Singh S, Kashyap N, Singla S, Bhadrecha P, Kaur P, Datta S, Kalia A, Singh J (2015) Bioremediation of heavy metals by employing resistant microbial isolates from agricultural soil irrigated with industrial waste water. Orient J Chem 31:357–361

Lee Y, Tebo BM (1994) Cobalt (II) oxidation by the marine manganese(II)-oxidizing *Bacillus* sp. strain SG-1. Appl Environ Microbiol 60(8):2949–2957

Lei D-Y, Liu Z, Y-h P, Liao S-b XH (2014) Biosorption of copper, lead and nickel on immobilized *Bacillus coagulans* using experimental design methodologies. Ann Microbiol 64:1371–1384

Liu Y-G, Liao T, He Z-B, Li T-T, Wang H, Hu X-J, Guo Y-M, He Y (2013) Biosorption of copper(II) from aqueous solution by *Bacillus subtilis* cells immobilized into chitosan beads. Trans Nonferrous Met Soc China 23:1804–1814

Lovely DR (1987) Organic matter mineralization with the reduction of ferric iron: a review. Geomicrobiol J 5(3–4):375–399

Lozano GA (1998) Parasitic stress and self-medication in wild animals. Adv Study Behav 27:291–317

Matthey J (2002) Platinum 2002. Platin Met Rev 46(3):116

Mishra J, Singh R, Arora NK (2017) Alleviation of heavy metal stress in plants and remediation of soil by rhizosphere microorganisms. Front Microbiol 8:1706

Naik MM, Pandey A, Dubey SK (2012) Microorganisms in environmental management: microbes and environment. In: Microorganisms in environmental management: microbes and environment. Springer Nature, Cham, pp 665–682

Olanow CW (2004) Manganese-induced parkinsonism and Parkinson's disease. Ann N Y Acad Sci 1012:209–223

Ottemann KM, Miller JF (1997) Roles for motility in bacterial-host interactions. Mol Microbiol 24:1109–1117

Oves M, Khan MS, Zaidi A (2013) Biosorption of heavy metals by Bacillus thuringiensis strain OSM29 originating from industrial effluent contaminated north Indian soil. Saudi J Biol Sci 20:121–129

Öztürk A (2007) Removal of nickel from aqueous solution by the bacterium *bacillus thuringiensis*. J Hazard Mater 147:518–523

Pakarinen J, Paatero E (2011) Recovery of manganese from iron containing sulfate solutions by precipitation. Miner Eng 24:1421–1429

Paraneeiswaran A, Shukla SK, Prashanth K, Rao TS (2015) Microbial reduction of [Co(III)–EDTA]− by *Bacillus licheniformis* SPB-2 strain isolated from a solar salt pan. J Hazard Mater 283:582–590

Pardo R, Herguedas M, Barrado E, Vega M (2003) Biosorption of cadmium, copper, lead and zinc by inactive biomass of *Pseudomonas putida*. Anal Bioanal Chem 376:26–32

Patterson JW (1985) Industrial waste water treatment technology, 2nd edn. Butterworths, USA

Pollmann K, Raff J, Schnorpfeil M, Radeva G, Selenska-Pobell S (2005) Novel surface layer protein genes in *Bacillus sphaericus* associated with unusual insertion elements. Microbiology 151:2961–2973

Pollmann K, Raff J, Merroun M, Fahmy K, Selenska-Pobell S (2006) Metal binding by bacteria from uranium mining waste piles and its technological applications. Biotechnol Adv 24:58–68

Pugazhendhi A, Ranganathan K, Kaliannan T (2018) Biosorptive removal of copper(II) by *Bacillus cereus* isolated from contaminated soil of electroplating industry in India. Water Air Soil Pollut 229:76

Que Q, Helmann JD (2000) Manganese homestasisin *Bacillus subtilis* is regulated by MntR, a bifunctional regulator related to the diphtheria toxin repressor family of proteins. Mol Microbiol 35:1454–1468

Ramírez V, Baez A, López P, Bustillos R, Villalobos MÁ, Carreño R, Contreras JL, Muñoz-Rojas J, Fuentes LE, Martínez J, Munive JA (2019) Chromium hyper-tolerant Bacillus sp. MH778713 assists phytoremediation of heavy metals by mesquite trees (*Prosopis laevigata*). Front Microbiol 10:1833

Ravindra K, Bencs L, Van Grieken R (2004) Platinum group elements in the environment and their health risk. Sci Total Environ 318:1–43

Sabae SZ, Hazaa M, Hallim SA, Awny NM, Daboor SM (2006) Bioremediation of Zn^{+2}, Cu^{+2} and Fe^{+2} using *Bacillus subtilis* D_{215} and *Pseudomonas putida* biovar A D_{225}. Biosci Res 3(1):189–204

Salehizadeh H, Shojaosadati SA (2001) Extracellular biopolymeric flocculants: recent trends and biotechnological importance. Biotechnol Adv 19:371–385

Salehizadeh H, Shojaosadati SA (2003) Removal of metal ions from aqueous solution by polysaccharide produced from *Bacillus firmus*. Water Res 37:4231–4235

Samarth DP Chandekar CJ, Bhadekar RK (2012) Biosorption of heavy metals from aqueous solution using *Bacillus Licheniformis*. Int J Pure Appl Sci Technol 10(2):12–19

Saranraj P, Stella D (2012) Bioremediation of sugar mill effluent by immobilized bacterial consortium. Int J Pure Appl Microbiol 2:43–48

Sari A, Mendil D, Tuzen M, Soylak M (2008) Biosorption of Cd(II) and Cr(III) from aqueous solution by moss (*Hylocomium splendens*) biomass: equilibrium, kinetic and thermodynamic studies. Chem Eng J 144:1–9

Sathyavathi S, Manjula A, Rajendhran J, Gunasekaran P (2013) Biosynthesis and characterization of mercury sulphide nanoparticles produced by *Bacillus cereus* MRS-1. Indian J Exp Biol 51:973–978

Shazia A, Yasmin A, Hasnain S (2002) Characterization of some indigenous mercury resistant bacteria from pollutant environment. Pak J Biol Sci 5:792–797

Sinha A, Singh VN, Mehta BR, Khare SK (2011) Synthesis and characterization of monodispersed orthorhombic manganese oxide nanoparticles produced by *Bacillus* sp. cells simultaneous to its bioremediation. J Hazard Mater 192:620–627

Sinha A, Pant KK, Khare SK (2012) Studies on mercury bioremediation by alginate immobilized mercury tolerant *Bacillus cereus* cells. Int Biodeterior Biodegrad 71:1–8

Sivakami G, Priyadarshini R, Baby V, Rajakumar S, Ayyasamy PM (2012) Bioremediation of ferric iron in synthetic metal oxide using *Bacillus* sp. (SO-10) 1:34–41

Sleytr UB, Beveridge TJ (1999) Bacterial S-layers. Trends Microbiol 7:253–260

Sorokin DY, Van Pelt S, Tourova TP (2008) Utilization of aliphatic nitriles under haloalkaline conditions by *Bacillus alkalinitrilicus* sp. nov. isolated from soda solonchak soil. FEMS Microbiol Lett 288:235–240

Stefanescu IA (2015) Bioaccumulation of heavy metals by *Bacillus megaterium* from phosphogypsum waste. Sci Study Res Chem Chem Eng Biotechnol Food Ind 16:93–97

Sukumar C, Janaki V, Vijayaraghavan K, Kamala-Kannan S, Shanthi K (2017) Removal of Cr(VI) using co-immobilized activated carbon and *Bacillus subtilis*: fixed-bed column study. Clean Techn Environ Policy 19:251–258

Taran M, Sisakhtnezhad S, Azin T (2015) Biological removal of nickel (II) by *Bacillus* sp. KL1 in different conditions: optimization by Taguchi statistical approach. Polish J Chem Technol 17:29–32

Upadhyay KH, Vaishnav AM, Tipre DR, Patel BC, Dave SR (2017) Kinetics and mechanisms of mercury biosorption by an exopolysaccharide producing marine isolate *Bacillus licheniformis*. 3. Biotech 7:1–10

Varghese R, Krishna MP, Arun Babu V, Mohamed Hatha AA (2012) Biological removal of lead by *Bacillus* sp. obtained from metal contaminated industrial area. J Microbiol Biotechnol Food Sci 2(2):756–770

Velásquez L, Dussan J (2009) Biosorption and bioaccumulation of heavy metals on dead and living biomass of *Bacillus sphaericus*. J Hazard Mater 167:713–716

Vijayakumar G, Tamilarasan R, Kumar MD (2011) Removal of Cd^{2+} ions from aqueous solution using live and dead *Bacillus subtilis*. Chem Eng Res Bull 15:18–24

Vishan I, Sivaprakasam S, Kalamdhad A (2017) Biosorption of lead using *Bacillus badius* AK strain isolated from compost of green waste (water hyacinth). Environ Technol (UK) 38:1812–1822

Volesky B (1994) Advances in biosorption of metals: selection of biomass types. FEMS Microbiol Rev 14:291–302

Wahl R, Mertig M, Raff J, Selenska-Pobell S, Pompe W (2001) Electron-beam induced formation of highly ordered palladium and platinum nanoparticle arrays on the S layer of *Bacillus sphaericus* NCTC 9602. Adv Mater 13:736–740

Wang J, Chen C (2009) Biosorbents for heavy metals removal and their future. Biotechnol Adv 27:195–226

Wang Y, Moore M, Levinson HS, Silver S, Walsh C, Mahler I (1989) Nucleotide sequence of a chromosomal mercury resistance determinant from a *Bacillus* sp. with broad-spectrum mercury resistance. J Bacteriol 171:83–92

Yilmaz EI (2003) Metal tolerance and biosorption capacity of *Bacillus circulans* strain EB1. Res Microbiol 154:409–415

Yilmaz EI, Ensari NY (2005) Cadmium biosorption by *Bacillus circulans* strain EB1. World J Microbiol Biotechnol 21:777–779

Yun-guo L, Bao-ying F, Ting F, Hai-zhou Z, Xin L (2008) Tolerance and removal of chromium(VI) by *Bacillus* sp. strain YB-1 isolated from electroplating sludge. Trans Nonferrous Met Soc China 18:480–487

Zhenggang X, Yi D, Huimin H, Liang W, Yunlin Z, Guiyan Y (2019) Biosorption characteristics of Mn (II) *by Bacillus cereus* strain HM-5 isolated from soil contaminated by manganese ore. Polish J Environ Stud 28:463–472

Chapter 2
Peptide Antibiotics Produced by *Bacillus* Species: First Line of Attack in the Biocontrol of Plant Diseases

Zobia Khatoon, Ma. del Carmen Orozco-Mosqueda, Suiliang Huang, Francisco X. Nascimento, and Gustavo Santoyo

Abstract *Bacillus* spp. contain a wide arsenal of weapons against plant pathogens such as bacteria, fungi, and oomycetes, including antibiotic peptides as part of the first line of attack and plant protection. In general, such peptides, based on their biosynthetic pathways, can be of ribosomal or non-ribosomal origin. In both cases, a complex genetic structure is required that codes for different enzyme complexes. Here, we review the capabilities and advantages of *Bacillus* species to antagonize and control diseases of different crops through the production of peptide antibiotics. Finally, this chapter discusses the potential of engineered microbes and different types and mechanisms of action of non-ribosomally or ribosomally synthesized peptide antibiotics in *Bacillus* spp., with the final goal of reducing the use of toxic chemicals in agriculture.

Zobia Khatoon and Ma. del Carmen Orozco-Mosqueda contributed equally with all other contributors.

Z. Khatoon · S. Huang
Key Laboratory of Pollution Processes and Environmental Criteria of the Ministry of Education, Key Laboratory of Urban Ecological Environment Rehabilitation and Pollution Control of Tianjin, Numerical Stimulation Group for Water Environment, College of Environmental Science and Engineering Nankai University, Tianjin, China

M. del Carmen Orozco-Mosqueda
Facultad de Agrobiología "Presidente Juárez", Universidad Michoacana de San Nicolás de Hidalgo, Uruapan, Michoacán, Mexico

F. X. Nascimento
IBET, Instituto de Biologia Experimental e Tecnológica, Oeiras, Portugal

G. Santoyo (✉)
Instituto de Investigaciones Químico Biológicas, Universidad Michoacana de San Nicolás de Hidalgo, Morelia, Michoacán, Mexico
e-mail: gsantoyo@umich.mx

M. T. Islam et al. (eds.), *Bacilli in Agrobiotechnology*, Bacilli in Climate Resilient Agriculture and Bioprospecting,
https://doi.org/10.1007/978-3-030-85465-2_2

Keywords Plant growth promoting bacteria · Plant pathogens · Antagonism · Bacilli

2.1 Introduction

Plant pathogens cause serious plant diseases that lead to great economic losses in conventional agricultural practices. The use of polluting agrochemicals is the immediate solution to control plant pathogens (Adesemoye and Kloepper 2009); however, the damage that these agrochemicals can cause to the environment, and human and animal health has been widely documented. As a result, there is a global trend to eliminate, or at least reduce, the use of agrochemicals in agriculture.

Organic agricultural production is a viable and sustainable alternative to conventional agriculture. These practices have allowed the production of grains, fruits, and vegetables in general, with reduced environmental damage (Bacenetti et al. 2016). On the other hand, the costs of the organic products are usually higher than those produced by conventional agriculture. In countries where per capita economic income is lower, such as Latin American countries, the increased cost of organic products limits their acceptance and the development of organic farming practices. Ultimately, novel solutions are necessary in order to decrease the production costs of sustainable organic agriculture.

Other options to control plant diseases in an eco-friendly way, and which have been suggested for decades, include the use of biological control agents such as beneficial fungi (i.e. *Trichoderma* spp.) and plant-growth-promoting bacteria (PGPB) such as *Bacillus spp.* (Santoyo et al. 2019).

PGPB can stimulate plant growth directly through the synthesis of phytohormones or substances that improve nutrition or nutrient uptake by the plant, or, indirectly, which involves antagonistic action towards potential pathogens of vegetable crops (Ghazanfar et al. 2018).

PGPB are part of the associated plant microbiome and are usually selected by the plant. PGPB can inhabit in environments such as the rhizosphere, which is defined as the part of the soil that surrounds the root system and is influenced by exudates from the roots or can colonize internal plant tissues without causing harm. The PGPB found in the rhizosphere are called rhizobacteria or plant-growth-promoting rhizobacteria (PGPR) (Vacheron et al. 2013). Likewise, PGPB found colonizing internal plant tissues are known as endophytic bacteria or plant-growth-promoting bacterial endophytes (PGPBE) (Santoyo et al. 2016).

Various PGPR and PGPBE have been studied for their direct or indirect abilities to promote plant growth and health. These include members of the Firmicutes such as, *Bacillus* spp. and *Paenibacillus* spp.; Actinobacteria such as *Streptomyces* spp. *and Arthrobacter* spp.; and several members of the Proteobacteria such as

Rhizobium spp., *Paraburkholderia* spp., *Pseudomonas* spp., *Pantoea* spp., and *Serratia* spp., to mention but a few (Glick 2012). The application of these PGPB have improved the yield of crops such as lettuce, cucumber, tomato, canola, wheat, maize, tomato, soybean, bean, sunflower, rice, cotton, beet, watermelon, sugar beet, blueberry, husk tomato, as well as various tree species like beech, scotch pine, black spruce, jack pine, white spruce, pine, mangrove, mango, apple, and spruce (Lucy et al. 2004).

The antagonistic effects of PGPB against plant pathogens can occur through a wide variety of mechanisms. For instance, Biocontrol PGPB can synthesize multiple volatile (i.e. hydrogen cyanide) and diffusible compounds (i.e. pyoluteorin, phenazines, pyrrolnitrin, siderophores, 2,4-diacetylphloroglucinol), lytic enzymes (i.e. proteases, cellulases, chitinases, and β-glucanases) as well as other compounds (i.e. bacteriocins, antibiotics) that inhibit the growth of potential plant pathogens, including bacteria, fungi, and oomycetes (Ongena and Jacques 2008).

In this chapter, we focus on the genus *Bacillus* and its different species with capacities to synthesize a group of peptides, of ribosomal and non-ribosomal origin, with antagonistic activity against fungal and bacterial pathogens of various plant crops.

2.2 Generalities of Antagonistic *Bacillus* Species

Some generalities of beneficial bacteria of the *Bacillus* genus have been described recently by Santoyo et al. (2012, 2019), as well as in some other chapters included in this book. Briefly, it is the genus *Bacillus* that was first reported by Cohn in 1872 (Cited in Santoyo et al. 2019). It was originally described as a species of bacteria that tolerated heat and produced endospores. Currently, just over 300 species have been described. *Bacillus* are ubiquitous and can be found in a wide variety of environments, including bulk and rhizospheric soil, where they are particularly abundant among the plant microbiome (Alcaraz et al. 2010; Bazinet 2017).

Each of the PGPB species or genera have different beneficial capacities, as well as advantages versus other genera. In the case of *Bacillus*, the ability to form spores allows them to be excellent candidates to be biological agents of commercial inoculants. It has been observed that *Bacillus* spores can be stored for a long time, remaining viable until inoculation in the field, and can survive even in adverse conditions such as saline or abiotic stress (Villarreal-Delgado et al. 2018). In fact, for several years, species of the genus *Bacillus*, such as *B. pumilus*, *B. subtilis*, *B. circulans*, *B. thuringiensis*, *B. megaterium*, *B. coagulans*, *B. amyloliquefaciens*, *B. licheniformis*, *B. methyltrophicus* or *B. toyonensis*, have stood out as effective bioinoculants (either as biofertilizers, biostimulants and biocides) due to their consistent results (Flores et al. 2020; Glick and Skof 1986; Lucy et al. 2004; Rojas-Solis et al. 2020).

2.3 *Bacillus* Peptide Antibiotics

Several species of the genus *Bacillus* are able to synthesize a wide range of antimicrobial peptides and therefore are viewed as a promising candidates for biotechnological applications and precursors in the search for new inhibitory compounds (Chen et al. 2019). These species have been the object of particular interest because of their safety, their ubiquity, and their ability to survive adverse environmental conditions (Al-Thubiani et al. 2018; Seel et al. 2018) (Fig. 2.1 and Table 2.1).

Bacillus antimicrobial compounds constitute a large collection of significant molecules that present an important ecological role and impact in the microbial community (Andersson and Hughes 2014). Their sub-inhibitory concentrations can trigger nonlethal physiological responses like induction of quorum sensing, antibiosis, defense responses in the host plants, production of virulence factors or biofilm formation, and competition for nutrient sources and space (Xu et al. 2019). Implication of peptide antibiotics by *Bacillus* species in biocontrol of phytopathogens causing aerial, soil, and post-harvest diseases in turn results in improved fitness in plant environment, thus ensuring sustainable agriculture. Among *Bacillus*, the production of peptide antibiotics have been recognized for *Bacillus subtilis*, *B. amyloliquefaciens, B. thuringiensis, B. cereus,* and *B. licheniformis,* but they were also reported in other Bacillus species (Saxena et al. 2019). These peptides fall into two categories, non-ribosomal peptides (NRP) (i.e. gramicidin, tyrocidine, bacitracin, glycopeptides, lipopeptides) and ribosomal peptides or natural peptides, including bacteriocins, and can be classified according to peptide biosynthesis, structure, and molecular weight. We will briefly summarize here with emphasis on three main families (surfactin, iturins, fengycins) and bacteriocins in detail.

2.4 Biosynthesis and Characteristics of *Bacillus* Peptide Antibiotics

Peptide antibiotics produced by *Bacillus* species are divided into two subgroups as mentioned above on the basis of synthesis pathway (Horwood et al. 2004; Myo et al. 2019). One of these subgroups includes small microbial peptides which are non-ribosomally produced by large enzymatic complexes, while the second subgroup constitutes ribosomally synthesized peptide (Fira et al. 2018).

2.4.1 Non-ribosomal Peptide Antibiotics

Bacillus antibiotics include non-ribosomal peptides (NRPs) such as gramicidin, tyrocidine, bacitracin, or lipopeptides such as surfactin, iturins, and fengycins, which are synthesized by non-ribosomal peptide synthetases via multistep

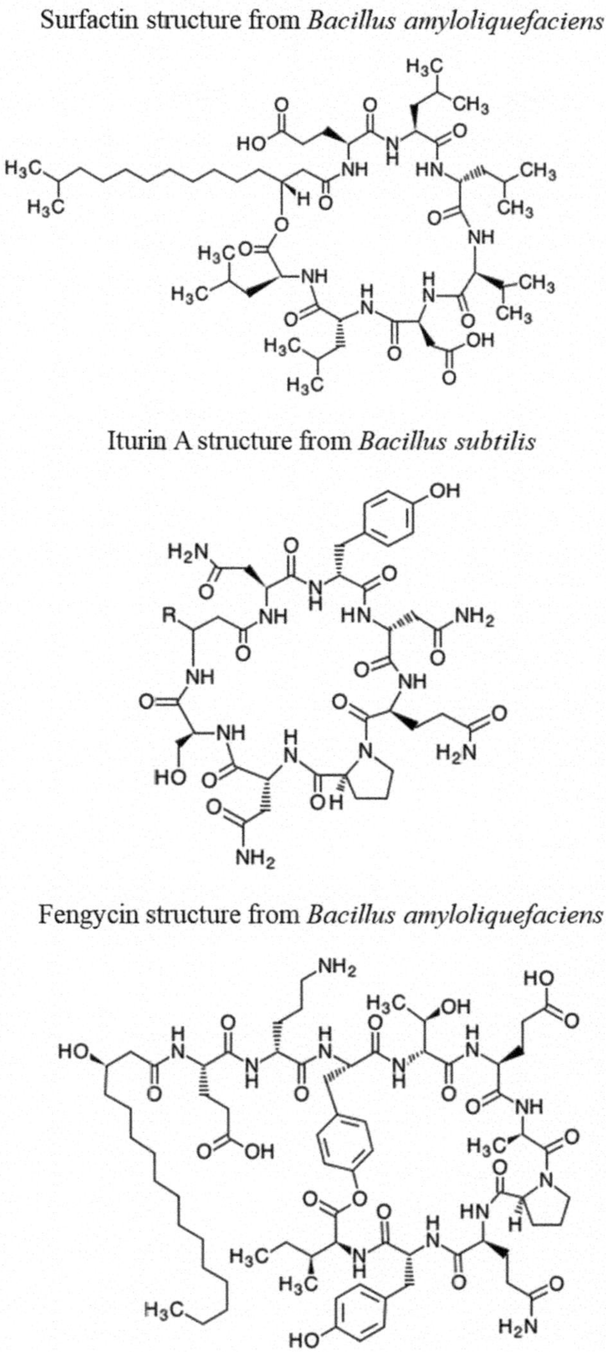

Fig. 2.1 Structures of surfactin, fengycin, and iturin A from *Bacillus* spp. Chemical structures were retrieved from KEGG COMPOUND Database (https://www.genome.jp/kegg/compound/)

Table 2.1 *Bacillus*-derived non-ribosomally or ribosomally synthesized peptide antibiotics. General description, types, and species of *Bacillus* that produces the antibiotic peptide

Surfactin family			
Description	Types	*Bacillus* spp.	References
Surfactin family is composed of heptapeptides with chiral carbon central sequence interlinked by *b*-hydroxy fatty acid to form cyclic lactone ring structure. Displays antiviral, antimycoplasma, antifungal, and antibacterial activities.	Surfactin, esperin, Hallobacillin, pumilacidin, linchenysin WH1fungin	*Bacillus subtilis,* *B. Amyloliquefaciens* *B. pumilus* *B. Thuringiensis* *B. Licheniformis* *B. firmus*	Hu et al. (2020) Penha et al. (2020) Kim et al. (2010) Roongsawang et al. (2011) Kakinuma et al. (1969)
Iturin family			
Description	Types	*Bacillus* spp.	References
Cyclic lipopetides contain a heptapeptide (L- AsnD-Tyr-D-Asn-L-Gln-L-pro-D-Asn-L-Ser) cyclized with a fatty amino acid, displaying a strong antimicrobial activity. Their efficiency against various phytopathogens is similar to available chemical pesticides. Iturin family has a wide spectrum of antibiotic activity, confers low toxicity, low allergic effects, high biodegradability, and are best candidates as environmentally safe biological pesticides	Iturin A, iturin C, Iturin D, iturin E, Bacillomycin D, bacillomycin F, bacillomycin L, bacillomycin Lc, mycosubtilin.	*Bacillus subtilis* *B. Amyloliquefaciens* *B. cereus* *B. Methyltrophicus* *B. Licheniformis*	Dang et al. (2019) Mizumoto et al. (2007) Shi et al. (2018) Yuan et al. (2012) Béchet et al. (2013)
Fengycin family			
Description	Types	*Bacillus* spp.	References
Peptides are synthesized by five synthetases (*FenC, FenD, FenE, FenA, and FenB)* that interlock to form a chain. Exhibit moderate surfactant activities. Regulation of fengycin genes is controlled by DegQ (enhancer of extracellular protease production), which strongly inhibits the growth of filamentous fungi	Fengycin, plipastatin, agrastatin1	*B. Amyloliquefaciens* *Bacillus subtilis* *B. Atrophaeus* *B. circulans*	Guo et al. (2014a) Roongsawang et al. (2011) Romero et al. (2007)
Bacteriocins			
Description	Types	*Bacillus* spp.	References
Heterologous group of proteinaceous antimicrobial substances produced by bacteria from every major lineage. They display a high degree of target specificity and are Ribosomally synthesized.	Lantibiotics Subclasses: Mersacidin, Sublancin Paenibacillin Haloduracin Lichenicidin Subtilosin A	*Bacillus subtilis* *Bacillus circulans* *B. Thuringiensis* *B. Megaterium* *B. coagulans*	Riley and Wertz (2002) Jack et al. (1995) Abriouel et al. (2011)

mechanisms that involve the selection and condensation of amino acid residues (Leães et al. 2016)(Tan et al. 2020). NRPs are assembled from more than 300 precursors and may assume linear or cyclic structures containing hydroxyl groups, *L*-amino acids, or *D*-amino acids. These peptides can be further modified by N-methylation, acylation, glycosylation, or heterocyclic ring formation (Tajbakhsh et al. 2017).

Bacilli NR lipopeptides are represented by three major families, namely surfactins, iturins, and fengycins (Dunlap et al. 2011). The surfactin family encompasses several structural variants (e.g. esperin, lichenysin, pumilacidin) but all members are heptapeptides associated with a *b*-hydroxy fatty acid and present a cyclic lactone ring structure (Tsuge et al. 1999).

Surfactins were the first members among the cyclic lipopeptides to be identified and characterized (Nair et al. 2016). These are stable in extreme environments such as high salinity, heat stress, and extreme pH conditions (Liu et al. 2015) and are involved in *Bacillus* motility and biocontrol functions in the rhizosphere (Sabaté et al. 2020). It was reported that surfactins are essential not only for efficient root colonization but also for reduction of infection caused by *Pseudomonas* species *(Pseudomonas syringae)* on *Arabidopsis* plants (Bais et al. 2004).

Iturins, which consist of iturin A-E, bacillomycin D, F, L, and mycosubtilin, are a group of antifungal cyclic heptapeptides mainly produced by *Bacillus* species that present a wide range of antagonistic effects against phytopathogens (Romero et al. 2007). Iturin A possesses potential antagonistic activity against several plant pathogens and is also used in the treatment of human and animal mycoses because of its low toxicity and lack of allergic effects on the host (Jin et al. 2014; Peng et al. 2014). It is synthesized by non-ribosomal peptide synthetases and its biosynthetic operon constitutes four genes, *ituD*, *ituA*, *ituB*, and *ituC* (Tsuge et al. 2001). The *ituD* gene plays a central role in the synthesis of Iturin A since it encodes a malonyl coenzyme A transacylase. A mutation in the *B. subtilis* RB14 *ituD* gene carries a phenotype where iturin A synthesis is lacking. With respect to the other genes, *ituA* encodes for a 449-kDA protein with three functional domains with high homology to fatty acid synthetase, amino acid transferase, and peptide synthetase. The other two genes, *ituB* and *ituC*, both encode for peptide synthetases with sizes of 609- and 297-kDa, respectively (Tsuge et al. 2001; Yao et al. 2003).

Work carried out by Yao and colleagues in 2003 characterized the role of the operon in the synthesis of iturin A, based on analyses of biocontrol strains of *Bacillus subtilis*. For example, *B. subtilis* strain B3 exhibits antagonism against pathogens like *Fusarium graminearum*, *Rhizoctonia solani*, *Rhizoctonia cerealis*, and *Pyricularia grisea* (Yao et al. 2003).

In the case of the *B. subtilis* RB14 strain, antagonism towards *Rhizoctonia solani* has been observed, being the production of iturin A and mycosubtilin the underlying mechanisms of inhibiting the growth of this pathogen causing the damping off disease in tomato (Asaka and Shoda. 1996). Iturin A and mycosubtilin are both members of the iturin group, whose structures are very similar (Ongena and Jacques 2008; Stein 2005). Overproduction of mycosubtilin by the biocontrol strain *Bacillus subtilis* BBG100 enhances hemolytic activity compared with that of the wild-type strain; however, this overexpression is also associated with an enhanced antifungal activity (Leclère et al. 2005). Iturin production has been reported in other several *Bacillus* species such as *B. amyloliquefaciens*, *B. licheniformis*, *B. thuringiensis*, and *B. methyltrophicus* (Kupper et al. 2020; Li et al. 2020a; Roslan et al. 2020; Yao et al. 2003).

Fengycin family contains B-hydroxy fatty acid that has a side chain length of 16–19 carbon atoms and is synthesized by five biosynthetic genes; *ppsA*, *ppsB*, *ppsC*, *ppsD*, and *ppsE* (Batool et al. 2011). It is composed of two isoforms, which differ in their amino acid residue at position 6, mainly fengycin A and fengycin B (Falardeau et al. 2013). Numerous Bacillus species are able to produce both isomeric forms of fengycin i.e. A and B (Wang et al. 2004). In *Bacillus amyloliquefaciens* Q426, fengycin A was more abundant than B, while, *B. subtilis* F29–3 generally produces twice as much fengycin B as A (Shu et al. 2002; Zhao et al. 2014). Fengycin reportedly has strong antifungal activity, particularly against filamentous fungal species like *Rhizoctonia solani*, *Monolinia laxa*, and *M. fruticola* (Alvarez et al. 2012; Guo et al. 2014b). (Deleu et al. 2005). Moreover, they play a functionally redundant role in defense against the inhibition of *R. solanacearum* as well as banana *Fusarium* wilt caused by *Fusarium* species (Cao et al. 2018).

2.4.2 *General Mechanism of Action of* Bacillus *Lipopeptides*

Due to their lipophilic nature, *Bacillus* lipopeptides present an increased affinity for biological membranes and their lipidic layers. As a result, *Bacillus* lipopeptides bind to the membranes of competing microbes and act as antimicrobials by disrupting membrane functions (Jourdan et al. 2009).

When present at increased concentrations, surfactins extensively bind to the membrane lipid layers (through its fatty acid moiety) inducing an increased membrane permeabilization and the formation of irreversible pores and mixed micelles. These effects ultimately lead to the disruption and solubilization of the membrane lipid bilayer. Curiously, the disruption and solubilizing effects of surfactins are decreased when sterols are present in the membrane (a trait mostly found in fungi), which may explain its low fungicidal activity (Ongena et al. 2007). Like many other natural antimicrobial peptides, its physicochemical characteristics, rather than its specific amino acid sequence, have been proposed to be the key to its antimicrobial activity (Findlay et al. 2010; Heerklotz et al. 2004).

Iturins and fengycins also bind to the membrane lipidic fraction; however, contrarily to surfactins, these seem to be unaffected by sterol components and present an increased fungitoxicity. Despite binding to membrane lipids, the action of iturins and fengycins results from a mechanism of action that does not solely rely on membrane solubilization. For example, Patel and colleagues (2011) observed that fengycins possessed a decreased ability to form micelles and solubilize membranes; however, they presented a strong pore-forming ability even at very low concentrations. Ultimately, the powerful pore-forming abilities of iturins and fengycins may impact the osmotic balance as well as the acquisition/maintenance of important ions/nutrients by cells, therefore, leading to an abnormal cell function and cell death (Ongena et al. 2007).

Bacillus lipopeptides can also induce plant defense responses; however, these do not seem to be caused by strong modifications in plant membrane functions (Ongena et al. 2007; Jourdan et al. 2009; Cawoy et al. 2014). Possibly, this occurs due to the fact that plants possess different sterols (phytosterols) in their membranes, as well as strong cell walls that limit the entrance of *Bacillus* lipopeptides. Still, lipopeptides may bind to specific receptors present in the plant cell and induce the activation of signaling cascades involved in plant defense. If so, *Bacillus* lipopeptides can be considered as Microbial Associated Molecular Patterns (MAMPs). Nevertheless, these molecular interactions remain to be conclusively proven.

2.4.3 Ribosomally Synthesized Peptide Antibiotics

Bacteriocins are microbicidal (or microbiostatic peptides), produced by a wide range of bacterial species, including *Bacillus* species (Abriouel et al. 2011). Bacteriocins have a wide variety of characteristics, such as a range of molecular masses (contain between 12 and 50 amino acid residues), heat stability, posttranslational modifications, and are typically cationic with greater structural diversity being largely distributed in nature (Marx et al. 2001). Bacteriocins belong to the group of peptides which are small, heat stable, amphiphilic protein components that act against target cells mainly via interaction with cell envelope (Papagianni 2003). *Bacillus*-originated bacteriocins like subtilosin A, plantozolicin, and subtilomycin are well studied as antibiotics (Li et al. 2020c; Nikiforova et al. 2016). They are classified into four major classes; Class I contains small ribosomally produced and posttranscriptionally modified peptides (lantibiotics); Class II contains heat-stable unmodified peptides; Class III contains large heat-labile antimicrobial peptides; and Class IV contains complex bacteriocins carrying lipids or carbohydrates moieties (Cotter et al. 2013; Zhao and Kuipers 2016). Bacteriocin production have been reported in several *Bacillus* species such as *B. pumilus*, *B. velezensis, B. altitudinis*, and *B. laterosporus* (Li et al. 2020b).

Most bacteriocins have been studied for their antibiotic activity against animal and human pathogens in lactic acid bacteria (Abriouel et al. 2011); however, there are some examples of bacteriocins isolated from *Bacillus* strains of rhizospheric soil origin. For example, *Bacillus subtilis* strain 14B was isolated from the rhizosphere of healthy bitter almond plants. Such strain, 14B, produces a bacteriocin-like inhibitory substance (BLIS) active against the tumor-producing *Agrobacterium tumefaciens*, reducing the percentage of infected plants (Hammami et al. 2009).

Thuricin 17 is a peptide belonging to the Class II bacteriocins produced by some strains of *B. thuringiensis* and *B. cereus* (Lee et al. 2009). Thuricin 17 has been characterized from the plant growth-promoting rhizobacterium *B. thuringiensis* NEB17, a strain isolated from soybean root nodules. Thuricin 17 is encoded by three copies in tandem of the same structural gene in the NEB17 genome and has a 3160-Da protease-sensitive peptide stable at a high temperature and within the pH range of 1.5–9.0 (Gray et al. 2006). Application of this bacteriocin to leaves or roots directly stimulated the growth of soybean and corn plants. Interestingly, the beneficial effect of Thuricin 17 was similar to the application of Nod factors in plants. (Lee et al. 2009).

Pseudomonas syringae is a bacterial plant pathogen that causes serious damages in diverse agronomically important plant crops, such as tomato, kiwifruit, pepper, olive, and soybean (Lamichhane et al. 2015). Recently, Rooney and colleagues (2019) demonstrated that one bacteriocin, putidacin L1, which was expressed at high levels in Arabidopsis and in *Nicotiana benthamiana in planta*, provided an effective resistance against multiple pathovars of the pathogen *Pseudomonas syringae*. The authors concluded that the expression of bacteriocins *in planta* might be an effective tool for managing bacterial diseases in crops.

2.4.3.1 General Mechanisms of Action of Bacteriocins

It has been proposed that different classes of bacteriocins exert their activity dependent on the target bacteria, if it is against Gram-positive or Gram-negative bacteria. Although it must be said that, in general, the action is mainly due to damage to the integrity of the cell wall. Also, it has been observed that there is an antibiotic action through the inhibition of DNA or protein synthesis. Additionally, it can be mentioned that bacteriocins are able to bind to different components of the cell wall (i.e. phospholipids). Likewise, bacteriocins bind to specific or nonspecific receptors, allowing the formation of pores. This leads to a loss of the membrane structure or cell lysis (Kumariya et al. 2019).

Moreover, they specifically target a particular subset of pathogenic strains through the formation of pores leading to the dissipation of membrane potential and the efflux of small metabolites from the sensitive cells (Lv et al. 2020). Antibiotic, antifungal, and anticancer activities have also been described for bacteriocins production by *Bacillus thuringiensis* strains (Raddadi et al. 2009).

2.5 Engineering of Peptide Antibiotics

Modifying what nature has built has always been attractive to science. Therefore, engineering or genetic modification of coding elements for non-ribosomal antibiotics is part of any biotechnological strategy. The objective may differ, but in general it is about improving the capacities or action spectra of such antibiotic peptides. Likewise, the genomic modification of the microorganisms that produce them is also part of a strategy to reduce production or purification costs at the industrial level, mainly. Exploiting the potential of *Bacillus* spp. to synthesize diverse antibiotic peptides with antibacterial and/or antifungal activity has been always important to control plant pathogens (Moyne et al. 2001). Unfortunately, efforts to engineer or modify *Bacillus* species or the production of non-ribosomal peptides are scarce. One possible explanation is the number of elements involved in the synthesis of such peptides, which requires several enzyme functions. However, some strategies have been proposed to modify such *Bacillus* cell factories. For example, the environmental strain widely studied as a pathogen biocontrol agent, *Bacillus amyloliquefaciens* FZB42, exhibits a broad range of action against phytopathogenic bacteria, fungi, and nematodes, through the production of several non-ribosomal cyclic lipopeptides and polyketides. Therefore, and although it has not been done, the modification of a global regulatory gene, *degU*, which controls the synthesis of bacillomycin D and bacilysin in the strain FZB42, has been proposed (Qiao et al. 2014).

Bioengineered bacteriocins have also been recommended as promising alternatives to already existing antibiotics due to their effectiveness and nontoxicity in both animals and humans. In part, because of the availability of narrow- and broad-spectrum peptides, their possibility in model membrane systems and their potential cytotoxicity reported against eukaryotes and erythrocytes (Abriouel et al. 2011).

Some genetic engineering methods have been listed by Ahmad and colleagues (2017), including mutagenesis, gene fusion, and/or specific amino acid inserts, in order to design or improve some functions of bacteriocins. For example, solubility at neutral pH has been improved by incorporating lysine residues or the stability and spectrum of some bacteriocins such as pyocin S-35 and microcin V have also been improved. Other molecular tools such as gene fusion have also achieved modifications in the spectrum of bacteriocins.

2.6 Conclusion

The antibiotic potential of peptides produced by bacteria of the *Bacillus* genus is broad and promising. However, more research is needed in the area of peptide stability for efficient action against pathogens prevalent in the field, where diverse environmental stresses affect the activity of such peptide antibiotics. Likewise, it is required to further explore its potential based on genetic modification techniques using *Bacillus* spp. like microbial factories. Lipopeptide extraction and purification techniques are important to obtain a product as pure as possible, which will result in rapid commercialization and effective application in different agricultural systems. Techniques such as acid precipitation, ultrafiltration, solid phase extraction, and chromatography are some of the tools most used in the study of lipopeptides, mainly those characterized and isolated from *Bacillus* strains (Valenzuela-Ruiz et al. 2020). The study and optimization of the aforementioned processes will allow increasing the field application of the different peptides with antimicrobial activity, reducing the use of agrochemicals and taking agriculture to terms of sustainability.

Acknowledgements G.S. thanks Consejo Nacional de Ciencia y Tecnología, México (Grant number: A1-S-15956) and Coordinación de la Investigación Científica-UMSNH (2020) for the financial support to research projects.

References

Abriouel H, Franz CMAP, Omar NB, Gálvez A (2011) Diversity and applications of *Bacillus* bacteriocins. FEMS Microbiol Rev 35:201–232

Adesemoye AO, Kloepper JW (2009) Plant–microbes interactions in enhanced fertilizer-use efficiency. Appl Microbiol Biotechnol 85(1):1–12

Alcaraz LD, Moreno-Hagelsieb G, Eguiarte LE, Souza V, Herrera-Estrella L, Olmedo G (2010) Understanding the evolutionary relationships and major traits of bacillus through comparative genomics. BMC Genomics 11(1):332

Al-Thubiani ASA, Maher YA, Fathi A, Abourehab MAS, Alarjah M, Khan MSA, Al- Ghamdi SB (2018) Identification and characterization of a novel antimicrobial peptide compound produced by Bacillus megaterium strain isolated from oral microflora. Saudi Pharm J 26:1089–1097

Alvarez F, Castro M, Principe A, Borioli G, Fischer S, Mori G, Jofre EJ (2012) The plant-associated Bacillus amyloliquefaciens strains MEP218 and ARP23 capable of producing the cyclic lipopeptides iturin or surfactin and fengycin are effective in biocontrol of sclerotinia stem rot disease. J Appl Microbiol 112:159–174

Andersson DI, Hughes DJ (2014) Microbiological effects of sublethal levels of antibiotics. Nat Rev Microbiol 12:465–478

Asaka O, Shoda M (1996) Biocontrol of *Rhizoctonia solani* damping-off of tomato with Bacillus subtilis RB14. Appl Environ Microbiol 62:4081–4085

Bacenetti J, Fusi A, Negri M, Bocchi S, Fiala M (2016) Organic production systems: sustainability assessment of rice in Italy. Agric Ecosyst Environ 225:33–44

Bahar AA, Ren D (2013) Antimicrobial Peptides 6: 1543–1575

Bais HP, Fall R, Vivanco JMJ (2004) Biocontrol of Bacillus subtilis against infection of Arabidopsis roots by Pseudomonas syringae is facilitated by biofilm formation and surfactin production. Plant Physiol 134:307–319

Batool M, Khalid MH, Hassan MN, Yusuf HFJ (2011) Homology modeling of an antifungal metabolite plipastatin synthase from the Bacillus subtilis. Bioinformation 168:384

Bazinet AL (2017) Pan-genome and phylogeny of *Bacillus cereus* sensu lato. BMC Evol Biol 17(1):176

Béchet M, Castéra-Guy J, Guez J-S, Chihib N-E, Coucheney F, Coutte F, Fickers P, Leclère V, Wathelet B, Jacques P (2013) Production of a novel mixture of mycosubtilins by mutants of Bacillus subtilis. Bioresour Technol 145:264–270

Cao Y, Pi H, Chandrangsu P, Li Y, Wang Y, Zhou H, Xiong H, Helmann JD, Cai Y (2018) Antagonism of two plant-growth promoting bacillus velezensis isolates against *Ralstonia solanacearum* and *fusarium oxysporum*. Sci Rep 8:4360

Chen Y, Kong Q, Liang Y (2019) Three newly identified peptides from Bacillus megaterium strongly inhibit the growth and aflatoxin B1 production of Aspergillus flavus. Food Control 95:41–49

Cotter PD, Ross RP, Hill CJ (2013) Bacteriocins—a viable alternative to antibiotics? Nat Rev Microbiol 11:95–105

Cawoy H, Mariutto M, Henry G, et al (2014) Plant defense stimulation by natural isolates of *Bacillus* depends on efficient surfactin production. Mol Plant-Microbe Interact 27:87–100

Dang Y, Zhao F, Liu X, Fan X, Huang R, Gao W, Wang S, Yang C (2019) Enhanced production of antifungal lipopeptide iturin A by *bacillus amyloliquefaciens* LL3 through metabolic engineering and culture conditions optimization. Microb Cell Factories 18:68

Deleu M, Paquot M, Nylander TJ (2005) Fengycin interaction with lipid monolayers at the air–aqueous interface—implications for the effect of fengycin on biological membranes. J Colloid Interface Sci 283:358–365

Dufour S, Deleu M, Nott K, Wathelet B, Thonart P, Paquot M (2005) Hemolytic activity of new linear surfactin analogs in relation to their physico-chemical properties. J Peptide Sci 1726:87–95

Dunlap CA, Schisler DA, Price NP, Vaughn SFJT (2011) Cyclic lipopeptide profile of three Bacillus subtilis strains; antagonists of Fusarium head blight. J Microbiol 49:603

Falardeau J, Wise C, Novitsky L, Avis TJ (2013) Ecological and mechanistic insights into the direct and indirect antimicrobial properties of *Bacillus subtilis* lipopeptides on plant pathogens. J Chem Ecol 39:869–878

Findlay B, Zhanel GG, Schweizer F (2010) Cationic amphiphiles, a new generation of antimicrobials inspired by the natural antimicrobial peptide scaffold. Antimicrob Agents Chemother 54(10):4049–4058

Flores A, Diaz-Zamora JT, Orozco-Mosqueda, MC, et al (2020) Bridging genomics and field research: draft genome sequence of *Bacillus thuringiensis* CR71, an endophytic bacterium that promotes plant growth and fruit yield in *Cucumis sativus* L. 3. Biotech 10:1–7

Fira D, Dimkić I, Berić T, Lozo J, Stanković S (2018) Biological control of plant pathogens by Bacillus species. J Biotechnol 285:44–55

Ghazanfar MU, Raza M, Raza W, Qamar MI (2018) Trichoderma as potential biocontrol agent, its exploitation in agriculture: a review. Plant Prot 2(3)

Glick BR (2012) Plant growth-promoting bacteria: mechanisms and applications. Scientifica 2012

Glick BR, Skof YC (1986) Environmental implications of recombinant DNA technology. Biotechnol Adv 4:261–277

Gray E, Di Falco M, Souleimanov A, Smith DL (2006) Proteomic analysis of the bacteriocin, thuricin-17 produced by *Bacillus thuringiensis* NEB17. FEMS Microbiol Lett 255:27–32

Guo Q, Dong W, Li S, Lu X, Wang P, Zhang X, Wang Y, Ma P (2014a) Fengycin produced by Bacillus subtilis NCD-2 plays a major role in biocontrol of cotton seedling damping-off disease. Microbiol Res 169:533–540

Guo Q, Dong W, Li S, Lu X, Wang P, Zhang X, Wang Y, Ma PJ (2014b) Fengycin produced by Bacillus subtilis NCD-2 plays a major role in biocontrol of cotton seedling damping-off disease. Microbiol Res 169:533–540

Hammami I, Rhouma A, Jaouadi B, Rebai A, Nesme X (2009) Optimization and biochemical characterization of a bacteriocin from a newly isolated *Bacillus subtilis* strain 14B for biocontrol of Agrobacterium spp. strains. Lett Appl Microbiol 48:253–260

Heerklotz H, Wieprecht T, Seelig JJT (2004) Membrane perturbation by the lipopeptide surfactin and detergents as studied by deuterium NMR. J Chem Physics 108:4909–4915

Horwood PF, Burgess GW, Jane Oakey H (2004) Evidence for non-ribosomal peptide synthetase production of cereulide (the emetic toxin) in Bacillus cereus. FEMS Microbiol Lett 236:319–324

Hu F, Liu Y, Lin J, Wang W, Li S (2020) Efficient production of surfactin from xylose-rich corncob hydrolysate using genetically modified Bacillus subtilis. Appl Microbiol Biotechnol 168:1–10

Jack RW, Tagg JR, Ray B (1995) Bacteriocins of gram-positive bacteria. Microbiol Rev 59:171–200

Jin H, Zhang X, Li K, Niu Y, Guo M, Hu C, Wan X, Gong Y, Huang F (2014) Direct bio-utilization of untreated rapeseed meal for effective iturin A production by Bacillus subtilis in submerged fermentation. Plos one 9:e111171

Jourdan E, Henry G, Duby F, Dommes J, Barthelemy J-P, Thonart P, Ongena M (2009) Insights into the defense-related events occurring in plant cells following perception of surfactin-type lipopeptide from Bacillus subtilis. Mol Plant-Microbe Int 22:456–468

Kakinuma A, Ouchida A, Shima T, Sugino H, Isono M, Tamura G, Arima K (1969) Confirmation of the structure of surfactin by mass spectrometry. Agric Biol Chem 33:1669–1671

Kim P-I, Ryu J, Kim YH, Chi Y-TJ (2010) Production of biosurfactant lipopeptides Iturin A, fengycin and surfactin A from Bacillus subtilis CMB32 for control of Colletotrichum gloeosporioides. J Microbiol Biotechnol 20:138–145

Kumariya R, Garsa AK, Rajput YS, Sood SK, Akhtar N, Patel S (2019) Bacteriocins: classification, synthesis, mechanism of action and resistance development in food spoilage causing bacteria. Microb Pathog 128:171–177

Kupper KC, Moretto RK, Fujimoto AJ (2020) Production of antifungal compounds by Bacillus spp isolates and its capacity for controlling citrus black spot under field conditions. World J Microbiol Biotechnol 36:7

Lamichhane JR, Messéan A, Morris CE (2015) Insights into epidemiology and control of diseases of annual plants caused by the *Pseudomonas syringae* species complex. J Gen Plant Pathol 81:331–350

Leães FL, Velho RV, Caldas DGG, Ritter AC, Tsai SM, Brandelli A (2016) Expression of essential genes for biosynthesis of antimicrobial peptides of Bacillus is modulated by inactivated cells of target microorganisms. Res Microbiol 167:83–89

Leclère V, Bechet M, Adam A, Guez JS et al (2005) Mycosubtilin overproduction by *Bacillus subtilis* BBG100 enhances the organism's antagonistic and biocontrol activities. Appl Environ Microbiol 71:4577–4584

Lee KD, Gray EJ, Mabood F, Jung WJ, Charles T, Clark SR, Ly A, Souleimanov A, Zhou X, Smith DL (2009) The class IId bacteriocin thuricin-17 increases plant growth. Planta 229:747–755

Li M, Mou H, Kong Q, Zhang T, Fu X (2020a) Bacteriostatic effect of lipopeptides from Bacillus subtilis N-2 on Pseudomonas putida using soybean meal by solid-state fermentation. Marine Life Sci Technol:1–9

Li Z, Song C, Yi Y, Kuipers OP (2020b) Characterization of plant growth-promoting rhizobacteria from perennial ryegrass and genome mining of novel antimicrobial gene clusters. BMC Genomics 21:157

Li Z, Song C, Yi Y, Kuipers OP (2020c) Characterization of plant growth-promoting rhizobacteria from perennial ryegrass and genome mining of novel antimicrobial gene clusters. BMC Genomics 21:1–11

Liu Q, Lin J, Wang W, Huang H, Li S (2015) Production of surfactin isoforms by Bacillus subtilis BS-37 and its applicability to enhanced oil recovery under laboratory conditions. Biochem Eng J 93:31–37

Lucy M, Reed E, Glick BR (2004) Applications of free living plant growth-promoting rhizobacteria. Antonie Van Leeuwenhoek 86(1):1–25

Lv J, Da R, Cheng Y, Tuo X, Wei J, Jiang K, Han B (2020) Mechanism of antibacterial activity of Bacillus amyloliquefaciens C-1 lipopeptide toward anaerobic Clostridium difficile. BioMed Res Int 2020

Marx R, Stein T, Entian KD, Glaser SJ (2001) Structure of the *Bacillus subtilis* peptide antibiotic subtilosin A determined by 1H-NMR and matrix assisted laser desorption/ionization time-of-flight mass spectrometry. J Protein Chem 20(6):501–506

Mizumoto S, Hirai M, Shoda M (2007) Enhanced iturin A production by *Bacillus* subtilis and its effect on suppression of the plant pathogen *Rhizoctonia solani*. Appl Microbiol Biotechnol 75:1267–1274

Moyne AL, Shelby R, Cleveland TE, Tuzun S (2001) Bacillomycin D: an iturin with antifungal activity against Aspergillus flavus. J Appl Microbiol 90(4):622–629

Myo EM, Liu B, Ma J, Shi L, Jiang M, Zhang K, Ge B (2019) Evaluation of Bacillus velezensis NKG-2 for bio-control activities against fungal diseases and potential plant growth promotion. Biol Control 134:23–31

Nair D, Vanuopadath M, Nair BG, Pai JG, Nair SS (2016) Identification and characterization of a library of surfactins and fengycins from a marine endophytic *Bacillus* sp. J Basic Microbiol 56(11):1159–1172

Nikiforova OA, Klykov S, Volski A, Dicks LM, Chikindas ML (2016) Subtilosin A production by Bacillus subtilis KATMIRA1933 and colony morphology are influenced by the growth medium. Annals Microbiol 66:661–671

Ongena M, Jacques PT (2008) *Bacillus* lipopeptides: versatile weapons for plant disease biocontrol. Trends Microbiol 16: 115–125

Ongena M, Jourdan E, Adam A, et al (2007) Surfactin and fengycin lipopeptides of *Bacillus subtilis* as elicitors of induced systemic resistance in plants. Environ Microbiol 9:1084–1090

Papagianni M (2003) Ribosomally synthesized peptides with antimicrobial properties: biosynthesis, structure, function, and applications. Biotechnol Adv 21(6):465–499

Peng W, Zhong J, Yang J, Ren Y, Xu T, Xiao S, Zhou J, Tan H (2014) The artificial neural network approach based on uniform design to optimize the fed-batch fermentation condition: application to the production of iturin A. Microbial Cell Factories 13:54

Penha RO, Vandenberghe LP, Faulds C, Soccol VT, Soccol CR (2020) Bacillus lipopeptides as powerful pest control agents for a more sustainable and healthy agriculture: recent studies and innovations. Planta 251:1–15

Qiao JQ, Wu HJ, Huo R, Gao XW, Borriss R (2014) Stimulation of plant growth and biocontrol by Bacillus amyloliquefaciens subsp. plantarum FZB42 engineered for improved action. Chem Biol Technol Agric 1(1):12

Raddadi N, Belaouis A, Tamagnini I, Hansen BM, Hendriksen NB, Boudabous A, Cherif A, Daffonchio D (2009) Characterization of polyvalent and safe Bacillus thuringiensis strains with potential use for biocontrol. J Phytopathol 49:293–303

Riley MA, Wertz JE (2002) Bacteriocin diversity: ecological and evolutionary perspectives. Biochimie 84:357–364

Rojas-Solis D, Vences-Guzmán MÁ, Sohlenkamp C et al (2020) Antifungal and plant growth–promoting bacillus under saline stress modify their membrane composition. J Soil Sci Plant Nutr. https://doi.org/10.1007/s42729-020-00246-6

Romero D, de Vicente A, Rakotoaly RH, Dufour SE, Veening J-W, Arrebola E, Cazorla FM, Kuipers OP, Paquot M, Pérez-García A (2007) The iturin and fengycin families of lipopeptides are key factors in antagonism of bacillus subtilis toward podosphaera fusca. Mol Plant-Microbe Interact 20:430–440

Roongsawang N, Washio K, Morikawa MJI (2011) Diversity of nonribosomal peptide synthetases involved in the biosynthesis of lipopeptide biosurfactants. Mol Diversity Preservation Int 12:141–172

Roslan HA, Husaini A, Lihan S, Kota MF (2020) Partial Purification and Characterization of Antifungal Peptides Produced by Bacillus amyloliquefaciens PEP3 Against Phytophthora capsici. Appl Sci Eng Progress 13:56–66

Ruiz VV, Gamboa GTG, Rodríguez EDV, Cota FIP, Santoyo G, de los Santos Villalobos S (2020) Lipopéptidos producidos por agentes de control biológico del género *Bacillus*: revisión de herramientas analíticas utilizadas para su estudio. Revista mexicana de ciencias agrícolas 11(2):419–432

Sabaté DC, Petroselli G, Erra-Balsells R, Audisio MC, Brandan CP (2020) Beneficial effect of Bacillus sp P12 on soil biological activities and pathogen control in common bean. Biol Control 141:104131

Santoyo G, Orozco-Mosqueda MDC, Govindappa M (2012) Mechanisms of biocontrol and plant growth-promoting activity in soil bacterial species of bacillus and pseudomonas: a review. Biocontrol Sci Tech 22(8):855–872

Santoyo G, Moreno-Hagelsieb G, del Carmen Orozco-Mosqueda M, Glick BR (2016) Plant growth-promoting bacterial endophytes. Microbiol Res 183:92–99

Santoyo G, Equihua A, Flores A, Sepulveda E, Valencia-Cantero E, Sanchez-Yañez JM, de los Santos-Villalobos, S. (2019) Plant growth promotion by ACC deaminase-producing bacilli under salt stress conditions. In: Bacilli and agrobiotechnology: phytostimulation and biocontrol. Springer, Cham, pp 81–95

Saxena AK, Kumar M, Chakdar H, Anuroopa N, Bagyaraj D (2019) Bacillus species in soil as a natural resource for plant health and nutrition. J Appl Microbiol. https://doi.org/10.1111/jam.14506

Seel W, Flegler A, Zunabovic-Pichler M, Lipski A (2018) Increased isoprenoid quinone concentration modulates membrane fluidity in listeria monocytogenes at low growth temperatures. J Bacteriol 200(13):e00148–e00118

Shi J, Zhu X, Lu Y, Zhao H, Lu F, Lu Z (2018) Improving iturin A production of Bacillus amyloliquefaciens by genome shuffling and its inhibition against saccharomyces cerevisiae in orange juice. Front Microbiol 9

Shu H-Y, Lin G-H, Wu Y-C, Tschen JS-M, Liu S-T (2002) Amino acids activated by fengycin synthetase FenE. Biochem Biophysical Res Commun 292:789–793

Stachelhaus T, Marahiel MA (1995) Modular structure of genes encoding multifunctional peptide synthetases required for non-ribosomal peptide synthesis. FEMS Microbiol Lett 125:3–14

Stein T (2005) Bacillus subtilis antibiotics: structures, syntheses and specific functions. Mol Microbiol 56:845–857

Tajbakhsh M, Karimi A, Fallah F, Akhavan M (2017) Overview of ribosomal and non-ribosomal antimicrobial peptides produced by Gram positive bacteria. Cellular Mol Biol 63:20–32

Tan K, Zhou M, Jedrzejczak RP, Wu R, Higuera RA, Borek D, Babnigg G, Joachimiak A (2020) Structures of teixobactin-producing nonribosomal peptide synthetase condensation and adenylation domains. Curr Res Structural Biol 2:14–24

Tsuge K, Ano T, Hirai M, Nakamura Y, Shoda M (1999) The genes degQ, pps, and lpa-8 (sfp) are responsible for conversion of Bacillus subtilis 168 to plipastatin production. Antimicrobial Agents Chemotherapy 43:2183–2192

Tsuge K, Akiyama T, Shoda M (2001) Cloning, sequencing, and characterization of the iturin A operon. Am Soc Microbiol 183:6265–6273

Vacheron J, Desbrosses G, Bouffaud ML, Touraine B, Moënne-Loccoz Y, Muller D, Prigent-Combaret C (2013) Plant growth-promoting rhizobacteria and root system functioning. Front Plant Sci 4:356

Villarreal-Delgado MF, Villa-Rodríguez ED, Cira-Chávez LA, Estrada-Alvarado MI, Parra-Cota FI, de los Santos-Villalobos, S. (2018) The genus Bacillus as a biological control agent and its implications in the agricultural biosecurity. Mex J Phytopathol 36:95–130

Wang J, Liu J, Wang X, Yao J, Yu Z (2004) Application of electrospray ionization mass spectrometry in rapid typing of fengycin homologues produced by Bacillus subtilis. Lett Appl Microbiol 39:98–102

Xu Z, Mandic-Mulec I, Zhang H, Liu Y, Sun X, Feng H, Xun W, Zhang N, Shen Q, Zhang R (2019) Antibiotic Bacillomycin D affects iron acquisition and biofilm formation in bacillus velezensis through a Btr-mediated FeuABC-dependent pathway. Cell Rep 29:1192–1202.e5

Yao S, Gao X, Fuchsbauer N, et al (2003) Cloning, sequencing, and characterization of the genetic region relevant to biosynthesis of the lipopeptides iturin A and surfactin in *Bacillus subtilis*. Curr Microbiol 47:272–277

Yuan J, Li B, Zhang N, Waseem R, Shen Q, Huang Q (2012) Production of Bacillomycin- and Macrolactin-type antibiotics by bacillus amyloliquefaciens NJN-6 for suppressing Soilborne plant pathogens. J Agric Food Chem 60:2976–2981

Zhao X, Kuipers OP (2016) Identification and classification of known and putative antimicrobial compounds produced by a wide variety of Bacillales species. BMC Genomics 17:882

Zhao P, Quan C, Wang Y, Wang J, Fan S (2014) Bacillus amyloliquefaciens Q-426 as a potential biocontrol agent against Fusarium oxysporum f sp spinaciae. J Basic Microbiol 54:448–456

Chapter 3
Discovery of Bioactive Natural Products from *Bacillus* Species: Chemistry, Biosynthesis and Biological Activities

Musrat Zahan Surovy, Shahinoor Rahman, Zerihun T. Dame, and Tofazzal Islam

Abstract *Bacillus* is one of the most widely distributed genera of bacteria. It produces various groups of bioactive compounds such as cyclic lipopeptides, surfactins, iturins, fengycins, various polyketide-derived nonribosomal peptides, linear lipopeptides, enzymes, and δ-endotoxins with strong surfactant, antimicrobial, and insecticidal properties. Biosynthesis and chemistry of these bioactive natural products are largely been known. Genes involved in the biosynthesis of some of the bioactive compounds produced by *Bacillus* spp. are used in the genetic engineering of plants and microorganisms. Among them, genetically engineered Bt-crops significantly improved yield and quality of several crops. Recently developed CRISPR-Cas genome editing technology would facilitates the large-scale application of new genes of *Bacillus* involved in biosynthesis of bioactive compounds for promoting crop production and drug industry. This chapter comprehensively reviews chemistry, biosynthetic pathways, and biological activities of natural products discovered

---

M. Z. Surovy
Institute of Biotechnology and Genetic Engineering (IBGE), Bangabandhu Sheikh Mujibur Rahman Agricultural University, Gazipur, Bangladesh

Division of Plant Pathology and Crop Protection, Georg-August Universität Göttingen, Göttingen, Germany

S. Rahman
Division of Agricultural Entomology, Georg-August Universität Göttingen, Göttingen, Germany

Z. T. Dame
Department of Biology, College of Natural and Computational Science, Ambo University, Ambo, Ethiopia

T. Islam (✉)
Institute of Biotechnology and Genetic Engineering (IBGE), Bangabandhu Sheikh Mujibur Rahman Agricultural University, Gazipur, Bangladesh
e-mail: tofazzalislam@bsmrau.edu.bd

M. T. Islam et al. (eds.), *Bacilli in Agrobiotechnology*, Bacilli in Climate Resilient Agriculture and Bioprospecting,
https://doi.org/10.1007/978-3-030-85465-2_3

in various strains of *Bacillus* spp., and discusses their application in biocontrol of plant diseases, medicines, and bioprospecting.

Keywords *Bacillus* spp. · Linear lipopeptides · Biocontrol agent · Bioremediation · CRISPR-Cas genome editing

3.1 Introduction

Microbes are the prospective source of bioactive natural compounds (Bhardwaj et al. 2017; Islam and Tahara 2005; Islam et al. 2005; Islam 2011; Tareq et al. 2015; Chakraborty et al. 2020a, b). The discovery of bioactive natural products from microbes has been an important scientific discovery of the early twentieth century (Fleming 1941; Stoica et al. 2019). The production of microbial secondary metabolites was regulated by the specific genetic constitution and growing conditions of microbes. *Bacillus* species are one of the leading sources of bioactive natural products (Matloub et al. 2019; Mondol et al. 2013). This genus of bacteria is ubiquitous as they can tolerate a wide range of environmental conditions (Rampelotto 2010; Mondol et al. 2013). The higher capability of biofilm formation and the quorum sensing mechanisms help them to survive even in extreme environmental conditions. Cohn first proposed the genus *Bacillus* in 1872 (Feng et al. 2016). From 2008 to 2018, 167 new species of *Bacillus* were described globally (https://www.bacterio.net/bacillus.html). However, a vast majority of the *Bacillus* species remain undiscovered. Many of the *Bacillus* spp. are used as model organisms in laboratory studies for many years. Biotechnological industries have employed these bacteria for synthesizing a broad range of biologically active compounds and enzymes. Due to the lack of outer membrane and natural competence, genetic manipulation of *Bacillus* is faster than many other bacteria (Drejer et al. 2018). The members of the *Bacillus* genus are known for their biological control efficiency. They are strong biocontrol agents against many notorious phytopathogens, promote plant growth by phytohormone production, and induce systemic resistance in the host plants (Islam et al. 2016a, b, 2019a, b; Rahman et al. 2018; Ansary et al. 2018; Tareq et al. 2015; Chakraborty et al. 2020a). Strains of *Bacillus* produce diverse metabolites such as lipopeptide, peptides, and polyene substances (surfactin, iturin, and fengycin) (Surovy et al. 2019). These metabolites suppress the growth and development of phytopathogens and thus are useful as a biopesticides (Kefi et al. 2015; Qiu et al. 2017). Bacteriocins and bacteriocin-like inhibitory substances, nonribosomal biosynthesized peptides, and nonpeptide-based antibiotics are also produced by the *Bacillus* species. Subtilin, ericin S (active against *Clavibacter michiganensis*), ericin A, subtilosin A, sublancin 168, and mersacidin were also isolated from this genus of bacteria. Extracellular polysaccharides with antioxidant capacity (Qiu et al. 2017) from *Bacillus* sp. protect them from starvation and drastic environmental conditions (Osta Oliveira et al. 2017). These polysaccharides are used as bio-thickener, texture stabilizers, or gelling agents to improve food texture and quality (Matloub et al. 2019).

Indiscriminate use of pesticides for crop production and toxic industrial waste results in the deposition of toxic heavy metals (Pd, Cu, Cd, Cr, Hg, Co, etc.) in the soil and water bodies. Detoxification of a trace amount of these heavy metals requires a huge amount of energy and biochemical activities. Interestingly, many strains of *Bacillus* spp. have the capacity to remediate toxic heavy metals and radionuclides from wastewater and natural waters (Bernard et al. 2018). They reduce the toxic heavy metals from the environment through accumulation, complexation, and reduction processes (Mondol et al. 2013).

The Bacilli are a promising group of bacteria producing diversified biologically active secondary metabolites with a novel mechanisms of action (Islam et al. 2016a, b, 2019a, b, Surovy et al. 2019; Dame and Islam 2019) (Fig. 3.1). This group of bacteria are also widely used in agriculture in controlling insect pests and pathogens. They can produce crystal proteins (δ-endotoxins) with specific activity against certain insect species, nematodes, mites, and protozoa. Moreover, they produce a good number of extracellular compounds including- chitinase, phospholipases, proteases, insecticidal proteins, and antibiotic compounds with antifungal activity (Nkiru and Franklin 2020; Rong et al. 2020). Intensive studies on the relationship between biological activities and molecular structures would help their industrial production as novel biopesticides, biofungicides, and drugs (Penha et al. 2020). The

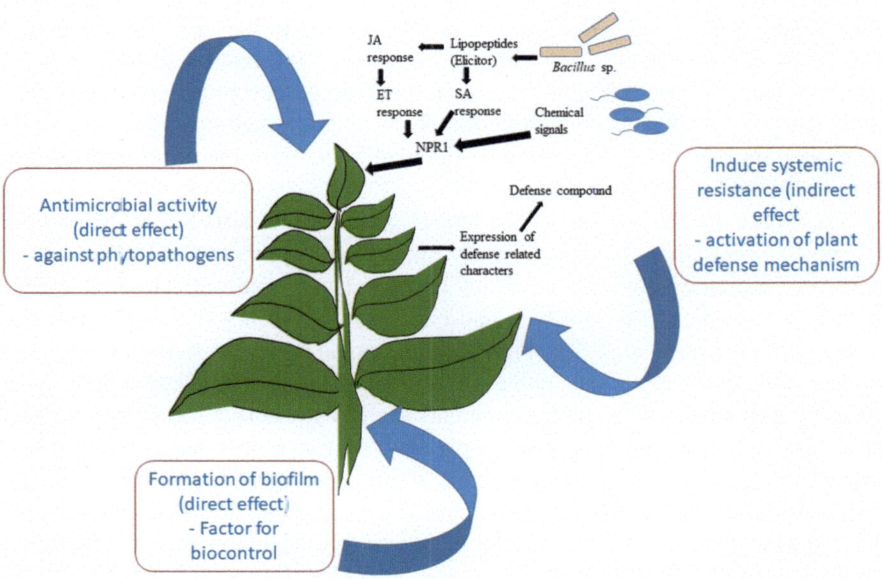

Fig. 3.1 Relationship between bioactive compounds from *Bacillus* species and their effects on plant. (JA jasmonic acid, ET ethylene, SA salicylic acid, NPR1 nonexpressor of pathogenesis-related genes 1)

molecular mechanisms of biosynthesis of bioactive metabolites from metabolites by *Bacillus* spp. are already been known and some of the responsible genes or biosynthesis pathways are exploited in the development of insect-resistant crop plants (like- corn, cotton, and brinjal) by genetic engineering. Further genomics and post-genomics studies are needed for better understanding the underlying molecular mechanisms involved in the production of diverse bioactive metabolites produced by *Bacillus* species. Recently developed CRISPR-Cas toolkits would facilitate to apply the genetic information in the practical field of crop improvement and industrial production of medicinally important compounds.

The application of Bacilli in agrobiotechnology and industry has been discussed in several books and reviews (Islam et al. 2016a, b, 2019a, b; Surovy et al. 2019; Dame and Islam 2019). A large body of literature describes the chemistry, biosynthesis, and biological activities of metabolites produced by *Bacillus* species (Mondol et al. 2013). This chapter comprehensively summarizes the biosynthesis, chemistry, and activities of diverse metabolites discovered from *Bacillus* species and discusses the application of this new knowledge in promoting sustainable agriculture and bioprospecting.

3.2 Isolation and Taxonomic Diversity of Bacilli

There are numerous protocols established for the isolation of *Bacillus* species from the environmental samples. Polanczyk and Alves (2004) proposed an effective modified isolation method for *Bacillus* species. According to their proposal, there are two steps in the isolation procedure *of Bacillus* - (i) bacterial growth in solid medium, and (ii) sporulation of bacteria in liquid medium supplemented with antibiotics under 12–48-h incubation period. This isolation method seems more efficient than the previously described World Health Organization developed method (World Health Organization (WHO) 1985).

The genus, *Bacillus* is one of the largest groups of bacteria under the phylum Firmicutes. More than 370 species and 8 subspecies with validly published names have been reported (www. Bacterio. net/ bacillus. Html, accessed: August 3, 2020). Bacilli are aerobic, Gram-positive, and endospore-producing rod-shaped bacteria (Cutting 2011; Cote et al. 2015). They are widespread in the environment and found in dust, soil, water, air, plants, and even in extreme environments. They have been isolated from a wide range of environments such as desert soil, forest soil, mineral pool, soda lake, alcohol fermentation pit mud, seawater, sediments, rice paddies, dried foods, milk, and honey (Zheng et al. 2020; Lee et al. 2016; Al-Thubiani et al. 2018). The endospores of *Bacillus* are generally oval, round, or cylindrical in shape. The exposure to air is not necessary for their sporulation (Bergey et al. 1975). Based on the morphology of the spore and sporangium, *Bacillus* sp. Are broadly divided into three subgroups (Savini 2016; Gillespie and Hawkey 2006). Group 1: Gram positive, produce central or terminal, cylindrical or ellipsoidal spore without distend sporangium. This group consists two subgroups. *B. anthracis, B. cereus,*

B. mycoides, *B. thuringiensis*, and *B. megaterium* under the large cell subgroup and *B. pumilus*, *B. subtilis*, and *B. licheniformis* under small cell subgroup. Group 2: Central or ellipsoidal spores with gram variable swollen sporangia. *B. circulans*, *B. coagulans*, *B. alvei*, *B. brevis*, and *B. marcerans* are under this group. Group 3: Gram variable, swollen sporangia with terminal or subterminal spores. *B. sphaericus* belongs this group. In recent years, the taxonomy of two selected groups of *Bacillus* species has been developed. They are *B. subtilis* group (*B. subtilis* subsp. *subtilis*, *B. subtilis* subsp. *spizizenii*, *B. mojavensis*, *B. vallismortis*, *B. clausii*, *B. atrophaeus*, *B. amyloliquefaciens*, *B. licheniformis*, *B. sonorensis*, *B. firmus*, *B. lentus*, and *B. sporothermodurans*) and *B. cereus* group (*B. anthracis*, *B. cereus*, *B. mycoides*, *B. pseudomycoides* *B. thuringiensis*, and *B. weihenstephanensis*) (Fritze 2004; PHEBacteriology-Identification 2018). Detailed phylogeny and taxonomy of agriculturally important *Bacillus* species have recently been reviewed by Dunlap (2019).

3.3 Purification and Identification of Bioactive Compounds

Bacillus species is one of the most studied species for the production of bioactive compounds (Table 3.1). Till date *B. subtilis* and *B. amyloliquefaciens* are the commonly studied groups of bacteria (Cawoy et al. 2015; Penha et al. 2020). Different media like brain heart infusion broth (Perez et al. 2017), nutrient broth (Soares et al. 2016), landy medium (Chen et al. 2016), and YPG medium (Gong et al. 2015) are used for culture and extraction of metabolites produced by *Bacillus* species. During fermentation, *Bacillus* sp. produce a huge amount of foam due to their surfactant nature. This intense foam production is the main challenge in the production of bioactive compounds that affect continuous recovery and identification of the compounds (Coutte et al. 2013). The use of rotating disk bioreactors can perform fermentation without producing foam and help in the industrial production of bioactive compounds from the strains of *Bacillus* species. Bubble-less membrane bioreactor, solid-state fermentation reactors, and biofilm reactors are also used for recovery of compounds but rotating disk bioreactors perform the best. The yield from the rotating disk bioreactor is significantly higher than any other reactor (Penha et al. 2020). The use of a bioreactor with continuous removal of foam from the fermentation chamber is good for the productivity of bioactive compounds (Barros et al. 2008).

Extraction of bioactive compounds from *Bacillus* sp. is generally performed by two methods. Firstly, adding ethyl acetate to the cell-free supernatant in addition with NaCl (30 g/L). The ethylacetate fraction is concentrated from the homogenized samples using a rotary evaporator (Dimkić et al. 2017). Secondly, use of acid precipitation method. In this method, the cell-free supernatant is adjusted in pH 2.0 with HCl. The precipitate from the suspension is resuspended in methanol. Methanol is more suitable to dissolve bioactive compounds (Asari et al. 2017; Dimkić et al. 2017). After the successful extraction of compounds, it is important to perform a finer purification process. A finer purification of extracts is essential

Table 3.1 List of bioactive compounds, their identification method, and biological activities from various strains of *Bacillus* spp

Species and strain	Bioactive compounds	Identification method	Inhibited pathogenic organisms	References
Bacillus methylotrophicus DR 08	Difficidin and oxydifficidin	Bioassay-guided chromatographic fractionation and spectroscopic analyses	*Ralstonia solanacearum*	Im et al. (2020)
B. amyloliquefaciens EZ1509	Surfactin, iturin, fengycin, and bacilysin	MALDI-TOF-MS analysis	*Sclerotinia sclerotiorum*	Farzand et al. (2020)
B. amyloliquefaciens	Fengycin	LC–MS.MS analysis	Phytopathogenic fungi	Kaki et al. (2020)
Bacillus sp.	Kurstakins, iturins, surfactins, and fengycins	HPTLC and MALDI-TOF MS analyses	*Xanthomonas arboricola* and *Pseudomonas syringae*	Dimkić et al. (2017)
B. subtilis DSM 10^T, *B. amyloliquefaciens* DSM 7^T and *B. methylotrophicus* DSM 23117	Iturin, surfactin, and fengycin	HPTLC (silica gel 60 plates—Merck)	Phytopathogens	Geissler et al. (2017)
B. amyloliquefaciens DSM 23117	Surfactin	HPLC (Luna C18 reversed phase column, equipped with a Luna C18 pre-column)	*Botrytis cinerea*	Pretorius et al. (2015)

for the identification of a bioactive compound. Thin layer chromatography, high performance thin layer chromatography, liquid column chromatography, and high pressure liquid chromatography are extensively been used for the purification of the compounds. Complete identification of compounds requires several spectroscopic methods such as various mass spectroscopy and ^{1}H and ^{13}C nuclear magnetic resonance (NMR) spectroscopy (Penha et al. 2020) (Fig. 3.2).

3.4 Bioactive Compounds Isolated from *Bacillus* spp.

3.4.1 Lipopeptides

Lipopeptides are small secondary metabolites that contain 7–10 cyclic amino acids (including 2–4 D amino acids) and 13–19 C atoms of beta-hydroxy fatty acid (Zhao et al. 2017). They are classified into two groups: cyclic and linear lipopeptides. Surfactin, iturin, fengycin, lacilomycin, and bacilotetrins are cyclic lipopeptides

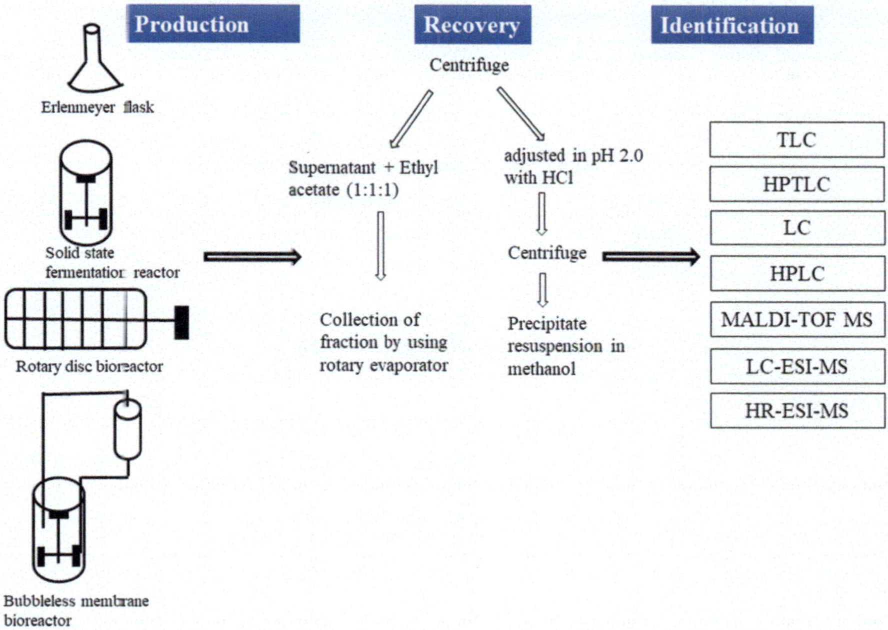

Fig. 3.2 Overview of production, recovery, and identification of bioactive compounds from *Bacillus* sp. (Adapted from Penha et al. 2020. *TLC Thin Layer Chromatography, HPTLC High Performance Thin Layer Chromatography, LC Liquid Chromatography, HPLC High performance Liquid Cromatography, MALDI-TOF MS Matrix Assistierte Laser Desorption/Ionisierung (MALDI) with the flight analysis (Time of Flight, TOF) released ions to mass spectrometry, LC-ESI-MS Liquid Chromatography/Electrospray Ionisation Tandem Mass Spectrometry, HR-ESI-MS High Resolution/ Electrospray Ionisation Tandem Mass Spectrometry

derived from *Bacillus* sp. On the other hand, gageotetrins, gagoepeptides, isocoumarins, amicoumacin, bacilosarcins, damxungmacin, hetiamacin, clorotetain, rhizocticins etc., are under linear polypeptides group. Lipopeptides directly act against fungi and bacteria and can induce systemic resistance (ISR) in plants against phytopathogens (Ongena et al. 2007) (Table 3.2). For example, *B. velezensis* SQR9 triggers ISR against *P. syringae* pv. tomato (Pst DC3000) and *Botrytis cinerea* phytopathogens in *Arabidopsis* plantlets (Wu et al. 2018). The surfactins, iturins, fengycins or plipastatins, and the kurstakins are the major families of lipopeptides produced by *Bacillus* species (Jacques 2011; Dimkić et al. 2017).

Surfactin (Fig. 3.3a) is a heptapeptide with LLDLLDL chiral central sequence which is internally linked by a β-hydroxy fatty acid. Surfactin, esperin, halobacillin, and pumilacidin are the main members of surfactin family (Penha et al. 2020). Surfactins are widely studied for their biological activities. They have antiviral, antimycoplasma, and antibacterial activities (Ongena and Jacques 2008). Amphiphilic structure of surfactin has led to its broad range of antibiotic treatment, cancer therapy, oil recovery, and biomedical applications. Under optimum conditions, wild type of *Bacillus* sp. can yield a large scale of surfactins for industrial

Table 3.2 List of lipopeptides produced by *Bacillus* sp. and their biological activities

Species and strain	Lipopeptide	Biological activity	Reference
Bacillus amyloliquefaciens FZB42	Surfactin	Support colonization and acquisition of nutrients through surface-wetting and detergent properties	Borriss (2020)
B. velezensis	Surfactin	Biofilm formation	Chen et al. (2007)
B. subtilis	Iturin	Antibiotic production	Dunlap et al. (2019)
B. safensis B21	Iturin A and Iturin B	Affect hyphal membrane permeability of blast fungus, *Magnaporthe oryzae*	Rong et al. (2020)
B. amyloliquefaciens HAB-2	Bacillomycin D	Roughening and damage of cell membrane of *Burkholderia pseudomallei*	Rajaofera et al. (2020)
B. megaterium (KC246043.1)	Bacitracin	Antimicrobial activities against *Micrococcus luteus, S. typhi, E. coli, P. aeruginosa*, and *S. aureus*	Al-Thubiani et al. (2018)
Bacillus sp. CND-914	Halobacillin	Anticancer activity	Trischman et al. (1994)
Bacillus sp.	Loloatin B	Antibacterial activity	Mondol et al. (2013)
B. laterosporus	Basiliskamides A and B	Antifungal activity	Barsby et al. (2002)
B. subtilis	Gageopeptides A-D (1–4) and Gageotetrin B (5)	Antifungal activity	Chakraborty et al. (2020a) and Mondol et al. (2013)

purposes. Biofilm produced by *Salmonella enterica* is destroyed by the lytic activity of *Bacillus* sp. (Kaspar et al. 2019).

Iturins (Fig. 3.3b) are also an important class of lipopeptides. They are monomeric or neutral lipopeptides with LDDLLDL chiral sequence with a restricted number of *Asx, Glx, Pro, Ser, Thr, Tyr* residues (Bonmatin et al. 2003). Iturin A, iturin C, bacillomycin D, bacillomycin F, bacillomycin L, bacillopeptin, and mycosubtilin belong to the iturin family (Penha et al. 2020). Iturin A composed of the heptapeptide NYNQPNS is linked to a β-amino fatty acid (Hiradate et al. 2002). They are widely studied for their antibiotic activities and are mostly produced by *B. subtilis* group (Ongena and Jacques 2008). Dunlap et al. (2019) found 330 iturinic lipopeptide clusters in *B. subtilis*. *B. amyloliquefaciens, B. siamensis,* and *B. velezensis*. Each isolate produces more than one lipopeptide (Dunlap et al. 2019). Cyclic lipopeptides (Iturin and surfactin) from *Bacillus* sp. have a broad range of antimicrobial activity against fungi (Phae et al. 1990). Both lipopeptides have the capacity to elicit induced systemic resistance (ISR) in the host plants (García-Gutiérrez et al. 2013; Jourdan et al. 2009; Kawagoe et al. 2015). They play an important role in protecting strawberry plant from the pathogens. They trigger plant

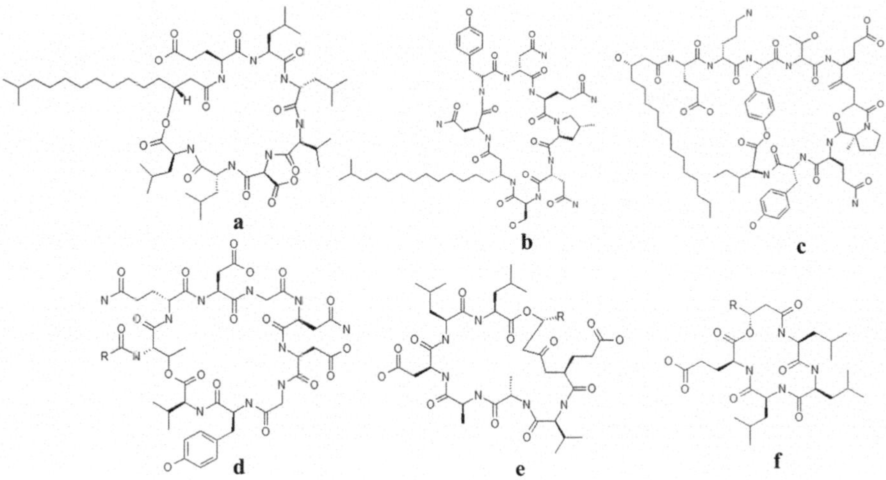

Fig. 3.3 Chemical structure of some lipopeptides (**a**) surfactin; (**b**) iturin; (**c**) fengycin; (**d**) locillomycin A (n =10), B (n =11) and C (n =12); (**e**) bacilotetrin; (**f**) gageopeptin

defense response against *Colletotrichum gloeosporioides* (Yamamoto et al. 2015). *Bacillus* LPs also induce defense responses in rice (Chandler et al. 2015), *Arabidopsis* (Kawagoe et al. 2015), tomato (Abdallah et al. 2017), and maize (Gond et al. 2015).

Nonribosomal peptides are a class of peptide secondary metabolites. Ten percent (10%) of *B. amyloliquefaciens* FZB42 genome possesses genes for NPRS. It produces antifungal lipopeptide bacillomycin D and the antibacterial polyketide difficidin, which are important for plant protection (Borriss 2020). Fengicin (Fig. 3.3c) was first discovered by Vanittanakom et al. in 1986. Fengycin A, fengycin B, plipastin A, and plipastin B are the main members of the fengycin family (Penha et al. 2020). This group is mainly responsible for antimicrobial activity against a wide range of filamentous fungi and yeasts (Zihalirwa et al. 2017). Fengycins A and B differ only in one amino acid residue. Fengycin A contains a D-Ala ring and Fengycin B contains a D-Val ring in their structure. They are less toxic to plants but have selective activity against filamentous fungus (Kaspar et al. 2019).

The plipastatins from *B. cereus* are related to fengycins. These are inhibitors of phospholipase A2 of *B. cereus* (Volpon et al. 2000). The structures of plipastatins are almost identical to fengycin counterparts, the only difference is the two *Tyr* stereocenters. Locillomycins (Fig. 3.3d) are cyclic lipopeptides isolated from a Chinese *B. subtilis* 916 strain. Gene cluster of *Bacillus* sp. encodes for *locA-locD* genes. This is potential in controlling *S. aureus* and *X. oryzae* but commercial applications are very limited due to their low fermentation yields (Luo et al. 2015).

Marine *B. subtilis* strain 109GGC020 produces bacilotetrin (Fig. 3.3e) and gageopeptin (Fig. 3.3f) (Tareq and Shin 2017; Tareq et al. 2015). They are moderately antibacterial and antifungal against *Colletotrichum acutatum, Botrytis cinerea, Staphylococcus aureus*, and *Pseudomonas aeruginosa* in broth dilution assay. Gageopetin A strongly inhibits the growth of *P. capsici* (Tareq et al. 2015).

Wheat blast is a destructive fungal disease caused by *Magnaporthe oryzae Triticum* (*MoT*) pathotype. It poses a serious threat to food security of South America, South Asia, and Africa (Islam et al. 2016a, b, 2019a, b, 2020). Recently, Chakraborty et al. (2020a) demonstrated that gageopeptides A–D and gageotetrin B isolated from a marine *B. subtilis* strain 109GGC020 inhibited the growth of *MoT* mycelia in a dose-dependent manner. Application of these natural compounds has also completely blocked formation of conidia in the *MoT* fungal mycelia in the agar medium. Further bioassay revealed that these compounds inhibited the germination of *MoT* conidia and, if germinated, induced deformation of germ tube and/or abnormal appressoria. Interestingly, application of these linear lipopeptides from strain 109GGC020 significantly suppressed wheat blast disease on detached wheat leaves. A further study is warranted to test the efficacy of wheat blast control at field conditions, and also to evaluate the mode of action of these natural compounds for considering them as biopesticides for managing this notorious cereal killer.

3.4.2 Polyketides/Lipoamides

Polyketides are a large association of multifunctional polypeptides with a series of catalytic domains (ketosynthase, KT; acyl transferase, AT; acyl carrier protein, ACP). They are an essential group of secondary metabolites, generally synthesized through decarboxylative condensation of carboxylic acids by polyketide synthases (PKSs) (Borriss 2020). *B. amyloliquefaciens* FZB42 has a special class of PKSs, which require a discrete AT enzyme and lack the cognate AT domain (Shen 2003). Type III polyketides catalyze the *B. amyloliquefaciens* FZB42 for priming, extension, and cyclization reactions to customize a huge array of various polyketide products (Yu et al. 2012). The *bspA* and *bspB* operons of *B. subtilis* are involved in the synthesis of triketide pyrones. The *bspA* synthesizes alkylpyrone and *bspB* operon acts on the alkylpyrones to produce alkylpyrone methyl ethers (Nakano et al. 2009). Aurantinin B (Fig. 3.4a), an antibacterial polyketide, was originally isolated from

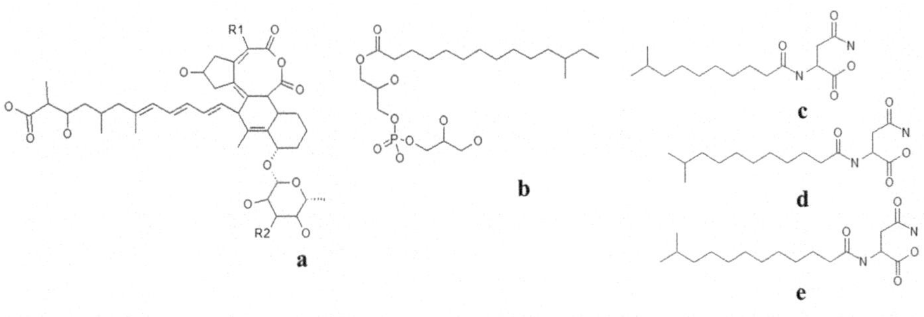

Fig. 3.4 Chemical structure of polyketides (**a**, **b**) and lipoamides (**c**, **d** and **e**) (**a**) aurantinin B (R1 = Me; R2 = O=), C (R1 = Me; R2 = OH), and D (R1 = H; R2 = O=); (**b**) bacilysocin; (**c**) lipoamide A (N = 5); (**d**) lipoamide B (N = 6); (**e**) lipoamide C (N = 7))

B. aurantinus (Kaspar et al. 2019). Aurantinins C and D (Fig. 3.4a) are two new analogs of aurantinin that are isolated from *B. subtilis* Fmb60. These two aurantinins possess very strong antibacterial activity against *M. luteus* and *B. pumilis*. These compounds disrupt cell wall and cell membrane with a highly selective mode of action (Yang et al. 2016).

Bacilysocin (Fig. 3.4b) is a unique example of modified phospholipid, which does not originate from non-ribosomal peptide synthetase or polyketide pathway of *B. subtilis*. This is very weak in controlling Gram-positive bacteria and its mode of action is still unrevealed (Tamehiro et al. 2002).

Furanoids and pyranoids were also isolated from a marine alga-associated *B. subtilis* (Chakraborty et al. 2017; Chakraborty et al. 2016). They are biosynthsized via hitherto polyketide synthase like pathways and moderately active against *Vibrio parahaemolyticus* (Chakraborty et al. 2017). Lipoamides A-C (Fig. 3.4c, d and e) were isolated from culture broth of *B. pumilus*. Lipoamide A is a weak antibacterial compound against *P. aeruginosa* and *S. aureus* with minimum inhibitory concentration >100 μg/mL (Mondol et al. 2013).

3.4.3 *Isocoumarins*

The isocoumarins comprise of a large diverse class of natural compounds (Saeed 2016). The first reported isocoumarins from *Bacillus* were amicoumacins A-C (Fig. 3.5a, b) and AI-77-F (Fig. 3.5c) (Shimojima et al. 1984; Kaspar et al. 2019). The *amiA-O* genes of *B. subtilis* utilize a genomic capture and with the help of expression vector enable to produce amicoumacins A-C in heterologous hosts. They are potential for drug development due to their potential antibacterial and anti-inflammatory properties (Li et al. 2015).

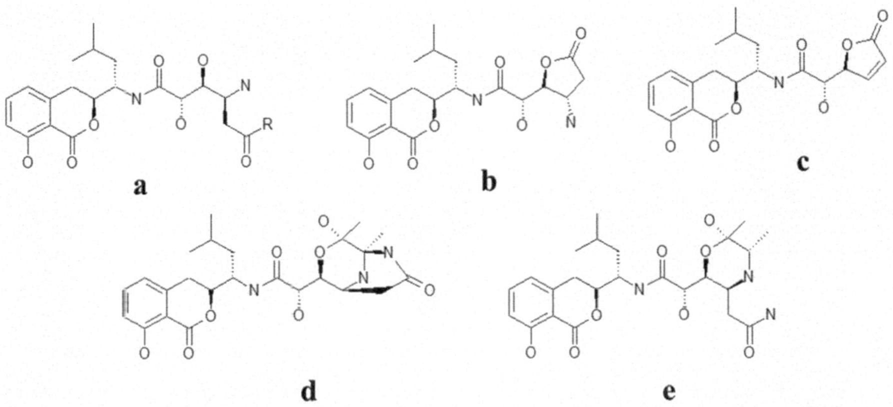

Fig. 3.5 Chemical structures of isocoumarins (**a**) amicoumacin A (R = NH_2), B (R = OH); (**b**) amicoumin C; (**c**) AI-7-F; (**d**) bacilosarcin A; (**e**) bacilosarcin B

Bacilosarcins A and B (Fig. 3.5d and e) are marine-derived bioactive compounds from *B. subtilis* TP-BO611. They showed herbicidal activity (Azumi et al. 2008).

The *B. subtilis* subsp. *subtilis* produces psoralen, 4, 6-dimethyl-3 coumarin, and xanthotoxin (all are member of coumarin group). The bioautophagy test showed that these compounds are capable of inhibiting the growth of *P. oryzae* race 173 (Wiraswati et al. 2020).

3.4.4 Fatty Acids

For synthesis of fatty acid, *Bacillus* sp. uses fatty acid synthase pathway (Mondol et al. 2013). Haplophilic *Bacillus* spp. produces iso-anteiso fatty acids (Fickers 2012; Li et al. 2011) with C_{11} to C_{19} but their biological activity still unknown (Carballeira et al. 2001). Ethanoic extract of marine *Bacillus* sp. 09ID194 contains ieodomycins A (Fig. 3.6a), B, C, and D (Tareq et al. 2013), the unsaturated hydroxyl fatty acids. These fatty acids are antimicrobial against *E. coli* and *S. cerevisiae* (Mondol et al. 2013).

3.4.5 Macrolactins

Polyene macrolactones with 24-membered rings of lactones are attached to glucose β-pyranoside. *B. amyloliquefaciens* FZB42 produces difficidin (59) (Fig. 3.6b) and oxydifficidin (60) (Fig. 3.6b). Difficidin inhibits the protein synthesis of *Erwinia amylovara.* This fungus causes blight diseases in apple, pear, and other rosaceous fruits (Mondol et al. 2013).

Macrolactin A is a potential compound to inhibit murine melanoma cancer cell and Herpes simplex virus *in vitro* (Mondol et al. 2013). Macrolactin G from marine *Bacillus* sp. PP19-H3 is active against *S. aureus,* whereas keto group (C=O) containing macrolactin K showed weak antibacterial activity (Nagao et al. 2001). *B. marinus* produces macrolactin T and macrolactin U. Macrolactin T exhibited

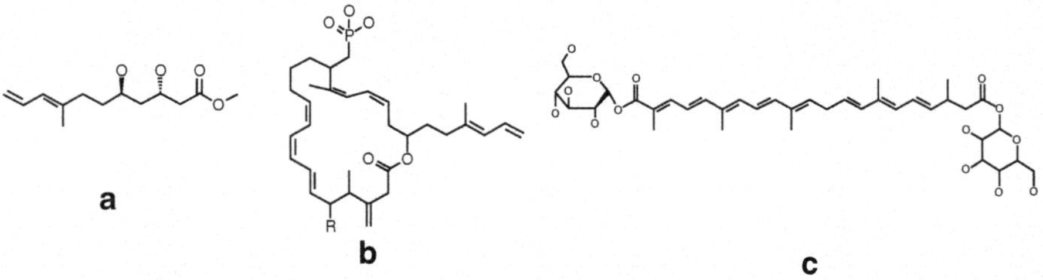

Fig. 3.6 Chemical structures of ieodomycin A (**a**), macrolactins (**b**) difficidin (R = H) and oxydifficidin (R = OH), and glycosylated carotenoid (**c**)

weak inhibitory activity towards *A. solani, P. oryzae*, and *S. aureus* (Xue et al. 2008). *B. amyloliquefaciens* SCSIO 00856 produces macrolactin V, a potential antibacterial agent against *E. coli* and *S. aureus* (Gao et al. 2010). Macrolactin W, X, Y, and Z were also isolated from *Bacillus* sp. 09ID194 in low saline medium (Mondol et al. 2011). Marine *Bacillus* sp. Sc026 produces macrolactin A, 7-0-succinyl macrolactin F and 7-0-succinyl macrolactin A. These compounds show antibacterial activity against *B. subtilis* and *S. aureus* (Jaruchoktaweechai et al. 2000).

3.4.6 Enzymes

Bacillus spp. produce different types of industrially important enzymes (Table 3.3), such as amylases, glucanases, and proteases. The *B. siamensis* also produces enzymes used in various Asian foods (Borriss 2020). *Bacillus* sp. C-4 produces serine protease, which is helpful in the degumming process in the silk industry (Prathumpai et al. 2020). Extracellular cellulase from *B. subtilis* helps in bio-refining (Reddy et al. 2016). Amylase production by *Bacillus* sp. is enhanced by the addition of ammonium nitrate in the culture medium. *B. amyloliquefaciens* produces amylase at pH 7.0 (Tanyildizi et al. 2007). *Bacillus* sp. FPF-1 is a potential source of keratinase. This enzyme is useful for degradation of chicken feathers and very important for agro chemical sectors (Nnolim et al. 2020). Some strains of *Bacillus* spp. are also capable of producing laccases enzymes from agro-waste. The laccases are important enzymes for bio-remediation, bio-pulping, and food industry (Kumar et al. 2020).

3.5 Detoxification of Heavy Metals

Heavy metal-resistant bacteria are gaining importance because of their prospective use in bioremediation. *Bacillus* spp. are one of the promising bacteria that are potential in detoxification of heavy toxic metals from soil and water bodies (Table 3.4). The negatively charged carbohydrates and protein moiety of the extracellular polymeric substances (EPS) of *B. licheniformis* strain KX657843 are capable of binding with heavy metals [Cu (II) and Zn (II)] through electrostatic interactions. This bacterial EPS can perform better at high pH. It is documented that, at pH 8, the EPS was able to remove 86 and 81% Cu (II) and Zn (II), respectively, from a metal solution (Biswas et al. 2020). Zheng et al. (2008) reported that *B. megaterium* strain causes the highest flocculation at pH 9, and the pH range for flocculation is 7–12. The *Bacillus* sp. S3 considerably enhances the adsorption and detoxification capacity of heavy metals [Sb (III), Cu (II), Cr (VI), Cd (II), Cr (VI) and Cu (II)] by the secretion of EPS (Zeng et al. 2019). *B. cereus* AJK3 is a potential natural renewable source for bacteriogenic synthesis of nanoparticle. This nanoparticle reduced selenite

Table 3.3 List of enzymes produced from *Bacillus* sp. and their role

Species and strain	Produced enzymes	Activities of the enzymes	References
Bacillus amyloliquefaciens and *B. subtilis*	Esterase, endoglucanase, β-glucanase, pectatelyase	Degrade fungal cell wall	Dutta et al. (2018)
Bacillus sp. VITL8	Lipase	Pretreatment of food industry effluents	Balaji et al. (2020)
B. australimaris P5	Xylanase	Prebleaching of bamboo pulp fiber treatment	Dutta et al. (2020)
Bacillus sp. MD-5	Chitosanase (Csn-BAC)	Good candidate for oligosaccharide production-based industries	Yang et al. (2020a)
Bacillus sp.	Xylanase	Degrade xylan into usable products such as xylooligosaccharides, xylotetrose, xylotriose, xylobiose, and xylose	Rosli et al. (2020)
Bacillus sp. Khoz1	Protease	Act as detergents, cheese-making, baking, and meat tenderization beside medical applications	Far et al. (2020)
Bacillus sp.	Protease, amylase and cellulase	Degrade alfatoxin	Pereyra et al. (2020)
Bacillus sp. CSB55	β-glucanase	Potential bio-industrial catalyst	Regmi et al. (2019)
Bacillussubtilis US191	β-mannanase	Potential candidate for poultry feed supplement	Blibech et al. (2020)
Bacillus sp. RN1	Pectate lyases	Effective in degumming ramie fibers	Zheng et al. (2020)
Bacillus sp. NR5 UPM	β-cyclodextrin glycosyltransferase	Catalyzes transglycosylation reactions, useful for food, agriculture and pharmaceutical industries	Nik-Pa et al. (2020)
Bacillus sp.	Mannanase	Degradation of mannan, important for shrimp farming	Sumardi et al. (2020)
Bacillus sp.	Keratinolytic enzymes	Chicken feather degradation	Nnolim et al. (2020)

accumulation and remove selenite from the contaminated environment through bioremediation process (Kora 2018). Li et al. (2020) found that exogenous sulphur stress increases the chemical composition of EPS of *B. vallismortis* sp. and it doubled at 20 mg/L concentration of Na_2S. This increased EPS concentration improves the adsorption capacity of Cu (II) (Li et al. 2020). The lipopeptides of *B. subtilis* form complexes and chelate with Pb^{2+} through -OH, C-O, O=C-O, and -NH- bonding. It follows pseudo-first-order model and adsorbs Pb^{2+} from water (Zhao et al. 2020). *Bacillus* sp. TZ5 immobilizes cadmium from the plant and reduces cadmium concentration in the plant. Some of the *Bacillus* spp. also reduces cadmium from the soil, increase microbial count and enzymatic activities. Therefore, they can improve soil health significantly (Ma et al. 2020).

Table 3.4 List of heavy metal detoxifying *Bacillus* spp. with their mode of action

Species and strain	Source of isolation	Contaminant heavy metal	Mode of action	Reference
B. subtilis	Brassica juncea (Indian Mustard)	Ni	Ni accumulation	Abou-Shanab et al. (2018); Chaudhary and Shukla (2019)
B. licheniformis strain KX657843	Earthworm	Cu (II) and Zn (II)	Sorption of Cu (II) and Zn (II) and in flocculation	Biswas et al. (2020)
Bacillus sp. S3	Soil from Xikuangshan, China	Sb (III), Cu (II), Cr (VI), Cd (II), Cr (VI) and Cu (II)	Stimulate secretion of EPS	Zeng et al. (2019)
B. cereus RC-1	–	Cd (II)	Absorb Cd (II) by bacterial cell membrane	Huang et al. (2013)
Bacillus sp.	Eloor-Edayar industrial area. Kerala, India	Pb	Bioaccumulation of Pb	Varghese et al. (2012)
Bacillus sp.	Eloor-Edayar industrial area, Kerala, India	Zn	Accumulation of Zn	Krishna et al. (2012)
B. cereus AJK3	Pedda cheruvu, India	Selenite	Bioremediation of selenite	Kora (2018)
B. cereus	Gut of C. gibelio	Cadmium	Reduce cadmium toxicity	Wang et al. (2020a)
B. vallismortis	Lijiao Sewage Treatment Plant, Guangzhou	Cu (II)	Adsorb Cu (II)	Li et al. (2020)
Bacillus sp. TZ5	Agricultural soil, Guangyuan, China	Cd	Immobilize cadmium	Ma et al. (2020)
B. licheniformis	–	Cr (VI)	Reduce Chromium toxicity	Kavitha et al. (2011)
Bacillus sp. strain SG1	–	Mn (II)	Oxidize Mn (II) to MnO_2	Mondol et al. (2013)

3.6 *Bacillus* Strains as a Source of Bioactive Compounds

3.6.1 *Antimicrobial Compounds*

Bacillus species are one of the largest sources of bioactive compounds, which exert a wide range of antimicrobial activities (Table 3.5). They produce low molecular weight polypeptides by non-ribosomal or ribosomal synthetic mechanisms. Bacteriocins and bacteriocin-like inhibitory substances (BLIS) are ribosomally synthesized peptides produced by *B. thuringiensis, B. subtilis, B. stearothermophilus,*

Table 3.5 List of antimicrobial activities shown by *Bacillus* spp. metabolites

Species and strain	Source of isolation	Compound(s) produced	Tested pathogenic organism	Reference
Bacillus amyloliquefaciens and *B. subtilis*	Rice seeds	Bacillaene, difficidin, macrolactin, bacilysin, plipastatin, bacillibactin	*Magnaporthe oryzae*	Dutta et al. (2018) and Surovy et al. (2017)
B. subtilis	Gageocho reef, Republic of Korea	Linear lipopeptides	*M. oryzae*	Chakraborty et al. (2020a)
B. velezensis	Aquaculture system	Anti-vibrio substances	*Vibrio* sp.	Gao et al. (2017)
B. licheniformis	Mussel	Anti-microbial compounds	Distrupt initial adhesion of *V. harveyi* and *P. aeruginosa*	Hamza et al. (2015)
B. pumilus	Sponge	Nonribosomal peptides	Antagonistic against *B. cereus*	Matobole et al. (2017)
B. pumilus	Collection of marine bacteria	Pumilacidin and surfactin	Antagonistic against *Listeria monocytogenes* and *S. aureus*	Saggese et al. (2018)
B. velezensis HC6	Intestinal tract of healthy piglet	Produce iturin, fengycin, and surfactin	Multiple pathogenic fungus of maize and reduce aflatoxin, ochratoxin	Liu et al. (2019)
B. amyloliquefaciens MTCC 12716	Gulf of Mannar region, India	Polyketides panned macrolides	Bactericidal activity against *Staphylococcus aureus, Enterococcus faecalis, P. aeruginosa and Klebsiella pneumoniae*	Kizhakkekalam et al. (2020)
Bacillus subsp. *subtilis*. strain A52	Marine sediment	Sublancin, surfactin	*Candida* sp.	Sharma et al. (2020)

Bacillus strain Bam22	Rhizosphere of oilseed rape	Cyclic lipopeptide antibiotics	*S. sclerotiorum*	Yang et al. (2020b)
B. velezensis BM21	Corn rhizosphere soil	Cyclic lipopeptide	*F. graminearum* strain YJH2	Wang et al. (2020b)
B. licheniformis W10	Plant rhizosphere	Serine protease	*B. cinerae*	Ji et al. (2020)
B. siamensis CBSMS07	Kimchi	Antimicrobial peptide	*E. coli, Alcaligenes faecalis*, and *P. aeruginosa*	Khan et al. (2020a)
B. velezensis strain AL7	Cotton soil	Antifungal antibiotics	*Verticillium dahlia* Kleb	Liu et al. (2020a)
B. subtilis	Rhizosphere soil	Surfactin	*Mucor* sp. and *Aspergillus niger*	Meena et al. (2020)
B. velezensis Ll-9	Bulbs of *Lilium leucanthum*	Diketopiperazines, cyclo-peptides, linear peptides, latrunculin A, 5α-hydroxy-6-ketocholesterol, (R)-S-lactoylglutathione, triamterene, rubiadin, moxifloxacin, 9-hydroxy-5Z, 7E, 11Z, 14Z-eicosatetraenoic acid, D-erythro-C18-sphingosine, citrinin, and 2-arachidonoyllysophosphatidylcholine.	*Botryosphaeria dothidea, F. oxysporum, B. cinerea* and *F. fujikuroi*	Khan et al. (2020b)
B. subtilis (SSL2)	Root of *S. surattense*	Citrulline, chloramphenicol and carnitine	*R. solanacearum* and *F. oxysporum*	Jinal and Amaresan (2019)
B. amyloliquefaciens	–	Cis-cyclo	*Candida albicans*	Adibi et al. (2017)
Bacillus sp. Sh10	–	Bacteriocin	*Proteus mirabilis*	Shayesteh et al. (2020)
B. tequilensis SDS21	Endo-rhizosphere of *Parthenium hysterophorus*	Surfactin	Planktonic cells	Singh and Sharma (2020)
B. subtilis	Gut of *Penaeus indicus*	Bacteriocin	*Pseudomonas* sp.	Kim et al. (2020)
B. safensis B21	*Osmanthus fragrans* Lour. Fruits	Iturin A2 and A6	*Magnaporthe oryzae*	Rong et al. (2020)

B lichenifirmis, B. megaterium, and *B. cereus*. Lanthionine and β-methyl lanthionine residues are post-translationally modified peptides. The lantibiotics and subtilin are single-peptide antibiotics with strong antibacterial activity, also produced by *Bacillus* species. Subtilin is a type of bacteriocin which is structurally similar to nisin. Subtilin is used as food preservatives and capable of inhibiting a broad range of Gram-positive bacteria (Gálvez et al. 2007). Fuchs et al. (2011) isolated entianin from *B. subtilis* subsp. *spizizenii* DSM 15029, which is potential in controlling *S. aureus, E. faecalis*, and other Gram-positive bacterial pathogens. The spore-forming *B. amyloliquefaciens* subsp. *plantarum* is able to enhance the yield of crop plants (plant growth promotion) by suppressing plant pathogens (biocontrol activity) (Borriss et al. 2011). Gene expression profiling of *B. subtilis* CBR05 induced vitamin B6 (VitB6) biosynthesis and salvage pathway against *Xanthomonas campestris* pv. *Vesicatoria* in tomato (Chandrasekaran et al. 2019). The *Bacillus* sp. Fcl1 inhibits a wide range of phytopathogens by producing iturin A and a blend of surfactin compounds (Jayakumar et al. 2018). *B. velezensis* F21 helps in the expression of numerous transcriptomic factors triggering ISR in watermelon plants and reduces the incidence of Fusarium wilt in watermelon (Jiang et al. 2019). *B. subtilis* NCIB 3610 shows antiviral, antimicrobial, antifungal, and cytotoxic activities. The 2, 6-di-t-butyl-4-methyl phenol, intracellular polysaccharide fraction (FC1), and 14-methyl hexadecanoic methyl ester are produced by this bacterium, and its intracellular polysaccharide (FC1) displays a promising antibacterial activity against *S. aureus* and *S. pneumoniae* (Matloub et al. 2019). A consortium of *B. subtilis* QST-713 and *B. pumilus* is effective against Asian soybean rust. It reduced 23% disease severity in field conditions (Dorighello et al. 2015). A cyclic polypeptide characterized from n-butanol extract of *B. megaterium* (KC246043.1) shows a wide range of antimicrobial activity against Gram-positive and Gram-negative bacteria. The lowest inhibitory concentrations were 0.25, 0.5, 1.0, 3.125, and 6.25 μg/ml against *M. luteus* ATCC10240, *S. typhi* ATCC19430, *E. coli* ATCC35218, *P. aeruginosa* ATCC27853, and *S. aureus* ATCC25923, respectively (Al-Thubiani et al. 2018). The *B. methylotrophicus* DR-08 also exhibits strong antibacterial activity against *R. solanacearum*. It produces two antibacterial metabolites: difficidin and oxydifficidin (Im et al. 2020). *B. amyloliquifaciens* EZ1509 produces surfactin, iturin, fenzycin, and bacilycin. These metabolites strongly inhibit the growth of *S. sclerotiorum*. These lipopeptides modulate the cell integration and cause shrinkage, plasmolysis, and breakdown of fungal hyphae through downregulation of endopolygalacturonase-3, oxalic acid hydrolase, and endopolygalacturonase-6 resistance genes of *S. sclerotiorum* (Farzand et al. 2020). *B. amyloliquifaciens* EZ1509 also secretes potential antibacterial material ß-1, 3–1, 4-glucanase and contains genes encoding *LCI* (47 residues cationic antimicrobial peptide), yellowfin tuna GAPDH-related antimicrobial peptide, and human GAPDH-related AMPs (Zhang et al. 2020).

3.6.2 Insecticidal Compounds

The use of bacterial agents to control insect pests is considered as an eco-friendly and safe approach to increase crop production (Dihazi et al. 2012). Application of plant-beneficial *Bacillus* spp. enhances the sustainable production of diverse crops in modern agriculture with minimum use of chemical fertilizers and pesticides (Myresiotis et al. 2014). Some strains of *Bacillus* spp. can kill the larvae of the insect pest and also enhance induced systemic resistance (ISR) in crop plants (Table 3.6). Maize root colonized with *B. subtilis* can be able to uptake insecticide (thiamethoxam) and control pest infestations (Myresiotis et al. 2014). However, the mechanism of *Bacillus*-induced insect control in crops varies with insect species as well as the crop varieties (Navon 2000; Paramasiva et al. 2014; Mnif and Ghribi 2015; Wielkopolan and Obrępalska-Stęplowska 2016). Generally, they colonize in plant organs, including the phyllosphere, and the larvae and/or adult insects are exposed to bacteria during ingestion of the *Bacillus*-colonized plant tissues during feeding. A well-known bio-insecticide, *B. thuringiensis*, can control a broad range of diverse lepidopteran insects and thus useful for pest management in the agricultural field (Navon 2000) without affecting other microbial organisms within the phyllosphere (Wang et al. 2014). Generally, the primary site of bacterial antagonism begins with a severe damage to the larval midgut epithelium by the bacterial crystal proteins, which interact with chitin and peritrophic membranes (Vachon et al. 2012; Feng et al. 2015). In later stages, the crystal protein endotoxin, lipopeptides, and polyketides (fengycin, macrolactin, iturin, surfactin, bacillomycin, bacillaene, and difficidin) modify the vacuolization of the cytoplasm, induce vesicle formation, lyse brush border membrane, and degenerate apical membranes, leading to damage of microvilli and finally resulting in the death of insect larvae (Khedher et al. 2015; Boukedi et al. 2016). Surfactin attaches to the Ca^{2+} receptor site and changes the peptide composition in the cellular phospholipid bilayer (Maget-Dana and Ptak 1995), while iturin increases cell membrane permeability via the formation of ion-conducting pores (Maget-Dana and Peypoux 1994). *Bacillus* spp. trigger the expression of the jasmonic acid (JA) pathway-related genes and simultaneously increase the gene expression for other secondary metabolites (allelochemicals, which inhibit the insect larval growth) in plants to defend against insects (Zebelo et al. 2016). A large body of literature suggests that *Bacillus* sp. controls the larval population of insects and triggers the ISR mechanism and allelochemicals in plants to prevent insect damage.

A protein elicitor, AMEP412, from *B. subtilis* can trigger plant defense and induce acquired systemic resistance. When AMEP412 protein is ingested by white fly (*Bemisia tabaci*), it gets digested by gut proteases and releases hydrophobic peptides in the white fly guts. This causes pores in guts and ultimately causes insect death (Liu et al. 2020b). Recombinant *B. subtilis* 26DCryChS showed aphicidal activity against green bug (*Schizaphis graminum*). Insertion of *Btcry11a* gene in *B. subtilis* 26D genome contributed higher aphicidal activity compared to wild-type *B. thuringiensis* B-5351 and *B. subtilis* 26D. These mutant *Bacillus* species

Table 3.6 List of insecticidal *Bacillus* sp. with their mode of action

Species and strain	Tested insect	LC_{50}/LD_{50} /Doses	Mode of action	Reference
B. thuringiensis	Larvae of mosquito	Highest mortality was found after 4 h exposure of 5 ml suspension of *Bt.*	Secretion of parasporal crystals with high polypeptide toxin	Nkiru and Franklin (2020)
B. thuringiensis var. kurstaki	Larvae of *Drosophila melanogaster*	5×10^7 to 1×10^9 Colony Forming Units (CFU) of *Btk.* spores/g of fly food medium	Disturb homeostasis of the larval intestine	Nawrot-Esposito et al. (2020)
B. subtilis, B. amyloliquefaciens and *B. thuringiensis var. kurstaki*	Tobacco whitefly (*Bemisia tabaci*)	–	Plant protection by antibiosis and/or ISR elicitation, parasporal inclusions produced during sporulation	Arabiat et al. (2018)
B. subtilis	Whitefly (Bemisia tabaci)	15.57 µg/ml AMEP412 protein	Causes gut cell lysis	Liu et al. (2020b)
B. subtilis 26DCryChS	*S. graminum*	Stimulate *TapI*, *TaChiI* genes and induce *Tapr1*, *TaPrx* genes	Inhibit the growth of green bug and aphid in wheat plants	Maksimov et al. (2020)
B. subtilis V26	*Tuta absoluta* larvae	278.78 ng/cm² surfactin	Causes histological damage of the larval midgut	Khedher et al. (2020)
B. amyloliquefaciens	*Tuta absoluta* larvae	180 ng/cm² surfactin	Damage larval midgut	Ben Khedher et al. (2015)
B. thuringiensis subsp. *israelensis*	Mosquito larvae	–	Damage larval gut	Dahmana et al. (2020)
B. clausii DTM1	Aphid and mosquito larvae	–	Causes cell lysis	Guo et al. (2015)
B. clausii BS02	Pulse beetle and mealybug	–	Causes cell lysis	Hazra et al. (2015)
B. amyloliquefaciens, *B. subtilis*, and *B. pumilus*	Silver whitefly (*Bremisia argentifolii*)	1×10^7 CFU/g of seed or planting medium	–	Murphy et al. (2000)
B. thuringiensis	*Spodoptera frugiperda* and *S. exigua*	–	Binding of *B. thuringiensis* toxins to the *Spodoptera* midgut	Herrero et al. (2016)

(continued)

Table 3.6 (continued)

Species and strain	Tested insect	LC_{50}/LD_{50} /Doses	Mode of action	Reference
B. thuringiensis and *B. sphaericus*	Mosquito larvae (*Culex pipiens* L.)	LC_{50} = for *B. thuringiensis* 5.49 × 10^6 spore/ml. and for *B. sphaericus* 6.70 × 10^6 spore/ml	–	Kelada and Shaker (1988)
B. cereus, B. subtilis, and *B. amyloliquefaciens*	Phloem-feeding aphid (*Brevicoryne brassicae*)	1 × 10^8 cfu/ml suspension	–	Gadhave and Gange (2015)
B. thuringiensis	Spotted stem borer (*Chilo partellus*)	2 × 10^7 spores/ml	–	Brownbridge (2001)
B. thuringiensis	Rice water weevil (*Oryzophagus oryzae*) and fall armyworm (*Spodoptera frugiperda*)	LC_{50} = for *O. oryzae* 5.40 μg/mL and for *S. frugiperda* 1 × 10^{10} CFU/mL resulting 100% mortality	–	Berlitz et al. (2012)
B. thuringiensis subsp. *kurstaki*	Second instar larvae of *Spodoptera littoralis*	*B. thuringiensis* spore-crystal mixture corresponding to 100 mg of delta-endotoxin	Delta-endotoxins that form the crystal of *B. thuringiensis kurstaki* in addition to *Cry1Aa* and *Cry2Aa*	Benfarhat-Touzri et al. (2013)
B. thuringiensis	Larvae of (*Helicoverpa armigera*)	Concentrations 10^9–10^{11} occlusion bodies (OBs) L^{-1} of *H. armigera* NPV resulting in 89–100% mortality	–	Arrizubieta et al. (2015)

synthesize surfactin and can attack green bug. Treatment of wheat seeds with this recombinant *Bacillus* species also promoted greater resistance against aphid in wheat plants. It stimulates the transcription of jasmonate-sensitive genes (*TapI* and *TaChiI*) and induces salicylate-dependent genes (*Tapr1* and *TaPrx*) (Maksimov et al. 2020). A complex of *B. cereus* contains *P19* and *P20* genes that are able to control mosquito larvae (Dahmana et al. 2020). African catfish diet supplemented with *Thymus vulgaris* (TV) essential oil and *B. subtilis* reduces the possible risk of thiamethoxam (TMX) insecticide. When this insecticide is applied in a waterbody to control fish insects, it causes gill, liver, head, and kidney necrosis of the African catfish (*Clarias garipenus*). Feeding with TV essential oil and *B. subtilis* significantly decrease splenocyte proliferation and percentage of melanomacrophage centers as well as increase the caspase-3 immunopositive cells. *B. subtilis* reduces lead toxicity by bioaccumulation of lead and increases the immunity of catfish (Euony et al. 2020). *Bacillus* spp. are potential alternatives to synthetic insecticides for controlling pests through an integrated pest management system.

3.6.3 Antinematicidal Compounds

Use of *Bacillus* spp. as biocontrol agents are safer for the environment and non-target organisms (humans, animals and other soil-dwelling organisms) (Gao et al. 2016). Under aerobic conditions, they form highly resistant endospores that are capable to survive in the environment for a longer period (Ongena and Jacques 2008; Padgham and Sikora 2007; Sansinenea and Ortiz 2011). These dormant spores have elevated levels of tolerance against extreme environmental conditions, such as high temperatures, nutrient and water deficiencies and unfavorable pH levels (Cawoy et al. 2011). They can colonize into plant roots (Jamal et al. 2017) and ubiquitous within rhizosphere (Sansinenea and Ortiz 2011). They promote plant growth (Abbasi et al. 2013; Lee and Kim 2015) and possess distinctive attributes as to control plant pathogenic nematodes (Table 3.7). They produce biologically active molecules (Moghaddam et al. 2014b) such as antibiotics, enzymes, exotoxins and other metabolites with nematicidal activity (Sansinenea and Ortiz 2011; Abbasi et al. 2013). From a technological viewpoint, the ability of *Bacillus* to produce heat and desiccation-resistant endospores is highly advantageous for the formulation of a stable product (Emmert and Handelsman 1999; Ongena and Jacques 2008) which has been used commercially under natural field conditions (Padgham and Sikora 2007). The endospores of nematicidal *Bacillus* spp. also contribute to the stability of bio-nematicides during storage (Rosas-Garcia 2009).

B. firmus is widely used as a bionematicidal agent (Tian et al. 2007; Wilson and Jackson 2013). It causes paralysis and mortality in the plant pathogenic nematode, *M. incognita*. It also controls burrowing nematode (*Radopholus similis*) and stem nematode (*Ditylenchus dipsaci*) (Mendoza et al. 2008). *B. firmus* reduces the infestation of sting nematode (*Belonolaimus longicaudatus*) on Bermuda grass (Crow 2014). Extracellular protease of *B. cereus* BCM2 has significant nematicidal activity against second-stage juveniles (J2) of *M. incognita.* It absorbs by the nematode body and degrade the body wall. The bacterial colonization into the plant induces systemic resistance in plants and expresses nematode defense-related genes (Hu et al. 2020a). Serine proteases and neutral protease from *B. nematocida* directly degrade the main chemical compositions of nematode cuticles and eggshell (Niu et al. 2006, 2011). However, the safety of using *Bacillus* spp. in bio-nematicides is evidenced by the US Food and Drug Administration (USFDA) granting *Bacillus* sp. the status of "generally regarded as safe" (Usta 2013). Therefore, *Bacillus* spp. are ideal bacteria for the biocontrol of nematodes. Discovery of novel species and strains of *Bacillus* spp. from diverse environments should be considered as a promising research approach for finding nematicidal bacteria with a novel mode of action.

Table 3.7 List of various nematicidal *Bacillus* spp. with their mechanisms of action

Species and strain	Tested nematode	Mode of action	Reference
B. cereus	*Meloidogyne incognita*	Produces extra cellular protease, sphingosine (a nematicidal toxin) and helps the metabolites	Hu et al. (2020a), Gao et al. (2016) and Xiao et al. (2018)
B. pumilus and *B. cereus*	*M. javanica*	Production of proteolytic enzyme	Moghaddam et al. (2014a, b)
B. firmus	*M. incognita*	Production of bioactive secondary metabolites	Jansen-Girgan et al. (2016) and Xiong et al. (2015)
B. *subtilis, B. firmus* and *B. coagulans*	*M. javanica*	Increases the enzymatic activity like superoxide dismutase, ascorbate peroxide, Guaiacol peroxidase, polyphenol oxidase etc.	Abbasi et al. (2013)
B. velezensis and *B. mojavensis*	*M. incognita*	–	Xiang et al. (2017)
B. nematocida	*Panagrellus redivius* and *Bursaphelenchus xylophilus*	Combined effect of two proteolytic enzyme, neutral protease and serine protease	Niu et al. (2006)
B. thuringiensis	*M. javanica*	Form pores in the cell membrane of gut epithelial cells	Ravari and Moghaddam (2016)
B. megaterium	*M. incognita*	Nematicidal volatiles (benzeneacetaldehyde, 2-nonanone, decanal, 2-undecanone and dimethyl disulphide) produced by the bacterium	Huang et al. (2009)
B. megaterium	*M. graminicola*	Production of secondary metabolites	Padgham and Sikora (2007)
B. pumilus	*M. arenaria*	Production of crude enzymes	Lee and Kim (2015)
B. coagulans and *B. cereus*	*M. javanica*	Production of secondary metabolites	Abbasi et al. (2017)
B. methylotrophicus	*M. incognita*	–	Zhou et al. (2016)
B. amyloliquefaciens	*M. incognita*	The dipeptide cyclo (d-pro-l-Leu) act as nematocidal activity	Jamal et al. (2017)
B. cereus and *B. pumilus*	*M. javanica*	Produces extracellular protease	Moghaddam et al. (2014a)
B. licheniformis	*M. incognita*	Degrade the nematode cuticle and/or to produce toxic metabolites	Colagiero et al. (2018)

(continued)

Table 3.7 (continued)

Species and strain	Tested nematode	Mode of action	Reference
B. subtilis, B. amyloliquefaciens, B. cereus, B. pumilus	*M. incognita*	–	Sarangi et al. (2014)
B. subtilis	*M. incognita*	Production of certain enzymes i.e. protease, chitinase and gelatinase	Metwally et al. (2015)
Bacillus spp.	M. *javanica* and *Ditylenchus* spp.	–	Turatto et al. (2018)
B. firmus	*Radopholus similis, M. incognita, D. dipsaci*	Production of bioactive compounds of secondary metabolites	Mendoza et al. (2008)

3.7 *Bacillus* Isolates as a Potential Source of Natural Carotenoids

Carotenoids are widely distributed in photosynthetic microorganisms (bacteria, fungi etc.) and plants. They are usually red, orange or yellow pigments that contain a poly C_{40} carbon skeleton with one or both ends having acyclic or cyclic group backbone (Nelis and Leenheer 1991; Mondol et al. 2013). Generally, plants, fungi and algae contain C_{40} carbon skeleton and bacteria produce C_{40} or C_{30} carbon skeleton. Carotenoids from natural sources are entirely "*trans*" form. They are highly recognized for their antioxidant, pro-vitamin A and other food or feed additives properties (Mondol et al. 2013). Duc et al. (2006) provides one of the first detailed reports of pigmented marine *Bacillus* species. The most commonly found pigments were yellow, orange and pink. The most related carotenoid producing species are *B. marisflavi, B. indicus, B. firmus, B. altitudinis* and *B. safensis.* The intensity of the colour varied depending on the strain (Khaneja et al. 2009). The presence of the pink carotenoids with the main absorbance maximum at 492 nm seems to be species specific, to the *B. firmus*-related isolates. Based on reference spectra in the literature, the predominant *Bacillus* carotenoids were putatively identified to be acyclic-carotenoids and potentially monocyclic, while other reports have stated the presence of astaxanthin (Pane et al. 1996). Other studies with the coat-associated melanin of *B. subtilis* spores have shown that in absence of the spore outer coat, the resistance to UV-C was increased (Riesenman and Nicholson 2000). *B. indicus* HU36 and *B. firmus* GB1 produce useful carotenoids (Fig. 3.6c) for humans. They produce glycosylated carotenoids, which is ca. 4.5 times more bioavailable than pure β-carotene (Duc et al. 2006).

3.8 Genetics and Biosynthesis Pathways

Bacillus spp. produce diverse bioactive compounds with different biosynthetic pathways. Surfactin genes in *Bacillus* species expressed with an increase of cell number. Fatty acid synthesis II system regulates fatty acid degradation or efficiency of fatty acid according to the requirement of *Bacillus* vegetative cells (Diomandé et al. 2015). Whereas biosynthesis of fengycins and iturins occur at the late stationary growth phase. Surfactin gene determinants are solely linked to the srf operon. Posttranscriptional activation of the Srf synthetase mainly depends on Spf transferease, which is encoded by *sfp* gene situated downstream of srfABCD (Steller et al. 2004; Raaijmakers et al. 2010). The surfactin and iturin of *Bacillus* spp. played a key role in the expression of chitinase and β-1, 3-glucanase in strawberry leaves (Yamamoto et al. 2015). For fengycin biosynthesis, degQ regulates ppsABCDE plipastatin operon (Tsuge et al. 2007). *Bacillus* spp. produce lipopeptides and regulate defense-related gene expression. Higher expression of the *PDF 1.2* bacterial gene encodes for defensin and induces host defense system (Chowdhury et al. 2015). Gene expression analysis also proposes that mycosubtilin (iturin family) triggered the SA and JA signaling pathways, and surfactin prompted a SA-regulated pathway in grapevine. As a result, grapevine leaves are more tolerant against *B. cinerea* (Farace et al. 2015). The expression of mycosubtilin gene in *B. subtilis* is influenced by *AbrB* transition state regulator (Duitman et al. 2007). *B. amyloliquefaciens* produces surfactin, iturin, and fengycin, which inhibit the growth of *Clostridium difficile*. The *srfABCD*, *ituABCD*, *fenABCDE*, *bmyABC*, *baeEDLMNJRS*, *difABCDEFGHIJ*, and *mlnABCDEFGHI* gene clusters of *B. amyloliquefaciens* contribute synergistically to control *C. difficile* (Lv et al. 2020). *B. amyloliquefaciens* FZB42 produces bacillomycin D. The *DegU* gene of *Bacillus* FZB42 regulates the cellular process of *Bacillus* and *bmy* operon helps in interacting with *DegU* regulator to produce bacillomycin D (Koumoutsi et al. 2007). It also contains three PKS operons (*pks1*, *pks2*, and *pks3*). These three gene clusters organize the type I PKS system. The *Pks1* and *Pks3* gene clusters produce difficidin and oxydifficide, respectively. On the other hand, *pks2* gene cluster is involved in macrolactin biosynthesis (Chen et al. 2006; Schneider et al. 2007). The production of surfactin by the *B. amyloliquefaciens* is enhanced with addition of Fe nanoparticle in the growth medium. It stimulates the key surfactin biosynthetic pathway (Yang et al. 2020c). By using genetic recombination of *B. subtilis* Symmank et al. (2002) produced lipohexapeptide for surfactin biosynthesis. In this process, deletion of individual A and T domain creates three novel plipastatin derivatives. Recombinant [Δ*Leu3*] and [Δ*Leu6*] surfactin offer reduced toxicity, and [Δ*Asp5*] surfactin displayed higher inhibition in comparison with intrinsic surfactin against *B. pumilus* and *M. luteus* (Jiang et al. 2016). The *bpsA-ypbQ* operon encodes type III polyketide synthases (PKSs) and a methyltransferase. The *bpsA* catalyzing the synthesis of acyl-coenzyme A (CoA) thioesters and malonyl-CoA and *ypbQ* involve in the biosynthesis of aliphatic polyketides (Nakano et al. 2009). The combination of *crtM* and *crtN* genes and *pHCMC04G* plasmid of *B. subtilis* carrying synthetic operon for MEP pathway produces carotenoids, diaponeurosporene, and diapolycopene (Xue et al. 2015). The hybrid PKS/NRPS gene cluster in marine *B. subtilis* TP-B0611 is responsible for the biosynthesis of amiocoumacins and bacilosarcins (Komaki et al. 2016). Surfactin A produced by *B. subtilis* upregulates the hypoxia-inducible factor-1α and vascular endothelial growth factor. These two factors accelerate keratinocyte signaling and regulate the pro-inflammatory cytokines and macrophage phenotypic switch of human body. As a result, it is used as a wound healing agent in medical science (Yan et al. 2020). Genetic engineering of *B. thuringiensis* overexpresses the *krsE* gene and produces kurskatin, which shows antimicrobial activity against *Galactomyces geotrichum* and *B. cinera* (Lereclus et al. 2017). Genetic engineering of lipopeptides synthetase polypeptide also produces higher amounts of iturin and surfactin, which effectively control insects and increase crop yield (Keenan et al. 2008). Ginsenoside *Rg3* is a bioactive compound that has notable biological properties. The final biosynthetic step of ginsenoside *Rg3* is glycosylation and glycosylation catalyzed by uridine diphosphate-dependent glycosyltransferase (*UGT*). A promiscuous *UGT Bs-YjiC* discovered from *B. subtilis* 168 and the C12 hydroxyl group of ginsenoside *Rg3* to unnatural ginsenoside *Rd12* selectively glycosylate by *Bs-Yjic* cloned from *B. subtilis* 168. Ginsenoside *Rd12* is effective against diverse cancer cell lines and might be a potential anticancer drug for cancer cell recovery (Hu et al. 2020b).

3.9 Conclusions and Future Perspectives

Bacillus species is a large group of bacteria that produces diverse classes of bioactive secondary metabolites (lipopeptides, polyketides, isocoumarins, carotenoids, fatty acids, lipoamides, etc.). These properties of *Bacillus* spp. make them an excellent option for using them as antimicrobial and biocontrol agents in agriculture. They also detoxify toxic heavy metals from the environment. Production of bioactive molecules from *Bacillus* sp. inexorably contribute for a clean and sustainable eco-friendly environment. The mechanisms of the biosynthesis of bioactive secondary metabolites, enzymes, and endotoxins by various species and strains of *Bacillus* are largely been known. Some of the genes involved in biosynthesis of these bioactive compounds are used in genetic engineering. Among them, genetically engineered Bt-crops such as cotton, corn, and brinjal not only increased productivity of these crops but also promoted green agriculture (reduced insecticide application). Discovery of the new biologically active strains of *Bacillus* spp. from diverse environment might lead to the development of new biologicals for promoting agriculture and industry. Recently developed CRISPR-Cas genome editing technology should facilitate to design commercially important microbes and plants using important genes discovered in the genomes of *Bacillus* species. To utilize the genome editing toolkit, further genomics and post-genomics studies are needed to dissect the underlying molecular mechanisms of biosynthesis of diverse bioactive compounds by the *Bacillus* species.

Acknowledgements This work is partially funded by the BAS-USDA (PALS CR-11) and KGF (TF 50-C/17). Bangladesh funded projects to MTI. The author (MZS) is thankful to the German Academy Exchange Service (DAAD) for offering PhD research fellowship under the Division of Plant Pathology and Crop Protection, Georg-August Universität Göttingen, Germany.

References

Abbasi MW, Ahmed N, Zaki MJ et al (2013) Potential of *Bacillus* species against *Meloidogyne javanica* parasitizing eggplant (*Solanum melongena* L.) and induced biochemical changes. Plant Soil 375:159–173. https://doi.org/10.1007/s11104-013-1931-6

Abbasi MW, Marium T, Khan MQ, Zaki MJ et al (2017) Assessment of extracellular metabolites from *Bacillus* species against root-knot nematodes and root-infecting fungi in *Abelmoschus esculentus* (L.) Moench. Pak J Bot 49:289–294

Abdallah RAB, Stedel C, Garagounis C et al (2017) Involvement of lipopeptide antibiotics and chitinase genes and induction of host defense in suppression of Fusarium wilt by endophytic *Bacillus* spp. in tomato. Crop Prot 99:45–58. https://doi.org/10.1016/j.cropro.2017.05.008

Abou-Shanab R, El-Sheekh M, Sadowsky M (2018) Role of rhizobacteria in phytoremediation of metal-impacted sites. In: Bharagava RN, Chowdhary P (eds) Emerging and eco-friendly approaches for waste management. Springer, Singapore, pp 299–328. https://doi.org/10.1007/978-981-10-8669-4_14

Adibi A, Rees ER, Mccarley S et al (2017) Characterization and isolation of peptide metabolites of an antifungal bacterial isolate identified as *Bacillus amyloliquefaciens* subspecies *plantarum* strain Fzb42. J Microbiol Biotechnol Food Sci 6:1309–1313. https://doi.org/10.15414/jmbfs.2017.6.6.1309-1313

Al-Thubiani AS, Maher YA, Fathi A, Abourehab MA, Alarjah M, Khan MS, Ghamdi SB (2018) Identification and characterization of a novel antimicrobial peptide compound produced by *Bacillus megaterium* strain isolated from oral microflora. Saudi Pharm J *26*(8):1089–1097. https://doi.org/10.1016/j.jsps.2018.05.019

Ansary MWR, Prince MFRK, Haque E, Sultana F, West HM, Rahman et al (2018) Endophytic *Bacillus* spp. from medicinal plants inhibit mycelial growth of *Sclerotinia sclerotiorum* and promote plant growth. Zeitschrift fur Naturforschung C 73:5–6. https://doi.org/10.1515/znc-2018-0002

Arabiat OWA, Araj SEA, Alananbeh KM, Antary TMA (2018) Effect of three *Bacillus* Spp. on tobacco whitefly *Bemisia tabaci* (Gennadius) (Homoptera: Aleyrodida). Fresenius Environ Bull 27:3706–3712

Arrizubieta M, Simón O, Torres-Vila LM et al (2015) Insecticidal efficacy and persistence of a co-occluded binary mixture of *Helicoverpa armigera* nucleopolyhedrovirus (HearNPV) variants in protected and field-grown tomato crops on the *Iberian Peninsula*. Pest Manag Sci 72:660–670. https://doi.org/10.1002/ps.4035

Asari S, Ongena M, Debois D et al (2017) Insights into the molecular basis of biocontrol of Brassica pathogens by *Bacillus amyloliquefaciens* UCMB5113 lipopeptides. Ann Bot 120:551–562. https://doi.org/10.1093/aob/mcx089

Azumi M, Ogawa K-I, Fujita T et al (2008) Bacilosarcins A and B, novel bioactive isocoumarins with unusual heterocyclic cores from the marine-derived bacterium *Bacillus subtilis*. Tetrahedron 64:6420–6425. https://doi.org/10.1016/j.tet.2008.04.076

Balaji L, Chittoor JT, Jayaraman G (2020) Optimization of extracellular lipase production by halotolerant *Bacillus* sp. VITL8 using factorial design and applicability of enzyme in pretreatment of food industry effluents. Prep Biochem Biotechnol 50:708–716. https://doi.org/10.1080/10826068.2020.1734936

Barros FFC, Ponezi AN, Pastore GM (2008) Production of biosurfactant by *Bacillus subtilis* LB5a on a pilot scale using cassava wastewater as substrate. J Ind Microbiol Biotechnol 35:1071–1078. https://doi.org/10.1007/s10295-008-0385-y

Barsby T, Kelly MT, Andersen RJ (2002) Tupuseleiamides and basiliskamides, new acyldipeptides and antifungal polyketides produced in culture by a *Bacillus laterosporus* isolate obtained from a tropical marine habitat. J Nat Prod 65:1447–1451. https://doi.org/10.1021/np0201321

Ben Khedher S, Boukedi H, Kilani-Feki O, Chaib I, Laarif A, Abdelkefi-Mesrati L, Tounsi S (2015) *Bacillus amyloliquefaciens* AG1 biosurfactant: histopathological effects and diversity of its putative receptors on *Tuta absoluta* midgut. J Invertbr Pathol 132:42–47

Benfarhat-Touzri D, Amira AB, Khedher SB et al (2013) Combinatorial effect of *Bacillus thuringiensis kurstaki* and *Photorhabdus luminescens* against *Spodoptera littoralis* (Lepidoptera: Noctuidae). J Basic Microbiol 54:1160–1165. https://doi.org/10.1002/jobm.201300142

Bergey DH, Buchanan RE, Gibbons NE (1975) Bergey's manual of determinative bacteriology. The Williams & Wilkins Company, Baltimore

Berlitz DL, Azambuja AOD, Sebben A et al (2012) Mortality of *Oryzophagus oryzae* (Costa Lima, 1936) (Coleoptera: Curculionidae) and *Spodoptera frugiperda* (J E Smith, 1797) (Lepidoptera: Noctuidae) larvae exposed to *Bacillus thuringiensis* and extracts of *Melia azedarach*. Braz Arch Biol Technol 55:725–731. https://doi.org/10.1590/s1516-89132012000500012

Bernard EI, Stanley IR, Grace OI, Ebere PA, Abraham OA, Ibe KE (2018) Toxicity and bioremediation of heavy metals contaminated ecosystem from tannery wastewater: a review. J Toxicol. https://doi.org/10.1155/2018/2568038

Bhardwaj A, Chaman S, Verma S (2017) Production of antibacterial agent from fungi isolated from pharmaceutical soil sample by fermentation under optimized conditions. Asian J Pharm Clin Res 10:110. https://doi.org/10.22159/ajpcr.2017.v10i7.18258

Biswas JK, Banerjee A, Sarkar B et al (2020) Exploration of an extracellular polymeric substance from earthworm gut bacterium (*Bacillus licheniformis*) for bioflocculation and heavy metal removal potential. Appl Sci 10:349. https://doi.org/10.3390/app10010349

Blibech M, Farhat-Khemakhem A, Kriaa M, Aslouj R, Boukhris I, Alghamdi OA, Chouayekh H (2020) Review for "Optimization of β-mannanase production by *Bacillus subtilis* US191 using economical agricultural substrates". https://doi.org/10.1002/btpr.2989/v2/review1

Bonmatin J-M, Laprévote O, Peypoux F (2003) Diversity among microbial cyclic lipopeptides: iturins and surfactins. Activity-structure relationships to design new bioactive agents. Comb Chem High Throughput Screen 6:541–556. https://doi.org/10.2174/138620703106298716

Borriss R (2020) Phytostimulation and biocontrol by the plant-associated *Bacillus amyloliquefaciens* FZB42: an update. In: Kumar M, Kumar V, Prasad R (eds) Phyto-microbiome in stress regulation, Environmental and microbial biotechnology. Springer, Singapore, pp 1–20. https://doi.org/10.1007/978-981-15-2576-6_1

Borriss R, Chen X-H, Rueckert C et al (2011) Relationship of *Bacillus amyloliquefaciens* clades associated with strains DSM 7T and FZB42T: a proposal for *Bacillus amyloliquefaciens* subsp. *amyloliquefaciens* subsp. nov. and *Bacillus amyloliquefaciens* subsp. *plantarum* subsp. *nov.* based on complete genome sequence comparisons. Int J Syst Evol Microbiol 61:1786–1801. https://doi.org/10.1099/ijs.0.023267-0

Boukedi H, Sellami S, Ktari S et al (2016) Isolation and characterization of a new *Bacillus thuringiensis* strain with a promising toxicity against Lepidopteran pests. Microbiol Res 186-187:9–15. https://doi.org/10.1016/j.micres.2016.02.004

Brownbridge M (2001) Greenhouse trials of new *Bacillus thuringiensis* isolates for control of Chilo partellus larvae in sorghum. Phytoparasitica 29:292–298. https://doi.org/10.1007/bf02981845

Carballeira NM, Miranda C, Lozano CM et al (2001) Characterization of novel methyl-branched chain fatty acids from a halophilic *Bacillus* species. J Nat Prod 64:256–259. https://doi.org/10.1021/np000494d

Cawoy H, Bettiol W, Fickers P, Onge M (2011) Bacillus-based biological control of plant diseases. In: Stoytcheva (ed) Pesticides in the modern world – pesticides use and management. InTech Academic Press, London, pp 273–302. https://doi.org/10.5772/17184

Cawoy H, Debois D, Franzil L et al (2015) Lipopeptides as main ingre-dients for inhibition of fungal phytopathogens by *Bacillus subtilis/amyloliquefaciens*. Microb Biotechnol 8:281–295. https://doi.org/10.1111/1751-7915.12238

Chakraborty M, Mahmud NU, Gupta DR et al (2020a) Inhibitory effects of linear lipopeptides from a marine *Bacillus subtilis* on the wheat blast fungus *Magnaporthe oryzae Triticum*. Front Microbiol. https://doi.org/10.3389/fmicb.2020.00665

Chakraborty M, Mahmud NU, Muzahid ANM, Rabby SMF, Islam T (2020b) Oligomycin inhibit *Magnaporthe oryzae Triticum* and suppress wheat blast disease. PLoS ONE 15(8):e0233665. https://doi.org/10.1371/journal.pone.0233665

Chakraborty K, Thilakan B, Chakraborty RD et al (2016) O-heterocyclic derivatives with antibacterial properties from marine bacterium Bacillus subtilis associated with seaweed, Sargassum myriocystum. Appl Microbiol Biotechnol 101:569–583. https://doi.org/10.1007/s00253-016-7810-3

Chakraborty K, Thilakan B, Raola VK (2017) Antimicrobial polyketide furanoterpenoids from seaweed-associated heterotrophic bacterium Bacillus subtilis MTCC 10403. Phytochemistry 142:112–125. https://doi.org/10.1016/j.phytochem.2017.06.019

Chandler S, Van HN, Coutte F et al (2015) Role of cyclic lipopeptides produced by Bacillus subtilis in mounting induced immunity in rice (Oryza sativa L.). Physiol Mol Plant Pathol 91:20–30. https://doi.org/10.1016/J.PMPP.2015.05.010

Chandrasekaran M, Paramasivan M, Chun S (2019) Bacillus subtilis CBR05 induces Vitamin B6 biosynthesis in tomato through the de novo pathway in contributing disease resistance against Xanthomonas campestris pv. vesicatoria. Scientific Reports. https://doi.org/10.1038/s41598-019-41888-6

Chaudhary T, Shukla P (2019) Bioinoculants for Bioremediation applications and disease resistance: innovative perspectives. Indian J Microbiol 59:129–136. https://doi.org/10.1007/s12088-019-00783-4

Chen X, Zhang Y, Fu X et al (2016) Isolation and characterization of Bacillus amyloliquefaciens PG12 for the biological control of apple ring rot. Postharvest Biol Technol 115:113–121. https://doi.org/10.1016/j.postharvbio.2015.12.021

Chen XH, Koumoutsi A, Scholz R, Eisenreich A, Schneider K et al (2007) Comparative analysis of the complete genome sequence of the plant growth-promoting bacterium *Bacillus amyloliquefaciens* FZB42. Nat Biotechnol 25:1007–1014

Chen X-H, Vater J, Jörn P et al (2006) Structural and functional characterization of three-polyketide synthase gene clusters in *Bacillus amyloliquefaciens* FZB 42. J Bacteriol 188:4024–4036. https://doi.org/10.1128/jb.00052-06

Chowdhury SP, Uhl J, Grosch R et al (2015) Cyclic Lipopeptides of Bacillus amyloliquefaciens subsp. plantarum colonizing the lettuce rhizosphere enhance plant defense responses toward the bottom rot pathogen Rhizoctonia solani. Mol Plant-Microbe Interact 28:984–995. https://doi.org/10.1094/MPMI-03-15-0066-R

Colagiero M, Rosso LC, Ciancio A (2018) Diversity and biocontrol potential of bacterial consortia associated to root-knot nematodes. Biol Control 120:11–16. https://doi.org/10.1016/j.biocontrol.2017.07.010

Cote CK, Heffron JD, Bozue JA, Welkos SL (2015) Bacillus anthracis and other Bacillus species. Mol Med Microbiol:1789–1844. https://doi.org/10.1016/b978-0-12-397169-2.00102-5

Coutte F, Leccuturier D, Leclère V et al (2013) New integrated bio-process for the continuous production, extraction and purification of lipopeptides produced by Bacillus subtilis in membrane bioreactor. Process Biochem 48:25–32. https://doi.org/10.1016/j.procbio.2012.10.005

Crow WT (2014) Effects of a commercial formulation of Bacillus firmus I-1582 on golf course bermudagrass infested with Belonolaimus longicaudatus. J Nematol 46:331–335

Cutting SM (2011) Bacillus probiotics. Food Microbiol 28:214–220. https://doi.org/10.1016/j.fm.2010.03.007

Dahmana H, Raoult D, Fenollar F, Mediannikov O (2020) Insecticidal activity of bacteria from larvae breeding site with natural larvae mortality: screening of separated supernatant and pellet fractions. Pathogens 9:486. https://doi.org/10.3390/pathogens9060486

Dame ZT, Islam MT (2019) Tapping the potential of metabolomics in new natural products discovery from Bacillus species. In: Islam M, Rahman M, Pandey P, Boehme M, Haesaert G (eds) Bacilli and agrobiotechnology: phytostimulation and biocontrol, Bacilli in climate resilient agriculture and bioprospecting. Springer, Cham, pp 201–216

Dihazi A, Jaiti F, Wafataktak et al (2012) Use of two bacteria for biological control of bayoud disease caused by Fusarium oxysporum in date palm (Phoenix dactylifera L) seedlings. Plant Physiol Biochem 55:7–15. https://doi.org/10.1016/j.plaphy.2012.03.003

Dimkić I, Stanković S, Nišavić M et al (2017) The profile and anti-microbial activity of Bacillus Lipopeptide extracts of five potential biocontrol strains. Front Microbiol 8:925. https://doi.org/10.3389/fmicb.2017.00925

Diomandé SE, Nguyen-The C, Guinebretièr M-H et al (2015) Role of fatty acids in Bacillus environmental adaptation. Front Microbiol. https://doi.org/10.3389/fmicb.2015.00813

Dorighello DV, Bettiol W, Maia NB, Leite RMVBDC (2015) Controlling Asian soybean rust (Phakopsora pachyrhizi) with Bacillus spp. and coffee oil. Crop Prot 67:59–65. https://doi.org/10.1016/j.cropro.2014.09.017

Drejer E, Hakvåg S, Irla M, Brautaset T (2018) Genetic tools and techniques for recombinant expression in thermophilic bacillaceae. Microorganisms 6:42. https://doi.org/10.3390/microorganisms6020042

Duc LH, Fraser PD, Tam NKM, Cutting SM (2006) Carotenoids present in halotolerant Bacillus spore formers. FEMS Microbiol Lett 255:215–224. https://doi.org/10.1111/j.1574-6968.2005.00091.x

Duitman EH, Wyczawski D, Boven LG et al (2007) Novel methods for genetic transformation of natural Bacillus subtilis isolates used to study the regulation of the mycosubtilin and surfactin synthetases. Appl Environ Microbiol 73:3490–3496. https://doi.org/10.1128/aem.02751-06

Dunlap CA, Bowman MJ, Rooney AP (2019) Iturinic lipopeptide diversity in the Bacillus subtilis species group – important antifungals for plant disease biocontrol applications. Front Microbiol. https://doi.org/10.3389/fmicb.2019.01794

Dunlap C (2019) Phylogeny and taxonomy of agriculturally important Bacillus species. In: Bacilli in agrobiotechnology: phytostimulation and biocontrol, Bacilli in climate resilient agriculture and bioprospecting. Springer, Cham, pp 143–150

Dutta PD, Neog B, Goswami T (2020) Xylanase enzyme production from Bacillus australimaris P5 for prebleaching of bamboo (Bambusa tulda) pulp. Mater Chem Phys 243:122227. https://doi.org/10.1016/j.matchemphys.2019.122227

Dutta S, Surovy MZ, Gupta DR, Mahmud NU, Chanclud E, Joe W et al (2018) Genomic analysis reveal that biocontrol of wheat blast by Bacillus spp. May be linked with production of antimicrobial compounds and induced systemic resistance in host plants. Figshare. https://doi.org/10.6084/m9.figshare.5852661.v1

Emmert EA, Handelsman J (1999) Biocontrol of plant disease: a (Gram-) positive perspective. FEMS Microbiol Lett 171:1–9. https://doi.org/10.1111/j.1574-6968.1999.tb13405.x

Euony OIE, Elblehi SS, Abdel-Latif HM et al (2020) Modulatory role of dietary Thymus vulgaris essential oil and Bacillus subtilis against thiamethoxam-induced hepatorenal damage, oxidative stress, and immunotoxicity in African catfish (Clarias garipenus). Environ Sci Pollut Res 27:23108–23128. https://doi.org/10.1007/s11356-020-08588-5

Far BE, Khosroushahi AY, Dilmaghani A (2020) In silico study of alkaline serine protease and production optimization in Bacillus sp. Khoz1 closed Bacillus safensis isolated from honey. Int J Pept Res Therap. https://doi.org/10.1007/s10989-020-10016-8

Farace G, Fernandez O, Jacquens L et al (2015) Cyclic lipopeptides from Bacillus subtilis activate distinct patterns of defense responses in grapevine. Mol Plant Pathol 16:177–187. https://doi.org/10.1111/mpp.12170

Farzand A, Moosa A, Zubair M et al (2020) Transcriptional profiling of diffusible lipopeptides and fungal virulence genes during Bacillus amyloliquefaciens EZ1509-mediated suppression of Sclerotinia sclerotiorum. Phytopathology 110:317–326. https://doi.org/10.1094/phyto-05-19-0156-r

Feng D, Chen Z, Wang Z et al (2015) Domain III of Bacillus thuringiensisCry1Ie toxin plays an important role in binding to peritrophic membrane of asian corn borer. PLoS ONE. https://doi.org/10.1371/journal.pone.0136430

Feng L, Liu D, Sun X et al (2016) Bacillus cavernae sp. nov. isolated from cave soil. Int J Syst Evol Microbiol 66:801–806. https://doi.org/10.1099/ijsem.0.000794

Fickers P (2012) Antibiotic compounds from Bacillus: why are they so amazing? Am J Biochem Biotechnol 8:38–43. https://doi.org/10.3844/ajbbsp.2012.38.43

Fleming A (1941, September 13) Penicillin. BMJ 2(4210):386

Fritze D (2004) Taxonomy of the genus Bacillus and related genera: the aerobic endospore-forming bacteria. Phytopathology 94:1245–1248. https://doi.org/10.1094/phyto.2004.94.11.1245

Fuchs SW, Jaskolla TW, Bochmann S et al (2011) Entianin, a novel subtilin-like lantibiotic from Bacillus subtilis subsp. spizizeniiDSM 15029T with high antimicrobial activity. Appl Environ Microbiol 77:1698–1707. https://doi.org/10.1128/aem.01962-10

Gadhave KR, Gange AC (2015) Plant-associated Bacillus spp. alter life-history traits of the specialist insect Brevicoryne brassicae L. Agric For Entomol 18:35–42. https://doi.org/10.1111/afe.12131

Gálvez A, Abriouel H, López RL, Omar NB (2007) Bacteriocin based strategies for food biopreservation. Int J Food Microbiol 120:51–60

Gao C-H, Tian X-P, Qi S-H et al (2010) ChemInform abstract: antibacterial and antilarval compounds from marine gorgonian-associated bacterium Bacillus amyloliquefaciens SCSIO 00856. ChemInform. https://doi.org/10.1002/chin.201036209

Gao H, Qi G, Yin R et al (2016) Bacillus cereus strain S2 shows high nematicidal activity against Meloidogyne incognita by producing sphingosine. Sci Rep. https://doi.org/10.1038/srep28756

Gao X-Y, Liu Y, Miao L-L et al (2017) Mechanism of anti-Vibrio activity of marine probiotic strain Bacillus pumilus H2, and characterization of the active substance. AMB Express. https://doi.org/10.1186/s13568-017-0323-3

García-Gutiérrez L, Zeriouh H, Romero D et al (2013) The antagonistic strain Bacillus subtilis UMAF6639 also confers protection to melon plants against cucurbit powdery mildew by activation of jasmonate and salicylic acid-dependent defense responses. Microb Biotechnol 6:264–274. https://doi.org/10.1111/1751-7915.12028

Geissler M, Oellig C, Moss K et al (2017) High-performance thin-layer chromatography (HPTLC) for the simultaneous quantification of the cyclic lipopeptides surfactin, iturin A and fengycin in culture samples of Bacillus species. J Chromatogr B 1044–1045:214–224. https://doi.org/10.1016/J.JCHROMB.2016.11.013

Genus Bacillus. In: Genus: Bacillus. https://www.bacterio.net/bacillus.html. Accessed 3 Aug 2020

Gillespie SH, Hawkey PM (2006) Principles and practice of clinical bacteriology. Wiley, Chichester

Gond SK, Bergen MS, Torres MS, White JF Jr (2015) Endophytic Bacillus spp. produce antifungal lipopeptides and induce host defence gene expression in maize. Microbiol Res 172:79–87. https://doi.org/10.1016/j.micres.2014.11.004

Gong AD, Li HP, Yuan QS et al (2015) Antagonistic mechanism of iturin a and plipastatin a from Bacillus amyloliquefaciens S76–3 from wheat spikes against Fusarium graminearum. PLoS ONE 10:1–18. https://doi.org/10.1371/journal.pone.0116871

Guo D-L, Wan B, Xiao S-J et al (2015) Cyclic lipopeptides with herbicidal and insecticidal activities produced by Bacillus clausii DTM1. Nat Prod Commun. https://doi.org/10.1177/1934578x1501001235

Hamza F, Kumar AR, Zinjarde S (2015) Antibiofilm potential of a tropical marine Bacillus licheniformis isolate: role in disruption of aquaculture associated biofilms. Aquac Res 47:2661–2669. https://doi.org/10.1111/are.12716

Hazra C, Kundu D, Chaudhari A (2015) Lipopeptide biosurfactant from Bacillus clausii BS02 using sunflower oil soapstock: evaluation of high throughput screening methods, production, purification, characterization and its insecticidal activity. RSC Adv 5:2974–2982. https://doi.org/10.1039/c4ra13261k

Herrero S, Bel Y, Hernández-Martínez P, Ferré J (2016) Susceptibility, mechanisms of response and resistance to Bacillus thuringiensis toxins in Spodoptera spp. Curr Opin Ins Sci 15:89–96. https://doi.org/10.1016/j.cois.2016.04.006

Hiradate S, Yoshida S, Sugie H et al (2002) Mulberry anthracnose antagonists (iturins) produced by Bacillus amyloliquefaciens RC-2. Phytochemistry 61:693–698. https://doi.org/10.1016/s0031-9422(02)00365-5

Hu H, Gao Y, Li X et al (2020a) Identification and nematicidal characterization of proteases secreted by endophytic bacteria Bacillus cereus BCM2. Phytopathology 110:336–344. https://doi.org/10.1094/phyto-05-19-0164-r

Hu Y, Li H, Qu Y et al (2020b) Biocatalytic synthesis of a novel bioactive ginsenoside using udp-glycosyltransferase from Bacillus subtilis 168. Catalysts 10:289. https://doi.org/10.3390/catal10030289

Huang F, Dang Z, Guo C-L et al (2013) Biosorption of Cd(II) by live and dead cells of Bacillus cereus RC-1 isolated from cadmium-contaminated soil. Colloids Surf B: Biointerfaces 107:11–18. https://doi.org/10.1016/j.colsurfb.2013.01.062

Huang Y, Xu C, Ma L et al (2009) Characterisation of volatiles produced from Bacillus megaterium YFM 3.25 and their nematicidal activity against Meloidogyne incognita. Eur J Plant Pathol 126:417–422. https://doi.org/10.1007/s10658-009-9550-z

Im SM, Yu NH, Joen HW et al (2020) Biological control of tomato bacterial wilt by oxydifficidin and difficidin-producing Bacillus methylotrophicus DR-08. Pestic Biochem Physiol 163:130–137. https://doi.org/10.1016/j.pestbp.2019.11.007

Islam MT, Gupta DR, Hossain A et al (2020) Wheat blast: a new threat to food security. Phytopathol Res 2:28. https://doi.org/10.1186/s42483-020-00067-6

Islam MT, Kim KH, Choi J (2019a) Wheat blast in Bangladesh: the current situation and future impacts. Plant Pathol J 35(1):1–10

Islam MT, Croll D, Gladieux P, Soanes DM, Persoons A, Bhattacharjee P et al (2016a) Emergence of wheat blast in Bangladesh was caused by a south American lineage of Magnaporthe oryzae. BMC Biol 14:84

Islam T, Rahman MM, Pandey P, Boehme MH, Haesaert G (2019b) Bacilli and agrobiotechnology: phytostimulation and biocontrol, Bacilli in climate resilient agriculture and bioprospecting, vol 2. Springer, Cham

Islam MT, Rahman PP, Jha CK, Aeron A (2016b) Bacilli and Agrobiotechnology. Springer

Islam MT (2011) Potentials of biological control of plant diseases by *Lysobacter* spp., with special reference to strain SB-K88. In: Maheshwari DK (ed) Bacteria in agrobiology: plant growth responses. Springer, Berlin/Heidelberg, pp 35–363

Islam MT, Hashidoko Y, Deora A, Ito T, Tahara S (2005) Suppression of damping-off disease in host plants by the rhizoplane bacterium Lysobacter sp. strain SB-K88 is linked to plant colonization and antibiosis against soilborne Peronosporomycetes. Appl Environ Microbiol 71:3786–3796

Islam MT, Tahara S (2005) Bioactive secondary metabolites related to life-cycle development of oomycete phytopathogens. In: Atta-ur-Rahman (ed) Studies in natural products chemistry, vol 32. Elsevier Science Publishers, Amsterdam, pp 1053–1122

Jacques P (2011) Surfactin and other Lipopeptides from Bacillus spp. In: Soberón-Chávez G (ed) Biosurfactants. Microbiology monographs, vol 20. Springer, Berlin/Heidelberg

Jamal Q, Cho J-Y, Moon J-H et al (2017) Identification for the first time of cyclo (d-Pro-l-Leu) produced by Bacillus amyloliquefaciens Y1 as a nematocide for control of Meloidogyne incognita. Molecules 22:1839. https://doi.org/10.3390/molecules22111839

Jansen-Girgan C, Claassens S, Fourie H (2016) In vitro evaluations to determine the effect of Bacillus firmus strains on the motility of Meloidogyne incognita second-stage juveniles. Trop Plant Pathol 41:320–324. https://doi.org/10.1007/s40858-016-0100-x

Jaruchoktaweechai C, Suwanborirux K, Tanasupawatt S et al (2000) New macrolactins from a marine Bacillus sp. Sc026. J Nat Prod 63:984–986. https://doi.org/10.1021/np990605c

Jayakumar A, Krishna A, Mohan M et al (2018) Plant growth enhancement, disease resistance, and elemental modulatory effects of plant probiotic endophytic Bacillus sp. Fcl1. Prob Antimicrob Prot 11:526–534. https://doi.org/10.1007/s12602-018-9417-8

Ji Z-L, Peng S, Chen L-L et al (2020) Identification and characterization of a serine protease from Bacillus licheniformis W10: A potential antifungal agent. Int J Biol Macromol 145:594–603. https://doi.org/10.1016/j.ijbiomac.2019.12.216

Jiang C-H, Yao X-F, Mi D-D et al (2019) Comparative transcriptome analysis reveals the biocontrol mechanism of Bacillus velezensis F21 against Fusarium wilt on watermelon. Front Microbiol. https://doi.org/10.3389/fmicb.2019.00652

Jiang J, Gao L, Bie X et al (2016) Identification of novel surfactin derivatives from NRPS modification of Bacillus subtilis and its antifungal activity against Fusarium moniliforme. BMC Microbiol 16:31. https://doi.org/10.1186/s12866-016-0645-3

Jinal HN, Amaresan N (2019) Characterization of medicinal plant-associated biocontrol Bacillus subtilis (SSL2) by liquid chromatography-mass spectrometry and evaluation of compounds by in silico and in vitro methods. J Biomol Struct Dyn 38:500–510. https://doi.org/10.1080/07391102.2019.1581091

Jourdan E, Henry G, Duby F et al (2009) Insights into the defense-related events occurring in plant cells following perception of surfactin-type lipopeptide from Bacillus subtilis. Mol Plant-Microbe Interact 22:456–468. https://doi.org/10.1094/mpmi-22-4-0456

Kaki AA, Smargiasso N, Ongena M et al (2020) Characterization of new fengycin cyclic lipopeptide variants produced by Bacillus amyloliquefaciens (ET) Originating from a Salt Lake of Eastern Algeria. Curr Microbiol 77:443–451. https://doi.org/10.1007/s00284-019-01855-w

Kaspar F, Neubauer P, Gimpel M (2019) Bioactive secondary metabolites from Bacillus subtilis: a comprehensive review. J Nat Prod 82:2038–2053. https://doi.org/10.1021/acs.jnatprod.9b00110

Kavitha V, Radhakrishnan N, Gnanamani A, Mandal AB (2011) Management of chromium induced oxidative stress by marine Bacillus licheniformis. Biol Med 3:16–26

Kawagoe Y, Shiraishi S, Kondo H et al (2015) Cyclic lipopeptide iturin A structure-dependently induces defense response in Arabidopsis plants by activating SA and JA signaling pathways. Biochem Biophys Res Commun 460:1015–1020. https://doi.org/10.1016/j.bbrc.2015.03.143

Keenan B, Reznik G, Jarrell K, Vishwanath P (2008) Lipopeptides and lipopeptide synthetases. International application published under the patent cooperation treaty (PCT) WO2008131014-A1

Kefi A, Slimene IB, Karkouch I et al (2015) Characterization of endophytic Bacillus strains from tomato plants (Lycopersicon esculentum) displaying antifungal activity against Botrytis cinerea Pers. World J Microbiol Biotechnol 31:1967–1976. https://doi.org/10.1007/s11274-015-1943-x

Kelada NL, Shaker N (1988) Toxicity of three chemical insecticides in combination with Bacillus spp. against mosquito larvae. Int J Trop Insect Sci 9:229–231. https://doi.org/10.1017/s1742758400006032

Khan MM, Kim YK, Cho SS et al (2020a) Response surface optimization of culture conditions for cyclic lipopeptide ms07 from Bacillus siamensis reveals diverse insights targeting antimicrobial and antibiofilm activity. Processes 8:744. https://doi.org/10.3390/pr8060744

Khan MS, Gao J, Chen X et al (2020b) Theendophytic bacteria Bacillus velezensis Lle-9, isolated from Lilium leucanthum, harbors antifungal activity and plant growth-promoting effects. J Microbiol Biotechnol 30:668–680. https://doi.org/10.4014/jmb.1910.10021

Khaneja R, Perez-Fons L, Fakhry S et al (2009) Carotenoids found in Bacillus. J Appl Microbiol. https://doi.org/10.1111/j.1365-2672.2009.04590.x

Khedher SB, Boukedi H, Kilani-Feki O et al (2015) Bacillus amyloliquefaciens AG1 biosurfactant: Putative receptor diversity and histopathological effects on Tuta absoluta midgut. J Invertebr Pathol 132:42–47. https://doi.org/10.1016/j.jip.2015.08.010
Khedher SB, Boukedi H, Laarif A, Tounsi S (2020) Biosurfactant produced by Bacillus subtilis V26: a potential biological control approach for sustainable agriculture development. Org Agric. https://doi.org/10.1007/s13165-020-00316-0
Kim YO, Mahboob S, Viayaraghavan P et al (2020) Growth promoting activity of Penaeus indicus by secondary metabolite producing probiotic bacterium Bacillus subtilis isolated from the shrimp gut. J King Saud Univ Sci 32:1641–1646. https://doi.org/10.1016/j.jksus.2019.12.023
Kizhakkekalam VK, Chakraborty K, Joy M (2020) Oxygenated elansolid-type of polyketide spanned macrolides from a marine heterotrophic Bacillus as prospective antimicrobial agents against multidrug-resistant pathogens. Int J Antimicrob Agents 55:105892. https://doi.org/10.1016/j.ijantimicag.2020.105892
Komaki H, Hosoyama A, Ichikawa N, Igarashi Y (2016) Draft genome sequence of marine-derived Bacillus subtilis tp-b0611, a producer of bacilosarcins and amicoumacins. Genome Announc. https://doi.org/10.1128/genomea.01134-16
Kora AJ (2018) Bacillus cereus, selenite-reducing bacterium from contaminated lake of an industrial area: a renewable nanofactory for the synthesis of selenium nanoparticles. Biores Bioproces. https://doi.org/10.1186/s40643-018-0217-5
Koumoutsi A, Chen X-H, Vater J, Borriss R (2007) DegU and YczE positively regulate the synthesis of bacillomycin D by Bacillus amyloliquefaciens strain FZB42. Appl Environ Microbiol 73:6953–6964. https://doi.org/10.1128/aem.00565-07
Krishna MP, Varghese R, Babu VA et al (2012) Bioremediation of zinc using Bacillus sp. isolated from metal-contaminated industrial zone. In: Sabu A, Augustine A (eds) Prospects in Bioscience: Addressing the Issues. Springer, New Delhi, pp 11–18. https://doi.org/10.1007/978-81-322-0810-5_2
Kumar A, Singh AK, Ahmad S, Chandra R (2020) Optimization of laccase production by Bacillus sp. strain akrc01 in presence of agro-waste as effective substrate using response surface methodology. J Pure Appl Microbiol 14:351–362. https://doi.org/10.22207/jpam.14.1.36
Lee M-H, Lee J, Nam Y-D et al (2016) Characterization of antimicrobial lipopeptides produced by Bacillus sp. LM7 isolated from chungkookjang, a Korean traditional fermented soybean food. Int J Food Microbiol 221:12–18. https://doi.org/10.1016/j.ijfoodmicro.2015.12.010
Lee YS, Kim KY (2015) Antagonistic potential of Bacillus pumilus L1 against root-knot nematode, Meloidogyne arenaria. J Phytopathol 164:29–39. https://doi.org/10.1111/jph.12421
Lereclus D, Jacques P, Gohar M, et al (2017) Genetically modified bacterial strain producing kurstakin in the culture medium. International application published under the patent cooperation treaty (PCT) WO2017125583-A1
Li D, Carr G, Zhang Y et al (2011) Turnagainolides A and B, cyclic depsipeptides produced in culture by a Bacillus sp.: isolation, structure elucidation, and synthesis. J Nat Prod 74:1093–1099. https://doi.org/10.1021/np200033y
Li Q, Song W, Sun M et al (2020) Response of Bacillus vallismortis sp. EPS to exogenous sulfur stress/ induction and its adsorption performance on Cu(II). Chemosphere 251:126343. https://doi.org/10.1016/j.chemosphere.2020.126343
Li Y, Li Z, Yamanaka K et al (2015) Directed natural product biosynthesis gene cluster capture and expression in the model bacterium Bacillus subtilis. Sci Rep. https://doi.org/10.1038/srep09383
Liu H, Zeng Q, Wang W et al (2020a) Complete genome sequence of Bacillus velezensis strain AL7, a biocontrol agent for suppression of cotton verticillium wilt. Microbiol Res Ann. https://doi.org/10.1128/mra.01595-19
Liu Q, Zhang B, Shen Y, Yin K (2020b) Effect of the protein elicitor AMEP412 from Bacillus subtilis artificially fed to adults of the whitefly, Bemisia tabaci (Genn.) (Hemiptera: Aleyrodidae). Egyp J Biol Pest Cont. https://doi.org/10.1186/s41938-019-0202-0

Liu Y, Teng K, Wang T et al (2019) Antimicrobial Bacillus velezensis HC6: production of three kinds of lipopeptides and biocontrol potential in maize. J Appl Microbiol 128:242–254. https://doi.org/10.1111/jam.14459

Luo C, Liu X, Zhou X et al (2015) Unusual biosynthesis and structure of locillomycins from Bacillus subtilis 916. Appl Environ Microbiol 81:6601–6609. https://doi.org/10.1128/aem.01639-15

Lv J, Da R, Cheng Y et al (2020) Mechanism of antibacterial activity of Bacillus amyloliquefaciens C-1 lipopeptide toward anaerobic Clostridium difficile. Biomed Res Int 2020:1–12. https://doi.org/10.1155/2020/3104613

Ma H, Wei M, Wang Z et al (2020) Bioremediation of cadmium-polluted soil using a novel cadmium immobilizing plant growth promotion strain Bacillus sp. TZ5 loaded on biochar. J Hazard Mat 388:122065. https://doi.org/10.1016/j.jhazmat.2020.122065

Maget-Dana R, Peypoux F (1994) Iturins, a special class of pore-forming lipopeptides: biological and physicochemical properties. Toxicology 87:151–174. https://doi.org/10.1016/0300-483x(94)90159-7

Maget-Dana R, Ptak M (1995) Interactions of surfactin with membrane models. Biophys J 68:1937–1943. https://doi.org/10.1016/s0006-3495(95)80370-x

Maksimov I, Blagova D, Veselova S et al (2020) Recombinant Bacillus subtilis 26DCryChS line with gene Btcry11a encoding Cry11a toxin from Bacillus thuringiensis promotes integrated wheat defense against pathogen Stagonospora nodorumBerk. and greenbug Schizaphis graminumRond. Biol Control 144:104242. https://doi.org/10.1016/j.biocontrol.2020.104242

Matloub AA, Gomaa EZ, Hassan AA et al (2019) Comparative chemical and bioactivity studies of intra- and extracellular metabolites of endophytic bacteria, Bacillus subtilis NCIB 3610. Int J Pept Res Ther 26:497–511. https://doi.org/10.1007/s10989-019-09856-w

Matobole R, Zyl LV, Parker-Nance S et al (2017) Antibacterial activities of bacteria isolated from the marine sponges Isodictya compressa and Higginsia bidentifera collected from algoa bay, South africa. Mar Drugs 15:47. https://doi.org/10.3390/md15020047

Meena KR, Sharma A, Kumar R, Kanwar SS (2020) Two factor at a time approach by response surface methodology to aggrandize the Bacillus subtilis KLP2015 surfactin lipopeptide to use as antifungal agent. J King Saud Univ Sci 32:337–348. https://doi.org/10.1016/j.jksus.2018.05.025

Mendoza AR, Kiewnick S, Sikora RA (2008) In vitro activity of Bacillus firmus against the burrowing nematode Radopholus similis, the root-knot nematode Meloidogyne incognita and the stem nematode Ditylenchus dipsaci. Biocontrol Sci Tech 18:377–389

Metwally WE, Mostafa FAM, Refaei AR (2015) In vitro Study on the antagonistic activity of different native isolates of rhizobacteria against Meloidogyne incognita. Egypt J Agronematol 14:1–9. https://doi.org/10.21608/ejaj.2015.57486

Mnif I, Ghribi D (2015) Potential of bacterial derived biopesticides in pest management. Crop Prot 77:52–64. https://doi.org/10.1016/j.cropro.2015.07.017

Moghaddam MR, Moghaddam EM, Ravari SB, Rouhani H (2014a) The first report of Bacillus pumilus influence against Meloidogyne javanica in Iran. J Crop Prot 3:105–112

Moghaddam MR, Moghaddam EM, Ravari SB, Rouhani H (2014b) The nematicidal potential of local Bacillus species against the root-knot nematode infecting greenhouse tomatoes. Biocontrol Sci Tech 24:279–290. https://doi.org/10.1080/09583157.2013.858100

Mondol M, Shin H, Islam M (2013) Diversity of secondary metabolites from marine Bacillus species: chemistry and biological activity. Mar Drugs 11:2846–2872. https://doi.org/10.3390/md11082846

Mondol MAM, Tareq FS, Kim JH et al (2011) Cyclic ether-containing macrolactins, antimicrobial 24-membered isomeric macrolactones from a marine Bacillus sp. J Nat Prod 74:2582–2587. https://doi.org/10.1021/np200487k

Murphy JF, Zehnder GW, Schuster DJ et al (2000) Plant growth-promoting rhizobacterial mediated protection in tomato against tomato mottle virus. Plant Dis 84:779–784. https://doi.org/10.1094/pdis.2000.84.7.779

Myresiotis CK, Vryzas Z, Papadopoulou-Mourkidou E (2014) Effect of specific plant-growth-promoting rhizobacteria (PGPR) on growth and uptake of neonicotinoid insecticide thiamethoxam in corn (Zea mays L.) seedlings. Pest Manag Sci 71:1258–1266. https://doi.org/10.1002/ps.391

Nagao T, Adachi K, Sakai M et al (2001) Novel macrolactins as antibiotic lactones from a marine bacterium. J Antibiot 54:333–339. https://doi.org/10.7164/antibiotics.54.333

Nakano C, Ozawa H, Akanuma G, Funa N, Horinouchi S (2009) Biosynthesis of aliphatic polyketides by type III polyketide synthase and methyltransferase in Bacillus subtilis. J Bacteriol 191(15):4916–4923. https://doi.org/10.1128/JB.00407-09

Navon A (2000) Bacillus thuringiensis insecticides in crop protection — reality and prospects. Crop Prot 19:669–676. https://doi.org/10.1016/s0261-2194(00)00089-2

Nawrot-Esposito M-P, Babin A, Pasco M et al (2020) Bacillus thuringiensis bioinsecticides induce developmental defects in non-target Drosophila melanogaster larvae. https://doi.org/10.1101/2020.04.30.071563

Nelis HJ, Leenheer APD (1991) Microbial sources of carotenoid pigments used in foods and feeds. J Appl Bacteriol 70:181–191. https://doi.org/10.1111/j.1365-2672.1991.tb02922.x

Nik-Pa NIM, Sobri MFM, Abd-Aziz S et al (2020) Combined optimization of codon usage and glycine supplementation enhances the extracellular production of a β-Cyclodextrin glycosyltransferase from Bacillus sp. NR5 UPM in Escherichia coli. Int J Mol Sci 21:3919. https://doi.org/10.3390/ijms21113919

Niu Q, Huang X, Zhang L et al (2006) A neutral protease from Bacillus nematocida, another potential virulence factor in the infection against nematodes. Arch Microbiol 185:439–448. https://doi.org/10.1007/s00203-006-0112-x

Niu Q, Tian Y, Zhang L et al (2011) Overexpression of the key virulence proteases Bace16 and Bae16 in Bacillus nematocida B16 to improve its nematocidal activity. J Mol Microbiol Biotechnol 21:130–137. https://doi.org/10.1159/000332805

Nkiru UC, Franklin OC (2020) The production of bioinsecticide using Bacillus thuringiensis. World J Adv Res Rev 5:1–12. https://doi.org/10.30574/wjarr.2020.5.3.0034

Nnolim NE, Okoh AI, Nwodo UU (2020) Bacillus sp. FPF-1 produced keratinase with high potential for chicken feather degradation. Molecules 25:1505. https://doi.org/10.3390/molecules25071505

Ongena M, Jacques P (2008) Bacillus lipopeptides: versatile weapons for plant disease biocontrol. Trends Microbiol 16:115–125. https://doi.org/10.1016/j.tim.2007.12.009

Ongena M, Jourdan E, Adam A, Paquot M, Brans A, Joris B et al (2007) Surfactin fengycin lipopeptides of Bacillus subtilis as elicitors of induced systemic resistance in plants. Environ Microbiol 9:1084–1090

Osta Oliveira BE, Cury JA, Filho APR (2017) Biofilm extracellular polysaccharides degradation during starvation and enamel demineralization. PLoS ONE 12(7):181

Padgham J, Sikora R (2007) Biological control potential and modes of action of Bacillus megaterium against Meloidogyne graminicola on rice. Crop Prot 26:971–977. https://doi.org/10.1016/j.cropro.2006.09.004

Pane L, Radin L, Franconi G, Carli A (1996) The carotenoid pigments of a marine Bacillus firmus strain. Boll Soc Ital Biol Sper 72:303–308

Paramasiva I, Krishnayya P, War A, Sharma H (2014) Crop hosts and genotypic resistance influence the biological activity of Bacillus thuringiensis towards Helicoverpa armigera. Crop Prot 64:38–46. https://doi.org/10.1016/j.cropro.2014.05.010

Penha RO, Vandenberghe LPS, Faulds C et al (2020) Bacillus lipopeptides as powerful pest control agents for a more sustainable and healthy agriculture: recent studies and innovations. Planta. https://doi.org/10.1007/s00425-020-03357-7

Pereyra MG, Giacomo AD, Lara A et al (2020) Aflatoxin-degrading Bacillus sp. strains degrade zearalenone and produce proteases, amylases and cellulases of agro-industrial interest. Toxicon 180:43–48. https://doi.org/10.1016/j.toxicon.2020.04.006

Perez KJ, Viana JD, Lopes FC et al (2017) Bacillus spp. isolated from puba as a source of biosurfactants and antimicrobial lipopeptides. Front Microbiol 8:1–14. https://doi.org/10.3389/fmicb.2017.00061

Phae CG, Shoda M, Kubota H (1990) Suppressive effect of Bacillus subtilis and its products on phytopathogenic microorganisms. J Ferment Bioeng 69:1–7. https://doi.org/10.1016/0922-338x(90)90155-p

PHEBacteriology-Identification | ID 9 | Issue no: 3.1 | Issue date: 04.04.18 | Page: 1 of 27 © Crown copyright 2018 UK Standards for Microbiology Investigations: Identification of Bacillus species. https://assets.publishing.service.gov.uk/government/uploads/system/uploads/attachment_data/file/697260/ID_9i3.1.pdf

Polanczyk RA, Alves SB (2004) Estudos de Bacillus thuringiensis Berliner visando ao controle de Spodoptera frugiperda (J.E. Smith). Thesis (Doutorado em Entomologia) - Escola Superior de Agricultura "Luiz de Queiroz", Universidade de São Paulo, Piracicaba, 2004. Available from http://www.teses.usp.br/teses/disponiveis/11/11146/tde-22092004-144726/pt-br.php. Accessed on 8 Aug 2020

Prathumpai W, Promboon A, Werapan B et al (2020) Pilot-Scale Protease Production by Bacillus sp. C4 for Silk Degumming Processes. J Nat Fiber:1–14. https://doi.org/10.1080/15440478.2020.1788482

Pretorius D, van Rooyen J, Clarke KG (2015) Enhanced production of antifungal lipopeptides by Bacillus amyloliquefaciens for bio-control of postharvest disease. New Biotechnol 32:243–252. https://doi.org/10.1016/j.nbt.2014.12.003

Qiu Z, Lu X, Li N et al (2017) Characterization of garlic endophytes isolated from the black garlic processing. MicrobiologyOpen. https://doi.org/10.1002/mbo3.547

Raaijmakers JM, Bruijn ID, Nybroe O, Ongena M (2010) Natural functions of lipopeptides from Bacillus and Pseudomonas: more than surfactants and antibiotics. FEMS Microbiol Rev 34:1037–1062. https://doi.org/10.1111/j.1574-6976.2010.00221.x

Rahman M, Sabir AA, Mukta JA, Khan MMA, Mohi-Ud-Din M, Miah MG, Rahman M, Islam MT (2018) Plant probiotic bacteria *Bacillus* and *Paraburkholderia* improve growth, yield and content of antioxidants in strawberry fruit. Sci Rep 8(1):2504. https://doi.org/10.1038/s41598-018-20235-1

Rajaofera MJN, Kang X, Jin P-F, Chen X et al (2020) Antibacterial activity of bacillomycin D-like compounds isolated from Bacillus amyloliquefaciens HAB-2 against Burkholderia pseudomallei. Asian Pac J Trop Biomed 10:183. https://doi.org/10.4103/2221-1691.280295

Rampelotto PH (2010) Resistance of microorganisms to extreme environmental conditions and its contribution to astrobiology. Sustainability 2:1602–1623. https://doi.org/10.3390/su2061602

Ravari SB, Moghaddam EM (2016) Efficacy of Bacillus thuringiensis Cry14 toxin against root knot nematode, Meloidogyne javanica. Plant Prot Sci 51:46–51. https://doi.org/10.17221/93/2013-pps

Reddy K, Lakshmi T, Reddy AK et al (2016) Isolation, screening, identification and optimized production of extracellular cellulase from Bacillus subtilis Sub.sps using cellulosic waste as carbon source. Int J Curr Microbiol App Sci 5:442–451. https://doi.org/10.20546/ijcmas.2016.504.052

Regmi S, Choi YS, Kim YK et al (2019) Industrial attributes of β-glucanase produced by Bacillus sp. CSB55 and its potential application as bio-industrial catalyst. Bioprocess Biosyst Eng 43:249–259. https://doi.org/10.1007/s00449-019-02221-7

Riesenman PJ, Nicholson WL (2000) Role of the spore coat layers in Bacillus subtilis spore resistance to hydrogen peroxide, artificial uv-c, uv-b, and solar UV radiation. Appl Environ Microbiol 66:620–626. https://doi.org/10.1128/aem.66.2.620-626.2000

Rong S, Xu H, Li L et al (2020) Antifungal activity of endophytic Bacillus safensis B21 and its potential application as a biopesticide to control rice blast. Pestic Biochem Physiol 162:69–77. https://doi.org/10.1016/j.pestbp.2019.09.003

Rosas-Garcia N (2009) Biopesticide production from Bacillus thuringiensis: an environmentally friendly alternative. Recent Pat Biotechnol 3:28–36. https://doi.org/10.2174/187220809787172632

Rosli SNA, Man RC, Masngut N (2020) Factorial experimental design for xylanase production by Bacillus sp. isolated from Malaysia landfill soil. IOP Conf Ser Mat Sci Eng 736:022074. https://doi.org/10.1088/1757-899x/736/2/022074

Saeed A (2016) Isocoumarins, miraculous natural products blessed with diverse pharmacological activities. Eur J Med Chem 116:290–317. https://doi.org/10.1016/j.ejmech.2016.03.025

Saggese A, Culurciello R, Casillo A et al (2018) A Marine isolate of Bacillus pumilus secretes a pumilacidin active against Staphylococcus aureus. Mar Drugs 16:180. https://doi.org/10.3390/md16060180

Sansinenea E, Ortiz A (2011) Secondary metabolites of soil Bacillus spp. Biotechnol Lett 33:1523–1538. https://doi.org/10.1007/s10529-011-0617-5

Sarangi T, Ramakrishnan S, Nakkeeran S (2014) Isolation and evaluation of indigenous isolates of Bacillus spp. against Meloidogyne incognita. Int J Trop Agric 32:15–18

Savini V (2016) Bacillus species outside the Bacillus cereus Group. In: Savini V (ed) The diverse faces of Bacillus cereus. Academic Press, London, pp 129–138. https://doi.org/10.1016/b978-0-12-801474-5.00011-6

Schneider K, Chen X-H, Vater J et al (2007) Macrolactin is the polyketide biosynthesis product of the pks2 cluster of Bacillus amyloliquefaciens FZB42. J Nat Prod 70:1417–1423. https://doi.org/10.1021/np070070k

Sharma D, Singh SS, Baindara P et al (2020) Surfactin like broad spectrum antimicrobial lipopeptide co-produced with sublancin from Bacillus subtilis strain A52: dual reservoir of bioactives. Front Microbiol. https://doi.org/10.3389/fmicb.2020.01167

Shayesteh F, Ahmad A, Usup G (2020) In vitro anti-biofilm activity of bacteriocin from a marine Bacillus sp. strain Sh10 against Proteus mirabilis. Iran J Microbiol. https://doi.org/10.18502/ijm.v12i1.2518

Shen B (2003) Polyketide biosynthesis beyond the type I, II and III polyketide synthase paradigms. Curr Opin Chem Biol 7:285–295

Shimojima Y, Hayashi H, Ooka T et al (1984) Studies on AI-77s, microbial products with gastroprotective activity. Structures and the chemical nature of AI-77s. Tetrahedron 40:2519–2527. https://doi.org/10.1016/s0040-4020(01)83504-3

Singh AK, Sharma P (2020) Disinfectant-like activity of lipopeptide biosurfactant produced by Bacillus tequilensis strain SDS21. Colloids Surf B: Biointerfaces 185:110514. https://doi.org/10.1016/j.colsurfb.2019.110514

Soares MA, Li HY, Bergen M et al (2016) Functional role of an endophytic Bacillus amyloliquefaciens in enhancing growth and disease protection of invasive English ivy (Hedera helix L.). Plant Soil 405:107–123. https://doi.org/10.1007/s11104-015-2638-7

Steller S, Sokoll A, Wilde C et al (2004) Initiation of surfactin biosynthesis and the role of the srfd-thioesterase protein. Biochemistry 43:11331–11343. https://doi.org/10.1021/bi0493416

Stoica R-M, Moscovici M, Tomulescu C et al (2019) Antimicrobial compounds of the genus Bacillus: a review. Roman Biotechnol Lett 24:1111–1119. https://doi.org/10.25083/rbl/24.6/1111.1119

Sumardi S, Farisi S, Ekowati CN et al (2020) Halotolerant Bacillus sp. for mannan degradation isolated from mangrove ecosystem at hanura beach lampung. J Pure Appl Microbiol 14:1237–1244. https://doi.org/10.22207/jpam.14.2.18

Surovy MZ, Gupta DR, Mahmud NU et al (2019) Genomics and post-genomics approaches for elucidating molecular mechanisms of plant growth-promoting Bacilli. In: Islam M, Rahman M, Pandey P, Boehme M, Haesaert G (eds) Bacilli and agrobiotechnology: phytostimulation and biocontrol, Bacilli in climate resilient agriculture and bioprospecting. Springer, Cham, pp 161–200. https://doi.org/10.1007/978-3-030-15175-1_10

Surovy MZ, Gupta DR, Chanclude E, Joe W, Kamoun S, Islam M (2017) Plant probiotic bacteria suppress wheat blast fungus Magnaporthe oryzae Triticum pathotype. Figshare. https://doi.org/10.6084/m9.figshare.5549278.v1

Symmank H, Franke P, Saenger W, Bernhard F (2002) Modification of biologically active peptides: production of a novel lipohexapeptide after engineering of Bacillus subtilis surfactin synthetase. Protein Eng Des Sel 15:913–921. https://doi.org/10.1093/protein/15.11.913

Tamehiro N, Okamoto-Hosoya Y, Okamoto S, Ubukata M, Hamada M, Naganawa H, Ochi K (2002) Bacilysocin, a novel phospholipid antibiotic produced by Bacillus subtilis168. Antimicrob Agents Chemother 46:315–320

Tanyildizi MS, Özer D, Elibol M (2007) Production of bacterial α-amylase by B. amyloliquefaciens under solid substrate fermentation. Biochem Eng J 37:294–297. https://doi.org/10.1016/j.bej.2007.05.009

Tareq FS, Shin HJ (2017) Bacilotetrins A and B, anti- staphylococcal cyclic-lipopeptides from marine derived Bacillus subtilis. J Nat Prod 80(11):2889–2892

Tareq FS, Hasan CM, Lee H-S et al (2015) Gageopeptins A and B, new inhibitors of zoospore motility of the phytopathogen Phytophthora capsici from a marine-derived bacterium Bacillus sp. 109GGC020. Bioorgan Med Chem Lett 25:3325–3329. https://doi.org/10.1016/j.bmcl.2015.05.070

Tareq FS, Kim JH, Lee MA et al (2013) Ieodoglucomides A and B from a marine-derived bacterium Bacillus licheniformis. Org Lett 15:2071–2071. https://doi.org/10.1021/ol4008603

Tian B, Yang J, Zhang K-Q (2007) Bacteria used in the biological control of plant-parasitic nematodes: populations, mechanisms of action, and future prospects. FEMS Microbiol Ecol 61:197–213. https://doi.org/10.1111/j.1574-6941.2007.00349.x

Trischman JA, Jensen PR, Fenical W (1994) Halobacillin: A cytotoxic cyclic acylpeptide of the iturin class produced by a marine Bacillus. Tetrahedron Lett 35:5571–5574. https://doi.org/10.1016/s0040-4039(00)77249-2

Tsuge K, Matsui K, Itaya M (2007) Production of the non-ribosomal peptide plipastatin in Bacillus subtilis regulated by three relevant gene blocks assembled in a single movable DNA segment. J Biotechnol 129:592–603. https://doi.org/10.1016/j.jbiotec.2007.01.033

Turatto MF, Dourado FDS, Zilli JE, Botelho GR (2018) Control potential of Meloidogyne javanica and Ditylenchus spp. using fluorescent Pseudomonas and Bacillus spp. Braz J Microbiol 49:54–58 https://doi.org/10.1016/j.bjm.2017.03.015

Usta C (2013) Microorganisms in biological pest control — a review (bacterial toxin application and effect of environmental factors). Curr Prog Biol Res. https://doi.org/10.5772/55786

Vachon V, Laprade R, Schwartz J-L (2012) Current models of the mode of action of Bacillus thuringiensis insecticidal crystal proteins: a critical review. J Invertebr Pathol 111:1–12. https://doi.org/10.1016/j.jip.2012.05.001

Vanittanakom N, Loeffler W, Koch U, Jung G (1986) Fengycin—a novel antifungal lipopeptide antibiotic produced by Bacillus subtilis F-29-3. J Antibiot (Tokyo) 39:888–901

Varghese R, Krishna MP, Babu VA, Mohamed Hatha AA (2012) Biological removal of lead by Bacillus sp. from metal contaminated industrial area. J Microbiol Biotechnol Food Sci 2(2):756–770

Volpon L, Besson F, Lancelin J-M (2000) NMR structure of antibiotics plipastatins A and B from Bacillus subtilis inhibitors of phospholipase A2. FEBS Lett 485:76–80. https://doi.org/10.1016/s0014-5793(00)02182-7

Wang N, Jiang M, Zhang P et al (2020a) Amelioration of Cd-induced bioaccumulation, oxidative stress and intestinal microbiota by Bacillus cereus in Carassius auratus gibelio. Chemosphere 245:125613. https://doi.org/10.1016/j.chemosphere.2019.125613

Wang S, Sun L, Zhang W et al (2020b) Bacillus velezensis BM21, a potential and efficient biocontrol agent in control of corn stalk rot caused by Fusarium graminearum. Egyp J Biol Pest Cont. https://doi.org/10.1186/s41938-020-0209-6

Wang X, Xue Y, Han M et al (2014) The ecological roles of Bacillus thuringiensis within phyllosphere environments. Chemosphere 108:258–264. https://doi.org/10.1016/j.chemosphere.2014.01.050

Wielkopolan B, Obrępalska-Stęplowska A (2016) Three-way interaction among plants, bacteria, and coleopteran insects. Planta 244:313–332. https://doi.org/10.1007/s00425-016-2543-1

Wilson MJ, Jackson TA (2013) Progress in the commercialisation of bionematicides. BioControl 58:715–722. https://doi.org/10.1007/s10526-013-9511-5

Wiraswati SM, Nawangsih AA, Rusmana I, Wahyudi AT (2020) Rice phyllosphere bacteria producing antifungal compounds as biological control agents of blast disease. Biodiv J Biol Div. https://doi.org/10.13057/biodiv/d210401

World Health Organization (WHO) (1985) Informal consultation on the development of Bacillus sphaericus as a microbial larvicide. WHO, Geneva, 24 p

Wu G, Liu Y, Xu Y, Zhang G, Shen Q, Zhang R (2018) Exploring elicitors of the beneficial Rhizobacterium Bacillus amyloliquefaciens SQR9 to induce plant systemic resistance and their interactions with plant signaling pathways. Mol Plant-Microbe Interact. https://doi.org/10.1094/MPMI-11-17-0273-R

Xiang N, Lawrence KS, Kloepper JW et al (2017) Biological control of Meloidogyne incognita by spore-forming plant growth-promoting rhizobacteria on cotton. Plant Dis 101:774–784. https://doi.org/10.1094/pdis-09-16-1369-re

Xiao L, Wan J-W, Yao J-H et al (2018) Effects of Bacillus cereus strain Jdm1 on Meloidogyne incognita and the bacterial community in tomato rhizosphere soil. 3 Biotech. https://doi.org/10.1007/s13205-018-1348-2

Xiong J, Zhou Q, Luo H et al (2015) Systemic nematicidal activity and biocontrol efficacy of Bacillus firmus against the root-knot nematode Meloidogyne incognita. World J Microbiol Biotechnol 31:661–667. https://doi.org/10.1007/s11274-015-1820-7

Xue C, Tian L, Xu M et al (2008) A New 24-membered lactone and a new polyene δ-lactone from the marine bacterium Bacillus marinus. J Antibiot 61:668–674. https://doi.org/10.1038/ja.2008.94

Xue D, Abdallah II, Haan IEMD et al (2015) Enhanced C30 carotenoid production in Bacillus subtilis by systematic overexpression of MEP pathway genes. Appl Microbiol Biotechnol 99:5907–5915. https://doi.org/10.1007/s00253-015-6531-3

Yamamoto S, Shiraishi S, Suzuki S (2015) Are cyclic lipopeptides produced by Bacillus amyloliquefaciens S13–3 responsible for the plant defence response in strawberry against Colletotrichum gloeosporioides? Lett Appl Microbiol 60:379–386. https://doi.org/10.1111/lam.12382

Yan L, Liu G, Zhao B et al (2020) Novel biomedical functions of surfactin A from Bacillus subtilis in wound healing promotion and scar inhibition. J Agric Food Chem 68:6987–6997. https://doi.org/10.1021/acs.jafc.0c01658

Yang G, Sun H, Cao R et al (2020a) Characterization of a novel glycoside hydrolase family 46 chitosanase, Csn-BAC, from Bacillus sp. MD-5. Int J Biol Macromol 146:518–523. https://doi.org/10.1016/j.ijbiomac.2020.01.031

Yang J, Zhu X, Cao M, Wang C, Zhang C, Lu Z et al (2016) Genomics-inspired discovery of three antibacterial active metabolites, aurantinins B, C, and D from compost-associated Bacillus subtilis fmb60. J Agric Food Chem 64:8811–8820

Yang N, Wu Q, Xu Y (2020c) Fe nanoparticles enhanced surfactin production in Bacillus amyloliquefaciens. ACS Omega 5:6321–6329. https://doi.org/10.1021/acsomega.9b03648

Yang X, Zhang L, Xiang Y et al (2020b) Comparative transcriptome analysis of Sclerotinia sclerotiorum revealed its response mechanisms to the biological control agent, Bacillus amyloliquefaciens. Sci Rep. https://doi.org/10.1038/s41598-020-69434-9

Yu D, Xu F, Zeng J, Zhan J (2012) Type III polyketide synthases in natural product biosynthesis. UBMB Life 64(4):285–229

Zebelo S, Song Y, Kloepper JW, Fadamiro H (2016) Rhizobacteria activates (+)-δ-cadinene synthase genes and induces systemic resistance in cotton against beet armyworm (Spodoptera exigua). Plant Cell Environ 39:935–943. https://doi.org/10.1111/pce.12704

Zeng W, Li F, Wu C et al (2019) Role of extracellular polymeric substance (EPS) in toxicity response of soil bacteria Bacillus sp. S3 to multiple heavy metals. Bioprocess Biosyst Eng 43:153–167. https://doi.org/10.1007/s00449-019-02213-7

Zhang W, Wei L, Xu R et al (2020) Evaluation of the antibacterial material production in the fermentation of Bacillus amyloliquefaciens-9 from whitespotted bamboo shark (Chiloscyllium plagiosum). Mar Drugs 18:119. https://doi.org/10.3390/md18020119

Zhao H, Shao D, Jiang C, Shi J, Li Q, Huang Q, Rajoka MSR, Yang H, Jin M (2017) Biological activity of lipopeptides from Bacillus. Appl Microbiol Biotechnol 101(15):5951–5960. https://doi.org/10.1007/s00253-017-8396-0

Zhao P, Xu X, Zhao X et al (2020) Capability of Bacillus subtilis to remove Pb2+ via producing lipopeptides. Sci Total Environ 730:138941. https://doi.org/10.1016/j.scitotenv.2020.138941

Zheng X, Zhang Y, Liu X et al (2020) High-level expression and biochemical properties of a thermo-alkaline pectate lyase from Bacillus sp. RN1 in Pichia pastoris with potential in ramie degumming. Front Bioeng Biotechnol. https://doi.org/10.3389/fbioe.2020.00850

Zheng Y, Ye Z-L, Fang X-L et al (2008) Production and characteristics of a bioflocculant produced by Bacillus sp. F19. Bioresour Technol 99:7686–7691. https://doi.org/10.1016/j.biortech.2008.01.068

Zhou L, Yuen G, Wang Y et al (2016) Evaluation of bacterial biological control agents for control of root-knot nematode disease on tomato. Crop Prot 84:8–13. https://doi.org/10.1016/j.cropro.2015.12.009

Zihalirwa ZK, Arias AA, Franzil L et al (2017) Stimulation of Fengycin-Type antifungal lipopeptides in Bacillus amyloliquefaciens in the presence of the maize fungal pathogen Rhizomucor variabilis. Front Microbiol 8:850. https://doi.org/10.3389/fmicb.2017.00850

Chapter 4
The Industrially Important Enzymes from *Bacillus* Species

Aurelio Ortiz and Estibaliz Sansinenea

Abstract *Bacillus* sp. genus produces several interesting extracellular enzymes, which have been used in a commercial way for detergent, textile, food, feed, and beverage industries. The improvement of strains and production strategies has recently been influenced or facilitated by the application of molecular biology techniques. *Bacillus* sp. species are attractive industrial organisms due to their short growth time, capacity to secrete proteins, and safety for humans. The enzymes that are secreted by many species of *Bacillus* can be used in several industrial processes. In this chapter, we revise the most important enzymes produced by *Bacillus* spp. species that are applied in industry reviewing the current status of their applications.

Keywords Enzymes · Industrial applications · *Bacillus* sp.

4.1 Introduction

Bacillus species have been used worldwide in several fields such as agriculture and biotechnology due to the production of antipathogenic chemicals, or triggering the host defense mechanism and predation against enemies and pathogens. *Bacillus* species have the ability to replicate rapidly, to generate spores when they do not find enough nutrients, and they have biocontrol ability, making them versatile and competitive weapons for pathogens.

They have been applied in several industries such as, textile, food, detergent, and beverages industries because they secrete extracellular proteins and enzymes, which are completely dissociated from the cell and found free in the surrounding medium. Genetic engineering has been the tool that has allowed the strain development to apply these biopesticides. Since *Bacillus* species have many advantages, including that they are classified as safe organisms, they are widely employed in biotechnology and industry. Besides, as they produce many compounds including enzymes, which have been applied in several industries, they are often considered microbial

A. Ortiz · E. Sansinenea (✉)
Facultad de Ciencias Químicas, Benemérita Universidad Autónoma de Puebla,
Puebla, México

M. T. Islam et al. (eds.), *Bacilli in Agrobiotechnology*, Bacilli in Climate
Resilient Agriculture and Bioprospecting,
https://doi.org/10.1007/978-3-030-85465-2_4

factories (Ortiz and Sansinenea 2019). This capacity allows a variety of industrial applications of this genus.

The industrial enzymes world market is 1.6 billion $US, (Schallmey et al. 2004); *Bacillus* spp. enzymes being the half of the total enzyme market. The enzymes produced by this genus have several applications in many industries. In this chapter we summarize the industrially important enzymes produced by *Bacillus* spp reviewing the current status of their applications.

4.2 Proteases

Proteolytic enzymes are one of the most interesting enzymes within the world market. These enzymes decompose proteins by hydrolyzing their peptide bonds having many applications in industrial processes, highlighting the search for new proteases as a research topic that it has been developing (Sharma et al. 2017). In this sense, the genus *Bacillus* produces neutral and alkaline proteolytic enzymes in high yields, with remarkable properties, such as their high stability to wide range of pHs and temperatures and their molecular weight between 27 and 71 KDa (Contesini et al. 2017). For example, some *Bacillus* species produce alkaline serine proteases which are applied as household detergents, while neutral zinc metalloproteinases are used in milk protein modification among others (Schallmey et al. 2004).

It is important to reduce their production cost, and the use of some agro-industrial wastes can be an alternative. However, the modern techniques such as genetic engineering have been studied and applied to produce proteases in good yields (Jaouadi et al. 2012).

One of the major applications of proteases is in the detergent industry to remove stains from fabrics. However, there are some features of proteases that are important to take into account, such as several compounds present in detergents that can inactivate the proteases (Annamalai et al. 2013). Besides, due to their stability in organic solvents they can be used in organic syntheses (Caille et al. 2002).

Proteases can be classified into four main groups such as serine-proteinase, cysteine proteinases, the aspartic-, carboxyl-, or acidic proteinases, and metalloproteases. Most extracellular proteases produced by *Bacillus* are serine proteases (Adinarayana et al. 2003), cysteine proteases (Rozs et al. 2001) and metalloproteases (Sookkheo et al. 2000).

For protease isolation and purification, they can be isolated using organic solvents or ammonium sulfate followed by chromatography purification (Anbu 2013). Their tertiary structure can be elucidated using X-ray diffraction of proteins from their crystal structure (Nonaka et al. 2004).

The production of these proteolytic enzymes should take into account the physiology of the producing microorganism and the genetic engineering to increase the production yield. Besides, some steps at industrial scale are very expensive therefore the cost of the fermentation process should be considered (Kirk et al. 2002). Efficient protease production also depends on the substrate concentration, physicochemical parameters, the initial pH, agitation, incubation time, and incubation temperature (Gupta et al. 2003). Genetic engineering can optimize the production by two ways: the improvement of biochemical and catalytic properties of these enzymes, altering specific amino acids in the protein structure, and the improvement of the proteases production enhancing secretion of homologous proteases.

Proteolytic enzymes and other enzymes such as lipases and amylases are used in detergent industry to remove proteinaceous stains. In food industry the proteases can generate bioactive peptides producing hydrolyzed peptides with important biological activities. Besides, *Bacillus* proteases have been applied on the food maturation process. In pharmaceutical industry the proteases are applied to some organic syntheses to obtain compounds with good yields reducing the use of organic solvents. In textile and leather industries, the proteases have been applied for hydrolyzing keratin, an insoluble fibrous protein (Contesini et al. 2017).

4.3 Amylases

This enzyme breaks down the starch into dextrin, maltose, and finally glucose units. They are applied in starch liquefaction, paper, desizing of textile fabrics, and preparing starch coatings of paints among others (Tanyildizi et al. 2005); therefore, the current amylase production implies 65% of enzyme market. Microbial amylase can be highly and easily produced by several microorganisms. *Bacillus* spp. amylases are applied in many industrial processes, such as the food, fermentation, textile, and paper industries (Pandey et al. 2000). α-Amylase (Fig. 4.1) from *Bacillus* spp. cleaves internal α-1,4-linkages, has a certain thermostability, and operates at 95 °C; therefore it is applied to liquefy starches, especially cornstarch, which only gelatinizes at 100 °C. On the other hand, *Bacillus* β-Amylases operate to remove maltose units from starch.

To improve the enzyme production the culture conditions can be optimized. Utilization of waste materials provides low-cost media and solves the environmental pollution problem (Kalyani and Rajesh 2018). Moreover, crude amylase can be applied in detergent industry and in starchy materials saccharification (Simair et al. 2017).

4.4 Galactosidases

α-Galactosidase hydrolyzes α-1,6-linked galactoside residues from simple oligosaccharides and the polysaccharides of galactomannans. It also catalyzes transglycosylation reactions (Carneiro et al. 2018) being applied in the beet sugar industry, pulp and paper industry, soya food processing, and animal feed processing. It is applied to remove oligosaccharides from soy-based foods avoiding the production of flatulence in the human small intestine (Patil et al. 2010). It is also used in blood

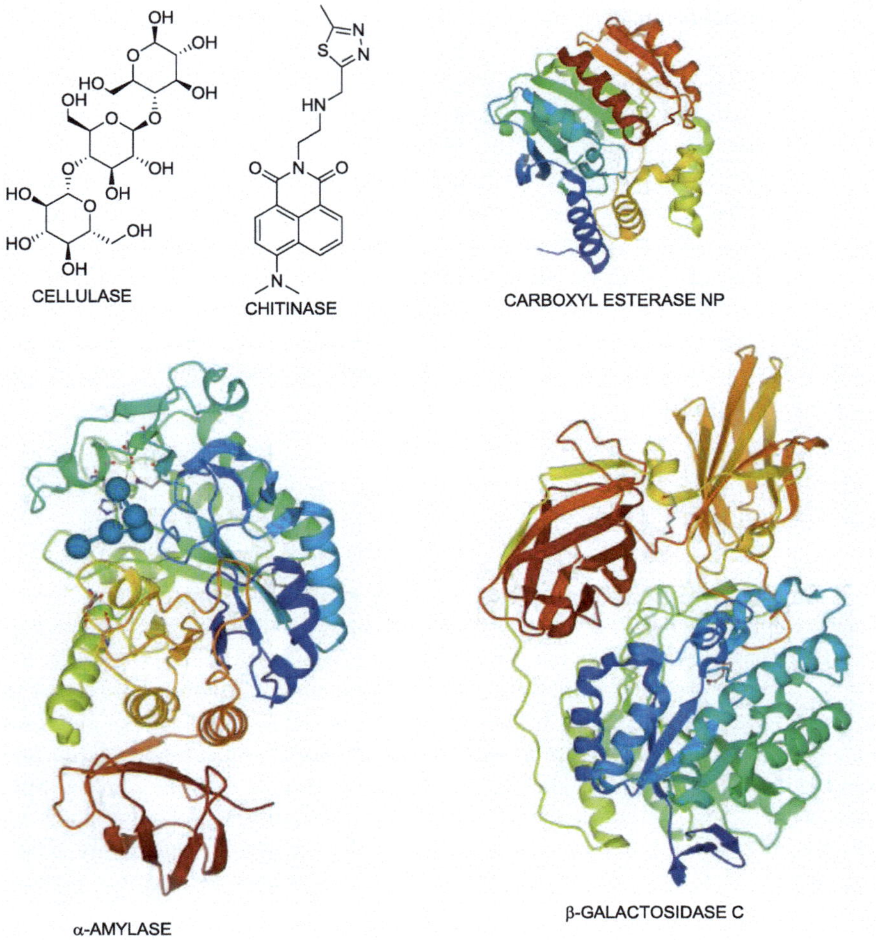

Fig. 4.1 Chemical crystal structures of some reviewed enzymes

group transformation, the treatment of Fabray's disease, and xenotransplantation (Fuller et al. 2004). New α-galactosidases have been described from *B. megaterium* in the last years (Xu et al. 2014; Huang et al. 2018).

On the other hand, the enzyme β-galactosidase (Fig. 4.1), found in microorganisms, plants, and animals, hydrolyzes β(1–3) and β(1–4) galactosyl bonds in oligosaccharides (Oliveira et al. 2011). In the food industry, it is used for the production of lactose-free milk for lactose-intolerant individuals (Ishikawa et al. 2015). Since lactose intolerance is a common problem, which affects over 70% of the world's adult population and causes symptoms such as abdominal pain, gas, nausea, and diarrhea, the market for lactose-free milk and dairy products is currently growing rapidly. This can be obtained by enzymatic hydrolysis using β-galactosidases.

4.5 Pullulanases

Bacillus spp. pullulanases, important debranching enzymes, hydrolyze the α-1,6 glucosidic linkages in starch, and related oligosaccharides, which convert the branched polysaccharides into small fermentable sugars (Hii et al. 2012). In the saccharification process pullulanase has been used to increase the final glucose concentration. To date, five groups of pullulanase enzymes have been reported: (1) pullulanase type I, (2) amylopullulanase, (3) neopullulanase, (4) isopullulanase, and (5) pullulan hydrolase type III. Some *Bacillus* sp. are pullulanase producers such as *Bacillus acidopullulyticus*, *Bacillus deramificans*, thermophilic *Bacillus* sp. AN-7 (Kunamneni and Singh 2006), *Bacillus cereus* FDTA-13 (Nair et al. 2007), and *Geobacillus stearothermophilus* (Zareian et al. 2010). However, pullulanases with appropriate features to industrially employ them is currently limited (Nisha and Satyanarayana 2016). However, molecular biotechnology may provide solutions to improve enzymatic properties for industrial application (Wang et al. 2019). *B. subtilis* is appropriate to produce pullulanase since this organism is recognized as safe.

4.6 β-glucanases

Glucanases break down the glucan polysaccharide hydrolyzing the glucosidic bond. They are used during the aging process of wine in enological practices improving the aroma of wines. Macerating enzymes also improve press-ability, settling, and juice yields of grapes used for wine fermentation.

It is used to reduce barley mash viscosity helping with the filtration process and improving wort yield. The *B. subtilis* enzyme is commercially available and used in the brewing industry (Niu et al. 2018), and hybrid enzymes have been developed using *B. macerans* and *B. amyloliquefaciens* enzymes (Borriss et al. 1988, 1989). The β-1,3-glucanase hydrolyzes the nonreducing end of the glucan, resulting in the formation of oligosaccharides and glucose. It has been seen that this enzyme delays the pathogenic fungi growth decreasing the damage caused by these fungi in fruits (Confortin et al. 2019).

4.7 Xylanases

Xylanase degrades the linear polysaccharide xylan into xylose (Beg et al. 2001), breaking down hemicellulose, one of the major components of plant cell walls. Thus, xylanase is applied for the chlorine-free bleaching of wood pulp prior to the papermaking process. They are also applied as food additives to poultry, in wheat flour for improving dough handling and quality of baked products (Ammoneh et al. 2014) among others (Subramaniyan and Prema 2002). These enzymes are widely produced by *Bacillus* spp. and applied for hydrolyzing cellulosic material such as cellulases and xylanases (Aizawa et al. 2010), for example, *B. pumilus* can produce a high quantity of xylanase (Nagar et al. 2011).

4.8 Cellulases

Microbial cellulases catalyze the bioconversion of cellulose to soluble sugars and glucose (Fig. 4.1) whose demand has grown rapidly (Li et al. 2009). Even though a variety of *Bacillus* species secrete cellulases, the study of these enzymes in this organism has remained behind (Kim et al. 2012). The explanation of this situation is due to that most *Bacillus* cellulases hydrolyze synthetic carboxymethyl cellulose, but barely hydrolyze the crystalline form of cellulose (Balasubramanian and Simões 2014). Even to improve the production cost of these enzymes *Bacillus* sp. can utilize the organic matter from several wastes (Ladeira et al. 2015). The cellulases has been applied in various industrial processes (Karmakar and Ray 2011; Kuhad et al. 2011), for example, they are used for deinking the paper wastes since they can remove the ink from the fiber surface hydrolyzing carbohydrate molecules. Cellulases are also applied for jeans biostoning and cotton biopolishing. The most popular application that is being investigated is enzymatic saccharification of lignocellulosic materials by cellulases for biofuel production (Sukumaran et al. 2005).

4.9 Chitinases

Chitin, an insoluble polysaccharide composed of linear chains of β-1,4-N-acetylglucosamine (GlcNAc), is abundant in crab, shrimp, and lobster shells. Each year, a vast amount of chitin waste material is released into the environment, which creates a serious environmental problem (Yan and Chen 2015). Therefore, a successful method for chitin recycling is very convenient for the ecosystem. Chitinase (Fig. 4.1) is the key enzyme in chitin degradation, since it hydrolyzes chitin polymer. The applications of chitinase includes bioconversion of chitin to useful products such as fertilizer and production of non-allergenic, non-toxic, biocompatible, and biodegradable materials (Hamid et al. 2013). Many *Bacillus* bacteria, such as *B. amyloliquefaciens, B. cereus, B. circulans, B. thuringiensis*, and *B. subtilis*, degrade chitin avoiding waste material accumulation (Wang et al. 2018). Many works have been done about chitinases including molecular engineering of chitinases (Pan et al. 2019).

B. thuringiensis produces small quantities of chitinases which are classified in four groups based on amino acid sequence alignments. Chitinases contain a catalytic domain and a putative chitin-binding domain, typical structure of enzymes that degrade biopolymers (Barboza-Corona et al. 2012). Chitinases have been employed for different environment purposes such as degradation of shrimp wastes, improving insecticidal activity, or controlling phytopathogenic fungi. Shrimps are part of traditional food in seafood restaurants. The wastes of them, which include heads and exoskeleton, represent more than the half of the volume. These wastes are composed of chitin, therefore chitinases can be employed to reduce shrimp wastes. On the other hand, chitinases destroy peritrophic membrane of the insects, increasing insecticidal activity of Cry proteins. Also, chitin is part of the cell wall of fungi, therefore, chitinases have activity against them (Barboza-Corona et al. 2012).

4.10 Esterases and Lipases

Esterases catalyze the cleavage and formation of ester bonds. Carboxyl ester hydrolases catalyze the cleavage or formation of carboxyl ester bonds, being often classified as esterases and lipases (De Vitis et al. 2018), which are the two major classes of hydrolases. Lipases prefer water-insoluble substrates, typically triglycerides composed of long-chain fatty acids, whereas esterases preferentially hydrolyze "simple" esters (e.g. ethyl acetate) and usually only triglycerides bearing fatty acids shorter than C6 (Bornscheuer 2002). They are used in the organic synthesis of optically pure compounds since they have high enantioselectivity, therefore they have been applied in the synthesis of chiral drugs employing them in reactions where chemo-or regioselectivity is an important matter. Probably the best studied enzyme is the so-called carboxyl esterase NP (Fig. 4.1) (NP from naproxen, a non-steroidal anti-inflammatory drug) originating from *B. subtilis* (Quax and Broekhuizen 1994). They can also be used in the mild removal of protecting groups. Esterases are employed in dairies and to produce wine, fruit juices, beer, and alcohol, besides they are used as trans-esterification catalysts to transform low value fats and oils into more valuable ones (Panda and Gowrishankar 2005).

4.11 Levansucroses

Levansucroses are the enzymes which catalyze transfer of the fructose moiety from sucrose to polyfructose (β-(2-6)-levan), releasing D-glucose. Levan-type fructooligosaccharides (FOSs) and β-(2-6)-levan selectively support the intestinal health having potential health benefits, besides they are applied in food and pharmaceutical fields. Levansucroses from microbial sources exhibit different oligomerization (FOSs) vs. polymerization (levans) ratios. In this sense, levansucroses from *Bacillus megaterium* and *Bacillus subtilis* were found to catalyze dominantly the synthesis of levan (Tian et al. 2011). Levansucroses effectivity is hindered by its hydrolytic activity, resulting in the release of glucose and fructose instead of desirable FOSs and levan. For this reason, the search of new levansucrase enzymes with improved catalytic activity to produce levan is desirable (Hill et al. 2019).

4.12 Keratinases

Keratinases are proteolytic enzymes that digest keratin, a strongly cross-linked structural polypeptide. Keratin wastes are increasingly accumulating in the environment mainly in the form of feathers or hair generated from various industries (Verma et al. 2017), for example in the leather industry dehairing process serves to eliminate the hair from the skin. The keratinases produced from hair degradation might be significantly applied due to their specificity (Hassan et al. 2020). Therefore, keratinases have many applications in several industries (Gupta and Ramnani, 2006). In the textile industry, keratinases serve to smooth the raw wool without damaging the internal parts of wool fibers, replacing the chemical methods (Ghaffar et al. 2018). The dermatologists prescribe various formulated drugs based on keratinases to treat hyperkeratosis which forms corns and calluses. For instance, Keratoclean®Hydra PB and Pure100 Keratinase have been produced by Proteos Biotech Company to treat the skin suffering from hyperkeratosis corns and calluses (Hassan et al. 2020). Keratinases are also used as fertilizers in organic farming to suppress nematodes and in Prion protein management (Verma et al. 2017). Therefore, these enzymes have multiple applications opening a new perspective to cover the unsatisfied demands in the pharmaceutical industries.

4.13 Conclusions and Perspectives

Bacillus enzymes are varied and with wide applications. This fact leads the great interest of industry to these metabolites. As we have seen, there are many enzymes produced by different strains of *Bacillus* with different industrial applications. In many of them, the major industrial application is oriented to green ecology minimizing the wastes produced by the human beings to avoid environment contamination. Besides, the enzymes produced by *Bacillus* have been applied to several industries and activities such as food, textile, detergents, and pharmaceutical. This chapter describes the enzymes produced by *Bacillus* species used in industries and their applications. The advances in recombinant DNA technologies can be highlighted, which help in improving enzymes production. By improving the production of the enzymes, they can be better applied in the industry. We expect that this review opens the door to new discoveries about these interesting enzymes which give the opportunity to approach the research to this topic. There is a need for a better production in almost all cases to take advantage and make better use of these amazing secondary metabolites from *Bacillus*.

References

Adinarayana K, Ellaiah P, Prasad DS (2003) Purification and partial characterization of thermostable serine alkaline protease from a newly isolated *Bacillus subtilis* PE-11. AAPS PharmSciTech 4:E56

Aizawa T, Urai M, Iwabuchi N, Nakajima M, Sunairi M (2010) *Bacillus trypoxylicola* sp. nov. xylanase-producing alkaliphilic bacteria isolated from the guts of Japanese horned beetle larvae (*Trypoxylus dichotomus septentrionalis*). Int J Syst Evol Microbiol 60:61–66

Ammoneh H, Harba M, Akeed Y, Al-Halabi M, Bakri Y (2014) Isolation and identification of local *Bacillus* isolates for xylanase biosynthesis. Iran J Microbiol 6:127–132

Anbu P (2013) Characterization of solvent stable extracellular protease from *Bacillus koreensis* (BK-P21A). Int J Biol Macromol 56:162–168

Annamalai N, Rajeswari MV, Thavasi R, Vijayalakshmi S, Balasubramanian T (2013) Optimization, purification and characterization of novel thermostable, haloalkaline, solvent stable protease from *Bacillus halodurans* CAS6 using marine shellfish wastes: a potential additive for detergent and antioxidant synthesis. Bioprocess Biosyst Eng 36:873–883

Balasubramanian N, Simões N (2014) *Bacillus pumilus* S124A carboxymethyl cellulase; a thermo stable enzyme with a wide substrate spectrum utility. Int J Biol Macromol 67:132–139

Barboza-Corona JE, Fuente Salcido NM, Leon-Galvan MF (2012) Future challenges and prospects of *Bacillus thuringiensis*. In: Sansinenea E (ed) *Bacillus thuringiensis* biotechnology. Springer, pp 367–384

Beg QK, Kapoor M, Mahajan L, Hoondal GS (2001) Microbial xylanases and their industrial applications: a review. Appl Microbiol Biotechnol 56:326–338

Bornscheuer U (2002) Microbial carboxyl esterases: classification, properties and application in biocatalysis. FEMS Microbiol Rev 26:73–81

Borriss R, Manteuffel R, Hofemeister J (1988) Molecular cloning of a gene coding for thermostable betaglucanase from *Bacillus macerans*. J Basic Microbiol 28:3–10

Borriss R, Olsen O, Thomsen KK, von Wettstein D (1989) Hybrid *Bacillus* endo-(1-3,1-4)-beta-glucanases: construction of recombinant genes and molecular properties of the gene products. Carlsb Res Commun 54:41–54

Caille J, Govindan CK, Junga H (2002) Hetero diels-alder biocatalysis approach for the synthesis of (*s*)-3-[2-{(methylsulfonyl)oxy}ethoxy]-4-(triphenylmethoxy)-1-butanol methanesulfonate, a key intermediate for the synthesis of the pkc inhibitor. Org Process Res Dev 6:471–476

Carneiro LABC, Yu L, Dupree P, Ward RJ (2018) Characterization of a β-galactosidase from *Bacillus subtilis* with transgalactosylation activity. Int J Biol Macromol 120:279–287

Confortin TC, Spannemberg SS, Todero I, Luft L, Brun T, Alves EA, Kuhn RC, Mazutti MA (2019) Microbial enzymes as control agents of diseases and pests in organic agriculture. New Future Dev Microbial Biotechnol Bioeng:321–332

Contesini FJ, Rodrigues de Melo R, Sato HH (2017) An overview of *Bacillus* proteases: from production to application. Crit Rev Biotechnol 1354354:1–14

De Vitis V, Nakhnoukh C, Pinto A, Contente ML, Barbiroli A, Milani M, Bolognesi M, Molinari F, Gourlay LJ, Romano D (2018) A stereospecific carboxyl esterase from *Bacillus coagulans* hosting non-lipase activity within a lipase-like fold. FEBS J 285:903–914

Fuller M, Lovejoy M, Brooks DA, Harkin ML, Hopwood JJ, Mickle PJ (2004) Immunoquantification of α- galactosidase evaluation for the diagnosis of Fabray's disease. Clin Chem 50:1979–1985

Ghaffar I, Imtiaz A, Hussain A, Javid A, Jabeen F, Akmal M, Qazi JI (2018) Microbial production and industrial applications of keratinases: an overview. Int Microbiol 21:163–174

Gupta R, Ramnani P (2006) Microbial keratinases and their prospective applications: an overview. Appl Microbiol Biotechnol 70:21–33

Gupta R, Beg QK, Khan S, Chauhan B (2003) An overview on fermentation, downstream processing and properties of microbial alkaline proteases. Appl Microbiol Biotechnol 60:381–395

Hamid R, Khan MA, Ahmad M, Ahmad MM, Abdin MZ, Musarrat J, Javed S (2013) Chitinases: An update. J Pharm Bioallied Sci 5:21–29
Hassan MA, Abol-Fotouh D, Omer AM, Tamer TM, Abbas E (2020) Comprehensive insights into microbial keratinases and their implication in various biotechnological and industrial sectors: A review. Int J Biol Macromol 154:567–583
Hii SL, Tan JS, Ling TC, Ariff AB (2012) Pullulanase: Role in Starch Hydrolysis and Potential Industrial Applications. Enzyme Res 2012:1–14
Hill A, Chen L, Mariage A, Petit J-L, de Berardinis V, Karboune S (2019) Discovery of new levansucrase enzymes with interesting properties and improved catalytic activity to produce levan and fructooligosaccharides. Catal Sci Technol 9:2931–2944
Huang Y, Zhang H, Ben P, Duan Y, Lu M, Li Z, Cui Z (2018) Characterization of a novel GH36 α-galactosidase from *Bacillus megaterium* and its application in degradation of raffinose family oligosaccharides. Int J Biol Macromol 108:98–104
Ishikawa K, Kataoka M, Yanamoto T, Nakabayashi M, Watanabe M, Ishihara S, Yamaguchi S (2015) Crystal structure of β-galactosidase from *Bacillus circulans* ATCC 31382 (BgaD) and the construction of the thermophilic mutants. FEBS J 282:2540–2552
Jaouadi NZ, Jaouadi B, Aghajari N, Bejar S (2012) The overexpression of the SAPB of *Bacillus pumilus* CBS and mutated sapB-L31I/T33S/N99Y alkaline proteases in *Bacillus subtilis* DB430: new attractive properties for the mutant enzyme. Bioresour Technol 105:142–151
Kalyani G, Rajesh EM (2018) Production and Purification of Amylase from *Bacillus subtilis* Isolated from Soil. Int J Eng Manage Res 8:246–254
Karmakar M, Ray RR (2011) Current trends in research and application of microbial cellulases. Res J Microbiol 6:41–53
Kim Y-K, Lee S-C, Cho Y-Y, Oh H-J, Ko YH (2012) Isolation of Cellulolytic *Bacillus subtilis* Strains from Agricultural Environments. ISRN Microbiol 2012:650563
Kirk O, Borchert TV, Fuglsang CC (2002) Industrial enzyme applications. Curr Opin Biotechnol 13:345–351
Kuhad RC, Gupta R, Singh A (2011) Microbial cellulases and their industrial applications. Enzyme Res 2011:1–10
Kunamneni A, Singh S (2006) Improved high thermal stability of pullulanase from a newly isolated thermophilic *Bacillus* sp. AN-7. Enzym Microb Technol 39:1399–1404
Ladeira SA, Cruz E, Delatorre AB, Barbosa JB, Martins MLL (2015) Cellulase production by thermophilic *Bacillus* sp. SMIA-2 and its detergent compatibility. Electron J Biotechnol 18:110–115
Li X, Yang H, Roy B, Wang D, Yue W-F, Jiang L-J, Park EY, Miao Y-G (2009) The most stirring technology in future: cellulase enzyme and biomass utilization. Afr J Biotechnol 8:2418–2422
Nagar S, Mittal A, Kumar D, Kumar L, Kuhad RC, Gupta VK (2011) Hyper production of alkali stable xylanase in lesser duration by *Bacillus pumilus* SV-85S using wheat bran under solid state fermentation. New Biotechnol 28:581–587
Nair SU, Singhal RS, Kamat MY (2007) Induction of pullulanase production in *Bacillus cereus* FDTA-13. Bioresour Technol 98:856–859
Nisha M, Satyanarayana T (2016) Characteristics, protein engineering and applications of microbial thermostable pullulanases and pullulan hydrolases. Appl Microbiol Biotechnol 100:5661–5679
Niu C, Liu C, Li Y, Zheng F, Wang J, Li Q (2018) Production of a thermostable 1,3-1,4-β-glucanase mutant in *Bacillus subtilis* WB600 at a high fermentation capacity and its potential application in the brewing industry. Int J Biol Macromol 107:28–34
Nonaka T, Fujihashi M, Kita A, Saeki K, Ito S, Horikoshi K, Miki K (2004) The crystal structure of an oxidatively stable subtilisin-like alkaline serine protease, KP-43, with a C-terminal beta-barrel domain. J Biol Chem 279:47344–47345
Oliveira C, Guimarães PMR, Domingues L (2011) Recombinant microbial systems for improved β-galactosidase production and biotechnological applications. Biotechnol Adv 29:600–609
Ortiz A, Sansinenea E (2019) Chemical compounds produced by sp. Factories and their role in nature. Mini Rev Med Chem 19:373–380

Pan M, Li J, Lv X, Du G, Liu L (2019) Molecular engineering of chitinase from *Bacillus* sp. DAU101 for enzymatic production of chitooligosaccharides. Enzym Microb Technol 124:54–62
Panda T, Gowrishankar BS (2005) Production and applications of esterases. Appl Microbiol Biotechnol 67:160–169
Pandey A, Nigam P, Soccol CR, Soccol VT, Singh D, Mohan R (2000) Advances in microbial amylases. Biotechnol Appl Biochem 31:135–152
Patil AG, Kumar PK, Mulimani VH, Veeranagouda Y, Lee K (2010) α-Galactosidase from *Bacillus megaterium* VHM1 and Its Application in Removal of Flatulence-Causing Factors from Soymilk. J Microbiol Biotechnol 20:1546–1554
Quax WJ, Broekhuizen CP (1994) Development of a new *Bacillus* carboxyl esterase for use in the resolution of chiral drugs. Appl Microbiol Biotechnol 41:425–431
Rozs M, Manczinger L, Vágvölgyi C, Kevei F (2001) Secretion of a trypsin-like thiol protease by a new keratinolytic strain of *Bacillus licheniformis*. FEMS Microbiol Lett 205:221–224
Schallmey M, Singh A, Ward OP (2004) Developments in the use of *Bacillus* species for industrial production. Can J Microbiol 50:1–17
Sharma KM, Kumar R, Panwar S, Kumar A (2017) Microbial alkaline proteases: optimization of production parameters and their properties. J Genet Eng Biotechnol 15:115–126
Simair AA, Qureshi AS, Khushk I, Ali CH, Lashari S, Bhutto MA, Mangrio GS, Lu C (2017) Production and Partial Characterization of α-Amylase Enzyme from *Bacillus* sp. BCC 01-50 and Potential Applications. Biomed Res Int 2017:1–9
Sookkheo B, Sinchaikul S, Phutrakul S, Chen ST (2000) Purification and characterization of the highly thermostable proteases from *Bacillus stearothermophilus* TLS33. Protein Expr Purif 20:142–151
Subramaniyan S, Prema P (2002) Biotechnology of microbial xylanases: enzymology, molecular biology, and application. Crit Rev Biotechnol 22:33–64
Sukumaran RK, Singhania RR, Pandey A (2005) Microbial cellulases-production, applications and challenges. J Sci Ind Res 64:832–844
Tanyildizi MS, Özer D, Elibol M (2005) Optimization of α- amylase production by *Bacillus* sp. using response surface methodology. Process Biochem 40:2291–2296
Tian F, Inthanavong L, Karboune S (2011) Purification and Characterization of Levansucrases from *Bacillus amyloliquefaciens* in Intra- and Extracellular Forms Useful for the Synthesis of Levan and Fructooligosaccharides. Biosci Biotechnol Biochem 75:1929–1938
Verma A, Singh H, Anwar S, Chattopadhyay A, Tiwari KK, Kaur S, Dhilon GS (2017) Microbial keratinases: industrial enzymes with waste management potential. Crit Rev Biotechnol 37:476–491
Wang D, Li A, Han H, Liu T, Yang Q (2018) A potent chitinase from *Bacillus subtilis* for the efficient bioconversion of chitin-containing wastes. Int J Biol Macromol 116:863–868
Wang X, Nie Y, Xu Y (2019) Industrially produced pullulanases with thermostability: Discovery, engineering, and heterologous expresion. Bioresour Technol 278:360–371
Xu H, Qin Y, Huang Z, Liu Z (2014) Characterization and site-directed mutagenesis of an α-galactosidase from the deep-sea bacterium *Bacillus megaterium*. Enzym Microb Technol 56:46–52
Yan N, Chen X (2015) Don't waste seafood waste: Turning cast -off shells into nitrogen-rich chemicals would benefit economies and the environment. Nature 524:155–158
Zareian S, Khajeh K, Ranjbar B, Dabirmanesh B, Ghollasi M, Mollania N (2010) Purification and characterization of a novel amylopullulanase that converts pullulan to glucose, maltose, and maltotriose and starch to glucose and maltose. Enzym Microb Technol 46:57–63

Chapter 5
Bacillus Species and Their Invaluable Roles in Petroleum Hydrocarbon Bioremediation

Cindy Baburam, Alfred Mitema, Tsepo Tsekoa, and Naser Aliye Feto

Abstract Exploration, production activities as well as improper waste disposal practices have led to major contamination of both aquatic and terrestrial ecosystems. Oil spills in particular have proven to be a global problem, with oil spills in the Gulf of Mexico, the coasts of Cape Town (South Africa) and Montara (CL, USA) being some of the landmarks. Efficient biodegradation of petroleum hydrocarbon (H-C) necessitates the presence of a stable biosurfactant and a microbe that can use the petroleum H-C as a sole carbon source. In light of this fact, *Bacillus* spp. have been reported to both produce a stable biosurfactant and tolerate high crude oil concentration and degrade aliphatics, monoaromatics and polyaromatics, which are the major constituents of crude petroleum. Such peculiarity makes the species the major source of potential candidates for H-C degradation. However, most of the oil-spill bioremediation products are presented as a magic cocktail of microorganisms, even though *Bacillus* spp. is the central genus if not the only one. Hence, this chapter aims to highlight the central and significant roles played by *Bacillus* spp. in oil-spill bioremediation and the mechanisms therein. Most importantly, highlighting the central roles played by *Bacillus* spp. and the mechanisms involved will contribute to targeted development of microbial and/or enzyme cocktail products and metabolic engineering to design next-generation oil-eaters in the future. Therefore, details of the major metabolic pathways, enzymes and biosurfactants deployed by *Bacillus* spp. during H-C biodegradation are addressed at possible depth.

Keywords *Bacillus* species · Biosurfactant · Enzyme cocktail · Hydrocarbon biodegradation · Oil-spill bioremediation · Petroleum

---

C. Baburam · A. Mitema · N. A. Feto (✉)
OMICS Research Group, Department of Biotechnology, Vaal University of Technology, Vanderbijlpark, South Africa
e-mail: naserf@vut.ac.za

T. Tsekoa
Council for Scientific and Industrial Research (CSIR), Pretoria, South Africa

M. T. Islam et al. (eds.), *Bacilli in Agrobiotechnology*, Bacilli in Climate Resilient Agriculture and Bioprospecting,
https://doi.org/10.1007/978-3-030-85465-2_5

5.1 Introduction

Our planet suffers greatly from various pollution problems, which include water, soil and air pollutions. Petroleum hydrocarbons are amongst the most toxic compounds released into oceans and spilled onto soil, thereby making them some of the major contributing factors to water and soil pollutions. Oil is introduced into the environment through natural seepage and human activities, which include pipeline and tanker leaks and spills. For instance, the British Petroleum (BP) /Deepwater Horizon discharge in 2010 stands out as the largest marine open water hydrocarbon discharge to date. The landmark incident caused a discharge of five million barrels of oil and at least 250, 000 metric tonnes of natural gas into the Gulf of Mexico (Joye et al. 2016).

The most important classes of organic pollutants in the environment include the mineral oil constituents and the halogenated products of petrochemicals (Chandra et al. 2013). Most petroleum compounds are reported to have carcinogenic properties (Souza et al. 2014). These compounds have a wide distribution and high toxicity hence hydrocarbons are considered as the principal organic markers of the anthropogenic activity in the ecosystems (Koshlaf and Ball 2017). Petroleum is a heterogeneous mixture of hydrocarbons (organic compounds containing carbon and hydrogen), which includes aliphatic (n-alkanes), alicyclic and aromatic hydrocarbons (polycyclic aromatic hydrocarbons), which are insoluble in water (Joye et al. 2016). Therefore there is an increasing need to develop a green technology to remove or transform hydrocarbons to a non-toxic or less-toxic version of the compound through application of hydrocarbon-degrading microbial or enzyme cocktails.

Bioremediation is a less-invasive and less-expensive process as compared to the classical decontamination methods (Zafra et al. 2016). Some microbes have been documented to inhabit and proliferate in hydrocarbon-contaminated environments, thus, using hydrocarbon as sole carbon and energy sources (Chebbi et al. 2017). The characterisation of certain metabolic pathways indicates that high enzymatic capacity allows the microbial communities to degrade complex hydrocarbons, and this therefore promotes the importance of enzymes in the bioremediation process (Peixoto et al. 2011). Bioremediation can be carried out through bioattenuation, biostimulation and/or bioaugmentation (Adams et al. 2015). The classical bioremediation approach that has been in use is bioaugmentation, which is application of a consortium of *in vitro*-cultivated inocula into the polluted site. However, there are multiple factors that lessen the effectiveness of this technique. This includes biotic factors such as predation and niche-exclusion, which could in turn affect the activity of the introduced microorganisms and result in the decline of the introduced populations. Moreover, bioaugmentation poses a potential risk with regard to the addition of microorganisms, in some cases a consortium that contains genetically modified ones, to the environment, which could be difficult to monitor and contain in the natural environment. Besides, relatively the time required for the microorganism to grow and be able to produce the enzyme required to biodegrade the hydrocarbons has negative implications on the rate of degradation (Adams et al. 2015).

There are a multitude of extreme environments on earth which are colonised by microorganisms called extremophiles, which can thrive under diverse harsh conditions (Mirete et al. 2016). Particularly those microorganisms with the evolved ability to grow in petroleum hydrocarbon-contaminated soils and water as highlighted in this chapter. Many technologies already developed through the years for remediation, rehabilitation and restoration of contaminated environments exploit the potential of microbial biological systems to use toxic compounds as substrates for growth and convert them into harmless by-products. Hence, the vast amount of research conducted to date concentrate on the capabilities of a single or a consortia of microbes exhibiting robust and effective hydrocarbon degradative properties (Mukherjee et al. 2017). The *Bacillus* spp. has been put in the spot light over the years for being an extremophile and has been studied in great extent over the years (Parthipan et al. 2017; Elumalai et al. 2019; Jin et al. 2019; Wang et al. 2019). More particularly, they have been found in hydrocarbon-polluted environments (Al-Dhabaan 2019; Christova et al. 2019; Lima et al. 2020) This provides promising potential for *Bacillus* spp. independent as well as in co-culture application for hydrocarbon bioremediation as reviewed extensively in this chapter.

5.2 The Chemistry of Hydrocarbons in Petroleum Crude Oil

Hydrocarbons are compounds that consist exclusively of carbon and hydrogen and lack functional groups. The lack of functional groups make hydrocarbons exhibit apolar properties, low chemical reactivity at room temperature and low water solubility in long-chain hydrocarbons. Hydrocarbons can be classified into the following groups (a) aliphatic groups which include straight-chain (n-alkanes), branched-chain and cyclic compounds and (b) aromatic groups which include mono- or polycyclic hydrocarbons and many important compounds which contain aliphatic hydrocarbon chains (e.g. alkylbenzenes) (Koshlaf and Ball 2017). Petroleum crude oil is a complex mixture of low and high molecular weight hydrocarbons like alkanes, aromatics, asphaltenes and resins associated with other organic compounds containing sulphur, nitrogen and oxygen (Fig. 5.1). Among these hydrocarbons mentioned, aromatic hydrocarbons constitute a major fraction (Arun et al. 2011). The structural complexity of the various hydrocarbon constituents from petroleum makes the process of degradation highly difficult.

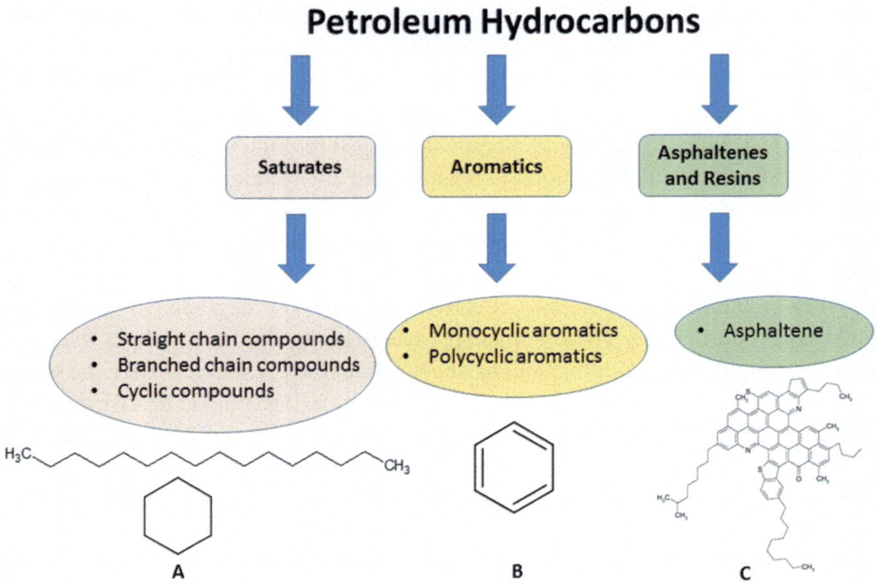

Fig. 5.1 The different classes of hydrocarbons that comprise crude oil are (**a**) saturated hydrocarbon; (**b**) aromatics and (**c**) asphaltenes and resins

5.2.1 Toxicity and Fate of Petroleum Hydrocarbons in the Environment

It is of utmost importance to expand our knowledge about the fate of hydrocarbons in the environment in order to have some level of pollution control. Due to their persistence and toxicity they cause permanent damage to the ecosystem (Atlas et al. 2015). The toxicity of the petroleum hydrocarbons depends on the solubility and bioavailability of the hydrocarbons. The water soluble fractions of the aromatics and polyaromatics are considered to be the most toxic (mutagenic, teratogenic and carcinogenic), though it is generally reported that most petroleum hydrocarbons have carcinogenic properties (Pandey et al. 2016). As a result of the reported toxicity, several hydrocarbons are classified by the Agency of Environmental Protection, the World Health Organisation and the European Union as top priority pollutants (Koshlaf and Ball 2017). Anthropogenic hydrocarbon contamination of soil and water is a global issue throughout the industrialised world (Umeh et al. 2017). Anthropogenic activity in different ecosystems are based on hydrocarbons as the principal organic markers due to their wide distribution and their high toxicity (Maikudi Usman et al. 2016).

Soil acts as a repository for many hydrocarbons; this poses a great concern due to the adverse effects on human health and the high levels of accumulation in the environment over long periods of time (Mukherjee et al. 2017). Throughout history there have been major global oil spills which have had catastrophic implications on the environment (Zampolli et al. 2014; Joye et al. 2016). The Deepwater Horizon oil spill also referred to as the Gulf of Mexico oil spill in 2010 was classified as a major oil spill . The US Government estimated the total discharge at 4.9 million barrels (210 million US gal; 780,000 m^3) (On Scene Coordinator Report on Deepwater Horizon Oil Spill 2011). In addition, in 2011, an Exxon Mobil pipeline spilled 63,000 gallons of crude oil into the Yellowstone River near Laurel, Montana. Following this, in January 2015 a ruptured oil pipeline leaked up to 50,000 gallons of crude oil into the Yellowstone River in Montana. Days after the spill, officials detected benzene, a cancer-causing agent, in the water supply downstream of the river (CBS news 2015). These are but a few of the major oil spills which have affected land and water ecosystems. Therefore, there is a great need for bioremediation methods and preventative measures to be put in place to bioremediate and prevent future incidences.

5.3 Bioremediation

Technologies commonly used for soil remediation of petroleum hydrocarbons include mechanical burying, evaporation, dispersion and washing. However, these measures for remediation are very costly, time-consuming and not effective in the long run. Bioremediation is the preferred method since it is cost-effective, efficient and eco-friendly as compared to other treatments (Varjani 2017). Bioremediation is applied after the use of physical and chemical methods and natural attenuation methods have not been successful (Roy et al. 2018). Bioremediation is the preferred method since it is cost effective, efficient and eco-friendly as compared to other treatments (Varjani 2017). Various species of archaea, bacteria, algae and fungi have been reported to play a significant role in treatment of oil spills in different environments (Sharma et al. 2018a). The details of different bioremediation approaches in general and of oil spills in particular are discussed in the following sections.

5.3.1 *Biostimulation and Bioaugmentation Methods for Bioremediation*

There are two main types of bioremediation methods, these include biostimulation and bioaugmentation (Wu et al. 2016). Many reports have shown that both methods have greatly enhanced the biodegradation of hydrocarbons in oil-polluted soils. The effects of both methods are case-specific and results could be inconsistent. Hence, the use of bioaugmentation and biostimulation needs to be carefully planned and studied for each type of contaminant and environmental conditions (Roy et al. 2018).

Biostimulation involves the modification of natural conditions of a site to increase the rates of biological degradation by stimulating indigenous microbial community endowed with catabolic pathways for hydrocarbon degradation. The environmental conditions can be modified through the addition of surfactants, nutrients, water and chemical species acting as donors or acceptors of electrons (Zrafi-Nouira and Saidane-Mosbahi 2012). Perfumo et al. (2007) added nitrogen, phosphate, inorganic potassium and surfactant on the bioremediation of a ground polluted with hexadecanes. Following treatment of the ground, there was a 10% increase in the degradation. Bragg et al. (1994) used fertilisation of the marine environment to stimulate the natural biomass in areas of Alaska following the accident of the Exxon Valdez. Recently Wu et al. (2016) compared the efficiency of bioaugmentation and biostimulation approaches for hydrocarbon biodegradation. Accordingly, after six weeks of treatment they reported that 60% and 34% of the total petroleum hydrocarbons biodegradation, respectively were achieved.

In some cases, the indigenous microbial populations of polluted ecosystems may not contain the metabolic tools to carry out complete degradation of the pollutants (Wu et al. 2013). In such cases bioaugmentation can be used to overcome discrepancy by incorporating microorganisms (*in vitro*-cultivated inoculum) with adapted metabolism for hydrocarbon degradation in contaminated sites (Zrafi-Nouira and Saidane-Mosbahi 2012). For instance, Da Silva et al. (2005) showed an improvement of anaerobic degradation of a mixture composed of benzene, toluene, ethylbenzene, xylene and ethanol after the addition of methanogen consortia.

It is important to emphasise that the success of the implementation of bioaugmentation, for example, depends on competitiveness of the inoculated strains in different environments (Adams et al. 2015). Genetically modified organisms can also prove to be beneficial in the bioremediation process but there are some limitations to their applications such as problems with government legislation and monitoring, and long-term assessment is therefore very difficult. In either case of introduction of genetically modified or wild-type strains, the potential impacts of introducing degrading microorganisms in the presence of indigenous microbes must be evaluated and monitored (Lim et al. 2016). Some examples of genetically modified *Bacillus* used for bioremediation studies include that of *Bacillus subtilis* developed for bioremediation of arsenic-contaminated organic waste (Huang et al. 2015).

Recent studies by Roy et al. (2018) demonstrated an effective strategy for the bioremediation of petroleum refinery sludge using a combined biostimulation–bioaugmentation approach. This nutrient amendment to the sludge resulted in a native microbial shift that promoted hydrocarbon degrading populations within the sludge microbiome. It resulted in 57–75% total petroleum hydrocarbon reduction in the sludge. Therefore, nutrient(s)-induced community dynamics and metabolic interplay might be involved in accelerating bioremediation applications.

5.3.2 *The Potential of Microorganisms for Hydrocarbon Bioremediation*

Microorganisms have the amazing ability to metabolise many organic contaminants, using them as an energy source or converting them to non-toxic products (carbon dioxide, water and biomass). Many petroleum-degrading strains are being characterised specifically for their metabolic pathways involved in biodegradation of petroleum. This natural biodegradation by microorganisms prevents the dispersion of petroleum hydrocarbons in the soil and water and minimises the pollution levels to some degree (Koshlaf and Ball 2017). Numerous microorganisms, such as bacteria, cyanobacteria, green algae and fungi, are capable of either degrading or producing hydrocarbons; this completely depends on the metabolic pathways involved and the different environmental conditions (e.g. aerobic, anaerobic, varied pHs and salinities (Varjani 2017). The use of microbes and their enzymes for the removal of pollutants has proven to be an effective, safe and a less-expensive method. Native microbes growing in a polluted environment are expected to be more robust in degradation than non-native species since they produce certain degrading metabolites, allowing them to out-perform non-native species (Parthipan et al. 2017). Alternatively, the method of co-cultivation of native microbes with efficient pollutant-degrading microbes is believed to be a valuable strategy to increasing contaminant removal in a short period of time (Patowary et al. 2016).

A large number of bacterial species have been classified as hydrocarbonoclastic bacteria and play key roles in the removal of hydrocarbons from polluted environments (Parthipan et al. 2017; Sharma et al. 2018a). *Bacillus* spp. are one such example of hydrocarbonoclastic alkane degrading bacteria found in literature (Freitas de Oliveira et al. 2013; Borah and Yadav 2014a; Liu et al. 2016; Parthipan et al. 2017). Some microorganisms have the ability to degrade aliphatics, monoaromatics or polyaromatics while others are involved in the degradation of resins (Varjani 2017). Some examples of *Bacillus* spp. involved in petroleum hydrocarbon compounds are listed in Table 5.1.

Although many bacteria are able to metabolise organic pollutants, a single bacterium is not capable of degrading all or even most of the compounds in polluted soils. The main reason for this stems from the fact that a single bacterium does not possess the enzymatic capability to complete such a task. Therefore, mixed microbial communities are most effective in degrading complex mixtures of hydrocarbons present in contaminated areas since the genetic information from all the microorganisms is present (Dombrowski et al. 2016). This process occurs gradually by sequential metabolism of its compounds. The genes involved in this degradation process can be located either on chromosomal or plasmid DNA. Therefore, it is crucial to assess biodegradation as a multi-domain community for one to completely appreciate the metabolic potential of the indigenous microbial community (Varjani 2017).

The importance of efficient bacterial consortiums for potential hydrocarbon remediation was reinforced by Patowary et al. (2016). Fourteen different bacterial

Table 5.1 *Bacillus* spp. involved in the degradation of specific classes of petroleum hydrocarbon compounds

Petroleum hydrocarbon	Microorganism responsible for biodegradation of the H-C	References
Aliphatics	*Bacillus thermoleovorans* *Bacillus subtilis* *Bacillus licheniformis Y-1* *Bacillus sp.* L26 *and* *Bacillus sp.* L30	(Ahmed et al. 2010, Darsa and Thatheyus 2014, Liu et al. 2016, Parthipan et al. 2017), (Lu et al. 2020), (Lima et al. 2020)
Monoaromatics	*Bacillus lentus* strain LP32 *Bacillus pumilus* MVSV3 *Bacillus amyloliquefaciens*	(Opere et al. 2013), (Surendra et al. 2017), (Wongbunmak et al. 2020)
Polyaromatics	*Bacillus cereus* Strain DRDU1 *Bacillus subtilis* 3KP *Bacillus subtilis* BL-27 *Bacillus cereus Strain* JMG-01	(Mittal and Singh 2009, Borah and Yadav 2014a, 2014b), (Vinothini et al. 2015), (Ni'Matuzahroh et al. 2017), (Wang et al. 2019), (Das et al. 2017)

consortia were designed involving both biosurfactant producing and non-producing isolates. The results obtained from this study showed that a consortium comprising two *Bacillus* stains namely, *B. pumilus* KS2 and *B cereus* R2, displayed the best result in the desired degradation of crude oil. This consortium showed degradation of up to 84.15% of TPH after 5 weeks of incubation using gravimetric analysis, further cementing the central and dominant role played by *Bacillus* spp. Fourier transform infrared (FTIR) and Gas chromatography-mass spectrometer (GC-MS) analyses correlated with the gravimetric results revealed that the consortium had removed a wide range of petroleum hydrocarbons including both aliphatic and aromatic hydrocarbons. Furthermore, analysis carried out by Dombrowski et al. (2016) through reconstruction of metabolic pathways of hydrocarbon-degrading bacteria from the Deepwater Horizon oil spill further supports the complexity of oil-degrading community and the co-ordination of its metabolic pathways to ensure complete degradation of petroleum hydrocarbons in the environment.

The enzymatic mechanisms are also crucial to the success of microbial degradation of hydrocarbons. Different microbial electron acceptors such as oxygen, nitrate, manganese, iron and sulphate are all involved in the biotransformation of aliphatic and aromatic hydrocarbons. For example, the breakdown of n-alkanes are catalysed by alkane activating enzymes, which are monooxygenases that lower chemical reactivity of the hydrocarbon by generating reactive oxygen species. The best alkane-degrading pathway identified is encoded by the OCT plasmid of *P. putida* GPo1 (van Beilen et al. 2001). The first enzyme in this pathway is an integral-membrane non-haem diiron monooxygenase (AlkB) that hydroxylates alkanes at the terminal position. This produces a primary alcohol which is further oxidised by alcohol and aldehyde dehydrogenases to form the corresponding aldehyde. The product is finally converted to fatty acid via oxidation, which is channelled into the β-oxidation pathway in the form of acetyl-CoA.

5.3.3 *The Mechanisms Employed by* **Bacillus** *spp. for Bioremediation of Hydrocarbons*

A number of bacterial species belonging to *Bacillus* have been identified as petroleum hydrocarbon degraders and have specifically been involved in naphthalene and pyrene degradation (Annweiler et al. 2000). Thamer et al. (2013) conducted studies showing that *Bacillus thuringiensis* has great capacity for the biodegradation of crude oil. This bacteria exhibited the ability to break down hydrocarbon compounds by 80% and total biomass reaches 5g/l, while the amount of emulsion reaches 2.3 g/l. Numerous studies have supported the implementation of *Bacillus* spp. in bioremediation activities. Sakthi Priya et al. (2015) showed that *Bacillus subtilis* isolated from a polymer dump site in India was used for degradation of crude oil. It was noted that the crude oil degradation and viscosity reduction was observed to be 80% and 60% respectively within a period of 10 days. It was concluded that high microbial adherence, surface tension reduction, emulsification activity, quantity of biosurfactant produced and stability provide a true indication that *Bacillus subtilis* is a potential microorganism for oil spill treatment. These results are compared with the other studies conducted (Borah and Yadav 2014a; Darsa and Thatheyus 2014; Jabeen et al. 2015; Vinothini et al. 2015; Parthipan et al. 2017; Tao et al. 2017).

Work done by Bujang et al. (2013) demonstrated a very good biodegradation capability of oily wastewater by *Bacillus cereus* from three different automotive workshops. Oily waste water has a complex composition of hydrocarbons and it was found that the degradation was up to 91% after 5 days of incubation. This finding was further supported by studies conducted (Borah and Yadav 2014b, 2017) on *Bacillus cereus* DRDU1 which was found to efficiently degrade 96% of kerosene. The results obtained for this biosurfactants emulsification index for kerosene, crude oil and used engine oil were in a good range thus making it an attractive future application for MEOR process. Other species of *Bacillus* recently found to have hydrocarbon degradation potential include *Bacillus thermoleovorans* (Annweiler et al. 2000), *Bacillus amyloliquefaciens* An6 (Ayed et al. 2015), *Bacillus licheniformis* Y-1 (Liu et al. 2016), *Bacillus pumilus* (Patowary et al. 2016) and *Bacillus methylotrophicus* USTBa (Chandankere et al. 2014).

5.3.3.1 Surfactants and Biosurfactants for Bioavailability of Pollutants

Surfactants are surface active substances which consist of hydrophilic (polar) and hydrophobic (nonpolar) portions on their molecules (Mulligan 2005). The polar part of the biosurfactant can be an amino acid, a carbohydrate and/or a phosphate group.

The long chain fatty acid makes up the nonpolar portion. Synthetic surfactants are widely used to treat oil spills by dispersing them and accelerating their mineralisation (Fernandes et al. 2016). The mechanism is the ability to reduce surface and interface tensions between liquids, solids and gases, allowing hydrocarbons to readily emulsify in water by forming aggregates called micelles (Maikudi Usman et al. 2016). Knowing the number of bacteria involved with biosurfactant production will enhance hydrocarbon degradation of hydrocarbons in two ways, the first is a direct method which favours the solubility of hydrophobic compounds and the second is by indirectly increasing the availability of hydrophobic compounds to the native microorganisms (Noparat et al. 2014). This bioavailability of petroleum crude pollutants to microbes is crucial for the process of bioremediation (Varjani and Upasani 2017a).

The demand for surfactants for many years has been met by those derived from petroleum which proved to be toxic to the environment and non-biodegradable. This has therefore led to using biosurfactants which serve the same purpose but are produced extracellularly or as part of the cell membrane by bacteria, fungi and yeast and utilise different substrates like simple sugars, oils and hydrocarbons from contaminated environments. Biosurfactants have advantages over their chemical counterparts with regard to their effective response to extreme environmental conditions, biodegradability, lower toxicity and the ability to be produced from cheap organic sources, which facilitate commercialisation (Díaz De Rienzo et al. 2016). Biosurfactants are produced extracellularly by a few microorganisms including bacteria, fungi and yeast. *Bacillus* spp. (Bezza and Chirwa 2015), *Streptomyce*s spp. (Ferradji et al. 2014) and *Sphingobacterium* spp. (Noparat et al. 2014) are but a few examples of biosurfactant producing bacteria. The addition of biosurfactant producing bacteria such as those mentioned above can resolve the problem of unavailability of hydrocarbons, which is due to the pollutants' lack of solubility (Varjani and Upasani 2016, 2017b).

The quality, quantity and type of biosurfactant produced are influenced by the concentration of carbon, nitrogen, iron and phosphorous iron sources. Environmental factors such as growth conditions (pH, agitation, temperature and oxygen accessibility) are also a major contributor and are also valuable in assessing biosurfactant production (Maikudi Usman et al. 2016). This was clearly demonstrated by Khan et al. (Ali Khan et al. 2017) with regard to the role of nutrients in bacterial biosurfactant production by four bacterial strains (*Pseudomonas poae* BA1, *Acinetobacter bouvetti* BP18, *Bacillus thuringiensis* BG3 and *Stenotrophomonas rhizophila* BG32). Further, pH and temperature were shown to influence biosurfactant production by *Pseudomonas* spp. as concluded by Müller et al. (2011). Kinetics of nutrient-enhanced crude oil degradation was also carried out by Chettri et al. (2016) for *Bacillus* spp. AKS2 isolated from a refinery in Guwahati, and further support work was done by Khan et al. (Ali Khan et al. 2017). The results showed a seven-fold increase in biodegradation due to nutrient enhancement.

Many recent reports have shown the application of biosurfactant producing microbes (Bezza and Chirwa 2015; Díaz De Rienzo et al. 2016; Joy et al. 2017). The remediation of petroleum-contaminated soils and water in the presence of microorganisms can be enhanced by the production of biosurfactants (Ferradji et al. 2014). It is therefore important to characterise biosurfactant producing bacteria. A vast amount of research have concluded that the *Bacillus* spp. are potential biosurfactant producers (Chandankere et al. 2013; Ferradji et al. 2014; Díaz De Rienzo et al. 2016; Maikudi Usman et al. 2016; Parthipan et al. 2017), biodegrading microbes and have been extensively used in microbial enhanced oil recovery (MEOR) (Gudiña et al. 2013; Youssef et al. 2013; Al-Wahaibi et al. 2014), bioremediation purposes (Bezza and Chirwa 2015; Liu et al. 2016) and biodegradation (Bujang et al. 2013; Thamer et al. 2013; Darsa and Thatheyus 2014).

There are pathways for biosurfactants synthesis that are well described by many studies for the *Bacillus* genera for the production of surfactin linked to the following genes (*spf, srfAA, srfAB, srfAC* and *srfAD* among others). In particular, two novel surfactin molecules isolated from cell-free cultures of *Bacillus subtilis* HSO121 were presented by Liu et al. (2016). Surfactins have the ability to form sphere-like micelles; their amphipathic and surface properties play a role in the minor polar and major hydrophobic domains of a surfactin molecule (Liu et al. 2015). Currently, a vast amount of research has been focused on the discovery of a novel hydrocarbon-degrading biosurfactant that would enhance biodegradation in hydrocarbon-polluted environments. Ayed et al. (2015) had successfully produced a biosurfactant from a bacterial strain *Bacillus amyloliquefaciens* An6. They reported that this biosurfactant displayed good stability over a wide range of temperatures, pH and salinity. The solubility of diesel was enhanced by up to 10% as compared to SDS or Tween 80 respectively.

Comparative studies have also been carried out between *B. thailandensis* E264 and *P. aeruginosa* ATCC 9027 with regard to rhamnolipid biosurfactant production (Díaz De Rienzo et al. 2016). Furthermore, Chandankere et al. (2014) carried out work to show the properties and characterise the biosurfactant in crude oil biodegradation by *Bacillus methylotrophicus* USTBa. The biosurfactant produced during the course of hydrocarbon degradation in this study was monitored by surface tension and cell hydrophobicity measurements. The results showed the ability of this biosurfactant to decrease the surface tension of water from 72 to 28 mN/m, with a critical micelle concentration of 35 mg/L. It exhibited 90% emulsification activity on the crude oil. The combination of these results made this biosurfactant an appropriate candidate for bioremediation of crude oil contaminants.

5.3.3.2 Bacterial Chemotaxis, Flagellar Motility and Biofilm Formation

The exposure of bacteria to pollutants such as hydrocarbon contaminants induces chemoattraction or chemorepellent reactions. Recent research suggests that the capacity to degrade harmful compounds have co-evolved in bacteria by reacting chemotactically to the pollutant therefore increasing their bioavailability which directly enhances biodegradation rates (Krell et al. 2013). Biosurfactant synthesis can be related to chemotaxis, flagellar motility and biofilm production (Chrzanowski et al. 2012). Structures used by microorganisms for locomotion such as flagella and pili have important roles in cell motility and chemotaxis. This helps the bacteria to move to relatively better niches as well as to attach to surfaces so as to start the biofilm formation through the complex cell-to-cell signalling. In hydrocarbon-polluted niches, the process of alkane degradation will take place in the oil–water interfaces. This makes more carbon available for growth, hence increasing cell density leading to the biosynthesis of biosurfactants which can disperse hydrophobic compounds and promote their bioavailability (Nie et al. 2012).

Research recently carried out by Vasconcellos et al. (2017) in which 16 genes associated with chemotaxis and flagellar motility in prokaryotes were found in fosmid clones from a constructed fosmid library using metagenomic DNA from a petroleum reservoir in Northeast Brazil, thus showing the link between hydrocarbon degradation and chemotaxis. Anthracene degradation studies carried out by Das et al. (2017) using *B. cereus* strain JMG-01 isolated from hydrocarbon-contaminated soils showed that the presence of anthracene and the process of degradation lead to the modification of the cell surface morphology and the formation of an exopolymeric matrix. The filamentous growth of the biomass in the form of biofilm reveals the chemotaxis behaviour of this *Bacillus* strain in enhanced anthracene degradation.

Further research by Dombrowski et al. (2016) showed the presence of major genes for motility and for use of scarce nutrients, which could suggest that microorganisms are well adapted for chemotactic motility towards a hydrocarbon substrate. These genes might be of great importance for both the survival and growth of these oil-degrading organisms found in the Gulf of Mexico.

5.3.3.3 Uptake and Trans-membrane Transport of Hydrocarbons

The uptake of hydrocarbons have been the subject of many studies through the years. In particular, biosurfactants discussed previously have received a lot of the limelight. However, the transport mechanisms by which hydrocarbons cross the cell membrane have received little attention. The membrane of bacterial cells is hydrophobic. This causes an issue with biodegradation due to the decreased availability of hydrocarbons for uptake by the cells. Most substrates need to undergo cellular attachment to become accessible. Bacteria need to have access to the substrate, hence they need to be either dissolved in the aqueous phase or the bacteria have to directly adhere to the hydrocarbon. Hydrocarbons that are adsorbed on the surface of the cell are transported across the membrane into the interior of the microorganism. These hydrocarbons are degraded in the presence of enzymes, which is a quick response (Hua and Wang 2014).

Following the physical interaction between bacterial cells and hydrocarbons, the next step is trans-membrane transport. Microbial cells have resources to physically access soluble, emulsified hydrocarbons and large oil droplets and transport these substrates across cell membranes and form inclusions before the hydrocarbons are metabolised (Alvarez et al. 1997; Bouchez-Naïtali and Vandecasteele 2008). Studies have been conducted on trans-membrane transport of phenanthrene, naphthalene and *n*-hexadecane (Kim et al. 2002; Kallimanis et al. 2007). For Gram-negative bacteria, outer membrane proteins have been reported to be involved in the transport of hydrocarbons across the cell membrane. Majority of the research has been carried out on *E. coli* with respect to long chain fatty acid transporter, FadL (Wiener and Horanyi 2011). This mechanism of trans-membrane transport is still poorly understood and needs to be further investigated to close the knowledge gap. However, it can be concluded that it is a huge player in the success for the ability of microorganisms to degrade petroleum hydrocarbons.

5.3.4 Enzymatic Approach for Bioremediation of Hydrocarbons

Enzymes are complex biological molecules with the sole purpose of catalysing a number of biochemical reactions involved in the degradation of hydrocarbons (Kalogerakis et al. 2017). Enzyme engineering has also been employed to improve the catalytic activity of isolated enzymes in different environments (Fig. 5.2). This is done by changing or modifying the basic amino acid structure of the enzymes

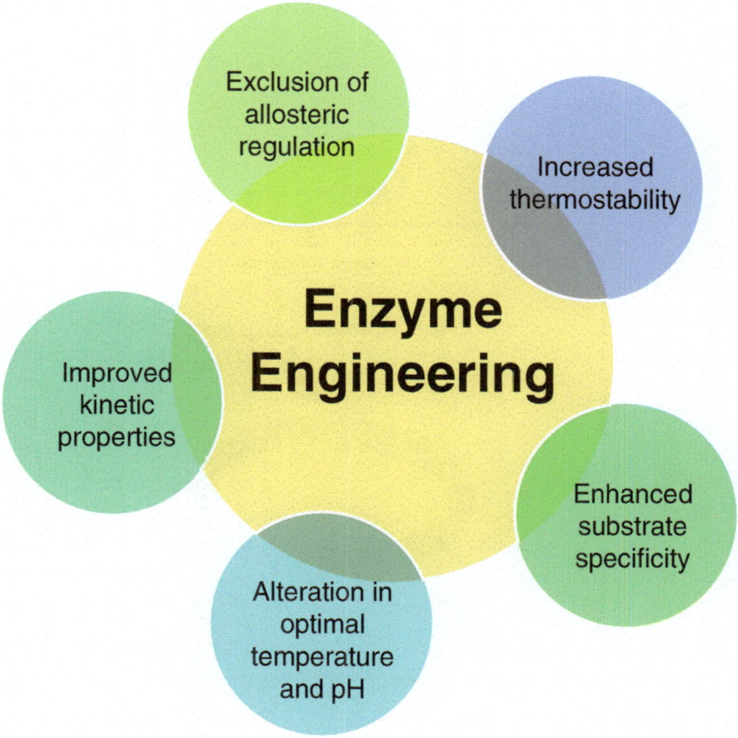

Fig. 5.2 Enzyme engineering to improve enzymatic properties for bioremediation

(Sharma et al. 2018a). Reseach carried out by Parthipan et al. (2017) showed that *Bacillus subtilis* A1 is able to produce degradative enzymes such as alkane hydroxylase and alcohol dehydrogenase. Recently Lu et al. (2020) showed the co-expression of alcohol dehydrogenase and aldehyde dehydrogenase in *Bacillus subtilis* for alcohol detoxification. Hence, these enzymes identified in *Bacillus* have the potential to be used in bioremediation applications.

Biodegradation of hydrocarbons including both aliphatic and aromatic compounds may occur under aerobic or anaerobic conditions (Koshlaf and Ball 2017). When considering aerobic conditions, oxygenase enzymes introduce oxygen atoms into hydrocarbons (monooxygenases introduce one oxygen atom to a substrate while dioxygenases introduce two (Parthipan et al. 2017). The most rapid and complete degradation of the majority of pollutants is brought about under aerobic conditions. Figure 5.3 shows the processes involved during the microbial degradation of hydrocarbons as a sole carbon source under aerobic conditions. Anaerobic degradation is catalysed by anaerobic bacteria; these include sulphate-reducing bacteria that use terminal electron acceptors (Van Hamme et al. 2003). The catabolism of hydrocarbons under aerobic conditions is faster than under anaerobic conditions due to the availability of oxygen as an electron acceptor (Cao et al. 2009).

There are several enzymes involved in the hydrocarbon bioremediation process which have been identified. These include oxidoreductases (laccases, oxygenases and peroxidases) and hydrolases, which are depicted in Table 5.2 with their catalytic function (Kadri et al. 2017).

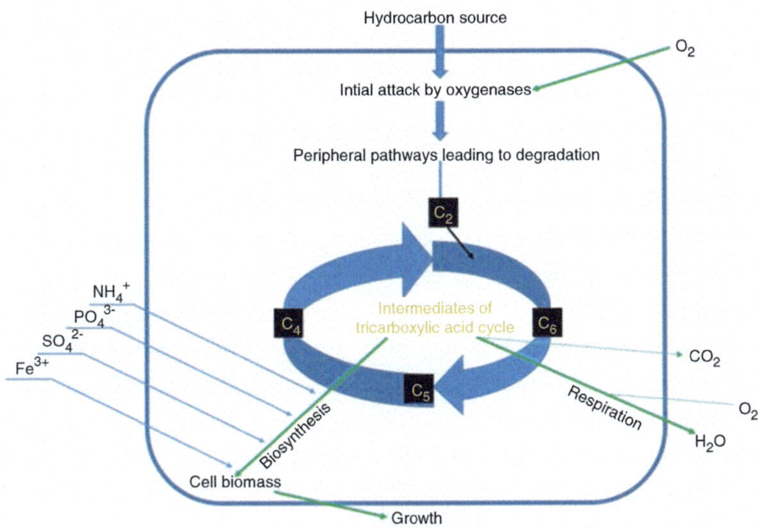

Fig. 5.3 Processes involved during the microbial degradation of hydrocarbons as a sole carbon source under aerobic conditions modified from (Fritsche and Hofrichter 2008)

Table 5.2 Enzymes involved in the bioremediation of hydrocarbons and their functions

Enzyme family	Examples	Functions	Reference
Oxidoreductases	Oxygenases	Catalyses the cleavage of the ring in aromatic compounds by adding one or two molecules of oxygen.	(Thomas et al. 2016, Sharma et al. 2018b)
	Laccases	Catalyses the oxidation of phenolic and aromatic compounds.	(Shekher et al. 2011)
	Peroxidases	Catalyses reduction reactions in the presence of hydrogen peroxide and generate reactive free radicals after oxidation of organic compounds.	(Sharma et al. 2018a)
Hydrolases	Haloalkane Dehalogenase	Catalyses the biodegradation of halogenated aliphatic compounds.	(Nagata et al. 2015)

5.3.4.1 Enzymatic Degradation of Aliphatic Hydrocarbons

When saturated aliphatic hydrocarbons are oxidised, the final product is acetyl-CoA, which is catabolised in the citric acid cycle as well as the production of electrons in the electron transport chain. This chain of events is repeated to further degrade the aliphatic hydrocarbons, which are then oxidised to carbon dioxide (Peixoto et al. 2011). The *n*-Alkanes are the main constituents of mineral oil contamination (Varjani 2017). Alkane metabolism is not a straightforward method since alkanes are very hydrophobic and less soluble in water (Rojo 2009). This poses a great challenge with regard to their uptake and hence facilitates their accumulation in the cytoplasmic membrane. Such accumulation could be detrimental as it might alter the membranes' fluidity. Despite such problems, many microorganisms have acquired the ability to degrade alkanes and use them as a carbon source (Wentzel et al. 2007). Four pathways have been identified for the initial attack on *n*-alkanes as shown in Fig. 5.4.

First is the terminal oxidation pathway, which has been well studied in *Geobacillus thermodenitrificans* NG80-2 (Li et al. 2008). In this pathway the alkanes are first attacked at their terminal methyl group to yield corresponding primary alcohols. These alcohols are further oxidised by alcohol dehydrogenases and aldehyde dehydrogenases to yield fatty acids which enter the β-oxidation pathway. The second pathway is called the biterminal oxidation, in which the termini of the *n*-alkane undergo oxidation to the corresponding fatty acid without rupturing of the carbon chain. The fatty acid produced undergoes ω-hydroxylation at the terminal methyl group. This yields a ω-hydroxy fatty acid that is further converted to a dicarboxylic acid which enters β-oxidation (Coon 2005; Li et al. 2008). Subterminal oxidation has also been identified in a number of microorganisms and takes place when alkanes are oxidised at the subterminal position to form a primary alcohol and a secondary alcohol with the same chain length as the substrate.

Microorganisms degrading short chain alkanes (C_2-C_4) have enzymes related to methane monooxygenase (Wang et al. 2017a). Those that degrade medium chain alkanes (C_5-C_{11}) or long chain alkanes ($>C_{12}$), contain integral membrane non-haem

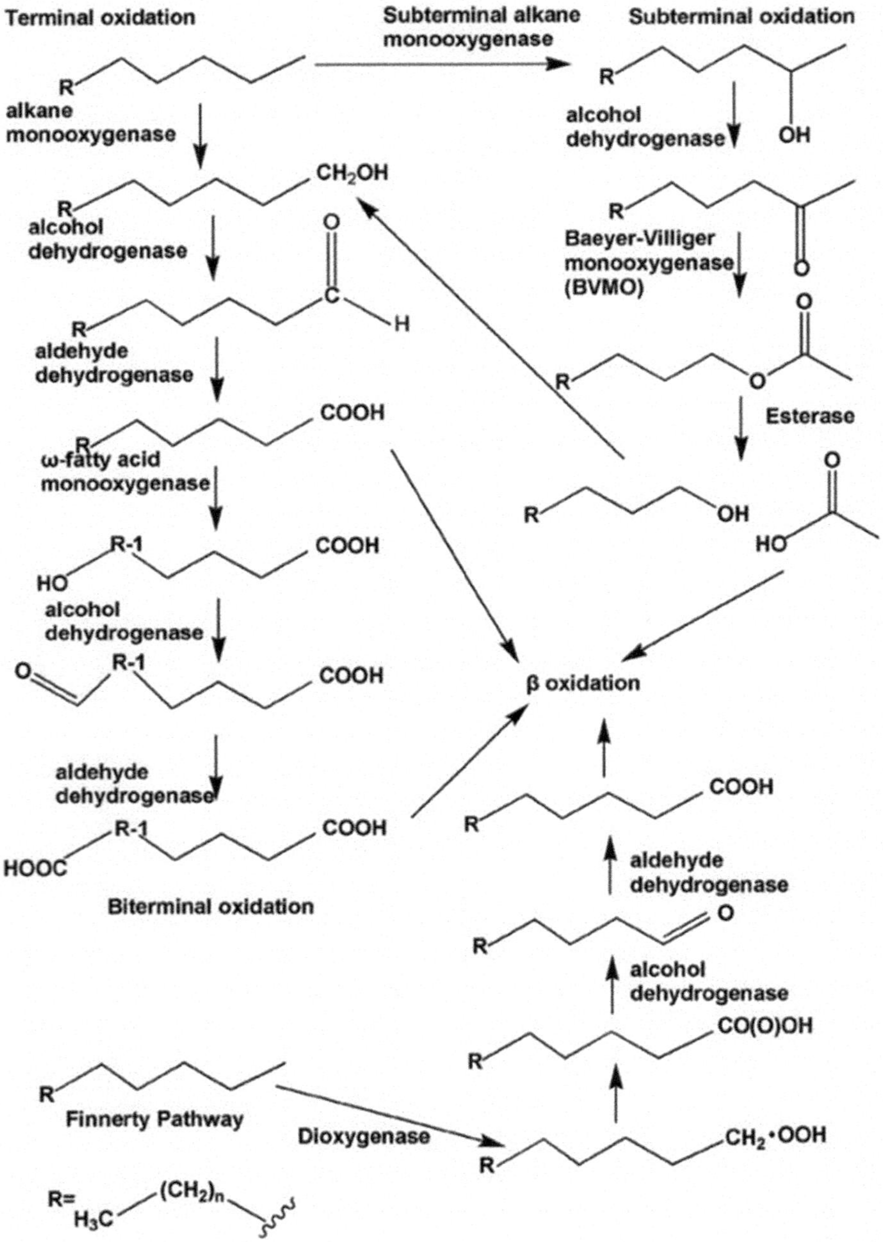

Fig. 5.4 Metabolic pathway for alkane degradation showing enzymes involved at different stages of degradation for both terminal oxidation and sub-terminal oxidation (Ji et al. 2013)

iron monooxygenases related to a well-characterised *Pseudomonas putida* GPo1 AlkB alkane hydroxylase (Rojo 2009). Several strains that assimilate alkanes greater than 18 carbon atoms contain alkane hydroxylases that are apparently unrelated to the former ones and have only recently been characterised (Wang et al. 2017b) Alkane hydroxylases are alkane-degrading enzymes that are distributed among many different species of bacteria, yeast, fungi and algae (Van Beilen and Funhoff 2007). The most extensively studied alkane degradation pathway is that for *Pseudomonas putida* GPo1, encoded by the OCT plasmid (van Beilen et al. 1994) and *Gordonia* sp. TF6 (FUJII et al. 2004). This pathway describes the conversion of an alkane into an alcohol using membrane monooxygenases, soluble rubredoxin and rubredoxin reductase (Van Hamme et al. 2003).

5.3.4.2 Enzymatic Degradation of Aromatic Hydrocarbons

Cyclic alkanes represent minor components of mineral oil and have been found to be relatively resistant to microbial attack. Only a few species are capable of using cyclohexane as sole carbon source. Alkyl side chains of cycloalkanes facilitate their enzymatic degradation. With regard to aromatic hydrocarbons e.g. benzene, toluene, ethylbenzene, xylenes and naphthalene, they belong to the large volume of petrochemicals which are used as fuels and industrial solvents. Many microorganisms have evolved catabolic pathways to degrade aromatic compounds. The compounds can be enzymatically converted to natural intermediates of the degradation: catechol and protecatechuate (Fritsche and Hofrichter 2008) (Fig. 5.5). Numerous studies have shown enzymes identified from *Bacillus* with the ability to degrade aromatic hydrocarbons (Bello-Akinosho et al. 2016; Das et al. 2017; Lu et al. 2020; Wongbunmak et al. 2020)

Benzene and related compounds are much more thermodynamically stable than aliphatics. They are very problematic since they are highly water soluble and toxic (Bello-Akinosho et al. 2016). Only a few reports on bacteria capable of degrading benzene have been reported but great progress has been made in this area through the years (Smith 1990; Vogt et al. 2011; Ren et al. 2015). The class of enzymes involved in the degradation of aerobic aromatic hydrocarbons is catechol dioxygenases, which are bacterial iron-containing enzymes. These enzymes are involved in aromatic ring cleavage since they are able to catalyse the addition of molecular oxygen atoms (Van Hamme et al. 2003).

Enzyme Bioprospecting Through Metagenomics

Over the past few years, the field of metagenomics has proven to be a very valuable tool for accessing the biosynthetic machinery of uncultured microbiota, that constitute for at least 80–99% of the total population, as opposed to the conventional culture-based method (Lam et al. 2015). Novel biocatalysts and small molecule biosynthetic genes can be identified in a very short span of time as compared to the conventional molecular biology methods (Duarte et al. 2017). The two complementary parallel approaches of metagenomic strategies for mining novel enzyme are as illustrated in Fig. 5.6.

Accordingly, there have been a great number of successes over the past few years with regard to the identification of various novel enzymes involved in hydrocarbon

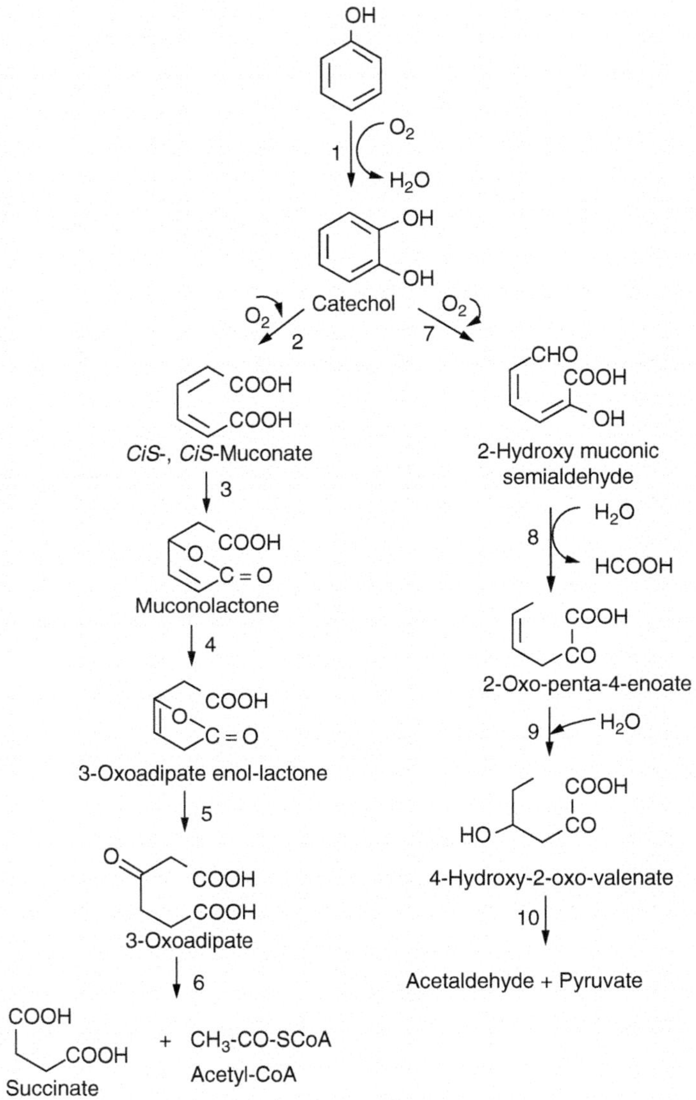

Fig. 5.5 Aerobic degradation pathway of aromatic compounds showing *o*- and *m*-cleavage (Rojo 2009)

degradation using metagenomic approaches. Recently, works carried out by Duarte et al. (2017) involving a metagenomic function-based screening approach were used to assess the microbial catabolome for polycyclic aromatic hydrocarbons from soils subjected to 12 years of *in situ* bioremediation. A total of 422 750 fosmid clones were screened for aromatic ring-cleavage activities using 2,3 dihydroxybiphenyl as a substrate. The positive fosmid clones that were sequenced showed that

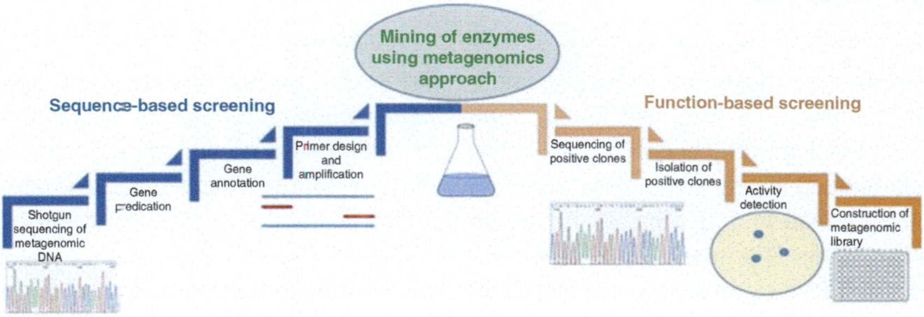

Fig. 5.6 Schematic diagram describing the general process of metagenomic strategies for mining novel genes

nearly 200 extradiol dioxygenase encoding genes of three different superfamilies were identified. Other genes including oxygenases were also identified and provided detailed information on enzymes activating aromatic compounds, thus providing insights into the complex microbial enzyme network. This was also shown by Dombrowski et al. (2016) working on reconstructing metabolic pathways of hydrocarbon-degrading bacteria from the Deepwater Horizon oil spill, where a broad gene set for degrading hydrocarbons was identified and found to work as a co-ordinated network for polycyclic aromatic degradation.

Moreover, in an ongoing experiment, Baburam and Feto (unpublished data) also used a function-driven metagenomic approach to identify diverse and potentially novel hydrocarbon biodegraders from petroleum-contaminated soils. A fosmid library was constructed using metagenomic DNA from contaminated soil samples. Hexadecane, octadecene and cyclohexane were used to screen the library for hydrocarbon-degrading fosmid clones. The growth of clones over a period of 5 days and the size of the fosmid clones on the different substrates (hexadecane, octadecene and cyclohexane) were an indication of positive clones with enzymatic-degrading capabilities for a specific hydrocarbon. Unpublished data from this research study showed a large number of clones with the ability to grow on multiple substrates. The ability of the positive fosmid clones to use the difficult-to-degrade substrate as a sole carbon source indicates the presence of the enzymatic machinery and the necessary catabolic pathways to degrade the candidate hydrocarbons. Therefore, the preliminary data show the potential of metagenomics approach to mine potent enzymes that could be part of an enzyme cocktail with a potential for an efficient bioremediation of oil spills. Most importantly, unravelling the role of an enzyme cocktail as an efficient oil-eater could lay solid ground for future development of next-generation oil-eaters through metabolic engineering.

5.4 Conclusion

Natural oil seepages and anthropogenic oil discharges continue to plague the water body including its flora and fauna. Some microorganisms have been found to be efficient in degrading hydrocarbons as they are capable of surviving in hydrocarbon-contaminated waters and soils across the world. Developing a deeper understanding of the regulation and capacity for microbial hydrocarbon remediation in various environments is critical. While much has been learned over the past few decades, there is still more that needs to be understood. From the review it is apparent that data and findings observed by various researchers in the field, the central roles of the *Bacillus* spp. cannot be denied and its future uses in the environment, be it natural or modified, will be of great value for bioremediation advances in the future. Research further points to the benefit of mining enzymes especially novel enzymes with hydrocarbon degrading characteristics from *Bacillus* spp. As a matter of fact the majority of the biomolecules discovered to date are developed using the culture-based approach, which accounts only for 1–10% of the microbial community. However, with the onset of the metagenomics approach, which literally deal with 100% of the population, scientists have been unravelling a number of novel biocatalysts with outstanding physiochemical as well as catalytic activities. Such approach will definitely contribute to discover novel biocatalysts with a potential to turn the tide against the oil spills across the globe. Nonetheless, the field is yet to develop, and thus, there are a number of limitations that should be addressed in the future research endeavour, from handling of huge data generated through Next Generation Sequencing (NGS) of metagenome, NGS-sequence analysis, accurate annotation, robust heterologous expression to a lack of modular high-throughput (HTP) screening techniques.

Acknowledgements The authors express their deepest gratitude to the research team at OMICS Research Group and Facility Laboratory at Vaal University of Technology for providing the comfort which stimulated the theoretical and practical research that contributed to the birth of this article.

Conflict of Interest The authors declare that they have no conflict of interest.

Ethical Approval This article does not contain any studies with human participants or animals performed by any of the authors.

References

Adams GO, Fufeyin PT, Okoro SE, Ehinomen I (2015) Bioremediation, biostimulation and bioaugmention: a review. *Int J Environ Bioremed Biodegradation* 3(1):28–39

Ahmed AM, Naif AH, Salem AD (2010) Hexadecane degradation by bacterial strains isolated from contaminated soils. *Afr J Biotechnol* 9(44):7487–7494

Al-Dhabaan FA (2019) Morphological, biochemical and molecular identification of petroleum hydrocarbons biodegradation bacteria isolated from oil polluted soil in Dhahran, Saud Arabia. *Saudi J Biol Sci* 26(6):1247–1252

Ali Khan AH, Tanveer S, Alia S, Anees M, Sultan A, Iqbal M, Yousaf S (2017) Role of nutrients in bacterial biosurfactant production and effect of biosurfactant production on petroleum hydrocarbon biodegradation. *Ecol Eng* 104:158–164

Alvarez HM, Pucci OH, Steinbüchel A (1997) Lipid storage compounds in marine bacteria. *Appl Microbiol Biotechnol* 47(2):132–139

Al-Wahaibi Y, Joshi S, Al-Bahry S, Elshafie A, Al-Bemani A, Shibulal B (2014) Biosurfactant production by Bacillus subtilis B30 and its application in enhancing oil recovery. *Colloids Surf B: Biointerfaces* 114:324–333

Annweiler E, Richnow HH, Antranikian G, Hebenbrock S, Garms C, Franke S, Francke W, Michaelis W (2000) Naphthalene degradation and incorporation of naphthalene-derived carbon into biomass by the thermophile Bacillus thermoleovorans. *Appl Environ Microbiol* 66(2):518–523

Arun K, Ashok M, Rajesh S, Vidyapith B, Sciences L, Pradesh H (2011) Crude oil PAH constitution , degradation pathway and associated bioremediation microflora: an overview. *Int J Environ Sci* 1(7):1420–1439

Atlas RM, Stoeckel DM, Faith SA, Minard-Smith A, Thorn JR, Benotti MJ (2015) Oil biodegradation and oil-degrading microbial populations in marsh sediments impacted by oil from the deepwater horizon well blowout. *Environ Sci Technol* 49(14):8356–8366

Ayed H, Jemil N, Maalej H, Bayoudh A, Hmidet N, Nasri M (2015) Enhancement of solubilization and biodegradation of diesel oil by biosurfactant from Bacillus amyloliquefaciens An6. *Int Biodeterior Biodegrad* 99(April):8–14

Bello-Akinosho M, Makofane R, Adeleke R, Thantsha M, Pillay M, Chirima GJ (2016) Potential of polycyclic aromatic hydrocarbon-degrading bacterial isolates to contribute to soil fertility. BioMed Res Int 2016

Bezza FA, Chirwa EMN (2015) Production and applications of lipopeptide biosurfactant for bioremediation and oil recovery by Bacillus subtilis CN2. *Biochem Eng J* 101:168–178

Borah D, Yadav RNS (2014a) Biodegradation of complex hydrocarbon by a novel bacillus cereus strain. J Environ Sci Technol 7:176–184

Borah D, Yadav RNS (2014b) Biodegradation of diesel, crude oil, kerosene and used engine oil by a newly isolated bacillus cereus strain DRDU1 from an automobile engine in liquid culture. *Arab J Sci Eng* 39(7):5337–5345

Borah D, Yadav RNS (2017) Bioremediation of petroleum based contaminants with biosurfactant produced by a newly isolated petroleum oil degrading bacterial strain. *Egypt J Pet* 26(1)

Bouchez-Naïtali M, Vandecasteele JP (2008) Biosurfactants, an help in the biodegradation of hexadecane? the case of Rhodococcus and Pseudomonas strains. *World J Microbiol Biotechnol* 24(9):1901–1907

Bragg JR, Prince RC, Harner EJ, Atlas RM (1994) Effectiveness of bioremediation for the Exxon Valdez oil spill. Nature 6470:413–418

Bujang M, Ibrahim NA, A. a/l E.R. (2013) Biodegradation of oily wastewater by pure culture of Bacillus cereus. *ARPN J Agric Biol Sci* 8(2):108–115

Cao B, Nagarajan K, Loh KC (2009) Biodegradation of aromatic compounds: Current status and opportunities for biomolecular approaches. *Appl Microbiol Biotechnol* 85(2):207–228

Chandankere R, Yao J, Choi MMF, Masakorala K, Chan Y (2013) An efficient biosurfactant-producing and crude-oil emulsifying bacterium Bacillus methylotrophicus USTBa isolated from petroleum reservoir. *Biochem Eng J* 74:46–53

Chandankere R, Yao J, Cai M, Masakorala K, Jain AK, Choi MMF (2014) Properties and characterization of biosurfactant in crude oil biodegradation by bacterium Bacillus methylotrophicus USTBa. *Fuel* 122:140–148

Chandra S, Sharma R, Singh K, Sharma A (2013) Application of bioremediation technology in the environment contaminated with petroleum hydrocarbon. *Ann Microbiol* 63(2):417–431

Chebbi A, Hentati D, Zaghden H, Baccar N, Rezgui F, Chalbi M, Sayadi S, Chamkha M (2017) Polycyclic aromatic hydrocarbon degradation and biosurfactant production by a newly isolated Pseudomonas sp. strain from used motor oil-contaminated soil. *Int Biodeterior Biodegrad* 122:128–140

Chettri B, Mukherjee A, Langpoklakpam JS, Chattopadhyay D, Singh AK (2016) Kinetics of nutrient enhanced crude oil degradation by Pseudomonas aeruginosa AKS1 and Bacillus sp. AKS2 isolated from Guwahati refinery, India. *Environ Pollut* 216:548–558

Christova N, Kabaivanova L, Nacheva L, Petrov P, Stoineva I (2019) Biodegradation of crude oil hydrocarbons by a newly isolated biosurfactant producing strain. *Biotechnol Biotechnol Equip* 33(1):863–872

Chrzanowski Ł, Ławniczak Ł, Czaczyk K (2012) Why do microorganisms produce rhamnolipids? *World J Microbiol Biotechnol* 28(2):401–419

Coon MJ (2005) Omega Oxygenases: Nonheme-iron enzymes and P450 cytochromes. *Biochem Biophys Res Commun* 338(1):378–385

Da Silva MLB, Ruiz-Aguilar GML, Alvarez PJJ (2005) Enhanced anaerobic biodegradation of BTEX-ethanol mixtures in aquifer columns amended with sulfate, chelated ferric iron or nitrate. *Biodegradation* 16(2):105–114

Darsa KV, Thatheyus AJ (2014) Biodegradation of Petroleum Compound Using Pseudomonas aeruginosa. *OALib* 01(05):1–9

Das M, Bhattacharya A, Banu S, Kotoky J (2017) Enhanced Biodegradation of Anthracene by Bacillus Cereus Strain JMG-01 Isolated from Hydrocarbon Contaminated Soils. *Soil Sediment Contam* 26(5):510–525

Díaz De Rienzo MA, Kamalanathan ID, Martin PJ (2016) Comparative study of the production of rhamnolipid biosurfactants by B. thailandensis E264 and P. aeruginosa ATCC 9027 using foam fractionation. *Process Biochem* 51(7):820–827

Dombrowski N, Donaho JA, Gutierrez T, Seitz KW, Teske AP, Baker BJ (2016) Reconstructing metabolic pathways of hydrocarbon-degrading bacteria from the Deepwater Horizon oil spill. *Nat Microbiol* 1(May):1–8

Duarte M, Nielsen A, Camarinha-Silva A, Vilchez-Vargas R, Bruls T, Wos-Oxley ML, Jauregui R, Pieper DH (2017) Functional soil metagenomics: elucidation of polycyclic aromatic hydrocarbon degradation potential following 12 years of in situ bioremediation. *Environ Microbiol*

Elumalai P, Parthipan P, Narenkumar J, Anandakumar B, Madhavan J, Oh BT, Rajasekar A (2019) Role of thermophilic bacteria (Bacillus and Geobacillus) on crude oil degradation and biocorrosion in oil reservoir environment. 3 Biotech 9(3):0

Fernandes PL, Rodrigues EM, Paiva FR, Ayupe BAL, McInerney MJ, Tótola MR (2016) Biosurfactant, solvents and polymer production by Bacillus subtilis RI4914 and their application for enhanced oil recovery. *Fuel* 180:551–557

Ferradji FZ, Mnif S, Badis A, Rebbani S, Fodil D, Eddouaouda K, Sayadi S (2014) Naphthalene and crude oil degradation by biosurfactant producing Streptomyces spp. isolated from Mitidja plain soil (North of Algeria). *Int Biodeterior Biodegrad* 86:300–308

Freitas de Oliveira DW, Lima França ÍW, Nogueira Félix AK, Lima Martins JJ, Aparecida Giro ME, Melo VMM, Gonçalves LRB (2013) Kinetic study of biosurfactant production by Bacillus subtilis LAMI005 grown in clarified cashew apple juice. *Colloids Surf B: Biointerfaces* 101:34–43

Fritsche W, Hofrichter M (2008) Aerobic degradation by microorganisms. *Biotechnol Second, Completely Rev Edition* 11–12:144–167

Fujii T, Narikawa T, Takeda K, Kato J (2004) Biotransformation of various alkanes using the *Escherichia coli* expressing an alkane hydroxylase system from *Gordonia* sp. TF6. *Biosci Biotechnol Biochem* 68(10):2171–2177

Gudiña EJ, Pereira JFB, Costa R, Coutinho JAP, Teixeira JA, Rodrigues LR (2013) Biosurfactant-producing and oil-degrading Bacillus subtilis strains enhance oil recovery in laboratory sand-pack columns. *J Hazard Mater* 261:106–113

Hua F, Wang HQ (2014) Uptake and trans-membrane transport of petroleum hydrocarbons by microorganisms. *Biotechnol Biotechnol Equip* 28(2):165–175

Huang K, Chen C, Shen Q, Rosen BP, Zhao FJ (2015) Genetically engineering Bacillus subtilis with a heat-resistant arsenite methyltransferase for bioremediation of arsenic-contaminated organic waste. *Appl Environ Microbiol* 81(19):6718–6724

Jabeen S, Khan MAS, Hassan QM, Ahmed MS, Nishat N, Zain H (2015) Comparative study of bioremediation of crude oil by Bacillus Subtilis and organic substances. J Sci J Chem Biol Phys Sci Sec B 55(33):2621–2633

Ji Y, Mao G, Wang Y, Bartlam M (2013) Structural insights into diversity and n-alkane biodegradation mechanisms of alkane hydroxylases. *Front Microbiol* 4(September 2014)

Jin M, Gai Y, Guo X, Hou Y, Zeng R (2019) Properties and applications of extremozymes from deep-sea extremophilic microorganisms: a mini review. *Marine Drugs* 17(12)

Joy S, Rahman PKSM, Sharma S (2017) Biosurfactant production and concomitant hydrocarbon degradation potentials of bacteria isolated from extreme and hydrocarbon contaminated environments. *Chem Eng J* 317:232–241

Joye S, Kleindienst S, Gilbert J, Handley K, Weisenhorn P, Overholt W, Kostka J (2016) Responses of microbial communities to hydrocarbon exposures. *Oceanography* 29(3)

Kadri T, Rouissi T, Kaur Brar S, Cledon M, Sarma S, Verma M (2017) Biodegradation of polycyclic aromatic hydrocarbons (PAHs) by fungal enzymes: a review. *J Environ Sci (China)* 51:52–74

Kallimanis A, Frillingos S, Drainas C, Koukkou AI (2007) Taxonomic identification, phenanthrene uptake activity, and membrane lipid alterations of the PAH degrading Arthrobacter sp. strain Sphe3. *Appl Microbiol Biotechnol* 76(3):709–717

Kalogerakis N, Fava F, Corvini PFX (2017) Bioremediation advances. *New Biotechnol* 38:41–42

Kim IS, Foght JM, Gray MR (2002) Selective transport and accumulation of alkanes by Rhodococcus erythropolis S+14He. *Biotechnol Bioeng* 80(6):650–659

Koshlaf E, Ball AS (2017) Soil bioremediation approaches for petroleum hydrocarbon polluted environments. *AIMS Microbiol* 3(1):25–49

Krell T, Lacal J, Reyes-Darias JA, Jimenez-Sanchez C, Sungthong R, Ortega-Calvo JJ (2013) Bioavailability of pollutants and chemotaxis. *Curr Opin Biotechnol* 24(3):451–456

Lam KN, Cheng J, Engel K, Neufeld JD, Charles TC (2015) Current and future resources for functional metagenomics. Front Microbiol 6(OCT):1–8

Li L, Liu X, Yang W, Xu F, Wang W, Feng L, Bartlam M, Wang L, Rao Z (2008) Crystal structure of long-chain alkane monooxygenase (LadA) in complex with coenzyme FMN: unveiling the long-chain alkane hydroxylase. *J Mol Biol* 376(2):453–465

Lim MW, Von Lau E, Poh PE (2016) A comprehensive guide of remediation technologies for oil contaminated soil — present works and future directions. *Mar Pollut Bull* 109(1):14–45

Lima SD, Oliveira AF, Golin R, Lopes VCP, Caixeta DS, Lima ZM, Morais EB (2020) Isolation and characterization of hydrocarbon-degrading bacteria from gas station leaking-contaminated groundwater in the Southern Amazon, Brazil. *Braz J Biol* 80(2):354–361

Liu JF, Mbadinga SM, Yang SZ, Gu JD, Mu BZ (2015) Chemical structure, property and potential applications of biosurfactants produced by Bacillus subtilis in petroleum recovery and spill mitigation. *Int J Mol Sci* 16(3):4814–4837

Liu B, Liu J, Ju M, Li X, Yu Q (2016) Purification and characterization of biosurfactant produced by Bacillus licheniformis Y-1 and its application in remediation of petroleum contaminated soil. *Mar Pollut Bull* 107(1):46–51

Lu J, Zhu X, Zhang C, Lu F, Lu Z, Lu Y (2020) Co-expression of alcohol dehydrogenase and aldehyde dehydrogenase in Bacillus subtilis for alcohol detoxi fi cation. *Food Chem Toxicol* 135(October 2019):110890

Maikudi Usman M, Dadrasnia A, Tzin Lim K, Fahim Mahmud A, Ismail S (2016) Application of biosurfactants in environmental biotechnology; remediation of oil and heavy metal. *AIMS Bioeng* 3(3):289–304

Mirete S, Morgante V, González-Pastor JE (2016) Functional metagenomics of extreme environments. *Curr Opin Biotechnol* 38:143–149

Mittal A, Singh P (2009) Isolation of hydrocarbon degrading bacteria from soils contaminated with crude oil spills. *Indian J Exp Biol* 47(9):760–765

Mukherjee A, Chettri B, Langpoklakpam JS, Basak P, Prasad A, Mukherjee AK, Bhattacharyya M, Singh AK, Chattopadhyay D (2017) Bioinformatic approaches including predictive metagenomic profiling reveal characteristics of bacterial response to petroleum hydrocarbon contamination in diverse environments. *Sci Rep* 7(1):1–22

Müller MM, Hörmann B, Kugel M, Syldatk C, Hausmann R (2011) Evaluation of rhamnolipid production capacity of Pseudomonas aeruginosa PAO1 in comparison to the rhamnolipid overproducer strains DSM 7108 and DSM 2874. *Appl Microbiol Biotechnol* 89(3):585–592

Mulligan CN (2005) Environmental applications for biosurfactants. *Environ Pollut* 133(2):183–198

Nagata Y, Ohtsubo Y, Tsuda M (2015) Properties and biotechnological applications of natural and engineered haloalkane dehalogenases. *Appl Microbiol Biotechnol* 99(23):9865–9881

Ni'Matuzahroh, Trikurniadewi N, Pramadita ARA, Pratiwi IA, Salamun F, Sumarsih S (2017) Biodegradation of naphthalene and phenanthren by Bacillus subtilis 3KP. AIP Conference Proceedings 1854

Nie Y, Tang YQ, Li Y, Chi CQ, Cai M, Wu XL (2012) The genome sequence of polymorphum gilvum SL003B-26A1 T reveals its genetic basis for crude oil degradation and adaptation to the saline soil. *PLoS ONE* 7(2)

Noparat P, Maneerat S, Saimmai A (2014) Application of biosurfactant from Sphingobacterium spiritivorum AS43 in the biodegradation of used lubricating oil. *Appl Biochem Biotechnol* 172(8):3949–3963

Opere BO, Obayori OS, Raji AA (2013) Degradation of Cyclohexane and Cyclohexanone by Bacillus lentus Strain LP32. *Afr J Biotechnol* 12(47):6632–6635

Pandey P, Pathak H, Dave S (2016) Microbial Ecology of Hydrocarbon Degradation in the Soil : A Review. *Res J Environ Toxicol* 10(1):1–15

Parthipan P, Preetham E, Machuca LL, Rahman PKSM, Murugan K, Rajasekar A (2017) Biosurfactant and Degradative Enzymes Mediated Crude Oil Degradation by Bacterium Bacillus subtilis A1. *Front Microbiol* 8(February):1–14

Patowary K, Patowary R, Kalita MC, Deka S (2016) Development of an efficient bacterial consortium for the potential remediation of hydrocarbons from contaminated sites. Front Microbiol 7(JUL):1–14

Peixoto RS, Vermelho AB, Rosado AS (2011) Petroleum-degrading enzymes: bioremediation and new prospects. *Enzyme Res* 2011(1)

Perfumo A, Banat IM, Marchant R, Vezzulli L (2007) Thermally enhanced approaches for bioremediation of hydrocarbon-contaminated soils. *Chemosphere* 66(1):179–184

Ren Y, Yang J, Chen S (2015) The fate of a nitrobenzene-degrading bacterium in pharmaceutical wastewater treatment sludge. *Chemosphere* 141:13–18

Rojo F (2009) Degradation of alkanes by bacteria: minireview. *Environ Microbiol* 11(10):2477–2490

Roy A, Dutta A, Pal S, Gupta A, Sarkar J, Chatterjee A, Saha A, Sarkar P, Sar P, Kazy SK (2018) Biostimulation and bioaugmentation of native microbial community accelerated bioremediation of oil refinery sludge. *Bioresour Technol* 253(November 2017):22–32

Sakthi Priya N, Doble M, Sangwai JS (2015) Bioremediation of costal and marine pollution due to crude oil using a microorganism Bacillus subtilis. *Proc Eng* 116(1):213–220

Sharma B, Dangi AK, Shukla P (2018a) Contemporary enzyme based technologies for bioremediation: a review. *J Environ Manag* 210:10–22

Sharma S, Tiwari S, Hasan A, Saxena V, Pandey LM (2018b) Recent advances in conventional and contemporary methods for remediation of heavy metal-contaminated soils. *3. Biotech* 8(4):1–18

Shekher R, Sehgal S, Kamthania M, Kumar A (2011) Laccase: microbial sources, production, purification, and potential biotechnological applications. Enzyme Res 2011

Smith MR (1990) The biodegradation of aromatic hydrocarbons by bacteria. *Biodegradation* 1(2–3):191–206

Souza EC, Vessoni-Penna TC, De Souza Oliveira RP (2014) Biosurfactant-enhanced hydrocarbon bioremediation: An overview. *Int Biodeterior Biodegrad* 89:88–94

Surendra SV, Mahalingam BL, Velan M (2017) Degradation of monoaromatics by Bacillus pumilus MVSV3. *Braz Arch Biol Technol* 60(December):1–18

Tao K, Liu X, Chen X, Hu X, Cao L, Yuan X (2017) Biodegradation of crude oil by a defined co-culture of indigenous bacterial consortium and exogenous Bacillus subtilis. *Bioresour Technol* 224:327–332

Thamer M, Al-Kubaisi AR, Zahraw Z, Abdullah HA, Hindy I, Khadium AA (2013) Biodegradation of Kirkuk light crude oil by Bacillus thuringiensis, Northern of Iraq. *Nat Sci* 5(7):865–873

Thomas F, Lorgeoux C, Faure P, Billet D, Cébron A (2016) Isolation and substrate screening of polycyclic aromatic hydrocarbon degrading bacteria from soil with long history of contamination. *Int Biodeterior Biodegrad* 107:1–9

Umeh AC, Duan L, Naidu R, Semple KT (2017) Residual hydrophobic organic contaminants in soil: Are they a barrier to risk-based approaches for managing contaminated land? *Environ Int* 98:18–34

Van Beilen JB, Funhoff EG (2007) Alkane hydroxylases involved in microbial alkane degradation. *Appl Microbiol Biotechnol* 74(1):13–21

van Beilen JB, Wubbolts MG, Witholt B (1994) Genetics of alkane oxidation by Pseudomonas oleovorans. *Biodegradation* 5(3–4):161–174

van Beilen JB, Panke S, Lucchini S, Franchini AG, Rothlis-berger M, Witholt B (2001) Analysis of Pseudomonas putida alkane degradation gene clusters and flanking insertion sequences: evolution and regulation of the alk-genes. *Microbiology (United Kingdom)* 147(May):1621–1630

Van Hamme JD, Singh A, Ward OP (2003) Recent Advances in Petroleum Microbiology Recent Advances in Petroleum Microbiology. *Microbiol Mol Biol Rev* 67(4):503–549

Varjani SJ (2017) Microbial degradation of petroleum hydrocarbons. *Bioresour Technol* 223:277–286

Varjani SJ, Upasani VN (2016) Biodegradation of petroleum hydrocarbons by oleophilic strain of Pseudomonas aeruginosa NCIM 5514. *Bioresour Technol* 222:195–201

Varjani SJ, Upasani VN (2017a) Critical review on biosurfactant analysis, purification and characterization using rhamnolipid as a model biosurfactant. *Bioresour Technol* 232:389–397

Varjani SJ, Upasani VN (2017b) A new look on factors affecting microbial degradation of petroleum hydrocarbon pollutants. *Int Biodeterior Biodegrad* 120:71–83

Vasconcellos SP, Sierra-Garcia IN, Dellagnezze BM, Vicentini R, Midgley D, Silva CC, Santos Neto EV, Volk H, Hendry P, Oliveira VM (2017) Functional and genetic characterization of hydrocarbon biodegrader and exopolymer-producing clones from a petroleum reservoir metagenomic library. *Environ Technol* 3330(August):1–12

Vinothini C, Sudhhakar S, Ravikumar R (2015) Biodegradation of petroleum and crude oil by Pseudomonas putida and Bacillus cereus. *Int J Curr Microbiol App Sci* 4(1):318–329

Vogt C, Kleinsteuber S, Richnow HH (2011) Anaerobic benzene degradation by bacteria. *Microb Biotechnol* 4(6):710–724

Wang VCC, Maji S, Chen PPY, Lee HK, Yu SSF, Chan SI (2017a) Alkane Oxidation: Methane Monooxygenases, Related Enzymes, and Their Biomimetics. *Chem Rev* 117(13):8574–8621

Wang Y, Nie M, Wan Y, Tian X, Nie H, Zi J, Ma X (2017b) Functional characterization of two alkane hydroxylases in a versatile Pseudomonas aeruginosa strain NY3. *Ann Microbiol* 67(7):459–468

Wang D, Lin J, Lin J, Wang W, Li S (2019) Biodegradation of petroleum hydrocarbons by bacillus subtilis BL-27, a strain with weak hydrophobicity. *Molecules* 24(17)

Wentzel A, Ellingsen TE, Kotlar HK, Zotchev SB, Throne-Holst M (2007) Bacterial metabolism of long-chain n-alkanes. *Appl Microbiol Biotechnol* 76(6):1209–1221

Wiener MC, Horanyi PS (2011) How hydrophobic molecules traverse the outer membranes of Gram-negative bacteria. *Proc Natl Acad Sci* 108(27):10929–10930

Wongbunmak A, Khiawjan S, Suphantharika M, Pongtharangkul T (2020) BTEX biodegradation by Bacillus amyloliquefaciens subsp. plantarum W1 and its proposed BTEX biodegradation pathways. *Sci Rep* 10(1):1–13

Wu M, Chen L, Tian Y, Ding Y, Dick WA (2013) Degradation of polycyclic aromatic hydrocarbons by microbial consortia enriched from three soils using two different culture media. *Environ Pollut* 178:152–158

Wu M, Dick WA, Li W, Wang X, Yang Q, Wang T, Xu L, Zhang M, Chen L (2016) Bioaugmentation and biostimulation of hydrocarbon degradation and the microbial community in a petroleum-contaminated soil. *Int Biodeterior Biodegrad* 107:158–164

Youssef N, Simpson DR, McInerney MJ, Duncan KE (2013) In-situ lipopeptide biosurfactant production by Bacillus strains correlates with improved oil recovery in two oil wells approaching their economic limit of production. *Int Biodeterior Biodegrad* 81:127–132

Zafra G, Taylor TD, Absalón AE, Cortés-Espinosa DV (2016) Comparative metagenomic analysis of PAH degradation in soil by a mixed microbial consortium. *J Hazard Mater* 318(55):702–710

Zampolli J, Collina E, Lasagni M, Di Gennaro P (2014) Biodegradation of variable-chain-length n-alkanes in Rhodococcus opacus R7 and the involvement of an alkane hydroxylase system in the metabolism. *AMB Express* 4(1):1–9

Zrafi-Nouira I, Saidane-Mosbahi D (2012) Crude Oil Metagenomics for Better Bioremediation of Contaminated Environments. 'Introduction to Enhanced Oil Recovery (EOR) Processes and Bioremediation of Oil-Contaminated Sites', 31

Chapter 6
Current Understanding and Future Directions of Biocontrol of Plant Diseases by *Bacillus* spp., with Special Reference to Induced Systemic Resistance

Sunita Mahapatra, Sunanda Chakraborty, Malay Samanta, Srikanta Das, and Tofazzal Islam

Abstract Plant pathogens represent one of the prime threats to sustainable crop production. Till date, synthetic agrochemicals are considered as effective tools for the management of various biotic stresses such as pathogenic microorganisms and insects in plants. Unfortunately, their injudicious and intensive usages in agriculture pose a serious threat to the environment and all living beings dwelling on the earth. Under such circumstances, the application of beneficial *Bacillus*-mediated management of plant pathogens has emerged as one of the most benevolent and sustainable options. A large number of *Bacillus* species has been identified as promising candidates for managing a number of plant pathogens through induction of systemic resistance in plants. Significant research progress has been attained in the characterisation and understanding of the role of *Bacillus*-induced systemic resistance (ISR) against a wide range of pathogens of crop plants. In this chapter, we aim to provide an overview of the mechanisms of *Bacillus*-induced ISR for instance, elici-

S. Mahapatra (✉) · S. Chakraborty · S. Das
Department of Plant Pathology, SSMP, Bidhan Chandra Krishi Viswavidyalaya, Nadia, West Bengal, India

M. Samanta
Hooghly Krishi Vigyan Kendra, Bidhan Chandra Krishi Viswavidyalaya, Chinsurah, West Bengal, India

T. Islam (✉)
Institute of Biotechnology and Genetic Engineering (IBGE), Bangabandhu Sheikh Mujibur Rahman Agricultural University, Gazipur, Bangladesh
e-mail: tofazzalislam@bsmrau.edu.bd

M. T. Islam et al. (eds.), *Bacilli in Agrobiotechnology*, Bacilli in Climate Resilient Agriculture and Bioprospecting,
https://doi.org/10.1007/978-3-030-85465-2_6

tors, phytoalexins, lipopeptides, antibiotics, hormones and enzymes to protect plants from various pests. Additionally, glimpses of the research progress in the identification of different *Bacillus* strains and their evaluation as a potential biocontrol agent have also been presented.

Keywords *Bacillus* · Induced resistance · Bio-control · Lipopeptides · Antibiotics

6.1 Introduction

Steady increase in crop yield to feed the ever-increasing population is the greatest challenge faced by the agriculturists worldwide. Apart from factors such as nutrient supply, soil conditions, moisture, degrading cultivable land, and cultivar quality, biotic stresses pose a huge threat to the food security of the ever-increasing global population (Wulff et al. 2011; Islam et al. 2016a, b, 2019a, 2020). Although practices like crop rotation, use of chemical pesticides, and development of better cultivar have been utilised for the management of emerging diseases, these approaches are no longer ecologically and economically sustainable. Thus, application of beneficial microorganisms have emerged as a suitable alternative to the pre-existing traditional approaches (Hardoim et al. 2015). Coordination between microbiota and root endodermis supports plant mineral nutrient homeostasis (Salas-González et al. 2020). Among the beneficial microorganisms studied so far, *Bacillus* spp. have been observed to perform especially well not only in managing diseases but also in increasing nutrient availability, enhancing plant growth, improving beneficial microbial community in the rhizosphere, and inducing sustained defence reactions in the plant (Islam et al. 2016a, b, 2019b).

Bacillus is a genus of Gram-positive and endospore-forming bacteria, which is cosmopolitan in nature and widely spread across a diverse range of environments. They occur naturally in crop fields and aid in crop productivity both directly and indirectly. Inherent traits like durable cell wall, formation of endospores, peptide and antibiotic secretion, and production of antimicrobial enzymes play a significant role in the survival of these ubiquitous organisms even in a highly adverse situation.

A large body of research on the *Bacillus* species has attributed its success in the production of antibiotics and lipopeptides, quorum quenching, lysis of pathogen hyphae, competition for space and nutrients, and induced systemic resistance (Li et al. 2013; Islam et al. 2016a, b, 2019a, b; Surovy et al. 2019; Rayanoothala et al. 2020). Furthermore, *Bacillus* spp. increase nutrient uptake, siderophore production, and promote plant growth. For example, *B. subtilis* increases N uptake, phosphate solubilisation, and siderophore and phytohormone production, and promotes plant growth. *Bacillus* species has also been observed to alter the microbial community in the rhizosphere which facilitated disease suppression (You et al. 2016; Mahapatra et al. 2020).

This chapter aims to review the current understanding of the induction of resistance mechanism in host plant by *Bacillus* species. It deals with the various mechanisms employed by *Bacillus* species to suppress the phytopathogens in the plant rhizosphere and phyllosphere. It also discusses the genomic and molecular bases of disease resistance in plants imparted by the *Bacillus* species. Finally, it summarises the field application of *Bacillus*-based formulation, its prospects and challenges. This report also tries to bridge the knowledge gap and help to develop *Bacillus* species as a reliable and effective strategy for the management of emerging pests and diseases.

6.2 *Bacillus* Diversity and Antagonism

The type of soil plays a crucial role in the management of phytopathogens. Suppressive soils contain a large number of beneficial microbes that are able to inhibit the growth of bacterial and fungal pathogens. Among the biocontrol agents, the *Pseudomonas* genus has been most extensively studied for its anti-fungal metabolites(AFMs)suchaspyrrolnitrin,phenazines,pyoluteorin,2,4-diacetylphloroglucinol (DAPG), and viscosinamide (Nielsen et al. 1999; Nielsen et al. 2000). However, *Bacillus* species has gained popularity in the recent years due to its production of a wide array of antimicrobial substances and the ability of certain strains to promote plant growth (Choudhary and Johri, 2008). A large number of cultivable strains of *B. cereus* and *B. subtilis* have been identified in different studies (Vargas-Ayala et al. 2000), while cultivation-independent studies have demonstrated the existence of a much more number of uncultivable strains of *Bacillus* spp. (McSpadden Gardener 2004).

Some studies have found *B. megatarium* to be the most abundantly available species, while some researchers observed *Paenibacillus*, previously known as *Bacillus polymyxa*, to dominate a variety of soils. *Paenibacillus* has the ability to fix atmospheric nitrogen and thus are helpful in meeting the N requirement of a large number of crop plants like Canadian wheat (Priest 1993). *Brevibacillus*, previously known as *Bacillus brevis*, is terrestrial and aquatic in nature (Panda et al. 2014). A member of the *Bacillus* species, *B. sphaericus* is a notable entomopathogen and thus thrives in the habitats of insect larvae like pools, ditches, and lakes (El-Bendary 2006).

Spatiotemporal analysis of the microbial communities in soil, rhizoplane, and rhizosphere revealed that soil type had greater effect than plant type in determining the microbial population diversity (Wieland et al. 2001). Arias et al. (1999) studied the distribution and diversity of *Bacillus* spp. in the soybean phylloplane. Their study revealed that *B. pumilus* is the most widely distributed species in the soybean phylloplane. Although other bacterial strains like *B. brevis*, *B. subtilis*, *B. circulans*, and *B. firmus* were also observed in the phylloplane, their population continued to decline with the growth of the crop and became completely undetectable at 85 days of the cropping season. As mentioned earlier, *Paenibacillus* genus contains many nitrogen fixing bacterial species, like *P. azotofixans*, *P. polymyxa*, and *P. macerans* (Ash et al. 1993). But amongst all the reported species, *P. azotofixans* has been found to be the most efficient, and hence, is abundant in the rhizosphere of wheat, maize, sorghum, banana, sugarcane, and some forage crops (Seldin 1992). *P. azotofixans* is more abundant in bulk soil than the rhizosphere, while its population varied largely with the soil type in the same crop rhizosphere (Rosado et al. 1998, Seldin et al. 1998).

The distribution and diversity of the *Bacillus* species is largely dependent on the interaction between the soil and the plant. The exudates from the plant roots, which play a significant role in determining the microbial distribution, are specific to the plant and are also correlated to a particular soil habitat (Crowley and Rengel 1999; Duineveld et al. 2001). The exudates vary with the stage of the crop as well, which is reflected in the findings of Jaegar and co-workers. The growth stage of the crop could be a crucial factor in determining the rhizobacterial community, as observed by Van Overbreek and van Elsas (2008) in potato. It has been observed that plants encourage the growth of specific microorganisms, from the indefinite microdiversity based on the chemical influences in the rhizosphere. Thus the manipulation of the chemical constituents of the rhizosphere could be done to encourage the growth of promising *Bacillus* strains among the rhizospheric community of microorganisms.

6.3 Mechanism of Induction of Resistance Against Plant Disease by *Bacillus* spp.

When *Bacillus* is introduced in the host-pathogen-environment system, it interacts with all the components in a number of complex mechanisms that affect the growth of the phytopathogen directly or indirectly. These interactions may lead to plant growth promotion, biofilm formation, induced systemic resistance (ISR), competition for nutrients, production of antibiotics, and cell lysis (Fig. 6.1). Strains of *Bacillus* spp. have been recorded to exhibit one or more of these traits that work synergistically with the plant and the environment in suppressing the phytopathogens. Furthermore, these interactions may lead to increased plant growth, vigour, and a shift in the rhizospheric microbial community, which would indirectly reduce the impact of the plant pathogens on the host plant.

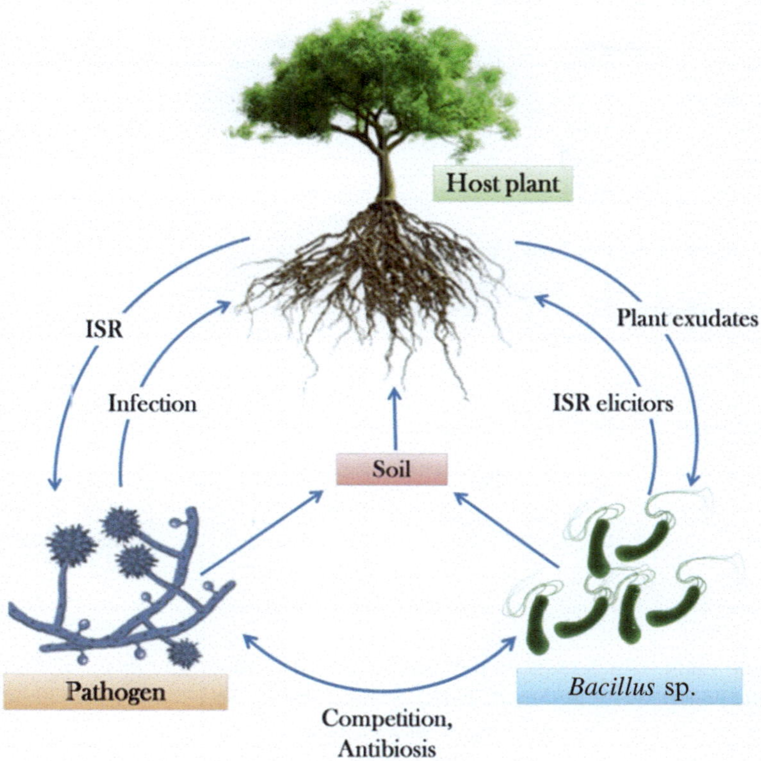

Fig. 6.1 Interactions between *Bacillus* spp., the pathogen and the host plant

6.3.1 Competition for Nutrients

Nutrients are always limiting factors for the growth and development of microorganisms and plants in the any crop fields. Biological control aims at utilizing this competition for nutrient and space to reduce the growth and productiveness of the pathogen by the non-pathogenic organisms. Generally, soil-borne pathogens like *Fusarium* and *Pythium* infect through mycelial contact, which makes them more vulnerable to competition from the non-pathogenic organisms. Micronutrients, such as iron, are extremely essential but its availability is limited and largely dependent on the soil conditions. In such situations, microbes produce siderophores, which are capable of attracting iron from the rhizosphere, thus meeting the iron requirement of the microorganism (Chakraborty et al. 2020a, b). In many instances, *Bacillus* has been found to compete with pathogenic microbes, such as *Erwinia carotovora*, by a similar mechanism, hence, restricting the growth and the development of the pathogen (Kloepper et al. 2004a, b). Studies also revealed that the inoculation of *Bacillus* sp. in the rhizosphere resulted in a shift of the pathogenic microflora from the site of infection. *Bacillus* species was reported to cause rapid colonisation in the tissues

Table 6.1 List of important *Bacillus* strains in respect to their competition, target pathogens, and host plants

Bacteria	Strain	Host	Target pathogen	References
Bacillus subtilis	HU5	Cotton	*Verticillium dahlia*	Li et al. (2013)
B. subtilis	SQR9	Cucumber	*Fusarium oxysporum*	Cao et al. (2011)
B. subtilis	E1R-j	Wheat	*Ustilago tritici*	Baker et al. (1983)
B. subtilis	SB24	Tomato	*Sclerotinia sclerotiorum*	Clayton and Hudelson (1991)
B. Amyloliquefaciens	CM-2 and T-5	Tomato	*Ralastonia solanacearum*	Tong-Jian et al. (2013)
B. Amyloliquefaciens	54	Watermelon	*Acidovorax avenae* subsp. *citrulli*	Jiang et al. (2015)
B. Megaterium	A6	Oilseed rape	*Sclerotinia sclerotiorum*	Hu et al. (2013)
B. Megaterium	B153-2-2	Soybean	*Rhizoctonia solani*	Zheng and Sinclair (2000)
B. pumilus	SE34	Pea	*F. oxysporum f.* sp. *pisi*	Benhamou et al. (1996)
B. cereus	UW 85	Tobacco	*Phytophthora parasitica*	Blagoeva-Nikolaeva et al. (1995)

of tomato plant, which reduced the disease severity and wilting index of *Fusarium* sp. in the host plant (Jangir et al. 2018). Similar results were obtained by Zhang and co-workers (2011) when working with *Fusarium* wilt of banana. They found that *Bacillus* was a rapid coloniser under hydroponic, sand and soil conditions, which inhibited the development of *Fusarium* sp. in the treated banana plant. Many studies have been conducted on the role of competition and colonisation in establishing *Bacillus* species as successful bio-control agents (Table 6.1).

6.3.2 Synthesis and Excretion of Lytic Enzymes

Aside from the production of lipoproteins and antibiotics, *Bacillus* has been known to produce a number of defense-related oxidative enzymes like peroxidase (PO) and polyphenol oxidase (PPO), which bring about structural changes in the host cell wall, thus strengthening the defense barriers against the invading pathogen. Several studies conducted so far have revealed that *Bacillus* synthesizes a large number of phytosanitary enzymes, which are enlisted in Table 6.2. Phenol oxidase enzymes produce quinones and release hydrogen peroxide. These enzymes trigger the release of toxic free radicals and polymerise the phenolic compounds into lignin like substances. The resulted metabolites are then deposited in the host cell wall, which

Table 6.2 List of enzymes activated by Bacillus species, along with the target pathogen and host plant

Bacteria	Strain	Host plant	Enzyme activated	Target pathogen	References
Bacillus subtilis	174	Tomato	PO,PPO, PAL	*Fusarium oxysporum*	Akram and Anjum (2011)
B. subtilis	AUBS1	Rice	PAL, PO and PR proteins	*Rhizoctonia solani*	Jayaraj et al. (2004)
B. subtilis	PTA-271	Grapevine	Lipoxygenase, PAL and chitinase	*Botrytis cinerea*	Trotel-Aziz et al. (2008)
B. subtilis	B4	Cucumber	Indole acetic acid	*Colletotrichum orbiculare*	Park et al. (2013a, b)
B. subtilis	SE34 and GBO3	Rice	PO and PAL, PPO	*Xanthomonas oryzae pv. oryzae*	Udayashankar et al. (2011)
B. pumilus	SE34	Pea	Phenolic compounds	*F. oxysporum f.* sp. *pisi*	Benhamou et al. (1996)
B. vallismortis	BS07	Chili pepper	Salicylic acid (SA)	*Phytophthora capsici and Colletotrichum acutatum*	Park et al. (2013a, b)
B. mycoides	Bac J	Sugar beet	Chitinase, b-1,3-glucanase and peroxidase	*Cercospora beticola*	Bargabus et al. (2002)
B. cereus	AR156	Loquat	PAL, PO, chitinase, β-1,3-glucanase, polyphenoloxidase and promoted accumulation of H_2O_2	*Colletotrichum acutatum*	Wang et al. (2014)
B. subtilis	BBG111	Rice	Jasmonic acid (JA) and ethylene (ET) as well as abscisic acid (ABA) and auxin signalling	*Rhizoctonia solani*	Chandler et al. (2015)

affect the growth of the pathogen. In many instances, *Bacillus* spp. have been found to be highly effective in the production of these oxidative enzymes. According to a study by Ramyabharathi et al. (2012), the liquid formulation of *B. subtilis* EPCO16 was capable of eliciting defense related enzymes like catalase and phenylalanine ammonia lyase (PAL) in tomato plants infected with *F. oxysporum* f. sp. *lycopersici*. Spectrophotometric analysis of plants revealed the activity of defense enzymes at their highest on the seventh day after inoculation by the pathogen.

Chitin is an important polysaccharide, which imparts structural rigidity and integrity to the fungal cell wall. But, the presence of hydrolysing enzymes like chitinase and glucanase breaks down the glycosidic bonds holding the constituting polysaccharides together, that leads to cell leakage and cell lysis. Many species of the *Bacillus* spp. are known to produce chitinolytic enzymes and suppress the phytopathogens (Podile and Prakash 1996). Wang et al. (2004) observed that two types of chitinase were produced by *B. amyloliquefaciens* V656, which were highly effective in inhibiting the growth of *F. oxysporum* growth. *B. thuringiensis* is known to inhibit the growth of *Sclerotium rolfsii* in soybean plant by a similar mechanism. Studies by Liu et al. (2010) revealed the production of chitinase from *B. thuringiensis* sub sp. *colmeri,* which was instrumental in preventing spore germination of various phytopathogenic fungi. Generally, bacteria possess a variety of chitinases for breaking down the wide array of chitin molecules that naturally exist in nature. Ruiz-Sanchez (2007) observed that five different types of chitinase (with 42, 49, 53, 62, and 66 kDa) were synthesised by *B. licheniformis*.

6.3.3 Production of Lipopeptides and Antibiotics

Bacillus spp. are considered to be the producers of a vast array of antimicrobial compounds such as bacillomycin and zwittermycin A (Mondol et al. 2013). According to studies, 4-5% of the *Bacillus* genome is devoted for the production of antimicrobial compounds (Stein 2005). Among these compounds, lipopepetides of the surfactins, iturins, and fengycin group are most widely studied. Surfactins are antiviral, antibacterial agents, while iturins and fengycins, also known as plipastatins, exhibit highly effective antifungal activities. The lipopeptides (LPs) interact with the each other in a synergistic way, and bring about an alteration of the cell membrane permeability of the microorganisms. The LPs produced by *Bacillus* are non-ribosomal in nature and are produced by non-ribosomal peptide synthetases (NRPSs) or hybrid polyketide synthases and non-ribosomal peptide synthetases (PKSs/NRPSs). The efficacy of biocontrol of plant diseases by lipopeptides produced by *Bacillus* spp. is well established (Chakraborty et al. 2020a, b).

The surfactins are excellent biosurfactants, with notable foaming ability. They attach themselves to the lipid layers, and thus hamper the integrity of the microbial membranes. The surfactins induce pore formation, which is followed by complete solubilisation and disruption of the membranes. Studies also reveal that the surfactins are unable to function in the presence of cholesterol in the phospholipid bilayer, which explains why they are inactive as antifungal compounds (Meena and Kanwar 2015).

Iturins are a group of heptapepetides linked to a β-amino fatty acid chain. They are named so after their place of discovery, Ituri, in the Democratic Republic of the Congo. The family of iturins mainly include iturins A, C, D, and E; bacillopeptin; bacillomycins D, F, L and LC; and mycosubtilin (Mnif and Ghribi 2015). They are highly fungitoxic in nature but with limited antiviral and antibacterial activity. The iturins are known to form ion-conducting pores, which lead to the permeablilisation of the membranes, leading to fungal toxicity.

The fengycins, also known as plipastatins, are a group of lipodecapeptides attached to a β hydroxy fatty acid chain (Wang et al. 2015). Fengycins are highly fungitoxic in nature, especially against filamentous fungi; however, they are less effective as compared to surfactins and iturins. The mechanism of fungitoxicity of the fengycins is elusive so far. However, these natural compounds have been shown to induce structural imbalance in the microbial membrane (Ongena and Jacques, 2008). Other antibiotics produced by *Bacillus* species include bacteriocins, bacillaene, difficidin, oxidifficidin, sporulenes A–C, bacisubin, and bacilysocin. All these antibiotics exhibit antifungal and antibacterial activity at varying levels of concentrations.

The antibiotics produced by *Bacillus* have a crucial role in its biocontrol activities. The LPs are attributed to biofilm formation by the bacteria in the soil and water surfaces. They bring about the flagella-driven motility of the bacteria, leading to biofilm spread over the plant and soil surfaces. They are also responsible for reducing the surface tension, which helps in further spread of the bacteria. Bais and collaborators (2004) reported that *B subtilis* strain 6051 produced surfactin molecules that resulted in biofilm formation in the roots of *Arabidopsis* plant.

It has been reported by Asaka and his co-workers (1996) that *B subtilis* strain RB14 produces iturin A, which is instrumental in inhibiting the growth of *Rhizoctonia solani* in tomato. Leclere and co-workers (2005) found that mycosubtillin, produced by *B. subtilis* ATCC, helps in the reduction of *Pythium aphanidermatum* infection. The combined role of fengycin and iturin in inhibiting the growth of *Podosphaera fusca* in melon leaves was demonstrated by Romero and co-workers (2007). Ongena and Jacques (2008) observed that *B. subtilis* S499 produces all three families of LPs that help in the management of plant diseases by this strain. This strain also inhibits the growth of *Botrytis cinerea* in wounded apple fruits by the production of lipopeptide-enriched extracts, which contained high doses of fengycins in it.

The production of antibiotics was also linked to the induction of defense reactions in a number of treated plants. In tomato and bean, fengycins and surfactins were directly attributed to induce ISR. *B subtilis* S499 inhibited the growth of *Colletotrichum lagenarium* in cucumber plants by the stimulation of systemic resistance. Potato tubers exhibited an accumulation of phenolics when being treated with purified fengycins, which activate the phenylpropanoid pathway. The activation of phenylpropanoid pathway leads to the accumulation of mRNAs, encoding phenylalanine ammonia lyase (PAL).

Although the antibiotics are highly potent in inhibiting the growth of harmful microorganisms, they have not been reported to produce any phytotoxicity in the host plants. These compounds are able to induce a cascade of biochemical reactions that stimulated the defense system of the plant against the pathogen, without causing any adverse effects on the cellular integrity of the plant. Researchers believed that the differing compositions of phytosterol from that of bacterial and fungal compositions are mainly responsible for the attenuation of the disruption of plant membranes.

6.4 Induction of Resistance in Plants

Several lines of evidence suggest that strains of *Bacillus* spp. elicit systemic resistance in the plants. Induced systemic resistance (ISR) was elicited in plants by *Bacillus* spp. in response to a number of biotic stresses like fungi, bacteria, viruses, nematodes, and insects. *Bacillus* spp. produce LPs which are considered to be the key components in ISR elicitation in plants (Rahman et al. 2015). The ISR modulates a number of cytological and biochemical processes in the host plant such as lignin deposition in the plant cell wall, production of phytoalexins, and synthesis of other antimicrobial substances like peroxidases and β-1, 3-glucanases (GarcíaGutiérrez et al. 2013). Among the LPs produced, surfactin is essential for ISR elicitation as observed by Chowdhury and his collaborators in 2015. As discussed earlier, the iturin and surfactin family of lipopepetides are highly effective in suppressing phytopathogens. After analysis of gene expression, it was observed that surfactin activated salicylic acid (SA) regulated pathway, while mycosubtillin of the iturin family, had the ability to activate both jasmonic acid (JA) and salicylic acid (SA) pathways in grapevine (Farace et al. 2015). Similar results were observed in strawberry, which helped in the successful inhibition of *Colletotrichum gleosporioides* (Yamamoto et al. 2015). The role of LPs was also demonstrated in maize (Gond et al. 2015), rice (Chandler et al. 2015), and tomato (Abdallah et al. 2017). The ISR generally leads to plant growth promotion as seen in strain *B. subtilis* 21-1. Lee and his co-workers (2014) observed that *B. subtilis* 21-1 helps in disease suppression of vegetable crops, by activating the plant defense system, as well as helps in plant growth promotion. It can be concluded that *B. subtilis* elicited plant defense system and growth promotion of the host, which collaborates to inhibit the phytopathogens.

The phytopathogens differ considerably in their morphology and cellular compositions. Hence, *Bacillus* employs different mechanisms against these phytopathogens (Fig. 6.2). It synthesises lipopeptides against some, while it produces enzymes for managing other pathogens. A brief description of the mechanisms used by *Bacillus* against phytopathogens is provided in the following section.

6.4.1 Bacillus *Against Fungi*

Extensive research has been conducted on the mechanism of *Bacillus* for the inhibition of fungal growth. Chitin forms a major constituent of the cell wall of all pathogenic fungi; hence, biocontrol agents such as *Trichoderma, Alteromonas,* and *Serratia* produce different type of chitinases (Elad et al. 1982). *Bacillus* has also been reported to produce chitinolytic enzymes for lysing the fungal cell wall (Mitchell and Alexander, 1962). *Bacillus* species was observed to reduce the radial growth *of Penicillium* sp. (72%), *Aspergillus niger* (64.58%), *Aspergillus*

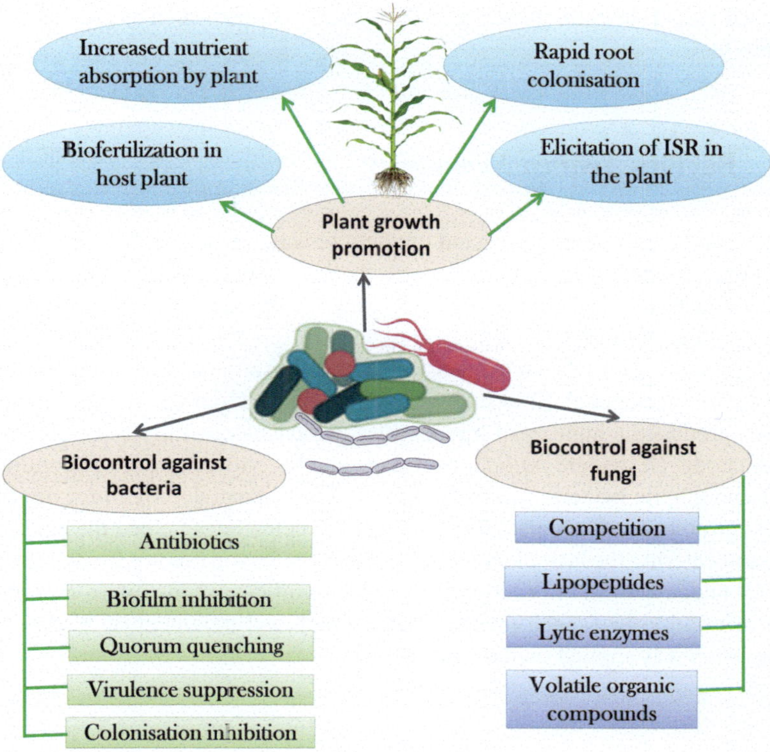

Fig. 6.2 Mechanisms used by *Bacillus* sp. as a growth promoter and biocontrol agent

fumigatus (51.61%), *Fusarium* sp. (39.13%), *Curvularia* sp. (37.50%), *Aspergillus flavus* (35.43%), and *Alternaria* sp. (31.82%), after 6 days of incubation. The inhibitory action was attributed to chitinolytic activity and hyperparasitism as concluded by Basha and Ulaganathan (2002). It was observed by Bargabus et al. (2002, 2004) that *B. mycoides* isolate BacJ and *B. pumilus* isolate 203–7 were highly effective in controlling the *Cercospora* leaf spot of sugar beet to a considerable extent.

Studies revealed that many *Bacillus* strains such as *B. lentimorbus* and *B. cereus* are able to inhibit the development of coffee rust. It has been observed that the bacterium prevented the pathogen from developing uredospores, which resulted in suppression of the pathogen up to 50% (Shiomi et al. 2006). The wide array of volatile compounds produced by *B. velezensis* ZSY-1 was studied by Gao et al. in 2017. They observed that these compounds possessed significant antifungal activities. *Bacillus* was able to inhibit the growth of *B. cinerea, Monilinia fructicola, Colletotrichum lindemuthianum, Alternaria solani,* and *F. oxysporum* f. sp. *capsicum*. It was also observed that benzothiazole, pyrazine, and phenolic compounds played a crucial role in inhibiting fungal diseases in tomato such as grey mould and early blight.

6.4.2 **Bacillus** ***Against Nematodes***

Limited studies have been conducted on the interaction between *Bacillus* and plant pathogenic nematodes. In some instances, the involvement of LPs has been correlated with the reduction in nematode infestation (Hallmann 2001). Hallmann (2001) observed that bacterial endophytes successfully inhibit the infestation of *Meloidogyne incognita* through extensive root colonisation and elicitation of ISR in the host plant. Reitz et al. (2000) observed that the bacterial endophytes, including *Bacillus*, produced a number of lipopeptides, which successfully controlled potato cyst nematode, *Globodera pallida*. Hallmann and his co-workers (1997) studied the role of endophytic bacteria in inhibiting root-knot nematode, *Meloidogyne incognita*, in cotton roots. They also observed that the nematodes helped the endophytic bacteria in entering the host plant which aided in successful colonisation of the plant roots. Mendoza and Sikora (2009) observed a significant reduction in *Radopholus similis* infestation in banana plants, when a combined treatment of *B. firmus* and *F. oxysporum* was applied on the banana plant.

6.4.3 **Bacillus** ***Against Bacterial Pathogens***

Phytopathogenic bacteria communicate among each other by the production of density-dependent autoinducers (AI), which is known as quorum sensing. Quorum sensing plays a crucial role in establishing the virulence factors in bacteria, which determines the disease severity. In the case of Gram-negative bacteria, *N*-acyl homoserine lactones (AHL) are mostly used as signalling molecules during quorum sensing. *Bacillus* species, however, have the ability to produce quorum quenching molecules that modify the AIs and hinder the bacterial signalling, thus affecting the virulence of the pathogen. *Bacillus* possesses *aiiA* gene encoding N-acyl homoserine lactonase (AiiA) enzyme, which brings about the hydrolysis of AHLs. *Bacillus* species, such as *B. cereus, B. subtilis, B. firmus,* and *B. toyonensis*, have been known to suppress a number of plant pathogenic bacteria by inhibiting quorum sensing, such as *Pseudomonas aeruginosa*, *Agrobacterium tumefaciens*, and *Pectobacterium carotovorum (Erwinia carotovora).* In addition to directly using the AiiA-producing *Bacillus* species to suppress the virulence of phytopathogenic bacteria (Dong et al. 2004), *aiiA* can be heterologously expressed in other bacterial strains (Molina et al. 2003) or in genetically modified plants (Ouyang and Li 2016).

6.4.4 Bacillus *Against Viral Phytopathogens*

Bacillus species has been found to provide resistance in plants against many viruses, such as Tomato Mottle Virus (ToMov) and cucumber mosaic virus. The application of *Bacillus* leads to the reduction in visible symptoms as well as reduced viral accumulation as is evident from ELISA analysis (Murphy et al. 2000). Some studies indicate that the *Bacillus* species results in an accumulation of IAA, leading to growth enhancement in the plants, that results in systemic protection against the virus as observed by Murphy and his co-workers (2000), in tomato against CMV. However, according to some researchers, *Bacillus* elicits ISR in the virus affected plants, that triggers the expression of PR genes, leading to an accumulation of PR proteins like chitinase and β 1,3-glucanase, as observed by Wang et al. (2009) in tobacco plant against TMV. It is known that NPR1 regulates SAR and ISR resistance pathways while Coi1 regulated ISR pathway. *Bacillus* species has been observed to induce the expression of both *NPR1* and *Coi1* genes in the TMV-resistant tobacco plants, indicating a correlation between the gene expression and disease resistance.

6.5 Genomics and Molecular Basis of Induction Resistance by *Bacillus* on Plant

Bacillus has been known to produce a number of antimicrobial compounds that aid in plant disease management. *B. subtilis* has been considered to be a model microorganism for the analysis of gene functions, as it encodes for a number of antibiotics and lytic enzymes in order to manage phytopathogens. Research on the genomics and molecular basis of resistance induction may help in the development of biologically feasible weapons against plant diseases. Fungal cell walls are made of chitin that provides mechanical strength to the fungus. According to a study, *B. subtilis* CHU26, isolated from a potato field in Taiwan, exhibited strong chitinase activity in vitro, that was successful in inhibiting the growth of *Rhizoctonia solani*. Further investigations revealed the presence of chitinase encoding gene *chi18* that was responsible for this antimicrobial activity (Yang et al. 2009).

Hydrogen peroxide is known to exhibit cytotoxic activity in organisms by the production of hydroxyl radicals that react to lipids, proteins, and nucleic acids. However, *B. subtilis* has been able to mitigate the H_2O_2 stress by the production of enzymes catalase and an alkyl hydroperoxide reductase, encoded by genes *katA* and *ahpC*, respectively (Broden et al. 2016).

For evaluating the role of *Bacillus* in stimulation of ISR, Lee and co-workers (2015) studied strain HK34 of *B. amyloliquefaciens* against *Phytophthora cactorum* by applying *Bacillus* on the leaves and roots of *Panax ginseng*. It was observed that the bacterium was able to reduce the pathogen growth up to 99.1%. The inhibition of the pathogen was found to correlate with the expression of *PgCAT*, *PgPR5*, and *PgPR10* genes. These outcomes showed the ISR-eliciting potential of strain HK34.

Table 6.3 List of genes of *Bacillus* species encoded for the production of various antimicrobial metabolites that help them in biocontrol of phytopathogens

Strain	Gene cluster	Products/function	Reference
Bacillus sp.	*srf*A	Surfactin	Hsieh et al. (2004)
Bacillus sp.	*fen*B	Fengycin	Ramarathnam et al. (2007)
Bacillus sp.	*bac*A	Bacilysin	Mora et al. (2011)
B. subtilis	*Eps* A-O operon	Synthesis of biofilm matrix	Vlamakis et al. (2013)
*B. subtilis*168	*deg*Q	Secretion of degradative enzymes	Parashar et al. (2013)
B. subtilis	*Pks* gene cluster	Bacillaene	Muller et al. (2014)
B. subtilis 916	locA, locB, locC and locD, gene cluster	Locillomycins	Luo et al. (2015)
B. subtilis	*tyr*Z	Growth and biofilm formation	Williams-Wagner et al. (2015)
B. subtilis	*tap*A	Surfactin synthesis	van Gestel et al. (2015)
B. subtilis	*Apr*E and *Npr*E	Extracellular proteases	Barbieri et al. (2016)
B. subtilis	*lta*S	Lipoteichoic acid (LTA) synthase	Kasahara et al. (2016)
B. subtilis	SWR01 *min*J	Swarming	Gao et al. (2016)

Ryu et al. (2004) worked on the production of volatile compounds, produced by *Bacillus* species. They observed that *B. subtilis* GB03 synthesised 2,3-butanediol and 3-hydroxy-2-butanone (acetoin) in abundance, which helped in the induction of ISR. Furthermore, the volatiles produced by the bacteria were controlled by the expression of genes *CHIB*, *GST2,* and *ERF1*, which were known for the biosynthesis of ethylene. Recent studies have been able to discover a number of gene clusters in *Bacillus* species, which are shown in Table 6.3.

6.6 Commercial Applications of *Bacillus* Species

Bacillus strains have been utilised commercially for the generation of a variety of products. It is used not only as a potent host for genetic modifications but also as a source of naturally obtained biocontrol compounds (Bunk et al. 2010). Furthermore, *Bacillus* species produces stress-resistant endospores that contribute towards better environmental stability and longer shelf-life of the products. Among the wide array of *Bacillus* species discovered so far, *B. thuringiensis* is the most widely exploited with more than 70% of the market share (Ongena and Jacques 2008). Different companies utilise different mechanisms of the biocontrol agent in developing its products. For example, Bio-Yield, produced by 3Bar Biologics Inc., USA, contains a combination of two *Bacillus* species, *B. amyloliquefaciens* GB99 which is responsible for the elicitation of ISR. The *B. subtilis* GB122 inhibits soil-borne pathogens by the production of lipopeptide iturin. Also the product contains chitosan, which is effective against insects and nematodes (Kloepper et al. 2004a, b). Another product,

Yield Shield produced by Bayer CropScience Inc., USA, is composed of *B. pumilus* GB34, which activates the plant defense system and promotes plant growth as well (Jeong et al. 2014). Similarly, many other *Bacillus*-based products have been developed commercially that can be utilised for the successful management of phytopathogens (Table 6.4).

Table 6.4 Commercial phytosanitary products obtained from *Bacillus*

Name of the product	*Bacillus* strains	Target pathogen	Crop	Company
Bio-yield	*B. subtilis* GB122, *B. Amyloliquefaciens* GB99	*Fusarium, Rhizoctonia, Pythium,*	Bedding plants	3Bar biologics, USA
Nacillus	*B. subtilis Antumávida, B. subtilis Vilcún, B. licheniformis Mallerauco Brevi, Bacillus brevis Maguellines Brevi, Bacillus brevis Maguellines I*	*Clavibacter pseudomonas, Xanthomonas, Acetobacter*	Vegetables, blueberry Cherry, hazelnut tree, tomato, pear, kiwi tree,	Bio Insumos Nativa, Chile
Yield shield	*B. pumilus* GB34	*Fusarium sp., Rhizoctonia solani,*	Soybean	Bayer CropScience, USA
Votivo	*B. firmus* I-1582	*Pratylenchus, Meloidogyne*	Corn, cotton, soybean,	Bayer CropScience, USA
Dipel	*B. thuringiensis* subsp. *kurstaki* HD-1	*Helicoverpa, Diatraea saccharalis, Grapholita molesta, Diaphania nitidalis, Plutella xylostella, Argyrotaenia sphaleropa*	Citrus, pineapple, tomato, cotton, apple tree, sugarcane, melon, cabbage,	Valent BioSciences, USA
AvoGreen	*B. subtilis*	*Colletotrichum gloeosporioides, Cercospora sp.*	Avocado	Ocean agriculture, South Africa
Ecoshot	*B. subtilis*	*Botrytis cinerea*	Citrus, legumes, vegetables, grape	Kumiai chemical industry, Japan
HiStick N/T, Subtilex	*B. subtilis* MB 1600	*Fusarium sp., aspergillus sp., Rhizoctonia sp., Pythium sp.*	Vegetables, ornamentals	Becker underwood, Ames, IA, USA
Kodiak	*B. subtilis* GB03	*Fusarium sp., aspergillus sp., Rhizoctonia sp., Alternaria sp.*	Legumes	Gustafson Inc., Dallas, Texas, USA
Biosafe	*B. subtilis*	*Xanthomonas axonopodis pv. Phaseoli*	Bean	Laboratorio de Biocontrole Farroupilha, Brazil

6.7 Mode of Application of *Bacillus* Species

The method of application of the biocontrol agent depends on its mode of action. Among the different methods of application, seed coating is the most extensively used, as it is easy to use and can be applied efficiently with a little quantity of inoculum. When the microbes are introduced as granular applications, the biocontrol agent is applied with a mixture of marble, peat, perlite, charcoal, and soil, with an aim of enhancing the contact between the plant roots and the biocontrol agent.

In case of *Bacillus*, both liquid and powdered formulations are available commercially, which are applied in the form of pellets, as well as soil drench, foliar spray, or as seed dressing. When *Bacillus* was applied as liquid formulation on tomato plant, it was able to successfully reduce the symptoms of wilt, as well as increase the shelf-life of *Bacillus* (Ramyabharti et al. 2016). Gao and his co-workers (2015) researched upon the most successful method of *B. subtilis* application against *Blumeria graminis* in wheat. They observed that all the preparations, i.e., cell-free culture, non-protein fermentation liquid, and crude proteins, were able to reduce the infection to some extent. However, fermentation liquid formulation was the most effective among all the preparations studied. Hsieh and his collaborators (2009) observed that talc-based formulation of *B. subtilis* and *P. fluorescens* was able to reduce the Banana Bunchy Top Virus incidence up to about 52% and helped in growth promotion of the plant under field conditions.

According to the studies conducted by Selim and his co-workers (2017), the application of endophytic bacteria as both soil drench and talc-based formulations was able to manage *Rhizoctonia solani* in cotton. But among the two formulations studied, soil drench was found to be more effective. However, Yamamoto and co-workers (2015) found that *Bacillus amyloliquefaciens* S13–3 was highly efficient in managing anthracnose in strawberry when applied as foliar spray. Spraying of fermentation liquid formulation of *Bacillus* sp. significantly improves growth, yield, and content of antioxidants in strawberry fruit (Rahman et al. 2015).

After the analysis of each method, Hallman and his co-workers (1997) observed that every method had its pros and cons. However, seed treatment was found to be the most efficient when ease of application, economic feasibility, and environmental stability was taken into consideration. They further stated that the combined application as seed treatment, soil drench, and foliar spray would be able to enhance the colonisation of the bacteria and also increase the benefits manifold.

6.8 Conclusion

Bacillus sp. is considered as the new tools for the promotion of sustainable agricultural system. A lot of information on biocontrol of plant diseases by *Bacillus* spp. has been generated so far by different researchers throughout the world. In this chapter, we summarised the details of the recent advances on mode of action, molecular basis of pathogenic antagonism, application in field level and formulations with their stability potential of *Bacillus* spp. In these aspects, this chapter is unique, informative and a good source of molecular insights for the future researchers to deal with the *Bacillus* biocontrol agent in the crop field without harming the nature.

In this chapter, we summarised the mode of actions involved in successful antagonism and the PGPR activities that include – secretion of lytic enzymes, production of lipoproteins, and antibiotics that are explained with examples. The mechanism of induced systemic resistance against plant pathogens depends on interactions of host-pathogen and environment. One or more *Bacillus* strains can work synergistically with plant and environment in suppressing the phytopathogens. The mechanism is varied according to the application of *Bacillus* sp. against fungus, bacteria, virus and nematodes. The most important part of this chapter is the genomics and molecular basis of induction of systemic resistance by the *Bacillus* spp. *Bacillus* is known to produce a number of antimicrobial compounds that deal with plant diseases resistance mechanism. In this case, the spore producing ability and genetically modified host with *Bacillus* genes may provide an effective solution in their colonisation simultaneously along with other beneficial microbes. But for successful colonisation, understanding ecological requirements are mandatory.

Among the wide functional assay of *Bacillus* sp. so far, *B. thuringiensis* is the most widely exploited even more than 70% of the market shares. But there are lots of conflicts on its commercialisation and acceptation country to country. So, successful application and commercialisation depends on the practical understanding on mode of action, methods of application, ecological distribution and interacting environment. But for more popularity among the farmers, a perfect demonstration of benefit:cost ratio would be mostly required.

References

Abdallah RAB, Stedel C, Garagounis C, Nefzi A, Jabnoun-Khiareddine H, Papadopoulou KK, Daami-Reamadi M (2017) Involvement of lipopeptide antibiotics and chitinase genes and induction of host defense in suppression of fusarium wilt by endophytic *Bacillus* spp. in tomato. Crop Prot 99:45–58

Adhikari TB, Joseph CM, Yang G, Phillips DA, Nelson LM (2001) Evaluation of bacteria isolated from rice for plant growth promotion and biological control of seedling disease of rice. Can J Microbiol 47:916–924

Akram W, Anjum T (2011) Quantitative changes in defense system of tomato induced by two strains of *Bacillus* against fusarium wilt. Indian J Fund Appl Life Sci 1:7–13

Arias RS, Sagardoy MA, van Vuurde JWL (1999) Spatio-temporal distribution of naturally occurring *Bacillus* spp. and other bacteria on the phylloplane of soybean under field conditions. J Basic Microbiol 39:283–292

Asaka O, Shoda M (1996) Biocontrol of Rhizoctonia solani damping-off of tomato with *Bacillus subtilis* RB14. Appl Environ Microbiol 62:4081–4085

Ash C, Farrow JAE, Priest FG, Collins MD (1993) Molecular identification of rRNA group 3 bacilli using a PCR probe test. Antonie Van Leeuwenhoek 64:253–260

Bacon CW, Hinton DM (2002) Endophytic and biological control potential of *Bacillus mojavensis* and related species. Biol Control 23:274–284

Bais HP, Park SW, Weir TL, Callaway RM, Vivanco JM (2004) How plants communicate using the underground information superhighway. Trends Plant Sci 9:26–32

Baker CJ, Stavely J, Thomas CA, Sasser M, Macfall JS (1983) Inhibitory effect of *Bacillus subtilis* on Uromyces phaseoli and on development of rust pustules on bean leaves. Phytopathology 73(8):1148–1152

Barbieri G, Albertini AM, Ferrari E, Sonenshein AL, Belitsky BR (2016) Interplay of CodY and ScoC in the regulation of major extracellular protease genes of *Bacillus subtilis*. J Bacteriol 198:907–920

Bargabus RL, Zidack NK, Sherwood JE, Jacobsen BJ (2002) Characterization of systemic resistance in sugar beet elicited by a nonpathogenic, phyllosphere-colonizing *Bacillus mycoides*, biological control agent. Physiol Mol Plant Pathol 61:289–298

Bargabus RL, Zidack NK, Sherwood JE, Jacobsen BJ (2004) Screening for the identification of potential biological control agents that induce systemic acquired resistance in sugar beet. Biol Control 30:342–350

Basha S, Ulaganathan K (2002) Antagonism of Bacillus species (strain BC121) towards *Curvularia lunata*. Curr Sci 82(12):1457–1462

Benhamou N, Kloepper JW, Quadt-Hallman A, Tuzun S (1996) Induction of defense-related ultrastructural modifications in pea root tissues inoculated with endophytic bacteria. Plant Physiol 112(3):919–929

Blagoeva-Nikolaeva V, Slavov S, Varsano L (1995) Possibilities for biological control of blackshank disease in tobacco with *Bacillus cereus*. Biotechnol Biotechnol Equip 9(1):36–39

Broden NJ, Flury S, King AN, Schroeder BW, Coe GD, Faulkner MJ (2016) Insights into the function of a second, nonclassical Ahp peroxidase, AhpA, in oxidative stress resistance in Bacillus subtilis. J Bacteriol 198:1044–1057

Bunk B, Schulz A, Stammen S, Munch R, Warren MJ, Jahn D, Biedendieck R (2010) A short story about a big magic bug. Bioeng Bugs 1:85–89

Cao Y, Zhang Z, Ling N, Yuan Y, Zheng X, Shen B, Shen Q (2011) Bacillus subtilis SQR 9 can control fusarium wilt in cucumber by colonizing plant roots. Biol Fertil Soils 47(5):495–506

Chakraborty S, Debnath D, Mahapatra S, Das S (2020a) Role of endophytes in plant disease management. In: Singh KP, Jahagirdar S, Sarma BK (eds) Emerging trends in plant pathology. Springer, Singapore. https://doi.org/10.1007/978-981-15-6275-4

Chakraborty M, Mahmud NU, Gupta DR, Tareq FS, Shin HJ, Islam T (2020b) Inhibitory effects of linear lipopeptides from a marine *Bacillus subtilis* on the wheat blast fungus *Magnaporthe oryzae Triticum*. Front Microbiol 11:665

Chandler S, Van Hese N, Coutte F, Jacques P, Monica H, De Vleesschauwer D (2015) Role of cyclic lipopeptides produced by *Bacillus subtilis* in mounting induced immunity in rice (*Oryza sativa* L.). Physiol Mol Plant Pathol 91:20–30

Choudhary DK, Johri BN (2008) Interactions of *Bacillus* spp. and plants – with special reference to induced systemic resistance (ISR). Microbiol Res 164:493–513

Chowdhury SP, Uhl J, Grosch R, Alqueres S, Pittroff S, Dietel K, Schmitt-Kopplin P, Borriss R, Hartmann A (2015) Cyclic Lipopeptides of *Bacillus amyloliquefaciens* subsp. *plantarum* colonizing the lettuce rhizosphere enhance plant defense responses toward the bottom rot pathogen *Rhizoctonia solani*. Mol Plant-Microbe Interact 28:984–995

Clayton MK, Hudelson BD (1991) Analysis of spatial patterns in the phyllosphere. In: Andrews JH, Hirano SS (eds) Microbial ecology of leaves. Springer, New York, pp 111–131

Crowley DE, Rengel Z (1999) Biology and chemistry of rhizosphere influencing nutrient availability. In: Rengel Z (ed) Mineral nutrition of crops: fundamental mechanisms and implications. The Haworth, New York, pp 1–40

Dong YH, Zhang XF, Xu JL, Zhang LH (2004) Insecticidal *Bacillus thuringiensis* silences *Erwinia carotovora* virulence by a new form of microbial antagonism, signal interference. Appl Environ Microbiol 70:954–960

Duineveld BM, Kowalchuk GA, Keijzer A, van Elsas JD, van Veen JA (2001) Analysis of bacterial communities in the rhizosphere of chrysanthemum via denaturing gradient gel electrophoresis of PCR amplified 16S rRNA as well as DNA fragments coding 16S rRNA. Appl Environ Microbiol 67:172–178

Elad Y, Chet I, Henis Y (1982) Parasitism of *Trichoderma* spp. on *Rhizoctonia solani* and *Sclerotium rolfsii* - scanning electron microscopy and fluorescence microscopy. Phytopathology 28:719–725

El-Bendary MA (2006) Bacillus thuringiensis and Bacillus sphaericus biopesticides production. J Basic Microbiol 46(2):158–170

Farace G, Fernandez O, Jacquens L, Coutte F, Krier F, Jacques P, Clement C, Barka EA, Jacquard C, Dorey S (2015) Cyclic lipopeptides from *Bacillus subtilis* activate distinct patterns of defence responses in grapevine. Mol Plant Pathol 16:177–187

Gao Y, Liu Q, Zang P, Li X, Ji Q, He Z, Zhao Y, Yang H, Zhao X, Zhang L (2015) An endophytic bacterium isolated from *Panax ginseng* CA meyer enhances growth, reduces morbidity, and stimulates ginsenoside biosynthesis. Phytochem Lett 11:132–138

Gao SF, Wu HJ, Yu XF, Qian LM, Gao XW (2016) Swarming motility plays the major role in migration during tomato root colonization by *Bacillus subtilis* SWR01. Biol Control 98:11–17

Gao Z, Zhang B, Llu H, Han J, Zhang Y (2017) Identification of endophytic *Bacillus velezensis* ZSY-1 strain and antifungal activity of its volatile compounds against *Alternaria solani* and Botrytis cinerea. Biol Control 105:27–39

García-Gutiérrez L, Zeriouh H, Romero D, Cubero J, de Vicente A, Perez-Garcia A (2013) The antagonistic strain *Bacillus subtilis* UMAF6639 also confers protection to melon plants against cucurbit powdery mildew by activation of jasmonate- and salicylic acid-dependent defence responses. Microb Biotechnol 6:264–274

Gond SK, Bergen MS, Torres MS, White JF Jr (2015) Endophytic *Bacillus* spp. produce antifungal lipopeptides and induce host defence gene expression in maize. Microbiol Res 172:79–87

Hallman J, Quadt-Hallmann A, Mahaffee WF, Kloepper JW (1997) Bacterial endophytes in agricultural crops. Can J Microbiol 43:895–914

Hallmann J (2001) Plant interactions with endophytic bacteria. In: Jeger MJ, Spence NJ (eds) Biotic interactions in plante pathogen associations. CABI Publishing, Wallingford, pp 87–119

Hallmann J, Rodríguez-Kábana R, Kloepper JW (1997) Nematode interactions with endophytic bacteria. In: Ogoshi A, Kobayashi K, Homma Y, Kodama F, Kondo N, Akino S (eds) Plant growth-promoting Rhizobacteria-present status and future prospects. Nakanishi Printing, Sapporo, pp 243–245

Hardoim PR, van Overbeek LS, Berg G, Pirttila AM, Compant S, Campisano A, Doring M, Sessitsch A (2015) The hidden world within plants: ecological and evolutionary considerations for defining functioning of microbial endophytes. Microbiol Mol Biol Rev 79:293–320

Hsieh FC, Li MC, Lin TC, Kao SS (2004) Rapid detection and characterization of surfactin- producing *Bacillus subtilis* and closely related species based on PCR. Curr Microbiol 49:186–191

Hsieh PW, Hsu LC, Lai CH, Wu CC, Hwang TL, Lin YK, Wu YC (2009) Evaluation of the bioactivities of extracts of endophytes isolated from Taiwanese herbal plants. World J Microbiol Biotechnol 25:1461–1469

Hu X, Roberts DP, Xie L, Maul JE, Yu C, Li Y, Zhang S, Liao X (2013) Bacillus megaterium A6 suppresses *Sclerotinia sclerotiorum* on oilseed rape in the field and promotes oilseed rape growth. Crop Protect 52:151–158

Islam MT, Croll D, Gladieux P, Soanes DM, Persoons A, Bhattacharjee P et al (2016a) Emergence of wheat blast in Bangladesh was caused by a south American lineage of Magnaporthe oryzae. BMC Biol 14:84

Islam MT, Rahman PP, Jha CK, Aeron A (2016b) Bacilli and Agrobiotechnology. Springer

Islam MT, Kim KH, Choi J (2019a) Wheat blast in Bangladesh: the current situation and future impacts. Plant Pathol J 35(1):1–10

Islam T, Rahman MM, Pandey P, Boehme MH, Haesaert G (2019b) Bacilli and agrobiotechnology: phytostimulation and biocontrol. Bacilli in climate resilient agriculture and bioprospecting, vol 2. Springer

Islam MT, Gupta DR, Hossain A et al (2020) Wheat blast: a new threat to food security. Phytopathol Res 2(2020):28. https://doi.org/10.1186/s42483-020-00067-6

Jaeger CH III, Lindow SE, Miller W, Clark E, Firestone MK (1999) Mapping of sugar and amino acids availability in soil around roots with bacterial sensors of sucrose and tryptophan. Appl Environ Microbiol 65:2685–2690

Jangir M, Pathak R, Sharma S, Sharma S (2018) Biocontrol mechanisms of Bacillus sp., isolated from tomato rhizosphere, against *Fusarium oxysporum* f. sp. lycopersici. Biol Control 123:60–70

Jayaraj J, Yi H, Liang G, Muthukrishnan S, Velazhahan R (2004) Foliar application of *Bacillus subtilis* AUBS1 reduces sheath blight and triggers defense mechanisms in rice. J Plant Dis Prot 111(2):115–125

Jeong H, Choi SK, Kloepper JW, Ryu CM (2014) Genome sequence of the plant endophyte *Bacillus pumilus* INR7, triggering induced systemic resistance in field crops. Genome Announc 2:e01093–e01014

Jiang CH, Wu F, Yu ZY, Xie P, Ke HJ, Li HW, Yu YY, Guo JH (2015) Study on screening and antagonistic mechanisms of *Bacillus amyloliquefaciens* 54 against bacterial fruit blotch (BFB) caused by *Acidovorax avenae* subsp. *citrulli*. Microbiol Res 170:95–104

Kasahara J, Kiriyama Y, Miyashita M, Kondo T, Yamada T, Yazawa K, Yoshikawa R, Yamamoto H (2016) Teichoic acid polymers affect expression and localization of DL-endopeptidase LytE required for lateral cell wall hydrolysis in *Bacillus subtilis*. J Bacteriol 198:1585–1594

Kloepper JW, Reddy MS, Rodríguez-Kabana R, Kenney DS, KokalisBurelle N, MartinezOchoa N (2004a) Application for rhizobacteria in transplant production and yield enhancement. Acta Hortic 631:219–229

Kloepper JW, Ryu CM, Zhang S (2004b) Induced systemic resistance and promotion of plant growth by *Bacillus* spp. Phytopathology 94(11):1259–1266

Leclere V, Bechet M, Adam A, Guez JS, Wathelet B, Ongena M, Thonart P, Gancel F, Chollet-Imbert M, Jacques P (2005) Mycosubtilin overproduction by *Bacillus subtilis* BBG100 enhances the organism's antagonistic and biocontrol activities. Appl Environ Microbiol 71:4577–4584

Lee SW, Lee SH, Balaraju K, Park KS, Nam KW, Park JW, Park K (2014) Growth promotion and induced disease suppression of four vegetable crops by a selected plant growth-promoting rhizobacteria (PGPR) strain *Bacillus subtilis* 21-1 under two different soil conditions. Acta Physiol Plant 36:1353–1362

Lee BD, Dutta S, Ryu H, Yoo SJ, Suh DS, Park K (2015) Induction of systemic resistance in *Panax ginseng* against *Phytophthora cactorum* by native *Bacillus amyloliquefaciens* HK34. J Ginseng Res 39(3):213–220

Li S, Zhang N, Zhang Z, Luo J, Shen B, Zhang R, Shen Q (2013) Antagonist *Bacillus subtilis* HJ5 controls *Verticillium* wilt of cotton by root colonization and biofilm formation. Biol Fertil Soils 49(3):295–303

Liu XY, Ruan LF, Hu ZF, Peng DH, Cao SY, Yu ZN, Liu Y, Zheng JS, Sun M (2010) Genome-wide screening reveals the genetic determinants of an antibiotic insecticide in *Bacillus thuringiensis*. J Biol Chem 285(50):39191–39200

Luo C, Zhou H, Zou J, Wang X, Zhang R, Xiang Y, Chen Z (2015) Bacillomycin L and surfactin contribute synergistically to the phenotypic features of *Bacillus subtilis* 916 and the biocontrol of rice sheath blight induced by *Rhizoctonia solani*. Appl Microbiol Biotechnol 99(4):1897–1910

Mahapatra S, Rayanoothala P, Solanki MK, Das S (2020) Wheat microbiome: present status and future perspective. In: Solanki M, Kashyap P, Kumari B (eds) Phytobiomes: current insights and future vistas. Springer, Singapore. https://doi.org/10.1007/978-981-15-3151-4_8

Mcspadden Gardener BB (2004) Ecology of Bacillus and *Paenibacillus* species in agricultural systems. Phytopathology 94:1252–1258

Meena KR, Kanwar SS (2015) Lipopeptides as the antifungal and antibacterial agents: applications in food safety and therapeutics. Biomed Res Int 2015:473050

Mendoza A, Sikora R (2009) Biological control of *Radopholus similis* in banana by combined application of the mutualistic endophyte *Fusarium oxysporum* strain 162, the egg pathogen *Paecilomyces lilacinus* strain 251 and the antagonistic bacteria *Bacillus firmus*. Biol Control 54(2):263–272

Mitchell R, Alexander M (1962) Lysis of soil fungi by bacteria. Can J Microbiol 9:689–696

Mnif I, Ghribi D (2015) Review lipopeptides biosurfactants: mean classes and new insights for industrial, biomedical, and environmental applications. J Pept Sci 104(3):129–147

Molina L, Constantinescu F, Michel L, Reimmann C, Duffy B, Defago G (2003) Degradation of pathogen quorum-sensing molecules by soil bacteria: a preventive and curative biological control mechanism. FEMS Microbiol Ecol 45:71–81

Mondol MA, Shin HJ, Islam MT (2013) Diversity of secondary metabolites from marine *Bacillus* species: chemistry and biological activity. Mar Drugs 11(8):2846–2872

Mora I, Cabrefiga J, Montesinos E (2011) Antimicrobial peptide genes in *Bacillus* strains from plant environments. Int Microbiol 14:213–223

Muller S, Strack SN, Hoefler BC, Straight PD, Kearns DB, Kirby JR (2014) Bacillaene and sporulation protect *Bacillus subtilis* from predation by *Myxococcus xanthus*. Appl Environ Microb 80:5603–5610

Murphy JF, Zehnder GW, Schuster DJ, Sikora EJ, Polston JE, Kloepper JW (2000) Plant growth-promoting rhizobacterial mediated protection in tomato against tomato mottle virus. Plant Dis 84:779–784

Nielsen TH, Christopheresen C, Anthoni U, Sørensen J (1999) Viscosinamide, a new cyclic depsipeptide with surfactant and antifungal properties produced by *Pseudomonas fluorescens* DR54. J Appl Microbiol 87:80–90

Nielsen TH, Thrane C, Christophersen C, Anthoni U, Sørensen J (2000) Structure, production characteristics and fungal antagonism of tensin — a new antifungal cyclic lipopeptide from *Pseudomonas fluorescens* strain 96.578. J Appl Microbiol 89:992–1001

Nihorimbere V, Ongena M, Cawoy H, Brostaux Y, Kakana P (2010) Beneficial effects of *Bacillus subtilis* on field-grown tomato in Burundi: reduction of local Fusarium disease and growth promotion. Afr J Microbiol Res 4:11–19

Ongena M, Jacques P (2008) Bacillus lipopeptides: versatile weapons for plant disease biocontrol. Trends Microbiol 16(3):115–125

Ouyang LJ, Li LM (2016) Effects of an inducible aiiA gene on disease resistance in *Eucalyptus urophylla* x *Eucalyptus grandis*. Transgenic Res 25:441–452

Panda AK, Bisht SS, DeMondal S, Kumar NS, Gurusubramanian G, Panigrahi AK (2014) Brevibacillus as a biological tool: a short review. Antonie Van Leeuwenhoek 105(4):623–639

Parashar V, Konkol MA, Kearns DB, Neiditch MB (2013) A plasmid-encoded phosphatase regulates *Bacillus subtilis* biofilm architecture, sporulation, and genetic competence. J Bacteriol 195:2437–2448

Park JW, Balaraju K, Kim JW, Lee SW, Park K (2013a) Systemic resistance and growth promotion of chili pepper induced by an antibiotic producing *Bacillus vallismortis* strain BS07. Biol Control 65(2):246–257

Park K, Park JW, Lee SW, Balaraju K (2013b) Disease suppression and growth promotion in cucumbers induced by integrating PGPR agent *Bacillus subtilis* strain B4 and chemical elicitor ASM. Crop Protect 54:199–205

Podile A, Prakash A (1996) Lysis and biological control of *Aspergillus niger* by *Bacillus subtilis* AF 1. Can J Microbiol 42(6):533–538

Priest F (1993) Systematics and ecology of Bacillus. In: Sonenshein AL, Hoch J, Losick R (eds) Bacillus subtilis and other gram positive bacteria, biochemistry, physiology and molecular genetics. American Society for Microbiology Press, Washington, DC, pp 3–16

Rahman A, Uddin W, Wenner NG (2015) Induced systemic resistance responses in perennial ryegrass against *Magnaporthe oryzae* elicited by semi-purified surfactin lipopeptides and live cells of *Bacillus amyloliquefaciens*. Mol Plant Pathol 16:546–558

Ramarathnam R, Bo S, Chen Y, Fernando WGD, Gao XW, de Kievit T (2007) Molecular and biochemical detection of fengycin- and bacillomycin D-producing Bacillus spp., antagonistic to fungal pathogens of canola and wheat. Can J Microbiol 53:901–911

Ramyabharathi S, Meena B, Raguchander T (2012) Induction of chitinase and b-1,3-glucanase PR proteins in tomato through liquid formulated Bacillus subtilis EPCO 16 against Fusarium wilt. J Today Biol Sci: Res Rev 1(1):50–60

Ramyabharathi S, Rajendran L, Karthikeyan G, Raguchander T (2016) Liquid formulation of endophytic *Bacillus* and its standardization for the management of Fusarium wilt in tomato. Bangladesh J Botany 45:283–290

Rayanoothala P, Divya M, Mahapatra S, Das S (2020) Microbial biofilm:formulation, quorum sensing and its application.In. In: Singh KP, Jahagirdar S, Sarma BK (eds) Emerging trends in plant pathology. Springer, Singapore. https://doi.org/10.1007/978-981-15-6275-4

Reitz M, Rudolph K, Schroder I, Hoffmann-Hergarten S, Hallmann J, Sikora RA (2000) Lippolsaccharides of *Rhizobium etli* strain G12 act in potato roots as an inducing agent of systemic resistance to infection by the cyst nematode *Globodera pallida*. Appl Environ Microbiol 66:3515–3518

Romero D, de Vicente A, Rakotoaly RH, Dufour SE, Veening JW, Arrebola E, Cazorla FM, Kuipers OP, Paquot M, Perez-Garcia A (2007) The iturin and fengycin families of lipopeptides are key factors in antagonism of *Bacillus subtilis* towards *Podosphaera fusca*. Mol Plant-Microbe Interact 20:430–440

Rosado AS, de Azevedo FS, da Cruz DW, van Elsas JD, Seldin L (1998) Phenotypic and genetic diversity of *Paenibacillus azotofixans* strains isolated from the rhizoplane or rhizosphere soil of different grasses. J Appl Microbiol 84:216–226

Ruiz-Sanchez R, Blake R, Castro-Gamez H, Sanchez F, Montaldo HH, Castillo-Juarez H (2007) Short communication: changes in the association between milk yield and age at first calving in Holstein cows with herd environment level for milk yield. J Dairy Sci 90(10):4830–4834

Ryu CM, Hu CH, Reddy M, Klopper JW (2003) Different signaling pathways of induced resistance by rhizobacteria in *Arabidopsis thaliana* against two pathovars of Pseudomonas syringae. New Phytol 160(2):413–420

Ryu CM, Farag MA, Hu CH, Reddy MS, Klopper JW, Pare PW (2004) Bacterial volatiles induce systemic resistance in Arabidopsis. Plant Physiol 134(3):1017–1026

Salas-González I, Reyt G, Flis P, Custódio V, Gopaulchan D, Bakhoum N, Dew TP, Suresh K, Franke RB, Dangl JL, Salt DE, Castrillo G. (2020) Coordination between microbiota and root endodermis supports plant mineral nutrient homeostasis. Science 19 Nov 2020: eabd0695. https://doi.org/10.1126/science.abd0695

Seldin L (1992) Primary characterization of the bacteriophage BA-4 from a nitrogen fixing *Bacillus azotofixans* strain. Microbios 71:167–177

Seldin L, Soares Rosado A, da Cruz DW, Nobrega A, van Elsas JD, Paiva E (1998) Comparison of *Paenibacillus azotofixans* strains isolated from rhizoplane, and non-root-associated soil from maize planted in two different Brazilian soils. Appl Environ Microbiol 64:3860–3868

Selim MM, Hend Nafisa Gomaa M, Essa AMM (2017) Application of endophytic bacteria for the biocontrol of *Rhizoctonia solani* (Cantharellales: ceratobasidiaceae) damping-off disease in cotton seedlings. Biocontrol Sci Tech 27(1):81–95

Shiomi HF, Silva HSA, de Melo IS, Nunes FV, Bettiol W (2006) Bioprospecting endophytic bacteria for biological control of coffee leaf rust. Scientia Agricola (Piracicaba, Braz) 63:32–39

Stein T (2005) Bacillus subtilis antibiotics: structures, syntheses and specific functions. Mol Microbiol 56(4):845–857

Surovy MZ, Gupta DR, Mahmud NU, et al (2019) Genomics and post-genomics approaches for elucidating molecular mechanisms of plant growth-promoting Bacilli. Bacilli in Climate

Resilient Agriculture and Bioprospecting Bacilli and Agrobiotechnology: Phytostimulation and Biocontrol 161–200. https://doi.org/10.1007/978-3-030-15175-1_10

Tong-Jian XI, Fang C, Chao GA, Quin-Yun ZH, Qi-Rong SH, Wei RA (2013) Bacillus cereus X5 enhanced bio-organic fertilizers effectively control rootknot nematodes (*Meloidogyne* sp.). Pedosphere 23(2):160–168

Trotel-Aziz P, Couderchet M, Biagianti S, Aziz A (2008) Characterization of new bacterial biocontrol agents *Acinetobacter, Bacillus, Pantoea* and *Pseudomonas* spp. mediating grapevine resistance against *Botrytis cinerea*. Environ Exp Bot 64(1):21–32

Udayashankar A, Nayaka SC, Reddy MS, Srinivas C (2011) Plant growth promoting rhizobacteria mediate induced systemic resistance in rice against bacterial leaf blight caused by *Xanthomonas oryzae* pv. *oryzae*. Biol Control 59(2):114–122

Van Gestel J, Vlamakis H, Kolter R (2015) New tools for comparing microscopy images: quantitative analysis of cell types in *Bacillus subtilis*. J Bacteriol 197:699–709

Van Overbeek L, van Elsas JD (2008) Effect of plant genotype and growth stage on the structure of bacterial communities associated with potato (*Solanum tuberosum* L.). FEMS Microbiol Ecol 64:283–296

Vargas-Ayala R, Rodriguez-Kaban R, Morgan-Jones G, McInroy JA, Kloepper JW (2000) Shifts in soil microflora induced by velvetbean (*Mucuna deeringiana*) in cropping systems to control root-knot nematodes. Biol Control 17:11–22

Vlamakis H, Chai Y, Beauregard P, Losick R, Kolter R (2013) Sticking together: building a biofilm the *Bacillus subtilis* way. Nat Rev Microbiol 11(3):157–168

Wang J, Liu J, Wang X, Yao J, Yu Z (2004) Application of electrospray ionization mass spectrometry in rapid typing of fengycin homologues produced by *Bacillus subtilis*. Lett Appl Microbiol 39(1):98–102

Wang Y, Ohara Y, Nakayashiki H, Tosa Y, Mayama S (2005) Microarray analysis of the gene expression profile induced by the endophytic plant growth-promoting rhizobacteria, *Pseudomonas fluorescens* FPT9601-T5 in Arabidopsis. Mol Plant-Microbe Interact 18:385–396

Wang S, Wu H, Qiao J, Ma L, Liu J, Xia Y, Gao X (2009) Molecular mechanism of plant growth promotion and induced systemic resistance of tobacco mosaic virus by *Bacillus* spp. J Microbiol Biotechnol 19(10):1250–1258

Wang X, Wang L, Wang J, Jin P, Liu H, Zheng Y (2014) Bacillus cereus AR156induced resistance to *Colletotrichum acutatum* is associated with priming of defense responses in loquat fruit. PLoS One 9(11):e112494

Wang T, Liang Y, Wu MB, Chen ZJ, Yang LR (2015) Natural products from *Bacillus subtilis* with antimicrobial properties. Chin J Chem Eng 23(4):744–754

Wieland G, Neumann R, Backhaus H (2001) Variation of microbial communities in soil, rhizosphere and rhizoplane in response to crop species, soil type and crop development. Appl Environ Microbiol 67:5849–5854

Williams-Wagner RN, Grundy FJ, Raina M, Ibba M, Henkin TM (2015) The Bacillus subtilis tyrZ gene encodes a highly selective Tyrosyl-tRNA synthetase and is regulated by a MarR regulator and T box riboswitch. J Bacteriol 197:1624–1631

Wulff BB, Horvath DM, Ward ER (2011) Improving immunity in crops: new tactics in an old game. Curr Opin Plant Biol 14(4):468–476

Yamamoto S, Shiraishi S, Suzuki S (2015) Are cyclic lipopeptides produced by *Bacillus amyloliquefaciens* S13–3 responsible for the plant defence response in strawberry against *Colletotrichum gloeosporioides*? Lett Appl Microbiol 60:379–386

Yang CY, Ho YC, Pang JC, Huang SS, Tschen JSM (2009) Cloning and expression of an antifungal chitinase gene of a novel *Bacillus subtilis* isolate from Taiwan potato field. Bioresour Technol 100:1454–1458

Yang L, Quan X, Xue B, Goodwin PH, Lu S, Wang J, Du W, Wu C (2015) Isolation and identification of *Bacillus subtilis* strain YB-05 and its antifungal substances showing antagonism against *Gaeumannomyces graminis* var. *tritici*. Biol Control 85:52–58

You C, Zhang C, Kong F, Feng C, Wang J (2016) Comparison of the effects of biocontrol agent *Bacillus subtilis* and fungicide metalaxyl–mancozeb on bacterial communities in tobacco rhizospheric soil. Ecol Eng 91:119–125

Zhang N, Wu K, He X, Li SQ, Zhang ZH, Shen B, Yang XM, Zhang RF, Huang QW, Shen QR (2011) A new bioorganic fertilizer can effectively control banana wilt by strong colonization with *Bacillus subtilis* N11. Plant Soil 344(1–2):87–97

Zheng X, Sinclair J (2000) The effects of traits of *Bacillus megaterium* onseed and root colonization and their correlation with the suppression of Rhizoctonia root rot of soybean. BioControl 45(2):223–243

Chapter 7
Enhanced Root Morphogenesis in Non-legumes as Induced by Rhizobacteria *Bacillus* spp.

M. A. Baset Mia

Abstract Rhizobacteria exert tremendous beneficial effects on various crop plants especially the non-legumes through multidimensional approaches namely biofertilizing, bioenhancing and biocontrolling activities. They create a conducive environment in the rhizosphere, endorhizosphere, apoplastic areas of tissue in the roots as well as stem, and form new organs by producing and secreting various types of phytohormones. Among the bacteria, species of the genera *Agrobacterium, Arthrobacter, Azotobacter, Azospirillum, Bacillus, Burkholderia, Herbaspirillum, Klebsiella, Pseudomonas, Rhizobium, Bradyrhizobium, Caulobacter, Chromobacterium, Erwinia, Flavobacterium, Micrococcus, Pseudomonas* and *Serratia* are main candidates for the beneficial effects on plants. Among them *Bacillus* spp. are gaining prominence; application of *B. sphaerichus* strain UPMB10 showed beneficial effects on banana, oil palm, rice, sweet potato and other non-legumes. The bacteria are also endophytic, and found in the apoplastic area of banana roots as have been observed by transmission electron micrography (TEM). Certain rhizobacteria isolated from rice seeds have shown beneficial effects on rice under field and laboratory conditions. Proliferation of root hairs in rice due to the application of *Rhizobium* and *Bacillus* has been documented in *Indica* type of rice under humid tropic condition. Plant juvenile hormones like auxin and gibberellin have been isolated from *Bacillus* spp.-inoculated roots of rice, banana and oil palm, which triggered induction of meristem organization in pericycle in the young roots and consequently forming the lateral roots as documented by various researchers. Inoculation process by *Bacillus* spp. resulted in more root hairs and seedling vigor index of rice and enhanced root growth in banana tissue-cultured plantlets under hydroponic condition.

M. A. B. Mia (✉)
Department of Crop Botany, Bangabandhu Sheikh Mujibur Rahman Agricultural University, Gazipur, Bangladesh
e-mail: miabaset@bsmrau.edu.bd

M. T. Islam et al. (eds.), *Bacilli in Agrobiotechnology*, Bacilli in Climate Resilient Agriculture and Bioprospecting,
https://doi.org/10.1007/978-3-030-85465-2_7

Keywords Legume-rhizobium symbiosis · Biofertilizer · Bioenhancer · *Bacillus* · Apoplastic area · Biocontrol

7.1 Introduction

Roots are the vital organ of higher plants for anchorage and absorption of nutrients. In dicotyledons, there are two types of roots, namely primary roots, which are produced from the embryo during germination phase, and secondary roots, which originate from the primary roots and also known as lateral roots. However, in monocotyledons, the embryonic roots are usually small, short-lived and only important at the early stage of seedling development. The primary roots are destroyed after few days of germination and new roots are formed from the base of the stem which are known as adventitious or fibrous roots (Bellini et al. 2014). It is interesting to note that root growth is continued until reproductive stage that ceases upon flowering of the plant. The architecture of roots reveals the spatial arrangement which are influenced by the growth and development of plant. The primary root produces the lateral (secondary) roots, and the ratio of both root types changes throughout the life of a plant (Waidmann et al. 2020). The growth, morphogenesis and development of roots are greatly influenced by various biotic and abiotic factors. The rhizobacteria are the most influential biotic factors affecting growth and development of roots by secreting various types of phytohormones. Biotic and abiotic stresses may stimulate morphogenic changes in roots viz. increased or decreased length, abnormal branching and hair formation. Some of these morphological changes of roots are correlated with altered auxin distribution (Potters et al. 2007; Péret et al. 2009). On the other hand, legumes have a strong symbiotic association between bacteria and roots by forming nodule for establishing a strong beneficial association. Non-legumes are potentially capable of establishing an association with diazotrophic bacteria consequently getting beneficial effects through diverse mechanisms (Mia and Shamsuddin 2013). They provide multiple beneficial effects to the host plant through enhanced root formation, acting as biological N_2 fixer (BNF), greater nutrient accumulation, rhizoremediation, serving as crop enhancer and functioning as biopesticides (Levanony and Bashan 1989; Mia and Shamsuddin 2013). They belong to various genera of bacteria like *Agrobacterium, Arthrobacter, Azotobacter, Azospirillum, Bacillus, Borkholderia, Herbaspirillum, Klebsiella, Pseudomonas, Rhizobium, Bradyrhizobium, Caulobacter, Chromobacterium, Erwinia, Flavobacterium, Micrococcus, Pseudomonas,* and *Serratia* (Bhattacharyya and Jha 2012). The bioenhancing vis-à-vis crop-enhancing activities is the most important avenue, which is mediated via morphological and physiological changes and enhancement of water and mineral nutrient uptake as observed in inoculated sorghum, banana, rice and oil palm roots (Sarig et al. 1988; Mia et al. 2010a, b; Amir et al. 2001). Murty and Ladha (1988) reported that inoculation by *A. lipoferum* resulted in reduced root length without showing any effect on root surface area in rice. These changes were attributed to the effect of auxin produced by the root-colonized bacterial strains of *B. subtilis*, which enhanced additional rooting in plants. Roots are the most sensitive organs to alter the IAA levels, resulting in the elongation of primary roots and forming the adventitious and lateral roots. The hormone produced by the bacteria can be used as a relation maker between roots and bacteria. The fact that *B. subtilis* improved rooting response in legumes like mung bean (*Vigna radiata* (L.) Wilczek) has been documented by various researchers. Recently, among the rhizobacteria, the genus *Bacillus* are gaining importance as bioenhancers and biofertilizers for non-legumes throughout the world.

The bacteria of genus *Bacillus*, are rod-shaped, Gram-positive, and facultative Gram-negative properties. They belong to the phylum Firmicutes containing 266 known species. They are motile, endospore-forming and have been found to be colonized in sugarcane roots, employing beneficial effects in association with VAM fungi, which has been documented by Bellone et al. (1997) in Argentina.

Inoculation of *B. licheniformis* along with other PGPR has been shown to enhance plant growth and nutrient uptake in *Salicornia bigelovii* (Bashan et al. 2000), and some species of *Bacillus* namely *B. macerans, B. polymyxa/Paenibacillus azotofixans* can fix N_2 and are also good as P-solubilizing agents, thus recognized as potential biofertilizers (Freitas et al. 1997). *B. azotofixans* are nitrate dependent for growth and are capable of fixing N_2 in the presence of combined N in the media (Subba Rao 1995). Wei (1997) reported that these bacteria can fix N_2 in association with different non-legumes; where inoculated plants had 15% higher yield compared to uninoculated control. Combined inoculation of *B. licheniformis* and *Phyllobacterium* sp. increased plant height and dry weight and consequently more P absorption and lipid content in *Salicornia bigelovii* Torr. (Bashan et al. 2000). Among the bioenhancing beneficial activities on non-legumes, the enhanced root stimulation is the champion by which numerous species of *Bacillus* showed significant effect as shown in the Table 7.1.

Table 7.1 List of *Bacillus* spp. capable of promoting root morphogenesis in different plant species

Bacillus species	Growth of roots	References
Bacillus amyloliquefaciens subsp. *plantarum* UCMB5113	Increases root growth, especially lateral root outgrowth and elongation and root-hair formation	Asari et al. (2017)
B. insolitus, B. subtilis, B. methylotrophicus	Increase root growth	Ashraf et al. (2004), Barnawal et al. (2013), and Radhakrishnan and Lee (2013)
B. subtilis, B. methylotrophicus	Synthesis of plant growth hormones (IAA, GAs, cytokinins and spermidines) trigger plant growth	Arkhipova et al. (2005), Xie et al. (2014), and Radhakrishnan and Lee (2013)
B. sphaericus strain UPMB10	Increases root growth	Mia et al. 2010b, and Amir et al. (2001)
B. Seropeaicae	Increases root growth	Weber et al. (1999)

The contrivances of root stimulation have been documented sporadically by a large body of researches throughout the globe. However, their chronological achievements by a thorough review and unravelling precise mechanisms are yet to be done. Keeping these views in mind, this chapter compiles available information and updates our knowledge on the root morphogenesis in non-legumes by rhizobacteria and their mechanism (s) in root morphogenesis.

7.2 Mode and Mechanism of Beneficial Effects

7.2.1 Colonization of Bacteria

Root colonization is recognized as a major factor in establishing a successful beneficial association between bacteria involving four phases, initial phase i.e. movement of microbes to the plant root surface, which can be passive, via soil water fluxes, or active, via specific induction of flagella activity by chemotaxis (Suslow 1982; Mia et al. 1999, 2010a, b). The second phase in colonization is the adsorption to the root surface followed by anchoring, specific and/or complex interaction between the bacterium and the host by ensuring of induction for bacterial gene expression (Brimecombe et al. 2001). It is an established phenomenon that PGPR could colonize the roots both ecto- and endophytically, which has been shown by scanning electron micrography (SEM) and transmission electron micrography (TEM) (Fig. 7.1) (Del Gallo and Fendrik 1994; Mia et al. 1999).

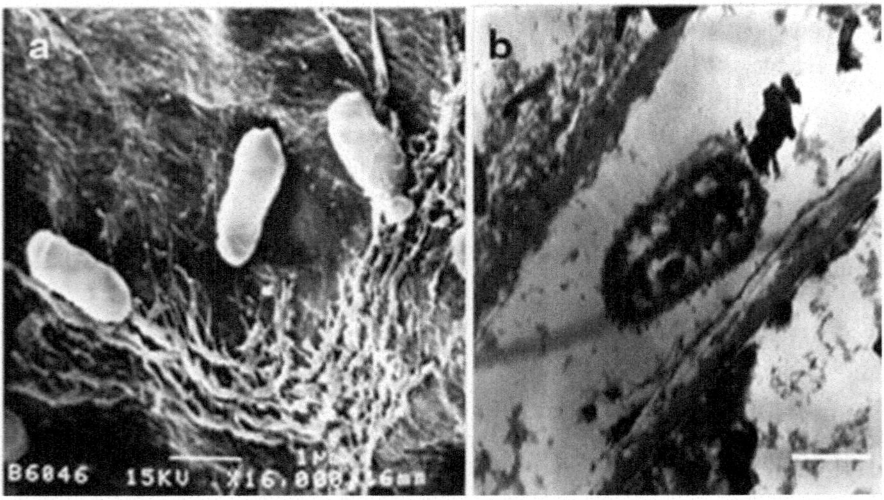

Fig. 7.1 (**a**) Scanning electron micrograph of banana roots shows bacterial colonization by *Azospirillum brasilense* strain Sp7; (**b**) Transmission electron micrograph (TEM) of banana roots shows bacterial colonization by *A. brasilense* strain Sp7as seen in the apoplastic area of root (horizontal bar shows 1 μm). (The images are adopted from Mia et al. 2013)

The root colonization study using *A. brasilense* and *B. sphareicus* has clearly demonstrated that bacterial colonization occurred mainly on root surface area and more cells were observed in the root hair proliferation zone of banana tissue-cultured plantlets. The latter might be due to the presence or availability of root exudates in this area.

Anchoring of bacterial cells on the root surface through a network of fibrillar materials was seen in Sp7 while fibrillar matrices were not observed in UPMB10 (Mia et al. 1999, 2010a, b). The formation of fibrillar matrix is a physiological process during the growth of Sp7 (Bashan and Levanony 1988). More number of Sp7 cells in the root hair proliferation zone is due to presence of fibrillar matrices, which restricted the movement of the bacterial cells to this zone.

The assured and smart attachment of rhizobacteria is essential for a sustainable long-term association with the host roots for having the following reasons:i) if the bacteria are not well attached to root epidermal cells, substances excreted by the bacteria will diffuse into the rhizosphere and are consumed by nutritionally versatile microorganisms before reaching the roots, ii) the bacteria might be washed out from the root zone to succumb to the surrounding soil and iii) association sites on roots with no attached PGPR are vulnerable to other virulent non-beneficial and pathogenic microbes (Bashan and Holguin 1997). The bacteria sometime or even frequently form a biofilm on the rhizoplane which interact between the root surface and the microbe consequently serve as a diffusion barrier of both nutrients and water (Vlamakis et al. 2013).

Bacillus and other rhizobacteria like *Azospirillum* are smart root colonizers, non-plant specific and able to colonize a wide array of plant species (Bashan and Holguin 1997; Mia et al. 2009a). They are also recognized as good colonizers of non-legumes (Bashan et al. 1986; Bashan et al. 1987; Murty and Ladha 1987), vegetables (Bashan et al. 1989), oil palm (Amir et al. 2001), banana (Mia et al. 1999, 2010a), and mangrove plants (Puente et al. 1999). However, the mode of colonization vary greatly among strains (Bashan et al. 1990) where a high level of bacterial colonization is found in the middle lamellae, an area of longitudinal contact between epidermal cells of the roots (Bowen and Rovira 1999), and generally, root caps have less or no bacterial colonizing capacity (Foster and Bowen 1982). However, the preferred site for colonization is the elongation and hairy zone forming an aggregated type of colonization supported by massive fibril material, and colonization sites in non-legumes correspond to the areas where root mucilage is present. The area around the point of emergence of lateral roots usually shows high colonization (Mia et al. 1999, 2010a, b; Bashan and Levanony 1990).

Recently, endophytic plant growth promoting bacteria have been introduced; these colonize roots, stems and leaves of cereals in the apoplastic area of tissue where they probably suffer less competition from other microorganisms for carbon substrates than rhizosphere bacteria, and possibly excrete part of their fixed N_2 and plant growth regulator directly into the host tissue (Baldani et al. 2000; Mia et al. 2007). It is clearly understood that rhizobacteria, especially the species of *Bacillus*, could form colonies on the roots both externally and endophytically i.e. in the apoplastic areas of the root cortical region.

Table 7.2 Root growth of banana plantlets inoculated with PGPR strains Sp7 and UPMB10 grown under hydroponic condition for 45 days (under N-free nutrient solution)

Treatments	Total no. of 1^0 roots (plant^{-1})	Total length of 1^0 roots (cm)	Root volume (cm^3 plant^{-1})	Root dry weight (g plant^{-1})
-PGPR	6.0 a	236 b	6.2 c	0.27 b
+Sp7	6.3 a	314 a	10.9 b	0.65 a
+UPMB10	7.0 a	341 a	16.6 a	0.64 a

Means having same letter (s) in a column do not differ significantly at 0.05 level by DMRT

7.2.2 Enhanced Root Growth and Development

Application of *A. brasilense* strain Sp7 and *B. spharecus* strain UPMB10 enhanced root growth in multiple facet in dessert-type banana under soilless culture. The pronounced effects have been found in root hair formation. Additionally, inoculation process greatly increased the production of primary, secondary and tertiary roots and total primary root length by 33-44% over un-inoculated control by inoculation process. Root volume was also significantly increased and effect of UPMB10 produced a greater volume than Sp7, an increase of 52%. The root dry weights were increased through inoculation by 137-141%. Strong positive results on root stimulation are shown in bananas under soilless culture condition (Table 7.2).

Results of a study on the application of rhizobacteria on tissue-cultured plantlets under hydroponics indicated that PGPR inoculation could increase root growth (root mass (137-141%), volume (76-168%) and length (33-44%)) with a consequent increase in N yield and higher growth of plant tops viz. plant height (42-50%), leaf area (128-134%) and dry matter (129-176%). It was demonstrated by scanning electron micrography (SEM) that *A. brasilense* strain Sp7 and *B. sphaericus* strain UPMB10 successfully formed colonies on the banana roots. Bacterial colonization by Sp7 and UPMB10 occurred mainly on the root surface and more cells were found in the root hair proliferation zone while the root hair per se is devoid of bacteria where Sp7 colonized mainly the root hair proliferation zone with a few cells found on the root elongation and root cap zone. Occurrence of more cells of Sp7 in the root hair zone was due to the presence of fibrillar matrix which aided in the attachment of bacterial cells on the root surface. But strain UPMB 10 performed better i.e. efficiently colonized all the three root zones viz. cap, elongation and hair proliferation areas. The presence of more UPMB10 cells throughout the whole roots indicated that this strain was able to utilize exudates produced by the three zones (Mia et al. 2010a, b). Similar fashions of root colonization have been documented by several researchers (Assmus et al. 1995; Bashan and Holguin 1997) who concluded that colonization of rhizobacteria corresponded to the areas where root mucilage was present. *A. brasilense* strain Sp7 preferred to colonize the root hair zone, and inoculation with different strains of *A. brasilense* to young wheat plants showed colonization both in rhizoplane and endorhizosphere and stimulated root and shoot growth. (Levanony et al. 1989; Saubidet and Barneix 1998) (Table 7.3).

Table 7.3 Root growth of banana plantlets inoculated with PGPR strains Sp7 and UPMB10 grown under low fertilizer-N regimes (33% of the requirement)

Treatments	No. of 1^0 roots $plant^{-1}$	Total 1^0 root lengths (cm)	Base diameter of 1^0 root (mm)		Root volume (cm^3 $plant^{-1}$)	Root dry wt. (g $plant^{-1}$)
			Feeder	Pioneer		
N_0-PGPR	6.0 b	236 b	0.13 b	0.25 c	11.3 d	0.27 c
N_0 + Sp 7	8.3 ab	314 ab	0.17 ab	0.27 bc	18.0 c	0.65 b
N_0 + UPMB 10	7.0 ab	341 a	0.23 a	0.35 abc	27.3 b	0.61 b
$N_{33\%}$ + Sp 7	9.6 a	308 ab	0.21 ab	0.38 b	55.0 a	1.02 a
$N_{33\%}$ + UPMB 10	9.6 a	355 a	0.19 ab	0.42 a	59.0 a	0.99 ab
$N_{100\%}$-PGPR	8.6 ab	372 a	0.15 ab	0.42 a	50.0 a	0.69 b

Means having same letter (s) in a column do not differ significantly at 0.05 level by DMRT

The growth enhancing effects by rhizobacteria are mainly derived from morpho-physiological attributes in roots of inoculated plant (Okon et al. 1988; Sarig et al. 1988), and the increased root growth occurred almost in all dimensions namely production of primary (1^0) and secondary (2^0) roots, longer roots (42%), and greater volume (48%) and mass (43%). Primary root elongation rate and 2^0 root initiation capacity of inoculated plant were increased due to the synergy of PGPR inoculation and minimal fertilizer-N application. Plants inoculated with strain UPMB10 and supplied with fertilizer-N showed more 1^0 root elongation while those with Sp7 and fertilizer-N showed greater 2^0 root initiation. The primary root elongation of UPMB10-inoculated plants might be due to more number of bacterial cells present throughout the whole root system which resulted in the root elongation from all the root zones, whereas the presence of more cells of Sp7 in the root hair zone stimulated the initiation of 2^0 roots. The higher 2^0 root initiation in Sp7-inoculated plants is due to the presence of more bacterial cells and their beneficial interaction in the hair proliferation zone *vis-a-vis* the zone of secondary root initiation in bananas.

It is interesting to note that as monocotyledon, banana roots are adventitious, arise in a group, white and fleshy when young, but corky when old. Most of the banana roots are not deep-statured, mainly present in the topsoil and grow horizontally (Sioussaram 1968). Swennen et al. (1986) classified banana roots as primary, secondary and tertiary roots which arise from the corm, primary roots and the secondary roots, respectively. The primary roots can further be classified into two groups namely feeders and pioneers. Feeders are significantly longer but thinner than pioneers and have a higher density of secondary roots. Every root has three different zones, namely root cap, elongation and hairy zone (Fig. 7.2).

Root surface area is considered as one of the important arena for absorption of nutrients through roots. The surface areas have been increased in maize by the inoculation of *Azospirillum*, which is reported by Jacoud et al. (1999). The surface area of inoculated plants were always greater than non-inoculated control plants. The positive results were remarkable during the early stage of the plants i.e. germination to seedling stage.

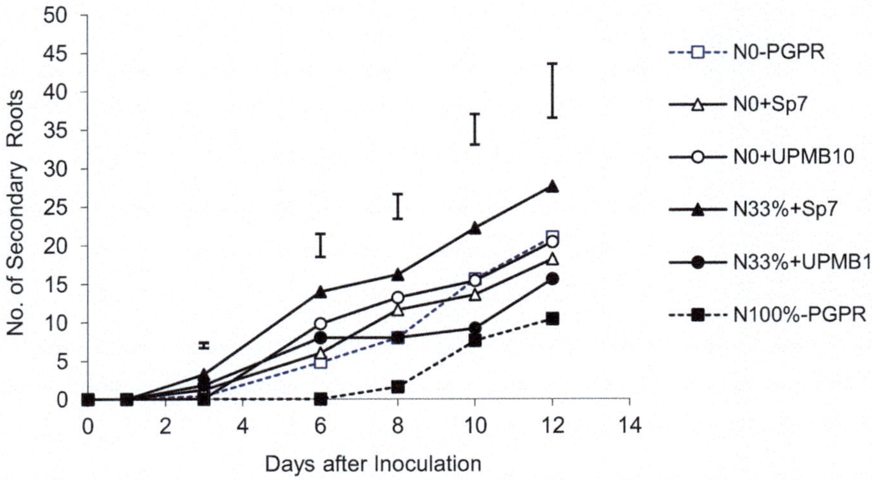

Fig. 7.2 Secondary root initiation of banana plantlets inoculated with PGPR strains Sp7 and UPMB10 (vertical bar represents LSD at 0.05 significant level) (Adopted from Mia et al. 2009b)

7.2.3 Root Hair

The epiblema generates distinctive outgrowths known as hair, which greatly increased the surface area of roots, consequently improving water and nutrient absorption. Generally, root hairs located 1-2 cm behind from the root cap, or tip, and hairs live for only a few days and never develop into multi-cell roots. Because of their short life, roots need to grow continually. Its development is particularly sensitive to both biotic and abiotic stimuli and is a useful marker of differentiation processes that take place in the root (López-Bucio et al. 2003). They perform significant roles in absorption of nutrients and water by increasing the surface area of root system. The formation of root hairs follows three distinct phases, namely selection of site, bulge formation and elongation phase, where the initiation phase is regulated by transcription factors and activity of certain enzymes, namely GTPase and wall modifier enzymes (Datta et al. 2011; Dittmer 1937).

Rhizosphere microorganisms contributed remarkable effects on the productivity of crop plants in agricultural and marginal soils by influencing growth and development of roots and improving nutrient availability in the rhizosphere (Mia 2015). These organisms have been recognized from a wide range of plant species of both dicot and monocot like sweet potato, tomato, chili, sugarcane, maize, wheat, rice, barley, canola, banana and bean. Their contribution can be exerted through direct and indirect methods. The direct methods include those by which bacteria directly influence root-system architecture and shoot development by secretion of plant growth-promoting substances such as auxins and cytokinins (Persello-Cartieaux et al. 2001).

Fig. 7.3 Increased root hairs of inoculated banana roots showed proliferated root hairs by the inoculation of (**a**) *A. brasilense* strain Sp7; (**b**) *B. sphaericus* strain UPMB10; (**c**) Control

Similarly, inoculation of *Bacillus* spp. influences root hair formation in banana (Fig. 7.3), rice, wheat, maize and oil palm, which has been observed by various researchers (Mia et al. 2013, 2016). Additionally, inoculation process of rhizobacteria on germinated low-land *Indica*-type rice showed profuse hair, elongated roots and consequently greater root mass production, which has been documented by Mia and Shamsuddin (2009).

The main mechanism of enhanced root hair formation is due to the production of phytohormone especially auxin and gibberellin, as they enable dynamic hormone-driven changes resulting in altered root hair growth, density, length and morphology (Vissenberg and Gonzalez 2020). The rhizobacteria influences on proliferation of hairs by secreting growth hormone like auxin and gibberellin, which has been documented by various researchers in both legumes and non-legumes. Similarly, production of plant growth regulators by the application of *Bacillus* spp. has been recognized by a large body of research findings in legumes and non-legumes (Hashem et al. 2019; Naher et al. 2013). Among the growth regulators, auxin is predominantly the secondary plant-friendly metabolite secreted by the bacteria, contributing in the enhancement of root hair in epiblema. This is a class of essential hormones which controls the positioning of the hair initiation where the alteration of density, length and morphology of hairs are governed by hormonal signalling interconnected with the hair signalling and the degree of morphological changes (Datta et al. 2011; Ishida et al. 2008; Péret et al. 2009).

7.2.4 Lateral Roots Formation

The roots branching out from the primary root are called secondary or lateral roots. In contrast to the taproot system, the majority of monocot roots form a fibrous 'homorhizic' root system, which is characterized by the development of many adventitious roots (Osmont et al. 2007; Bellini et al. 2014; Levanony and Bashan 1989). These are very important in absorbing nutrients, and at the tip of the root, there is an area where new cells develop, called the apical meristem. Root cap over the top of roots protect the meristem from damage when it grows quickly through soil particles, and the surface of the root is covered with a layer of cells called the epidermis which is devoid of cuticle. The rhizobacteria influence the stimulation of lateral roots and enhance proliferating of hairs by secreting growth hormones like auxin and gibberellin and have been documented by various researchers both legumes and non-legumes. As it is known that there are three parts of roots, namely root tip, root elongation and root hairy zones which are responsible for varied levels of nutrients absorption. Nevertheless, rhizosphere, an adjacent layer which is very close and outside of the surface and influenced by the root activity, is the hub of nutrients which are taken up by the epiblema of roots. The epiblema contains root hair, the extension of the epidermal cell is directly involved in nutrient absorption. The increased number of lateral roots along with root hair formation have been observed in bean by the inoculation of *B. megaterium*. However, primary-root growth was inhibited by the process of inoculation which might be happened due to reduction in cell elongation and by reduction of cell proliferation in the root meristem. Nevertheless, lateral roots originate from the pericycle beneath the endodermis. These tissues are very important to initiate lateral or secondary root formation, and they are located at the periphery of vascular cylinder. The layer of those tissues are located just beneath the endodermis in front of vascular tissue especially the phloem, composed of parenchymatous cell and surrounds the stele, and are heterogeneous in nature (Beeckman et al. 2001; Himanen et al. 2004; Laplaze et al. 2005; Mähönen et al. 2006; Parizot et al. 2008). The pericycle may be dedifferentiated into meristem by the activity of certain growth regulators. The inoculated roots resulted in the production of more lateral roots, which is mediated by the secretion of phytohormone by the PGPR. The lateral roots are initiated by the induction of hormone, especially auxin, which is accumulated in the pericycle, and it triggers the asymmetric meristematic activity. The endodermis plays a significant role in supplying auxin to the pericycle (Marhavy et al. 2012), and this accumulation of auxin in a selective number of pericycle cells is one of the earliest events during founder cell specification that precedes lateral roots initiation (Dubrovsky et al. 2008), although the molecular mechanism of auxin-mediated lateral root formation is not clear (Fukaki et al. 2007). However, it is recognized that the process progress is mediated by at least four recognizable phases like priming, initiating, patterning, and emerging (Malamy 2009; Péret et al. 2009; Joseph et al. 2012), and each of these phases is either controlled or at least influenced by auxin. The steps have been shown in the schematic diagram below:

The IAA can initiate the developmental program for lateral roots formation via a model where it requires at least two steps viz. initiation of cell division in the pericycle and promote cell division and maintain cell viability in the developing lateral roots (Celenza et al. 1995). The overall mechanism has been shown in the schematic diagram below (Figs. 7.4 and 7.5):

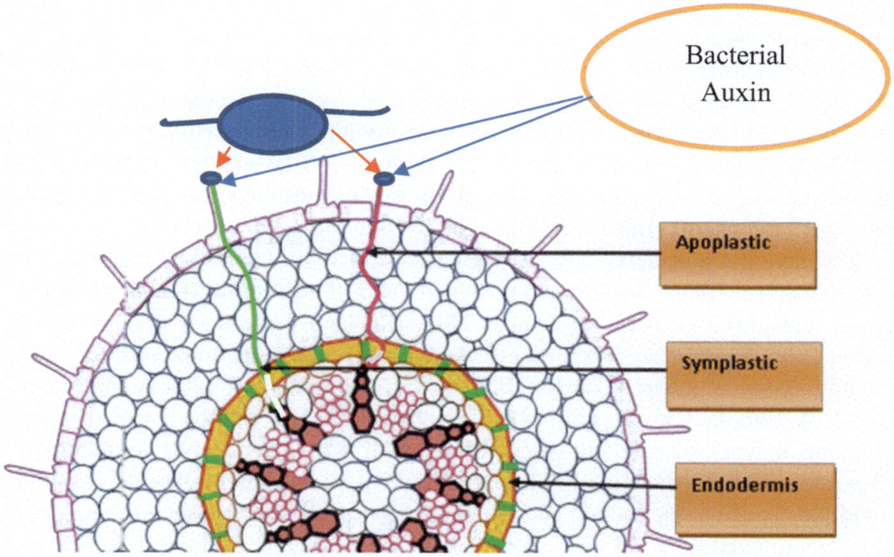

Fig. 7.4 Diagram shows movement of bacterial auxin to the endodermis through symplastic and apoplastic pathways in the root (Adapted from Mia 2015)

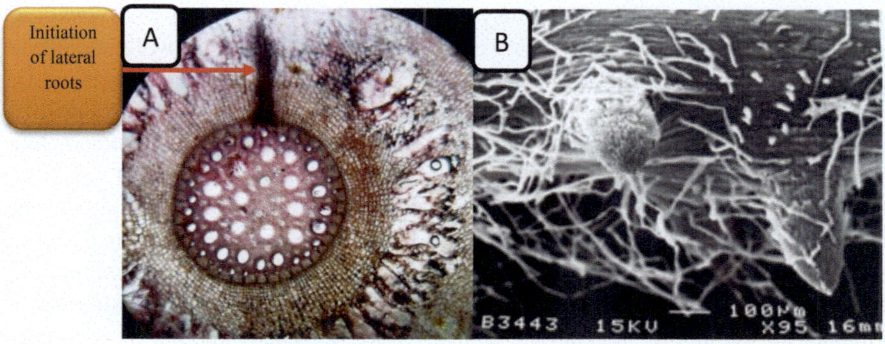

Fig. 7.5 (**a**) Transverse section of banana roots showing initiation of lateral roots from pericycle observed by compound light microscope (X-100); (**b**) Emerging of lateral roots as shown by electron micrography (SEM) of banana roots (horizontal bar shows the micrometre)

7.2.5 Role of Auxin in Lateral Root Initiation

Root systems show variable architectures that contribute to functionality of plants, and the phytohormone auxin fulfils a large body of roles throughout lateral root development. The recent advancement clarified four phases of lateral root formation: (i) positioning of lateral roots, which determines the spatial distribution of lateral root primordia and lateral roots along primary roots; (ii) lateral root initiation, encompassing the activation of nuclear migration in specified lateral root founder cells (LRFCs) up to the first asymmetric cell division; (iii) lateral root outgrowth, the 'primordium-intrinsic' patterning of de novo organ tissues and a meristem; and (iv) lateral root emergence, an interaction between lateral root primordia and draping tissues for allowing passage through cell layers. The auxin plays a significant role through signalling and changing the developmental cortex for completing all the phases (Bao et al. 2014).

The phytohormone auxin is a key regulator of various features in the growth and development of plant including cell division and elongation, differentiation, tropisms, apical dominance, senescence, abscission and flowering (Woodward and Bartel 2005; Teale et al. 2006), where Naxillan, a non-auxin compound, enhances the conversion of auxin precursor IBA (indole-3-butyric acid) to the active auxin-like IAA (indole-3-acetic acid) (De Rybel et al. 2012). The root cap turnover dynamic may co-ordinate primary root growth with lateral root formation in response to environmental stimuli, namely gravity, water and nutrients, and this highlights possible interaction sites in plants to transform extrinsic environmental signals into intrinsic developmental strategies where rhizobacterial influence is yet to be elucidated (Xuan et al. 2016).

The plant growth regulators like auxin and cytokinin are involved in several stages of plant growth and development, namely division and elongation of cell, tissue differentiation and apical dominance. However, the biosynthesis and the underlying mechanism of auxin action on root formation need to be continued. The genetic mechanism of auxin biosynthesis and regulation by *Pseudomonas, Agrobacterium*, *Rhizobium*, *Bradyrhizobium* and *Azospirillum* are well studied; in these bacteria several physiological effects have been correlated to the bacterial phytohormones biosynthesis. However, they do also follow chromosomally encoded indole-3-pyruvic acid pathway; additionally, they have genes that can conjugate free auxins or hydrolyze conjugated forms of auxins and cytokinins. Several genes are located near the auxin and cytokinin biosynthetic genes which are involved in the regulation of auxins and cytokinins sensibility of the transformed plant tissue in *Agrobacterium*. On the other hand, symbiotic bacteria viz. *Rhizobium* and *Bradyrhizobium* synthesize IAA via indole-3-pyruvic acid; also the genetic determinants for the indole-3-acetamide pathway have been detected, but their activity has not been demonstrated. In the plant growth-promoting bacterium *Azospirillum*, as in *Agrobacterium* and *Pseudomonas*, both the indole-3-pyruvic acid and the indole-3-acetamide pathways are present, although in *Azospirillum* the indole-3-pyruvic acid pathway is of major significance.

In case of *Bacillus* strains *B. cereus* (So3II) and *B. subtilis*(Mt3b) showed variable pathways for the production of IAA under different set of growth and environmental conditions where the IAA production potential can be enhanced by affecting optimum growth conditions (Wagi and Ahmed 2019).

7.2.6 Secretion of Hormone and Translocation Through Root System

Plant growth hormones are biochemical substances which are produced in one place of a plant, translocate to another site and perform distinct functions on growth and development. They are produced in the shoot of the plant and translocate basipetally up to the base of the stem, thereafter moving acropetally in the root system (Skoog 1938). Most of the beneficial rhizobacteria, endophytic and phyllospheric bacteria are able to synthesize IAA, and it is a potent signaling molecule essential for plant–microbe interactions and improving plant growth directly (Matsuda et al. 2018).

Nearly 80% of the rhizobacteria are capable of synthesizing IAA, and the majority of the IAA produced from tryptophan is released by root exudates (Khalid et al. 2009). It is proved that higher amount of IAA could be produced by applying tryptophan in the culture media. In the symbiotic bacteria *Rhizobium fredii* and *Bradyrhizobium japonicum*, the production of IAA via IAM pathway was suggested, as bacterial cultures can convert an IAM analog. A gene (*bam*) with sequence similarity to *iaaH* could be isolated from *B. japonicum* (Peyush et al. 2000). Later, the presence of the IAM pathway was also confirmed for *Rhizobium* sp. strain NGR234 by LC-MS detection of intermediates (Theunis 2004).

7.2.7 Translocation of Phytohormone Through Root System

Whatever the production of hormone via various pathways, bacteria secrete these in the epiblema or in the apoplastic area of root cortical regions. The secreted hormones are translocated through apoplastic or symplastic movement most probably by the diffusion process. However, endogenous IAA in the root cap could move to the growing zone and cause a unilateral inhibition of growth. In the case of exogenously applied IAA, the pathway of transport pool is also similar.

Nevertheless, both rhizospheric and endophytic bacterium-released IAA change the plant auxin pool to either optimal or supraoptimal levels and induce root stimulation (Iqbal et al. 2017). The IAA is a signalling molecule comprising heterogeneous carboxylic acid group responsible for controlling diversified physiological processes especially the additional root formation in the plants. Generally, it is produced in aerial parts like shoot apex and is transported through phloem to sub-aerial parts of the plants actively by utilizing two common transportation pathways towards the root via non-directional passive pathway or via cell to bi-directional active pathway known as polar auxin transport (Park et al. 2017).

Anyway, IAA production by bacteria is one of the most important direct mechanisms utilized by plant growth-promoting bacteria (PGPB) for the betterment of plants naturally because auxin is a plant-friendly secondary metabolite synthesized naturally by bacteria, and hence improves the growth of associated plants. Bacterial IAA changes the auxin pool to either supraoptimal or optimal, thereby improving root growth, especially the development of lateral roots resulting in greater surface area that promotes plant nutrient uptake (Ahmed and Hasnain 2010).

7.3 Conclusions and Future Perspective

The application of plant beneficial bacteria is increasing tremendously throughout the world for its environment-friendly uses. They provide advantages in counteracting adverse soil condition where enhanced root growth is required for better productivity. The use of these bacteria for increasing productivity is gaining eminence. Keeping the plants in juvenility is important, as it is vital for proper morphogenesis of plant organ which is mediated by the action of phytohormone like IAA. Plant beneficial microbes viz. rhizobacteria, endophytic, and phyllosphere, provide significant contribution by producing hormone and subsequently secreting to the host plant for providing beneficial effects. Root morphogenesis through hair and lateral roots formation is mediated by this hormone, and the mechanism has been discussed here comprehensively. The mechanism of root morphogenesis by rhizobacteria especially the *Bacillus* involves the following steps: synthesis of IAA by using tryptophan, release of IAA to the epiblema or apoplastic region of root cortical zone of root, translocation to the endodermis and release to pericycle cells for induction, forming founder cell followed by lateral root primordia development and finally emerging the lateral roots. In addition, this root hair is formed by the induction of IAA through differentiation of epiblema to epiblemal outgrowth formation. However, the molecular mechanism for bacterial synthesis of hormones, their secretion and utilization by the root tissue were not investigated comprehensively, which needs to be elucidated in future studies.

References

Ahmed A, Hasnain S (2010) Auxin-producing *bacillus* sp.: auxin quantification and effect on the growth of *Solanum tuberosum*. Pure Appl Chem 82(1):313–319

Amir HG, Shamsuddin ZH, Halimi MS, Ramlan MF, Marziah M (2001) Effects of *Azospirillum* inoculation on N_2 fixation and growth of oil palm plantlets at nursery stage. J Oil Palm Res 13(1):42–49

Arkhipova TN, Veselov SU, Melentiev AI, Martynenko EV, Kudoyarova GR (2005) Ability of bacterium *Bacillus subtilis* to produce cytokinins and to influence the growth and endogenous hormone content of lettuce plants. Plant Soil 272:201–209

Asari S, Tarkowská D, Rolčík J, Novák O, Velázquez Palmero BS, Meijer J (2017) Analysis of plant growth-promoting properties of *Bacillus amyloliquefaciens* UCMB5113 using *Arabidopsis thaliana* as host plant. Planta 245(1):15–30

Ashraf M, Hasnain S, Berge O, Mahmood T (2004) Inoculating wheat seedlings with exopolysaccharide-producing bacteria restricts sodium uptake and stimulates plant growth under salt stress. Biol Fertil Soils 40:157–162

Assmus B, Hutzler P, Kirchhof G, Amann R, Lawrence JR, Hartmann A (1995) *In situ* localization of *Azospirillum brasilense* in the rhizosphere of wheat with florescently labelled, rRNA-targeted oligonucleotide probes and scanning confocal laser microscopy. Appl Environ Microbiol, 61:1013–1019

Baldani VLD, Baldani JI, Dobereiner J (2000) Inoculation of rice plants with the endophytic diazotrophs *Herbaspirillum seropedicae* and *Burkholderia* spp. Biol Fertil Soils 30:485–491

Bao Y, Aggarwal P, Robbins NE, Sturrock CJ, Thompson MC, Tan HQ, Tham C, Duan L, Rodriguez PL, Vernoux T, Mooney SJ, Bennett MJ, Dinneny JR (2014) Plant roots use a patterning mechanism to position lateral root branches toward available water. Proc Natl Acad Sci U S A 111(25):9319–9324

Barnawal D, Maji D, Bharti N, Chanotiya CS, Kalra A (2013) ACC deaminase-containing *Bacillus subtilis* reduces stress ethylene-induced damage and improves mycorrhizal colonization and rhizobial nodulation in *Trigonella foenum-graecum* under drought stress. J Plant Growth Regul 32:809–822

Bashan Y, Holguin G (1997) *Azospireillum*/plant relationship: environmental and physiological advances (1990–1996). Can J Microbiol, 43:103–121

Bashan Y, Levanony H (1988) Interaction between *A. brasilense* Cd and wheat root cells during early stages of root colonization. In: Klingmuller W (ed) *Azospirillum IV: genetics, physiology, ecology*. Springer, Berlin, pp 166–173

Bashan Y, Levanony H (1990) Current status of *Azospirillum* inoculation technology: *Azospirillum* as a challenge for agriculture. Can J Microbiol 36:591–608

Bashan Y, Levanony H, Klein E (1986) Evidences for a weak active external adsorption of *A. brasilense* Cd to wheat roots. J Gen Appl Microbiol 132:3069–3073

Bashan Y, Levanony H, Ziv-Vecht O (1987) The fate of field-inoculated *Azospirillum brasilense* cd in wheat rhizosphere during the growing season. Can J Microbiol 33:1074–1079

Bashan Y, Levanony H, Mitiku G (1989) Changes in proton efflux in intact wheat roots inoculated by *A. brasilense* Cd. Can J Microbiol 35:691–697

Bashan Y, Moreno M, Troyo E (2000) Growth promotion of seawater-irrigated oilseed halophyte *Salicorniabigelovii* inoculated with mangrove rhizosphere bacteria and halotolerant Azospirillum spp. Biol Fertil Soil 32:265–272

Bashan Y, Harrison SK, Whit Moyer RE (1990) Enhanced growth of wheat and soybean plants inoculated with *A. brasilense* is not necessarily due to general enhancement of mineral uptake. Appl Environ Microbiol 56:769–775

Beeckman T, Burssens S, Inzé D (2001) The peri-cell-cycle in Arabidopsis. J Exp Bot 52:403–411

Bellini C, Pacurar DI, Perrone I (2014) Adventitious roots and lateral roots: similarities and differences. Annu Rev Plant Biol 65(1):639–666

Bellone CH, de Bellone SDVC, Pedraza RO, Monzon MA (1997) Cell colonization and infection thread formation in sugar cane roots by *Acetobacter diazotrophicus*. Soil Biol Biochem 29(56):965–967

Bhattacharyya PN, Jha DK (2012) Plant growth-promoting rhizobacteria (PGPR): emergence in agriculture. World J Microbiol Biotechnol 28(4):1327–1350

Bowen GD, Rovira AD (1999) The rhizosphere and its management to improve plant growth. Adv Agron 66:1–102

Brimecombe, J. M., De Leij, F. A. & Lynch, J. M. (2001). The effect of root exudates on rhizosphere microbial population. In *The Rhizosphere*. Ed: Pinton, R., Varanini, Z, and Nannipieri, P pp. 95-140. Marcel Dekker Inc, New York

Celenza JL, Grisafi PL, Fink GR (1995) A pathway for lateral root formation in Arabidopsis thaliana. Genes Dev 19(17):2131–2142

Datta S, Kim CM, Pernas, M.& Dolan, L. (2011) Root hairs: development, growth and evolution at the plant-soil interface. Plant Soil 346(1):1–14

De Rybel B, Audenaert D, Xuan W et al (2012) A role for the root cap in root branching revealed by the non-auxin probe naxillin. Nat Chem Biol 8(9):798–805

Del Gallo M, Fendrik I (1994) The Rhizosphere and *Azospirillum*. In: Okon Y (ed) Azospirillum/ plant associations. CRC Press, London, pp 57–85

Dittmer HJ (1937) A quantitative study of the roots and root hairs of a winter rye plant (*Secale creale*). Am J Bot 24:417

Dubrovsky JG, Sauer M, Napsucialy-Mendivil S, Ivanchenko MG, Friml J, Shishkova S, Celenza J Benková E (2008) Auxin acts as a local morphogenetic trigger to specify lateral root founder cells. PNAS, 105(25):8790–8794

Foster RC, Bowen GD (1982) In: Lacy GH (ed) Plant surfaces and bacterial growth: the rhizosphere, phytopathogenic prokaryotes. Academic, New York, pp 159–185

Freitas JRD, Banerjee MR, Germida JJ (1997) Phosphate-solubilizing rhizobacteria enhance the growth and yield but not phosphorus uptake of canola (*Brassica napus* L.). Biol Fertil Soils 24:358–364

Fukaki H, Yoko H, Masao T (2007) Auxin-mediated lateral root formation in higher plants. Int Rev Cytol 256:111–137

Hashem A, Tabassum B, Fathi E, Allah A (2019) *Bacillus subtilis*: A plant-growth promoting rhizobacterium that also impacts biotic stress. Saudi J Biol Sci 26(6):1291–1297

Himanen K, Vuylsteke M, Vanneste S, Vercruysse S, Boucheron E, Alard P, Chriqui D, Van Montagu M, Inze D, Beeckman T (2004) Transcript profiling of early lateral root initiation. Proc Nat Acad Sci USA 101:5146–5151

Iqbal M, Wagi S, Ahmed A (2017) Phyllospheric bacterial treatments improve growth in *Helianthus annuus* L. RADS J Biol Res Appl Sci 9(1):2521–8573

Ishida T, Kurata T, Okada K, Wada T (2008) A genetic regulatory network in the development of trichomes and root hairs. Annu Rev Plant Biol 59:365–386

Jacoud C, Job D, Wadoux P, Bally R (1999) Initiation of root growth stimulation by *Azospirillum* lipoferumCRT1 during maize seed germination. Can J Microbiol 45:339–342

Joseph G, Dubrovsky JG, Rost TL (2012) Pericycle. In: Encyclopedia in life science. Facts On File, New York

Khalid A, Arshad M, Shaharoona B, Mahmood T (2009) Plant growth promoting rhizobacteria and sustainable agriculture. In: Khan M, Zaidi A, Musarrat J (eds) Microbial strategies for crop improvement. Springer, Berlin/Heidelberg

Laplaze L, Parizot B, Baker A, Ricaud L, Martinière A, Auguy F, Franche C, Nussaume L, Bogusz, D.& Haseloff, J. (2005) GAL4-GFP enhancer trap lines for genetic manipulation of lateral root development in *Arabidopsis thaliana*. J Exp Bot 56:2433–2442

Levanony H, Bashan Y (1989) Enhancement of cell division in wheat root tips and growth of root elongation zone induced by *Azospirillum brasilense* Cd. Can J Bot 67:2213–2216

Levanony H, Bashan Y, Romano B, Klein E (1989) Ultra-structural colonization and identification of *A. brasilense* Cd on and within wheat root by immuno-gold labelling. Plant Soil 117:207–218

López-Bucio J, Cruz-Ramírez A, Herrera-Estrella L (2003) The role of nutrient availability in regulating root architecture. Curr Opin Plant Biol 6:280–287

Mähönen AP, Bishopp A, Higuchi M, Nieminen KM, Kinoshita K, Törmäkangas K, Ikeda Y, Oka A, Kakimoto T, Helariutta Y (2006) Cytokinin signaling and its inhibitor AHP6 regulate cell fate during vascular development. Science 311:94–98

Malamy JE (2009) Lateral root formation. Ann Plant Rev 37: Root development, 37:83–126, Beeckman eds

Marhavy P, Vanstraelen M, Rybell B, Zhaojun D, Bennett MJ, Beeckman T, Benkova E (2012) Auxin reflux between the endodermis and pericycle promotes lateral root initiation. EMBO J:1–10

Matsuda R, Handayani ML, Sasaki H, Takechi K, Takano H, Takio S (2018) Production of indoleacetic acid by strains of the epiphytic bacteria *Neptunomonas* spp. isolated from the red alga *Pyropia yezoensis* and the seagrass *Zostera marina*. Arch Microbiol 200(2):255–265

Mia MAB (2015) Nutrition of crop plants. Nova Science Publisher, New York
Mia MAB, Shamsuddin ZH (2009) Enhanced emergence and vigour seedling production of rice through plant growth promoting bacterial inoculation. Res J Seed Sci 2:96–104
Mia MAB, Shamsuddin ZH (2013) Biofertilizer for banana production. Lambert Academic Publisher, Germany
Mia MAB, Shamsuddin ZH, Marziah M (1999) External and internal root colonization of Azospirillumbrasilenseon tissue-cultured banana plantlets. In: Vidyadaran MK, Tengku ASTM (eds) Proceedings of the eighth scientific conference of Electron Microscopy Society Malaysia. Genting Highlands, Pahang, pp 173–174
Mia MAB, Shamsuddin ZH, Marziah M (2007) Associative nitrogen fixation by *Azospirillum* and *Bacillus* spp. in bananas. Infomusa 16(1&2):11–15
Mia MAB, Shamsuddin ZH, Maziah M (2009a) Use of plant growth promoting bacteria in banana: A new insight for sustainable banana production. Int J Agric Biol 12(3):459–467
Mia MAB, Shamsuddin, ZH, Zakaria W, Marziah M (2009b) The effect of rhizobacterial inoculation on growth and nutrient accumulation of tissue-cultured banana plantlets under low N-fertilizer regime. Afr. J. Biotechnol, 8(21):5855–5866 https://doi.org/10.5897/AJB09.882
Mia MAB, Shamsuddin ZH, Zakaria W, Marziah M (2010a) Rhizobacteria as bioenhancer for growth and yield of banana (*Musa* spp. cv. "Berangan"). Sci Hortic 126(2):80–87
Mia MAB, Shamsuddin ZH, Marziah M (2010b) Rhizobacterial inoculation on growth and nitrogen incorporation in tissue-cultured *Musa* plantlets under nitrogen-free hydroponics condition. Aust J Crop Sci 4(2):85–90
Mia MAB, Hossain MM, Shamsuddin ZH, Islam MT (2013) Plant-associated bacteria in nitrogen nutrition in crops, with special reference to rice and banana. In: Maheshwari DK (ed) Bacteria in agrobiology: crop productivity. Springer, Berlin/Heidelberg
Mia MAB, Naher UA, Panhwar QA, Islam MT (2016) Growth promotion of non-legumes by the inoculation of Bacillus species. In: Islam MT et al (eds) Bacillus and Agrobiotechnology. Springer
Murty MG, Ladha, JK (1987) Differential colonization of *Azospirillum lipoferum* on roots of two varieties of rice (*Oryza sativa* L.). Biol Fertil Soil 4:3–7
Murty MG, Ladha JK (1988) Influence of *Azospirillum* inoculation on the mineral uptake and growth of rice under hydroponics conditions. Plant Soil 108:281–285
Naher UA, Radziah O, SitiZuraidah Y (2013) Effect of urea-N on growth and indoleacetic acid production of *Stenotrophomonas maltophilia* (Sb16) isolated from rice growing soils in Malaysia. Chilean J Agricult Res 73(2):187–192
Okon Y, Kapulnik Y, Sarig S (1988) Field inoculation studies with *Azospirillum* in Israel. In: Suba Rao NS (ed) Biological nitrogen fixation, recent developments. IBH, Oxford/New Delhi, p 175
Osmont KS, Sibout R, Hardtke CS (2007) Hidden branches: developments in root system architecture. Annu Rev Plant Biol 58:93–113
Parizot B, Laplaze L, Ricaud L (2008) Diarch symmetry of the vascular bundle in Arabidopsis root encompasses the pericycle and is reflected in distich lateral root initiation. Plant Physiol 146:140–148
Park YG, Mun BG, Kang SM, Hussain A, Shahzad R, Seo CW et al (2017) *Bacillus aryabhattai* SRB02 tolerates oxidative and nitrosative stress and promotes the growth of soybean by modulating the production of phytohormones. PLoS One 12:e0173203
Péret B, De Rybel B, Casimiro I, Benková E, Swarup R, Laplaze L, Beeckman T, Benne MJ (2009) *Arabidopsis* lateral root development: an emerging story. Trends Plant Sci 14(7):399–408
Persello-Cartieaux F, David P, Sarrobert C, Thibaud MC, Achouak W, Robaglia C, Nussaume L (2001) Utilization of mutants to analyze the interaction between *Arabidopsis thaliana* and its naturally root-associated *Pseudomonas*. Planta 212:190–198
Peyush P, Moriyama S, Takahashi A (2000) Molecular cloning of growth hormone complementary DNA in Barfin Flounder (*Verasper moseri*). Mar Biotechnol 2:21–26
Potters G, Pasternak T, Guisez Y, Jansen MAK (2007) Stress-induced morphogenic responses: growing out of trouble? Trends Plant Sci 12(3):98–105

Puente ME, Holguin G, Bernard RG, Bashan Y (1999) Root-surface colonization of black mangrove seedlings by *Azospirillum halopraeferens* and *Azospirillum brasilense* in sea water. FEMS Microbiol Ecol 29:283–292

Radhakrishnan R, Lee IJ (2013) Regulation of salicylic acid, jasmonic acid and fatty acids in cucumber (*Cucumis sativus* L.) by spermidine promotes plant growth against salt stress. Acta Physiol Plant 35:3315–3322

Sarig S, Blum A, Okon Y (1988) Improvement of water status and yield of field-grown grain sorghum (*Sorghum bicolor*) by inoculation with *Azospirillum brasilense*. J Agric Sci 110:271

Saubidet MI, Barneix AJ (1998) Growth stimulation and nitrogen supply to wheat plants inoculated with *Azospirillum brasilense*. J Plant Nutr 21(12):2565–2577

Sioussaram D (1968) Observacions preliminaries sur Ienracinement des bananiers dans les sols de la station de neufchateau, Guadeloupe. Fruits 23:473–479

Skoog F (1938) Absorption and translocation of auxin. Am J Bot 25(5):361–372

Subba Rao NS (1995) Nitrogen fixation in free-living and associative symbiotic bacteria. In: Soil microorganisms and plant growth. IBH Publishing Co. Pvt. Ltd, Oxford

Suslow TV (1982) Role of root-colonizing bacteria in plant growth. In: Mount MS, Lacy GH (eds) Phytopathogenic prokaryotes, vol 1. Academic, New York, pp 186–223

Swennen R, De Langhe E, Janssen J, Decoene D, D. (1986) Study of the root development of some *Musa* cultivars in hydroponics. Fruits 41(9):515–524

Teale WD, Paponov IA, Palme K (2006) Auxin in action: signalling, transport and the control of plant growth and development. Nat Rev Mol Cell Biol 7:847–859

Theunis M (2004) Flavonoids, NodD1, NodD2, and *Nod*-Box NB15 modulate expression of the y4wEFG locus that is required for indole-3-acetic acid synthesis in *Rhizobium* sp. strain NGR234. Mol Plant-Microbe Interact 17:1153–1161

Vissenberg K, Gonzalez N (2020) Plant organ and tip growth. J Exp Bot 71(8):2363–2364

Vlamakis H, Chai Y, Beauregard P, Losick R, Kolter R (2013) Sticking together: building a biofilm the *Bacillus subtilis* way. Nat Rev Microbiol 11:157–168

Wagi S, Ahmed A (2019) *Bacillus* spp.: potent microfactories of bacterial IAA. PeerJ 7:7258

Waidmann S., Sarkel E, & Kleine-Vehn J. (2020). Same same, but different: growth responses of primary and lateral roots *Journal of Experimental Botany*, 23, 71(8):2397-2411

Weber OB, Baldani VLD, Teixeira KRS, Kirchhof G, Baldani JI, Dobereiner J (1999) Isolation and characterization of diazotrophic bacteria from banana and pineapple plants. Plant Soil 210:103–113

Wei CT (1997) Development and application of biofertilizer in China. In Proceedings Biological Nitrogen Fixation: The Global and Future Needs, Italy, pp 71–72

Woodward AW, Bartel B (2005) Auxin: regulation, action, and interaction. Ann Bot 95:707–735

Xie S, Wu HJ, Zang H, Wu L, Zhu Q, Gao X (2014) Plant growth promotion by spermidine-producing *Bacillus subtilis* OKB105. Mol Plant Microbe Interact 27:655–663

Xuan W, Band LR, Kump RP, Damme DV, Parizot B, De Ro G, Opdenacker D, Barbara Möller K, Skorzinsk N, Njo MF, Rybel BD, Audenaert D, Nowack MK, Vanneste S, Beeckman T (2016) Cyclic programmed cell death stimulates hormone signalling and root development in *Arabidopsis*. Science 351(6271):384–387

Chapter 8
Mechanisms Involved with Bacilli-Mediated Biotic and Abiotic Stress Tolerance in Plants

Mahfuz Rahman, Md Nur Alam Miah, and Whitney Dudding

Abstract Due to increased demand for food and feed, plants are being grown in marginal lands dominated by abiotic stresses. These abiotic stresses predispose plants to biotic stresses compromising the yield and quality. Mitigation efforts of these stresses with synthetic chemicals further complicated the situation. However, use of beneficial microbes opened a new horizon for managing these stresses in the agricultural ecosystem. To date, an appreciable amount of research elucidated the underlying mechanisms how these microbes, especially numerous species belonging to the genus *Bacillus*, play a positive role in mitigating these stresses. Colonization of plant rhizosphere or phyllosphere by these microbes contributes to alleviating these stresses through up- or downregulation of major metabolic pathways in plants. Regulation of metabolic pathway helps in reducing/neutralizing the level of stressors or inducing plants to overproduce stress-mitigating biochemicals. This chapter compiles all the major mechanisms pertaining to biotic and abiotic stress alleviation in plants by *Bacilli* to aid in elucidating more complex mechanisms by future research endeavors.

Keywords Stress alleviation · Metabolic pathway · Molecular mechanisms · Gene expression · Peptides · Climate change · Biotic factors · Agro-ecosystem · Plant health · Toxic effluent · Innate immune system · *Bacillus* sp.

M. Rahman (✉) · M. N. A. Miah · W. Dudding
WVU Extension Service, West Virginia University, Morgantown, WV, USA
e-mail: mm rahman@mail.wvu.edu

M. T. Islam et al. (eds.), *Bacilli in Agrobiotechnology*, Bacilli in Climate Resilient Agriculture and Bioprospecting,
https://doi.org/10.1007/978-3-030-85465-2_8

8.1 Introduction

Crop productivity is influenced by a combination of environmental and genetic factors (Kleinwechter et al. 2016). Genetic potential of many crop varieties is not achieved since crops in most cases are grown in stressful or less-than ideal growing conditions, which include unfavorable edaphic and climatic environments. It is universally proven that various stress factors adversely influence the physiology of a plant starting from germination through growth and yield. These stresses are primarily classified into two major groups: abiotic and biotic. Our agro-ecosystem is subjected to continuous exposure from a series of ever-changing abiotic and biotic factors, many of which eventually create an unfavorable environment for soil health, fertility, crop growth, and productivity. It is now well documented that these abiotic and biotic stresses can affect all aspects of plant growth including biomass production and intended yield (Pandey et al. 2017). While all biotic stresses are of biological origin like diseases, insects, and weeds; abiotic stresses are generally physical or chemical, and imposed on plants by their environment. Disease-causing organisms include fungi, bacteria, viruses, nematodes, and phytoplasmas. Abiotic stresses on plants are on the rise due to human activities that drive global warming and climate change, releasing toxic effluent to arable land. Despite the consequences, people attempt to grow crops in saline or nutrient-deficient, unsuitable land, due to the increased demand for food and feed worldwide. Exposure of plants to abiotic stresses predispose them to biotic stresses such as pathogens, insects, and may also reduce their ability to compete with weeds. Combined abiotic and biotic stresses can severely affect crop productivity. Even if these stresses occur separately, 50% and 30% of worldwide agricultural productivity could be lost to abiotic and biotic stresses, respectively. However, plants have evolved a sophisticated defense network, also called innate immune system, in response to fluctuating environmental conditions that provide protection to some extent against these stress factors. There are no simple solutions to fully counteract these stresses with modern agricultural technologies discovered over the last few centuries. However, many plant-beneficial microbes living in the soil and microbes existing on or inside plants as endophytes or epiphytes have been found in several studies to trigger a plant's innate defense system, thereby promoting growth and protecting plants from biotic and abiotic stress factors (Glick 1995; Radhakrishnan et al. 2014, 2017; Tonelli et al. 2010). These beneficial microbes contributed to boosting crop yield by minimizing either abiotic or biotic or both types of stresses. Among the microbes, *Bacillus* and *Pseudomonas* species are the major plant stress mitigators and growth-promoting bacteria (Kang et al. 2015a). However, the ability of *Bacilli* to form spores distinguishes them from that of *Pseudomonas*. The survival capacity of the members of this genus for a long time under unfavorable environmental conditions make them suitable to be used for prevention of both biotic and abiotic stresses encountered by plants. That is why *Bacillus* spp. is ranked at the top as a microbial agent for mitigating both abiotic and biotic plant stresses due to their unparalleled genetic fitness. Dihazi et al. (2012) noted that organic and sustainable farming practices prefer application of *Bacillus* or other bacterial biocontrol agents as an eco-friendly option to enhance disease resistance in crops, which is also considered as a safer method of increasing crop productivity. Use of synthetic pesticides can be reduced in modern agriculture by utilization of *Bacillus*-derived formulations or products due to their proven efficacy (Myresiotis et al. 2015). In addition, an appreciable number of studies have been directed towards understanding *Bacillus*-mediated protection of crop against adverse biotic and environmental stresses at the physiological, biochemical, and molecular levels. Similar advancements through numerous studies have also taken place in the area of *Bacillus*-based remediation or detoxification of metals and other pollutants from the edaphic environments. As a result, remarkable development has occurred in elucidating the underlying mechanisms pertaining to biotic stress tolerance in plants by the use of *Bacillus* (Choudhary and Johri 2009; Ongena and Jacques 2008). However, information on *Bacillus*-mediated abiotic stress tolerance mechanisms are still very inadequate (Arkhipova et al. 2007; Wolter and Schroeder 2012) as most of the studies on *Bacillus*-based abiotic stress tolerance focused on evaluating plant growth promoting effects rather than unraveling specific mechanisms (Dimkpa et al. 2009). *Bacillus*-mediated stress mitigation mechanisms have been explored by many researchers to improve efficacy by using a consortium of beneficial microbes and combining them for diverse multiple modes of action. Many potential stress mitigation mechanisms that have been observed indicated that *Bacillus* spp.-induced actions on the adverse environmental factors and host plants are diverse, numerous, and may be indirect or direct (Berg 2009; Numan et al. 2018; Perez-Garcia et al. 2011; Turan et al. 2012). It is not uncommon for some species of *Bacillus* to mitigate multiple stresses if present at the right time in ample populations. This chapter highlights the major *Bacillus*-mediated plant abiotic and biotic stress tolerance mechanisms by comparing with plants' innate responses to stresses.

8.2 Major Abiotic Stresses and Their Impacts on Crop Growth and Yield

Abiotic stressors are a major cause that puts obstacle against enhancement of worldwide crop production (Bray et al. 2000; Wang et al. 2003). Crop plants have to cope with adverse environmental and edaphic factors with biological mechanisms that are intrinsic to them like hormonal signaling (Nguyen et al. 2016) and interaction with beneficial microbes (Numan et al. 2018). Plant growth, development, and productivity can suffer immensely if they fail to counteract those stresses (Pereira 2016). As more arable land is lost to urbanization and non-agricultural use, crops are being grown in less-suitable areas where abiotic stresses are common. As a result, 50–82% of major crops encounter losses that pose a serious threat to agriculture and food security due to adverse environmental conditions like drought, salinity, extreme temperature, UV radiation, heavy metals, or various oxidative stresses

(Benedetto et al. 2017; FAO 2016). By impairing biochemical/physiological and molecular processes, these stress factors (individually or combined) may induce numerous hostile effects in plants, eventually reducing plant growth, development, and productivity. Stresses from abiotic factors are now considered as one of the biggest potential threats all over the world to agricultural productivity that may affect yields up to 70% for staple food crops as projected by many studies (Kaur et al. 2008; Mantri et al. 2012). Among many different types of abiotic stresses, accumulation of various heavy metals in agricultural soil has created a major concern to the agriculture system and health of many millions of people in countries such as Zambia, Ukraine, Russia, Peru, China, and India (ENS 2006). Wang et al. (2003) estimated that 30% arable land may be lost by the end of 2028, and at this rate it may reach 50% by mid-twenty-first century due to high salinity. Other environmental factors such as the projected increase of mean temperature by 3 °C due to the rise in CO_2 concentration by the end of twenty-first century by about 500–1000 ppm, which will cause heat stress to crops (Khan et al. 2013). Thus, salinity of arable land, nutrient deficiency, drought, metal toxicity, and unforeseen impact of climate change are likely to significantly worsen the problem (Anjum et al. 2014). The combined effect of these stresses may result in losses of soil microbial diversity, soil fertility, and availability of nutrients (Chodak et al. 2015). Plants require a favorable growing site for their physiological and developmental processes. Less than favorable growing condition induced abiotic stress factors can also predispose plants to biotic stresses that interfere with normal growth and development thereby reducing productivity. Plants use their intrinsic recognition mechanism to detect and respond to stresses to some extent by activating the defense pathways to support their nourishment (Jiang et al. 2016; Ahmad et al. 2015; Crane et al. 2011). However, this response may not be enough to overcome stresses if stress levels exceed a certain threshold. When beneficial microbes such as *Bacilli* are present in or on plants within a stressful environment, they modulate the environment and provide enhanced stress tolerance capacity to plants through several procedures described below.

8.2.1 Mechanisms of Abiotic Stress Alleviation

Plants have developed an efficient innate defense network that is turned on by external stimuli representing abiotic (environmental) and biotic stresses. Depending on the genetic makeup of plants, they can employ various self-defense mechanisms to prevent these stresses through up- or downregulation of major metabolic pathways, such as TCA cycle, photosynthesis, accumulation of secondary metabolites, and specific sugars or amino acids (Lotfi et al. 2010; Rai 2002; Lewis et al. 2001). This wide array of different metabolites may play a significant role in plant stress tolerance. For example, phytohormones profoundly influence control of specific molecular mechanisms in plants, and thus optimizing plant responses against stresses of abiotic nature (Nguyen et al. 2016). However, presence of plant beneficial microbes such as *Bacillus* alleviate these stress factors by either reducing/neutralizing the level of stressors or inducing plants to overproduce stress-mitigating biochemicals. During low water availability, salt and heavy metals can accumulate on top soil layers. Heavy metal accumulation in agricultural soil has also been increasing due to the release of industrial effluent and spread with water. *Bacillus* spp. if present in an environment can produce ample amount of exopolysaccharides and siderophores to bind Fe, which usually prevent the movement of toxic ions and adjust the ionic balance and water transport in plant tissues. Plants can also rapidly sense changing environmental conditions and defend themselves with their innate defense mechanism. Plant responses to abiotic stress involve an induced metabolic cross talk within various biosynthetic pathways. Root system can also sense abiotic stress signals and respond accordingly to the stresses originating from soils (khan et al. 2016). These responses are results of an intricate mechanism that involves changes at genetic, cellular, metabolic, and physiological levels (Atkinson and Urwin 2012). Major impact of abiotic stress is water-deficient conditions created within cells, which is followed by a series of biochemical, molecular, and phenotypic defense action (Xu and Zhou 2006; Almoguera et al. 1995). As the number of stressors experienced by a plant increases, so does the complexity of their responses when compared to plants with a single stress. The complex nature of such responses is due to the activation of a specific gene along with metabolic programming in cells against individual stresses in a specific growing condition. Stress tolerance is a vital phenomenon that may vary with different stages of plant development. Abiotic stress responses may reduce or increase the susceptibility of plants toward biotic stress caused by pests or pathogens (Rizhsky et al. 2004). This becomes more important in agricultural crops because, in various agricultural systems, most crops grow in unfavorable environmental conditions that restrict genetic growth and development potential of plants (Bray et al. 2000). Metabolic regulations including wide changes in the composition, concentration, and distribution of secondary and primary metabolites are among the common responses of plants to abiotic stresses. While secondary metabolites such as alkaloids and flavonoids produced by plants in response to stresses are utilized mostly for their defense, primary metabolites such as carbohydrates and amino acids play a crucial role in the plant's growth and development. Major *Bacillus*-mediated abiotic stress mitigation mechanisms are presented below:

8.2.1.1 Mechanisms to Mitigate Drought Stress on Plants

Global climate change is arguably one of the most pressing ecological concerns of our lifetime. Future changes in water availability and increasing drought stress will likely alter the crop production system, ultimately leading to exploring sustainable mitigation measures and necessary modifications to the system. Drought was identified as a key factor that limits crop productivity among various abiotic stresses plants encounter in noncontrolled environments or vast crop growing environments that lack irrigation facilities. Incidence and effect of drought is predicted to get worse in the coming years as a result of global warming and adverse climate change.

Crop breeders continue their efforts to overcome drought problems with resistant varieties (Araus et al. 2002). However, it is true that many of these varieties can only tolerate and survive under drought stress for a short period of time. As such, any other natural agent such as *Bacillus* spp. capable of alleviating drought stress is considered one of the significant agricultural inputs as it relates to sustainable agriculture. Lack of water availability primarily restricts nutrient uptake and photosynthesis as plants curtail water loss by increased diffusive resistance and make morpho-physiological adaptations, such as closing their stomates. As a result, plants lose turgidity, and leaf size expansion, stem extension, and root proliferation are halted. Overall, innate plant adaptation strategies against drought stress include either or all of physiological, biochemical (increased production of multiple phytohormones, accumulation of compatible osmolytes, antioxidant and other secondary metabolites or signaling molecules), and molecular (upregulation of multiple drought stress related genes) mechanisms (Chaves et al. 2003; Krasensky and Jonak 2012). These changes help plants maintaining turgor pressure of cells, protect cell membranes, macromolecules and enzymes from oxidative injury (Krasensky and Jonak 2012; Gill and Tuteja 2010). Unfavorable environmental stimuli such as drought can affect normal plant metabolism leading to suppression of crop growth and yield. Association of *Bacillus* with plant rhizosphere or phyllosphere stimulates plant immunity against drought stress by altering stress-responsive genes that optimize production of proteins, phytohormones, and other related metabolites.

8.2.1.2 *Bacillus*-Mediated Mechanisms of Mitigating Drought Stress

As most of the studies conducted to evaluate efficacy of *Bacillus* in providing abiotic stress tolerance and unraveling the associated mechanisms had to exert drought stress on plants, effect of *Bacillus* could be confounded with innate drought response of plants. However, additive effects from *Bacillus* application and quantitative expression of responses were indicative of the role of beneficial microbes. For example, exposure of *Bacillus* inoculated timothy (*Phleum pratense* L.) plants to drought stress for 8 weeks provided enhanced shoot (26.6%) and root (63.8%) biomass, stomatal conductance (214.9%), and photosynthetic activity (55.2%) compared to plants grown under similar drought stress conditions without inoculation (Gagne-Bourque et al. 2016). Underlying mechanisms that supported increased stress tolerance in this study by *Bacillus* included enhanced accumulation of osmolytes, such as many different types of amino acids and sugars in roots and shoots. Another documented mechanism of drought stress avoidance in wheat by *Bacillus* is production of growth-promoting phytohormones such as indole-3-acetic acid (IAA) and 1-aminocyclopropane-1-carboxylate deaminase (ACCD) to counteract the stress-induced increase of abscisic acid (Barnawal et al. 2017). *Bacillus* spp. treatment of plants exposed to drought were found to increase the uptake of water and macronutrients like N, P, and K (Barnawal et al. 2013), which is the direct influence of increased root growth by enhanced IAA production. A higher production of scavenging enzymes of reactive oxygen species (ROS) was observed by Saikia et al.

(2018) when mung bean was inoculated with a group of plant growth enhancing bacteria including *Bacillus subtilis* RJ46 followed by exposure to drought stress. Quantity of ROS scavenging enzymes (ascorbate peroxidase, SOD, PO, CAT, glutathione reductase) and relevant compounds (glutathione, ascorbic acid, and cysteine) that prevent buildup of ROS is usually found to be higher in *Bacillus*-inoculated plants, and they lessen the oxidative damage that occurs at the time of drought stress (Kaushal and Wani 2015). Thus, production of antioxidants is the most important known mechanism of *Bacillus*-mediated drought stress management.

8.2.1.3 Extent and Impact of salinity Stress on Plants

Climate change and global warming has been affecting regular rainfall and distribution over recent decades. More specifically, productivity of agricultural land has been worsening worldwide due to accumulation of salt that resulted from low rainfall, improper irrigation practices and high water evaporation rates due to dry weather condition (Al-Karaki 2006). The high concentration of salts in saline soils cause oxidative, osmotic, ionic, and water stress in plants. Accumulation of salt in upper soil layers eventually affects nutrient and water uptake by plant roots due to reduced soil water potential (Porcel et al. 2012). This mimics a drought stress situation that induces plant response to accumulate compounds with osmo-protective properties such as special types of sugars, amino acids, and other secondary metabolites that were previously absent or present in lower quantity. Under conditions of salinity, plants are subjected to nutrient imbalance like K^+ deficiency and Na^+ toxicity and may also suffer from oxidative stress. Plants may eventually face metabolic pathway disorders related to respiration, photosynthesis, homeostasis in redox system, and phytohormone regulation. As a result, carbohydrate and amino acid syntheses are affected leading to reduced seed germination, plant vigor, growth, and yield (Radhakrishnan and Lee 2013, 2014; Rady 2011; Munns and Tester 2008). However, induction of osmoregulators may counteract the above disorders as well as protect structures of different cellular organelles including membranes (Hare et al. 1998). These osmoregulators also work as scavengers of free radicals to prevent their damaging effects to DNA (Ashraf and Foolad 2007).

8.2.1.4 *Bacillus*-Based Mechanism of Salinity Stress Tolerance

Inoculation of plants with beneficial microbes including *Bacillus* spp. enhances plant growth and development during salt stress compared with non-inoculated. This is considered an eco-friendly approach to counteract abiotic stresses and make agriculture more sustainable (Hashem et al. 2015, 2016a, b; Radhakrishnan et al. 2014). Production of growth hormones such as IAA has been found by Bochow et al. (2001) as one of the mechanisms to compensate salt stress-induced growth and yield loss. Authors found significant yield increase in pepper and eggplant due to inoculation with *B. subtilis* FZB24 compared to the nontreated, despite irrigating the plots with saline groundwater. For further investigation to unravel the mode of action, pepper seedlings were pretreated with millimolar amounts of auxin precursors (indole-3-acetic aldehyde or indole-3-pyruvic acid, tryptophan) followed by exposure to saline water. Results revealed that auxin precursors treatment had compensated 75% growth loss of seedlings after 1 week that would occur due to salt stress. This finding further supports that one of the *Bacillus*-based modes of action of alleviating salinity-induced stress in plants is production of IAA. However, other investigators found that growth hormones like gibberellins were also produced by *Bacillus* and played a role in addition with enhanced uptake of N, P, K^+, $Mg2^+$, and $Ca2^+$(Mohamed and Gomaa 2012). They also reported decreased abscisic acid and Na^+ and Cl^- content compared to the non-inoculated, which may have occurred due to the relatively high uptake of required nutrients as facilitated by *Bacillus* to alleviate stress induced by salinity. Egamberdieva et al. (2017) also reported improved uptake of N, P, K, and Mg by chickpea, grown under saline soil conditions that were inoculated with a consortium of *B. subtilis* and *Mesorhizobium ciceri* compared with non-inoculated plants. Salinity stress significantly increases reactive oxygen species level and H_2O_2 in plants resulting in peroxidation of membrane lipid (Yazici et al. 2007; Koca et al. 2007). Thus, reduction of membrane lipid peroxidation or oxidation of other cellular organelles due to *Bacillus* inoculation under salinity is a proof of *Bacillus*-based stress mitigation. Lastochkina et al. (2017) reported reduction of stress-induced lipid peroxidation (MDA) in wheat due to inoculation of plants by *B. subtilis* 10–4 followed by exposure to water containing 2% NaCl salt compared with non-inoculated plants. If *Bacilli* colonize plants under salt stress, they trigger general antioxidant defense process resulting in synthesis of various anti-oxidant enzymes namely peroxidase, catalase (CAT), nitrate reductase (NR), superoxide dismutase (SOD), polyphenol oxidase (PO), glutathione reductase (GR), and guaiacol peroxidase (GP) that may vary with plant species (Chawla et al. 2013). However, halotolerant (HT) PGPR *Bacillus licheniformis* A2 recovered from saline soil showed higher growth promoting effect when plants were inoculated with the organism and grown in saline soil. Although several plant growth-promoting features such as production of IAA, phosphate solubilization, and siderophore production were considered as potential *Bacillus*-based mechanisms in alleviating the salt stress-induced detrimental effects and increasing plant growth of stressed peanuts (Goswami et al. 2014), study conducted by Zhang et al. (2008) with Arabidopsis in salt-affected soil revealed that HT *Bacillus subtilis* reduced the uptake of Na^+ by roots that involved downregulation of high-affinity potassium transporter (HKT1). However, IAA synthesis being transcriptionally related to ethylene production in plants via the expression of ACC synthase gene can halt their growth and development despite ethylene's capacity of providing salt tolerance to plants. Thus, after the initial ethylene production step, HT Bacilli start synthesizing ACC deaminase (ACCD) enzyme to stop ACC and ethylene production, supporting resumption of growth of stressed plants (Nabti et al. 2015). In addition, *Bacillus* can excrete exopolysaccharides (EPS) that facilitate binding of Na^+ in root cells thereby preventing their translocation to foliage. Similar binding of Na^+ by EPS may take place in the rhizosphere by *Bacillus* excreted EPS to prevent its absorption like a physical barrier around the roots (Arora et al. 2020). Thus, association of *Bacillus* spp. with salt-stressed plants has the potential to alter their metabolism to sustain or promote plant growth. Halotolerant *Bacillus*-mediated expression of genes to alleviate salinity stress is presented in Table 8.1.

Table 8.1 Bacillus-mediated gene expression in plants that were identified to be responsible for salinity stress amelioration

Plant species	*Bacillus* species included in the study	Genes responding to stresses	Role/Results
Triticum aestivum	*Bacillus safensis* W10	Upregulation of multiple genes including sulfur-rich thionin, S adenosylmethionine decarboxylase precursor, expansins, endotransglucosylase/ hydrolase and metallothionines Downregulation of protein phosphatases, flavonones hydroxylases, oxalates, and oxidases	Mitigation of salt stress
Solanum lycopersicum	*Bacillus megaterium*	Expression of MT2 receptor and gamma response (GR1)	Synthesis of metallothionin and glutathione reductase enzyme
Glycine max	*Bacillus firmus* SW5	Upregulation of salt tolerance genes GmVSP, CHS, GmPHD2, GmbZIP62, GmOLPb, and GmWRKY54 Expression of antioxidant enzyme-encoding APX, Fe-SOD, CAT and POD	Production of antioxidant enzyme, salinity tolerance, flavonoid biosynthesis
Zea mays	*Bacillus amyloliquefaciens* SQR9	Upregulation of several genes including NHX1, NHX2, RBCS, RBCL, H + -PPase, and NHX3	Enhanced photosynthesis, Na^+ sequestration, and export
Puccinellia tenuiflora	*Bacillus subtilis (GB3)*	Upregulation of *PtSOS1* and *PtHKT1;5* Downregulation of NCED	Modulation of Na^+ homeostasis
Oryza sativa	*B. Amyloliquefaciens* SN13	Upregulation of EREBP, *NADP-Me2, SOSI, SERK1*, and *BADH* Suppression of *GIG* and *SAPK4*	Na^+/H^+ reverse porter system, ion homeostasis; abiotic stress response and oxidative decarboxylation of L-malate
Arabidopsis thaliana	*Bacillus subtilis* (GB03)	Expression of HKT1	Na^+ transport in roots

Adapted from Arora et al. (2020)

8.2.1.5 Impact of Heavy Metal Stress on Plants

Due to unplanned industrialization all over the world, arable lands are inadvertently getting contaminated with metal toxicants that are part of industrial effluents released. These contaminants are affecting ecology of food chain by altering microbial communities and crop cultivation (Hu et al. 2009; Ashraf et al. 2017). Accumulation of Cu, Mn, Zn, Pb, Cr, and other heavy metals are listed as major pollutants in soil and water that are not degraded into harmless substances easily (Ma et al. 2009; Arthur et al. 2012). These metal-contaminated soil and water are not only toxic to the flora and fauna in a certain area but also create a huge risk to human health if contaminated soils are used for crop production and metal is taken up and transferred into food chain at higher-than-acceptable concentrations (Oves et al. 2016). Heavy metal content can affect activities of microorganisms, such as respiration and metabolism (metabolic entropy response), thereby affecting soil respiration (Blagodatskaya et al. 2006). Thus, microbes such as *Bacillus* that can survive these harsh conditions are more suitable for heavy metal remediation. Traditionally, chelators have been used to reduce metal toxicity from soil; however, chelators can be harmful to organisms in the edaphic environment (Tandy et al. 2006). In contrast, beneficial microorganisms like *Bacillus* spp. solubilize and change toxic metals to nontoxic forms. This method can be utilized in management of heavy metal phytoremediation when used in the integrated approach with hyperaccumulator plants (Kang et al. 2015c, Bosecker 1997). However, this remediation method of heavy metals is also known as bioremediation when microbes are used solely, which is considered the most sustainable, environmentally friendly, and cost-effective without any adverse effect to any component of the environment (Dixit et al. 2015). The release of *Bacillus* spp. into soil contaminated with heavy metals can enhance reduction of toxic effects of these metals on plant growth when combined in a phytoremediation effort. Brunetti et al. (2012) found that *B. licheniformis* enhanced Cu, Zn, Cd, Pb, especially Cr accumulation in *Brassica* plants that were grown to test their capacity as hyperaccumulators of metals for phytoremediation of heavy metal-contaminated soil. This eventually led to reduced levels of toxic metals in soil compared to nontreated plants. However, due to the low bioconcentration factors (>1), investigators could not conclude the suitability of the species for the phytoextraction of toxic metals from polluted soils. However, these species can be utilized successfully for low metal polluted soil. Besides accumulation and uptake of heavy metals or supporting enhanced uptake by plants, *Bacillus* and similar microbes mitigate plant stress from heavy metal through other mechanisms described below.

8.2.1.6 *Bacillus*-Based Heavy Metal Stress Alleviation in Plants

Rhizospheric and endophytic bacteria including *Bacillus* enhance growth and development of plants in metal polluted soils by two major methods: i) These microbes can remove heavy metals from soil or modify capacity of metal accumulation through efflux of ions outside the cells, transformation of metal ions to lower forms of toxicity, sequestration of metal ions on the cell surface or in polymers inside cells, and biomethylation, precipitation, adsorption or desorption; ii) these microbes can also alleviate heavy metal-induced plant stress through production of beneficial plant growth enhancing substances that may include solubilization/transformation of mineral nutrients, such as phosphate, nitrogen, and potassium, and production of plant beneficial enzymes, siderophores, and phytohormones (Ma et al. 2011).

8.2.1.6.1 Adsorption and Absorption of Heavy Metal by *Bacillus*

Bacillus can remove large amount of soil heavy metals by both adsorption and absorption, although Wang et al. (2013, 2001) reported that the principal mechanism of heavy metal ion accumulation is adsorption, which normally is independent of energy metabolism. However, absorption, a closely related mechanism depends largely on energy metabolism and occurs mostly in living cells. Active export of heavy metals via an ATPase efflux P-type pump was found by Shin et al. (2012) in an endophytic bacterial strain *Bacillus* sp. MN3–4. *Bacillus* has evolved a well-defined metal-resistance mechanism, which is capable of transporting metal ions against the concentration gradient across cell membranes to enhance hyperaccumulator plants' capacity of removing heavy metals from soil. This process uses ATP hydrolysis-released energy. In this regard, Wang et al. (2001) found that *Bacillus* could saturate 60% of its Cu^{2+} adsorption capacity within the first minute and reach an equilibrium as early as within 10 min. However, Wierzba (2015) found that addition of ethylenediaminetetraacetic acid (EDTA) and lemon oil in the media could improve removal rates by 31.5 and 26.3%, respectively.

8.2.1.6.2 Bioleaching

The process of extracting metals from waste or ores by using microorganisms is defined as bioleaching as microbes oxidize the metals and produce soluble compounds in the form of organic acids. Thus, production of organic acids and chelating or complexing compounds that are excreted into the environment are used for metal extraction. Bacteria belonging to the genus *Thiobacilli* are considered the most active group in bioleaching of metal ions; they generally grow under aerobic conditions. It was shown that bacteria leach a higher amount of heavy metals from sludge deposited when elementary sulfur was added as a stimulant for the activity of thionic acidophilic *Bacilli*. It took place by way of bacterial oxidation of elementary sulfur releasing sulfuric acid (Marchenko et al. 2015). As a result, the pH of sludge deposits goes down, which is an important factor for effective leaching of heavy metal.

8.2.1.6.3 Other Mechanisms of *Bacillus*-Based Heavy Metal Remediation

Microbial cells are capable of converting metals from one oxidation state to another, which can reduce their toxicity. Microbial enzymes and other secretions from their metabolic activities can dissolve heavy metals that are stuck in soil particles. Thus, precipitation, biosorption, and enzymatic transformation are the processes used by *Bacillus* and similar microbes to degrade, detoxify, or transform heavy metals to more stable, less mobile, or inert forms (Kumar and Bharadvaja 2020). *B. subtilis* inoculation of rice under Cd stress showed that roots and shoots at 45 days after inoculation (DAI) and grains at 120 DAI had lower Cd accumulation compared with non-inoculated (Treesubsuntorn et al. 2017). Authors hypothesized that the mechanism by which *B. subtilis* reduced Cd accumulation was its capacity to effectively absorb Cd from the medium. Their findings also suggested that *B. subtilis* was more effective in absorbing Cd compared to *B. cereus*. Ahmad et al. (2014) showed that *Bacillus* and a few other bacterial genera such as *Klebsiella*, *Stenotrophomonas*, and *Serratia* had supported plant growth under Cd stress by increasing water uptake and reducing electrolyte leakage in maize and wheat. As *Bacillus* increases availability and uptake of essential nutrients, heavy metal uptake by plants is reduced due to competitive exclusion. For example, Naseem et al. (2016) found that *Bacillus* sp. AMP2-inoculated wheat seedlings took up less Cr of different chromium salts (CrCl3, K2CrO4, K2Cr2O7) both at 10 and 20 ppm concentrations than control. Similar to other *Bacillus*-based abiotic stress mitigation mechanisms, when *Bacillus*-inoculated plants were grown in high concentration of heavy metal containing medium, it was found that tissues increased antioxidant enzyme activity including CAT, PA, APX, and SOD and concurrent decrease of metal concentration in the roots compared to non-inoculated plants (Hao et al. 2015). Inoculation of *B. subtilis* could also increase dry weight of rice plant as well as protect them from Cd stress. Mechanisms included ability of *B. subtilis* producing IAA, solubilizing phosphate, and controlling ethylene levels by the activity of ACC deaminase (Treesubsuntorn et al. 2017).

8.2.1.7 Mechanism of Nutrient Stress Alleviation by *Bacillus*

Multiple species of *Bacillus* were found to improve plant uptake of P and K by solubilizing fixed soil nutrients (Chen et al. 2006). For example, use of insoluble K sources together with *Bacillus edaphicus* NBT strain to treat soil for growing cotton and rape could increase K content in plants by 30% and improve growth compared with no *Bacillus* inoculation (Sheng and He 2006). Plant growth promotion in this study was attributed to K solubilization by *B. edaphicus* strain. In a similar study, Wu et al. (2005) found improved N, P, and K assimilation in maize due to *B. megaterium* and *B. mucilaginous* inoculation. Several other investigators also reported similar mechanism of nutrient stress alleviation such as solubilization of tricalcium phosphate (Calvo et al. 2010; Almoneafy et al. 2012) and zinc (Ajilogba and Babalola 2013) by *B. amyloliquefaciens* isolated from potato. Phosphate solubilization abilities of multiple *Bacillus* species, *B. thuringiensis*, *B. sphaericus*, and *B. megaterium*, were found by Akgul and Mirik (2008).

8.3 Biotic Stress Mitigation in Plants by *Bacillus* spp.

8.3.1 Extent and Impact of Biotic Stress on Crop Growth and Productivity

Like abiotic stresses, plants encounter many different types of biotic agents such as fungi, viruses, bacteria, nematodes, and insects during their life cycle. These entities may exert biotic stresses when they invade plants to use them as their host resulting in disruption of plant's normal metabolism. This interrupts plant growth, and in some cases, biotic stresses become the cause of plant mortality (Hashem et al. 2019). Many of these biotic agents can also cause post-harvest losses of crop yield (Singla and Krattinger 2016). However, in most cases plants do well even under these stressful situations because some beneficial microbes, if present in the microenvironment, interact with their host plants symbiotically or synergistically to counteract harmful effects from pathogenic microbes. These beneficial microbes can play similar role as synthetic fertilizers or pesticides to minimize adverse effect from biotic stresses and promote plant growth. Due to their significant role in plant growth enhancement they are often termed as plant growth promoting rhizobacteria (PGPR), where *Bacillus* is considered a prominent member. PGPR, including *Bacillus,* have been used in agricultural, especially crop production, systems to alleviate biotic stresses and develop environmentally friendly, sustainable management options (Grover et al. 2011, Vejan et al. 2016) for a long time, which is now gaining momentum. Application of beneficial microbes to the rhizosphere in an augmentative approach can be utilized for improving biotic stress tolerance. A few *Bacillus* species including *B. megaterium*, *B. subtilis*, and *B. cereus* were found by GroBkinsky et al. (2016) to produce cytokinins and other growth hormones. Synergism is another known mechanism of effect of Bacilli when used with other beneficial organisms for plant growth and development. Alam et al. (2011) applied *B. subtilis* together with arbuscular mycorrhizal (AM) fungi to rose-scented geranium. Results indicated a synergistic effect of *B. subtilis* that increased >10% yield (59.5%) compared to AM fungi alone (49.4%). Total oil yield from the harvested biomass was increased significantly as biomass production went up, although oil content percentage did not increase on a dry weight basis. Underlying mechanisms of *B. subtilis* providing synergistic effect to greater promotion of plant growth included increased growth supporting enzyme production, higher antioxidants and P solubilization, root nodulation, nitrogen fixation, and biocontrol activity.

8.3.2 *Mechanism of Biotic stress Mitigation in Plants by* Bacilli

Members of the genus *Bacillus* alleviate biotic stress of plants through a variety of mechanisms that include depriving biotic agents by outcompeting them for nutrients and space on plant rhizosphere or phyllosphere; inhibition of biotic agents by producing a variety of inhibitory biochemicals; induction of plant resistance; and facilitating plant growth by producing hormones, so that plant pathogens fail to infect plants. These can further be divided into two groups: direct and indirect mechanisms. The direct mechanism refers to production of biochemicals by cells of *Bacillus* spp. such as synthesis of several secondary metabolites (antibiotics, cell wall-degrading enzymes, hormones, antioxidants) that stimulate plant growth and assist plants to defend against pathogenic attack (Goswami et al. 2016). Stimulation of plant growth and induction of acquired systemic resistance are among the major indirect mechanisms involved with *Bacillus*-based biotic stress mitigation in plants.

8.3.2.1 Depriving Biotic Agents by Outcompeting them for Nutrients and Space

Early colonization of plant roots by *Bacillus* can provide a barrier against pathogenic microorganism invasion (Bais et al. 2004). Root exudates provide required nutrients to rhizosphere microbes to thrive. That is why microbial density in proximity of roots is higher compared to areas away from them, and there is always a competition for space on the root surface. Due to the competitive advantage of *Bacilli* for nutrients and space in plant rhizosphere, they can suppress pathogen proliferation, which is an important feature of the mode of action of *Bacillus*-based biotic stress alleviation in plants (Cawoy et al. 2011). It was found that many members of *Bacilli* can form biofilms, a densely packed surface or multicellular interface of associated aggregates under unfavorable environmental factors (Morikawa 2006) occupying most of the root surface. Formation of a biofilm by *Bacillus* on plant root can be very quick, requiring only a few hours (Allard-Massicotte et al. 2016), thus preventing attachment of pathogenic microbes on roots (Morikawa 2006). In general, bacteria utilize chemotaxis for finding root to colonize at early stages of plant growth (Allard-Massicotte et al. 2016). Due to the presence of many chemoreceptors involved in root colonization, *Bacilli* can efficiently colonize root system in soil (Allard-Massicotte et al. 2016). In the case of lack of availability of nutrients to *Bacilli*, major competition occurs for carbon, and is considered an important factor in biological control. *Bacillus* also competes for several micronutrients with other biotic disease-causing biotic agents that include copper, manganese, iron, and zinc. Among these elements, iron is highly important because of its limited presence in available form in soil (Loper and Henkels 1997). Due to their ability of producing siderophores that have a strong affinity for iron, *Bacillus* can solubilize and acquire ferric ions efficiently. Thus, depriving other soil microbes, including pathogens, from acquiring iron (Cawoy et al. 2011; Haas and Defago 2005; Loper and Henkels 1997).

8.3.2.2 Production of Inhibitory Biochemicals by *Bacillus*

Production of compounds of low-molecular-weight including antibiotics is common to many *Bacillus* species (B. subtilis, *B. megaterium*, *B. amyloliquefaciens*, *B. cereus*, *B. licheniformis*, *B. mycoides*, and *B. pumilus*) that directly affect other microorganisms through a process known as antibiosis (Weller and Thomashow 1993; Handelsman and Stabb 1996; Weller 1988). Some of these species can produce multiple antibiotics. For example, *B. subtilis* can produce more than two dozens of antimicrobial compounds with diverse structures (Stein 2005). These include polymyxin, subtilin, difficidin, and mycobacillin usually possessing broad-spectrum antimicrobial activity. *Bacilli* can produce antibiotic molecules during most of their life cycle, but production reaches very high levels during sporulation. Most of these compounds belong to the peptide class despite very high diversity in their sizes and structures. These antibiotics mostly contain amino acids and, in some cases, other residues that help them forming linear, cyclic or basic aminoglycoside type antibiotics (Stein 2005). Almost all the antimicrobial cyclic peptides can directly affect the integrity of fungal cell membranes through lysis and change their structure, thereby inhibiting their growth and development. *Bacillus*-produced antimicrobial peptides can reliably be used as fungicides due to their direct interaction with fungal cell membrane or interference in biosynthesis of chitin, glucan, and sphingolipid that are essential components of fungal cell wall and membrane. In this regard, positive results have been obtained by investigators from numerous studies. For example, two *B. brevis*- and *B. polymyxa*-produced peptide antibiotics gramicidin S and polymyxin B, respectively, inhibited gray mold-causing fungus *Botrytis cinerea* both in vitro and in vivo (Haggag 2008). The three main *Bacillus* spp.-produced families of cyclic lipopeptides (CLPs) are surfactins, iturins, and fengycins. It was reported that all three may although be present in bacterial secretion/culture filtrate but may not be needed for antimicrobial activity. In a study by Waewthongrak et al. (2015) on the inhibitory effect of all three CLPs, they found that growth of *Penicillium digitatum* was inhibited by iturin A and fengycin to control green mold on mandarin, but surfactins had no direct effect. However, Surfactin likely supports bacterial colonization of root tissues and establishment in the rhizosphere of plants through biofilm layer formation (Mihalache et al. 2017; Cawoy et al. 2014), and is considered essential component of the mechanisms by which *Bacilli* become successful in alleviating biotic stresses in plants. Investigators from different studies with different strains of *Bacilli* found diverse peptide antibiotics such as bacillomycin D, subtilisin, and xanthobaccin. However, due to the consistency and frequency of its occurrence, bacilysin is regarded taxonomically the most related and significant for the genus *Bacillus* (Loeffler et al. 1990). In addition, chitinase, amylase, protease, cellulase, pectinase, glucanase, and similar cell wall-degrading enzymes or substances like HCN can damage pathogens and pests to lower their population on plants.

8.3.2.2.1 Specific Mechanism Associated with Activity of Cyclic Peptides

Some cyclic peptide molecules disrupt membrane structure by binding only to the surface without passing through. However, other cyclic peptides can bind with specific membrane-associated structures such as ion channels, transporters and different types of receptors by traversing membranes. After successful binding, cyclic peptide molecules aggregate in a specific site on membrane to form variable size aqueous pores. Ions and other solutes pass through the channels of these pores in an uncontrolled way that ultimately leads to cell death. Debono and Gordee (1994) found that impairment of biosynthesis microbial cell wall macromolecular components such as glucan, chitin, and mannoproteins was the underlying mechanism of antifungal activity of several cyclic peptides. These cyclic peptides have the capacity to form a complex with the precursor molecule of microbial cell wall macromolecules that eventually promote pore formation and cell wall disruption. While pore formation on membrane or degradation of cell walls is the major mode of action, alternation of ribosome function is also considered a potential mechanism to suppress growth of pathogenic microbes. For example, Katz and Demain (1977) found that bacitracin A produced by *B. licheniformis* inhibited cell wall synthesis but the butirosin complex belonging to amino glycosides produced by *B. circulans* altered ribosome function of diverse microbial groups (Defuria and Claridge 1976).

8.3.2.3 Induction of Host Defense against Biotic Stress

In addition, with above-mentioned mechanisms, some strains of *Bacilli* activate defense systems in host plants that results in an enhanced level of resistance against pathogen attack (Conrath et al. 2006). This can be explained as supporting plant's immune defense arsenal through sensitization and priming to defeat invading pathogens. This process initiated by biotic stimuli helps in scaling up expression of a plant's defense-related genes to accumulate antifungal biochemicals. Strains of *Bacilli* or their metabolites can turn on plant's defense system when pathogens attack host plants or can be triggered by pre-inoculation (Schonbeck et al. 1993) that results in an enhanced resistance level (Conrath et al. 2006). Higher level of induced resistance in plants is superior over other modes of action of *Bacilli* such as antibiosis or competition, as it provides protection to plants usually for a long period of time even when bacterial population subsides. This happens as the response has already transduced to the distal organ of the plant from the point of instigation. Induction of systemic resistance occurs through a well-orchestrated sequence of biological events. Beneficial rhizobacteria such as *Bacilli* can trigger a defense response by stimulating the plant through activation of a variety of cellular defense responses in a well-coordinated manner. These responses result in oxidative burst, defense-related enzymes buildup (Rahman et al. 2015), cell-wall strengthening (Heil and Bostock 2002), and production of secondary metabolites (Yedidia et al. 2003). The presence of the microbe is detected by pattern recognition receptors (PRRs) of a plant cell membrane through microbe-associated molecular patterns (MAMPs), which may include lipopolysaccharides, flagellin, glycoproteins, and chitin (Jones and Dangl 2006). These are also known as elicitors. *Bacillus* spp. are also known to produce lipopeptides and volatile compounds, which can play similar role as elicitors for inducing systemic resistance in plants. Surfactin produced by *B. subtilis* strain S499 was found to induce systemic resistance (Ongena et al. 2007). Complex interaction between PRRs and MAMPs (elicitors) subsequently results in immunity (PTI) through defense-related gene expression, oxidative burst, and callose deposition (Schwessinger and Ronald 2012; Altenbach and Robatzek 2007) from ISR-type defense signaling. Different *Bacillus* spp. may induce different but relevant signaling pathways. For example, ISR mediated by *Bacillus cereus* AR156 required both SA and JA/ET signaling pathways together with NPR1 (Niu et al. 2011). Defense response against pathogens involve several molecules such as phytoalexins, pathogenesis-related proteins (proteinase inhibitors, chitinases, β-1,3-glucanases), and lignin (Van Loon 2007). Fungal hyphal growth during infection process is prevented by thickened cell wall in combination with PR proteins (Lugtenberg et al. 2001).

8.3.3 Mechanism of Bacterial Disease Prevention by Bacillus *Spp.*

Together with fungi, viruses, and nematodes, plant disease-causing pathogenic bacteria pose major challenges to plant health and yield in agricultural production systems (Hussey and McGuire 1987; Guo et al. 2013; Narasimhan and Shivakumar 2015). Major pathogenic bacteria that are known to infect plant with detrimental effect on plant growth and development include but not limited to *Pseudomonas, Xanthomonas, Erwinia,* and *Ralstonia*. Although *Bacillus* spp. belong to microbial group bacteria, upon inoculation of plant or plant growing media they not only counteract and suppress pathogen growth but also promote plant growth (Krid et al. 2012; Yi et al. 2013). Mechanisms of *Bacillus*-based protection of plant from bacterial infections include biofilm formation around the root surface and their secretion of several toxins such as surfactin, iturin, macrolactin, bacillomycin, and fengycin that destroy pathogenic populations of bacteria resulting in plant disease control (Hinarejos et al. 2016; Chen et al. 2013; Huang et al. 2014; Elshakh et al. 2016). Pathogenic bacterial cell walls can be degraded quickly by the secretions of *Bacillus* spp., eventually killing the pathogen (Elshakh et al. 2016).

8.3.4 Mechanism of Fungal Disease Control

Mycelial growth of many different fungi is inhibited by the antagonistic activity of *Bacillus* spp. to control fungal diseases (Handelsmann and Stabb 1996; Aydi-Ben-Abdallah et al. 2016; Abdalla 2015; Chowdhury et al. 2015a; Akram et al. 2016;), thereby enhancing plant growth and yield (Narasimhan and Shivakumar 2015).

Underlying mechanism of *Bacillus* spp.-mediated fungal disease control include either or all of the events and interactions occur when *Bacillus* come in contact with fungal pathogens. Immediately after attachment of *Bacillus* to the mycelial cell walls, production of fungal cell wall-degrading enzymes (chitosanase, protease, glucanase, cellulase siderophores) and HCN takes place from the bacteria, which crack and deform the hyphae leading to altered cell functions and structures such as protoplast leakage and vacuolation (Ben-Khedher et al. 2015; Han et al. 2015). Mitigation of pathogen-induced biotic stress may also occur via *Bacillus* spp.-mediated physiological changes in plants. These include alteration of respiratory and photosynthetic pathways in diseased plants together with regulation of phenyl-propanoid, carbohydrates, defense-related proteins, and nitrogen metabolism (Jain et al. 2015). Plant beneficial *Bacillus* spp. increase antioxidant enzymes (APX, CAT, GR, GPX, POD, PPO) and reduce lipid peroxidation in plants. They also enhance production of other defense enzymes such as PAL, β-1,3-glucanase, chitinase, and phenolic acids that lessen the hostile effects of plant infection by pathogens (Chowdappa et al. 2013; Jain et al. 2013; Solanki et al. 2012).

8.3.5 Mechanism of Nematode and Virus Disease Control by **Bacillus**

Viruses, the second largest group of plant pathogens cause most plant diseases after fungi. Many *Bacillus* spp., produce antiviral compounds that minimize the adverse effects of these pathogens on plants (Esawy et al. 2011). Prevention of viral diseases by *Bacillus* spp. has been reported in a few cases. However, in most cases it was due to reduction of disease rate as a consequence of *Bacillus* spp. induced systemic resistance (ISR) in plants. Zhang et al. (2004) reported enhanced plant growth of cucumber during cucumber mosaic virus infection when plants were inoculated with *Bacillus* compared to non-*Bacillus* treated. Formation of biofilm and surfactin production by *B. amyloliquefaciens plantarum* was found to subvert the viral disease in plants by Chowdhury et al. (2015b). Underlying mechanism in this case was due to triggering of ISR machinery. Similarly, viral disease caused by tobacco mosaic virus was suppressed by *Bacillus* spp. because of inhibition of viral coat protein synthesis from induced systemic resistance. Additional mechanisms involved were increased expression of plant defense genes (PR-1a and PR-1b), disease-resistant signaling genes (*Coil* and *NPR1*), and cell wall expansion (*NtEXP2* and *NtEXP6*) genes (Wang 2009).

Nematodes, which are known as microscopic worms, can also damage plants by being parasitic and feeding on roots. Among many plant parasitic nematodes, root-knot causing ones are most damaging worldwide. Nearly 5500 plant species are within the host range of this nematode (Trudgill and Blok 2001). Prevention of root-knot nematode infection in crops by *Bacillus* spp. includes resistance development and reduction of gall and egg masses (Adam et al. 2014). Chowdhury et al. (2015b) reported that *Bacillus* spp. synthesized antimicrobial peptides such as bacteriocins that inhibit pathogenic nematode growth. Genes encoding nematicidal activity were identified by Liu et al. (2013) as PZN gene cluster in *B. amyloliquefaciens*. *Bacillus* spp. secreted crystal proteins (Cry5B and Cry6A) were found to control the growth of plant-parasitic and free-living nematodes *Meloidogyne hapla* and *Caenorhabditis elegans*, respectively (Yu et al. 2015).

8.4 Mechanism of Insect Stress Alleviation in Plants by *Bacillus*

A broad range of insect control in plants is provided by *Bacillus thuringiensis* (Bt)-produced Bt toxin (Navon 2000). In addition, Bt was also found to inhibit the growth of insect larvae, thereby decreasing plant damage and increasing growth (Boukedi et al. 2016; Arrizubieta et al. 2014). Upon sporulation, *B. thuringiensis* forms crystals of proteinaceous insecticidal δ-endotoxins also known as crystal proteins or Cry proteins (Roh et al. 2007). In most strains of *Bacillus*, genes located on a plasmid instead of a chromosome that are known as cry genes encode these toxins. Cry toxins show precise activities against insect species belonging to several orders such as Coleoptera (beetles), Diptera (flies and mosquitoes), Lepidoptera (moths and butterflies), and Hymenoptera (wasps, bees, ants and sawflies) (Schnepf et al. 1998). Upon ingestion of toxin crystals by insects, insoluble crystals get denatured in their alkaline digestive tracts. Proteases from the insect gut then cut these soluble crystals to liberate the toxin. Inside the insect gut, the Cry toxin at this stage is injected into the cell membranes, which paralyzes the digestive tract including formation of pores. The insect starves to death as it stops eating. Live Bt bacteria from the environment may also colonize the insect causing death. Additional findings on the mechanism suggested that bacteria in the midgut of susceptible larvae stimulate insecticidal activity of *B. thuringiensis* (Broderick et al. 2006). Some relevant studies suggested that due to its insecticidal and plant growth promotion properties, *B. thuringiensis* could be used as a biological control agent (Compant et al. 2005).

8.5 Conclusion and Future Perspectives

Field grown plants are constantly exposed, either sequentially or simultaneously, to many abiotic or biotic stresses. These biotic and abiotic stresses significantly affect crop yield, food quality and ultimately global food security. Plants must cope with these stressful conditions to thrive and complete their life cycle. Among many beneficial microbes, *Bacillus* species are exceptional, as members of this group can form endospores that are extremely robust under harsh environmental conditions, suppress harmful microbes, and can also secrete secondary metabolites that stimulate plant growth. Thus, beneficial microbes such as *Bacillus* provide a model for enhancing stress tolerance upon successful application in plant growing environment. *Bacillus* spp. were found in numerous studies to alleviate biotic and abiotic stresses and improve plant growth either directly or indirectly. Both biotic and abiotic stresses can cause physiological, biochemical, and molecular changes in plants affecting normal growth and development. *Bacillus* spp. either prevent or counteract these negative changes through a series of biological and biochemical mechanisms that have been interpreted from an appreciable number of studies. *Bacilli*-mediated abiotic and biotic stress tolerance in plants includes biological, physiological, biochemical, metabolic, and molecular mechanisms triggered in response to stresses. Promotion of plant growth by these underlying mechanisms usually involve regulation of plant hormones, improved nutrition acquisition, siderophore production, and enhanced antioxidant activity. Secretion of exopolysaccharides and siderophores by *Bacillus* spp. inhibit or stop movement of toxic ions and assist in uptaking water by roots as well as maintaining ionic balance. These compounds were also found to inhibit pathogenic microbial populations through multiple mechanisms that have been unraveled by numerous studies. However, many of the possible mechanisms either remain unclear or not compiled in a systematic method to make them available to researchers of *Bacillus* spp. This review compiled most of the processes unraveled by various studies to date and provides a reliable source of information for designing relevant research plan to further explore the mechanisms that have not been fully elucidated to the molecular level. As new and more effective strains of *Bacilli* to counteract plant abiotic and biotic stresses are discovered on a regular basis, future studies should focus on discerning mechanisms associated with enhanced efficacy.

References

Abdalla MY (2015) Biological control and induction of systemic resistance against cucumber Fusarium wilt by plant growth promoting rhizo-organisms. Egypt J Biol Pest Cont 25:407–413

Adam M, Heuer H, Hallmann J (2014) Bacterial antagonists of fungal pathogens also controls root-knot nematodes by induced systemic resistance of tomato plants. PLoS One 9:e90402. https://doi.org/10.1371/journal.pone.0090402

Ahmad I, Akhtar MJ, Zahir ZA, Naveed M, Mitter B, Sessitsch A (2014) Cadmium-tolerant bacteria induce metal stress tolerance in cereals. Environ Sci Pollut Res 21:11054–11065. https://doi.org/10.1007/s11356-014-3010-9

Ahmad P, Hashem A, Abd-Allah EF, Alqarawi AA, John R, Egamberdieva D (2015) Role of Trichoderma harzianum in mitigating NaCl stress in Indian mustard (*Brassica juncea* L) through antioxidative defense system. Front Plant Sci 6:868

Ajilogba CF, Babalola OO (2013) Integrated management strategies for tomato Fusarium wilt. Biocontrol Sci 18:117–127

Akgul DS, Mirik M (2008) Biocontrol of *Phytophthora capsici* on pepper plants by *Bacillus megaterium* strains. J Plant Pathol 90:29–34

Akram W, Anjum T, Ali B (2016) Phenylacetic acid is ISR determinant produced by *Bacillus fortis* IAGS162, which involves extensive re-modulation in metabolomics of tomato to protect against Fusarium wilt. Front Plant Sci 7:498. https://doi.org/10.3389/fpls.2016.00498

Alam M, Khaliq A, Sattar A, Shukla RS, Anwar M, Dharni S (2011) Synergistic effect of *arbuscular mycorrhizal* fungi and *Bacillus subtilis* on the biomass and essential oil yield of rose-scented geranium (*Pelargonium graveolens*). Arch Agron Soil Sci 57(8):889–898

Almoneafy AA, Xie GL, Tian WX, Xu LH, Zhang GQ, Ibrahim M (2012) Characterization and evaluation ofB acillus isolates for their potential plant growth and biocontrol activities against tomato bacterial wilt. African Journal of Biotechnology 11:7193-7201

Al-Karaki GN (2006) Nursery inoculation of tomato with *arbuscular mycorrhizal* fungi and subsequent performance under irrigation with saline water. Sci Hortic 109:1–7. https://doi.org/10.1016/j.scienta.2006.02.019

Allard-Massicotte R, Tessier L, Lecuyer F, Lakshmanan V, Lucier J, Garneau D, Caudwell L, Vlamakis H, Bais HP, Beauregard PB (2016) *Bacillus subtilis* early colonization of *Arabidopsis thaliana* roots involve multiple chemotaxis receptors. mBio 7: e01664-16. https://doi.org/10.1128/mBio.01664-16

Almoguera C, Coca MA, Jordano J (1995) Differential accumulation of sunflower tetraubiquitin mRNAs during zygotic embryogenesis and developmental regulation of their heat-shock response Plant Physiol 107:765–773

Altenbach D, Robatzek S (2007) Pattern recognition receptors: from the cell surface to intracellular dynamics. Mol Plant-Microbe Interact 20:1031–1039

Anjum NA Gill SS, Gill R (2014) Plant adaptation to environmental change: significance of amino acids and their derivatives, 1st edn. CABI, Wallingford. https://doi.org/10.1079/9781780642734.0000

Arrizubieta M, Williams T, Caballero P, Simón O (2014) Selection of a nucleopolyhedrovirus isolate fromH elicoverpa armigera as the basis for a biological insecticide. Pest Manag Sci 70:967–976

Araus JL, Slaffer GA, Reynolds SMP, Royo C (2002) Plant breeding and drought in C3 cereals: what should we breed for? Ann Bot 89:925–940. https://doi.org/10.1093/aob/mcf049

Arkhipova TN, Prinsen E, Veselov SU, Martynenko EV, Melentiev AI, Kudoyarova GR (2007) Cytokinin producing bacteria enhances plant growth in drying soil. Plant Soil 292:305–315

Arora NK, FatimamT MJ, Mishra I, Verma S, Verma R, Verma M, AnkitaBhattacharya A, Verma P, Mishra P, Bharti C (2020) Halo-tolerant plant growth promoting rhizobacteria for improving productivity and remediation of saline soils. J Adv Res 26:69–82

Arthur E, Moldrup P, Holmstrup M, Schjonning P, Winding A, Mayer P et al (2012) Soil microbial and physical properties and their relations along a steep copper gradient. Agricult Ecosys Environ 159:9–18. https://doi.org/10.1016/j.agee.2012.06.021

Ashraf M, Foolad MR (2007) Roles of Glycine Betaine and Proline in Improving Plant Abiotic Stress Resistance. Environmental and Experimental Botany 59:206-216 https://doi.org/10.1016/j.envexpbot.2005.12.006

Ashraf MA, Hussain I, Rasheed R, Iqbal M, Riaz M, Arif MS (2017) Advances in microbe-assisted reclamation of heavy metal contaminated soils over the last decade: a review. J Environ Manag 198:132–143. https://doi.org/10.1016/j.jenvman.2017.04.060

Atkinson NJ, Urwin PE (2012) The interaction of plant biotic and abiotic stresses: from genes to the field. J Exp Bot 63:3523–3543

Aydi-Ben-Abdallah R, Jabnoun-Khiareddine H, Nefzi A, Mokni-Tlili S, Daami-Remadi M (2016) Biocontrol of Fusarium wilt and growth promotion of tomato plants using endophytic bacteria isolated from *Solanum elaeagnifolium* stems. J Phytopathol 164:811–824. https://doi.org/10.1111/jph.12501

Bais HP, Fall R, Vivanco JM (2004) Biocontrol of *Bacillus subtilis* against infection of arabidopsis roots by *Pseudomonas syringae* is facilitated by biofilm formation and surfactin production. Plant Physiol 134:307–319

Barnawal D, Maji D, Bharti N, Chanotiya CS, Kalra A (2013) ACC deaminase-containing Bacillus subtilis reduces stress ethylene-induced damage and improves mycorrhizal colonization and rhizobial nodulation in Trigonella foenum-graecum under drought stress. J Plant Growth Regul 32: 809–822

Barnawal D, Bharti N, Pandey SS, Pandey A, Chanotiya CS, Kalra A (2017) Plant growth-promoting rhizobacteria enhance wheat salt and drought stress tolerance by altering endogenous phytohormone levels and TaCTR1/TaDREB2 expression. Physiol Plant 161(4):502–514. https://doi.org/10.1111/ppl.12614

Benedetto NA, Corbo MR, Campaniello D, Cataldi MP, Bevilacqua A, Sinigaglia M, Flagella Z (2017) The role of plant growth promoting bacteria in improving nitrogen use efficiency for sustainable crop production: a focus on wheat. AIMS Microbiol 3(3):413–434

Ben-Khedher S, Boukedi H, Kilani-Feki O, Chaib I, Laarif A, Abdelkefi-Mesrati L et al (2015) *Bacillus amyloliquefaciens* AG1 biosurfactant: putative receptor diversity and histopathological effects on Tuta absoluta midgut. J Invertebr Pathol 132 42–47. https://doi.org/10.1016/j.jip.2015.08.010

Berg G (2009) Plant-microbe interactions promoting plant growth and health: perspectives for controlled use of microorganisms in agriculture. Appl Microbiol Biotechnol (1):84, 11–88

Blagodatskaya EV, Pampura TV, Myakshina TN, Dem Yanova EG (2006) The influence of lead on the respiration and biomass of microorganisms in gray forest soil in a long-term field experiment. Eurasian Soil Sci 39:498–506

Bochow H, El-Sayed SF, Junge H, Stavropoulou A, Schmiedeknecht G (2001) Use of *Bacillus subtilis* as biocontrol agent IV salt-stress tolerance induction by *Bacillus subtilis* FZB24 seed treatment in tropical vegetable field crops, and its mode of action. Z Pflanzenkrankh. Pflanzenschutz 108(1):21–30

Bosecker K (1997) Bioleaching: metal solubilization by microorganisms. FEMS Microbiol Rev 20:591–604. https://doi.org/10.1111/j.1574-6976.1997.tb00340.x

Boukedi H, Sellami S, Ktari S, Belguith-Ben-Hassan N, Sellami-Boudawara T, Tounsi S et al (2016) Isolation and characterization of a new *Bacillus thuringiensis* strain with a promising toxicity against Lepidopteran pests. Microbiol Res 186–187:9–15. https://doi.org/10.1016/j.micres.2016.02.004

Bray EA, Bailey-Serres J, Weretilnyk E (2000) Responses to abiotic stresses. In: Gruissem W, Buchannan B, Jones R (eds) Biochemistry and molecular biology of plants. American Society of Plant Physiologists, Maryland, pp 1158–1203

Broderick NA, Raffa KF, Handelsman J (2006) Midgut bacteria required for *Bacillus thuringiensis* insecticidal activity. Proc Natl Acad Sci U S A 103(41):15196–15199

Brunetti G, Farrag K, Soler-Rovira P, Ferrara M, Nigro F, Senesi N (2012) The effect of compost and *Bacillus licheniformis* on the phytoextraction of Cr, Cu, Pb and Zn by three brassicaceae species from contaminated soils in the Apulia region, Southern Italy. Geoderma 170:322–330. https://doi.org/10.1016/j.geoderma.2011.11.029

Calvo P, Ormeno-Orrillo E, Martinez-Romero E, Zuniga D (2010) Characterization of Bacillus isolates of potato rhizosphere from andean soils of Peru and their potential PGPR characteristics. Braz J Microbiol 41:899–906

Cawoy H, Debois D, Franzil L, Pauw ED, Thonart P, Ongena M (2014) Lipopeptides as main ingredients for inhibition of fungal phytopathogens by *Bacillus subtilis/amyloliquefaciens*. Microb Biotechnol 8:281–295

Cawoy H, Bettiol W, Fickers P, Ongena M (2011) Bacillus-based biological control of plant diseases. In pesticides in the modern world – pesticides use and management. Available: https://www.intechopen.com/books/pesticides-in-the-modern-world-pesticides-use-and-management/bacillus-based-biological-control-of-plant-diseases

Chaves MM, Maroco JP, Pereira JS (2003) Understanding plant responses to drought-from genes to the whole plant. Funct Plant Biol 30:239–264. https://doi.org/10.1071/FP02076

Chawla S, Jain S, Jain V (2013) Salinity induced oxidative stress and antioxidant system in salt-tolerant and salt-sensitive cultivars of rice (Oryza sativa L.). J Plant Biochem Biotechnol 22:27–34

Chen YP, Rekha PD, Arun AB, Shen FT, Lai WA, Vand Young CC (2006) Phosphate solubilizing bacteria from subtropical soil and their tricalcium phosphate solubilizing abilities. Appl Soil Ecol 34:33–41

Chen Y, Yan F, Chai Y, Liu H, Kolter R, Losick R et al (2013) Biocontrol of tomato wilt disease by *Bacillus subtilis* isolates from natural environments depends on conserved genes mediating biofilm formation. Environ Microbiol 15:848–864. https://doi.org/10.1111/j.1462-2920.2012.02860.x

Chodak M et al (2015) Soil chemical properties affect the reaction of forest soil bacteria to drought and rewetting stress. Annal Microbiol 65:1627–1637

Choudhary DK, Johri BN (2009) Interactions of Bacillus spp. and plants – with special reference to induced systemic resistance (ISR). Microbiological Research 164:493–513. https://doi.org/10.1016/j.micres.2008.08.007

Chowdappa P, Mohan-Kumar SP, Jyothi-Lakshmi M, Upreti KK (2013) Growth stimulation and induction of systemic resistance in tomato against early and late blight by *Bacillus subtilis* OTPB1 or *Trichoderma harzianum* OTPB3. Biol Control 65:109–117. https://doi.org/10.1016/j.biocontrol.2012.11.009

Chowdhury SP, Hartmann A, Gao X, Borriss R (2015b) Biocontrol mechanism by root-associated *Bacillus amyloliquefaciens* FZB42-a review. Front Microbiol 6:780. https://doi.org/10.3389/fmicb.2015.00780

Chowdhury SP, Uhl J, Grosch R, Alqueres S, Pittroff S, Dietel K et al (2015a) Cyclic lipopeptides of *Bacillus amyloliquefaciens* subsp. plantarum colonizing the lettuce rhizosphere enhance plant defence responses towards the bottom rot pathogen *Rhizoctonia solani*. Mol Plant Microb Interact 28:17–18. https://doi.org/10.1094/MPMI-03-15-0066-R

Compant S, Duffy B, Nowak J, Clement C, Barka EA (2005) Use of plant growth-promoting bacteria for biocontrol of plant diseases: principles, mechanisms of action, and prospects. Appl Environ Microbiol 71(9):4951–4959

Conrath U, Beckers GJM, Flors V, Garcia-Agustin P, Jakab G, Mauch F, Newman MA, CMJ P, Poinssot B, Pozo MJ, Pugin A, Schaffrath U, Ton J, Wendehenne D, Zimmerli L, Mauch-Mani B, Prime A (2006) Priming: getting ready for battle. Mol Plant-Microbe Interact 19:1062–1071

Crane TA, Roncoli C, Hoogenboom G (2011) Adaptation to climate change and climate variability: the importance of understanding agriculture as performance. NJAS Wag J Life Sci 57:179–185

Debono M, Gordee RS (1994) Antibiotics that inhibit fungal cell wall development. Annu Rev Microbiol 48:471–497

Defuria, MD, Claridge CA (1976) Aminoglycoside antibiotics produced by the genus Bacillus. In: Schlessinger M (ed) Microbiology. American Society of Microbiology, Washington, D.C, pp 421–436

Dihazi A, Jaiti F, Jaoua S, Driouich A, Baaziz M, Daayf F, Serghini MA (2012) Use of two bacteria for biological control of bayoud disease caused by *Fusarium oxysporum* in date palm (*Phoenix dactylifera* L) seedlings. Plant Physiol Biochem 55:7–15

Dimkpa C, Weinand T, Asch F (2009) Plant-rhizobacteria interactions alleviate abiotic stress conditions. Plant Cell Environ 32:1682–1694

Dixit R, Wasiullah Malaviya D, Pandiyan K, Singh UB, Sahu A, Shukla R, Singh BP, Rai JP, Sharma PK, Lade H, Paul D (2015) Bioremediation of heavy metals from soil and aquatic environment: an overview of principles and criteria of fundamental processes. Sustainability (Switzerland) 7(2):2189–2212. https://doi.org/10.3390/su7022189

Egamberdieva D, Wirth SJ, Shurigin VV, Hashem A, Abd Allah EF (2017) Endophytic bacteria improve plant growth, symbiotic performance of chickpea (*Cicer arietinum* L.) and induce suppression of root rot caused by *Fusarium solani* under salt stress. Front Microbiol 8:1887. https://doi.org/10.3389/fmicb.2017.01887

Elshakh ASA, Anjum SI, Qiu W, Almoneafy AA, Li W, Yang Z et al (2016) Controlling and defence-related mechanisms of *bacillus* strains against bacterial leaf blight of rice. J Phytopathol 164:534–546. https://doi.org/10.1111/jph.12479

ENS (2006) Environment news service: report list world's 10 worst pollution spots. Available at: http://www.google.pt/url?sa=t&rct=j&q=&esrc=s&source=web&cd=2&cad=rja&uact=8&ved=0CCgQFjAB&url=http%3A%2F%2Fwww.chem.unep.ch%2Fpb_and_cd%2FDocuments%2FNews%2F10%2520most%2520polluted%2520cities.pdf&ei=1zZgVb2iOKKR7AaMoIAo&usg=AFQjCNHOAyar_69TBPA143KXTt-cAvkP3g&sig2=_lTZpa2YQnsF0slt0Xm7ew&bvm=bv.93990622,d.ZGU

Esawy MA, Ahmed EF, Helmy WA, Mansour NM, El-Senousy WM, El-Safty MM (2011) Production of levansucrase from novel honey *Bacillus subtilis* isolates capable of producing antiviral levans. Carbohydr Polym 86:823–830. https://doi.org/10.1016/j.carbpol.2011.05.035

FAO (2016) Available online: http://www.fao.org/3/a-i6030e.pdf (2016)

Gagne-Bourque F, Bertrand A, Claessens A, Aliferis KA, Jabaji S (2016) Alleviation of drought stress and metabolic changes in timothy (*Phleum pratense* L.) colonized with *Bacillus subtilis* B26. Front Plant Sci 7:584. https://doi.org/10.3389/fpls.2016.00584

Gill SS, Tuteja N (2010) Reactive oxygen species and antioxidant machinery in abiotic stress tolerance in crop plants. Plant Physiol Biochem 48:909–930. https://doi.org/10.1016/j.plaphy.2010.08.016

Glick BR (1995) The enhancement of plant growth by free-living bacteria. Can J Microbiol 41:109–117. https://doi.org/10.1139/m95-015

Goswami D, Dhandhukia P, Patel P, Thakker JN (2014) Screening of PGPR from saline desert of Kutch: growth promotion in *Arachis hypogea* by *Bacillus icheniformis* A2. Microbiol Res 169:66–75

Goswami D, Thakker JN, Dhandhukia PC, Tejada MM (2016) Portraying mechanics of plant growth promoting rhizobacteria (PGPR): a review. Cogent Food Agric 2:1127500

GroBkinsky DK, Tafner R, Moreno MV, Stenglein SA, De Salamone IEG, Nelson LM, Roitsch T (2016) Cytokinin production by *Pseudomonas fluorescens* G20–18 determines biocontrol activity against *Pseudomonas syringae*. Arabidopsis Sci Rep 6:23310

Grover M, Ali SZ, Sandhya V, Rasul A, Venkateswarlu B (2011) Role of microorganisms in adaptation of agriculture crops to abiotic stresses. World J Microbiol Biotechnol 27(5):1231–1240

Guo L, Rasool A, Li C (2013) Antifungal substrates of bacterial origin and plant disease management in bacteria. In: Maheshwari DK (ed) Agrobiology disease management. Springer, Heidelberg, pp 473–485

Haas D, Defago G (2005) Biological control of soil-borne pathogens by fluorescent pseudomonads. Nat Rev Microbiol 3:307–319

Haggag WM (2008) Isolation of bioactive antibiotic peptides from *Bacillus brevis* and *Bacillus polymyxa* against Botrytis grey mould in strawberry. Biocontrol Sci Tech 41:477–491

Han Y, Zhang B, Shen Q, You C, Yu Y, Li P et al (2015) Purification and identification of two antifungal cyclic peptides produced by *bacillus amyloliquefaciens* L-H15. Appl Biochem Biotechnol 176:2202–2212. https://doi.org/10.1007/s12010-015-1708-x

Handelsmann J, Stabb EV (1996) Biocontrol of soilborne pathogens. Plant Cell 8:1855–1869

Hao Y, Wu H, Liu Y, Hu Q (2015) Mitigative effect of Bacillus subtilis QM3 on root morphology and resistance enzyme activity of wheat root under lead stress. Adv Microbiol 5:469–478

Hare PD, Cress WA, van Staden J (1998) Dissecting the roles of osmolyte accumulation during stress. Plant, Cell and Environment 21:535–553

Hashem A, Abd-Allah EF, Alqarawi AA, Al-Huqail AA, Alshalawi SRM, Wirth S et al (2015) Impact of plant growth promoting *Bacillus subtilis* on growth and physiological parameters of *Bassia indica* (Indian bassia) grown under salt stress. Pak J Bot 47:1735–1741

Hashem A, Abd Allah EF, Alqarawi AA, Al-Huqail AA, Wirth S, Egamberdieva D (2016a) The interaction between arbuscular mycorrhizal fungi and endophytic bacteria enhances plant growth of *Acacia gerrardii* under salt stress. Front Microbiol 7:1089. https://doi.org/10.3389/fmicb.2016.01089

Hashem A, Abd Allah EF, Alqarawi AA, Al-Huqail AA, Shah MA (2016b) Induction of osmoregulation and modulation of salt stress in *Acacia gerrardii* benth. By arbuscular mycorrhizal fungi and *Bacillus subtilis* (BERA 71). Bio Med Res Int 2016:6294098. https://doi.org/10.1155/2016/6294098

Hashem A, Tabassum B, Fathi Abd Allahd E (2019) *Bacillus subtilis*: a plant-growth promoting rhizobacterium that also impacts biotic stress. Saudi J Biol Sci 26(6):1291–1297

Heil M, Bostock RM (2002) Induced systemic resistance (ISR) against pathogens in the context of induced plant defences. Ann Bot 89:503–512. https://doi.org/10.1093/aob/mcf076

Hinarejos E, Castellano M, Rodrigo I, Belles JM, Conejero V, Lopez-Gresa MP et al (2016) *Bacillus subtilis* IAB/BS03 as a potential biological control agent. Eur J Plant Pathol 146: 597–608. https://doi.org/10.1007/s10658-016-0945-3

Hu N, Zheng JF, Ding DX, Liu J, Yang LQ, Yin J et al (2009) Metal pollution in Huayuan river in Hunan province in China by manganese sulphate waste residue. Bull Environ Contamin Toxicol 83:583–590. https://doi.org/10.1007/s00128-009-9802-9

Huang J, Wei Z, Tan S, Mei X, Shen Q, Xu Y (2014) Suppression of bacterial wilt of tomato by bioorganic fertilizer made from the antibacterial compound producing strain *Bacillus amyloliquefaciens* HR62. J Agric Food Chem 62:10708–10716. https://doi.org/10.1021/jf503136a

Hussey RS, McGuire JM (1987) Interactions with other organisms. In: Brown RH, Kerry BR (eds) Principles and practice of nematode control in crops. Academic, Marrickville, pp 294–320

Jain A, Singh A, Singh S, Singh HB (2013) Microbial consortium-induced changes in oxidative stress markers in pea plants challenged with *Sclerotinia sclerotiorum*. J. Plant Growth Regul 32:388–398. https://doi.org/10.1007/s00344-012-9307-3

Jain A, Singh A, Singh S, Singh V, Singh HB (2015) Comparative proteomic analysis in pea treated with microbial consortia of beneficial microbes reveals changes in the protein network to enhance resistance against *Sclerotinia sclerotiorum*. J Plant Physiol 182:79–94. https://doi.org/10.1016/j.jplph.2015.05.004

Jiang QY, Zhuo F, Long SH, Zhao HD, Yang DJ, Ye ZH (2016) Can arbuscular mycorrhizal fungi reduce cd uptake and alleviate cd toxicity of Lonicera japonica grown in cd-added soils? Sci Rep 6:21805

Jones JD, Dangl JL (2006) The plant immune system. Nature 444:323–329. https://doi.org/10.1038/nature05286

Kang SM, Radhakrishnan R, Lee KE, You YH, Ko JH, Kim JH et al (2015a) Mechanism of plant growth promotion elicited by Bacillus sp. LKE15 in oriental melon. Acta Agric Scand Sect B Soil Plant Sci 65:637–647. https://doi.org/10.1080/09064710.2015.1040830

Kang SM, Radhakrishnan R, You YH, Khan AL, Lee KE, Lee JD et al (2015c) *Enterobacter asburiae* KE17 association regulates physiological changes and mitigates the toxic effects of heavy metals in soybean. Plant Biol 17: 1013–1022. doi: https://doi.org/10.1111/plb.12341

Katz E, Demain AL (1977) The peptide antibiotics of *Bacillus*. Bacteriol Rev 41:449–474

Kaur G, Kumar S, Nayyar H, Upadhyaya HD (2008) Cold stress injury during the pod-filling phase in Chickpea (*Cicer arietinum* L.): effects on quantitative and qualitative components of seeds. J Agron Crop Sci 194:457–464

Kaushal M, Wani SP (2015) Plant-growth-promoting rhizobacteria: drought stress alleviators to ameliorate crop production in drylands. Ann Microbiol 66(1):1–8. https://doi.org/10.1007/s13213-015-1112-3

Koca H, Bor L, Ozdemir F, Turkan I (2007) The effect of salt stress on lipid peroxidation, antioxidative enzymes and proline content of sesame cultivars. Environ Exp Bot 60:344–351

Khan MA, Gemenet DC, Villordon A (2016) Root system architecture and abiotic stress tolerance: current knowledge in root and tuber crops. Front Plant Sci 7:1584

Khan MIR, Asgher M, Khan NA (2013) Rising temperature in the changing environment: a serious threat to plants. Climate Change Environ Sustain 1:25–36. https://doi.org/10.5958/j.2320-6411.1.1.004

Kleinwechter U, Gastelo M, Ritchie J, Nelson G, Asseng S (2016) Simulating cultivar variations in potato yields for contrasting environments. Agric Syst 145:51–63. https://doi.org/10.1016/j.agsy.2016.02.011

Krasensky J, Jonak C (2012) Drought, salt, and temperature stress-induced metabolic rearrangements and regulatory networks. J Exp Bot doi. https://doi.org/10.1093/jxb/err460

Krid S, Triki MA, Gargouri A, Rhouma A (2012) Biocontrol of olive knot disease by *Bacillus subtilis* isolated from olive leaves. Ann Microbiol 62:149–154. https://doi.org/10.1007/s13213-011-0239-0

Kumar L, Bharadvaja N (2020) Microbial remediation of heavy metals. In: Shah M (ed) Microbial bioremediation and biodegradation. Springer, Singapore, pp 49–72

Lastochkina O, Pusenkova L, Yuldashev R, Babaev M, Garipova S, Blagova D, Khairullin R, Aliniaeifard S (2017) Effects of Bacillus subtilis on some physiological and biochemical parameters of *Triticum aestivum* L. (wheat) under salinity. Plant Physiol Biochem 121:80–88

Lewis BD, Hirsch RE, Sussman MR, Spalding EP (2001) Functions of AKT1 and AKT2 Potassium channels determined by studies of single and double mutants of Arabidopsis. Plant Physiol 127:1012–1019

Loeffler W, Kratzer W, Kremer S, Kugler M, Petersen F, Jung G, Rapp C, Tschen JSM (1990) Gegen Pilze wirksame Antibiotika der Bacillus subtilis-Gruppe. Forum Mikrobiologie 3:156–163

Liu Z, Budiharjo A, Wang P, Shi H, Fang J, Borriss R et al (2013) The highly modified microcin peptide plantazolicin is associated with nematicidal activity of *Bacillus amyloliquefaciens* FZB42. Appl Microbiol Biotechnol 97:10081–10090. https://doi.org/10.1007/s00253-013-5247-5

Loper JE, Henkels MD (1997) Availability of iron to *Pseudomonas fluorescens* in rhizosphere and bulk soil evaluated with an ice nucleation reporter gene. Appl Environ Microbiol 63:99–105

Lotfi N, Vahdati K, Amiri R, Kholdebarin B, Mcneil DL (2010) Drought-induced accumulation of sugars and proline in radicle and plumule of tolerant walnut varieties during germination phase. Acta Hortic 861:289–296

Lugtenberg BJJ, Dekkers L, Bloemberg GV (2001) Molecular determinants of rhizosphere colonization by Pseudomonas. Annu Rev Phytopathol 39:461–490

Ma Y, Prasad MNV, Rajkumar M, Freitas H (2011) Plant growth promoting rhizobacteria and endophytes accelerate phytoremediation of metalliferous soils. Biotechnol Adv 29(2):248–258. https://doi.org/10.1016/j.biotechadv.2010.12.001

Ma Y, Rajkumar M, Freitas H (2009) Inoculation of plant growth-promoting bacterium *Achromobacter xylosoxidans* strain Ax10 for the improvement of copper phytoextraction by *Brassica juncea*. J Environ Manag 90:831–837. https://doi.org/10.1016/j.jenvman.2008.01.014

Mantri N, Patade V, Penna S, Ford R (2012) Abiotic stress responses in plants: present and future. In: Ahmad P, Prasad MNV (eds) Abiotic stress responses in plants. Springer, New York, pp 1–19. https://doi.org/10.1007/978-1-4614-0634-1_1

Marchenko AM, Pshinko GN, Demchenko VY, Goncharuk VV (2015) Leaching heavy metal from deposits of heavy metals with bacteria oxidizing elemental Sulphur. J Water Chem Technol 37:311–316

Mihalache G, Balaes T, Gostin I, Stefan M (2017) Lipopeptides produced by Bacillus subtilis as new biocontrol products against fusariosis in ornamental plants. Environ Sci Pollut Res 25:29784–29793

Mohamed HI, Gomaa EZ (2012) Effect of plant growth promoting *Bacillus subtilis* and *Pseudomonas fluorescens* on growth and pigment composition of radish plants (*Raphanus sativus*) under NaCl stress. Photosynthetica 50:263–272. https://doi.org/10.1007/s11099-012-0032-8

Morikawa M (2006) Beneficial biofilm formation by industrial bacteria *Bacillus subtilis* and related species. J Biosci Bioeng 101:1–8

Munns R, Tester M (2008) Mechanisms of salinity tolerance. Annu Rev Plant Biol 59:651–681. https://doi.org/10.1146/annurev.arplant.59.032607.092911

Myresiotis CK, Vryzas Z, Papadopoulou-Mourkidou E (2015) Effect of specific plant-growth-promoting rhizobacteria (PGPR) on growth and uptake of neonicotinoid insecticide thiamethoxam in corn (*Zea mays* L.) seedlings. Pest Manag Sci 71:1258–1266. https://doi.org/10.1002/ps.3919

Nabti E, Schmid M, Hartmann A (2015) Application of halotolerant bacteria to restore plant growth under salt stress. In: Maheshwari DK, Saraf M (eds) Halophiles biodiversity and sustainable exploitation. Springer, Cham, pp 235–259

Narasimhan A, Shivakumar S (2015) Evaluation of *Bacillus subtilis* (JN032305) biofungicide to control chilli anthracnose in pot-controlled conditions. Biocontrol Sci Tech 25:543–559. https://doi.org/10.1080/09583157.2014.996737

Naseem S, Yasin M, Faisal M, Ahmed A (2016) Comparative study of plant growth promoting bacteria in minimizing toxic effects of chromium on growth and metabolic activities in wheat (*Triticum aestivum*). J Chem Soc Pak 38(3):509–516

Navon A (2000) Bacillus thuringiensis insecticides in crop protection — reality and prospects. Crop Protection 19:669–676

Niu DD, Liu HX, Jiang CH, Wang YP, Wang QY, Jin HL et al (2011) The plant growth-promoting rhizobacterium Bacillus cereus AR156 induces systemic resistance in Arabidopsis thaliana by simultaneously activating salicylate- and jasmonate/ethylene-dependent signaling pathways. Mol Plant-Microbe Interact 24:533–542. https://doi.org/10.1094/MPMI-09-10-0213

Nguyen D, Rieu I, Mariani C, Dam NM (2016) How plants handle multiple stresses: hormonal interactions underlying responses to abiotic stress and insect herbivory. Plant Mol Biol 91:727–740

Numan M, Bashir S, Khan Y, Mumtaz R, Shinwari ZK, Khan AL, Khan A, Al-Harrasi A (2018) Plant growth promoting bacteria as an alternative strategy for salt tolerance in plants: a review. Microbiol Res 209:21–32. https://doi.org/10.1016/j.micres.2018.02.003

Ongena M, Jacques P (2008) *Bacillus* lipopeptides: versatile weapons for plant disease biocontrol. Trends Microbiol 16: 115–125. pmid:18289856

Ongena M, Jourdan E, Adam A et al (2007) Surfactin and fengycin lipopeptides of *Bacillus subtilis* as elicitors of induced systemic resistance in plants. Environ Microbiol 9(4):1084–1090

Oves M, Saghir Khan M, Huda Qari A, Nadeen Felemban M, Almeelbi T (2016) Heavy metals: biological importance and detoxification strategies. J Bioremed Biodegr 7:334. https://doi.org/10.4172/2155-6199.1000334

Pandey P, Irulappan V, Bagavathiannan MV, Senthil-Kumar M (2017) Impact of combined abiotic and biotic stresses on plant growth and avenues for crop improvement by exploiting physio-morphological traits. Frontiers Plant Sci 8(art. no. 537). https://www.frontiersin.org/articles/10.3389/fpls.2017.00537/full

Pereira A (2016) Plant abiotic stress challenges from the changing environment. Front Plant Sci 7:–1123. https://doi.org/10.3389/fpls.2016.01123

Perez-Garcia A, Romero D, Vicente A (2011) Plant protection and growth stimulation by microorganisms: biotechnological applications of *Bacilli* in agriculture. Curr Opin Biotechnol Plant Anal 43(12):1658–1673

Porcel R, Aroca R, Ruiz-Lozano JM (2012) Salinity stress alleviation using arbuscular mycorrhizal fungi, a review. Agron Sustain Dev 32:181–200. https://doi.org/10.1007/s13593-011-0029-x

Radhakrishnan R, Lee IJ (2013) Regulation of salicylic acid, jasmonic acid and fatty acids in cucumber (*Cucumis sativus* L.) by spermidine promotes plant growth against salt stress. Acta Physiol Plant 35(12):3315–3322. https://doi.org/10.1007/s11738-013-1364-0

Radhakrishnan R, Lee IJ (2014) Effect of low dose of spermidine on physiological changes in salt stressed cucumber plants. Russ J Plant Physiol 61:90–96. https://doi.org/10.1134/S1021443714010129

Radhakrishnan R, Hashem A, Abd Allah EF (2017) Bacillus: a biological tool for crop improvement through bio-molecular changes in adverse environments. Front Physiol 8:667

Radhakrishnan R, Kang SM, Baek IY, Lee IJ (2014) Characterization of plant growth-promoting traits of *Penicillium* species against the effects of high soil salinity and root disease. J Plant Interact 9:754–762. https://doi.org/10.1080/17429145.2014.930524

Rady MM (2011) Effect of 24-epibrassinolide on growth, yield, antioxidant system and cadmium content of bean (*Phaseolus vulgaris* L.) plants under salinity and cadmium stress. Sci Horticult 129:232–237. https://doi.org/10.1016/j.scienta.2011.03.035

Rahman A, Uddin W, Wenner NG (2015) Induced systemic resistance responses in perennial ryegrass against *Magnaporthe oryzae* elicited by semi-purified surfactin lipopeptides and live cells of *Bacillus amyloliquefaciens*. Mol Plant Pathol 16:546–558. https://doi.org/10.1111/mpp.12209

Rai VK (2002) Role of amino acids in plant responses to stresses. Biol Plantarum 45:481–487

Rizhsky L, Liang H, Shuman J, Shulaev V, Davletova S, Mittler R (2004) When defense pathways collide. The response of arabidopsis to a combination of drought and heat stress. Plant Physiol 134:1683–1696

Roh JY, Choi JY, Li MS, Jin BR Je YH (2007) *Bacillus thuringiensis* as a specific, safe, and effective tool for insect pest control. J Microbiol Biotechnol 17(4):547

Saikia J, Sarma R K, Dhandia R, Yadav A, Bharali R, Gupta V K et al. (2018) Alleviation of drought stress in pulse crops with ACC deaminase producing rhizobacteria isolated from acidic soil of Northeast India. S. cRi ep. 8:3560. https://doi.org/10.1038/s41598-018-21921-w

Sheng XF, He LY (2006) Solubilization of potassium-bearing minerals by a wild type strain of *Bacillus edaphicus* and its mutants and increased potassium uptake by wheat. Can J Microbiol 52:66–72

Shin M, Shim J, You Y, Myung H, Bang KS, Cho M, Kamala-Kannan S, Oh BT (2012) Characterization of lead resistant endophytic Bacillus sp. MN3-4 and its potential for promoting lead accumulation in metal hyperaccumulator *Alnus firma*. J Hazard Mater 199:314–320

Schnepf E, Crickmore N, Van Rie J, Lereclus D, Baum J, Feitelson J, Zeigler DR, Dean DH (1998) Bacillus thuringiensis and its pesticidal crystal proteins. Microbiol Mol Biol Rev 62(3):775–806

Schonbeck FU, Steiner U, Kraska T (1993) Induced resistance: criteria, mechanisms, practicalapplications and estimation. Journal of Plant Diseases and Protection 100:541–557

Schwessinger B, Ronald PC (2012) Plant innate immunity: perception of conserved microbial signatures. Annu Rev Plant Biol 63:451–482. https://doi.org/10.1146/annurev-arplant-042811-105518

Singla J, Krattinger SG (2016) Biotic stress resistance genes in wheat. In: Wrigley CW, Faubion J, Corke H, Seetharaman K (eds) Encyclopedia of food grains. Elsevier, Oxford, pp 388–392

Solanki MK, Robert AS, Singh RK, Kumar S, Pandey AK, Srivastava AK et al (2012) Characterization of mycolytic enzymes of *Bacillus* strains and their bio-protection role against *Rhizoctonia solani* in tomato. Curr Microbiol 65:330–336. https://doi.org/10.1007/s00284-012-0160-1

Stein T (2005) *Bacillus subtilis* antibiotics: structures, syntheses and specific functions. Mol Microbiol 56:845–857

Tandy S, Schulin R, Nowack B (2006) The influence of EDDS on the uptake of heavy metals in hydroponically grown sunflowers. Chemosphere 62:1454–1463. https://doi.org/10.1016/j.chemosphere.2005.06.005

Tonelli ML, Taurian T, Ibanez F, Angelini J, Fabra A (2010) Selection and in vitro characterization of biocontrol agents with potential to protect peanut plants against fungal pathogens. J Plant Pathol 92:73–82

Treesubsuntorn C, Dhurakit P, Khaksar G, Thiravetyan P (2017) Effect of microorganisms on reducing cadmium uptake and toxicity in rice (Oryza sativa L.). Environ Sci Pollut Res:1–12

Trudgill DL, Blok VC (2001) Apomictic, polyphagous root-knot nematodes: exceptionally successful and damaging biotrophic root pathogens. Annu Rev Phytopathol 39:53–77. https://doi.org/10.1146/annurev.phyto.39.1.53

Turan M, Gulluce M, Şahin F (2012) Effects of plant-growth-promoting rhizobacteria on yield, growth, and some physiological characteristics of wheat and barley plants. Commun Soil Sci Plant Anal 43(12):1658–1673

Van Loon LC (2007) Plant responses to plant growth-promoting rhizobacteria. Eur J Plant Pathol 119:243–254

Vejan P, Abdullah R, Khadiran T, Ismail S, Nasrulhaq Boyce A (2016) Role of plant growth promoting rhizobacteria in agricultural sustainability-a review. Molecules 21(5):573

Waewthongrak W, Pisuchpen S, Leelasuphakul W (2015) Effect of Bacillus subtilis and chitosan applications on green mold (*Penicilium digitatum* Sacc.) decay in citrus fruit. Postharvest Biol Technol 99:44–49

Wang H, Xu R, You L, Zhong G (2013) Characterization of Cu-tolerant bacteria and definition of their role in promotion of growth, Cu accumulation and reduction of Cu toxicity in *Triticum aestivum* L. Ecotoxicol Environ Saf 94:1–7. https://doi.org/10.1016/j.ecoenv.2013.04.005

Wang S (2009) Molecular mechanism of plant growth promotion and induced systemic resistance to tobacco mosaic virus by *Bacillus spp*. J Microbiol Biotechnol 19:1250–1258. https://doi.org/10.4014/jmb.0901.008
Wang W, Vinocur B, Altman A (2003) Plant responses to drought, salinity and extreme temperatures: towards genetic engineering for stress tolerance. Planta 218:1–14. https://doi.org/10.1007/s00425-003-1105-5
Wang Y, Guo J, Liu R (2001) Biosorption of heavy metals by bacteria isolated from activated sludge. Appl Biochem Biotechnol 91–93:171–184
Weller DM (1988) Biological control of soilborne plant pathogens in the rhizosphere with bacteria. Annu Rev Phytopathol 26:379–407
Weller DM, Thomashow LS (1993) Microbial metabolites with biological activity against plant pathogens. In: Lumsden RD, Vaughn JL (eds) Pest management: biologically based technologies, Amer Chem Soc, Washington, DC, pp173–180
Wierzba S (2015) Biosorption of lead(II), zinc(II) and nickel(II) from industrial wastewater by Stenotrophomonas maltophilia and Bacillus subtilis. Pol. J. Chem. Technol 17:79–87
Wolter and Schroeder (2012) Effect of drought stress on the productivity of ivy treated with rhizobacterium Bacillus subtilis. In: Proceedings of the international symposium on soilless cultivation 1004, Shanghai, pp107–113
Wu S, Cao Z, Li Z, Cheung K, Wong M (2005) Effects of biofertilizer containing N-fixer, P and K solubilizers and AM fungi on maize growth: a greenhouse trial. Geoderma 125:155–166. https://doi.org/10.1016/j.geoderma.2004.07.003
Xu ZZ, Zhou GS (2006) Combined effects of water stress and high temperature on photosynthesis, nitrogen metabolism and lipid peroxidation of a perennial grass *Leymus chinensis*. Planta 224:1080–1090
Yazici I, Turkan I, Sekmen AH, Demiral T (2007) Salinity tolerance of purslane (*Portulaca oleracea* L.) is achieved by enhanced antioxidative system, lower level of lipid peroxidation and proline accumulation. Environ Exp Bot 62:49–57. https://doi.org/10.1016/j.envexpbot.2007.02.010
Yedidia I, Shoresh M, Kerem Z, Benhamou N, Kapulnik Y, Chet I (2003) Concomitant induction of systemic resistance to *Pseudomonas syringae* pv. *lachrymans* in cucumber by *Trichoderma asperellum* (T-203) and accumulation of phytoalexins. Appl Environ Microbiol 69:7343–7353. https://doi.org/10.1128/AEM.69.12.7343-7353.2003
Yi HS, Yang JW, Ryu CM (2013) ISR meets SAR outside: additive action of the endophyte *Bacillus pumilus* INR7 and the chemical inducer, benzothiadiazole, on induced resistance against bacterial spot in field-grown pepper. Front Plant Sci 4:122. https://doi.org/10.3389/fpls.2013.00122
Yu Z, Xiong J, Zhou Q, Luo H, Hu S, Xia L et al (2015) The diverse nematicidal properties and biocontrol efficacy of *Bacillus thuringiensis* Cry6A against the root-knot nematode *Meloidogyne hapla*. J Invertebr Pathol 125:73–80. https://doi.org/10.1016/j.jip.2014.12.011
Zhang H, Kim MS, Sun Y, Dowd SE, Shi H, Pare PW (2008) Soil bacteria confer plant salt tolerance by tissue-specific regulation of the sodium transporter HKT1. Mol Plant-Microbe Interact 21(6):737–744
Zhang S, Reddy MS, Kloepper JW (2004) Tobacco growth enhancement and blue mold disease protection by rhizobacteria: relationship between plant growth promotion and systemic disease protection by PGPR strain 90–166. Plant Soil 262:277–288

Chapter 9
Amelioration of Salinity Stress by *Bacillus* Species as Promoters of Plant Growth in Saline Soil

Mohammad Tofajjal Hossain and Tofazzal Islam

Abstract Salinity is a major factor of osmotic stress of the high salt in the rhizosphere, which limits plant growth and development. The uptake of salt inhibits the physiological and metabolic processes of plants and severely impacts agricultural productivity. About 62 million hectares (20%) of the world's irrigated land are affected by salinity. *In vitro* and *in vivo* bioassays reveal that some salt-tolerant plant-associated *Bacillus* spp. enhance plant tolerance to salinity. The roots and fresh weight of plant seedlings showed better growth promotion significantly by the application of *Bacillus* species in the saline environment. However, the precise mechanism of higher tolerance of plants to salinity by *Bacillus* spp. is still poorly understood. Salt-tolerant plant growth-promoting *Bacillus* species, viz., *B. oryzicola* YC7007, *B. velezensis*, *B. licheniformis* SA03, *B. safensis* ZY16, *B. megaterium*, *B. pumilus*, *B. firmus* SW5, *B. subtilis* SU-12, *B. amyloliquefaciens*, *B. aryabhattai* MS3, *B. cereus*, and *B. aquimaris* DY-3 enhance salinity tolerance of plants through inducing plant gene/protein/pattern recognition receptors against salt stress. Out of different genera of salt-tolerant bacteria such as *Pseudomonas*, *Enterobacter*, *Agrobacterium*, *Streptomyces*, *Klebsiella*, and *Ochromobacter*, *Bacillus* show the highest performances in enhancing plant tolerance to salinity. This chapter summarizes current knowledge on the amelioration of salinity stress in plants by plant growth-promoting *Bacillus* species.

Keywords Abiotic stress tolerance · Climate change · Food security · PGPR · Genomics

M. T. Hossain (✉)
Plant Pathology Division, Regional Agricultural Research Station, Bangladesh Agriculture Research Institute (BARI), Chittagong, Bangladesh

T. Islam
Institute of Biotechnology and Genetic Engineering (IBGE), Bangabandhu Sheikh Mujibur Rahman Agricultural University, Gazipur, Bangladesh

M. T. Islam et al. (eds.), *Bacilli in Agrobiotechnology*, Bacilli in Climate Resilient Agriculture and Bioprospecting,
https://doi.org/10.1007/978-3-030-85465-2_9

9.1 Introduction

Plants face various abiotic stresses that significantly impact crop production (Bao et al. 2009; Kumar et al. 2019; Mahajan and Tuteja 2005). Abiotic stresses such as salinity, extreme temperature, heavy metals, heat, and cold stresses can cause up to 80% of yield reduction in crops (Bharti and Barnawal 2019). Among these stresses, salt adversely affects plant health, which is a serious threat to the reduction of yield losses between 20 and 50% (Shrivastava and Kumar 2015; Vij and Tyagi 2007). In plants, salinity induces oxidative stresses and causes a deficiency of essential nutrients, such as K^+, which enhances the toxicity of Na^+ inside the plants, which subsequently hinders plant photosynthesis, lipid metabolism, protein synthesis, and biomass accumulation (Munns and Tester 2008). It also causes water deficit, osmotic stress, stomatal closer, and reduction of leaf expansion. The salinity is a serious threat to crop production in coastal areas worldwide. About 62 million hectares (20%) of the world's irrigated land is affected by salinity. The coastal region of Bangladesh also covers almost 29,000 Km^2 or about 20% of the country that is almost 30% of the cultivable land area (Chinnusamy et al. 2006; Egamberdieva et al. 2019; Haque 2006; Zhang et al. 2008). To avoid salinity, now, it is challenging how it should be controlled over the world. In these circumstances, to mitigate the salinity from the soil, many strategies have been imposed already, and one of the promising ways is the use of plant growth-promoting rhizobacteria (PGPR) (Gray and Smith 2005). The PGPR are naturally soil microorganisms that colonize roots and stimulate plant growth. More than 33,000 bacterial species under proteobacteria, firmicutes, and actinobacteria are present out of one million microbes in one gram of bulk soils in plant rhizosphere which was discovered by phyloChip analysis. To influence bacterial diversity and activities, root exudates contain about 21% photosynthetically fixed carbon in the root-soil interface (Lynch et al. 2012; Mendes et al. 2011). These types of beneficial bacteria can be either free-living in the soil or colonized in the rhizosphere, phyllosphere, or plant tissue interior (endophytes). They enhance plant growth and induce systemic resistance in host plants to a broad spectrum of pathogens (Chung et al. 2015; Hossain et al. 2016; Kloepper et al. 2004; Khan et al. 2016a) and also increase tolerance of plants to abiotic stresses including salinity (Baek et al. 2020; Chinnusamy et al. 2006). The PGPR include the strains in the genera *Bacillus*, *Acinetobacter*, *Alcaligenes*, *Arthrobacter*, *Azospirillum*, *Azotobacter*, *Beijerinckia*, *Burkholderia*, *Erwinia*, *Enterobacter*, *Rhizobium*, *Flavobacterium*, and *Serratia* (Bashan and de-Bashan 2005). Out of these PGPR, *Bacillus* consortium has been utilized since long with an eco-friendly manner for promoting sustainable agriculture (Rahman et al. 2018; Dame et al. 2021). Application of salt-tolerant *Bacillus* spp. to alleviate salinity stress in crop production is a new approach in crop production. The beneficial *Bacillus* species play a prominent role in the plant's development in many ways. Among them, resistance induction in host plants (Niu et al. 2016) against various biotic and abiotic stress including soil salinity is prominent. *Bacillus subtilis* GB03 and *B. oryzicola* YC7007 are soil-borne plant symbionts that promote plant growth promotion and protect host plants from salinity stress (Baek et al. 2020; Hossain et al. 2016). *Bacillus* species confer plant tolerance by tissue-specific regulation of the sodium transporter *HKT1* (Zhang et al. 2008). *Bacillus subtilis* GBO3 induces growth promotion and enhances tolerance in the halophyte grass *Puccinellia tenuiflora* to salinity (Han et al. 2014). Furthermore, *Bacillus* species induce physical and chemical changes that increase tolerance against abiotic stresses especially salinity (Yang et al. 2009). Although improvement of salinity tolerance in plants by various plant growth-promoting bacteria has been reviewed (Numan et al. 2018). No review has so far been published on salinity tolerance by *Bacillus* species (Paul and Lade 2014). This chapter describes the potential of *Bacillus* strains to increase plant tolerance to salt stress and discusses the utilization of these beneficial bacteria in the promotion of sustainable agriculture.

9.2 Diversity of Salt-Tolerant PGPR Bacteria

Bacteria from diverse taxonomic groups show tolerance to salinity stress. A large body of literature describes the tolerance of plant growth-promoting rhizobacteria (PGPR) to salinity. Since the discovery of salt-tolerant PGPR in high-salt environments, many researchers explore novel strains of bacteria that enhance plant growth under soil salinity. The PGPR belonging to different genera such as *Bacillus*, *Microbacterium*, *Pantoea*, *Achromobacter*, *Rhizobium*, *Pseudomonas*, *Paenibacillus*, *Enterobacter*, *Burkholderia*, *Methylobacterium*, *Azospirillum*, *Variovorax*, etc. induce tolerance in host plants to abiotic stresses, including salinity (Vacheron et al. 2013). During the last few decades, the beneficial effects of these bacterial genera have been shown to enhance the tolerance of crop plants (e.g., rice, wheat, maize, cotton, lettuce, tomato, and pepper) to drought, salinity, heat stress, and chilling injury (Jha and Subramanian 2014). Based on plant growth-promoting traits, seed germination and seedling vigor index of plants in salt-affected soils were analyzed to identify the salt-tolerant PGPR strains. NaCl uptake and tolerance at the different concentrations of 0–20% were also examined in different studies (Upadhyay et al. 2009; Zhang et al. 2008). Diversity of salt-tolerant PGPR bacteria representative cluster groups based on taxonomic and phylogenetic placement was investigated by analysis of the 16S rRNA gene sequencing (Bibi et al. 2012). Out of the different strains, the findings belong to three phyla (Proteobacteria, Firmicutes, Actinobacteria), four orders (Bacillales, Actinomycetales, Rhizobiales, Oceanospirillales), nine families, eighteen genera, and thirty-five species. The 18 genera were as follows (number of strains in parentheses): *Bacillus* (39), *Streptomyces* (9), *Fictibacillus* (5), *Halomonas* (3), *Nocardia* (3), *Microbacterium* (2), *Nesterenkonia* (2), *Oceanobacillus* (1), *Arthrobacter* (1), *Rhizobium* (1), *Jeotgalibacillus* (1), *Bhargavaea* (1), *Chryseomicrobium* (1), *Paenisporosarcina* (1), *Rhodococcus* (1), *Brevibacterium* (1), *Arsenicicoccus* (1), and *Solibacillus* (1). Bacillales, Rhizobiales, and Oceanospirillales were 88%, and out of them, Bacillales belonged 80% which increased shoot and root lengths 63% and 87%, respectively

(Zhang et al. 2018). The novel PGPR strain *Bacillus* species of the Firmicutes group display various beneficial effects in the host plants including abiotic stress tolerance. Among the endophytic bacteria, *Bacillus* spp. exhibit the highest abundance of 52.7% of the strains followed by *Streptomyces* and *Fictibacillus* of 12.2% and 6.8%, respectively (Damodaran et al. 2019; Upadhyay et al. 2009). *Bacillus pumilus* (Jha and Subramanian 2013), *B. amyloliquefaciens* (Nautiyal et al. 2013), *B. thuringiensis* (Raheem and Ali 2015), and *B. oryzicola* YC7007 (Baek et al. 2020) are most effective in enhancing plant growth in the saline environment. Chen et al. (2010) screened 114 strains of halophiles and salt-tolerant bacteria from orchards, paddy, sandy soil, and forest soils using five media and different NaCl concentrations (5–20%). Bacillaceae (33 strains; 54.1%) was the most frequently occurring family that was the most functional to salt stress. Similarly, Echigo et al. (2005) isolated 176 strains of salt-tolerant bacteria from orchards, lawn, pasture, and woodlands around Tokyo, Japan, using one medium (20% NaCl concentration). From all together of our studies, salt-tolerant PGPR are *Bacillus* species (80%) in the saline environment; therefore, we should focus salt-tolerant *Bacillus* species to salt stress.

9.3 The Genus *Bacillus* Is a Good Source for Making a Green Revolution in the Saline Area

Considering the importance of PGPR for mitigating the salinity in plants, we reviewed literature mainly related to the *Bacillus* species that play a key role in the saline environment. The genus *Bacillus* was first described by Cohn in 1872 (Claus and Berkeley 1986), and since then, many scientific works have been conducted on many aspects of this genus in relation to agriculture (McSpadden Gardener 2010). Numerous *Bacillus* species have been reported as biocontrol agents against phytopathogens. Some of them enhance plant tolerance to various abiotic stresses including salinity. Recently, an endophytic *B. oryzicola* YC7007 has been well-versed as a novel species, which successfully controlled the rice bakanae and bacterial blast diseases and mitigated salinity by inhibiting malondialdehyde and Na^+ in the dicot model plant *Arabidopsis thaliana,* cabbage, and radish (Baek et al. 2020; Chung et al. 2015; Hossain et al. 2016). A strain of another species, *B. subtilis* GB03, effectively enhances salinity stress tolerance in *A. thaliana* (Zhang et al. 2008). The cocultivation of wheat seeds with *B. amiloliquefaciens* BcA12 and *B. laevolactivus* BcL28 increases root and shoot growth (15–50%) in the saline environment compared with untreated plants (Egamberdiyeva and Höflich 2003). In another study, *Bacillus* strains BcP26 and BcM33 increase root and shoot growth of pea (18%) and maize (27%) and increase the uptake of N and P uptake (55%) under arid saline soil conditions compared with untreated plant (Egamberdiyeva and Höflich 2004). *Bacillus aryabhattai* RS341 increases up to 40% in root elongation and dry weight compared with untreated salt-stressed canola seedlings (Siddikee et al. 2010). Application of the rhizobacterium *B. subtilis* FZB24 at the dose of 1×10^8 CFU/ml has been increased 6.5 and 5.3 times more yield in the eggplant and pepper plant, respectively, compared to untreated control under the saline environments. The bacterization by the strain FZB24 resulted in 50% and 25% reduction in salinity effects on the yield of eggplant and pepper plant, respectively (Bochow et al. 2001). *Bacillus pumilus* stimulates plant growth and enhances stress tolerance of rice under salt stress by inducing the antioxidative enzymes and lowering Na^+ accumulation in the leaves (Khan et al. 2016a, b). Under saline soils, *B. subtilis* significantly increases in fresh and dry masses of roots and leaves, photosynthetic pigments, proline, total free amino acids, crude protein, and N, P, K^+, Ca^{2+}, and Mg^{2+} uptake compared to untreated radish plants (Mohamed and Gomaa 2012). In fact, *B. megaterium* is an absolute majority of the Bacillaceae family. *Bacillus megaterium* was found to be the most dominant phosphate solubilizer in saline conditions (Nakade 2020). *Bacillus megaterium* F-58 exhibited significant improvement in fresh weight 42% of plants compared to untreated control under the saline soil at 200 mM concentration (Aslam and Ali 2018). *Bacillus subtilis* EY2, *B. atrophaeus* EY6, and *B. spharicus* enhance plant growth and ultimately increase yield in radish and strawberry plants in saline conditions (Karlidag et al. 2010). These findings indicate that *Bacillus* species is an ideal candidate for plant growth promotion under saline soils.

9.4 Induction of Salt Tolerance in Plant by the *Bacillus* Species

Plants have their own mechanisms to bear salinity stress at a certain level depending on species and genotypes. Most of the plants release different hormones, activate antioxidant enzymes, and regulate their ion uptake and their metabolic and genetic pathways to grow under harsh saline conditions (Perez-Alfocea et al. 2010). The soil biota may improve salt tolerance efficiency in crop plants through altering hormonal root-shoot signaling, various classes of plant hormones that respond to changes in salinity. Plant growth-promoting bacteria (PGPB), especially *Bacillus* species are playing an important role in mitigating salinity by triggering biosynthesis of various plant growth hormones such as auxin, cytokinin, and gibberellin and by inducing the pattern recognition receptors (PRRs) or volatile organic compounds (VOCs) (Arkhipova et al. 2005; Ortiz-Castro et al. 2008; Hossain et al. 2016). The colonization of plants by *Bacillus* spp. also triggers the accumulation of abscisic acid (ABA) and antioxidants in the plant cells. Antioxidant defense enzymes such as ascorbate peroxidase (APOX), superoxide dismutase (SOD), peroxidase (POD), and catalase (CAT) have been induced by the *Bacillus* and other PGPR to detoxify reactive oxygen species (ROS) (Chiquito-Contreras et al. 2019; Hossain et al. 2016: Zhao et al. 2019). These PGPR *Bacillus* species also produce siderophores to enhance plants' iron uptake and also promote atmospheric nitrogen fixation and solubilization of organic and inorganic phosphates in the soils. The elite strains of *Bacillus* spp. might be used as an effective measure for inducing salinity tolerance genes in the plant that could help the plant to survive in a saline environment. Several studies have been reported that stress-tolerant genes or proteins or transcription factor or pattern recognition receptors in plants are induced by these beneficial bacteria through various mechanisms (Munns and Tester 2008). Somewhat, they produce indole acetic acid and some unidentified elements that increase root architecture, viz., root length, lateral roots, and root tips, and most importantly enhance nutrient content, by improving the health of plants under salt stress conditions (Idris et al. 2007: Khan et al. 2016a, b; Sayyed 2019). The ACC (1-aminocyclopropane-1-carboxylic acid) deaminase produced by bacteria degrades plant ACC for acquiring energy. Moreover, it also decreases the harmful impact of ethylene, by enhancing stress tolerance and promoting the growth of plant. Some *Bacillus* spp. also produce exo-poly saccharides (EPS) that mitigate the salinity of soil. Inoculations of specific Bacilli PGPR help to enhance salt stress tolerance in plants via induced systemic tolerance (IST), which causes many functional and biochemical changes (Yang et al. 2009). Recently, an endophytic *B. oryzicola* YC7007 has been found to mitigate salinity stress by inducting *SOS1* in the dicot model plant *A. thaliana* (Baek et al. 2020; Rodríguez et al. 2005). The salinity of soils can be attributed to cations such as magnesium (Mg^{2+}), calcium (Ca^{2+}), and sodium (Na^{+}) and also anions like bicarbonate ($HCO3^{-}$), sulfate ($SO4^{2-}$), and chloride (Cl^{-}) (Tester and Davenport 2003). However, under stressed conditions, K^{+} and Na^{+} are two major contributors of osmotic pressure and ionic strength. Salinity increases the uptake of sodium ion (Na^{+}), which eventually results in decrease in the uptake of calcium (Ca^{2+}) and potassium (K^{+}) ions (Yildirim et al. 2006). The PGPB improves plant growth under salt stress by maintaining a favorable K^{+}/Na^{+} ratio, by enhancing the production of certain osmolytes such as proline, total soluble sugar, and total protein content, or by both these methods (Singh and Jha 2016). Another strain *B. subtilis* GB03 induces tissue-specific regulation called *HKT1* (High-affinity K^{+} transporter) that controls salinity in *A. thaliana* (Zhang et al. 2008). The *AVP1* (*Arabidopsis Vacuolar H^{+}-pyrophosphatase*) gene transformed into alfalfa (*Medicago sativa*) enhanced salt tolerance in the alfalfa up to 200 mM NaCl (Bao et al. 2009). *Pisum sativum* calcineurin B-like protein (*PsCBL*) and *P. sativum* calcineurin B-like (CBL)-interacting protein kinase (*PsCIPK*) genes into rice (*Oryza sativa* L.) have been transformed, and therefore, rice was able to survive at 150 mM NaCl (Sikdar et al. 2015). Research would be focused on how to induce salt tolerance genes in the plant by the application of PGPR *Bacillus* species.

9.5 Conclusion and Future Trends

Bacillus spp. plays an important role in regulating the PGP traits, tolerance to salt stress which enhances agricultural crop production. They could induce tissue-specific regulation of *HKT1* and *SOS* signaling in the plant system to mitigate the salinity stress. The application of *Bacillus* spp. enhance the accumulation of abscisic acid (ABA) and antioxidant defense enzymes such as ascorbate peroxidase (APOX), superoxide dismutase (SOD), peroxidase (POD), and catalase (CAT) that detoxify ROS and protect the host plant from salt stress toxicity. Endophytic *Bacillus* species are rich in bioactive compounds that induce systemic resistance in host plants to tolerate harsh environments. The good symbiotic relationship of Bacilli with the plant host could be utilized for increasing crop yield under various abiotic stresses including salinity. However, underlying molecular mechanisms of beneficial effects of various strains *Bacillus* spp. on plants are largely unknown. The advancement of genomics and post-genomics analytical methods and recently developed CRISPR-Cas genome editing technology would be helpful for dissecting molecular cross-talks between *Bacillus* spp. and their plant hosts during exerting their beneficial effects. A better understanding of molecular plant-Bacilli interactions would immensely promote the development of *Bacillus*-based technology for alleviating salinity and other abiotic stresses for promoting sustainable crop production under the changing climate of the world.

References

Arkhipova TN, Veselov SU, Melentiev AI, Martynenko EV, Kudyarova GR (2005) Ability of bacterium *Bacillus subtilis* produce cytokinins and to influence the growth and endogenous hormone content of lettuce plants. Plant Soil 272:201–209

Aslam F, Ali B (2018) Halotolerant bacterial diversity associated with Suaeda fruticosa (L.) forssk. Improved growth of maize under salinity stress. Agronomy 8:131

Baek D, Rokibuzzaman M, Khan A, Kim MC, Park HJ, Yun DJ, Chung YR (2020) Plant-growth-promoting *Bacillus oryzicola* YC7007 modulates stress-response gene expression and provides protection from salt stress. Front Plant Sci 10:1646

Bao AK, Wang SM, Wu GQ, Xi JJ, Zhang JL, Wang CM (2009) Overexpression of the *Arabidopsis* H+-PPase enhanced resistance to salt and drought stress in transgenic alfalfa (*Medicago sativa* L.). Plant Sci 176:232–240

Bashan Y, De-Bashan LE (2005) Plant growth-promoting. Encyclopedia Soils Environ 1:103–115

Bharti N, Barnawal D (2019) Amelioration of salinity stress by PGPR: ACC deaminase and ROS scavenging enzymes activity. In: Singh AK, Kumar A, Singh PK (eds) PGPR amelioration in sustainable agriculture, food security and environmental management. Woodhead Publishing, pp 85–106

Bibi F, Yasir M, Song GC, Lee SY, Chung YR (2012) Diversity and characterization of endophytic bacteria associated with tidal flat plants and their antagonistic effects on oomycetous plant pathogens. Plant Pathol J 28:20–31

Bochow H, El-Sayed SF, Junge H, Stavropoulou A, Schmiedeknecht G (2001) Use of *Bacillus subtilis* as biocontrol agent. IV. Salt-stress tolerance induction by *Bacillus subtilis* FZB24 seed treatment in tropical vegetable field crops, and its mode of action. J Plant Dis Prot 108:21–30

Chen QH, Zhang L, Liu ZX, Huang K, Xiong LZ, He JW, Chen YG (2010) Screening and biological characteristics of antibiotic active halophilic and halo-tolerant bacteria from Xiaoxi National Natural Reserve. J Microbiol 4

Chinnusamy V, Zhu J, Zhu JK (2006) Salt stress signaling and mechanisms of plant salt tolerance. In: Setlow JK (ed) Genetic engineering. Genetic engineering: principles and methods, vol 27. Springer, Boston. https://doi.org/10.1007/0-387-25856-6_9

Chiquito-Contreras RG, Murillo-Amador B, Carmona-Hernandez S, Chiquito-Contreras CJ, Hernandez-Montiel LG (2019) Effect of marine bacteria and Ulvan on the activity of antioxidant defense enzymes and the bio-protection of papaya fruit against *Colletotrichum gloeosporioides*. Antioxidants 8:580

Chung EJ, Hossain MT, Khan A, Kim KH, Jeon CO, Chung YR. (2015) *Bacillus oryzicola* sp. nov., an endophytic bacterium isolated from the roots of rice with anti-microbial, plant-growth-promoting, and systemic resistance-inducing activities in rice. Plant Pathol J 31: 152–164

Claus D, Berkeley RCW (1986) Genus *Bacillus* Cohn 1872. In: PHA S, Mair NS, Sharpe ME, Holt JG (eds) Bergey's manual of systematic bacteriology, vol 2. Williams and Wilkins, Baltimore, pp 1105–1139

Dame ZT, Rahman M, Islam T (2021) Bacilli as sources of agrobiotechnology: recent advances and future directions. Green Chem Lett Rev, 14(2): 246–271

Damodaran T, Mishra VK, Jha SK, Pankaj U, Gupta G, Gopal R (2019) Identification of rhizosphere bacterial diversity with promising salt tolerance, PGP traits and their exploitation for seed germination enhancement in sodic soil. Agric Res 8:36–43

Echigo A, Hino M, Fukushima T, Mizuki T, Kamekura M (2005) Endospores of halophilic bacteria of the family Bacillaceae isolated from non-saline Japanese soil may be transported by kosa event (Asian dust storm). Aquat Biosyst 1:1–13

Egamberdieva D, Wirth S, Bellingrath-Kimura SD, Mishra J, Arora NK (2019) Salt-tolerant plant growth promoting rhizobacteria for enhancing crop productivity of saline soils. Front Microbiol 10

Egamberdiyeva D, Höflich G (2003) Influence of growth-promoting bacteria on the growth of wheat in different soils and temperatures. Soil Biol Biochem 35:973–978

Egamberdiyeva D, Höflich G (2004) Effect of plant growth-promoting bacteria on growth and nutrient uptake of cotton and pea in a semi-arid region of Uzbekistan. J Arid Environ 56:293–301

Gray EJ, Smith DL (2005) Intracellular and extracellular PGPR: commonalities and distinctions in the plant–bacterium signaling processes. Soil Biol Biochem 37:395–412

Han QQ, Lü XP, Bai JP, Qiao Y, Paré WP, Wang SM, Zhang LJ, Wu YN, Pang XP, Xu WB, Wang ZL (2014) Beneficial soil bacterium *Bacillus subtilis* (GB03) augments salt tolerance of white clo ver. Front Plant Sci 5:525

Haque SA (2006) Salinity problems and crop production in coastal regions of Bangladesh. Pak J Bot 38:1359–1365

Hossain MT, Khan A, Chung EJ, Rashid HO, Chung YR (2016) Biological control of rice bakanae by an endophytic *Bacillus oryzicola* YC7007. Plant Pathol J 32:228–242

Idris EE, Iglesias DJ, Talon M, Borriss R (2007) Tryptophan dependent production of indole-3-acetic acid (IAA) affects level of plant growth promotion by *bacillus amyloliquefaciens* FZB42. Mol Plant-Microbe Interact 20:619–626

Jha Y, Subramanian RB (2013) Paddy physiology and enzymes level is regulated by rhizobacteria under saline stress. J Appl Bot Food Qual 85:168

Jha Y, Subramanian RB (2014) PGPR regulate caspase-like activity, programmed cell death, and antioxidant enzyme activity in paddy under salinity. Physiol Mol Biol Plants 20:201–207

Karlidag H, Esitken A, Yildirim E, Donmez MF, Turan M (2010) Effects of plant growth promoting bacteria on yield, growth, leaf water content, membrane permeability, and ionic composition of strawberry under saline conditions. J Plant Nutr 34:34–45

Khan A, Hossain MT, Park HC, Yun DJ, Shim SH, Chung YR (2016a) Development of root system architecture of Arabidopsis thaliana in response to colonization by *Martelella endophytica* YC6887 depends on auxin signaling. Plant Soil 405:81–96. https://doi.org/10.1007/s11104-015-2775-z

Khan A, Zhao XQ, Javed MT, Khan KS, Bano A, Shen RF, Masood S (2016b) *Bacillus pumilus* enhances tolerance in rice (Oryza sativa L.) to combined stresses of NaCl and high boron due to limited uptake of Na+. Environ Exp Bot, 124, 120–129

Kloepper JW, Ryu CM, Zhang S (2004) Induced systemic resistance and promotion of plant growth by *bacillus* spp. Phytopathology 94:1259–1266

Kumar A, Patel JS, Meena VS, Ramteke PW (2019) Plant growth-promoting rhizobacteria: strategies to improve abiotic stresses under sustainable agriculture. J Plant Nutr 42:1402–1415

Lynch J, Marschner P, Rengel Z (2012) Effect of internal and external factors on root growth and development. In: Marschner H (ed) Mineral nutrition of higher plants. Academic press, London, pp 331–345

Mahajan S, Tuteja N (2005) Cold, salinity and drought stresses: an overview. Arch Biochem Biophys 444:139–158

McSpadden Gardener B (2010) Biocontrol of plant pathogens and plant growth promotion by *bacillus*. In: Gishi U, Chet I, Lodovica Gullino M (eds) Recent developments in Management of Plant Diseases, plant pathology in the 21st century. Springer, Amsterdam, pp 71–79

Mendes R, Kruijt M, de Bruijn I, Dekkers E, van der Voort M, Schneider JHM, Piceno YM, DeSantis TZ, Andersen GL, Bakker PAHM, Raaijmakers JM (2011) Deciphering the rhizosphere microbiome for disease-suppressive bacteria. Science 332:1097–1100

Mohamed HI, Gomaa EZ (2012) Effect of plant growth promoting *Bacillus subtilis* and *Pseudomonas fluorescens* on growth and pigment composition of radish plants (*Raphanus sativus)* under NaCl stress. Photosynthetica 50:263–272

Munns R, Tester M (2008) Mechanisms of salinity tolerance. Annu. Rev Plant Biol 59:651–681

Nakade D (2020) Biodiversity and plant growth promoting potential of bacteria from soybean rhizosphere of saline soil. Am J Agric For 8:77–80

Nautiyal CS, Srivastava S, Chauhan PS, Seem K, Mishra A, Sopory SK (2013) Plant growth-promoting bacteria *bacillus amyloliquefaciens* NBRISN13 modulates gene expression profile of leaf and rhizosphere community in rice during salt stress. Plant Physiol Biochem 66:1–9

Niu SQ, Li HR, Paré PW, Aziz M, Wang SM, Shi H, Guo Q (2016) Induced growth promotion and higher salt tolerance in the halophyte grass *Puccinellia tenuiflora* by beneficial rhizobacteria. Plant Soil 407(1–2):217–230

Numana M, Bashir S, Khan Y, Mumtaz R, Shinwari ZK, Khan AL, Khan A, AL-Harrasi A (2018) Plant growth promoting bacteria as an alternative strategy for salt tolerance in plants: a review. Microbiol Res 209:21–32

Ortiz-Castro R, Valencia-Cantero E, Lopez-Bucio J (2008) Plant growth promotion by *bacillus megaterium* involves cytokinin signalling. Plant Signal Behav 3:263–265

Paul D, Lade H (2014) Plant-growth-promoting rhizobacteria to improve crop growth in saline soils: a review. Agron Sustain Dev 34:737–752. https://doi.org/10.1007/s13593-014-0233-6

Pérez-Alfocea F, Albacete A, Ghanem ME, Dodd IC (2010) Hormonal regulation of source–sink relations to maintain crop productivity under salinity: a case study of root-to-shoot signalling in tomato. Funct Plant Biol 37:592–603

Raheem A, Ali B (2015) Halotolerant rhizobacteria: beneficial plant metabolites and growth enhancement of *Triticum aestivum* L. in salt-amended soils. Arch Agron Soil Sci 61:1691–1705

Rahman M, Sabir AA, Mukta JA, Khan MMA, Mohi-Ud-Din M, Miah MG, Rahman M, Islam MT (2018) Plant probiotic bacteria *Bacillus* and *Paraburkholderia* improve growth, yield and content of antioxidants in strawberry fruit. Sci Rep, 8, 2504

Rodríguez M, Canales E, Borrás-Hidalgo O (2005) Molecular aspects of abiotic stress in plants. Biotecnol Aplic 22:1–10

Sayyed RZ (Ed) (2019) Plant growth promoting Rhizobacteria for sustainable stress management: volume 2: Rhizobacteria in biotic stress management (Vol. 13). Springer Nature

Shrivastava P, Kumar R (2015) Soil salinity: a serious environmental issue and plant growth promoting bacteria as one of the tools for its alleviation. Saudi J Biol Sci 22:123–131

Siddikee MA, Chauhan PS, Anandham R, Han GH, Sa T (2010) Isolation, characterization, and use for plant growth promotion under salt stress, of ACC deaminase-producing halotolerant bacteria derived from coastal soil. Microbiol Biotechnol J 20:1577–1584

Sikdar SU, Zobayer N, Nasrin S, Prodhan SH (2015) Agrobacterium-mediated *PsCBL* and *PsCIPK* gene transformation to enhance salt tolerance in indica rice (*Oryza sativa*). In Vitro Cellular Dev Biol Plant 51:143–151

Singh RP, Jha PN (2016) The multifarious PGPR Serratia marcescens CDP-13 augments induced systemic resistance and enhanced salinity tolerance of wheat (*Triticum aestivum* L.). PLos one 11(6):e0155026

Tester M, Davenport R (2003) Na+ tolerance and Na+ transport in higher plants. Ann Bot 91:503–527
Upadhyay SK, Singh DP, Saikia R (2009) Genetic diversity of plant growth promoting rhizobacteria isolated from rhizospheric soil of wheat under saline condition. Curr Microbiol 59:489–496
Vacheron J, Desbrosses G, Bouffaud ML, Touraine B, Moënne-Loccoz Y, Muller D, Prigent-Combaret C (2013) Plant growth-promoting rhizobacteria and root system functioning. Front Plant Sci 4:356
Vij S, Tyagi AK (2007) Emerging trends in the functional genomics of the abiotic stress response in crop plants: review article, plant biot. J 5(3):361–380
Yang J, Kloepper JW, Ryu CM (2009) Rhizosphere bacteria help plants tolerate abiotic stress. Trends Plant Sci. https://doi.org/10.1016/j.tplants.2008.10.004
Yildirim E, Taylor AG, Spittler TD (2006) Ameliorative effects of biological treatments on growth of squash plants under salt stress. Sci Hortic, 111, 1–6
Zhang H, Kim MS, Sun Y, Dowd SE, Shi H, Paré PW (2008) Soil bacteria confer plant salt tolerance by tissue-specific regulation of the sodium transporter *HKT1*. Mol Plant-Microbe Interact 21:737–744
Zhang S, Fan C, Wang Y, Xia Y, Xiao W, Cui X (2018) Salt-tolerant and plant-growth-promoting bacteria isolated from high-yield paddy soil. Can J Microbiol 64:968–978
Zhao P, Li P, Wu S, Zhou M, Zhi R, Gao H (2019) Volatile organic compounds (VOCs) from *Bacillus subtilis* CF-3 reduce anthracnose and elicit active defense responses in harvested litchi fruits. AMB Express 9:119

Chapter 10
Bacillus spp. of Ruminant Origin as Major Sources of Potential Industrial Amylases

Kgodiso Judith Rabapane, Alfred Mitema, Karen Nelson, and Naser Aliye Feto

Abstract The microbiome of ruminant's gastrointestinal tract origin is dominated by two major bacterial phyla *Firmicutes* and *Bacteroidetes*. Meanwhile, *Bacillus* spp. belonging to *Firmicutes*, have been reported to be potent producers of thermostable amylases and other biocatalysts that are of considerable commercial interest. Hence, they are primarily exploited as a significant source of thermostable biomolecules. Among some industrial enzymes, amylases are highly efficient commercial biocatalysts with an extensive range of utilization in textile, paper and pulp, detergent, leather, waste management, biofuel production and pharmaceutical and food industry. The ruminant's GIT microbiome is a relatively less exploited niche despite being the potentially rich source of biocatalysts of therapeutic, environmental and industrial importance. Therefore, in this chapter, we highlight the need to unravel the ruminant's GIT microbiome with particular emphasis on exploiting *Bacillus* spp. of ruminant origin as potential producers of amylases for utilization on a commercial level.

Keywords *Bacillus* spp. · Amylases · Ruminant · GIT · Microbiome

10.1 Introduction

The use of chemical catalysts to drive several processes in a wide range of sectors has been a significant contributor to environmental pollution. Besides the harmful effects they pose to the environment, chemical catalysts are carried in adverse

---

K. J. Rabapane · A. Mitema · N. A. Feto (✉)
OMICS Research Group, Department of Biotechnology, Vaal University of Technology, Vanderbijlpark, South Africa
e-mail: nasef@vut.ac.za

K. Nelson
J. Craig Venter Institute (JCVI), Rockville, MD, USA

M. T. Islam et al. (eds.), *Bacilli in Agrobiotechnology*, Bacilli in Climate Resilient Agriculture and Bioprospecting,
https://doi.org/10.1007/978-3-030-85465-2_10

reaction conditions, thus leading to elevated costs due to high energy consumptions (Singh et al. 2016a). Microorganisms are the preferred source of many industrial enzymes because of their generation time, ease for genetic manipulation towards desired traits, high yield at low production cost and the relatively stable enzymes they produce, thus resulting in less energy being consumed during industrial applications. In addition, microorganisms can be reproducible within a reasonable time frame and can produce more than one enzyme when used as whole cells during biocatalysis and bioprocessing of biomass.

Among major microbial enzymes, amylases are highly efficient commercial biocatalysts that are in high demand due to their broad range of usage from textile, paper and pulp, detergent, leather, waste management, biofuel production and pharmaceutical to food industry (Gopinath et al. 2017). Although amylases from fungal strains are employed extensively in the starch processing industry, most of the amylases that are applied in the industry have been reported to be produced by *Bacillus* species (Gopinath et al. 2017; Schallmey et al. 2004). *Bacillus* spp. are renowned as bacterial workhorses in microbial fermentation, and their proficiency of certain strains to produce large quantities (20–25 g/l) of extracellular enzymes and to withstand harsh environmental conditions such as various pH and temperature ranges among others has rated them among the most industrial enzyme producers (Schallmey et al. 2004). The ruminant's GIT is a potentially rich, under-exploited source of consortia of microorganisms including *Bacillus* spp., and their presence in the GIT can contribute to our knowledge of microbial responses to environmental stress and their function in complex and fastidious microbial ecologies. *Bacillus* is a broad and diverse genus of bacteria originating from the family of *Bacillaceae*. *Bacillus subtilis* and *B. mesentericus* were the first species to be used for the production of amylase on a commercial scale by Boidin and Effront during the early nineteenth century (Boidin and Effront 1917). This elucidation paved a way for other *Bacillus* and bacterial species and have since been used for the production of enzymes on a commercial scale. Among the primary enzymes produced by *Bacillus* spp. are amylases (Gopinath et al. 2017). Amylase enzymes are required for the complete hydrolysis of starch into fermentable sugars such as glucose and maltose (Liu and Kokare 2017; Sundarram and Murthy 2014). The breakdown process of starch and starch-related polymers by amylase enzymes in food industries is carried out under harsh conditions at very low or high pH and temperature, which involves a two-step process, liquefaction and saccharification that run at 50–60 °C, and thus, the process requires stable amylases that can also tolerate these conditions (Jiang et al. 2015). Furthermore, the detergent industry requires amylases which are stable under alkaline and oxidative, consistent and active at a broad range of temperatures (Fitter 2005). Therefore, there is a high demand to develop enzymes with improved or enhanced operational parameters such as pH and temperature among others that will meet the requirements set by specific applications.

Accordingly, the potential use of microbial enzymes on a commercial scale continues to encourage the development of novel extracellular enzymes with enhanced characteristics from unusual sources of microorganisms (Gopinath et al. 2017). The ruminant's GIT is among one of the sources of microbiomes that haven't been explored comprehensively for industrial use due to certain limitations; hence, in this chapter, we highlight the importance of unravelling the GIT microbiome with particular emphasis on exploiting *Bacillus* spp. of ruminant GIT origin as a potential source of industrial amylases.

10.2 Mechanism of Action and Structure Variation of Amylases

As mentioned in the previous section, among many industrial enzymes, amylases are highly efficient commercial biocatalysts that are in high demand globally. This class of enzymes is responsible for the degradation of starch and other related polymers into simple fermentable sugars (Sundarram and Murthy 2014). The monomeric sugars in starch are held together by glycosidic bonds, which are targeted by amylase during the breakdown of the polymer. Consequently, the manner in which various amylases act on the glycosidic bonds assists in their classification (Liu and Kokare 2017).

In order to fully understand how amylases act on the glycosidic bonds mentioned briefly in the previous section, it is crucial to first elaborate on the metabolism of their primary substrate, starch, a glucose polymer that serves as a major component of food as well as feed, providing a source of energy and carbon to living organisms. This polysaccharide is composed of glucose molecules that are linked together through glucosidic bonds which are required to be hydrolyzed in order for the glucose monomers to be assimilated as fuel. Amylases are responsible for the breakdown down of these glucosidic bonds to produce glucose monomers, maltodextrin, modified starches or glucose syrups (El-Fallal et al. 2012).

Furthermore, starch represents the most abundant form of polysaccharide stored in plants and next to cellulose is the most abundant polysaccharide on earth (Walsh 2014). The two significant constituents that contribute to the complexity of starch are amylose which is a linear molecule and the branched amylopectin (Fig. 10.1). Frequently, amylose makes up approximately 20–30% of the starch, while amylopectin makes 70–80% of the starch. The structure and composition of these molecules depend highly on the source of the carbohydrate. Examples include amylo-maizes which contain over 50% of amylose whereas 'waxy' maize has almost none (~3%) (JJM 1985; Singh et al. 2003).

Starch and starch-containing substrates are utilized in food, textiles, alcohol and other applications; hence, its industry has a significant stake in the market. These substrates are widely available from cheap plant sources, rendering the potential uses of the enzyme more plentiful regarding costs (Gopinath et al. 2017). According to market research future®, the global industrial starch market is projected to amass a revenue of $112 billion by 2024 with a compound annual growth rate of approximately 5.9%. These projections ultimately stimulate the demand for amylase enzymes (Gopinath et al. 2017). The complete breakdown of starch requires a consortium of amylolytic enzymes which break down the $\alpha 1 \rightarrow 4$ and $\alpha 1 \rightarrow 6$ glycosidic linkages into glucose.

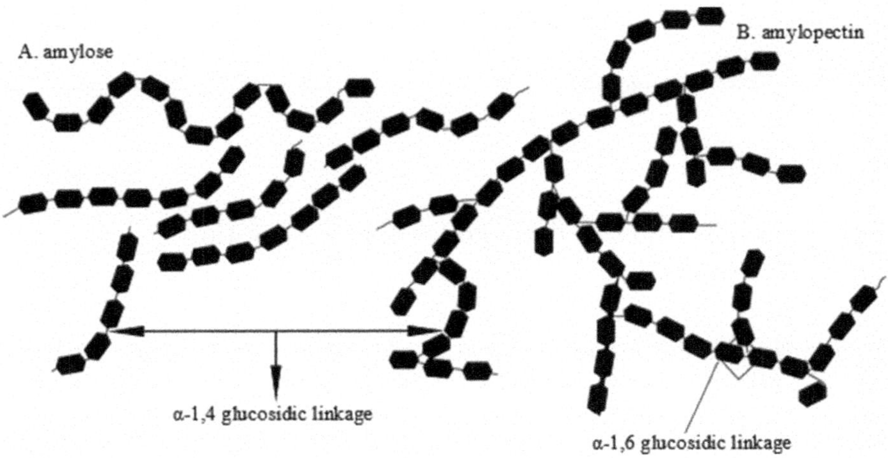

Fig. 10.1 Starch consists of two polymers. (**a**) Amylose is a linear polymer that comprises approximately 6000 glucose units (Gopinath et al. 2017) that are attached by α1 → 4 glycosidic linkages and (**b**). Successive glucose residues are linked via α1 → 4 glycosidic linkages along the linear portion of the molecule, with branch points consisting of α1 → 6 glycosidic linkages. Amylopectin has short α1 → 4 bonded to linear chains of 10–60 glucose units that are branched by α1 → 6 tied to side chains with 15–45 glucose units

Consequently, they are broadly classified into four distinct groups based on the manner in which they act on the substrate, mainly as endoamylases, exoamylases, debranching amylases and glucotransferases. Accordingly, their mechanism of action and structural variations are outlined in the following subsection outlines.

10.2.1 Endoamylases

Endo-acting amylases such as α-amylase (EC 3.2.1.1), also known as endo-1,4-α-D-glucan glucohydrolase, cleave the α-1,4 glycosidic linkages in the internal part of the linear amylose and branched amylopectin chain to yield oligosaccharides with varying lengths and α-limit dextrins, respectively, that ultimately contributes to the branched oligosaccharides (Fig. 10.2) (Walsh 2014; Aiyer 2005). It is also crucial to note that generally, endo-acting amylases cannot cleave the α-1 → 6 of amylopectin, one exception being the α-amylase produced by *Thermoactinomyces vulgaris*, which has been reported to hydrolyze both α-1 → 6 and α-1 → 4 glycosidic linkages in trisaccharides (Sakano et al. 1983). Most α-amylases are metalloenzymes and thus require a calcium ion for their function and stability. Compared to the other classes of amylases, α-amylases are faster rate as they degrade the starch at random locations (Tiwari et al. 2015b; Sarian 2016).

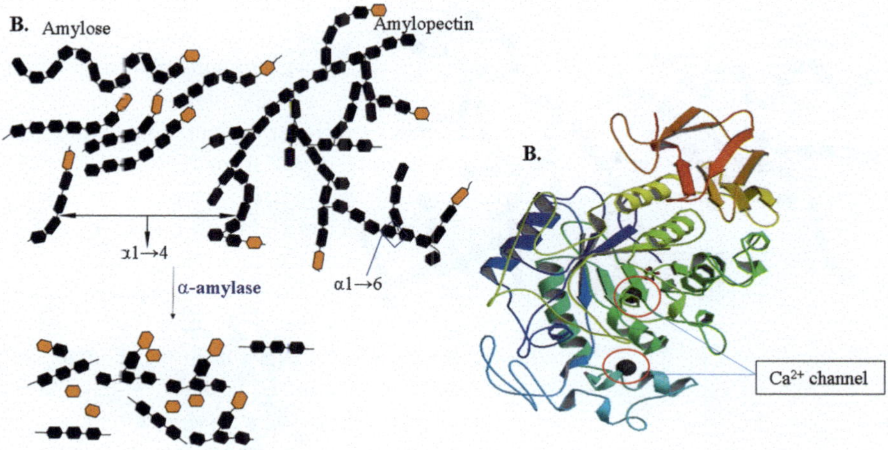

Fig. 10.2 (c) Schematic diagram of the action of alpha amylase on starch and (**b**) the three-dimensional structure of α-amylase with a Ca^{2+} in the middle (Ramasubbu et al. 1996)

10.2.2 Exoamylases

The second group of amylases act on the substrate from the non-reducing end to generate small oligo- and monosaccharides such as maltose by β-amylase or the smallest monomer glucose by glucoamylase. There are three well-known types of exo-acting amylases in the starch degradation process. (**1**) β-amylase (EC 3.2.1.2) exclusively acts on the α-1 → 4 glycosidic bonds at non-reducing ends of the substrate to ultimately generate maltose and β-limit. Animal tissues do not contain this type of amylase; however, it has been reported to be present in microbes inhabiting the digestive tract (Gurung et al. 2013), further cementing the potential of the digestive tract of animals as a source of potent enzymes. (**2**) Glucoamylase (3.2.1.3) hydrolyzes the α-1 → 4 glucosidic bonds at the non-reducing end of amylose to generate glucose monomers in the β-configuration and even though, at a slower rate, it also acts on the α-1 → 6 glucosidic bonds at the branch points of amylopectin to yield glucose molecules (Fig. 10.3). Glucoamylases are capable of reversing hydrolysis reactions to produce maltose and isomaltose, which in turn has a great significance in the industrial process where the sugar content is present (Negi and Vibha 2017). The most distinguishing feature of glucoamylase is its ability to operate under acidic conditions (active around pH <3). The third type of amylase in the exo-acting group is (**3**) α-glucosidase (3.2.1.20); this type of amylase acts on the external glucose residues of either amylose or amylopectin and both to ultimately yield glucose (Sarian 2016).

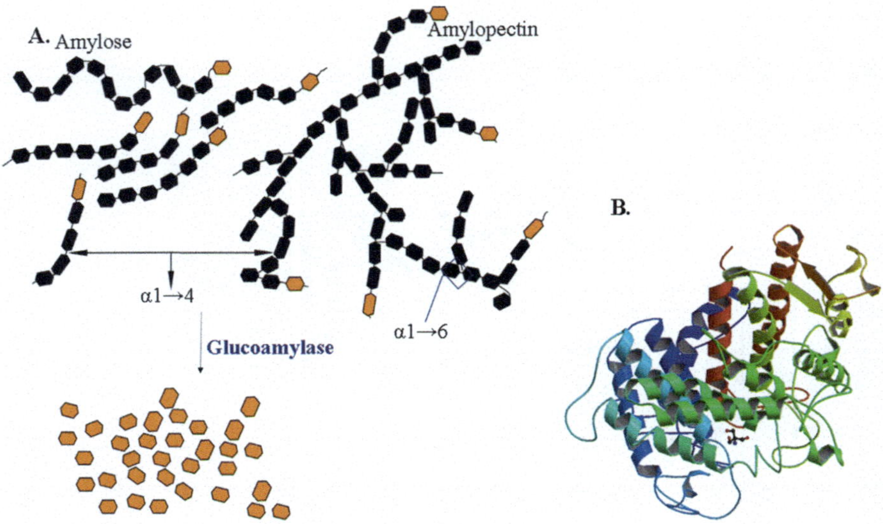

Fig. 10.3 (**a**) Schematic diagram of the action of glucoamylase on starch and (**b**) the three-dimensional structure of glucoamylase from *Saccharomycopsis fibuligera* (Sevcik et al. 1998)

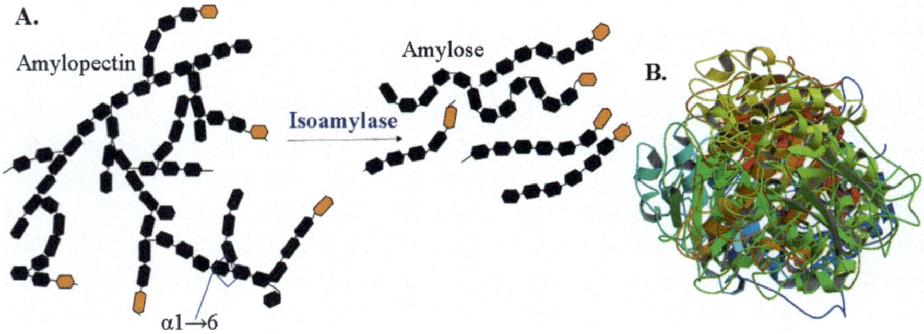

Fig. 10.4 (**a**) Schematic diagram of the action of isoamylase on starch and (**b**) the three-dimensional structure of isoamylase 1 from *Chlamydomonas reinhardtii* (Sim et al. 2014)

10.2.3 Debranching Enzymes

This group of amylases catalyzes the hydrolysis of α1 → 6 glucosidic bonds in amylopectin and or glycogen and other related polymers to yield linear molecules (Fig. 10.4). An example of this group is isoamylase (EC 3.2.1.68).

10.2.4 Glucotransferases

The last group of amylases cleaves the α-1 → 4 glycosidic bond of the donor molecule and transfer a portion of the donor to a glycosidic acceptor with the formation of the new glycosidic bond. Enzymes such as amylomaltase (EC 2.4.1.25) and

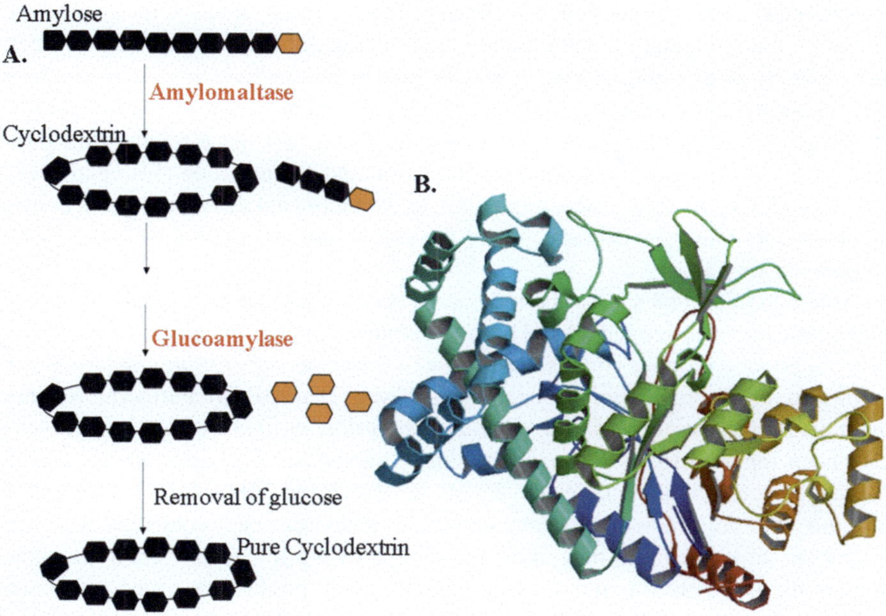

Fig. 10.5 (**a**) Schematic diagram of the effect of amylomaltase on starch (Fujii et al. 1998) and (**b**) the three-dimensional structure of amylomaltase from *Thermus aquaticus* (Przylas et al. 2000)

cyclodextrin glycosyltransferase (EC 2.4.1.19) form a new α-1 → 4 glycosidic bond while branching enzyme (EC. 2.4.1.18) establishes a new α1 → 6 glycosidic bond (Van Der Maarl et al. 2002). A unique feature of amylomaltase is its ability to catalyze the cyclization reaction of cyclic amylose to generate cyclodextrins (Rachadech et al. 2015). Once the cyclodextrins have been created, the amylomaltase is then inactivated, and glucoamylase is introduced to catalyze the hydrolysis of the linear oligosaccharides to produce glucose monomers, Fig. 10.5 shows the action of amylomaltase.

10.3 Source of Microbial Amylases

The breakdown of the glycosidic bonds by amylases is a very crucial process that occurs within biological systems to produce and store energy. Thus, amylases are diversified among various sources ranging from plant, to animal, to microbial kingdoms where they play a crucial role in carbohydrate metabolism (Gopinath et al. 2017). Though plants and animals produce amylases, amylases from both sources don't meet industrial demands, mainly because plant sources are seasonal, and therefore, inconsistencies in enzyme production can be experienced, and yield may be less (Belorkar n.d.). Also, enzymes from plant sources are not stable under adverse conditions. On the other hand, using animal cells to produce enzymes involves sacrificing animals in the process, and the overall operation of animal cell culture makes the production of enzymes from animal sources uneconomical (Belorkar n.d.; Mahalakshmi and Jayalakshmi 2016). As a result, microbial sources are generally dominating driving forces for industrial applications due to enzyme yield, stability and specificity as these attributes in conjunction with additional benefits such as cost-effectiveness, consistency, less time and space required for production, as well as ease of genetic manipulation and optimization (El-Fallal et al. 2012; Pandey et al. 2000), which continue to be significant advantages or requirements to meet industrial demands. Thus, microbial sources are preferred over plant and animal sources.

Although microbial sources include yeast, bacteria and fungi, majority of industrial applications utilize fungal and bacterial strains. Among fungal strains that have been reported to produce amylases are those of the genus *Aspergillus* and have been employed for the preparation of Asian foods (Tiwari et al. 2015a). Furthermore, filamentous fungi, namely, *Rhizopus oryzae*, *Aspergillus awamori* and *A. niger*, have also been reported to be the major producers of industrial glucoamylases which operate at a temperature range of 55–60 °C (Negi and Vibha 2017). Generally, *Bacillus* spp. are renowned to have enhanced attributes when compared to other microbial strains because they are capable of growing in extreme environmental conditions, and therefore, it is expected that they will produce enzymes which are stable in adverse conditions (Feto 2016). Accordingly, some of the species that have been reported to overproduce amylases include *B. cereus (*Sundarram and Murthy 2014*)*, *B. circulans (*Singh and Rani 2014*)*, *B.subtilis* (El-Banna et al. 2007) and others (Gopinath et al. 2017). *Bacillus* spp. including *B. licheniformis*, *B. amyloliquefaciens* and *B. stearothermophilus* are reported to be prolific producers of different types of amylase particularly α-amylases (Liu and Kokare 2017; Pandey et al. 2000; Saini et al. 2017) and also have been widely exploited for the production of many other industrial enzymes due to the inheritance of these thermostable attributes (Schallmey et al. 2004; Tiwari et al. 2015b). *Bacillus subtilis*, *B. licheniformis*, *B. amyloliquefaciens* and *B. stearothermophilus* are reported to provide approximately 60% of commercially available enzymes (Dash et al. 2015). Moreover, *Bacillus* spp. have been approved as Generally Recognized As Safe (GRAS) by the Food Drug and Administration (FDA) (Irajie et al. 2016).

Emerging research studies are continuously focusing on developing *Bacillus* spp. for the production of amylases with improved characteristics to meet industrial demands. Li et al. (2013) purified, characterized and cloned a thermotolerant isoamylase produced from *Bacillus* sp. CICIM 304. This isoamylase displayed its optimal activity at a remarkably high temperature of 70 °C and pH 6.0, with thermostability between 30 °C and 70 °C and an alkaline pH range from 5.5 to 9.0 (Li et al. 2013). Gurumurthy and Neelagund (year) characterized *Geobacillus* sp. iso5, an industrially viable extreme thermostable novel alpha-amylase. The strain was isolated from the thermal water of a geothermal spring for industrial application. This bacterium showed attributes of thermotolerance and alkali-resistance, and it is reported as novel due to its ability to withstand an optimum activity of temperature 90 °C and pH 8.0 (Gurumurthy and Neelagund 2012). The primary interest in exploring thermophilic amylases from unusual sources is their use in elevated temperatures as this increases the reaction rate and thus accelerates the overall process. Moreover, higher temperatures are beneficial because they reduce the risk of microbial contamination and also reduce the viscosity of the reaction mixture, thus providing a considerable energy saving process (Sarian 2016; Pandey et al. 2000). For instance, in the production of glucose from starch, the liquefaction process of starch is carried out by a thermostable α-amylase produced by *B. licheniformis*, which is then followed by a saccharification process achieved by the use of a fungal glucoamylase. The liquefaction process is completed faster because it is carried out by an α-amylase that can operate at 95–105 °C (Sundarram and Murthy 2014; Sarian 2016; Ward 1991). However, because fungal glucoamylases are stable generally up to a temperature range of 55–60 °C (Pavezzi et al. 2008), they delay the overall process because the reaction mixture must be cooled to a temperature that is within that range. Therefore, potential thermostable glucoamylases from thermophilic bacteria can be used at a higher temperature for saccharification processes and thereby reduce the production cost of glucose from the high energy required for cooling (Negi and Vibha 2017).

A suitable organism is crucial for the production of amylases that will meet industrial demands. Depending on the type of application, thermotolerance is a desirable trait for some amylase enzymes and other types of enzymes to meet industrial needs (Gurung et al. 2013). Table 10.1 shows some of the enzymes and parameters produced by fungal and bacterial strains. From these findings, it is clear that *Bacillus* spp. sources have a higher thermostability and higher pH range, and this, among other attributes, has resulted in their utilization in industrial applications that are carried out at higher temperatures and alkaline conditions. Hence, there is a continuous need to explore bacterial species with enhanced characteristics, in particular those of *Bacillus*, for the production of amylases for use on a commercial scale.

A suitable organism is crucial for the production of amylases that will meet industrial demands. For an amylase enzyme and other types of enzymes to meet

Table 10.1 Optimum pH and temperature of enzymes produced by fungal and bacterial strains

Microbe	Enzyme	pH	Temperature °C	References
Bacillus flavocaldarius	Type I pullulanase	7.0	75–80	Suzuki et al. (1991)
Bacillus subtilis	α-Amylase	7.0	135	Asgher et al. (2007)
Halomonas meridian	α-Amylase	7.0	37	Coronado et al. (2000)
Aspergillus flavus	α-Amylase	6.0	55	Khoo et al. (1994)
Aspergillus awamori	Glucoamylase	4.5	60	Yamasaki et al. (1977)
Clostridium	Glucoamylase	4.5	65	Ohnishi et al. (1991)
Thermomyces	Glucoamylase	–	70	Rao et al. (1981)
Bacillus clausii	Protease	11.5	60	Horikoshi (1971)
Aspergillus awamori	Glucoamylase	5–7	55	Pavezzi et al. (2008)
Geobacillus sp. Iso5	α-Amylase	8.0	90	Gurumurthy and Neelagund (2012)

industrial needs, it has to be thermotolerant because thermotolerance is a desirable attribute for significant groups of enzymes that are exploited at the industrial level (Gurung et al. 2013). Table 10.1 shows some of the enzymes and parameters produced by fungal and bacterial strains. From these findings, it is clear that *Bacillus* spp. sources have a higher thermostability and higher pH range, and this, among other attributes, has resulted in their utilization in industrial applications that are carried out at higher temperatures and various pH ranges. Hence, there is a continuous need to explore bacterial strains with enhanced characteristics more especially those of *Bacillus* for the production of amylases for use on a commercial scale.

10.4 Amylase Enzyme Production by *Bacillus* spp. in the Ruminant's GIT

The digestive system in ruminants is represented by a four-stomach compartment that is divided into the paunch (rumen), reticulum, omasum and abomasum (Fig. 10.6) (Hungate 1947). This digestive system is uniquely designed to process plant-based feed such as phytate and lignocellulosic material, which sets them apart from monogastric mammals. Unlike monogastrics, which have specialized enzymes, ruminants depend on consortia of microorganisms within their GIT to bring about feed conversion through a fermentation process (Ribeiro et al. 2016). Thus, the ruminant's GIT is an anaerobic ecosystem that is inhabited by an internetwork of various microorganisms working in synergy to bring about feed conversion. Feed intake by ruminants triggers the release of large volumes of saliva which facilitate the mechanical pretreatment of feed from the oesophagus to the rumen, reticulum and/or omasum until it reaches the abomasum which is also called the true stomach because it is similar to that of monogastrics (Moran 2005). The actual microbial fermentation of feed starts in the rumen, and the rumino-reticulum permits regurgitation of partially fermented ingesta (cud) back to the mouth for rechewing (rumination) which further breaks down the ingested feed into a small particle size that is

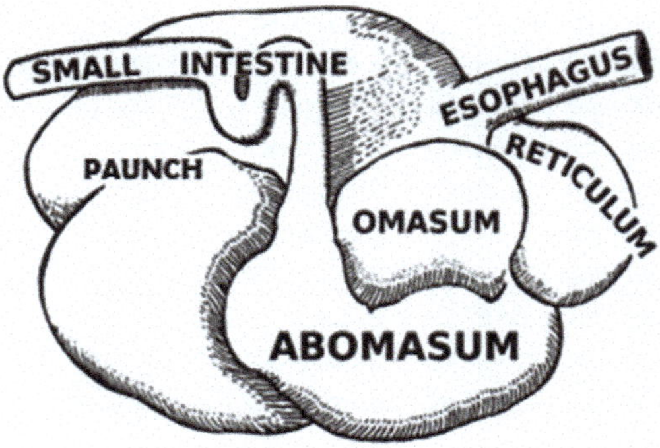

Fig. 10.6 A digestive system of a ruminant animal (Ruminant n.d.)

able to pass through the other compartments for complete digestion (Moran 2005) and further goes processed through the intestines as faecal matter.

The secreted saliva also releases salts such as bicarbonate and ultimately acts as a buffering agent regulating the physiological pH range between 5.5 and 6.9 (Choudhury et al. 2015). In addition to the maintained pH, the temperature range in the ruminant's GIT is between 38 °C and 42 °C (Hill et al. 2016), creating a conducive environment which potentially allows a vast array of anaerobic and facultative anaerobic microorganisms to grow. Accordingly, the microorganisms in the GIT include bacteria (10^{11} cells/mL), methanogenic archaea (10^6 cells/mL), protozoa (10^4–10^6 cells/mL), fungi (10^3–10^6 zoospore/mL) and viruses (10^9 particles/mL), with bacteria being the most abundant and extensively studied [54, 59]. Although diverse and operate synergistically to facilitate feed conversion sequentially, anaerobic fungi being the most efficient fibre degraders by acting on the cuticle of lignocellulosic material which then becomes easier for bacteria and ciliate protozoa to gradually act upon. With a biomass turnover of approximately 50%, bacteria in the GIT are also actively involved in both the degradation of complex and simple carbohydrates. The protozoa make use of the complex and simple carbohydrates by means of engulfment to bring about end products including ammonia and amino acids from microbial protein as they also engulf bacteria in addition to feed (Kamra 2005). Ultimately, the interaction between bacteria, fungi and protozoa results in the production of gases such as carbon dioxide (CO_2) and hydrogen (H_2) which are then utilized by methanogens to produce methane gas (CH_4). Hence, methanogens are accountable for CH_4 belching in ruminant animals as they actively reduce the buildup of H_2, which would interfere with bacterial dehydrogenases in the GIT, consequently reducing the fermentation rate, digestibility and eventually feed intake (Patra et al. 2017; Cieslak et al. 2013). Meanwhile, the viruses in the GIT are reported to be responsible for the biomass turnover through cell lysis which leads to direct availability of amino acids from microbial protein for the ruminants as a host (Kamra 2005), and some of these phages are under investigation for the abatement of methane in the GIT (Leahy et al. 2010).

Overall, the microorganisms in the GIT all play unique roles including pectinolytic (Paster and Canale-Parola 1985), cellulolytic (Fujimoto et al. 2011), amylolytic (Hungate et al. 1952), xylanolytic (Seo et al. 2013), proteolytic (Wallace 1996), lipolytic (Hobson and Mann 1961) and hemicellulolytic (Seo et al. 2013; Nyonyo et al. 2014; Šimůnek et al. 2018), among others, during feed fermentation which results in the production of VFAs, amino acids and various gases among other by-products, thus making the ruminant's GIT a potentially rich source of not only amylase enzymes but also other biocatalysts for industrial applications, such as cellulase, protease, lipase and xylanase. Although the GIT microbiome is well adapted to convert various ingested feed into nutrients, it is still mostly unknown and represents an untapped microbial consortium. This limitation is supported by the fact that the GIT is a fastidious environment in that it is anoxic, the temperature and pH are naturally maintained and mimicking this type of environment in the lab remains a challenge (Zehavi et al. 2018). Accordingly, conventional culture techniques have resulted as a major limitation as they have been reported to cover about 11%

(Zehavi et al. 2018) of the representative sample which ultimately results in insufficient functional genomic studies on the ruminant's GIT species.

Bacteriodetes, *Firmicutes*, *Proteobacteria* and *Actinobacteria* are the most reported dominant phyla in the GIT (Patel et al. 2014; Firkins and Yu 2006). The population of the GIT microbiome is highly dependent on the host species, age, feeding habits (diet) and geographical localizations with some being common among ruminants worldwide (Ogata et al. 2019; Kim et al. 2018; Henderson et al. 2015). Cunha et al. (2011) reported the characterization of the bacterial and archaeal community structure in the rumen microbiome of goats (*Capra hircus*) from the semiarid region of Brazil, and accordingly, sequences of the bacteria from the phyla *Bacteroidetes* and *Firmicutes* were predominant (Cunha et al. 2011). Patel et al. (2014) explored the buffalo rumen metagenome where they outlined carbohydrate-active enzymes (CAZymes) and provided insight on the abundance of other enzyme families and the alteration of microbial diversity in response to variation in the diet (Patel et al. 2014). They described the analysis of (CAZymes) from 3.5 gigabase sequences of metagenomic data from rumen samples of Mehsani buffaloes fed on different proportions of green or dry roughages to concentrate ration. A total of 2597 contigs encoding putative CAZymes were identified, and the phylogenetic analysis of these contigs by MG-RAST revealed the predominance of *Bacteroidetes*, followed by *Firmicutes*, *Proteobacteria* and *Actinobacteria* phyla (Patel et al. 2014). Consequently, with *Firmicutes* being among the dominant phyla in the GIT, chances of obtaining *Bacillus* spp. are potentially higher. This, however, stands to be supported by research that is focused on the objective of exploring *Bacillus* spp. of which is currently insufficient in this niche due to the primary limitation mentioned previously. Thus, although the sequence and structural community studies reveal knowledgeable insight on the abundance of microbial population in the GIT, this has resulted in limited information on the individual species because there is no link between the individual species and the predicted genes.

Interestingly, there are developments on the improvement of the reference databases such as the Hungate 1000 project which aims to increase the microbial culture collection of rumen origin and to culture as-yet uncultured microorganisms from the ruminant's GIT in order to fully understand how the GIT system operates (Seshadri et al. 2018). In a study by Seo et al. (2013) (Seo et al. 2013), a facultative anaerobic bacteria *Bacillus licheniformis* JK7 was isolated and was reported to be capable of secreting endoglucanase, β-glucosidase and xylanase in the rumen of a native Korean goat, which can survive on harsh condition, such as provision of low-quality roughage as a sole feed source. According to the authors, the isolation of cellulolytic and xylanolytic *Bacillus* sp. from the rumen of goats has not previously been reported. Among the Hungate 1000 project is a study by Nathani et al. (2014) (Nathani et al. 2014), which reported the whole genome sequencing and functional annotation of amylolytic anaerobic spore-forming *Bacillus nealsonii strain* AAU1 isolated from the rumen of Surti buffalo. The optimal growth of *B. nealsonii strain* AAU1 was observed at 40 °C (range: 30 °C to 45 °C) and pH 6.5 (range: 5.5 to 7.5) under anaerobic conditions. Three unpublished strains *B. cereus* KPR-7A and *Bacillus* sp. MB2021 from cow rumen as well as *B. licheniformis* VTM3R78 from moose rumen are also reported (Seshadri et al. 2018). Among these three strains, only the latter is capable of utilizing starch as a substrate. According to literature reviews, research on the production of amylase enzymes by *Bacillus* spp. of ruminant's GIT origin is insufficient and remains a field to be explored in depth.

10.5 Industrial Application of Amylases

The industrial enzyme market is estimated to reach US $ 6.2 billion by 2020 (Mehta and Satyanarayana 2016). One of the significant reasons for the continuous increase in the global sales of microbial enzymes is the rise in demand for consumer goods and biofuels. The use of amylase enzymes has replaced acid hydrolysis of starch, and currently, amylases have a great significance in present-day biotechnology having approximately 25–30% of the world enzyme market to date (Dash et al. 2015). These vast potentials of amylases to be employed in a wide range of industries have placed a more significant challenge for researchers. As a result, there is a continuous exploration for more active amylase producing *Bacillus* strains with enhanced potential characteristics for industrial utilization (Table 10.2).

As mentioned previously, amylases are one of the primary enzymes that are exploited in the industry. The history of the industrial production of enzymes dates back to the time when Dr. Jhokichi Takamine began the production of digestive enzyme preparation by wheat bran koji culture of *Aspergillus oryzae* in 1894.

Table 10.2 Industrial applications of amylases from *Bacillus* spp.

Industry	Enzyme	Function	Source	References
Baking	Amylase Maltogenic α-Amylase	Flour adjustment Enhance the shelf life of bread	*Bacillus* sp. *Bacillus stearothermophilus*	Singh et al. (2016a)
Beverage	α-Amylase β-Amylase	Starch hydrolysis Starch hydrolysis	*Bacillus* sp. *Bacillus* sp.	Singh et al. (2016a) and Rejzek et al. (2011)
Paper and pulp	Amylase	Deinking, drainage improvement	*Bacillus licheniformis*	Mojsov (2012)
Detergent	Amylase	Carbohydrate strain removal	*Bacillus subtilis*	Schallmey et al. (2004)
Leather	Amylase	Fibre splitting	*Bacillus subtilis*	Singh et al. (2016a)
Waste management	Amylase	Bioremediation of vegetable wastes	*Bacillus licheniformis*	Singh et al. (2016a)
Textile	Amylase	Desizing	*Bacillus* sp., *B. licheniformis*	Schallmey et al. (2004)
Therapeutic	α-Amylase	As a digestive disorder treatment	*Bacillus* spp.	Singh et al. (2016a)
Ethanol production	α-Amylase	Starch hydrolysis	*Bacillus licheniformis*	Gurung et al. (2013)

Industrial production of dextrose powder and dextrose crystals from starch using α-amylase and glucoamylase then began in 1959. From then on, amylases are being utilized for various purposes (Aiyer 2005). Moreover, amylases have recently found their use in other applications such as fuel production, removal of starch sizer from textile, detergent industry, paper and pulp, brewing and pharmaceutical industry among others (Saini et al. 2017).

10.5.1 Biofuel Production

The focus on biofuels to secure future energy solutions promotes a high demand for enzymes that are able to produce biofuels in an eco-friendly manner while proficiently utilizing abundant polysaccharides such as lignocellulosic materials and starch polymers, for instance, ethanol, which is among the most utilized liquid biofuel that can be derived from renewable resources such as waste generated from crops and by-products. Accordingly, starch is the most used substrate for the production of bioethanol due to its low price and readily available raw material in most regions of the world (Gopinath et al. 2017). Amylases, in particular alpha amylases and glucoamylases, are crucial to producing fermentable sugars for the production of ethanol (Saini et al. 2017).

10.5.2 Therapeutic

Mayo Clinic refers to malabsorption syndrome as many disorders in which the small intestine can't absorb enough specific nutrients and fluids. As a result of this syndrome, individuals can experience symptoms such as weight loss, bloating and diarrhoea, and in most cases, this condition can eventually affect the brain, nervous system, bones, liver and other organs. One of these disorders caused by malabsorption is a digestive disorder which generally occurs due to lack of or insufficient digestive enzymes in the body. In cases like these, amylases are used as therapeutic drugs in health issues related to enzymatic deficiency and gastrointestinal disorders (Singh et al. 2016a; Mane and Tale 2015).

10.5.3 Brewing

In the brewing industry, once the starch has been solubilized, it is subjected to two enzymatic steps including saccharification, where starch is converted into sugar using an amylolytic microorganism or enzymes such as glucoamylase and α-amylase to obtain fermentable sugars, followed by fermentation, where sugar is converted into ethanol using an ethanol fermenting organism such as yeast *Saccharomyces cerevisiae* (Tiwari et al. 2015a; Mobini-Dehkordi and Javan 2012). Accordingly, in the conversion of crushed starch to fermentable sugars, such as maltose, amylase enzymes are used to improve malting and mashing as well as enhancing the aroma and quality of beer together with the primary fermentation (Kuhad et al. 2011).

10.5.4 Paper and Pulp

The paper and pulp industry is among the practices that utilize a large amount of resources such as water, wood and energy. The activities in this industry result in a significant amount of solid wastes and waste water that require extensive treatment. Thus, the increasing awareness of sustainability concerns the use of microbial amylases in this industry has grown steadily to minimize the adverse effect on the environment. The use of amylase enzymes improves draining and enzymatic deinking which then result in reduced energy and chlorine requirement. Furthermore, the paper and pulp use amylases for the reduction of starch viscosity to achieve the appropriate coating of paper for application to fibres.

10.5.5 Textile

Starch is incorporated into the yarn before fabric production for a fast and weaving process where amylase is then employed to hydrolyze and solubilize starch, which then washes out the cloth increasing the stiffness of the finished material. Fabrics are further sized with starch, and alpha-amylase is then used as a desizing agent for removing starch from the grey cloth before it is further processed in bleach and dye (Saini et al. 2017).

10.5.6 Bioremediation

The execution of strict protocols for waste disposal into the environment has compelled the urge to reinforce research that is based on finding alternative processes in particular enzymes for waste water treatment. Starch occurs extensively in waste materials produced from the processing of plant raw material. Starch processing waste is produced in large quantities, and this results in environmental pollution. A cocktail of microbial amylases or a consortium of microbes producing amylases are then used for bioremediation of starch pollutant materials (Tiwari et al. 2015a; Mobini-Dehkordi and Javan 2012).

10.5.7 Detergent

The detergent industry is among the primary consumers of enzymes as it relates to both volume and value. It dates back to 1913 when the pancreatic extract was used for the first time in the enzyme-detergent development, and then later, a microbial enzyme under the trade name BIO-40 was commercialized. Amylases are the second group of enzymes used in the formulation of enzymatic detergent, and 90% of all liquid soaps contain these enzymes. They catalyze the hydrolysis of glucosidic linkages in starch stains, commonly found in foods such as pasta, fruit, chocolate, baby food and barbecue sauce, as well as gravy, and consequently provide an effective cleaning action without damaging fibres while making the detergent environmentally safe. As coloured stains, their removal is of interest in both detergent and dishwashing contexts. Amylases used in the detergent industry are derived from *Bacillus* or *Aspergillus* (Sahni and Goel 2015; Mobini-Dehkordi and Javan 2012).

10.5.8 Medicine and Analytical

Enzyme utilization in medicine is extensive as in the industry and is overgrowing. Plasma serum amylase is measured for medical diagnosis. A standard concentration of amylase for adults in the blood is in the range of 21–101 U/L. Higher than normal levels may speculate one of several medical conditions including acute inflammation of the pancreas, perforated peptic ulcer, strangulation of ileus in humans, torsion of an ovarian cyst, macroamylasemia and mumps. Amylase concentration may be measured in other body fluids, including urine and peritoneal fluid (Tiwari et al. 2015a).

In recombinant DNA technology, the presence of amylase serves as an additional approach for selecting successful recombinants of a reporter construct in addition to antibiotic resistance. As homologous regions of the structural gene flank reporter genes for amylase, successful integration will disrupt the amylase gene and prevent starch degradation, which is easy to detect when stained with iodine (Singh et al. 2011).

10.5.9 Baking

The global market for baking enzymes is estimated at US $754.1 million for the current year, 2020, and is projected to reach a revised size of US $1 Billion by 2027, growing at a compound annual growth rate of 4.7%. Among the major baking enzymes, amylases play a pivotal role in the hydrolysis of starch and produce small dextrins for the yeast to act upon. Additionally, amylases also degrade the damaged starch in the wheat flour to produce small dextrins, which allows the yeast to be continually active during dough fermentation, resulting in improved bread volume and crumb texture (Saini et al. 2017). Although amylases can be produced by these yeast cells, it takes a while for the yeast to produce enough of these amylases to hydrolyze substantial amounts of starch in the bread. Hence, amylases are included in the bread improver to accelerate the process, thereby making the bread-making process economical. Currently, a thermostable maltogenic amylase of *Bacillus stearothermophilus* is used commercially in the bakery industry (Mobini-Dehkordi and Javan 2012).

10.5.10 Other Applications

The starch hydrolysis industry has the most widespread applications of amylases that are used during starch hydrolysis in the starch liquefaction process which converts starch into fructose and glucose syrups (Nielsen and Borchert 2000). In the therapeutic industry, amylases are utilized as therapeutic drugs in health issues related to enzymatic deficiency and gastrointestinal disorders (Mane and Tale 2015; Singh et al. 2016b) such as malabsorption, a digestive disorder which generally occurs due to lack of or insufficient digestive enzymes in the body (Singh et al. 2016a; Mane and Tale 2015). In the candy industry, amylases are used for the production of desired softness in candy. They are also used for partial or total hydrolysis of cornstarch to produce large quantities of sweeteners in the glucose and syrup industry. Moreover, they are applied for hydrolysis of starch causing turbidity due to insolubility in the fruit juice industry (Saini et al. 2017).

10.6 Future Research Outlooks

1. Evaluate the cultural possibilities of capturing more *Bacillus* strains including other microorganisms from the ruminant's GIT by tampering with the culturing conditions and exploring the depth of OMICS technology, in particular, culturomics and functional genomics, among others, in order to bridge the gap between potential *Bacillus* spp. and the predicted sequences.
2. It may also be advisable to conduct more studies on ruminants that are free grazing as this will potentially increase the chances of obtaining more strains as compared to studies whereby animals are under a specified diet due to a constant evolution of microorganisms adapting to various feed.
3. Although the utilization of amylases, in starch processing, has been playing a major role for decades, the application of amylases or enzymes in general as pure extracts are not competent as industrial biocatalysts in their natural state because their catalytic efficiency is low, and they are prone to denaturation, thus affecting their performance and shelf-life. Furthermore, even when they are used as whole microbial cells for bioprocessing of biomass, overproduction or high

titer values are near impossible to obtain in the native microorganism due to feedback inhibition because the organism can only produce the required amount of enzyme for its survival. Therefore, it is advisable to overcome these limitations with molecular techniques such as genetic engineering and recombinant DNA technology that improve or enhance the attributes of the potential *Bacillus* strains once obtained.

4. Another alternative and sustainable approach which contributes towards the economic use of enzymes and to enhance operational activity and continuous usage of enzymes is immobilization, which involves physical adsorption (an enzyme can be attached to an inert, organic or inorganic or insoluble support), entrapment, covalent bonding and cross-linking as some of the most used methods to facilitate enzyme immobilization. This technique should be employed to ensure specificity and stability of the biocatalysts for continuous applications on a large scale under a wide range of adverse conditions, such as pH and temperature. Additionally, with the advent of frontiers, it may also be worthy to explore the use of nanomaterials to immobilize and stabilize amylases.

10.7 Concluding Remarks

When considering all the research that has been done so far, it can be concluded that the starch industry is on a continuous rise due to its high demand in different industries from food, to detergent, to textile industries, which in turn implies the increased demand for amylolytic enzymes. Amylases continue to prove their importance in the industrial and environmental sectors and for medicinal purposes. To this effect, efforts are being made to develop amylases with an extreme range of pH, temperature and better catalytic efficiency to meet the industrial requirements. As far as the literature reviewed is concerned, biomolecules of *Bacillus* spp. largely fulfil such requirements, further affirming the central role of the genus as potential microbial cell factories. The utilization of the technique progression can also contribute in many significant ways towards the improvement and enhancement of amylase enzymes for a constant supply to the high demand and to meet future challenges. Despite the potential of ruminal microorganisms as a source of industrial enzymes, knowledge about these microbial diversity remains relatively unknown.

References

Aiyer PV (2005) Amylases and their applications. Afr J Biotechnol 4:1525–1529

Asgher M, Asad MJ, Rahman S, Legge R (2007) A thermostable α-amylase from a moderately thermophilic Bacillus subtilis strain for starch processing. J Food Eng 79(3):950–955

Belorkar SA (n.d.). Current research in microbiology

Boidin A, Effront J. Process for treating amylaceous substances. 1917, US Patent 1,227,374:1–5

Choudhury PK, Salem AZM, Jena R, Kumar S, Singh R, Puniya AK (2015) Rumen microbiology: an overview. In: Puniya AK, Singh R, Kamra DN (eds) Rumen microbiology: from evolution to revolution. Springer, New Delhi, pp 3–16

Cieslak A, Szumacher-Strabel M, Stochmal A, Oleszek W (2013) Plant components with specific activities against rumen methanogens. Animal 7(s2):253–265

Coronado M-J, Vargas C, Mellado E, Tegos G, Drainas C, JNJ N, Ventosa A (2000) The α-amylase gene amyH of the moderate halophile Halomonas meridiana: cloning and molecular characterization. Microbiology 146(4):861–868

Cunha IS, Barreto CC, Costa OYA, Bomfim MA, Castro AP (2011) Bacteria and archaea community structure in the rumen microbiome of goats (*Capra hircus*) from the semiarid region of Brazil. Anaerobe 17:118–124

Dash BK, Rahman MM, Sarker PK (2015) Molecular identification of a newly isolated *Bacillus subtilis* BI19 and optimization of production conditions for enhance production of extracellular amylase. Biomed Res Int:1–9

El-Banna TE, Abd-Aziz AA, Abou-Dobara MI, Ibrahim RI (2007) Production and immobilization of alpha-amylase from Bacillus subtilis. Pak J Biol Sci 10(12):2039–2047

El-Fallal A, Dobara MA, El-Sayed A, Omar N (2012) Starch and microbial alpha-amylase: from concepts to biotechnological applications. In Chuan-Fa Chang (ed) Carbohydrates- comprehensive studies on glycobiology and glycotechnology. Intech Open, United Kingdom, pp 461–488

Feto NA (2016) Bacillus spp. and their biotechnological roles in green industry. In: Islam MT, Rahman M, Pandey P, Jha CK, Aeron A (eds) Bacilli and agrobiotechnology. Springer, Cham, pp 143–162

Firkins JL, Yu Z (2006) Characterisation and quantification of the microbial populations of the rumen. In: Ruminant physiology, digestion, metabolism and impact of nutrition on gene expression, immunology and stress. Wageningen Academic Publishers, Wageningen, pp 19–54

Fitter J (2005) Structural and dynamical features contributing to thermostability in alpha-amylases. Cell MolLife Sci 62:1925–1937

Fujii K, Okada S, Takaha T, Takata H, Terada Y, Yanase M (1998) Heat-resistant amylomaltase. Google Patents

Fujimoto N, Kosaka T, Nakao T, Yamada M (2011) Bacillus licheniformis bearing a high cellulose-degrading activity, which was isolated as a heat-resistant and micro-aerophilic microorganism from bovine rumen. Open Biotechnol J 5(1)

Gopinath SCB, Anbu P, Md-Arshad MK, Lakshmipriya T, Voon-Hong C, Hashim U, Chinni SV (2017) Biotechnological processes in microbial amylase production. Biomed Res Int 2017:1–9

Gurumurthy DM, Neelagund SE (2012) Molecular characterization of industrially viable extreme thermostable novel alpha-amylase of gebacillus sp Iso5 isolated from geothermal spring. J Pure Appl Microbiol 6:1759–1773

Gurung N, Ray S, Bose S, Rai V (2013) A broader view: microbial enzymes and their relevance in industries, medicine and beyond. Biomed Res Int:1–18

Henderson G, Cox F, Ganesh S, Jonker A, Young W, Collaborators GRC, Abecia L, Angarita E, Aravena P, Arenas GN (2015) Rumen microbial community composition varies with diet and host, but a core microbiome is found across a wide geographical range. Sci Rep 5:14567

Hill J, McSweeney C, Wright A-DG, Bishop-Hurley G, Kalantar-Zadeh K (2016) Measuring methane production from ruminants. Trends Biotechnol 34(1):26–35

Hobson P, Mann S (1961) The isolation of glycerol-fermenting and lipolytic bacteria from the rumen of the sheep. Microbiology 25(2):227–240

Horikoshi K (1971) Production of alkaline enzymes by Alkalophilic microorganisms: part I. alkaline protease produced by Bacillus no. 221. Agric Biol Chem 35(9):1407–1414

Hungate RE (1947) Studies on cellulose fermentation: III. The culture and isolation for cellulose-decomposing bacteria from the rumen of cattle. J Bacteriol 53(5):631–645

Hungate R, Dougherty R, Bryant M, Cello R (1952) Microbiological and physiological changes associated with acute indigestion in sheep. Cornell Veterinarian 42:423–449

Irajie C, Mohkam M, Nezafat N, Hosseinzadeh S, Aminlari M, Ghasemi Y (2016) In silico analysis of glutaminase from different species of Escherichia and Bacillus. Iran J Med Sci 41(5):406

Jiang T, Cai M, Huang M, He H, Lu J, Zhou X, Zhang Y (2015) Characterization of a thermo-stable raw-starch hydrolyzing alpha amylase from deep-sea thermophile *Geobacillus sp*. Protein Expr Purif 114:15–22

JJM S (1985) Composition and properties of commercial native starches. Wiley Online Library 37(1)

Kamra DN (2005) Rumen microbial ecosystem. Curr Sci:124–135

Khoo S, Amirul A-A, Kamaruzaman M, Nazalan N, Azizan M (1994) Purification and characterization of α-amylase from *Aspergillus flavus*. Folia Microbiol 39(5):392–398

Kim Y-H, Nagata R, Ohkubo A, Ohtani N, Kushibiki S, Ichijo T, Sato S (2018) Changes in ruminal and reticular pH and bacterial communities in Holstein cattle fed a high-grain diet. BMC Vet Res 14(1):310

Kuhad RC, Gupta R, Singh A (2011) Microbial cellulases and their industrial applications. Enzyme Res 2011

Leahy SC, Kelly WJ, Altermann E, Ronimus RS, Yeoman CJ, Pacheco DM, Li D, Kong Z, McTavish S, Sang C (2010) The genome sequence of the rumen methanogen Methanobrevibacter ruminantium reveals new possibilities for controlling ruminant methane emissions. PLoS One 5(1):e8926

Li Y, Niu D, Zhang L, Wang Z, Shi G (2013) Purification, characterization and cloning of a thermotolerant isoamylase produced from Bacillus sp. CICIM 304. J Ind Microbiol Biotechnol 40:437–446

Liu X, Kokare C (2017) Microbial enzymes of use in industry. In: Biotechnology of microbial enzymes. Elsevier, pp 267–298

Mahalakshmi N, Jayalakshmi S (2016) Research article Cellulase production by *Aspergillus niger* under solid state fermentation using agro industrial wastes. Int J Adv Multidiscip Res 3(2):78–83

Mane P, Tale V (2015) Overview of microbial therapeutic enzymes. Int J Curr Microbiol App Sci 4(4):17–26

Mehta D, Satyanarayana T (2016) Bacterial and archaeal α-amylases: diversity and amelioration of the desirable characteristics for industrial applications. Front Microbiol 7:1129

Mobini-Dehkordi M, Javan FA (2012) Application of alpha-amylase in biotechnology. J Biol Today's World 1(1):15–20

Mojsov K (2012) Microbial alpha-amylases and their industrial applications: a review. International Journal of Management, IT and Engineering (IJMIE) 2(10):583–609

Moran J (2005) Tropical dairy farming: feeding management for small holder dairy farmers in the humid tropics. Csiro Publishing

Nathani NM, Duggirala SM, Bhatt VD, KaPatel J, Joshi CG (2014) Genomic analysis of a novel strain of *Bacillus nealsonii*, isolated from Surti buffalo rumen. Adv Biosci Biotechnol 2014

Negi S, Vibha K (2017) Amylolytic enzymes: Glucoamylases. In: Current developments in biotechnology and bioengineering: production and isolation of industrial products. Elsevier Science, Saint Louis, pp 25–46

Nielsen JE, Borchert TV (2000) Protein engineering of bacterial α-amylases. Biochimica et Biophysica Acta (BBA)-Prot Struct Mol Enzymol 1543(2):253–274

Nyonyo T, Shinkai T, Mitsumori M (2014) Improved culturability of cellulolytic rumen bacteria and phylogenetic diversity of culturable cellulolytic and xylanolytic bacteria newly isolated from the bovine rumen. FEMS Microbiol Ecol 88(3):528–537

Ogata T, Makino H, Ishizuka N, Iwamoto E, Masaki T, Ikuta K, Kim Y-H, Sato S (2019) Long-term high-grain diet altered the ruminal pH, fermentation, and composition and functions of the rumen bacterial community, leading to enhanced lactic acid production in Japanese black beef cattle during fattening. PLoS One 14(11):e0225448

Ohnishi H, Sakai H, Ohta T (1991) Purification and some properties of a glucoamylase from Clostridium sp. G0005. Agric Biol Chem 55(7):1901–1902

Pandey A, Nigam P, Soccol CR, Soccol VT, Singh D, Mohan R (2000) Advances in microbial amylases. Biotechnol Appl Biochem 31(Pt 2):135–152

Paster BJ, Canale-Parola E (1985) Treponema saccharophilum sp. nov., a large pectinolytic spirochete from the bovine rumen. Appl Environ Microbiol 50(2):212–219
Patel DD, Patel AK, Parmar NR, Shah TM, Patel JB, Pandya PR, Joshi CG (2014) Microbial and carbohydrate active enzyme profile of buffalo rumenmetagenome and their alteration in response to variation in the diet. Gene 545:88–94
Patra A, Park T, Kim M, Yu Z (2017) Rumen methanogens and mitigation of methane emission by anti-methanogenic compounds and substances. J Anim Sci Biotechnol 8(1):13
Pavezzi FC, Eleni G, da Silva R (2008) Production and characterization of glucoamylase from fungus *Aspergillus Awamori* expressed in yeast saccharomyces cerevisiae using different carbon sources. Bra J Microbiol 39:108–114
Przylas I, Tomoo K, Terada Y, Takaha T, Fujii K, Saenger W, Strater N (2000) Crystal structure of amylomaltase from thermus aquaticus, a glycosyltransferase catalysing the production of large cyclic glucans. J Mol Biol 296(3):873–886
Rachadech W, Nimpiboon P, Naumthong W, Nakapong S, Krusong K, Pongsawasdi P (2015) Identification of essential tryptophan in amylomaltase from Corynebacterium glutamicum. Int J Biol Macromol 76:230–235
Ramasubbu N, Paloth V, Lou Y, Brayer GD, Levine MJ (1996) Structure of human salivary alpha amylase at 1.6 angstrom resolution: implications for its role in the oral cavity. Acta Crystallogr
Rao VB, Sastri N, Rao PS (1981) Purification and characterization of a thermostable glucoamylase from the thermophilic fungus *Thermomyces lanuginosus*. Biochem J 193(2):379–387
Rejzek M, Stevenson CE, Southard AM, Stanley D, Denyer K, Smith AM, Naldrett MJ, Lawson DM, Field RA (2011) Chemical genetics and cereal starch metabolism: structural basis of the non-covalent and covalent inhibition of barley β-amylase. Mol BioSyst 7(3):718–730
Ribeiro G, Gruninger R, Badhan A, McAllister T (2016) Mining the rumen for fibrolytic feed enzymes. Anim Front 6(2):20–26
Ruminant (n.d.). https://en.wikipedia.org/w/index.php?title=Ruminant&oldid=847549212
Sahni TK, Goel A (2015) Microbial enzymes with special reference to α–amylase. Bio Evolution 2(1):19–25
Saini R, Sing-Saini H, Dahiya A (2017) Amylases: characteristics and industrial applications. J Pharmacol Phytochem 6(4):1865–1871
Sakano Y, Fukushima J, Kobayashi T (1983) Hydrolysis of alpha-1,4- and alpha-1,6-glucosidic linkages in trisaccharides by the Thermoactinomyces vulgaris alpha-amylase. Agric Biol Chem 47(10):2211–2216
Sarian FD (2016) Unique features of several microbial α-amylases active on soluble and native starch. University of Groningen
Schallmey M, Singh A, Ward OP (2004) Developments in the use of *Bacillus* species for industrial production. Ca J Microbiol 50:1–17
Seo J, Park T, Kwon I, Piao M, Lee C, Ha JK (2013) Characterization of cellulolytic and xylanolytic enzymes of *Bacillus licheniformis* JK7 isolated from the rumen of a native Korean goat. Asian Australas J Anim Sci 26(1):50
Seshadri R, Leahy SC, Attwood GT, Teh KH, Lambie SC, Cookson AL, Eloe-Fadrosh EA, Pavlopoulos GA, Hadjithomas M, Varghese NJ (2018) Cultivation and sequencing of rumen microbiome members from the Hungate1000 collection. Nat Biotechnol 36(4):359
Sevcik J, Solovicova A, Hostinova E, Gasperik J, Wilson KS, Dauter Z (1998) Structure of glucoamylase from Saccharomycopsis fibuligera at 1.7 a resolution. Acta Crystallogr D Biol Crystallogr 54(Pt 5):854–866
Sim L, Beeren SR, Findinier J, Dauvillee D, Ball SG, Henriksen A, Palcic MM (2014) Crystal structure of the Chlamydomonas starch debranching enzyme isoamylase ISA1 reveals insights into the mechanism of branch trimming and complex assembly. J Biol Chem 289(33):22991–23003
Šimůnek J, Killer J, Sechovcová H, Pechar R, Rada V, Švec P, Sedláček I (2018) Characterization of a xylanolytic bacterial strain C10 isolated from the rumen of a red deer (Cervus elaphus) closely related of the recently described species *Actinomyces succiniciruminis*, *A. glycerinitolerans*, and *A. ruminicola*. Folia Microbiol 63(3):391–399

Singh P, Rani A (2014) Isolation and partial characterization of amylase producing Bacillus sp. from soil. Int J Pharm Technol Res 6:2064–2069
Singh N, Singh J, Kaur L, Singh-Sodhi N, Singh-Gill B (2003) Morphological, thermal and rheological properties of starch from different botanical sources. Food Chem 81(2):219–231
Singh S, Sharma V, Soni M, DAS S (2011) Biotechnological applications of industrially important amylase enzyme. Int J Pharma Bio Sci 2(1):486–495
Singh R, Kumar M, Mittal A, Mehta P (2016a) Microbial enzymes: industrial Progress in 21st century. 3 Biotech 6(174):1–15
Singh G, Verma A, Kumar V (2016b) Catalytic properties, functional attributes and industrial applications of β-glucosidases. 3 Biotech 6(1):3
Sundarram A, Murthy TPK (2014) α-amylase production and applications: a review. J Appl Environ Microbiol 2(4):166–175
Suzuki Y, Hatagaki K, Oda H (1991) A hyperthermostable pullulanase produced by an extreme thermophile, Bacillus flavocaldarius KP 1228, and evidence for the proline theory of increasing protein thermostability. Appl Microbiol Biotechnol 34(6):707–714
Tiwari SP, Srivastava R, Singh CS, Shukla K, Singh RK, Singh P, Singh R, Singh NL (2015a) Amylases: an overview with special reference to alpha-amylase. J Global Biosci 4(1):1886–1901
Tiwari S, Srivastava R, Singh C, Shukla K, Singh R, Singh P, Singh R, Singh N, Sharma R (2015b) Amylases: an overview with special reference to alpha amylase. J Global Biosci 4:1886–1901
Van Der Maarl M, +J+E+C, Van Der Veen B, Uitdehaag JCM, Leemhuis H, Dijkhuizen L (2002) Properties and applications of starch-converting enzymes of the alpha amylase family. J Biotechnol 94(2):137–155
Wallace RJ (1996) Ruminal microbial metabolism of peptides and amino acids. J Nutr 126(Suppl_4):1326S–1334S
Walsh G (2014) Proteins: biochemistry and biotechnology, 2nd edn. Wiley. ISBN: 978-0-470-66986-0
Ward OP (1991) Bioprocessing. Open University Press, Buckingham
Yamasaki Y, Suzuki Y, Ozawa J (1977) Three forms of α-glucosidase and a glucoamylase from *Aspergillus awamori*. Agric Biol Chem 41(11):2149–2161
Zehavi T, Probst M, Mizrahi I (2018) Insights into culturomics of the rumen microbiome. Front Microbiol 9:1999

Chapter 11
Bacilli and Sustainable *Jhum* Agrobiotechnology

Aishiki Banerjee, S. K. Barik, and S. R. Joshi

Abstract *Bacillus,* the name that rekindles a lot of inquisitiveness in agricultural research, has been known for its multifaceted impacts on microbial processes. Mostly soil-dwelling, the genera of *Bacillus* has been studied for various biotechnological usage in the environment and industry. This group of eubacteria undergo various external environmental influences which helped them develop adaptive mechanisms at the genetic level related to survival strategies, making them tough performers in biotechnological applications.

In the northeastern region of India, *Jhumming,* a primitive form of agriculture, is still in practice where the natural environmental set-up is taken under the control of man, employing several tools including different stressors, fire being the main. The *jhum* fields in their coordinated cropping and fallow cycles involving slash and burn undergo a net change in nutrients along with other important components which results in a competitive environment for the soil micro-dwellers. Eco-restoration and making the entire *jhum* system sustainable is a key goal of many research strategies. Organic and environmental-friendly farming being the trend of modern-day agriculture, the usage of Bacilli in the rehabilitation of degraded ecosystems look promising to be exploited for eco-restoration goals. The well-documented roles of *Bacillus* spp. in plant protection and plant growth-promoting (PGP) properties along with other beneficial parameters is presented in this chapter, with reference to their usage in *jhum* agroecosystems prevalent in the north-eastern part of India.

A. Banerjee
Microbiology Laboratory, Department of Biotechnology & Bioinformatics, North-Eastern Hill University, Shillong, Meghalaya, India

Department of Environmental Studies, Siksha-Bhavana, Visva Bharati, Santiniketan, West Bengal, India

S. K. Barik
CSIR-National Botanical Research Institute, Lucknow, Uttar Pradesh, India

S. R. Joshi (✉)
Microbiology Laboratory, Department of Biotechnology & Bioinformatics, North-Eastern Hill University, Shillong, Meghalaya, India

M. T. Islam et al. (eds.), *Bacilli in Agrobiotechnology*, Bacilli in Climate Resilient Agriculture and Bioprospecting,
https://doi.org/10.1007/978-3-030-85465-2_11

Keywords *Bacillus* species · *Jhumming* · Soil-micro dwellers · PGP · Sustainable agriculture · Eco-restoration

11.1 Introduction

Soil, an excellent culture medium pulsating with life, harbours an enormous population of micro dwellers and actively contributes to the physico-chemical equilibriums of the earth. Microorganisms, the reservoir of resources, represent by far the richest array of molecular and chemical diversity in nature, providing huge space for innovative research useful to man (Riesenfeld et al. 2004). In this dynamic environment, *Bacillus* has been one of the major micro-workers with a long history of importance, that is much thoroughly studied to date, both from an economic point of view and as a source of investigational microbes. Mostly occurring as spores, *Bacillus* was reported by Waksman as 'saprophytic organisms whose natural habitat is the soil' (Vilain et al. 2006).

The major species of *Bacillus* belong to *B. cereus*, *B. mycoides*, *B. pseudomycoides*, *B. weihenstephanensis*, and *B. thuringiensis* (Jensen et al. 2003; Priest et al. 2004; Rasko et al. 2005). Although these species of *Bacillus* report high genetic closeness, their phylogenetic and taxonomic relationships are still argued (Helgason et al. 2004). *B. cereus* is one of the widely reported soil *Bacillus* groups (Guinebretière et al. 2008; Liu et al. 2015). This genus is characterised by the intricate cell wall structure, stress-resistant endospores, synthesis of secondary metabolites like peptide antibiotics, various peptide signal molecules, and extracellular enzymes, with the ability to enrich plant growth, suppression of diseases of the rhizosphere, and perform a plethora of beneficial functions important to natural life (Gardener 2004; Beneduzi and Passaglia 2011; Bahadir et al. 2018). While *Bacillus* remains most studied and reported bacteria isolated from soils, our knowledge of the ecology of these *Bacillus* species in soil remains rudimentary and far from complete.

11.1.1 **Bacillus** *and Its Distribution in Soil*

This genus of aerobic gram-positive spore-formers comprising of almost 200–300 bacterial species are essentially ubiquitous in nature with their habitats ranging from the extreme hot springs (Pandey et al. 2015; Mohammad et al. 2017) to the polar deep sea sediments (Rűger et al. 2000; da Silva et al. 2013) existing in composts like *Bacillus composti* and *Bacillus thermophilus* (Yang et al. 2013) and high pH conditions like *Bacillus firmus* (Pikuta et al. 2007) apart from residing in its regular environments like soil, water, and plant surfaces (Ichimatsu et al. 2001; Motta et al. 2004). The aerobic spore-forming members of the genera *Bacillus* have been isolated from many soil types depending on depths and altitudes (Wang et al. 2006; Lyngwi et al. 2013), with the population varying to different extents depending on the soil organic content, with strains being isolated from the extremes of the desert and Arctic or tundra type climatic conditions (Stabb et al. 1994; Garbeva et al. 2003; Matsui et al. 2016; Yadav et al. 2011) (Table 11.1). Culture-based isolation methods from soil samples recorded a CFU (colony forming unit) range from log 3 to log 6 cells with multiple incidences of related species of *B. subtilis* and *B. cereus* (Vargas-Ayala et al. 2000), while the CFU counts characteristically exceeding in the case of rhizospheric soil (Mahaffee and Kloepper 1997; Seldin et al. 1998). Soils with low organic content are normally dominated by *B. subtilis*, *B. licheniformis*, and *B. cereus* but with change in the nutrient input, a wide-ranging population emerge (Ranjard and Richaume 2001; Yang et al. 2014). Microbes are the key components in soil that determine life support functions, but the functional redundancy in the microbiota of most soils has long been thought to engulf microbial diversity-function relationships (van Elsas et al. 2012).

The significant soil-microbe-plant communications are the underlying aspects behind plant health and soil fertility (Glick 2012), and these microbes tend to play a central role in the biogeochemical cycling of both organic and inorganic nutrients in the soil, maintaining its health and quality (Jeffries et al. 2003; Madsen 2011). Soil is considered to be biologically unhealthy with little or no microbial growth in it, which in turn makes it not fertile to support the growth of plants. Several species of *Bacillus* along with the gram-negative *Pseudomonas* fall in the group of potent growth-promoting bacteria (Tilak et al. 2005; Islam et al. 2016). Many species have been studied for their characteristic phosphate solubilisation, phytohormone production, and antagonistic properties with respect to their possible use for plant growth and biocontrol (Lugtenberg and Kamilova 2009; Banerjee et al. 2017). In sight of the ecological specificity, isolation, identification, and characterisation of these native beneficial microbial organisms from diverse environmental habitats has become predominantly essential (Wang et al. 2007). Most farm fields have been found to be inhabited by different *Bacillus* spp., associated with a variety of crop plants playing a significant role in enhancing crop productivity by its direct or indirect mechanisms (Grayston et al. 1998; Bais et al. 2006; Berendsen et al. 2012). The rhizosphere region of soil has been one of the most favourable habitats for a number of *Bacillus* spp., attracting a huge number of microbiologists to study their role in plant growth promotion and biocontrol (Ramani and Patel 2011; Tallapragada and Seshachala 2012). The plant-microbe exchanges in the rhizosphere carry out crucial roles in the transformation, mobilisation, and solubilisation of nutrients from a confined pool in soil that is uptaken by the crop plants. Usage of PGPRs in agriculture is gradually increasing which offers a potential way of replacing the unfavourable chemicals in fertilizers and pesticides (Dey et al. 2004; Herman et al. 2008; Minorsky 2008).

As the genus *Bacillus* sp. exhibits diverse physiological and biochemical properties, such as the ability to produce enzymes like cellulase (Ray et al. 2012; Nair et al. 2018), phytase (Choi et al. 1999), tannase (Mondal et al. 2001), chitinase (Chang et al. 2003), xylanase (Sharma et al. 2015), protease (Liu et al. 2009), and lipase (Ghosh et al. 2017) as well as certain bacteriocins and peptide and lipopeptide antibiotics (Stein 2005; Abriouel et al. 2011), they are at the growing stage for bioprospecting.

Table 11.1 Recent reports on *Bacillus* from soil and unusual terrains with its physiological features

Species of *Bacillus*	Sampling site	Sampled source	Features	References
Bacillus subtilis	Assam, India	Petroleum contaminated soil	Capable of degrading petroleum	Das and Mukherjee (2007)
Geobacillus thermodenitrificans.	Northern China	Soil from depths of Subterranean oil reservoir	Isolate capable of degrading long-chain n-alkanes	Wang et al. (2006)
Geobacillus stearothermophilus, Geobacillus stearothermophilus, Bacillus oceanisediminis, Bacillus pichinotyi, Bacillus alcaliinulinus, Bacillus firmus, Bacillus sporothermodurans, Geobacillus subterraneus and *Bacillus licheniformis*	Panarea Island, Italy	Hydrothermal vents		Gugliandolo et al. (2012)
Bacillus, Oceanobacillus, Paenibacillus	Eastern South Atlantic Ocean	Deep sea sediment soil		da Silva et al. (2013)
Bacillus ectoiniformans	South China Sea	Deep sea sediment soil	Halotolerant	Zhu et al. (2016)
Cohnella kolymensis sp. nov.	Siberian permafrost	Deep sea sediment soil	Novel *Bacillus*	Kudryashova et al. (2018)
Paenibacillus borealis, P. donghaensis, P. macquariensis subsp. *macquariensis* and *B. circulans*	Arctic tundra	Arctic soil	could grow at 10°C- 20°C	Kim et al. (2013)
Bacillus licheniformis B. altitudinis, B. cereus, B. halmapalus, B. thuringiensis, B. isronensis, B. pseudomycoides, Lysinibacillus xylanilyticus, L. fusiformis, L. sphaericus Paenibacillus taichungensis, P. nanensis, P. amylolyticus, P. alkaliterrae and *P. pabuli*	Domiasiat, Meghalaya	Uranium ore deposited soil	Uranium Metal tolerant bacteria	Kumar et al. (2013)
Bacillus sp.	Tarbalo hot springs, Odisha, India	Muddy soil of hot spring		Panda et al. (2013)

(continued)

Table 11.1 (continued)

Species of *Bacillus*	Sampling site	Sampled source	Features	References
B. licheniformis, B. sonorensis and *B. tequilensis, Paenibacillus ehimensis,*	Ringigad hot springs, Uttarakhand, India	Soil from hot spring	(80 °C)	Pandey et al. (2015)
Bacillus sp.	Ganeshpuri, Maharashtra, India.	Soil from hot spring	Extracellular enzymatic activities	Lele and Deshmukh, (2016)
Bacillus licheniformis	5 thermal hot springs in Jordan (Hammamat Ma'in, Zara Dead Sea, Hammamat Afra, Al-Burbita, and Al-Hemma)	Thermal hot springs		Mohammad et al. (2017)
Bacillus zhangzhouensis sp. nov. and *Bacillus australimaris* sp. nov.	South China sea, China	Soil from surface sediment		Liu et al. (2016)
Bacillus thuringiensis, B. marisflavi, B. aryabhattai, B. psychrosaccharolyticus, B. weihenstephanensis, L. xylanilyticus, L. parviboronicapiens, Viridibacillus arenosi, V. arvi, P. taichungiensis and *P. tylopili*	Meghalaya, India	Pristine soils of sacred groves	Uninhabited soil source potential for bioprospection	Lyngwi et al. (2016)
Bacillus clausii, B. licheniformis, B. megaterium, B. subtilis, and *B. thuringiensis*	Itanagar, Arunachal Pradesh, India	'Fired Plots' Under Shifting Cultivation in Lobi of Papumpare	Plant growth promoters and biocontrol agents	Pandey et al. (2011)
Bacillus thuringiensis (Bt) strains	Mizoram, India	*Jhum* agroforestry fallow soil samples	Novel Bt entomopathogens may help in regulating mosquito vectors	Zothansanga et al. (2016)
Bacillus cereus and *B methylotrophicus*	Northeast India	Fallow soil samples of shifting cultivation fields	Plant growth promotion enhancing rice	Banerjee et al. (2017)

11.2 *Bacillus* as a Potential PGPR for Agriculture: Roles and Reports

Soil *Bacillus* plays a significant role in improving soil health in both cultivated and natural environments. Bacterial populations colonizing the root region in the rhizosphere improving and sustaining plant growth are designated as plant growth-promoting rhizobacteria (PGPR) (Prashar et al. 2014). The rhizo-microbiome is a diverse community of microorganisms, (Hinsinger et al. 2009; Chaparro et al. 2013) influenced by the plant root exudates released in the rhizosphere, which act as chemical signals (Venturi and Keel 2016), allowing microorganisms to thrive and flourish (Kaul et al. 2018). *Bacillus* and its relative genera are important performers in the most studied PGPRs (Kumar et al. 2011; Sivasakthi et al. 2014). Phosphate solubilisation, auxin synthesis, siderophore production, synthesis of antagonistic substances, and suppression of fungal pathogens are some of the important traits of *Bacillus* (Mehta et al. 2010; Kumar et al. 2011; Goswami et al. 2014;). They exhibit an extensive interaction with plant roots (Sharma et al. 2014) delivering the inaccessible nutrients available to the plants affecting plant growth and also reducing both biotic and abiotic stresses (Grover et al. 2011; Shameer and Prasad 2018).

11.2.1 Bacillus*, a PGPR*

The PGP role of *Bacillus* has been studied to positively stimulate plant growth by two distinguished behaviours: directly or indirectly, forming long-lasting, stress-tolerant spores, secreting metabolites that stimulate plant growth, and aid in crop productivity (Radhakrishnan et al. 2017). The soil nutrients, utilised by the saprophytic *Bacillus* and converted to molecules directly utilized by the plants, help to alleviate stress and entail the direct mode of plant growth promotion (Goswami et al. 2016). Solubilisation of inorganic phosphorus and mineralisation of organic phosphates, increasing phosphorus availability to plants (Zaidi et al. 2009); production of phytohormones, such as IAA, GAA (Idris et al. 2007), fixation of atmospheric N_2 (Hernandez et al. 2009); reduction of root membrane potential (Choudhary and Johri 2009); and synthesis of ACC-deaminase, which modulates the level of plant hormone (Penrose and Glick 2003), are a few of the direct roles exerted by *Bacillus* and other PGPRs. The indirect mode of the action takes place when bacteria reduce or prevent some toxic effect of a phytopathogenic organism by synthesizing antibiotics (Cazorla et al. 2007; Kumar et al. 2011), produce induced systemic resistance against the number of plant diseases (Tripathi et al. 2012; Pingping et al. 2017; Wang et al. 2018), and by siderophores (Gray and Smith 2005; Hashem et al. 2019). Application of these identified PGPR *Bacillus* possessing exclusive features, diversity, and relationship to plants could be used to improve various biotic and abiotic stress tolerance encountered by plants. These bacilli could be deployed in agricultural production systems as environmentally-friendly management tools to advance sustainable forms of agriculture (Grover et al. 2011; Vejan et al. 2016).

11.2.1.1 *Bacillus* as a P Solubiliser

Most soils have an insufficient amount of total available phosphorus required for a significant increase in plant growth and also have limited numbers of phosphate solubilising organisms in soil which fail to compete with the existing bacterial population thriving in the rhizosphere. Hence, inoculating the phosphate solubilizing bacteria (PSB) in the soil at a higher concentration than ordinarily found is essential for eliciting plant growth promotion. A number of studies reported bacterial plant growth promotion by solubilising in organic and/or organic phosphorus after being inoculated in soil or plant seeds (Chaiharn and Lumyong 2011; Qureshi et al. 2012). Synthesis of plant favourable metabolites such as phytohormones, antibiotics, or siderophores (Kloepper et al. 1989), among the many roles of PGPRS, the specific contribution of phosphate solubilizing PGPRs in plant stimulation is still under research (Glick 2012). The fundamental contribution of PSBs to plant nutrition suggests that phosphorus uptake by plants is a limited step in plant nourishment (Rengel and Marschner 2005), hence, highlighting the profound contribution of PSBs to plant nutrition and, consequently, in enhancing plant growth (Beneduzi et al. 2012). Considering phosphorus availability as a limiting step in plant nutrition (Rengel and Marschner 2005), suggests a fundamental contribution of PSBs to plant nutrition and consequently, in enhancing plant growth (Beneduzi et al. 2012). Phosphate solubilising bacteria is always under some environmental stress with reports on *Bacillus* sp. isolates present in tomato rhizosphere able to solubilise higher amounts of $AlPO_4$ and $FePO_4$ (Banerjee et al. 2010). *Bacillus* has been found in the *Zea mays* (Szilagyi-Zecchin et al. 2014), rice (Chaiharn and Lumyong 2009), wheat (Ahmad et al. 2008), soybean (Wahyudi et al. 2011), mustard (Kang et al. 2014), aubergine, and chilli (Pena-Yam et al. 2016) rhizosphere enhancing the respective plant growth. Tricalcium phosphate solubilizing *B. cereus* and *B. methylotrophicus* from acidic soil environments of a shifting cultivated field have shown to enhance rice growth *in vitro* (Banerjee et al. 2017).

Bacillus brings about the phosphate solubilisation by the production of various organic acids which in one study showed to improve with an increase in the presence of acids like malic acid or quinic acid resulting in increased chlorophyll, sucrose, glucose, fructose, and amino acids in the treated plant (Kang et al. 2014). Mineral phosphate solubilisation is well documented in many gram-negative bacteria (*Pseudomonas* sp. *Acinetobacter* sp.) with active participation in direct oxidation pathways (Sashidhar and Podile 2010) which is lacking in the case of the gram-positive bacilli, but enzymes like phytase and phosphatase have been reported in many plant growth-promoting *Bacillus* carrying out phosphate solubilisation (Yip et al. 2003; Matos et al. 2017). Detailed and systematic studies on the mode of action of phosphate solubilisation in the gram-positive *Bacillus* to better understand the biological process and optimisation of the nutrient cycling is way scarce, and the genetic studies to better improve the trait is required for developing the strains as effective biological tools for increasing crop productivity.

11.2.1.2 Role in IAA

Phyto stimulation by most *Bacillus* PGPRs occurs due to the biochemical synthesis of bacterial growth regulators such as indole-3-acetic acid (IAA) (Idris et al. 2007), gibberellins (Bottini et al. 2004), and cytokinins (Bloemberg and Lugtenberg 2001; Karadeniz et al. 2006). Different species of *Bacillus* have been tested to produce IAA *in vitro* but reports on the gram-positive soil-living bacteria are only a few. IAA production by *B. amyloliquefaciens* utilising tryptophan dependant pathways was confirmed by Idris et al. (2007) and *B. pumilus* and *B. licheniformis* produced high amounts of gibberellins in the alder rhizosphere (Gutiérrez-Mañero et al. 2001). The different phytohormones required and produced by plants to control their metabolism and growth in very minute concentrations are also secreted by the PGPRs as metabolic end products in various biochemical pathways (Glick 2010; Müller and Munné-Bosch 2011). Tryptophan is the major precursor molecule for the biosynthesis of IAA in bacteria which helps in root development, tissue differentiation, lateral root development, root anchoring, and root hair positioning (Lewis et al. 2013). These secreted molecules act as mechanical signals, transducing information among bacteria, thus, synchronizing their activities in the rhizosphere (Spaepen et al. 2007). *B. megaterium* and *B. cereus* along with other gram-negative microorganisms was reported to produce more than one phytohormone (auxin, gibberellin, cytokinin and abscisic acid) simultaneously (Karadeniz et al. 2006); however, this can occur as the production of one hormone and degradation of the other, which may act as precursor in the chain of molecular pathways (Boiero et al. 2007; Leveau and Lindow 2005). Multifaceted *Bacillus* sp. PGPRs like *B. methylotrophicus* have been shown to increase seed germination, shoot, and root elongation that are potent IAA producers (Nain et al. 2012; Banerjee et al. 2017).

11.2.1.3 *Bacillus* in the Indirect Mechanism of Plant Growth Promotion

Several literatures target mostly a single trait in the indirect mechanism of action by the plant growth promoters which mainly include the biocontrol ability by siderophore production, ammonia production, secretion of hydrolytic enzymes, antibiosis, induced systemic resistance (ISR) etc. (Choudhary and Johri 2009; Kumar et al. 2011). The indirect ways of bringing about plant growth by bacteria is a composite action of the many interlinked biochemical pathways carried out by the cells that result in stress reduction or provide local and systemic host resistance. *Bacillus* is reported in numerous studies as a biocontrol agent along with *Trichoderma* and *Pseudomonas* (McSpadden Gardener and Driks 2004). The biocontrol ability of *Bacillus* is mostly attributed to antibiosis, quorum quenching, siderophore production, and ISR (Ryu et al. 2004) and this has been well portrayed for *Bacillus* in the review by Kumar et al. (2011). Results of extensive research and field testing of different species of *Bacillus* biocontrol agents made it widely available as formulations or immobilised cakes for *in situ* usage, thus making *Bacillus* a promising candidate as a biocontrol agent in agricultural fields (Borriss 2011; Pérez-García et al. 2011; Yánez-Mendizábal et al. 2012).

11.2.2 **Bacillus** *in Agricultural Use*

Bacillus, as described in the above sections, has been identified as a species with abundant benefits and very few drawbacks, finds ample use in agriculture mainly due to the many growth-promoting properties, and disease resistance in plants. The most researched *Bacillus thuringiensis*, a proven biopesticide, whose formulations of spores and crystals have been conventionally used for integrated insect and pest management in several agricultural and horticultural crops (George and Crickmore 2012). The setback in the use of the same has led to further research towards improving their efficacy with nanoscience in crop protection, opening new boundaries in the field of nanobiotechnology (Mahadeva Swamy and Asokan 2013; Vimala Devi et al. 2019). Many studies have targeted to enhance the growth of various plants using a single strain of *Bacillus* or in consortia. In one study, 12 strains of *Bacillus* from wheat rhizospheres were evaluated for their effects on the growth of staple crops like corn, soybean, and wheat. *Bacillus megaterium*, *B. safensis*, *B. simplex*, and *Paenibacillus graminis* isolates simultaneously increased the growth in pot experiments but no single PGPR trait could be significantly predicted for growth promotion efficacy (Akinrinlola et al. 2018). In field studies, with an isolated strain of *B. methylotrophicus,* increased plant growth and height was observed which showed potential commercial application as a biofertilizer or biocontrol agent (Ge et al. 2016). Trials using consortia of *Bacillus cereus*, *Bacillus subtilis*, and *Serratia* sp. application significantly reduced the incidence of *Phytophthora* blight disease, improving fruit quality and soil properties in sweet pepper as compared to the control. Interestingly, soil microflora transformed with the dominance of species like *Comamonas, Burkholderia* and *Ramlibacter*, which are negatively associated with disease severity (Guo et al. 2019). Thus, the consortium of *Bacillus* treatment-controlled soil-borne disease and improved the soil chemical properties (Guo et al. 2019). A very recent study reported a consortium of indigenous *Bacillus* sp., consisting of 2–3 well-suited species, which were able to increase field emergence in chilli seeds compared to the controls. Increased growth of chilli both in the nursery stage and after planting was noted while simultaneous controlling of wilting was found to be effective (Yanti et al. 2020).

Due to its multipurpose use and ubiquitous nature, research in *Bacillus* has received ardent interest from the scientific community with more work focussing on its advancement and application. Govindasamy et al. (2010) highlighted a considerable amount of work on *Bacillus* and *Paenibacillus spp.* and their potential PGPR for sustainable agriculture. A review on the secondary metabolites of the soil

Bacillus of agricultural and industrial importance, mostly of *B. thuringiensis*, help in describing the role of *Bacillus* spp. in the soil ecosystem (Sansinenea and Ortiz Sansinenea and Ortiz 2011). Plant-associated *Bacillus* strains used as biocontrol agents and biofertilizers in agriculture was a cumbersome review work published by Borriss (2011). This article aims to disseminate the environmental views and the importance of the PGPR *Bacillus* species with emphasis and relevance to its role in shifting cultivated (*Jhum*) fields of north-eastern India and its possible implications in constructing a sustainable *jhum* agrobiotechnology. Some published and unpublished work has been summarised in this chapter to highlight the role of *Bacillus* as a PGPR and soil inoculant.

11.3 *Bacillus* Diversity in a *Jhum* System of Farming

Jhum, a primitive shifting cultivation practice predominant in the most biologically diverse northeastern part of India, is a source of livelihood in several parts of the Indian subcontinent concentrated in the states of Assam, Arunachal Pradesh, Manipur, Meghalaya, Mizoram, and Nagaland in the north-east and some belts in Andhra Pradesh, Madhya Pradesh, Kerala, and Orissa (Jha 1997; Freeman 1999). The *jhum* system, as reported by Ramakrishnan and Toky (1981) in the lower altitude of Meghalaya, is characteristic of that practised in several parts of the north-eastern region. The *jhum* cycle varies from 4–5 years to 10–30 years, depending on the length of the fallow period between two successive croppings. The agrarian practices of shifting cultivation and horticulture significantly impact temporal and spatial changes in microbial biomass and several enzyme activities in soil (dehydrogenase and urease). These factors were reported to be greater in the soils of the undisturbed forest ecosystem than in the profoundly affected *jhum* soils (Ralte et al. 2005). Recurrent slashing and burning processes on the hilly slopes accompanied by heavy rainfall result in the loss of soluble organic matter and nutrients in the topsoil. This appears to be responsible for low enzyme activity in the *jhum* fallows (Maithani et al. 1998; Reza et al. 2014). Low input of detrital matter and nutrients through litter-fall in the *jhum* field results in decreased microbial biomass nitrogen and enzyme activity (Saplalrinliana et al. 2016). With increasing age of the plant growth on the *jhum,* fallows result in increased microbial biomass carbon and nitrogen, which may be attributed to greater input of plant detritus that is eventually incorporated into the soil, improving the nutrient pool (Arunachalam et al. 1996; Maithani et al. 1998; Saplalrinliana et al. 2016).

In the past few years, research has been directed towards the PGPR efficiency of *Bacillus*, but only recently, the importance of the application of a consortium of bacteria has been realised. Studies on microbes and microbial diversity with additional emphasis on *Bacillus* sp. from this ancient type of agriculture is rare. Shifting agriculture has its own biological merits and demerits. The uncultivated or fallow stages in *jhum* help in regeneration and enrichment of diverse soil properties along with increased soil microbial biomass and diversity (Pandey et al. 2011). The survival of the plant growth-promoting microbes under such stressful environmental conditions and their regeneration during the fallow period is essential for the health and yield of the cultivated crops. Depleting nutrients like reduction in organic carbon and nitrogen content (Ramakrishnan and Toky 1981) with a decrease in soil phosphatase activity, and β-glucosidase activity is the characteristic of *jhum* soils observed after burning (Boerner et al. 2000). In context to such degrading soil conditions, it has become immensely vital to study the existence of PGPRs inherent to these soils and their function in improving the different features of soil conditions, which in turn will enhance the plant growth, making the system a sustainable one (Fig. 11.1).

The species of *Bacillus* isolated from the various shifting cultivated fired plots or fallows of north-east India are provided in Table 11.1. In a study conducted by Pandey et al. (2011), bacterial species of *Bacillus* and *Pseudomonas* were recovered from the 'fired plots' soil samples in Arunachal Pradesh which revealed lower amounts of total organic carbon and nitrogen as compared to the fallowed soils. A report on *B. thuringiensis* (Bt) strains characterised by diverse endotoxin genes and crystal proteins, isolated by acetate selection method from *jhum* fallow soil of Mizoram suggested it to be novel species capable of controlling mosquito vectors (Zothansanga et al. Zothansanga et al. 2016). In an investigation for microbial isolates with combined PGP properties from *jhum* fallows, an inadequate number of isolates were obtained but the isolated organisms present were able to enhance plant growth. This work reported two isolates of *Bacillus* sp. out of a number of bacteria and fungi studied. The nearest homologs of bacterial isolates reported were *B. cereus* and *B. methylotrophicus* (Banerjee et al. 2017). In present days, with increased population, the span of fallows has considerably decreased (up to 3 years) which has led to severe problems including reduced soil fertility, crop productivity, and increased soil erosion. In this scenario, plant growth-promoting bacteria can be utilised in the nutrient-depleted soils of the *jhum* fields This expands the need for study in the field of isolation of varied *Bacillus* species concentrated in *jhum* fallows for their unique properties to be exploited in bioprospection and sustainable agriculture.

Fig. 11.1 (**a**) Fired plot in a *Jhum* field in Mizoram; (**b**) fallow land in Nagaland

11.4 *Bacillus* and Their PGP Role in *Jhum* Agroecosystem

The use of beneficial microbes is a rising trend in the field of agriculture and is required for maintaining its sustainability. Uneven distribution of microbial inhabitants has been demonstrated in soil with reports suggesting that shifting cultivated fields depict a significant diminution in the microbial population, along with loss of certain species of bacteria and fungi, as compared to that of uncultivated forestlands (Miah et al. 2010). Characterisation of soil microbial communities is difficult due to their vast phenotypic and genotypic diversity. This is applicable to different ecosystems, populations, and even to different micro-individuals (Fierer and Jackson 2006). The soil bacterial community ratio, composition, density, and activity, with respect to the wide range of organic root exudates, vary with the stage of the aboveground plant vegetation and an improved understanding of facts affecting the ability of bacterial colonisation the associated plant population has to be taken into account. The study of microbes from agricultural soils is not new. However, a study on the effect of the activities of *jhum,* mainly on the soil bacterial community, the differences in the ages of fallow phases and their effect on soil restoration would help to assess the diversity of bacteria associated with these soils. In the different states of north-east India, shifting cultivation is practised in their own indigenous way where the biological diversity of ecosystems is used and preserved by ethnic communities through numerous informal institutions using traditional knowledge. The practice varies in terms of principles, actions, culture and types of crops grown among *jhum* fields of the lower altitudes of Assam, Mizoram, Meghalaya, and Nagaland.

The cultivation-based methods shown in Fig. 11.2 used to study the gram-positive rod-shaped cells (Fig. 11.3) of the aerobic chemoheterotrophic bacteria from different soil samples from fallows of *jhum* revealed the occurrence of *Bacillus* isolates (Figs. 11.2, 11.3, and 11.4), identified and accession numbers were received for the sequences of marker gene from GenBank. Biochemical characters along with the plant growth-promoting properties and molecular identification with 16S rRNA gene sequences of these isolates revealed their phylogenetic relationship with 12 different genera. *Bacillus* and *Actinobacteria* being the most populous (Banerjee et al. unpublished) (Fig. 11.5).

11.5 *Bacillus* Bioinoculants in *Jhum* Fields: *In Vitro* and *In Situ* Applications

Jhum cultivation is based on a two-step cropping system i.e. sequential cropping and intercropping. The rich agro-biodiversity results in a sustainable cropping pattern. The environmental load on using the recalcitrant chemical fertilizers have become intertwined with the food chain with the phenomenon of biomagnification or biological amplification. Cutting down usages of these chemicals replacing with biofertilizers can be considered an important strategy for ecological management

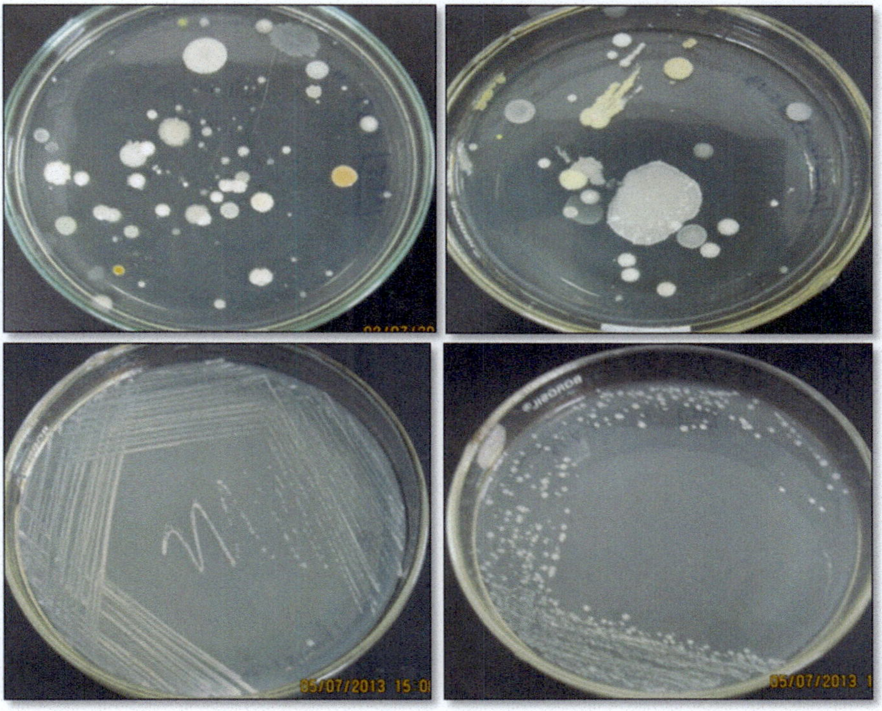

Fig. 11.2 Isolated bacteria from the fallow soil samples by spread plating and a pure culture grown in Agar plates

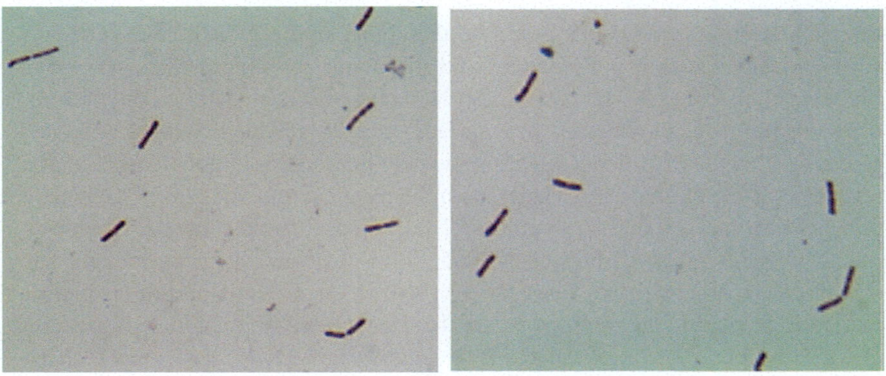

Fig. 11.3 Gram-stained rod-shaped cells of *Bacillus* sp. isolated from the fallow fields of north-east India observed in 100X OIL (Oil Immersion Lens)

and for reducing environmental hazards (Balasubramanian et al. 2004; Hungria et al. 2010, 2013). Bacterial inoculants parameterised by their PGPR traits determine the ability to be used as successful inoculants. The effectiveness and usage of

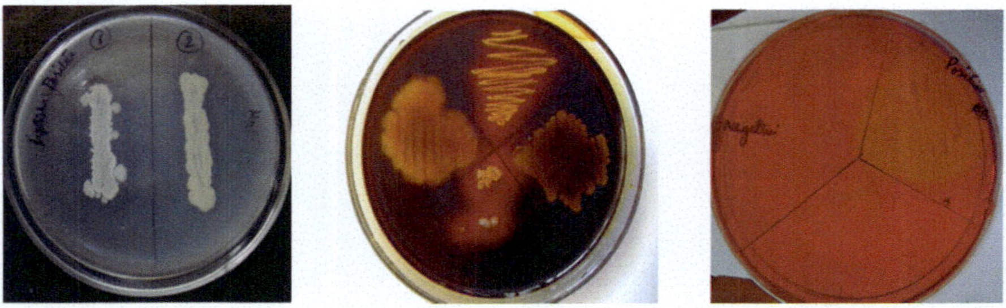

Fig. 11.4 Biochemical properties of the *Bacillus* isolated from the *jhum* fallows in northeast India. (**a**) Lipase assay; (**b**) Amylase assay; (**c**) Cellulase assay

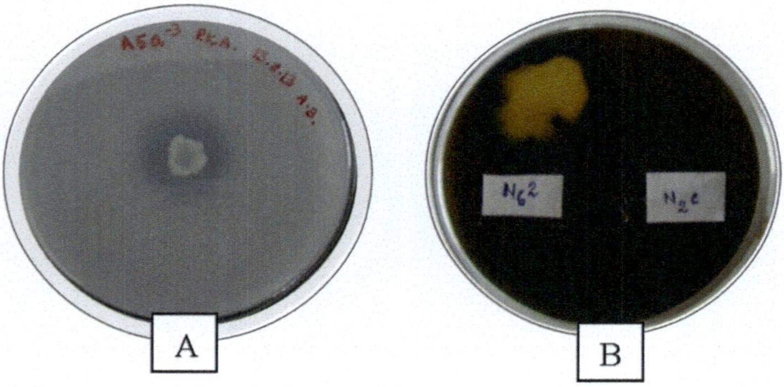

Fig. 11.5 Plant growth-promoting characters of the *Bacillus* isolated from the *jhum* fallows in north-east India. (**a**) Phosphate solubilizing test; (**b**) Siderphore production

these PGPR microbes inoculants rest on the ability of these bacteria to interact in the rhizosphere rich in root exudates and to inhabit and colonise the plant roots (Meneses et al. 2011; Alquéres et al. 2013; Beauregard et al. 2013). The plant-root secretions influence the bacteria-root surface interaction which changes with varying environmental conditions, affecting the potentiality of the inoculants (Cai et al. 2012; Carvalhais et al. 2013). Their role in nutrient cycling and ability to boost up plant growth, make them a dynamic tool in the sustainable form of modern-day agriculture (Rengel and Marschner 2005). Phosphate solubilizing bacteria have been in use as bioinoculants since the 90's and in the Soviet Union 'phosphobacterin' was prepared and commercialised for agricultural applications (Khan et al. 2006). This contained the isolate of *B. megaterium* var. *phosphaticum* which was later introduced in India.

An *in vitro* study in the primitive phase was conducted to isolate and identify the native PGPRs inherent to the shifting cultivated fields of north-eastern India. The traits of phosphate solubilisation, IAA production, along with siderophore, ammonia, and catalase production tests for the isolates were used to determine the PGPR ability out of the 55 bacterial isolates and 32 fungal isolates obtained from the different fallows of *jhum*. Three bacterial isolates reported, showed plant growth promotion in the local upland variety of rice along with its fungal counterparts. The rice seeds treated with the microbial inoculum exhibited positive plant growth-promoting properties in the plate assay (Fig. 11.6a). Early induced seed germination was observed in the plate assay by the bacterial isolates SB3, SB5, and SB9 than that in the control. Pot experiments with the microbial isolates exhibiting PGP abilities under greenhouse conditions were carried out with a more definitive outcome (Fig. 11.6b). The rice seeds treated with SB3 induced the formation of the longest roots as it was the highest IAA producer while the isolate SB9 induced the highest growth as it was both phosphate solubiliser and a moderate IAA producer. The closest homologs of bacterial isolates were identified as *B. cereus*, *B. methylotrophicus*, and *Curtobacterium oceanosedimentum* (Banerjee et al. 2017).

These were considered as potent bioinoculants which was further tested in field applications in a similar environment of the *jhum* cultivation fields (Fig. 11.7). The same variety of rice was used which was treated with the bacterial inoculum (*Bacillus* sp was used) and sowed intermittently with each isolate treatment divided into plots by permutation and combination. Weekly weeding and watering of the seeds were performed until the onset of the rains. Results showed increased seed germination rate and harvesting index was high. The potent isolates when tested in actual field trials gave satisfactory outcomes in accordance with the pot

Fig. 11.6 *In vitro* study of the upland variety rice seed germination by microbial isolates from *Jhum* fields. (**a**) plate assay; (**b**) pot experiments under greenhouse conditions

Fig. 11.7 *In situ* application of the microbial isolates under field conditions showing germination of rice and full-grown rice plant

Fig. 11.8 Field sites of *Jhum* in Nagaland (**a**) and Meghalaya(**b**)

experiments. There was significant variation in rice growth from the treated to that of control (unpublished) (Fig. 11.8).

11.6 Conclusion

In order to achieve efficient PGPR strains using *Bacillus* and its consortia, with utmost growth-promoting association between the microbe and the plant root, it is important to identify the various factors affecting the interaction and also the presence of other microbes (Bent et al. 2001). The possible way to find efficient strains is to carry out a diversity study on the inherent soil microbes that are native to the particular soil conditions and can perform amidst the stresses present in the soil environment. In bioinoculant production, formulating a competent microbial strain depending on its performance is crucial in determining the achievement of the biological agent (Bashan 1998). Often an isolate with optimised *in vitro* reports deliberately fail in field trials. Thus, research is still in its budding stage to provide an optimal efficient performing bioinoculant PGPR.

Jhum has been a different situation with its complicated system of cropping intricated to the cultural practices native to the various ethnic human populations of north-eastern India which still perform this indigenous type of farming. The use of modern-day agricultural practices is not prominent in *jhum* type of farming. Hence, the application of native PGPs for plant growth and yield enhancement is a way of supporting the eco-restoration and maintaining the sustainability of an already degraded *jhum* ecosystem. The study carried out with the *Bacillus* sp. isolates obtained from *jhum* fields was the first report of its kind showcasing phosphate solubilizing and IAA producing PGPRs from the natural *jhum* conditions which possibly will aid in plant growth promotion in the nutrient-depleted soils of the *jhum* fields. This will provide a better understanding of the soil microbiota while enabling the *jhum* farmers to become technologically enriched with emphasis on the use of the biological tools to optimise nutrient cycling and crop productivity and simultaneously, in eco-restoration of the *jhum* fallows.

Acknowledgement The present chapter forms a part of the study under the National Mission on Himalayan Studies (NMHS), Ministry of Environment, Forest and Climate Change, Government of India, and sanctioned project (MoEFCC-NMHS/LG-2016/005 Dated: 31/03/2016) to the authors.

References

Abriouel H, Franz CMAP, Ben Omar N, Galvez A (2011) Diversity and applications of *Bacillus* bacteriocins. FEMS Microbiol Rev 5:201–232

Ahmad F, Ahmad I, Khan MS (2008) Screening of free-living rhizospheric bacteria for their multiple plant growth promoting activities. Microbiol Res 163:173–181

Akinrinlola RJ, Yuen GY, Drijber RA, Adesemoye AO (2018) Evaluation of Bacillus strains for plant growth promotion and predictability of efficacy by in vitro physiological traits. Int J Microbiol 2018:1–11

Alquéres S, Meneses C, Rouws L, Rothballer M, Baldani I, Schmid M, Hartmann A (2013) The bacterial superoxide dismutase and glutathione reductase are crucial for endophytic colonization of rice roots by Gluconacetobacter diazotrophicus PAL5. Mol Plant Microbe Interact 26:937–945

Arunachalam A, Pandey HN, Tripathi RS, Maithani K (1996) Fine root decomposition and nutrient mineralization patterns in a subtropical humid forest following tree cutting. For Ecol Manage 86:141–150

Bahadir PS, Liaqat F, Eltem R (2018) Plant growth promoting properties of phosphate solubilizing Bacillus species isolated from the Aegean Region of Turkey. Turk J Bot 42:1–14

Bais HP, Weir TL, Perry LG, Gilroy S, Vivanco JM (2006) The role of root exudates in rhizosphere interactions with plants and other organisms. Annu Rev Plant Biol 57:233–266

Balasubramanian V, Alves B, Aulakh M, Bekunda M, Cai Z, Drinkwater L, Mugendi D, van Kessel C, Oenema O (2004) Crop, environmental, and management factors affecting nitrogen use efficiency. In: Mosier AR, Syers JK, Freney J (eds) Agriculture and the nitrogen cycle, vol 65. Scope, pp 19–33

Banerjee S, Palit R, Sengupta C, Standing D (2010) Stress induced phosphate solubilization by *Arthrobacter* sp. and *Bacillus* sp. isolated from tomato rhizosphere. Aus J Crop Sci 4:378

Banerjee A, Bareh DA, Joshi SR (2017) Native microorganisms as potent bioinoculants for plant growth promotion in shifting agriculture (*Jhum*) systems. J Soil Sci Plant Nutr 17:127–140

Bashan Y (1998) Inoculants of plant growth-promoting bacteria for use in agriculture. Biotechnol Adv 16:729–770

Beauregard PB, Chai Y, Vlamakis H, Losick R, Kolter R (2013) *Bacillus subtilis* biofilm induction by plant polysaccharides. PNAS 110:E1621–E1630

Beneduzi A, Passaglia LM (2011) Genetic and phenotypic diversity of plant growth promoting bacilli. In: Bacteria in agrobiology: plant growth responses. Springer, Berlin/Heidelberg, pp 1–20

Beneduzi A, Ambrosini A, Passaglia LM (2012) Plant growth-promoting rhizobacteria (PGPR): their potential as antagonists and biocontrol agents. Genet Mol Biol 35:1044–1051

Bent E, Tuzun S, Chanway CP, Enebak S (2001) Alterations in plant growth and in root hormone levels of lodgepole pines inoculated with rhizobacteria. Can J Microbiol 47:793–800

Berendsen RL, Pieterse CM, Bakker PA (2012) The rhizosphere microbiome and plant health. Trends Plant Sci 17:478–486

Bloemberg GV, Lugtenberg BFJ (2001) Molecular basis of plant growth promotion and biocontrol by rhizobacteria. Curr Opin Plant Biol 4:343–350

Boerner REJ, Decker KLM, Sutherland EK (2000) Prescribed burning effects on soil enzyme activity in a southern Ohio hardwood forest: a landscape scale analysis. Soil Biol Biochem 32:899–908

Boiero L, Perrig D, Masciarelli O, Penna C, Cassan F, Luna V (2007) Phytohormone production by three strains of *Bradyrhizobium japonicum* and possible physiological and technological implications. Appl Microbiol Biotechnol 74:874–880

Borriss R (2011) Use of plant-associated Bacillus strains as biofertilizers and biocontrol agents in agriculture. In: Bacteria in agrobiology: plant growth responses. Springer, Berlin/Heidelberg, pp 41–76

Bottini R, Cassan F, Picolli P (2004) Gibberellin production by bacteria and its involvement in plant growth promotion and yield increase. Appl Microbiol Biotechnol 65(5):497–503

Cai Z, Kastell A, Knorr D, Smetanska I (2012) Exudation: an expanding technique for continuous production and release of secondary metabolites from plant cell suspension and hairy root cultures. Plant Cell Rep 31:461–477

Carvalhais LC, Dennis PG, Fan B, Fedoseyenko D, Kierul K, Becker A, von Wiren N, Borriss R (2013) Linking plant nutritional status to plant-microbe interactions. PLoS One 8:e68555

Cazorla F, Romero D, Pérez-García A, Lugtenberg B, Vicente AD, Bloemberg G (2007) Isolation and characterization of antagonistic *Bacillus subtilis* strains from the avocado rhizoplane displaying biocontrol activity. J Appl Microbiol 103:1950–1959

Chaiharn M, Lumyong S (2009) Phosphate solubilization potential and stress tolerance of rhizobacteria from rice soil in Northern Thailand. W J Microbiol Biotechnol 25:305–314

Chaiharn M, Lumyong S (2011) Screening and optimization of indole-3-acetic acid production and phosphate solubilization from rhizobacteria aimed at improving plant growth. Curr Microbiol 62:173–181

Chang WT, Chen CS, Wang SL (2003) An antifungal chitinase produced by Bacillus cereus with shrimp and crab shell powder as a carbon source. Curr Microbiol 47:0102–0108

Chaparro JM, Badri DV, Bakker MG, Sugiyama A, Manter DK, Vivanco JM (2013) Root exudation of phytochemicals in Arabidopsis follows specific patterns that are developmentally programmed and correlate with soil microbial functions. PLoS One 8:e55731

Choi YM, Noh DO, Cho SH, Lee HK, Suh HJ, Chung SH (1999) Isolation of a phytase-producing Bacillus sp. KHU-10 and its phytase production. J Microbiol Biotechnol 9:223–226

Choudhary DK, Johri BN (2009) Interactions of Bacillus spp. and plants–with special reference to induced systemic resistance (ISR). Microbiol Res 164:493–513

da Silva MAC, Cavalett A, Spinner A, Rosa DC, Jasper RB, Quecine MC, Bonatelli ML, Pizzirani-Kleiner A, Corção G, de Souza Lima AO (2013) Phylogenetic identification of marine bacteria isolated from deep-sea sediments of the eastern South Atlantic Ocean. SpringerPlus 2:1–10

Das K, Mukherjee AK (2007) Crude petroleum-oil biodegradation efficiency of Bacillus subtilis and Pseudomonas aeruginosa strains isolated from a petroleum-oil contaminated soil from North-East India. Bioresour Technol 98:1339–1345

Dey R, Pal KK, Bhatt DM, Chauhan SM (2004) Growth promotion and yield enhancement of peanut (*Arachis hypogaea* L.) by application of plant growth-promoting rhizobacteria. Microbiol Res 159:371–394

Fierer N, Jackson RB (2006) The diversity and biogeography of soil bacterial communities. PNAS USA 103:626–631

Freeman JR (1999) Gods, groves and the culture of nature in Kerala. Mod Asian Stud 33:257–302

Gardener BBM (2004) Ecology of Bacillus and Paenibacillus spp. in agricultural systems. Phytopathol 94:1252–1258

Garbeva P, Van Veen JA, Van Elsas JD (2003) Predominant Bacillus spp. in agricultural soil under different management regimes detected via PCR-DGGE. Microb Ecol 45:302–316

Ge B, Liu B, Nwet TT, Zhao W, Shi L, Zhang K (2016) *Bacillus methylotrophicus* strain NKG-1, isolated from Changbai Mountain, China, has potential applications as a biofertilizer or biocontrol agent. PloS one 11:e0166079

George Z, Crickmore N (2012) *Bacillus thuringiensis* applications in agriculture. In: Sansinenea E (ed) *Bacillus thuringiensis* biotechnology. Springer, Brighton, pp 19–39

Ghosh K, Banerjee S, Moon UM, Khan HA, Dutta D (2017) Evaluation of gut associated extracellular enzyme-producing and pathogen inhibitory microbial community as potential probiotics in Nile tilapia, *Oreochromis niloticus*. Int J Aquacult 7:143–158

Glick BR (2010) Using soil bacteria to facilitate phytoremediation. Biotechnol Adv 28:367–374

Glick BR (2012) Plant growth-promoting bacteria: mechanisms and applications. Scientifica 2012:963401

Goswami D, Dhandhukia P, Patel P, Thakker JN (2014) Screening of PGPR from saline desert of Kutch: growth promotion in Arachis hypogea by *Bacillus licheniformis* A2. Microbiol Res 169:66–75

Goswami D, Thakker JN, Dhandhukia PC (2016) Portraying mechanics of plant growth promoting rhizobacteria (PGPR): a review. Cogent Food Agric 2:1–19

Govindasamy V, Senthilkumar M, Magheshwaran V, Kumar U, Bose P, Sharma V, Annapurna K (2010) *Bacillus* and *Paenibacillus* spp.: potential PGPR for sustainable agriculture. In: Plant growth and health promoting bacteria. Springer, Berlin/Heidelberg, pp 333–364

Gray EJ, Smith DL (2005) Intracellular and extracellular PGPR: commonalities and distinctions in the plant–bacterium signaling processes. Soil Biol Biochem 37:395–412

Grayston SJ, Wang SQ, Campbell CD, Edwards AC (1998) Selective influence of plant species on microbial diversity in the rhizosphere. Soil Biol Biochem 30:369–378

Grover M, Ali SZ, Sandhya V, Rasul A, Venkateswarlu B (2011) Role of microorganisms in adaptation of agriculture crops to abiotic stresses. World J Microbiol Biotechnol 27:1231–1240

Gugliandolo C, Lentini V, Spanò A, Maugeri TL (2012) New bacilli from shallow hydrothermal vents of Panarea Island (Italy) and their biotechnological potential. J Appl Microbiol 112:1102–1112

Guo JH, Zhang L, Wang D, Hu Q, Dai X, Xie Y, Li Q, Liu H (2019) Consortium of plant growth-promoting Rhizobacteria strains suppresses sweet pepper disease by altering the rhizosphere microbiota. Front Microbiol 10:1668–1677

Guinebretière MH, Thompson FL, Sorokin A, Normand P, Dawyndt P, Ehling-Schulz M, Svensson B, Sanchis V, Nguyen-The C, Heyndrickx M, De VP (2008) Ecological diversification in the Bacillus cereus group. Environ Microbiol 10:851–865

Gutiérrez-Mañero FJ, Ramos-Solano B, Probanza AN, Mehouachi J, Tadeo FR, Talon M (2001) The plant-growth-promoting rhizobacteria *Bacillus pumilus* and *Bacillus licheniformis* produce high amounts of physiologically active gibberellins. Physiol Plant 111:206–211

Hashem A, Tabassum B, Abd_Allah EF (2019) *Bacillus subtilis*: a plant-growth promoting rhizobacterium that also impacts biotic stress. Saudi J of Biol Sci 26:1291–1297

Helgason E, Tourasse NJ, Meisal R, Caugant DA, Kolstø AB (2004) Multilocus sequence typing scheme for bacteria of the Bacillus cereus group. Appl Environ Microbiol 70:191–201

Herman MAB, Nault BA, Smart CD (2008) Effects of plant growth promoting rhizobacteria on bell pepper production and green peach aphid infestations in New York. J Crop Prot 27:996–1002

Hernandez JP, de Bashan LE, Rodriguez DJ, Rodriguez Y, Bashan Y (2009) Growth promotion of the freshwater microalga Chlorella vulgaris by the nitrogen-fixing, plant growth-promoting bacterium *Bacillus pumilus* from arid zone soils. Eur J Soil Biol 45:88–93

Hinsinger P, Bengough AG, Vetterlein D, Young IM (2009) Rhizosphere: biophysics, biogeochemistry and ecological relevance. Plant and Soil 321:117–152

Hungria M, Campo RJ, Souza EM, Pedrosa FO (2010) Inoculation with selected strains of *Azospirillum brasilense* and *A. lipoferum* improves yields of maize and wheat in Brazil. Plant and Soil 331:413–425

Hungria M, Nogueira MA, Araujo RS (2013) Co-inoculation of soybeans and common beans with rhizobia and azospirilla: strategies to improve sustainability. Biol Fertil Soils 49:791–801

Ichimatsu T, Mizuki E, Nishimura K, Akao T, Saitoh H, Higuchi K, Ohba M (2001) Occurrence of *Bacillus thuringiensis* in fresh waters of Japan. Curr Microbiol 40:217–220

Idris EE, Iglesias DJ, Talon M, Borriss R (2007) Tryptophan-dependent production of indole-3-acetic acid (IAA) affects level of plant growth promotion by *Bacillus amyloliquefaciens* FZB42. Mol Plant-microbe Int 20:619–626

Islam S, Akanda AM, Prova A, Islam MT, Hossain MM (2016) Isolation and identification of plant growth promoting rhizobacteria from cucumber rhizosphere and their effect on plant growth promotion and disease suppression. Front Microbiol 6:1–12

Jeffries P, Gianinazzi S, Perotto S, Turnau K, Barea JM (2003) The contribution of arbuscular mycorrhizal fungi in sustainable maintenance of plant health and soil fertility. Biol Fertil Soils 37:1–16

Jensen GB, Hansen BM, Eilenberg J, Mahillon J (2003) The hidden lifestyles of *Bacillus cereus* and relatives. Environ Microbiol 5:631–640

Jha LK (1997) Shifting cultivation. APH Publishing, Delhi, pp 1–189

Kang SM, Radhakrishnan R, You YH, Joo GJ, Lee IJ, Lee KE, Kim JH (2014) Phosphate solubilizing *Bacillus megaterium* mj1212 regulates endogenous plant carbohydrates and amino acids contents to promote mustard plant growth. Indian J Microbiol 54:427–433

Karadeniz A, Topcuoğlu ŞF, Inan S (2006) Auxin, gibberellin, cytokinin and abscisic acid production in some bacteria. World J Microbiol Biotechnol 22:1061–1064

Kaul S, Gupta S, Sharma T, Dhar MK (2018) Unfolding the role of rhizomicrobiome toward sustainable agriculture. In: Root biology. Springer, Cham, pp 341–365

Khan MS, Zaidi A, Wani PA (2006) Role of phosphate solubilizing microorganisms in sustainable agriculture-a review. In: Lichtfouse E, Navarrete M, Debaeke P, Veronique S, Alberola C (eds) Sustainable agriculture, vol 5. Springer, Netherlands, pp 551–570

Kim HM, Chae N, Jung JY, Lee YK (2013) Isolation of facultatively anaerobic soil bacteria from Ny-Ålesund, Svalbard. Polar Biol 36:787–796

Kloepper JW, Lifshitz R, Zablotowicz RM (1989) Free-living bacterial inocula for enhancing crop productivity. Trends Biotechnol 7:39–44

Kudryashova EB, Karlyshev AV, Ariskina EV, Streshinskaya GM, Vinokurova NG, Kopitsyn DS, Evtushenko LI (2018) *Cohnella kolymensis* sp. nov., a novel bacillus isolated from Siberian permafrost. Int J Syst Evol Micr 68:2912–2917

Kumar A, Prakash A, Johri BN (2011) *Bacillus* as PGPR in crop ecosystem. In: Bacteria in agrobiology: crop ecosystems. Springer, Berlin/Heidelberg, pp 37–59

Kumar R, Nongkhlaw M, Acharya C, Joshi SR (2013) Bacterial ccommunity structure from the perspective of the uranium ore deposits of domiasiat in India. Proc Natl Acad Sci India Sect B Biol Sci 83:485–497

Lele O, Deshmukh P (2016) Isolation and characterization of thermophilic Bacillus sp. with extracellular enzymatic activities from hot spring of Ganeshpuri, Maharashtra, India. Int J Appl Res 2:427–430

Leveau JHJ, Lindow SE (2005) Utilization of the plant hormone indole-3-acetic acid for growth by *Pseudomonas putida* strain 1290. Appl Environ Microbiol 71:2365–2371

Lewis DR, Olex AL, Lundy SR, Turkett WH, Fetrow JS, Muday GK (2013) A kinetic analysis of the auxin transcriptome reveals cell wall remodeling proteins that modulate lateral root development in Arabidopsis. Plant Cell 25:3329–3346

Liu CH, Chiu CS, Ho PL, Wang SW (2009) Improvement in the growth performance of white shrimp, *Litopenaeus vannamei*, by a protease-producing probiotic, Bacillus subtilis E20, from natto. J Appl Microbiol 107:1031–1041

Liu Y, Lai Q, Göker M, Meier-Kolthoff JP, Wang M, Sun Y, Wang L, Shao Z (2015) Genomic insights into the taxonomic status of the Bacillus cereus group. Sci Rep 5:1–11

Liu Y, Lai Q, Du J, Shao Z (2016) *Bacillus zhangzhouensis* sp. nov. and *Bacillus australimaris* sp. nov. Int J Syst Evol Micr 66:1193–1199

Lugtenberg B, Kamilova F (2009) Plant-growth-promoting rhizobacteria. Annu Rev Microbiol 63:541–556

Lyngwi NA, Koijam K, Sharma D, Joshi SR (2013) Cultivable bacterial diversity along the altitudinal zonation and vegetation range of tropical Eastern Himalaya. Rev Biol Trop 61:467–490

Lyngwi NA, Nongkhlaw M, Kalita D, Joshi SR (2016) Bioprospecting of plant growth promoting Bacilli and related genera prevalent in soils of pristine sacred groves: biochemical and molecular approach. PLoS One 11:e0152951

Madsen EL (2011) Microorganisms and their roles in fundamental biogeochemical cycles. Curr Opin Biotechnol 22:456–464

Mahaffee WF, Kloepper JW (1997) Temporal changes in the bacterial communities of soil, rhizosphere, and endorhiza associated with field grown cucumber (*Cucumis sativus* L.). Microb Ecol 34:210–223

Maithani K, Arunachalam A, Tripathi RS, Pandey HN (1998) Influence of leaf litter quality on N mineralization in soils of subtropical humid forest regrowths. Biol Fertil Soils 27:44–50

Matsui K, Yoshinami S, Narita M, Chien MF, Phung LT, Silver S, Endo G (2016) Mercury resistance transposons in Bacilli strains from different geographical regions. FEMS Microbiol 363:1–8

Matos AD, Gomes IC, Nietsche S, Xavier AA, Gomes WS, Dos Santos Neto JA, Pereira MC (2017) Phosphate solubilization by endophytic bacteria isolated from banana trees. An Acad Bras Ciênc 89:2945–2954

McSpadden Gardener BB, Driks A (2004) Overview of the nature and applications of biocontrol microbes: Bacillus spp. Phytopathology 94:1244

Mehta P, Chauhan A, Mahajan R, Mahajan PK, Shirkot CK (2010) Strain of *Bacillus circulans* isolated from apple rhizosphere showing plant growth promoting potential. Curr Sci 98(4):538–542

Meneses CHSG, Rouws LFM, Simões-Araújo JL, Vidal MS, Baldani JI (2011) Exopolysaccharide production is required for biofilm formation and plant colonization by the nitrogen-fixing endophyte *Gluconacetobacter diazotrophicus*. Mol Plant Microbe Interact 24:1448–1458

Miah S, Dey S, Haque SMS (2010) Shifting cultivation effects on soil fungi and bacterial population in Chittagong Hill Tracts, Bangladesh. J Forest Res 21:311–318

Minorsky PV (2008) On the inside. Plant Physiol 146:1455–1456

Mohammad BT, Al Daghistani HI, Jaouani A, Abdel-Latif S, Kennes C (2017) Isolation and characterization of thermophilic bacteria from Jordanian hot springs: *Bacillus licheniformis* and *Thermomonas hydrothermalis* isolates as potential producers of thermostable enzymes. Int J Microbiol 2017:1–12

Mondal KC, Banerjee D, Banerjee R, Pati BR (2001) Production and characterization of tannase from *Bacillus cereus* KBR9. J Gen Appl Microbiol 47:263–267

Motta AS, Cladera-Olivera F, Brandelli A (2004) Screening for antimicrobial activity among bacteria isolated from the Amazon basin. Braz J Microbiol 35:307–310

Müller M, Munné-Bosch S (2011) Rapid and sensitive hormonal profiling of complex plant samples by liquid chromatography coupled to electrospray ionization tandem mass spectrometry. Plant Methods 7:37–48

Nain L, Yadav RC, Saxena J (2012) Characterization of multifaceted Bacillus sp. RM-2 for its use as plant growth promoting bioinoculant for crops grown in semi-arid deserts. Appl Soil Ecol 59:124–135

Nair AS, Al-Battashi H, Al-Akzawi A, Annamalai N, Gujarathi A, Al-Bahry S, Dhillon GS, Sivakumar N (2018) Waste office paper: a potential feedstock for cellulase production by a novel strain *Bacillus velezensis* ASN1. J Waste Manag 79:491–500

Panda MK, Sahu MK, Tayung K (2013) Isolation and characterization of a thermophilic Bacillus sp. with protease activity isolated from hot spring of Tarabalo, Odisha, India. Iran J Microbiol 5:159–165

Pandey A, Chaudhry S, Sharma A, Choudhary VS, Malviya MK, Chamoli S, Rinu K, Trivedi P, Palni LMS (2011) Recovery of Bacillus and Pseudomonas spp. from the 'fired plots' under shifting cultivation in northeast India. Curr Microbiol 62:273–280

Pandey A, Dhakar K, Sharma A, Priti P, Sati P, Kumar B (2015) Thermophilic bacteria that tolerate a wide temperature and pH range colonize the Soldhar (95°C) and Ringigad (80°C) hot springs of Uttarakhand, India. Ann Microbiol 65:809–816

Pena-Yam LP, Ruíz-Sánchez E, Barboza-Corona JE, Reyes-Ramírez A (2016) Isolation of *Mexican bacillus* species and their effects in promoting growth of chili pepper (*Capsicum annuum* L. cv Jalapeño). Indian J Microbiol 56:375–378

Penrose DM, Glick BR (2003) Methods for isolating and characterizing ACC deaminase-containing plant growth-promoting rhizobacteria. Physiol Plant 118:10–15

Pérez-García A, Romero D, De Vicente A (2011) Plant protection and growth stimulation by microorganisms: biotechnological applications of Bacilli in agriculture. Curr Opin Biotechnol 22:187–193

Pikuta EV, Hoover RB, Tang J (2007) Microbial extremophiles at the limits of life. Crit Rev Microbiol 33:183–209

Pingping SU, Jianchao CU, Xiaohui JI, Wenhui WA (2017) Isolation and characterization of *Bacillus Amyloliquefaciens* L-1 for biocontrol of pear ring rot. Hortic Plant J 3:183–189

Prashar P, Kapoor N, Sachdeva S (2014) Rhizosphere: its structure, bacterial diversity and significance. Rev Environ Sci Biotechnol 1:363–377

Priest FG, Barker M, Baillie LW, Holmes EC, Maiden MC (2004) Population structure and evolution of the *Bacillus cereus* group. J Bacteriol 186:7959–7970

Qureshi MA, Ahmad ZA, Akhtar N, Iqbal A, Mujeeb F, Shakir MA (2012) Role of phosphate solubilizing bacteria (PSB) in enhancing P availability and promoting cotton growth. J Anim Plant Sci 22:204–210

Radhakrishnan R, Hashem A, Abd_Allah EF (2017) Bacillus: a biological tool for crop improvement through bio-molecular changes in adverse environments. Front Physiol 8:667–681

Ralte V, Pandey HN, Barik SK, Tripathi RS, Prabhu SD (2005) Changes in microbial biomass and activity in relation to shifting cultivation and horticultural practices in subtropical evergreen forest ecosystem of north-east India. Acta Oecologica 28:163–172

Ramakrishnan PS, Toky OP (1981) Soil nutrient status of hill agroecosystems and recovery pattern after slash and burn agriculture (jhum) in northeastern India. Plant and Soil 60:41–64

Ramani V, Patel HH (2011) Phosphate solubilization by *Bacillus sphaericus* and *Burkholderia cepacia* in presence of pesticides. J Agric Technol 7:1331–1337

Ranjard L, Richaume A (2001) Quantitative and qualitative microscale distribution of bacteria in soil. Res Microbiol 152:707–716

Rasko DA, Altherr MR, Han CS, Ravel J (2005) Genomics of the *Bacillus cereus* group of organisms. FEMS Microbiol Rev 29:303–329

Ray AK, Ghosh K, Ringø E (2012) Enzyme-producing bacteria isolated from fish gut: a review. Aquacult Nutr 18:465–492

Rengel Z, Marschner P (2005) Nutrient availability and management in the rhizosphere: exploiting genotypic differences. New Phytol 168:305–312

Reza SK, Baruah U, Nath DJ, Sarkar D, Gogoi D (2014) Microbial biomass and enzyme activity in relation to shifting cultivation and horticultural practices in humid subtropical North-Eastern India. Range Manag Agrofor 35:78–84

Riesenfeld CS, Goodman RM, Handelsman J (2004) Uncultured soil bacteria are a reservoir of new antibiotic resistance genes. Environ Microbiol 6:981–989

Rűger HJ, Fritze D, Sproer C (2000) New psychrophilic and psychrotolerant *Bacillus marinus* strains from tropical and polar deep-sea sediments and emended description of the species. Int J Syst Bacteriol 50:1305–1313

Ryu CM, Farag MA, Hu CH, Reddy MS, Kloepper JW, Pare PW (2004) Bacterial volatiles induce systemic resistance in Arabidopsis. Plant Physiol 134:1017–1026

Sansinenea E, Ortiz A (2011) Secondary metabolites of soil Bacillus spp. Biotechnol Lett 33:1523–1538

Saplalrinliana H, Thakuria D, Changkija S, Hazarika S (2016) Impact of shifting cultivation on litter accumulation and properties of Jhum soils of North East India. J Indian Soc Soil Sci 64:402–413

Sashidhar B, Podile A (2010) Mineral phosphate solubilization by rhizosphere bacteria and scope for manipulation of the direct oxidation pathway involving glucose dehydrogenase. J Appl Microbiol 109:1–12

Seldin L, Soares Rosado A, da Cruz DW, Nobrega A, van Elsas JD, Paiva E (1998) Comparison of *Paenibacillus azotofixans* strains isolated from rhizoplane, and non-root-associated soil from maize planted in two different Brazilian soils. Appl Environ Microbiol 64:3860–3868

Shameer S, Prasad TNVKV (2018) Plant growth promoting rhizobacteria for sustainable agricultural practices with special reference to biotic and abiotic stresses. Plant Growth Regul 84:603–615

Sharma A, Shankhdhar D, Shankhdhar SC (2014) Growth promotion of the rice genotypes by PGPRs isolated from rice rhizosphere. J Soil Sci Plant Nutr 14:505–517

Sharma P, Sood C, Singh G, Capalash N (2015) An eco-friendly process for biobleaching of eucalyptus kraft pulp with xylanase producing *Bacillus halodurans*. J Clean Prod 87:966–970

Sivasakthi S, Usharani G, Saranraj P (2014) Biocontrol potentiality of plant growth promoting bacteria (PGPR)-*Pseudomonas fluorescens* and *Bacillus subtilis*: a review. African J Agri Res 9:1265–1277

Spaepen S, Vanderleyden J, Remans R (2007) Indole-3-acetic acid in microbial and microorganism-plant signaling. FEMS Microbiol Rev 31:425–448

Stabb EV, Jacobson L, Handelsman J (1994) Zwittermicin A-producing strains of *Bacillus cereus* from diverse soils. Appl Environ Microbiol 60:4404–4412

Stein T (2005) *Bacillus subtilis* antibiotics: structures, syntheses and specific functions. Mol Microbiol 56:845–857

Swamy HMM, Asokan R (2013) *Bacillus thuringiensis* as 'Nanoparticles'—a perspective for crop protection. Nanosci Nanotechnol Asia 3:102–105

Szilagyi-Zecchin VJ, Ikeda AC, Hungria M, Adamoski D, Kava-Cordeiro V, Glienke C, Galli-Terasawa LV (2014) Identification and characterization of endophytic bacteria from corn (*Zea mays* L.) roots with biotechnological potential in agriculture. Amb Express 4:1–9

Tallapragada P, Seshachala U (2012) Phosphate-solubilizing microbes and their occurrence in the rhizospheres of Piper betel in Karnataka. India Turk J Biol 36:25–35

Tilak K, Ranganayaki N, Pal KK, De R, Saxena AK, Nautiyal CS, Mittal S, Tripathi AK, Johri BN (2005) Diversity of plant growth and soil health supporting bacteria. Curr Sci 89:136–150

Tripathi DK, Singh VP, Kumar D, Chauhan DK (2012) Impact of exogenous silicon addition on chromium uptake, growth, mineral elements, oxidative stress, antioxidant capacity, and leaf and root structures in rice seedlings exposed to hexavalent chromium. Acta Physiol Plant 34:279–289

Van Elsas JD, Chiurazzia M, Cyrus AM, Dana E, Krištu VF, Sallesa JF (2012) Microbial diversity determines the invasion of soil by a bacterial pathogen. Proc Natl Acad Sci U S A 109:1159–1164

Vargas-Ayala R, Rodriguez-Kaban R, Morgan-Jones G, McInroy JA, Kloepper JW (2000) Shifts in soil microflora induced by velvetbean (*Mucuna deeringiana*) in cropping systems to control root-knot nematodes. Biol Control 17:11–22

Vejan P, Abdullah R, Khadiran T, Ismail S, Nasrulhaq Boyce A (2016) Role of plant growth promoting rhizobacteria in agricultural sustainability-a review. Mol Ther 21:573–590

Venturi V, Keel C (2016) Signaling in the rhizosphere. Trends Plant Sci 21:187–198

Vilain S, Luo Y, Hildreth MB, Brözel VS (2006) Analysis of the life cycle of the soil saprophyte *Bacillus cereus* in liquid soil extract and in sdoil. Appl Environ Microbiol 72:4970–4977

Vimala Devi PS, Duraimurugan P, Chandrika KSVP (2019) *Bacillus thuringiensis*-based nanopesticides for crop protection. In: Nano-biopesticides today and future perspectives, pp 249–260

Wahyudi AT, Astuti RP, Widyawati A, Mery A, Nawangsih AA (2011) Characterization of *Bacillus* sp. strains isolated from rhizosphere of soybean plants for their use as potential plant growth for promoting rhizobacteria. J Microbiol Antimicrob 3:34–40

Wang L, Tang Y, Wang S, Liu RL, Liu MZ, Zhang Y, Liang FL, Feng L (2006) Isolation and characterization of a novel thermophilic Bacillus strain degrading long-chain n-alkanes. Extremophiles 10:347–356

Wang Q, Garrity GM, Tiedje JM, Cole JR (2007) Naïve Bayesian classifier for rapid assignment of rRNA sequences into the new bacterial taxonomy. Appl Environ Microbiol 73:5261–5267

Wang X, Zhao D, Shen L, Jing C, Zhang C (2018) Application and mechanisms of *Bacillus subtilis* in biological control of plant disease. In: Role of rhizospheric microbes in soil. Springer, pp 225–250

Yadav S, Kaushik R, Saxena AK, Arora DK (2011) Diversity and phylogeny of plant growth-promoting bacilli from moderately acidic soil. J Basic Microbiol 51:98–106

Yánez-Mendizábal V, Viñas I, Usall J, Torres R, Solsona C, Teixidó N (2012) Production of the postharvest biocontrol agent *Bacillus subtilis* CPA-8 using low cost commercial products and by-products. Biol Conserv 60:280–289

Yang G, Chen M, Yu Z, Lu Q, Zhou S (2013) Bacillus composti sp. nov. and Bacillus thermophilus sp. nov., two thermophilic, Fe(III)-reducing bacteria isolated from compost. Int J Syst Evol Microbiol 63:3030–3036

Yang JK, Zhang JJ, Yu HY, Cheng JW, Miao LH (2014) Community composition and cellulase activity of cellulolytic bacteria from forest soils planted with broad-leaved deciduous and evergreen trees. Appl Microbiol Biotechnol 98:1449–1458

Yanti Y, Hamid H, Habazar T (2020) The ability of indigenous Bacillus spp. consortia to control the anthracnose disease (*Colletrotricum capsici*) and increase the growth of chili plants. Biodivers J Biol Diversity 21

Yip W, Wang L, Cheng C, Wu W, Lung S, Lim BL (2003) The introduction of a phytase gene from *Bacillus subtilis* improved the growth performance of transgenic tobacco. Biochem Biophys Res Commun 310:1148–1154

Zaidi A, Khan M, Ahemad M, Oves M (2009) Plant growth promotion by phosphate solubilizing bacteria. Acta Microb Imm H 56:263–284

Zhu D, Zhang P, Niu L, Xie C, Li P, Sun J, Hang F (2016) *Bacillus ectoiniformans* sp. nov., a halotolerant bacterium isolated from deep-sea sediments. Int J Syst Evol Microb 66:616–622

Zothansanga R, Senthilkumar N, Gurusubramanian G (2016) Diversity and toxicity of *Bacillus thuringiensis* from shifting cultivation (Jhum) habitat. Biocontrol Sci 21:99–111

Chapter 12
Bacillus Species of Ruminant Origin as a Major Potential Sources of Diverse Lipolytic Enzymes for Industrial and Therapeutic Applications

Grace Mujinga Mukendi, Alfred Mitema, Karen Nelson, and Naser Aliye Feto

Abstract Lipases are biocatalysts that catalyze a wide range of reactions such as the hydrolysis of triglycerides or lipids and esterification of fatty acids in a nonaqueous medium. This versatility renders lipases to be potential biocatalysts for the food, detergent, paper and pulp, leather and textile industries, biodiesel production, and therapeutic applications. Lipases are naturally sourced from either plants, animals, or microorganisms. Microbial lipases are quite stable, selective, substrate-specific, and, thus, classified as one of the most extensively used industrial enzymes. The rumen is a four-chamber stomach of the ruminant animal representing the fermentation-vat where microorganisms, *Bacillus* species, work in a symbiotic relationship with the host to digest complex ingested feed. The ruminant microbiome is a relatively less exploited yet potentially rich source of biocatalysts of industrial, environmental, and therapeutic importance. However, no or minimal comprehensive review has been reported on biocatalysts of ruminant origin. Therefore, this review presents an in-depth analysis of biocatalysts of rumen microbiome with particular emphasis on *Bacillus* spp. of ruminant origin as an untapped source of diverse lipase-isoforms with potential industrial and therapeutic applications.

Keywords *Bacillus* species · Biocatalyst · Lipase · Microbiome · Rumen · Ruminant · Therapeutics

---

G. M. Mukendi · A. Mitema · N. A. Feto (✉)
OMICS Research Group, Department of Biotechnology, Vaal University of Technology, Vanderbijlpark, South Africa
e-mail: naserf@vut.ac.za

K. Nelson
J. Craig Venter Institute (JCVI), Rockville, MD, USA

M. T. Islam et al. (eds.), *Bacilli in Agrobiotechnology*, Bacilli in Climate Resilient Agriculture and Bioprospecting,
https://doi.org/10.1007/978-3-030-85465-2_12

12.1 Introduction

The exponential growth of the global population in the last centuries saw an increase in demand and consumption of daily life products: food, potable water, chemicals, cosmetics, medicine, transportation, and pharmaceutical (Sarmah et al. 2018). The industrial world has been revolutionized to meet the demand for daily life products, and it played a significant role in the improvement of the economy. The production of such consumables was deemed to be highly dependent on raw materials and require too much energy, thus, generating a considerable amount of waste which substantially affects the environment (Sarmah et al. 2018). To save the environment from industrial pollution and effects of climate change, death of aquatic and vegetable life, ozone layer destruction, and fossil depletion, industries globally have drifted their focus towards the development of technologies of less environmental effect and that would continue to meet the increasing demand for products while using fewer resources (Sarmah et al. 2018).

Enzyme catalysis or enzymatic processing among biotechnologies are promising alternatives for greener, cleaner, safer, and sustainable in comparison to the traditional or conventional processes (Sarmah et al. 2018). Enzymes or biocatalysts are proteins that are produced by living organisms for the synthesis of critical sustainable biomolecules and the digestion of foods. They accelerate the rate of various biochemical reactions that can also be mimicked in various industries (Gurung et al. 2013).

Enzymes possess significant advantages over harsh chemicals used in conventional processing due to their specificity, reusability, and biodegradability (Andualema and Gessesse 2012). Enzymes are highly specific and target some particular bonds to produce the necessary products. Thus, the enzymatic processes eliminate the generation of unwanted products and are reputed to withstand competitive conditions. Additionally, enzymes do not leave any ecological footprints because they are readily degraded by microscopic life. The resulting decomposed materials are always recycled back to nature (Leow et al. 2007).

The diverse functions that enzymes perform on different substrates during chemical reactions greatly influence their classification. They have been classified into six groups; oxidoreductase (EC 1), transferase (EC 2), hydrolase (EC 3) currently being the most used in biotechnology, lyase (EC 4), isomerase (EC 5), and ligase (EC 6), per the Enzyme Commission (Fig. 12.1). Hydrolase is comprised of carbohydrase, proteases, and Lipolytic enzymes which represent more than 70% of the enzyme market sales (Haki and Rakshit 2003).

12.2 Rumen Microbiota as a *Bacillus* Habitat

The genus *Bacillus* is composed of many species that are gram-positive, rod-shaped, motile, aerobic and facultative anaerobic, and gram-variable endospore-forming bacteria (Guncheva and Zhiryakova 2011). *Bacillus* genera and other

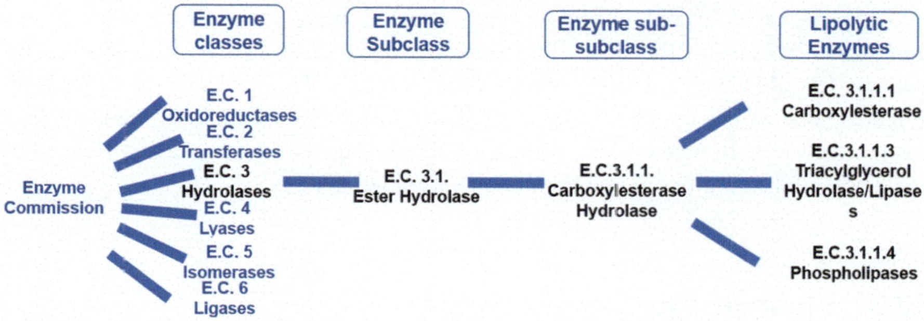

Fig. 12.1 A representation of the numerical classification of lipolytic enzymes by the Enzyme Commission (EC) (Casas-Godoy et al. 2012; Robinson 2015). The classification is based on the chemical reactions that enzymes catalyze. As a system of enzyme nomenclature, every EC number is associated with a recommended name for the respective enzyme (Robinson 2015)

bacteria-producing lipases inhabit their nature from the temperature ecological environment to the extreme environment such as the soil, hot springs, marine sediments, volcanic water, industrial effluents, human gut, and the rumen (Guncheva and Zhiryakova 2011).

The rumen is a four-chambered stomach of the ruminants [the rumen, reticulum, omasum, and abomasum). herbivorous animals digesting their feeds through rumination. The rumination process starts by consuming or engulfing the raw material, regurgitating it into a semi-digested form known as cud, and by chewing the resulting cud (Oyeleke and Okusanmi 2008).

The rumen, which also qualifies as the unique digestive vessel, is inhabited by a group of microbes, namely protozoa, archaea, fungi, and bacteria, which work in a symbiotic relationship (Choudhury et al. 2015). Bacteria consist of 50-75% of the rumen microbiome (Choudhury et al. 2015). Ishaq et al. (2015) isolated 31 bacteria from the rumen of North American Moose (*Alces alces*) whereby, 26 were identified as *Bacillus* spp. thus, confirmed the genus *Bacillus* as one of the predominant bacterial genera inside the rumen. Similarly, Oyeleke and Okusanmi (Oyeleke and Okusanmi 2008) isolated microorganisms from the rumen of the cow, sheep, and goat and discovered that *Bacillus* spp. represented 37.8% of the bacterial population of the rumen microbiome.

The rumen is a harsh environment that is prevailed by a strict anaerobic condition, a physiological 5.5-6.0 pH range, and a temperature of about 40 °C (Hungate 1966; Miyagi et al. 1995). *Bacillus* genus survives in such a complex environment, and their biocatalysts, especially lipases, are believed to demonstrate exceptional qualities. The rumen microbiome plays a pivotal role in the digestion processes by synthesizing enzymes that are applied in the breaking down of complex ingested feeds into simpler compounds (Hungate 1966; Miyagi et al. 1995). Thus, the rumen, a very complex microbiota, can be explored for its microbiome sources of enzymes like lipases with potential in biotechnology, research, and industrial applications (Hess et al. 2011).

Lipolytic enzymes notably lipase have attracted the attention of the biotechnology world because of their excellent features (Javed et al. 2018). The subsequent section describes in detail different Lipolytic enzymes, their common and divergent features such as the lipase isoforms, the three-dimensional structure, their mechanism of action, different sources, mining methods of novel lipases, stabilization, and their potential in industrial and therapeutic applications.

12.3 Lipolytic Enzymes

Lipolytic enzymes are the third large subclass of the hydrolases after carbohydrases and proteases being the first and second respectively (Gurung et al. 2013). They both make up to 70% of the enzyme sales (Gurung et al. 2013; Ouyang et al. 2013). Lipolytic enzymes are further characterized by their chemo-, stereo-, enantio- and regio-specificity characteristics that render them extensively useful in different industries, like food, detergent, paper and pulp, leather, and textile industries (Haki and Rakshit 2003), biodiesel production, and therapeutic applications (Sarmah et al. 2018).

Lipolytic enzymes are comprised of esterases also called carboxylesterase (EC.3.1.1.), phospholipases [E.C. 3.1.1.4), and "true" lipases also called triacylglycerol hydrolases (EC.3.1.1.3) (Fig. 12.1) (Bornscheuer 2002; Casas-Godoy et al. 2012; Fan et al. 2012; Hudlicky and Reed 2009). Although the three listed lipolytic enzymes share the same functionality, mechanism of action, and α/β hydrolase canonical fold patterns, they differ in the three-dimensional structure or amino acid sequences of the active site or the substrate preference (Bornscheuer 2002; Casas-Godoy et al. 2012; Fan et al. 2012; Hudlicky and Reed 2009).

12.3.1 Esterases

Esterases, as phospholipases and lipases, catalyze the hydrolysis reactions (Bornscheuer 2002; López-López et al. 2014). They mainly fasten the hydrolysis of water-soluble esters which are usually made up of a short acyl chain, precisely less than or equivalent to twelve carbons. Esterases do not only catalyze the hydrolysis of esters, but they also catalyze a plethora of reactions under micro-aqueous conditions to form ester bonds. Such reactions are mainly called esterification, alcoholysis, and transesterification (Bornscheuer 2002; López-López et al. 2014).

Being a superfamily of the α/β hydrolase enzyme, esterases share a similar 3D structure with the enzymes of the α/β hydrolase family (Bornscheuer 2002). The structure is mainly determined by the presence of eight stranded β-sheets that are parallel and antiparallel at the center of the enzyme molecules. The stranded β-sheets are connected to the surrounded helices and connecting loops (Bornscheuer 2002). The β-sheets and α-helices are the support of the active site which is made up of signatures of three amino acids residues namely serine, histidine, and aspartic acid or glutamic residue (Bornscheuer 2002). Serine residue is mainly located in the center of the pentapeptide motif (G-X-S-X-G-) with X representing any residue, and the nucleophile elbow is located on the pentapeptide motif between α-helix and β-sheet (Bornscheuer 2002). Unlike other Lipolytic enzymes, the active site of the esterases is always opened, and the binding pocket for the acid moiety of the substrate is remarkably large. Therefore, esterases have broad substrate profiles because they are specific for a broad range of substrates (Bornscheuer 2002).

Esterases act similarly to the enzymes of α/β hydrolase during hydrolysis. In order to release fatty acid and alcohol from an ester in the presence of water, the hydroxyl (-OH) group of the serine residue from the catalytic triad attacks the carbonyl carbon of the ester bond to be broken down. Upon attacks, an alcohol metabolite and an acylated enzyme (tetrahedral intermediate) are released (Bornscheuer 2002). The acylated enzyme is formed by the formation of a covalent bond between the enzyme and the fatty acid moiety of the substrate. The stabilization step then occurs by first adding hydrogen to the nitrogen atom of histidine residue thereafter, adding a carboxylic group of the aspartic or glutamic acid residue of the oxyanion hole to the histidine. Lastly, the stabilized histidine residue then gets attracted to the water molecules and form a bond with water to release an acid moiety and enzyme at its active state (Bornscheuer 2002). Therefore, esterases obey the Michaelis-Menten classical kinetics because they only catalyze water-soluble substrates. Their broad substrate specificity and versatility features make them useful for industrial applications. However, their limitation only acts on short acyl-chain substrate that reduces their application compared to phospholipases and lipases (Borrelli and Trono 2015; Javed et al. 2018).

12.3.2 Phospholipases

Just like esterases, phospholipases catalyze the hydrolysis of a classical substrate called phospholipids (Borrelli and Trono 2015). Phospholipids are water-insoluble triglycerides having a phosphate group at the third carbon of the glycerol group. Apart from acting on phospholipids, phospholipase can also cleave the ester bonds of the triacylglycerols at the first and second carbon of the glycerol backbone. Contrary to esterases, phospholipases catalyze the hydrolysis of the water-insoluble substrate that is simply esterases having a long acyl chain. Therefore, the reactions catalyzed by phospholipases take place at the water/lipids interphase (Borrelli and Trono 2015).

Phospholipases are not only involved in the hydrolysis of phospholipids or triglycerides (exceptional conditions), but they also catalyze diverse reactions (Borrelli and Trono 2015). Such diverse reactions include acyl transesterification during which free fatty acids are added to lysophospholipids, transphospahtidylation during which the phosphatidyl group from the phospholipid is transferred to the alcohol to give phosphatidyl alcohol under low water activity in the presence of organic substrates or solvents. Phospholipases are a group of enzymes that are classified into six classes that are Phospholipases A, Phospholipases A1, Phospholipases A2, Phospholipases B, Phospholipases C, and Phospholipases D (Borrelli and Trono 2015). The classification of Phospholipases A and B is entirely based on the carbon position of the glycerol backbone at which the hydrolysis of the ester bond occurred, while the classification of phospholipase C and D entirely depends on the cleavage of glycerophosphate and phosphodiester bonds, respectively (Borrelli and Trono 2015).

Phospholipases have the 3D structure of the α/β hydrolase family i.e., they have the same amino acid signature on their active site. The active site is comprised of the catalytic triad Ser-, Serine the central or nucleophilic residue of the pentapeptide motif (GXSXG), Asp- Aspartic acid the catalytic acid residue, and His- Histidine the catalytic base residue (Borrelli and Trono 2015). Some phospholipases (PLA) have their catalytic acid Aspartic acid replaced with Asn- Asparagine while other phospholipases (PLA2) possess an additional Ca^{+2} binding loop (Borrelli and Trono 2015). Compared to esterases, phospholipases have shorter β5 and β9 sheets and a lid domain to cover the active site when it is in a closed conformation (Fig. 12.2); thus, the two mentioned differences render phospholipases substrate-specific (Borrelli and Trono 2015).

It is made up of two domains (C2 A in orange, indicated with a black arrow, and C2 B in blue, indicated with a red arrow). The two domains are linked to the α/β hydrolase core (in blue sky color, wrapped in the red box). The hydrolase core consists of 10 β sheets represented as blue tiny ribbons. The β sheets are connected to

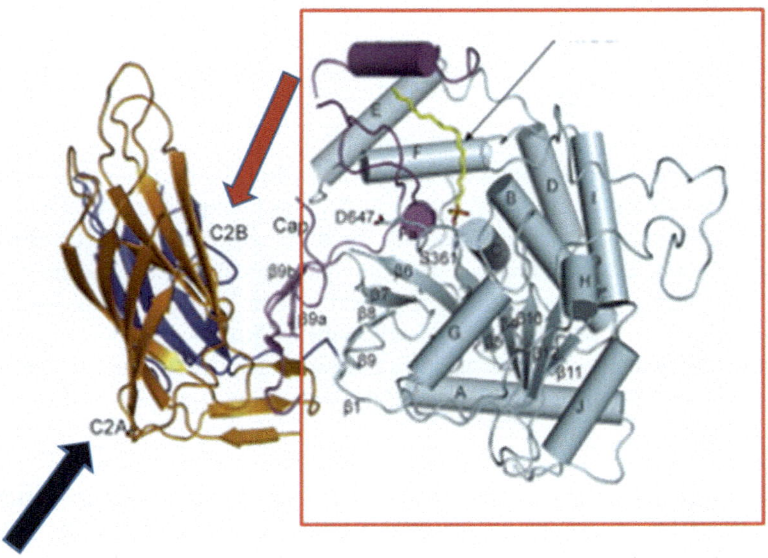

Fig. 12.2 The three-dimensional structure of phospholipase A2 from a human GIVD (cytosolic Group IV D) cytosol (Wang et al. 2016)

the 10 α helices (represented in long cylinders A-J in blue color) in the form of a can (Wang et al. 2016).

The mode of action of phospholipases is like the classical mechanism of action of esterases. However, some phospholipase A2 starts their nucleophilic attack with the Histidine residue playing the role of the nucleophilic residue (Borrelli and Trono 2015). The substrate specificity and versatility make phospholipases to be a great potential for application in various industries and therapeutic domains. Nevertheless, their limitation of reactions to the glycerol carbon position or sites leads to a search for more versatile and broad substrate-specific enzymes such as lipases (Borrelli and Trono 2015).

12.3.3 *Lipases*

Lipases are serine hydrolases that catalyze the cleavage of ester bonds between the glycerol group and fatty acids moiety to yield free fatty acids or acyl-glycerides and alcohol (Ramnath et al. 2017). Unlike esterases, lipases have an affinity for water-insoluble long-chain acyl triglycerides notably more than twelve carbons. Lipases are always activated by the occurrence of water and lipids interphase for lipids hydrolysis (Ramnath et al. 2017).

12.3.3.1 Lipase-Catalyzed Reaction

Hydrolysis of triglycerides usually takes place under natural conditions in the presence of excess water. In limited water conditions, lipases can catalyze the reverse reaction designated as esterification (Ramnath et al. 2017). Moreover, under deficient water activity, lipases are still capable of catalyzing diverse transesterification reactions such as alcoholysis, acidolysis, aminolysis, and interesterification. Therefore, based on the diverse reactions that lipases catalyzed (Table 12.1), they are always regarded as one of the versatile enzymes found in nature (Borrelli and Trono 2015). As a member of the serine hydrolase family, lipases rely on their serine residue to cleave the ester bonds of the triglyceride substrate as described in the subsequent section.

12.3.3.2 Lipase Structure and Mechanism of Action

12.3.3.2.1 Lipase Structure

As mentioned previously, lipases are serine hydrolases and are generally characterized based on the α/β folding signature. Lipases have a backbone that is made up of eight different strands constituting the central β sheet (Gupta et al. 2015). The eight strands are designated as β_1, β_2, β_3, β_4, β_5, β_6, β_7, and β_8. Studies performed on the three-dimensional structure of lipases from various sources except for the lipases

Table 12.1 List of reactions catalyzed by microbial lipases

Type of reaction	Industrial application	Source of enzyme	References
Hydrolysis	Bioorganic synthesis, Food industry Detergent industry	*Bacillus pumilus*	Casas-Godoy et al. (2012), Chaplin et al. (2004), Laachari et al. (2015), and Litantra et al. (2013)
	Oil and fat industry	*B. thermoleovorans*	Patil et al. (2011)
	Solid waste treatment	*Pseudomonas aeruginosa*	Bose and Keharia (2013)
	Medical and healthcare	*Staphylococcus aureus*	Vaquero et al. (2016)
Esterification	Chemical industry	*Acinetobacter* sp.	Thakur (2012)
	Food processing industry	*B. coagulans*	Thakur (2012)
	Medical and healthcare	*B. licheniformis*	Sharma et al. (2015)
	Flavor industry	*Staphylococcus epidermidis*	Patil et al. (2011)
Transesterification	Biodiesel	*B. subtilis*	Bajaj et al. (2010) and Treichel et al. (2010)
	Jet fuel	*Burkhodelia cepacia*	Bajaj et al. (2010), Salum et al. (2010)
	Light hydrocarbon oil	*Enterobacter aerogenes*	Bajaj et al. (2010)
	Fuel oil	*Pseudomonas cepacia*	Bajaj et al. (2010)
	Incorporation of different acids	*B. stearothermophilus*	Thakur (2012)
	Therapeutic agents	*Candida antarctica*	Hernández-Fernández et al. (2010)
Alcoholysis	Quick-drying oil	*Pseudomonas* sp.	Salis et al. (2009)
	Cosmetics	*Pseudomonas fluorescens*	Patil et al. (2011)
	Biodiesel production	*Pseudomonas* sp.	Salis et al. (2009)
Enantioselective hydrolysis	Medical and healthcare products	*Serratia marcescens*	Su et al. (2014)

from the pancreas revealed the presence of the eight β sheets that are sequentially connected to about six α helices. The helices are named as α1, α2, α3, α4, α5, and α6 (Fig. 12.3b) (Gupta et al. 2015).

The backbone holds the active site of the lipases, and the site is made up of some amino acids that are involved in the hydrolysis mechanism (Gupta et al. 2015). The

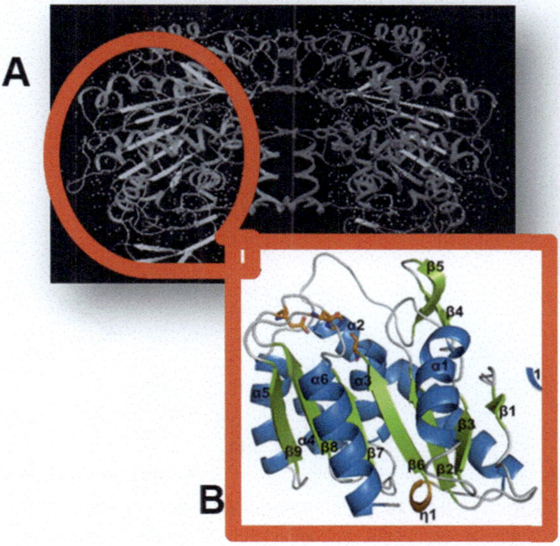

Fig. 12.3 (**a**) The Quaternary structure of lipase; (**b**) The α/β folding signature of the lipase formed of eight different strands (β_1, β_2, β_3, β_4, β_5, β_6, β_7, and β_8) in green that are connected to the six α helices (α1, α2, α3, α4, α5, and α6) in blue

three most essential amino acids called the catalytic triad, of the active site, are like those of esterases and phospholipases (i.e., the serine and nucleophilic residues respectively). The serine residue is the central residue of the pentapeptide motif and the other two amino acids of the active site are the Aspartic or Glutamic acid, catalytic acid, and Histidine residues (Gupta et al. 2015).

12.3.3.2.2 Mechanism of Action

Lipases possess the nucleophilic elbow containing the Serine residue that forms the pentapeptide motif and a lid covering the active site (Jaeger et al. 1999). Lipases are generally known to have a closed conformation in their inactive state i.e., the lid covers the active site when the lipases do not perform their catalytic activity (Ito et al. 1998, Jaeger et al. 1999). During activation, the lid is removed from the active site for ease of serine accessibility to the carbonyl carbon atom of the substrate (either fats or lipids). Triglycerides of long acyl-chain carbon are insoluble in water as mentioned previously, (Jaeger et al. 1999) but during their hydrolysis in the presence of lipases, they tend to dissolve partially in the water. Thus, the observation and formation of a water-oil emulsion activate the lipases by having their lid moved away from the active site leaving it open for catalysis. Therefore, one of the characteristics of the lipases is the activation by the water-lipids interphase (Jaeger et al. 1999; Sharma et al. 2001).

The hydrolysis of triglycerides by lipases has three steps of mechanism (Fig. 12.4).

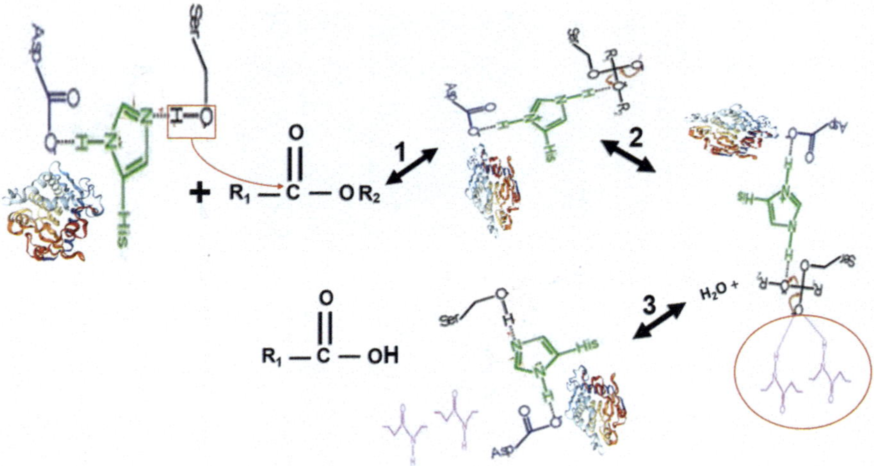

Fig. 12.4 The steps involved in lipid or triglycerides hydrolysis. (1), shows the covalent bonding of hydroxyl group from the serine residue and the carbonyl carbon of the triglyceride to form a tetrahedral intermediate. (2), tetrahedral intermediate stabilization step. (3) the water molecule hydrolysis of the covalent bonds between the enzyme and acyl product and stabilization of the enzyme in a closed conformation and the acyl (diacylglycerol) (Ribeiro et al. 2011)

When the active site of the lipases is opened, the –OH group of the serine residue binds to the carbonyl carbon of the substrate (Fig. 12.4, reaction 1) leading to the formation of tetrahedral intermediate (enzyme-acyl complex). The resultant intermediate is then stabilized by two hydrogen bonding between the negatively charged carbonyl oxygen atoms of the intermediate to nitrogen atoms of the amino acid residues of the oxyanion hole (Fig. 12.4, reaction 2) leading to the release of alcohol as a product (Ribeiro et al. 2011). The water molecule hydrolyzes the covalent bond between the enzyme and acyl product to release the stabilized enzyme in a closed conformation and the acyl can be the diacylglycerol (Fig. 12.4, reaction 3) (Ribeiro et al. 2011).

After hydrolysis, the lipases reclaim the closed conformation by having their helices and strands rearranging themselves in such a way to form a helical segment, the lid, to protect the active site of the lipases93. The active site of the lipases, with its catalytic triad, is also found in the serine proteases, thus, lipases are classified as serine hydrolase.

12.3.3.3 Properties of Lipases

Lipases are known to catalyze a wide range of reactions (Borrelli and Trono 2015; Javed et al. 2018; Thakur 2012). They are said to be versatile, a feature having great potential for industrial applications. The versatility of lipases originates from various properties that lipases possess (Javed et al. 2018). Lipases are said to be solvent saline, high and low temperature tolerant, protease-resistant, and substrate specificity (Javed et al. 2018).

12.3.3.3.1 Stability in Organic Solvents

Organic solvents mainly used in industries improve the solubility of substrates, reduce microbial contamination, and side reactions. Therefore, the stability of lipases in organic solvents makes lipases a potential for industrial processes. An example of an organism producing such lipases is the *Pseudomonas* genera (Borrelli and Trono 2015; Bose and Keharia 2013; Thakur 2012).

12.3.3.3.2 Tolerance to High and Low Temperatures

Many industrial processes operate at high temperatures (Borrelli and Trono 2015). Such processing requires thermophilic enzymes to avoid loss of activity during the process. Some household chores such as laundry require high temperatures for effective removal of stains and thus, thermophilic enzymes are of great importance (Borrelli and Trono 2015).

12.3.3.3.3 Tolerance to Alkaline and Acidic pHs

Alkalinity condition is one of the primary requirements by the detergent industry (Chauhan et al. 2013); thus, alkaline enzymes attract such sectors. Since washing conditions are harsh, lipases, as one of the main components of the detergents for the removal of stubborn greasy stains, are required to be active and stable in the alkaline condition (Chauhan et al. 2013).

Some other industrial applications take place under acidic conditions (Andualema and Gessesse 2012). Therefore, lipases of great potential for industrial applications should not only be alkaline, but those required for acidic operations should be active at acidic pH (Andualema and Gessesse 2012).

12.3.3.4 Classification of Lipases Based on Substrate and Region-Specificity

Lipases having the same mechanism of action are said to differ among themselves in the order of amino acids or residue sequences making them up (Sarmah et al. 2018). There has been no formal classification of lipases as far as literature is concerned (Javed et al. 2018). However, many classifications (Fig. 12.5) have been made based on different properties such as their sources, substrate specificity (Fig. 12.6), and region-specificity (Sarmah et al. 2018).

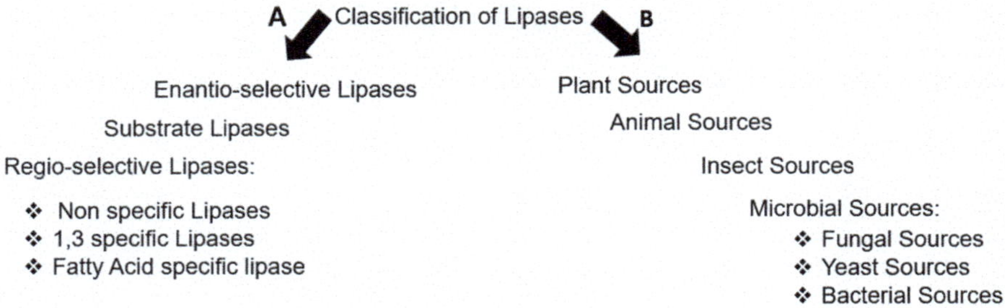

Fig. 12.5 Classification of lipases based on (**a**) properties or specificity and (**b**) their sources

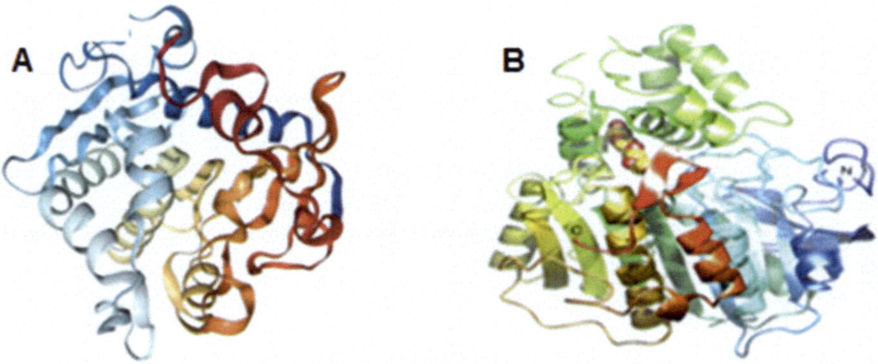

Fig. 12.6 The illustration of different lipase isoforms based on their folding pattern obtained from the X-ray crystallography of lipase from *Rhizomucor miehei* (**a**) from *Candida antartica* (**b**). They are all substrate specific because they target triglycerides with specific fatty acids (Gupta et al. 2015; Gurung et al. 2013)

The first class of lipases includes the nonspecific lipases which catalyze triglycerides into free fatty acids and glycerol. During the reaction, the intermediates that are produced comprise the mono- and diglycerides. Such lipases attack the carbonyl atom irrespective of their position in the molecule (Kapoor and Gupta 2012; Ribeiro et al. 2011). The second class of lipases includes the 1, 3- specific lipases which catalyze triglycerides into monoglycerides and fatty acids (Kapoor and Gupta 2012; Ribeiro et al. 2011). The third class of lipases comprises the fatty-acid specific lipases which catalyze triglycerides based on selective fatty acid. This isoform has an affinity for long-chain acyl group having a double bond between carbon 9 and 10 in cis position.

The ability of lipases to perform multiple types of reactions in heterogeneous media, large-scale substrate specificity stability under extreme conditions, readily available, are the primary factors influencing the choice of lipases isoforms for industrial and therapeutic applications over other chemical catalysts which are not eco-friendly and highly toxic (Ribeiro et al. 2011).

12.4 Sources of Lipase

Lipases are major enzymes employed in the turnover of the lipid's biomass in nature. They are mainly sourced from plants, animals, and microorganisms; thus, they are ubiquitous (Sarmah et al. 2018).

12.4.1 Plant Lipase

Plant lipases are in various parts of the plants such as the germinating seeds and cereals, and they appear to attract substantial attention in biotechnology due to their low cost, uniqueness, and specificity in their ability to cleave the three ester bonds of triglycerides and to synthesize designed esters (Sarmah et al. 2018). Plant lipases find their application in the food and beverage industries, pharmaceutical, and therapeutic applications (Sarmah et al. 2018). Plant lipases are not extensively studied and explored because their purification and bulk production are confronted with some difficulties (Sarmah et al. 2018).

12.4.2 Animal Lipase

On the other hand, animal and insect lipases, which are mainly composed of pancreatic lipase, hepatic lipase, and hormone-sensitive lipase, are primarily used for digestion inside the hosts (Wong and Schotz 2002). Animal lipases are used in medicine and diagnosis (Sarmah et al. 2018). Animal lipases are less studied because they are bulky and difficult to purify. The use of animal lipases has some disadvantages such as health implications and defective products (Sarmah et al. 2018).

12.4.3 Microbial Lipase

Lipases from microorganisms are the most industrially used owing to their high abundance, ease of handling, amenability to genetic manipulation, ease of purification, bulk production, stability, and substrate specificity (Ray 2012). The above-listed features influence the attractiveness of lipases for industrial applications and make lipases to be a potential tool for biotechnological applications. The majority of microbial lipases are extracellularly produced by fungi, yeast, and bacteria (Thakur 2012).

12.4.3.1 Fungal Lipase

Fungal lipases are produced by fungi and yeast either outside or inside the fungal cells depending on the composition of the crude enzyme production medium (Thakur 2012). Fungal lipases are obtained through Solid Substrate Fermentation (SSF) or Submerged Fermentation (SmF), and they can catalyze a range of reactions such as hydrolysis and transesterification, etc... Fungal lipases are the widely used lipases thanks to their attributes including thermal stability, low-cost extraction and production, alcohol and other organic solvents tolerance, and pH stability (Patil et al. 2011). Despite such qualities, an isoform of lipase named Lipase B from *Candida antarctica* (CALB) was noticed to be cold-active although widely applied in various industries (Sarmah et al. 2018). Some of the major genera involved in the production of commercially important fungal lipases include *Aspergillus* spp., *Mucor* spp., *Penicillium* spp., *Rhizomucor* spp., *Rhizopus* spp., *Candida* spp., *Yarrowia* spp., *Pichia* spp., and *Saccharomyces* spp. (Miyagi et al. 1995). Such genera are ubiquitous and have so long been isolated from diverse microbial niches such as soil contaminated with oil, hot springs, compost, industrial wastes and effluents, dairy plants, and vegetable oil processing factories (Sarmah et al. 2018).

12.4.3.2 Bacterial Lipase

Bacterial lipases are mostly produced extracellularly depending on the composition of the production medium and are deemed to be glycoproteins while some are lipoproteins (Javed et al. 2018). They can also be intracellular or attached to the membrane (Javed et al. 2018). Additionally, bacterial lipases are considered more suitable for industrial applications because of their ability to withstand the harsh conditions imposed by the industrial environment, alkaline pH and organic solvent tolerance, substrate nonspecific, and abundance (Sarmah et al. 2018). Some of the important genera reputed to produce lipases of commercial importance include *Staphylococcus* spp. *Achromobacter* spp., *Micrococcus* spp. *Burkholderia* spp., *Pseudomonas* spp., *Chromobacterium* spp., and *Bacillus* spp., the most useful lipase producer (Jaeger et al. 1994).

Among the bacterial genera listed, *Bacillus* is identified as a much safer bacterium to use for industrial production, and it is generally regarded as safe (GRAS) (Rashid et al. 2013). Bacillus lipases with their fascinating feature (Table 12.2), which is thermos-stability, are considered to be potential lipases for biotechnological applications (Rashid et al. 2013). *Bacillus* lipases differ from other lipases groups despite their diversified features or properties as demonstrated by the analysis of many *Bacillus* lipases (Table 12.2) that have been isolated to date.

The pentapeptide conserved motif of *Bacillus* lipases have their amino sequences as indicated: Ala-His-Ser-[Met or Gln)-Gly. Alanine residue substituted the Glycine residue present in the canonical pentapeptide conserved motif of other microbial lipases. Such substitution conveys the increased thermos-stability to the *Bacillus* lipases (Jeong et al. 2002). *Bacillus* lipases are divided into two groups: I.4

Table 12.2 Biochemical properties of bacterial lipases

Enzymes	Microorganisms	Sources	pH	Temp. (°C)	Tolerance	References
Lipases	*Bacillus alcalophillus*	Mangrove detritus	10-11	60 °C	7.5% NaCl tolerance	Ghanem et al. (2000)
	Bacillus sp. *A30-1*	Hot spring	9.5	60	H_2O_2, alkaline protease tolerant	Wang et al. (1995)
Monoacylglycerol lipase	*Bacillus* sp. *H-257*	Soil	6.0-8.0	75	–	Kitaura et al. (2001)
Lipase	*Geobacillus* sp. *T1*	–	9.0	70	Ions tolerant	Leow et al. (2007)
	Geobacillus thermocatenulatus BTL2	–	8.0-9.0	65–75	–	Eggert et al. (2003)
	Acinetobacter sp.	–	10.0	50	–	Ahmed et al. (2010)
	Ralstonia sp.	–	8.0-9.5	50–55	Methanol tolerant	Chouhan and Sarma (2011)
	Talaromyces thermophiles	–	9.5	50	–	–

subfamily and I.5 subfamily *Bacillus* lipases (Guncheva and Zhiryakova 2011). I.4 subfamily *Bacillus* lipases are produced by *B. subtilis, B. pumilus,* and *B. licheniformis* and have a low molecular weight of 19-20 kDa. Additionally, I.4 subfamily *Bacillus* lipases are extremely alkaline tolerant, not thermostable, and do not possess a lid to protect the active site. Therefore, their active Serine residue is opened to the organic solvent and the polar (Guncheva and Zhiryakova 2011; Van Pouderoyen et al. 2001).

I.5 subfamily *Bacillus* lipases are synthesized by *Geobacillus thermocatenulatus* and have a molecular weight doubled the size of the I.4 subfamily *Bacillus* lipases (Ahmed et al. 2010; Eggert et al. 2003; Ghanem et al. 2000; Kitaura et al. 2001; Leow et al. 2007; Wang et al. 1995) kDa. They have neutral to moderate alkaline pH and are thermostable. They possess a lid to cover the active site. *Bacillus* lipases are mostly produced by submerged culture fermentation (Guncheva and Zhiryakova 2011).

12.5 Mining of Bacterial Lipase

Despite the diversity, specificity, uniqueness versatility, and the ease of bulk production of microbial lipases, the quest for lipases with improved and novel catalytic and stability under competitive conditions are still intense. To facilitate the hunt for novel lipases, different approaches (Table 12.3) have been used to answer the above-stated issues.

Table 12.3 List of approaches used to mine Lipases

Enzymes	Sources	Methods of mining	References
Moderately thermostable (and thermally activated) lipase	Soil	Metagenomics	Faoro et al. (2012)
Protease insensitive feruloyl esterase	Cow rumen	Metagenomics	Cheng et al. (2012)
Two lipases	Cattle rumen	Metagenomics	Liu et al. (2009)
Two esterases	Sheep rumen	Metagenomics	Bayer et al. (2010)
Two esterases	Soil and water	Metagenomics	Ouyang et al. (2013)
Three carboxylic ester hydrolases	Soil and drinking water	Metagenomics	Elend et al. (2006)
Thermostable esterase	Mud-sediment rich water	Metagenomics	Rhee et al. (2005)
Lipase	Palm oil-contaminated waste	Culture-based methods	Hasan et al. (2018)
Lipase	Degrading oil cakes	Culture-based methods	Sarkar and Chatterji (2018)
Lipase	–	Culture-based methods	Navvabi et al. (2018)

12.5.1 Conventional Method

The conventional approach (Fig. 12.7), which consists of cultivation and isolating the potential lipases producing strains, has been used for a century to get access to the lipases of interest. The traditional methods or culture-dependent methods consist of growing the organism of interest on an appropriate substrate, usually the emulsified synthetic oil (Tributyrin) under laboratory conditions like the natural habitat of the organisms to be studied.

The detection of positive lipases producing strains is based on the formation of a transparent or clear zone of halos around the colony. Nevertheless, it is stated that the culture-dependent methods managed to tap 1% of the microbial diversity of an ecosystem leaving about 99%, the vast natural variety untapped (Liu et al. 2009). The 99% which are made of the viable but non-culturable microbes are believed to produce a novel family of lipase isoforms (Miyagi et al. 1995). Such findings implied that the exploration of sources of novel lipases could be limited and restricted. To explore the vast intrinsic diversity, the culture-independent method called metagenomics is applied (Fig. 12.7).

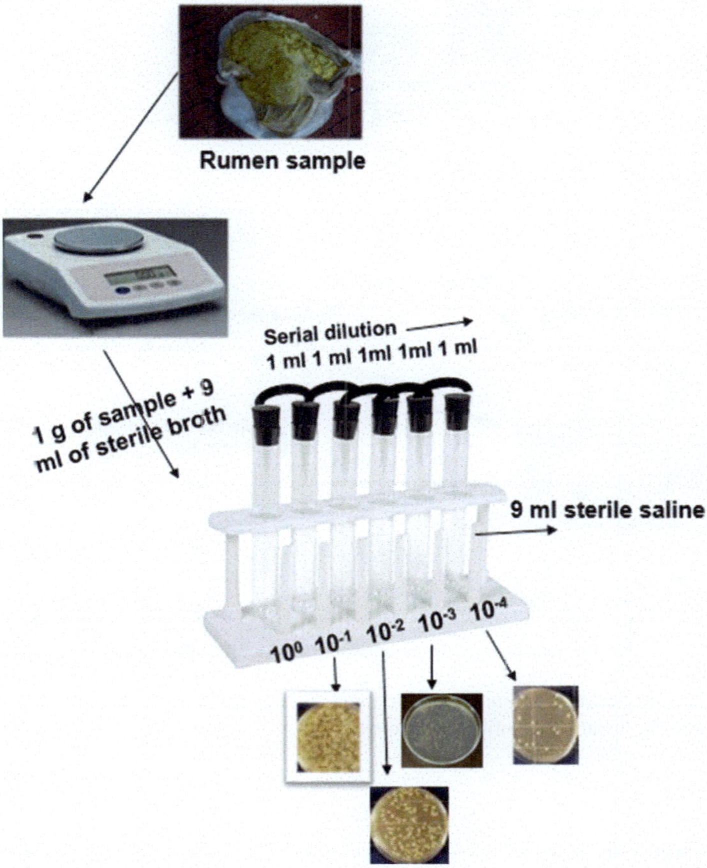

Fig. 12.7 The conventional culture-based technique used for the isolation of culturable organisms

12.5.2 Metagenomic Approach

The Metagenomic technique consists of three steps. The first step entirely relies on the extraction of the pooled genomes from an environmental microbiome, followed by the genomic libraries construction and the sequencing steps (Miyagi et al. 1995). The genomic library construction step helps in storing the genomes of microbes that are viable but non-culturable. Such genomes could be sources of genes coding for unknown or novel enzymes notably lipases possessing novel and unique properties. The progress that was made recently in the field of metagenomics favored the accessing of enzymes from unaccommodating habitats such as hot springs, compost, Antarctic soil, flat intertidal sediments, seawater, coastal environment, deep-sea marine sediments, marine sponges, mangrove sediments, and fat-contaminated soils (Miyagi et al. 1995). Treichel et al. (Treichel et al. 2010) reported the isolation of novel lipolytic enzymes that had variation in a protein sequence as compared to those

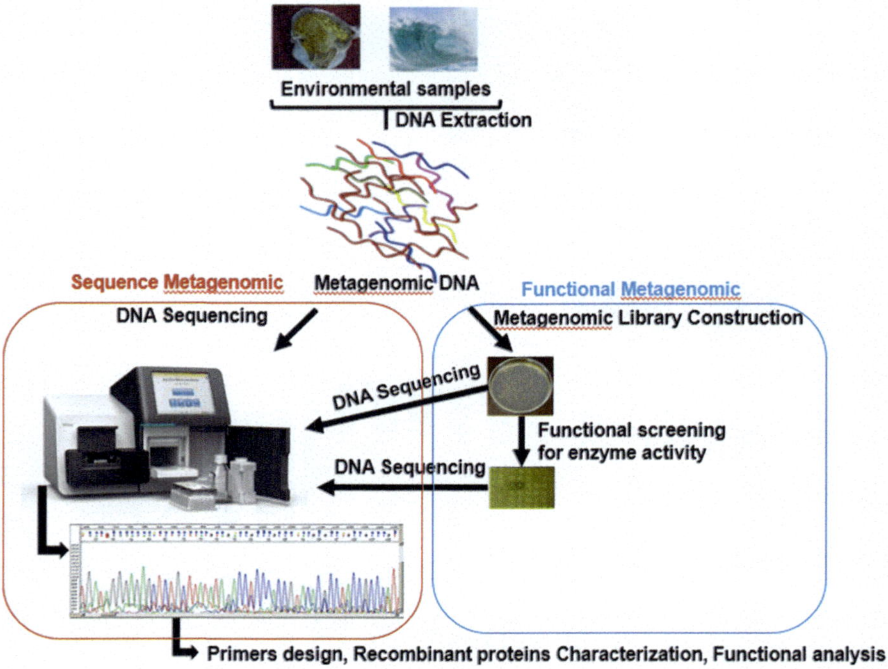

Fig. 12.8 The illustration of the metagenomic approach (Sequence and Function-based screening) used to exploit the unculturable microbial ecology

from the commonly reported classes of enzymes. Thus, the most significant accomplishment of the metagenomic approach in the discovery of novel families of biocatalysts, notably lipases, is the isolation of biocatalysts possessing unique protein sequences from deeply divergent lineages without exhibiting any close relativeness and representing brand new families. Metagenomic studies are revealed to continuously adding enzymes to the growing number of already existing industrial biocatalysts. It is hypothesized that biocatalysts discovered through metagenomics are on the verge to surpass biocatalysts determined through the conventional approach (López-López et al. 2014). There are two types of metagenomic approach: the sequence-based strategy and the function or phenotype-based strategy (Fig. 12.8).

12.5.2.1 Sequence-Based Metagenomic Approach

New enzymes are discovered through the analysis of the sequences obtained via the sequencing using platforms such as MiSeq and HiSeq Illumina from the Next-Generation Sequencing (Fig. 12.9). The study of sequences relies on the alignment and annotation of the metagenomic data against known and available sequences data

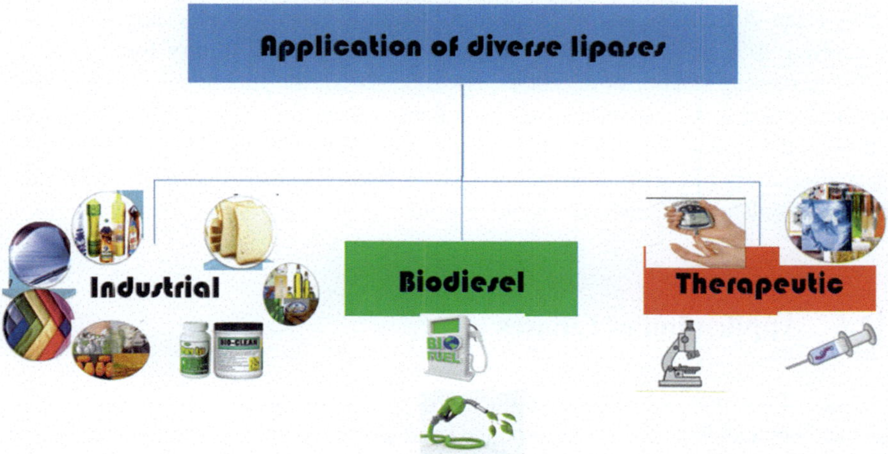

Fig. 12.9 Different industrial Applications of diverse lipases in terms of white, green, and red biotechnology

for enzymes homologous to known biocatalysts such as lipases/esterases (López-López et al. 2014). The sequence-based approach metagenomic is not readily used because this approach only detects genes coding for enzymes that are related to previously reported families. Consequently, the new sequences are overlooked because of a lack of similarity with known sequences (López-López et al. 2014).

12.5.2.2 Function-Driven Metagenomic Approach

Alternatively, with the function-based metagenomic approach, metagenomic libraries are screened based on their activity such as lipolytic activity reflecting on their phenotypes and many factors are influencing the success of this approach (López-López et al. 2014). The cloned genes should be compatible with the transcription and translation machinery of the heterologous host which is usually *Escherichia coli (*López-López et al. 2014*)*. The expression of lipases dramatically depends on the requirement for specific chaperones for the correct folding of the enzyme or can be hindered by its toxicity to the host cells (López-López et al. 2014). Using the function-driven metagenomic approach is advantageous because it allows the discovery of entirely new genes coding for novel biocatalysts. Currently, new families of lipases are still discovered through both traditional and metagenomic approaches (Nagarajan 2012).

Metagenomics like the conventional approach also experiences some challenges and bottlenecks such as the underrepresentation of genes of interest that are quickly picked through inefficient screening of metagenomic libraries. In this case, genes coding for enzymes of interest is represented in the minority in the metagenome

(Nagarajan 2012). The above bottlenecks made the metagenomic approach only recovered 40% of desirable enzymes from a metagenomic library through functional screening (Gabor et al. 2004). By 2009–2010, the functional metagenomic approach had so far only identified two lipases from the cattle rumen (Liu et al. 2009) and two esterases from the sheep rumen (Bayer et al. 2010). The isolated genes were less like genes coding for lipases from other environments. The resulting lipases also had their function unidentified.

12.6 Application of Lipases

Nature is known as one of the manufacturers of biocatalysts that are so searched for and such biocatalysts have been produced and used in vivo for centuries (Nagarajan 2012). Lipases, one of the biocatalysts, are said to contribute to the industrial world through the catalysis of a wide range of chemical reactions taking place in the transformation of chemicals (Biotransformation) (Sarmah et al. 2018). In virtue of their versatility, stability, regiospecificity, enantioselectivity, ease of production, catalytic properties, and microbial lipases constitute one of the critical groups of enzymes valuable to biotechnology and therapeutic applications (Nagarajan 2012) (Fig. 12.9).

12.6.1 Industrial Applications

12.6.1.1 Food Industry

In the food industry, lipases are used for many applications as discussed in the following sections:

12.6.1.1.1 Fat and Oil Industry

The value of triglycerides dramatically depends on their structure. Therefore, triglycerides of low cost can be transformed into valuable triglycerides using lipase. In the fat or oil industry, lipase is used for the modification of lipids by either changing the position of one fatty acid moiety from one carbon of the glycerol backbone to another or replacing one or more fatty acids moieties with new ones (Pabai et al. 1995; Undurraga et al. 2001). The newly produced lipids usually have the form of vegetable oils with nutritional importance, low calories, and oleic acid-enriched. An example of the use of lipase in the oil industry is the use of lipase for the interesterification of the palm (cheap) oil to produce cocoa butter equivalents. The palmitic acid of the palm oil at the second carbon position is replaced with oleic acid, and the cocoa butter is then used in chocolate manufacturing (Nakajima et al. 2000).

12.6.1.1.2 Dairy Industry

In the dairy industry, lipase is used for flavor modification by synthesizing esters of short-chain fatty acids moiety plus alcohols. The resulting esters are compounds of developed flavor for dairy products and fragrance (Alves Macedo et al. 2003), whereas the fatty acids moiety acts as flavor compounds or flavor precursors in cheese making of milk production (Andualema and Gessesse 2012). In the production of mayonnaise, phospholipases are used to treat the egg yolk (Sarmah et al. 2018).

12.6.1.1.3 Meat Processing Industry

In the meat processing industry, lipase is used to remove the fat from the meat products to obtain lean meat. Additionally, lipase is used to remove fat from fatty fish products through the bio-lipolysis process (Kazlauskas and Bornscheuer 1998).

12.6.1.1.4 Bakery Industry

Lipase aids in the extension of bread shelf-life in the baking industry, and as an enhancer that controls the nonenzymatic browning while increasing the loaf volume and improving the crumb structure. An example of such is the lipase from *Bacillus subtilis* (Ray 2012; Sangeetha et al. 2011).

12.6.1.2 Detergent Industry

Detergent is one of the cleaning agents used worldwide for household chores. It is used for either automatic or hand-wash laundry, dishwashing, and household cleaning. The addition of lipase in the detergent industry constitutes one of the most commercially essential applications of lipase. Lipase has been used in combination with other enzymes such as amylase and protease to increase the performance of the detergent (Ito et al. 1998). However, during the improvement of the detergent, some factors are said to be taken into consideration (Chauhan et al. 2013). Lipase is incorporated in the detergent to help in the removal of fatty stains like fried oil, sauce oil, and lipstick (Jaeger and Reetz 1998). Some of the lipase used in the detergent industry is sourced from *Staphylococcus arlettae*, *Burkholderia cepacia*, *P. fluorescens*, and *Candida* spp. (Ahmed et al. 2010; Phuah et al. 2015; Su et al. 2014; Su et al. 2015). The chelating agents in commercial detergents inactivate the incorporated enzyme, but the lipase from *B. licheniformis* had its activity restored through the addition of calcium chloride to the enzyme-detergent complex (Romdhane et al. 2010). Similarly, recent studies on anionic surfactants reported the retention of lipase stability and activity in the presence of sodium lauryl ether sulfate with two ethylene oxide units (Magalhaes et al. 2016). Novel lipases from *Talaromyces thermophilus*, *P. aeruginosa*, and many more have been studied, and the findings gave a promising insight into their potential use in the detergent industry (Grbavčić et al. 2011; Kanjanavas et al. 2010).

12.6.1.3 Pulp and Paper Industry

Lipase is mainly used in the paper industry (Califano et al. 2014) as one of the solutions provided to the problem of pitch formation. Pitch is mostly made of hydrophobic triglycerides or waxes coming from the wood or lignocellulosic biomass components (Jaeger and Reetz 1998). Pitch makes its way through the paper manufacture process and is deposited in the paper mill machines ending up making holes and spots in the final paper. The reason why pitch control in the pulp and paper industry should be implemented is to enhance the quality of the paper to be produced, prolong the equipment shelf life, save energy, and reduce the level of pollution caused by the chemical process. Lipase does solve the problem by being applied to the wood to hydrolyze 90% of the pitch into monoglycerides and fatty acids. The hydrolysis renders the pitch less sticky, thus easily washable (Jaeger and Reetz 1998).

12.6.1.4 Leather Industry

Leather is a sturdy material produced from hides and skins of animals such as bovine. Hides and skins that are transformed into leather through tanning usually possess proteins and fats that pose a major problem in the tanning industry (Andualema and Gessesse 2012). To remediate the problem, Lipase is applied in the leather industry to decrease and remove fat from the skin prior to the initiation of the tanning process. The leather processing operation undergoes three steps: pre-tanning, tanning, and post-tanning (Thanikaivelan et al. 2004). During the stages of soaking, bating, and degreasing to remove subcutaneous fat, hair, and stuffing; lipases are applied to break down lipids found in the skins. The use of lipases has been deemed eco-friendly compared to the conventional use of solvents (Andualema and Gessesse 2012).

12.6.1.5 Textile Industry

Lipase has been utilized in the textile industry as a tool for sizing and improving features of clothing materials (Rowe 1999). Lipase is mainly used in the industry to render the polyester fabric capable of uptaking chemical compounds such as cationic compounds, fabric finishing compositions, dyes, antistatics, anti-staining, antimicrobial, antiperspirant, and deodorant compounds (Rowe 1999).

12.6.1.6 Cosmetics

In the cosmetic industry, lipase is used to yield esters such as isopropyl myristate, isopropyl palmitate, and 2-Ethylhexyl palmitate which is applied as a moisturizer or soothing ingredients in personal skin care products (Sarmah et al. 2018). It was discovered that the replacement of conventional acid catalysts with lipase generates higher quality products than those obtained from the traditional procedures (Andualema and Gessesse 2012). The downstream refining process was also minimized (Andualema and Gessesse 2012). Lipases from *P. fluorescens* and *P. cepacia* were used in the beauty industry to yield the menthol esters which provided a peppermint flavor in mouthwashes and shaving creams (Chaplin et al. 2004).

12.6.1.7 Agrochemical Industry

Lipase has been incorporated in the agrochemical industry to synthesize organic compounds used as herbicides or pesticides in crops (Barbosa et al. 2011). An example is the production of insecticides and fungicides by transesterification using *Pseudomonas* lipase as a catalyst. Notably, one of the herbicides (S)-indanofan, which was produced using lipases as a catalyst, is proven successful and useful against wild grass and weeds (Sangeetha et al. 2011). Lipases from *B. subtilis* was employed for the synthesis of optically active forms of pesticides from a racemic mixture of alcohol and carboxylic esters.

12.6.1.8 Waste Management

Lipase, a fat-splitting molecule, is incorporated in chemicals that are widely used to continuously remove thin layers of fats formed on the surface of aerated tanks used for aerobic waste processes of many industries (Sarmah et al. 2018). Lipase from organisms like *Bacillus, Pseudomonas, Candida,* etc. is also used to treat oily effluent from abattoirs, poultry waste processing, food processing, restaurants, and domestic sewage (Sarmah et al. 2018) Lipases from *Bacillus subtilis* COM-B6 has been used as one of the effective decontaminants of wastes from lipid-processing factories. Lipase was also found to be useful in the degradation of diesel oil in freshly contaminated, unfertilized, and fertilized soils.

12.6.2 Therapeutic Applications

Lipases constitute the primary enzyme involved in the metabolism of lipids in the body (Thanikaivelan et al. 2004). In virtue of its hydrophobicity, lipids are not assimilated by the body because they are insoluble in the blood which is majorly made of water. Therefore, lipase needs to break down lipids into small fatty acids to be absorbed by the body system. Additionally, lipases are used as digestive aids to stimulate the digestion process of lipids ingested. Bacterial lipases are currently used to treat cystic fibrosis and pancreatitis (Ye et al. 2011). They replaced pancreatic lipases for such treatment (Ye et al. 2011). Bacterial lipases are also used in the treatment of malignant tumors by activating the tumor necrosis factor. Moreover, monoglyceride lipase is a promising drug target for cancer, neurodegenerative, and inflammatory diseases (Grabner et al. 2017).

Lipases that are generally isolated from the wax moth (*Galleria mellonella*) do possess a bactericidal action on *Mycobacterium tuberculosis* (MBT) H37Rv (Javed et al. 2018; Nagarajan 2012)). Infection initiated by *Mycobacterium tuberculosis* is diagnosed through the detection of lipases produced by *M. tuberculosis*. Therefore, lipases were deemed suitable for health care (Nagarajan 2012).

Lipases can also be used in the treatment of skin scalp disease and hair loss, while *Bacillus* lipase is used for the synthesis of enantio-pure compounds to be used as pharmaceutical products (Javed et al. 2018).

As biochemical biosensors, lipases are used in the immobilized form in combination with other enzymes to determine the concentration of lipids/ triglycerides or cholesterol through the quantification of glycerol (Javed et al. 2018). Lipases are used as biosensors in diagnosis to determine the concentration of lipids in the blood, in food, and in pharmaceutical products. Lipases can also be used as a diagnostic tool because it is one of the enzyme markers in medicine. Since some microbes produce lipase as a means of defense or as a poison for the adverse organisms, the presence of lipase at a concentration can indicate the initiation of an infection or disease. For further illustration, the presence and quantity of lipase in the blood serum prognosticate (Aronne et al. 2015; Califano et al. 2014; Solanki et al. 2016; Zehani et al. 2015).

12.6.3 Biodiesel Production

Biodiesel has been one of the alternatives to fuel since the raising of the environmental conservation issue (Sarmah et al. 2018). Biodiesel is a liquid produced from the transesterification of lipids with methanol or ethanol, short-chain alcohol. Industrial production is said to provide less amount of greenhouse gases [sulfur oxide) as compared to the burning of fossil fuels to produce diesel or petrol (Sarmah et al. 2018). Biodiesel production is additionally seen as a replacement for fossil fuel because of the depletion of the resources of fossil fuels. The transesterification process requires lipase as a catalyst because the use of lipase renders the production of biodiesel less harmful to the environment (Lukovic et al. 2011). Another advantage that the enzymatic transesterification has over the chemical one is the easy recovery of the purified glycerol which is a byproduct of the transesterification reaction (Lukovic et al. 2011).

The enzymatic production of biofuel consumes less energy and produces waste in a reduced quantity as compared to the chemical process. Examples of raw materials used in the production of biodiesel are nonedible oils. The use of nonedible oils does not affect food production (Ahmia et al. 2014). Furthermore, edible oils can also be used for the production of biodiesel. The properties and features of lipases determine their relevancy in the industry; Aguieiras et al. (2015) reported on the current status and perspectives of the use of fungal lipases for biodiesel production. Aguieiras et al. (2015) tested the lipase activity and stability against variant raw materials, alcohols, the concentration of solvents, etc.. The type of lipase to be used for the transesterification of lipids/fats with alcohol is highly dependent on the variety of raw materials. For example, lipase from *M. miehi* (Lipozyme IM-20) was deemed more effective on animal fat, while lipase from *Geotrichum* spp. worked best for the transesterification of waste cooking oil. Whereas, lipase from *P. fluorescens* was suitable for the transesterification of waste sunflower oil. Further studies are also conducted to develop strategies to reduce the cost of biodiesel production (Ghaly et al. 2010).

12.7 Conclusion and Future Prospects

Lipolytic enzymes notably lipases, being a versatile biocatalysts in virtue of their catalytic activities towards a wide range of reactions are one of the biocatalysts that attract the attention of many industries. The lipase-catalysis reactions make them a potential candidate for industrial, biotechnological, and research applications. Therefore, lipases with improved features such as selectivity, region-specificity, enantio-selectivity, thermos-stability, pH stability, and organic solvent tolerance are necessary for such applications. The rumen, as one of the ecosystems possessing thermophilic microbes (data unpublished), should be explored for sources of potential biocatalysts with enhanced features discussed above, for biorefinery applications and other industrial applications. The recently developed "metagenomics" technique would be of great help for mining novel lipases. The mined lipases are to be characterized to serve industrial purposes.

References

Aguieiras EC, Cavalcanti-Oliveira ED, Freire DM (2015) Current status and new developments of biodiesel production using fungal lipases. Fuel 159:52–67

Ahmed EH, Raghavendra T, Madamwar D (2010) An alkaline lipase from organic solvent tolerant *Acinetobacter* sp. EH28: application for ethyl caprylate synthesis. Bio Resource Technol 101(10):3628–3634

Ahmia A, Danane F, Bessah R, Boumesbah I (2014) Raw material for biodiesel production. Valorization of used edible oil. Rev Energ Renouv 17(2):335–343

Alves Macedo G, Soberón Lozano MM, Pastore GM (2003) Enzymatic synthesis of short chain citronellyl esters by a new lipase from *Rhizopus* sp. Electron J Biotechnol 6(1):3–4

Andualema B, Gessesse A (2012) Microbial lipases and their industrial applications. Biotechnology 11(3):100–118

Aronne A, Bloisi F, Calabria R, Califano V, Depero LE, Fanelli E et al (2015) Lipase biofilm deposited by matrix assisted pulsed laser evaporation technique. Appl Surf Sci 336:196–199
Bajaj A, Lohan P, Jha PN, Mehrotra R (2010) Biodiesel production through lipase catalyzed transesterification: an overview. J Mol Catal B Enzym 62(1):9–14
Barbosa AM, Messias JM, Andrade MM, Dekker RF, Venkatesagowda B (2011) Soybean oil and meal as substrates for lipase production by *Botryosphaeria ribis*, and soybean oil to enhance the production of Botryosphaeran by *Botryosphaeria rhodina*, Soybean – Biochemistry, Chemistry and Physiology, Tzi-Bun Ng, IntechOpen. https://doi.org/10.5772/15623
Bayer S, Kunert A, Ballschmiter M, Greiner-Stoeffele T (2010) Indication for a new lipolytic enzyme family: isolation and characterization of two esterases from a metagenomic library. J Mol Microbiol Biotechnol 18(3):181–187
Bornscheuer UT (2002) Microbial carboxyl esterases: classification, properties and application in biocatalysis. FEMS Microbiol Rev 26(1):73–81
Borrelli GM, Trono D (2015) Recombinant lipases and phospholipases and their use as biocatalysts for industrial applications. Int J Mol Sci 16(9):20774–20840
Bose A, Keharia H (2013) Production, characterization and applications of organic solvent tolerant lipase by *Pseudomonas aeruginosa* AAU2. Biocatal Agric Biotechnol 2(3):255–266
Califano V, Bloisi F, Aronne A, Federici S, Nasti L, Depero LE et al (2014) Biosensor applications of MAPLE deposited lipase. Biosensors 4(4):329–339
Casas-Godoy L, Duquesne S, Bordes F, Sandoval G, Marty A (2012) Lipases: an overview. Springer, Lipases and phospholipases, pp 3–30
Chaplin J, Gardiner N, Mitra R, Parkinson C, Portwig M, Mboniswa B, et al (2004) Process for preparing (−)-menthol and similar compounds. Google Patents
Chauhan M, Chauhan RS, Garlapati VK (2013) Evaluation of a new lipase from *Staphylococcus* sp. for detergent additive capability. BioMed research international. 2013
Cheng F, Sheng J, Cai T, Jin J, Liu W, Lin Y et al (2012) A protease-insensitive feruloyl esterase from China Holstein cow rumen metagenomic library: expression, characterization, and utilization in ferulic acid release from wheat straw. J Agric Food Chem 60(10):2546–2553
Choudhury PK, Salem AZM, Jena R, Kumar S, Singh R, Puniya AK (2015) Rumen microbiology: an overview. Rumen microbiology: from evolution to revolution, pp 3–16
Chouhan APS, Sarma AK (2011) Modern heterogeneous catalysts for biodiesel production: a comprehensive review. Renew Sust Energ Rev 15(9):4378–4399
Eggert T, Brockmeier U, DrÃ¶ge MJ, Quax WJ, Jaeger K-E (2003) Extracellular lipases from *Bacillus subtilis*: regulation of gene expression and enzyme activity by amino acid supply and external pH. FEMS Microbiol Lett 225(2):319–324
Elend C, Schmeisser C, Leggewie C, Babiak P, Carballeira JD, Steele H et al (2006) Isolation and biochemical characterization of two novel metagenome-derived esterases. Appl Environ Microbiol 72(5):3637–3645
Fan X, Niehus X, Sandoval G (2012) Lipases as biocatalyst for biodiesel production. In: Lipases and phospholipases. Springer, pp 471–483
Faoro H, Glogauer A, Couto GH, de Souza EM, Rigo LU, Cruz LM et al (2012) Characterization of a new Acidobacteria-derived moderately thermostable lipase from a Brazilian Atlantic Forest soil metagenome. FEMS Microbiol Ecol 81(2):386–394
Gabor EM, Alkema WB, Janssen DB (2004) Quantifying the accessibility of the metagenome by random expression cloning techniques. Environ Microbiol 6(9):879–886
Ghaly A, Dave D, Brooks M, Budge S (2010) Production of biodiesel by enzymatic transesterification. Am J Biochem Biotechnol 6(2):54–76
Ghanem EH, Al-Sayed HA, Saleh KM (2000) An alkalophilic thermostable lipase produced by a new isolate of *Bacillus alcalophilus*. World J Microbiol Biotechnol 16(5):459–464
Grabner GF, Zimmermann R, Schicho R, Taschler U (2017) Monoglyceride lipase as a drug target: at the crossroads of arachidonic acid metabolism and endocannabinoid signaling. Pharmacol Ther 175:35–46

Grbavčić S, Bezbradica D, Izrael-Živković L, Avramović N, Milosavić N, Karadžić I et al (2011) Production of lipase and protease from an indigenous Pseudomonas aeruginosa strain and their evaluation as detergent additives: compatibility study with detergent ingredients and washing performance. Bioresour Technol 102(24):11226–11233

Guncheva M, Zhiryakova D (2011) Catalytic properties and potential applications of *Bacillus* lipases. J Mol Catal B Enzym 68(1):1–21

Gupta R, Kumari A, Syal P, Singh Y (2015) Molecular and functional diversity of yeast and fungal lipases: their role in biotechnology and cellular physiology. Prog Lipid Res 57:40–54

Gurung N, Ray S, Bose S, Rai V (2013) A broader view: microbial enzymes and their relevance in industries, medicine, and beyond. Biomed Res Int 2013:1–18. https://doi.org/10.1155/2013/329121

Haki G, Rakshit S (2003) Developments in industrially important thermostable enzymes: a review. Bioresour Technol 89(1):17–34

Hasan N, Nawahwi M, Yahya N, Othman N (2018) Identification and optimization of lipase producing bacteria from palm oil contaminated waste. J Fundament Appl Sci 10(2S):li300–li310

Hernández-Fernández FJ, de los Ríos AP, Lozano-Blanco LJ, Godínez C (2010) Biocatalytic ester synthesis in ionic liquid media. J Chem Technol Biotechnol 85(11):1423–1435

Hess M, Sczyrba A, Egan R, Kim T-W, Chokhawala H, Schroth G et al (2011) Metagenomic discovery of biomass-degrading genes and genomes from cow rumen. Science 331(6016):463–467

Hudlicky T, Reed JW (2009) Applications of biotransformations and biocatalysis to complexity generation in organic synthesis. Chem Soc Rev 38(11):3117–3132

Hungate R (1966) The rumen and its microbes. Academic, New York/London

Ishaq SL, Kim CJ, Reis D, Wright AD (2015) Fibrolytic bacteria isolated from the Rumen of North American Moose (*Alces alces*) and their use as a probiotic in neonatal lambs. PLoS One 10(12):e0144804

Ito S, Kobayashi T, Ara K, Ozaki K, Kawai S, Hatada Y (1998) Alkaline detergent enzymes from alkaliphiles: enzymatic properties, genetics, and structures. Extremophiles 2(3):185–190

Jaeger K-E, Reetz MT (1998) Microbial lipases form versatile tools for biotechnology. Trends Biotechnol 16(9):396–403

Jaeger KE, Ransac S, Dijkstra BW, Colson C, van Heuvel M, Misset O (1994) Bacterial lipases. FEMS Microbiol Rev 15(1):29–63

Jaeger K, Dijkstra B, Reetz M (1999) Bacterial biocatalysts: molecular biology, three-dimensional structures, and biotechnological applications of lipases. Ann Rev Microbiol 53(1):315–351

Javed S, Azeem F, Hussain S, Rasul I, Siddique MH, Riaz M et al (2018) Bacterial lipases: a review on purification and characterization. Prog Biophys Mol Biol 132:23–34

Jeong S-T, Kim H-K, Kim S-J, Chi S-W, Pan J-G, Oh T-K et al (2002) Novel zinc-binding center and a temperature switch in the *Bacillus stearothermophilus* L1 lipase. J Biol Chem 277(19):17041–17.57

Kanjanavas P, Khuchareontaworn S, Khawsak P, Pakpitcharoen A, Pothivejkul K, Santiwatanakul S et al (2010) Purification and characterization of organic solvent and detergent tolerant lipase from thermotolerant bacillus sp. RN2. Int J Mol Sci 11(10):3783–3792

Kapoor M, Gupta MN (2012) Lipase promiscuity and its biochemical applications. Process Biochem 47(4):555–569

Kazlauskas RJ, Bornscheuer UT (1998) Biotransformations with lipases. Biotechnol Biotransform I:36–191

Kitaura S, Suzuki K, Imamura S (2001) Monoacylglycerol lipase from moderately thermophilic *bacillus* sp. strain H-257: molecular cloning, sequencing, and expression in *Escherichia coli* of the gene. J Biochem 129(3):397–402

Laachari F, El Bergadi F, Sayari A, Elabed S, Mohammed I, Harchali EH et al (2015) Biochemical characterization of a new thermostable lipase from *Bacillus pumilus* strain/*Bacillus pumilus* suşundan elde edilen yeni termostabil lipazın biyokimyasal karakterizasyonu. Turkish J Biochem 40(1):8–14

Leow TC, Rahman RNZRA, Basri M, Salleh AB (2007) A thermoalkaliphilic lipase of *Geobacillus* sp. T1. Extremophiles 11(3):527–535
Litantra R, Lobionda S, Yim JH, Kim HK (2013) Expression and biochemical characterization of cold-adapted lipases from Antarctic *Bacillus pumilus* strains. J Microbiol Biotechnol 23(9):1221–1228
Liu K, Wang J, Bu D, Zhao S, McSweeney C, Yu P et al (2009) Isolation and biochemical characterization of two lipases from a metagenomic library of China Holstein cow rumen. Biochem Biophys Res Commun 385(4):605–611
López-López O, Cerdan ME, Gonzalez Siso MI (2014) New extremophilic lipases and esterases from metagenomics. Curr Prot Peptide Sci 15(5):445–455
Lukovic N, Knežević-Jugović Z, Bezbradica D (2011) Biodiesel fuel production by enzymatic transesterification of oils: recent trends, challenges and future perspectives. Alternative fuel: Intech
Magalhaes SS, Alves L, Sebastiao M, Medronho B, Almeida ZL, Faria TQ et al (2016) Effect of ethyleneoxide groups of anionic surfactants on lipase activity. Biotechnol Prog 32(5):1276–1282
Miyagi T, Kaneichi K, Aminov RI, Kobayashi Y, Sakka K, Hoshino S et al (1995) Enumeration of transconjugated *Ruminococcus albus* and its survival in the goat rumen microcosm. Appl Environ Microbiol 61(5):2030–2032
Nagarajan S (2012) New tools for exploring "old friends—microbial lipases". Appl Biochem Biotechnol 168(5):1163–1196
Nakajima M, Snape J, Khare S, Gupta M (2000) Method in non-aqueous enzymology. In: Gupta MN (ed) Biochemistry. Birkhauser Verlag, Basel, pp 52–69
Navvabi A, Razzaghi M, Fernandes P, Karami L, Homaei A (2018) Novel lipases discovery specifically from marine organisms for industrial production and practical applications. Process Biochem 70:61–70
Ouyang L-M, Liu J-Y, Qiao M, Xu J-H (2013) Isolation and biochemical characterization of two novel metagenome-derived esterases. Appl Biochem Biotechnol 169(1):15–28
Oyeleke S, Okusanmi T (2008) Isolation and characterization of cellulose hydrolysing microorganism from the rumen of ruminants. Afr J Biotechnol 7(10)
Pabai F, Kermasha S, Morin A (1995) Lipase from Pseudomonas fragi CRDA 323: partial purification, characterization and interesterification of butter fat. Appl Microbiol Biotechnol 43(1):42–51
Patil KJ, Chopda MZ, Mahajan RT (2011) Lipase biodiversity. Indian J Sci Technol 4(8):971–982
Phuah E-T, Tang T-K, Lee Y-Y, Choong TS-Y, Tan C-P, Lai O-M (2015) Review on the current state of diacylglycerol production using enzymatic approach. Food Bioprocess Technol 8(6):1169–1186
Ramnath L, Sithole B, Govinden R (2017) Classification of lipolytic enzymes and their biotechnological applications in the pulping industry. Can J Microbiol 63(3):179–192
Rashid FAA, Rahim RA, Ibrahim D, Balan A, Bakar NMA (2013) Purification and properties of thermostable lipase from a Thermophilic bacterium, *Bacillus licheniformis* IBRL-CHS2. J Pure Appl Microbiol 7:1635–1645
Ray A (2012) Application of lipase in industry. Asian J Pharm Technol 2(2):33–37
Rhee J-K, Ahn D-G, Kim Y-G, Oh J-W (2005) New thermophilic and thermostable esterase with sequence similarity to the hormone-sensitive lipase family, cloned from a metagenomic library. Appl Environ Microbiol 71(2):817–825
Ribeiro BD, AMD C, MAZ C, DMG F (2011) Production and use of lipases in bioenergy: a review from the feedstocks to biodiesel production. Enzyme Res 2011:2011
Robinson KP (2015) Enzymes: principles and biotechnological applications. Essays Biochem 59:1–41. https://doi.org/10.1042/BSE0590001
Romdhane IB-B, Fendri A, Gargouri Y, Gargouri A, Belghith H (2010) A novel thermoactive and alkaline lipase from *Talaromyces thermophilus* fungus for use in laundry detergents. Biochem Eng J 53(1):112–120

Rowe HD (1999) Biotechnology in the textile/clothing industry–a review. J Consumer Stud Home Econom 23(1):53–61

Salis A, Bhattacharyya MS, Monduzzi M, Solinas V (2009) Role of the support surface on the loading and the activity of *Pseudomonas fluorescens* lipase used for biodiesel synthesis. J Mol Catal B Enzym 57(1-4):262–269

Salum TFC, Villeneuve P, Barea B, Yamamoto CI, Côcco LC, Mitchell DA et al (2010) Synthesis of biodiesel in column fixed-bed bioreactor using the fermented solid produced by *Burkholderia cepacia* LTEB11. Process Biochem 45(8):1348–1354

Sangeetha R, Arulpandi I, Geetha A (2011) Bacterial lipases as potential industrial biocatalysts: an overview. Res J Microbiol 6(1):1–24

Sarkar S, Chatterji A (2018) Characterization of lipase-producing bacteria isolated from degrading oil cakes. Utilization and management of bioresources. Springer, pp 253–260

Sarmah N, Revathi D, Sheelu G, Yamuna Rani K, Sridhar S, Mehtab V et al (2018) Recent advances on sources and industrial applications of lipases. Biotechnol Prog 34(1):5–28

Sharma R, Chisti Y, Banerjee UC (2001) Production, purification, characterization, and applications of lipases. Biotechnol Adv 19(8):627–662

Sharma S, Kanwar SS, Dogra P, Chauhan GS (2015) Gallic acid-based alkyl esters synthesis in a water-free system by celite-bound lipase of *bacillus licheniformis* SCD 11501. Biotechnol Prog 31(3):715–723

Solanki S, Pandey CM, Soni A, Sumana G, Biradar AM (2016) An amperometric bienzymatic biosensor for the triglyceride tributyrin using an indium tin oxide electrode coated with electrophoretically deposited chitosan-wrapped nanozirconia. Microchim Acta 183(1):167–176

Su E, Xu J, You P (2014) Functional expression of Serratia marcescens H30 lipase in *Escherichia coli* for efficient kinetic resolution of racemic alcohols in organic solvents. J Mol Catal B Enzym 106:11–16

Su J, Zhang F, Sun W, Karuppiah V, Zhang G, Li Z et al (2015) A new alkaline lipase obtained from the metagenome of marine sponge *Ircinia* sp. World J Microbiol Biotechnol 31(7):1093–1102

Thakur S (2012) Lipases, its sources, properties and applications: a review. Int J Sci Eng Res 3(7):1–29

Thanikaivelan P, Rao JR, Nair BU, Ramasami T (2004) Progress and recent trends in biotechnological methods for leather processing. Trends Biotechnol 22(4):181–188

Treichel H, de Oliveira D, Mazutti MA, Di Luccio M, Oliveira JV (2010) A review on microbial lipases production. Food Bioprocess Technol 3(2):182–196

Undurraga D, Markovits A, Erazo S (2001) Cocoa butter equivalent through enzymic interesterification of palm oil midfraction. Process Biochem 36(10):933–939

Van Pouderoyen G, Eggert T, Jaeger K-E, Dijkstra BW (2001) The crystal structure of *Bacillus subtilis* lipase: a minimal α/β hydrolase fold enzyme1. J Mol Biol 309(1):215–226

Vaquero ME, Barriuso J, Martínez MJ, Prieto A (2016) Properties, structure, and applications of microbial sterol esterases. Appl Microbiol Biotechnol 100(5):2047–2061

Wang Y, Srivastava KC, Shen G-J, Wang HY (1995) Thermostable alkaline lipase from a newly isolated thermophilic *Bacillus*, strain A30-1 (ATCC 53841). J Ferment Bioeng 79(5):433–438

Wang H, Klein MG, Snell G, Lane W, Zou H, Levin I et al (2016) Structure of human GIVD cytosolic phospholipase A2 reveals insights into substrate recognition. J Mol Biol 428(13):2769–2779

Wong H, Schotz MC (2002) The lipase gene family. J Lipid Res 43(7):993–999

Ye L, Zhang B, Seviour EG, K-x T, X-h L, Ling Y et al (2011) Monoacylglycerol lipase (MAGL) knockdown inhibits tumor cells growth in colorectal cancer. Cancer Lett 307(1):6–17

Zehani N, Kherrat R, Dzyadevych SV, Jaffrezic-Renault N (2015) A microconductometric biosensor based on lipase extracted from *Candida rugosa* for direct and rapid detection of organophosphate pesticides. Int J Environ Anal Chem 95(5):466–479

Chapter 13
Bacillus spp. Facilitated Abiotic Stress Mitigation in Rice

Meenu Thampi, Edna Mary Varghese, and M. S. Jisha

Abstract Rice (*Oryza sativa* L.) is the main source of staple food for human population. One of the significant challenges currently is to obtain higher crop yield. Environmental conditions, cultivar quality, and plant diseases enormously affect plants productivity. Understanding rice responses to stress may help breeding for more tolerant varieties. On the other hand, several endophytic *Bacillus* species have emerged as a complementary, efficient, and safe alternative to current crop management practices. This study focuses on the critical role of endophytic *Bacillus* spp. in plant health and their stimulatory different mechanisms to tolerance against abiotic stress in rice. Bacterial endophytes have the ability to act as plant growth–promoting agents through producing phytohormones and also enable plants to grow in contaminated soils through breakdown of hazardous compounds. These endophytes manage plant growth under adverse conditions such as salinity, drought, temperature, heavy metal stress, and nutrient stress through different mechanisms. This chapter presents new approaches for the utilization of endophytic *Bacillus* spp. to battle abiotic stresses in agricultural fields, which increments global crop production.

Keywords *Bacillus* spp. · Endophyte · Abiotic stress · Climate change

13.1 Introduction

Rice (*Oryza sativa*) is the most widely consumed cereal grain in the world's human population. Of all the continents, Asia is the chief producer of rice (FAO 2007). Studies show that rice is a prehistoric food source among other crops. Therefore, any stress on this crop will have a devastating effect on its production which can eventually lead to huge economic losses. Abiotic stresses, including drought,

---

M. Thampi · E. M. Varghese · M. S. Jisha (✉)
School of Biosciences, Mahatma Gandhi University, Kottayam, Kerala, India
e-mail: jishams@mgu.ac.in

M. T. Islam et al. (eds.), *Bacilli in Agrobiotechnology*, Bacilli in Climate Resilient Agriculture and Bioprospecting,
https://doi.org/10.1007/978-3-030-85465-2_13

salinity, low temperatures, high temperatures, and toxic metals, can make rice production extremely problematic and limit agricultural productivity. The response to abiotic stress depends on the duration and progression of the stress and the different plant growth and development stages. (Feller and Vaseva 2014). Drought is the most important abiotic stress on crop productivity around the world. Salinity is one of the second most severe factors in crop survival, biomass, and yield (Mariani and Ferrante 2017). Major abiotic stresses adversely affect crop survival, biomass, and yield of up to 70%, mainly reducing plant growth and profitability (Thakur et al. 2010). Stressors affect plant morphology, physiology, and biochemistry, as well as the gene regulations by developmental stage, resulting in the loss of microbial diversity of soil, fertility of soil, and competition for nutrients (Chodak et al. 2015).

The internal metabolic abilities help plants to cope with rapid fluctuations and adverse environmental conditions (Simontacchi et al. 2015). Plants have acquired a range of conservation mechanisms to facilitate their regular biophysical-chemical processes regardless of the external conditions (Mickelbart et al. 2015). Therefore, plants can resist abiotic stresses with the aid of their associated microbial communities. Mycorrhizal fungus, a microbial association with plants, promotes plant growth and development under various abiotic stresses. The plant-microbial associations can be classified into beneficial, deleterious, and neutral microbes. *Pseudomonas* (Grichko and Glick 2001; Ali et al. 2009), *Bacillus* (Sorty et al. 2016), *Azospirillum* (Creus et al. 2004; Omar et al. 2009), *Azotobacter* (Sahoo et al. 2014a, b), *Bradyrhizobium* (Swaine et al. 2007; Tittabutr et al. 2013), *Enterobacter* (Grichko and Glick 2001; Nadeem et al. 2007; Sorty et al. 2016), *Methylbacterium* (Madhaiyan et al. 2007; Meena et al. 2012), *Rhizobium* (Alami et al. 2000; Remans et al. 2008; Sorty et al. 2016), *Trichoderma* (Ahmad et al. 2015) and cyanobacteria (Singh et al. 2011) help to promote plant growth and alleviate multiple types of abiotic stresses.

Plant growth-promoting bacteria (PGPB) can be free-living or colonize the phyllosphere, rhizosphere, or interior of plant tissues (endophytes) to produce phytohormones, solubilize mineral phosphate, antagonize plant pathogen (Beneduzi et al. 2012), and alleviate abiotic stresses. The colonization areas of PGPB in the plant are generally classified as rhizosphere and endosphere, and the inhabitants are called rhizobacteria and endophytes, respectively (Timm et al. 2015). PGPB stimulates plant growth through a wide range of mechanisms. These include the availability of nutrients to plants derived from processes such as biological nitrogen fixation, phosphate solubilization, stress mitigation through ACC deaminase expression modulation, and phytohormone and siderophore production. These create a favourable environment for growth and functioning. Non-symbiotic endophytic microbes colonize the intercellular spaces of plant cells with high amino acids, carbohydrates, and volatile nutrient levels (Bacon and Hinton 2007).

Among the various PGPB, the dominant species are *Bacillus* and *Pseudomonas* spp. (Kang et al. 2015a). Owing to their high endurance in varied abiotic and biotic environments, a few of the PGPB have been commercialized. Thus, organic fertilizers could be employed as a substitute for harmful chemical fertilizers and pesticides, as well as offer new insights into plant growth and development, even in the presence of pathogenic organisms (Choudhary 2011). Alinit, developed from *Bacillus spp.,* was the first commercial bacterial fertilizer that displayed a 40% rise in crop production (Kilian et al. 2000). Kodiak (*B. subtilis* GB03), Quantum-400 (*B. subtilis* GB03), Rhizovital (*B. amyloliquefaciens* FZB42), Serenade (*B. subtilis* QST713), and YIB *(Bacillus spp.)* have been commercialized to improve crop production (Brannen and Kenney 1997; Ngugi and Nyariki 2005; Cawoy et al. 2011). The fertilizer properties of *Bacillus* spp. are high compared to *Pseudomonas*-based fertilizers owing to their effective spore-forming ability and metabolite production, which enhance cell viability in commercial formulations (Haas and Défago 2005). *Bacillus sp.* are ubiquitous, gram-positive organisms that can be found in all environment niches. They have been used in various agricultural, medicinal, and industrial products (Lyngwi and Joshi 2014). *Bacillus* spp. interact with roots and produce biofilms to enhance plant growth (Beauregard et al. 2013). Therefore, *Bacillus*-based fertilizers can be applied to the soil. Gingivitis can increase the available nutrients of plants in the rhizosphere, regulate the growth of pathogenic microorganisms, and stimulate pests (García-Fraile et al. 2015; Kang et al. 2015b). Therefore, this chapter focuses on the growth-enhancing potential of *Bacillus* spp. in paddy plants to cope with abiotic stresses in agricultural sectors that increase global crop production.

13.2 Abiotic Stresses

Abiotic stress is any factor that disrupts a plant's growth, development, or metabolism rate. It is any substance that decreases plant growth and productivity compared to that of its actual genotype (Krishnan et al. 2011). Therefore, in this chapter, we deliver an elucidation of different types of stresses—drought, salinity, heavy metal, cold, and heat stresses—and their effect on paddy cultivation, as well as their bias. In mitigating these stresses, the typical abiotic stresses affecting paddy cultivation are illustrated in Fig. 13.1.

13.2.1 Drought

Drought is a chief factor limiting the production of drought-affected crops. This has been described as a situation that brings plant water potential and turgor to the point where the plant encounters difficulties in performing normal physio-biochemical functions (Noel et al. 2018). Compared to other crops such as maize and wheat, crops such as paddy have a much higher drought challenge because it has a relatively high water requirement (Todaka et al. 2015). In rice, drought stress increases or prolongs the growth period and causes a series of physical and chemical changes in the plant and leaves, *viz.* decrease in size and root growth reduction, leaf drying, and leaf rolling (Lafitte et al. 2004; Lima et al. 2015; Prakash et al. 2016), increased ethylene production, changes in chlorophyll content, impairment of photosynthetic devices leading to inhibition of photosynthesis (Lata and Prasad 2011),

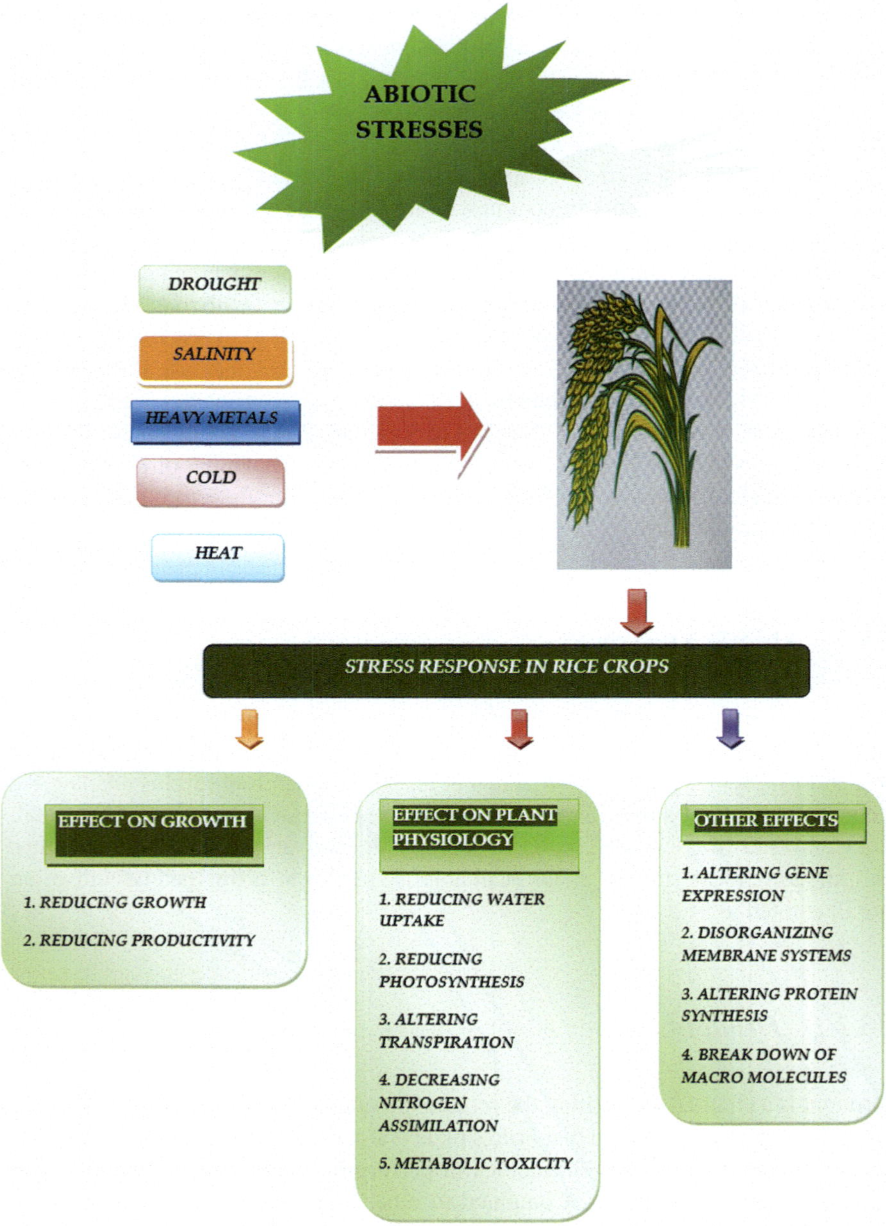

Fig. 13.1 Common abiotic stresses affecting rice

accumulation of free radicals altering membrane function, protein formation, lipid peroxidation, and eventually cell death (Tiwari et al. 2016).

13.2.2 Effect of Drought Stress in Rice Crop

There have been several investigations into the behaviour of various plants in drought conditions, including paddy cultivation (Chen et al. 2011; Gao et al. 2008). Effects of drought on different growth and development stages, vegetation stage, grain filling stage, and yield were studied. Drought conditions (Ahmad et al. 2009) affect the early developmental stages like seed germination and seedling development, which adversely affects germination rate, especially in polyethylene glycol concentrations (Kaya et al. 2006). Zeid and Shedeed (2006) reported that due to water scarcity initiated by polyethylene glycol, seed germination capacity, hypocotyl length, and root/shoot biomass decreased at the same time as root measurement was developed. Drought stress causes disabled mitosis and cell proliferation, leading to stunted growth and growth attributes of rice (Hussain et al. 2008). Swain et al. 2010 reported a decrease in panicle number (72%) and grain yield (12%) during the evaluation of eighteen rice genetic variants. Liu et al. (2006) and Rang et al. (2011) reported that drought negatively affects the production of viable pollen, panicle initiation exertion, pollen shed and germination, and embryo development, which are parts of fertilization and initiation of grain filling. This causes the spikelet fertility to decrease and the dry weight of the fertile spikelets to cause a loss of grain yield. Dehydration is considered multidimensional stress as it affects a wide variety of cells, tissues, and plants as a whole (Rahdari and Hoseini 2012). Nair et al. (2008) reported that drought stress produces free radicals that affect the antioxidant immune system and reactive oxygen species (ROS) and therefore cause oxidative stress which adversely affect the various levels associated with the paddy plant. Lack of water reduces photosynthetic pigmentation, resulting in photo-oxidation (Anjum et al. 2011, 2016a), and also causes decreased NO_3^- uptake, which restricts plant growth (Ali et al. 2014; Awais et al. 2017a, b).

During the grain filling stage, sucrose, ADP-glucose pyrophosphorylase, starch, and starch branched compounds are formed (Taiz and Zeiger 2002). In rice, the lack of water adversely affected the rearrangement of carbohydrates stored during the grain filling phase into rice grains (Yang et al. 2012; Naseem et al. 2016). This is due to low sucrose synthesis activity in drought conditions and ADP-glucose-pyrophosphate inactivation (Ahmadi and Baker 2001; Nasim et al. 2016). Lack of water disrupts the nature of the leaves by controlling the amount of source and sink tissues. In addition, foam stacking and drying tissues were disabled (Akram et al. 2019). Dry conditions cause stomatal closure, which regulates the entry of CO2, resulting in photosynthesis (Flexas et al. 2004), stunted vegetative growth and development, flower formation, and grains leading to reduced grain yield (Akram et al. 2019). Rice responses to drought stress through various physiological, morphological, and molecular modifications are illustrated in Fig. 13.2.

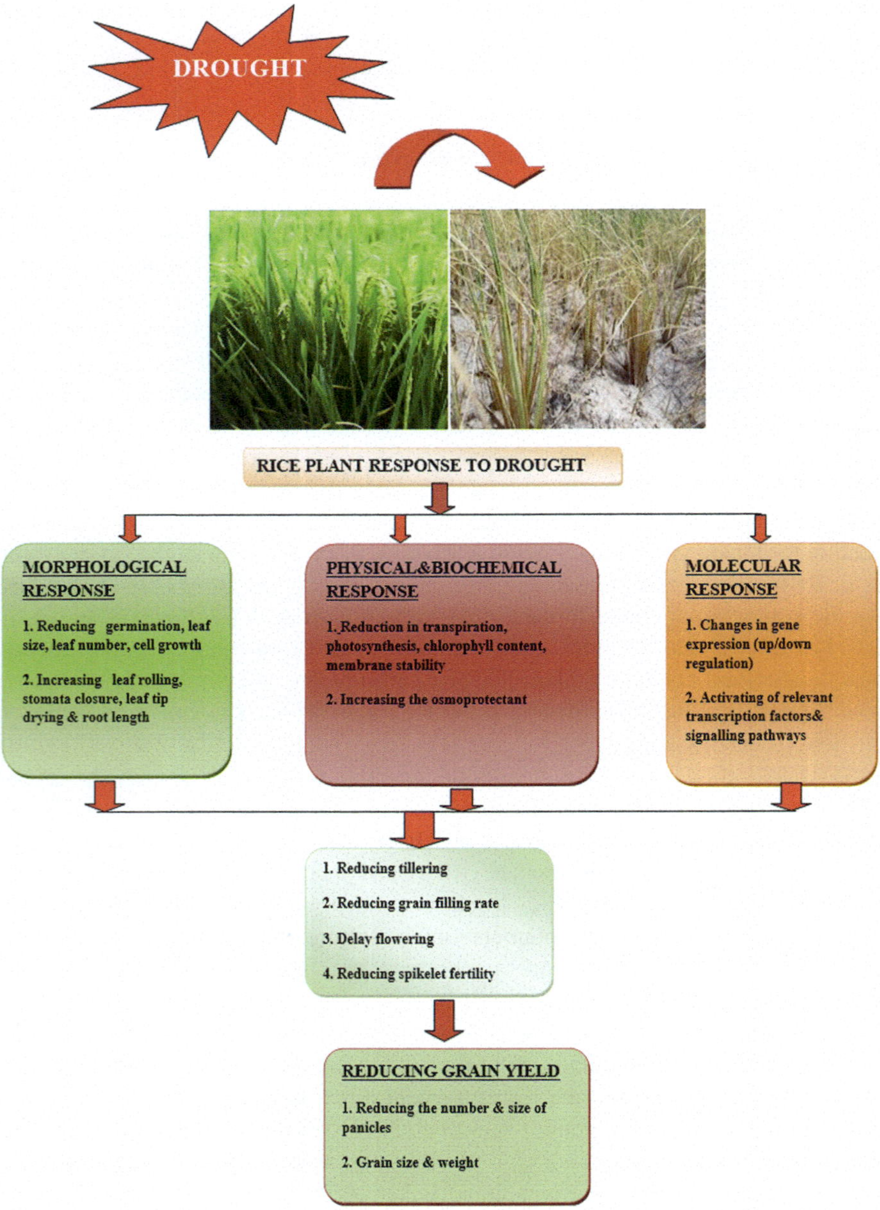

Fig. 13.2 Rice response to drought stress

Normal gene expression pattern of plants changes during drought conditions. The expression of OsDREB1A (dehydration-responsive element-binding protein 1A), OsWRKY11 (WRKY transcription factor 11), OsGAPDH (glyceraldehyde-3-phosphate dehydrogenase), OsDIL (drought-induced lipid transfer protein), P4H (prolyl-4-hydroxylase), sHSP (small heat shock protein), Cadhn (dehydrin-like protein), VA (vacuolar H+-ATPase), cAPX (cytosolic ascorbate peroxidase), CaPR-10 (pathogenesis-related protein 10), rbcL (ribulose-1,5-bisphosphatecarboxy/oxygenase large subunit), and rbcS (ribulose-1,5-bisphosphate carboxy/oxygenase small subunit) genes of crop plants are affected in drought stress conditions (Radhakrishnan et al. 2017).

13.3 Salt Stress

Excessive accumulation of salt in soil is salt stress, which inhibits crop growth and eventually leads to crop death. In more than 100 countries, 1–10 billion hectares of salt-affected land are reported to exist, accounting for more than 7–10% of the total land area (Panta et al. 2014; Jesus et al. 2015; Zhang et al. 2017). Munns (2002) and Ruan et al. (2010) estimated that 20% of irrigated land is inclined to high salinity and 50% to moderate or secondary salinization. Rahman et al. (2019) suspected that salinity is related to irrigation, its quality, and saltwater use. Salt pressure affects the plant in many ways. Nutrient deficiencies, oxidative stress, ion toxicity, alteration of metabolic processes, genetics, cell division reduction, development, lack of essential nutrients, *viz.* K^+, and the toxicity of Na^+ within the plants hamper plant processes like lipid metabolism, photosynthesis, biomass accumulation and protein synthesis (Munns and Tester 2008; Asgari et al. 2012).

13.3.1 Effect of Salt Stress in Rice Crop

Paddy is considered a salt-sensitive crop where production is affected by the spread of salt-affected land in the croplands. Salt pressures applied to plants are of two types: osmotic pressure due to increased rhizosphere osmotic potential and ionic pressure due to enhanced ionic concentration (Ghosh et al. 2016). In rice, early vegetation and later reproductive stages are more susceptible to saltwater. Due to additive gene effects, rice genetic variants differ in salinity tolerance (Shannon et al. 1998; Mass and Hoffman 1977; Sahi et al. 2006). Salt causes mutations and anatomical changes in roots, stems, and leaf cells that inhibit cell division and development, reducing the size of the agricultural meristem, cortex, and vascular cylinder (Islam et al. 2019). The effects of salinity have been reported. They include the growth and establishment of seedling and grain yield components such as spikelet and tiller number. Increased salinity caused a decrease in the number of spikelets per panicle, 1000 grain weight, and infertility, irrespective of season and stage of development (Ghosh et al. 2016; Zeng et al. 2003). Rice response to salt stress is shown in Fig. 13.3.

Salinity affects ion status, leading to excessive accumulation of Na^+/Cl^- inside and outside the plant, which in turn affects ion homeostasis of essential elements causing nutrient imbalances or deficiencies (Naeem et al. 2010). And also saline

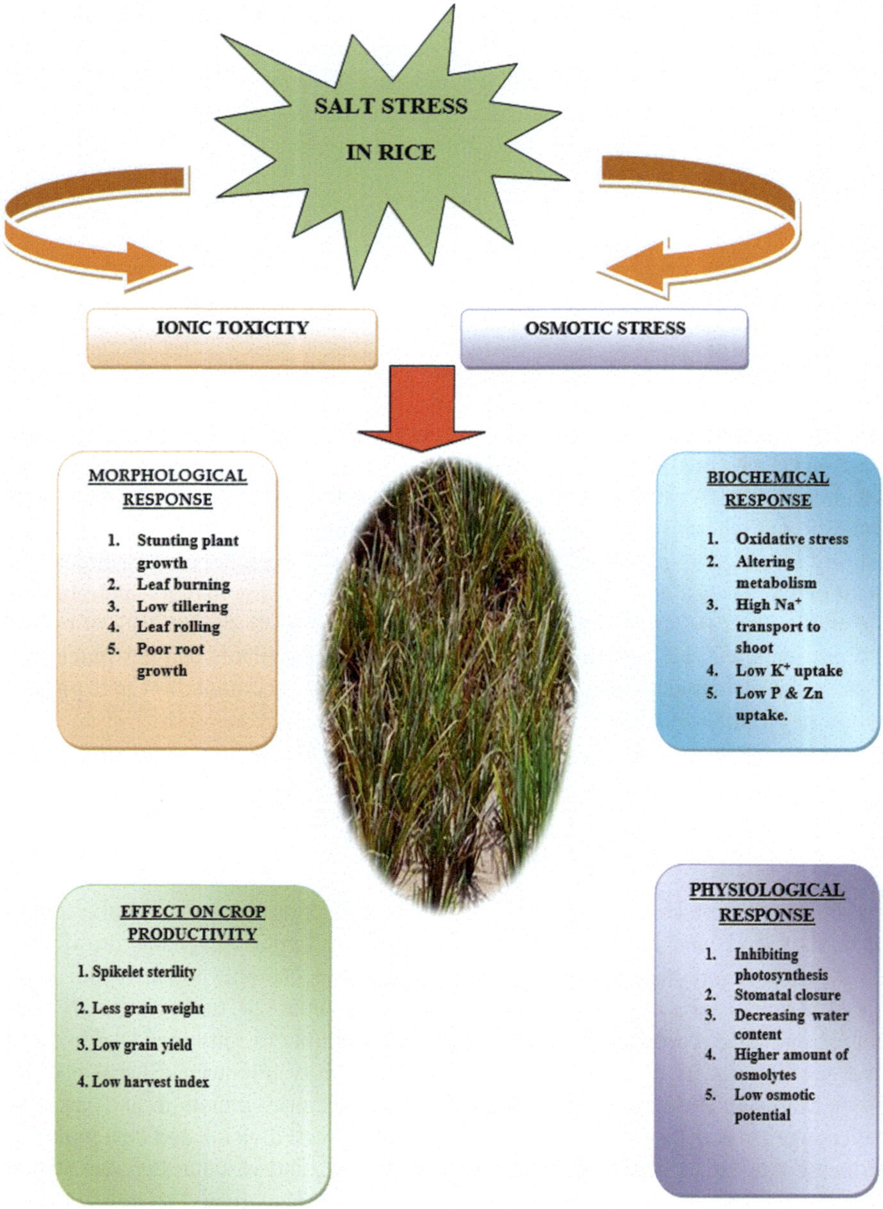

Fig. 13.3 Rice response to salt stress

stressed soil is greatly susceptible to Fe deficit affecting the morphology of the root and root/shoot biomass ratio due to the increased pH, insoluble Fe complex formation, and Fe chelator uptake in rice plants (Islam et al. 2019). Salt pressure creates oxidative stress through the production of ROS in rice cells. Lipid peroxide-derived short-chain carbon mediators are thought to lead to salt-producing cell death in plants (Biswas and Mano 2015). Golldack et al. (2014) reviewed that signalling mediated by lipids in rice plant is activated in response to abiotic stresses, which leads to the maintenance and regeneration of membrane properties and affects membrane-related protein function.

13.4 Heavy Metal Stress

Intensified industrialization processes lead to heavy metal pollution causing problems in environmental and agricultural sectors apart from human health deterioration problems (Fahad et al. 2015a, b, c, d, e). The major heavy metal pollutants are cobalt (Co), cadmium (Cd), mercury (Hg), lead (Pb), copper (Cu), chromium (Cr), and manganese (Mn) (Goyer 1997; Adrees et al. 2015; Anjum et al. 2015, 2016a, b; Bakhat et al. 2017; Shahzad et al. 2018). Some of these are essential metals such as Zn, Ni, and Mn, which are critical for plant physiological processes in small quantities (Goyer 1997; Kabata-Pendias 2010). Most metals are available in the rhizosphere (Wenzel et al. 2003), which pass through the root cell plasma membrane and form a wide substrate specificity through the transport of other cations (Manara 2012).

13.4.1 Effect of Heavy Metal Stress in Rice Crop

Mining, metal processing, wastewater, sludge, and fertilizer application are the major metalloid soil contaminants caused by rapid industrialization and urbanization (Zheng et al. 2010; Niu et al. 2013; Anjum et al. 2016a, b). Metal poisoning mainly affects plant growth by disrupting proteins, lipids, thylakoids, and cellular structures (Fahad et al. 2019).

Heavy metal toxicity is increasing in arable lands, particularly in paddy (Yu et al. 2016). Du et al. (2013) found that two-thirds of the soil in Hunan province was contaminated with Cd (above 0.3 ppm), in rice with 0.2 ppm density. In a study of paddy fields in the Yangtze River region of China by Liu et al. (2016), heavy metal transfer in paddy is mainly through agricultural involvements, *viz.* fertilizers, pesticides, irrigation, etc. Heavy metal poisoning first appears on young leaves with dark green rib formation, where the leaves develop chlorosis and eventually turn white. Cu, Pb, and Zn toxins in rice reduce germination rate and seedling growth characteristics (Fahad et al. 2019). Ashraf et al. (2015) reported that lead poisoning in rice can cause variations in physiological and biochemical characteristics. The first mechanical reaction to heavy metal poisoning is the production of ROS. Cd poisoning increased oxidative stress due to ROS accumulation, MDA content, and electrolyte leakage in rice seedlings (Srivastava et al. 2014; Yu et al. 2013). High Cd toxicity is regulated by the ups and downs of many genes (Lee et al. 2013) and the overexpression of the glutamine synthesis (GS) gene in rice causes plant death (Lee et al. 2013). Cd pressure slows seed germination and seedling growth. Pb mainly affects seed germination, root/shoot ratio, and the dry and fresh weight of rice (Sethy and Ghosh 2013). Cd and Cu pressure accumulate high thiobarbituric acid reactive substance (TBRS) in plant biomass. Stress of excessive Cu is minimized by the control of antioxidant and stress-related protein genes such aldos reductase, peroxyroxine, glyoxalase I, allopi proteins, DNAc-type molecular chaperone, and receptor-like kinase, as well as other metabolic processes. Chatterjee et al. (2004) reported that PB toxins reduced levels of chlorophyll, carotene, nitrogen, and protein. Singh et al. (2010) estimated that lead poisoning adversely affected photosynthesis, chloroplast structure, its pigments, reduction of CO_2 fixation, and electron transport network. The reaction of rice to heavy metal toxins is shown in Fig. 13.4.

Nickel reduces rice seedling roots and is pH-dependent, although it also contributes to the activity of ferulic acid (FPOX) and syringaldazine peroxidase, and the lignin content is observed in $NiSO_4$ application (Lin and Kao 2005). Lead toxicity unfavourably affects photosynthesis by decomposition of chloroplast structure, biosynthesis of photosynthetic pigments, reduction of CO_2 fixation, and activity of enzymes responsible for carbon metabolism. Cd /Cu heavy metal accumulation causes lipid peroxidation, modulation of salicylic acid, abscisic acid, and jasmonic acid leading to decreased plant biomass (Kim et al. 2014a, b). Cultivation of paddy in Cd contaminated soil poses a serious challenge to food production quality (Jallad 2015). Heavy metal poisoning not only affects rice yield but also leads to human health problems due to the intake of heavy metal contaminated rice.

13.5 Temperature Stress

The main effect of the plasma membrane and water level (transpiration) is that during the growing season, plants can be exposed to one or more environmental stress factors; one is the intensity and frequency of temperature stress. It is a major detrimental factor in leaf emergence in plants (Ritchie 1993). Thermal pressure (HS) and cold pressure have become significant abiotic stress conditions for crop productivity and food security worldwide as they impair photosynthetic activity, enzyme activity, cell division, and plant growth (Kumar et al. 2018).

If not handled properly, heat stress can cause a lot of physical and biochemical effects on plants. According to Qu et al. (2013) and Kotak et al. (2007), plants undergo several mechanisms to surpass thermal stress, including the production and accumulation of osmolytes and enzymes. The main proteins are heat shock proteins (HSP20, HSP60, HSP70, HSP90, HSP100) and ROS-scavenging enzymes (catalase and ascorbate peroxidase).

Under cold pressure, ROS accumulate (Song et al. 2011) causing serious harm to the various cellular components, changing membrane lipid composite, and increasing the accumulation of malondialdehyde (MDA) (Sato et al. 2011).

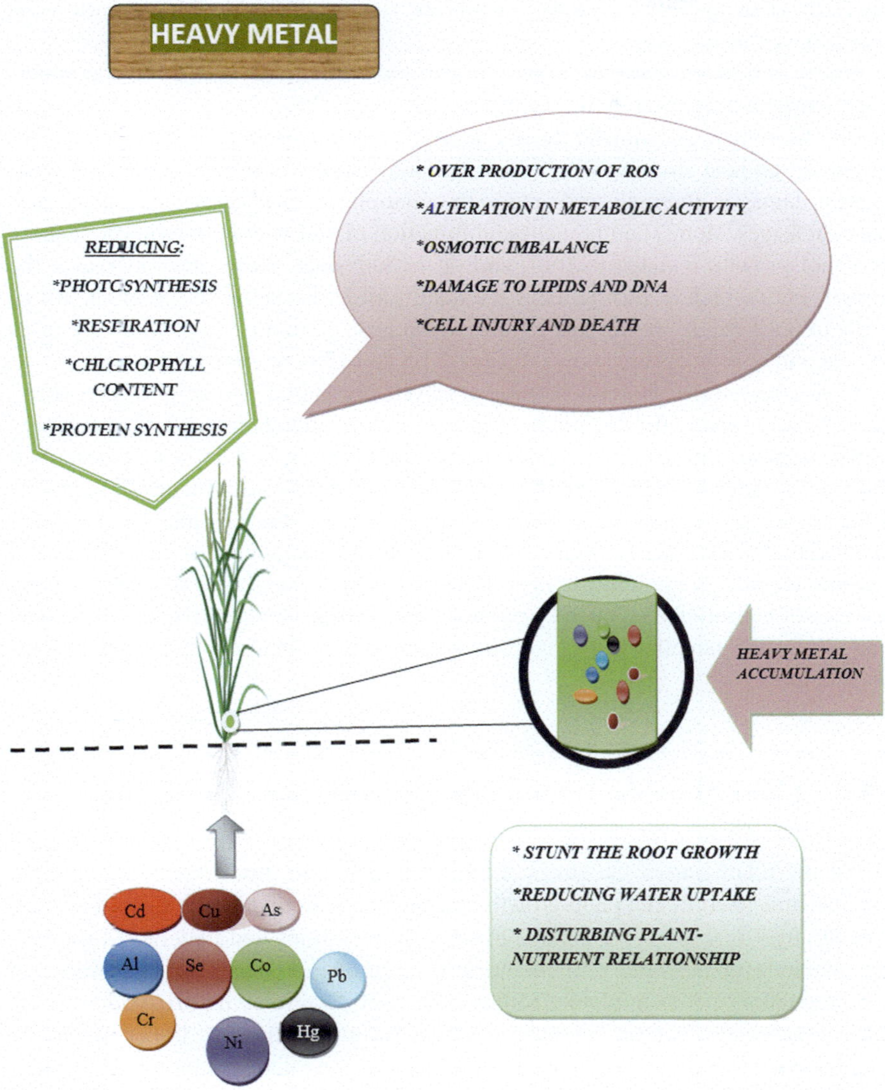

Fig. 13.4 Rice response to heavy metal toxicity

13.5.1 *Effect of Temperature (Heat and Cold) Stress in Rice Crop*

Rice (*Oryza sativa* L.) is an important grain and heat-sensitive plant (Jagadish et al. 2010; Fahad et al. 2015a). According to Bahuguna et al. (2015), the quality of the reproductive or grain filling stages is adversely affected by high temperature/

pressure. Heat stability of the cell membrane is a key factor determining the yield performance of the crop.

Paddy is now grown in areas with higher than growth (28/22 °C) temperatures. High temperature reduces photosynthesis (Oh-e et al. 2007; Fahad et al. 2016b, c, a) and is also reported to transfer the dry matter to roots and shoots. Fahad et al. (2018) reported that heat stress and high radiation may cause considerable pre- and post-harvest damages, thermal stress and high radiation, burns of leaves and twigs, sunburn on leaves, stems, and branches, elimination of leaf ageing, inhibition of shoot and root growth, damage to fruit, and decreased yield. Heat stress increases the growth rate, which in turn reduces the quality of assimilation in the grain, which leads to poor-quality millet rice grains (Kobata and Uemuki 2004). Rice responses to heat and cold stress are shown in Figs. 13.5 and 13.6, respectively.

Cold-stressed rice shows yellowing of leaves, low height, and crop failure (Zhang et al. 2014), causing the depletion of energy sources and photosynthetic machinery impairment, specifically to the chloroplasts modifying light-harvesting chlorophyll antenna complexes (Huner et al. 2012), altering thylakoid structures (Santarius 1992) leading to a decline in plant energy resources. Hughes and Dunn (1996) showed that the plasma layer alters sowing to the rearrangement of unsaturated and unsaturated fatty acids at low temperatures. Enerstvedt et al. (2017) have shown that the lack of hydration instigated by cooling can damage the plasma layer. These may include lysine, lamellar-to-hexagon-phase advances, additional lesions, and ROS presentation.

13.6 Plant-Microbe Interaction Assisting Stress Tolerance

When the plant senses the stress stimuli microbes, use both direct and indirect mechanisms to stimulate plant growth and development under stressful conditions. One strategy is to perceive signals in stressful situations and understand the best molecular level machinery. It includes descriptions of metabolic pathways and their regulatory genes in plants (Miller et al. 2010; Dahro et al. 2016), identifying the characteristic features involved in stress responses, and the linker marker of such genes. The other strategy we focused on was the micro-mediation determination of abiotic stresses. Based on stress stimuli, plants show a complex stress-specific signal cascade (Chinnusamy et al. 2004; Andreasson and Ellis 2010), accumulation of acid and ethylene, flavonoids, and phenolic acids (Singh et al. 2011; Tiwari et al. 2011), and activation of transcription factors (TFs), initiating the expression of defence-specific genes (Koussevitzky et al. 2008; Atkinson et al. 2013; Prasch and Sonnewald 2013). Plants could be made capable to withstand these adverse conditions by the utilization of stress-tolerant plant growth-promoting microorganisms (PGPM) and arbuscular mycorrhizal (AM) fungi (Nadeem et al. 2014). Others include strains belonging to the genera *Acinetobacter*,

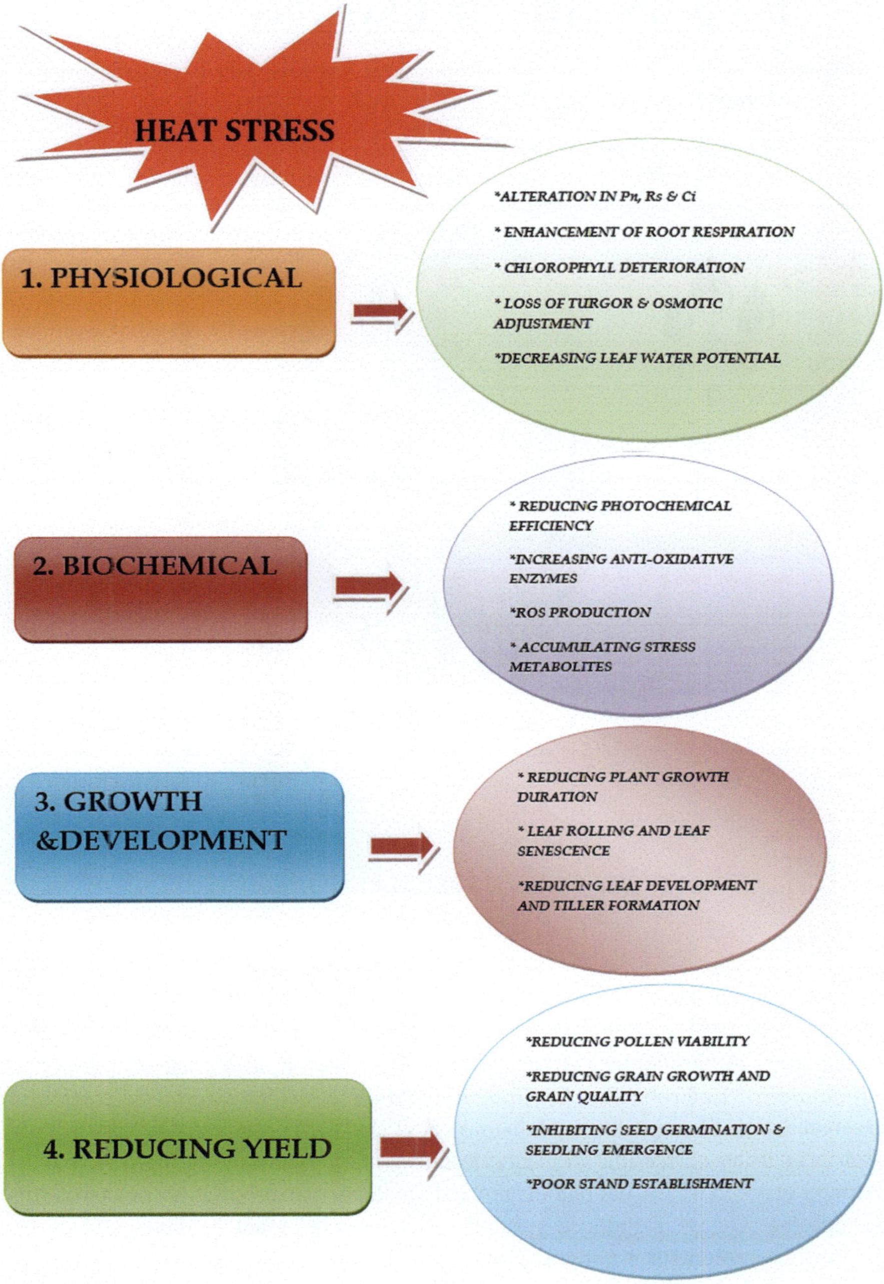

Fig. 13.5 Rice response to heat stress

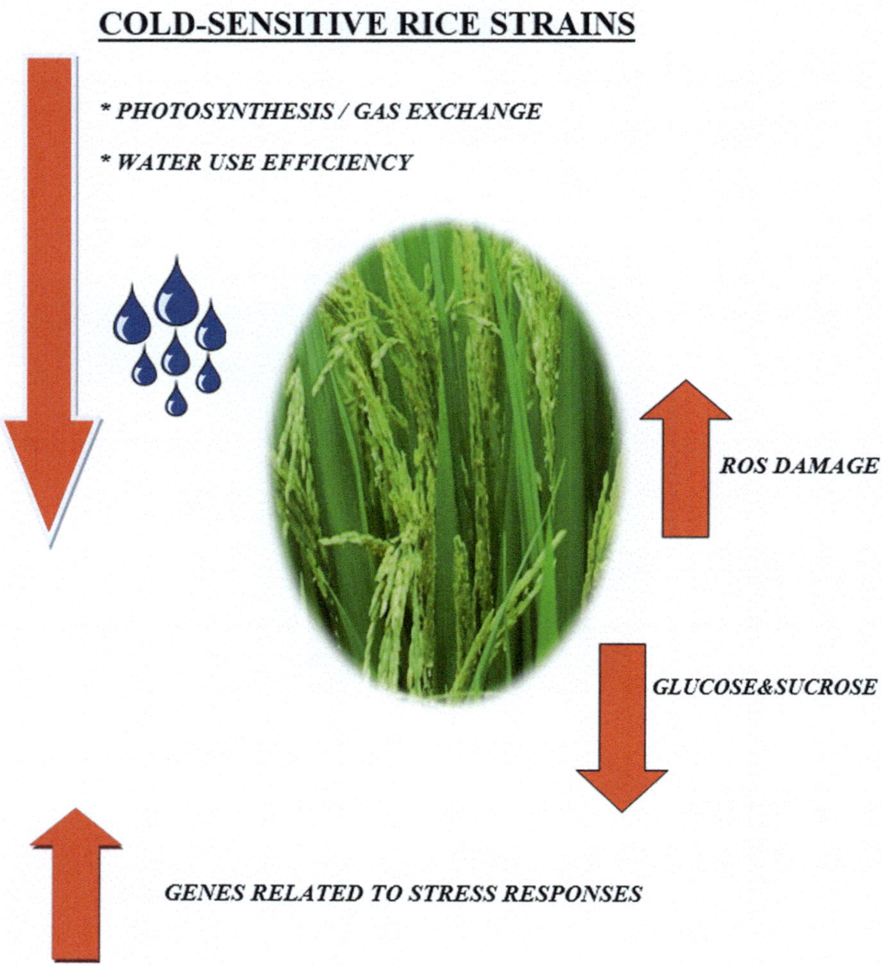

Fig. 13.6 Rice response to cold stress

Alcaligenes, *Arthrobacter*, *Azospirillum*, *Azotobacter*, *Bacillus*, *Beijerinckia*, *Burkholderia*, Enterobacter, *Erwinia*, *Rhizobium*, *Flavobacterium,* and *Serratia* (Bashan and de-Bashan 2005; Maksimov et al. 2015). These microorganisms help plant growth by correcting atmospheric nitrogen by synthesizing plant hormones (Ahmad et al. 2011). Chauhan et al. (2015) stated that production of cytokinins through the process of induced systemic tolerance (IST), antioxidants, and bacterial ethylene precursor 1 aminocyclopropane-1-carboxylate (ACC) by plant-associated bacteria can help to tolerate abiotic stresses more proficiently. Microbes have systems for nutrient accumulation, exopolysaccharide production, and rhizobitoxine, which help the plant cope with adverse conditions (Vardharajula et al. 2011). These microorganisms develop important enzymes such as glucanase and chitinase under stress circumstances (Farooq et al. 2009). In sustainable agriculture, the application of PGPM increases the yield and nutritional value of food grains. Fatnassi et al. (2015) reported that more than 1 mm of copper (Cu) inhibits the vegetative growth of *Vicia faba*, but with *Rhizobia* and PGPR, its effects subsided. Harmful effects of cadmium stress on the reduction of malonaldehyde and hydrogen peroxide in crops are investigated (Hashem et al. 2016). Ghosh et al. (2015) reported that arsenic-resistant bacteria (ARB) from *Pteris vittata* is an effective siderophore manufacturer, enhancing plant nutrition through phosphate solubilization.

Andrews (1992) and Pandey et al. (2012) stated that endophytic bacteria were found to be more potent in providing strong immune responses against stress than rhizospheric or bulk soil microorganisms. Endophytes, *viz.* bacteria and fungi, reside within healthy plant cells and can stimulate plant growth under stress without causing any harmful effects. Depending on the pathogen, they may or may not be pathogenic microorganisms in the endophytic relationship, and may be pathogenic to another host, or non-pathogenic, but can still be colonized by selection methods or genetics (Backman and Sikora 2008). An individual plant is considered to be a host for more than one endophyte (Strobel et al. 2004; Ryan et al. 2008). Environmental changes alter the metabolism of plants from homeostasis, forcing the plant to adapt to systems within the cellular system (Gill and Tuteja 2010). The significance of endophytes is greatly enhanced by the fact that they help plants grow by increasing nutrients, water efficiency (osmotic adjustment, stomata regulation), and reducing abiotic stresses. Instead, endophytes gain access to the nutrients of the host plants and their distribution to the next generation. However, *Bacillus spp.* has a special interest in, has been identified as a dominant community, and has been commercialized as it exists within the boundaries of biotic and abiotic environments (Kilian et al. 2000). *Bacillus*-based biofertilizers, such as Kodiak (*B. subtilis* GB03), Quantum-400 (*B. subtilis* GB03), Rhizovital (*B. amyloliquefaciens*FZB42), Serenade (*B. subtilis* QST713), and YIB (*Bacillus spp.*) have been commercialized for improved crop yield due to high metabolic production (Radhakrishnan et al. 2017). Endophytic bacteria are resistant to metals, thereby promoting plant growth under metal stress by providing a direct supply of mineral nutrients, plant growth regulators, and enzymes. Some PGPMs produce low-molecular-weight organic acids, which play a major role in phytoremediation. Ullah et al. (2015) and Janoušková et al. (2006) reported that citric, oxalic, and gluconic acids are the most effective in accumulating and supplying heavy metals to plants.

Numerous studies and statistics have attempted to identify bacteria belonging to *Bacillus* (Yu et al. 2011; Gaind and Gaur 1991; Zaidi et al. 2006) genera that can help to alleviate the abiotic stresses of many crops, especially rice (Bhambure et al. 2018). Not all PGPBs can withstand different environmental pressures. Therefore, research that incorporates endophytic bacteria as stress alleviators is a good job of combating the adverse effects of abiotic factors.

13.7 *Bacillus* spp.: A Multifunctional Toolbox

Many microorganisms are proved to be naturally beneficial to plants and helpful for maintaining plant growth and yield under abiotic and biotic pressures. PGPB and fungus, which inhabit the endophytic part, phyllosphere part, and rhizospheric part, cause plant growth and protect the plants from diseases and abiotic factors (Tonelli et al. 2010; Radhakrishnan et al. 2014). It includes many PGPB such as *Bacillus* spp., *Agrobacterium, Pseudomonas*, and Azo*spirillum.* Among them, genus *Bacillus* spp. is the most naturally occurring PGPB which produces phytohormones which are able to regulate the endogenous level of phytohormones in plants, thereby balancing the overall hormonal balance of the plant and its reaction to stress (Glick et al. 2007; Kundan et al. 2015).

Most of the PGP microorganisms have multifunctional features (e.g. indole acetic acid production, siderophore production, potassium, phosphate, zinc solubilization), abiotic stress tolerance, and resistance to plant pathogens and withstand extreme environmental conditions (Hassan 2017; Gond et al. 2015). In agricultural field, *B. pumilus*, *B. thuringiensis*, *B. amyloliquefaciens*, *B. subtilis*, *B. velezensis*, *B. cereus*, and *B. licheniformis* are the most studied *Bacillus* spp. (Lopes et al. 2018). Endophytic *Bacillus* spp. has been mainly documented to produce a vast variety of antimicrobials, including compounds that block a wide variety of pathogens that exhibit broad-spectrum activity (Lopes et al. 2015; Jasim et al. 2016a, b; Hu et al. 2017). They are also reported to control quorum sensing-regulated virulence in phytopathogenic bacteria (Lopes et al. 2018). They act as an elicitor of plant immunity and have the ability to induce plants to an improved immune system condition called induced systemic resistance (ISR). Kang et al. (2015a) and Kuan et al. (2016) investigated that many *Bacillus* spp. can convert the complex/complicated form of essential nutrients P and N into the simplest available form. *Bacillus* spp. releases ammonia from nitrogen organic matter (Hayat et al. 2010). Ding et al. (2005) stated that few *Bacillus* spp. can undergo nitrogen fixation and produce a gene that contains the enzyme nitrogenase, which delays senescence by increasing plant growth (Kuan et al. 2016). Shabanamol et al. (2018) reported the production of phytohormones (IAA, GA, cytokinin), synthesis of siderophores, ACC deaminase activity, biological N_2 fixation, phosphate solubilization, ammonia production, and biological resistance by rice endophytic diazotrophic bacteria *Lysinibacillus sphaericus*. The use of endophytic nitrogen-fixing bacteria with gibberellic acid (GA) production leads to effective and direct plant growth (Shabanamol et al. 2018).

Under controlled field conditions, *Bacillus* spp. adopt various mechanisms to promote plant growth such as synthesis of plant growth hormones (e.g. indole acetic acid, gibberellins, cytokinins, and spermidines) (Arkhipova et al. 2005; Xie et al. 2014; Radhakrishnan and Lee 2016; Hassan 2017, Shabanamol et al. 2018), secretion of 1-aminocyclopropane-1-carboxylate (ACC) deaminase for inhibition of ethylene synthesis (Yaish et al. 2015), solubilization of minerals through biofortification, nitrogen fixation or ammonia release from nitrogenous organic matter (Zhao et al. 2011; Hassan 2017; Yousaf et al. 2017; Shabanamol et al. 2020), and production of volatile organic compounds (VOCs) which regulate the metabolism of phytohormones (Tahir et al. 2017; Rath et al. 2018). For, e.g. *B. amyloliquefaciens*, which produces gibberellins and regulates endogenous phytohormones (Shahzad et al. 2016), conformational nitrogen fixation ability by nif gene analysis by *L. sphaericus* (Shabanamol et al. 2018), *B. atrophaeus*, which produces IAA and siderophores (Zhao et al. 2015; Yaish et al. 2015), and *B. aryabhattai*, which produces IAA, ammonia, and ACC deaminase (Yaish 2017), Zhao et al. (2015) investigated that *B. megaterium* produces IAA, ACC deaminase, and siderophores, as well as have the ability to solubilizes phosphate, and also reported *B. mojavensis* produces plant growth-stimulating VOCs (2,3-butanediolandacetoin) (Rath et al. 2018).

13.8 *Bacillus* spp. Mediated Alleviation of Abiotic Stress in Rice

Plant growth and development are restricted due to environmental pressures and stresses created by living communities. Two mechanisms by which plants can survive the abiotic stress are (i) activation after stress exposure (Meena et al. 2017) and (ii) endophytic synthesis of biochemical agents, working as anti-stress agents (Schulz et al. 2002) to regulate abiotic stressors. As a result, plants undergo similar transformations, such as physiochemical, biochemical, and molecular changes (Potters et al. 2007; Huang et al. 2008). Numerous studies by researchers proved that various strains of endophytic *B. subtilis* protect host plants from a variety of abiotic stresses and contribute to increased plant growth and yield (Turan et al. 2012; Ahmad et al. 2017; Martin et al. 2017; Egamberdieva et al. 2017; Martins et al. 2018; Wani et al. 2018, Karthika et al. 2020).

13.8.1 Drought Tolerance by **Bacillus** *spp.*

Under drought conditions, *Bacillus* spp. colonized plants take more water (Marulanda et al. 2009) and are reported to take nutrients such as N, P, K, Mg^{2+}, Zn^{2+}, Ca^{2+}, Mn^{2+}, Fe^{2+}, and Cu^{2+} (Zawadzka et al. 2009; Barnawal et al. 2013; Armada et al. 2014). Sandhya et al. (2011) investigated the application of *Bacillus* spp. in soil, which increases the population of bacteria in roots resulting in the stimulation of root exudation promotes both bacterial and plant growth and also migitates the disturbances in normal cellular physiological and biochemical processes and prevents the formation of ROS (Zgallaï et al. 2005; Miller et al. 2010), which results in the degradation of proteins, lipids, and nucleic acids. Barnawal et al. (2013). In drought-affected plants, PGPB such as *Bacillus* spp. accumulate endogenous amino acids and synthesis of chlorophylls pigments such as chlorophyll a and b and carotenoids, thereby triggering the production of metabolites, which causes oxidative stress reduction and increased photosynthesis. Under adverse conditions, EPS suppresses plant drought resistance by combining suitable solvents and fertilizers (Chitrasree et al. 2011). Bacteria retain membrane permeability by the accumulation of glycine, proline, betaine, and trehalose in drought conditions. Ortiz et al. (2015) stated that mycorrhizal interactions of *Pseudomonas aeruginosa* with specific bacteria, such as *B. thuringiensis*, reduce stomatal conductivity and electrolyte leakage as proline accumulates in the shoot and root.

Gusain et al. (2015) investigated that *Bacillus* spp. cultured rice was found to be mitigating oxidative damage by improving plant growth and activating antioxidant immune systems, which increases the membrane stability in plant cells. Inoculation of crops with *B. altitudinis* FD48 increased the accumulation of relative water content (RWC) and proline which suggested to contribute to osmoregulation during stress tolerance in plants (Kumar et al. 2017) and also showed increased catalase activity than the control plants. *Bacillus altitudinis* FD48 treated rice plant reported less ethylene emission, which usually increases when rice plant was subjected to drought conditions. *Bacillus altitudinis* FD48 was also reported to show inhibitory effect induced by drought stress through the ACC deaminase activity (Zahir et al. 2008). Gusain et al. (2014) stated that inoculation with *B. cereus* shows better parameters and prevention of forming ROS such as superoxide dismutase, catalase, and peroxidase as compared to uninoculated plants under drought stress.

During environmental stressed conditions, production of ROS such as H_2O_2 and catalase is regulated by bacteria, which makes them act as scavenging enzymes, thereby protecting oxidative damage of cells (Shukla et al. 2012, Sandhya et al. 2011 and Gusain et al. 2015), such as catalase and exopolysaccharides; producing ability of rhizobacteria has been already reported (Hussain et al. 2014). It has been recently reported in rice crop plants inoculated with *B. amyloliquefaciens* SN13 (Nautiyal et al. 2013; Kim et al. 2014a; Tiwari et al. 2016) improved the presence of stress-induced symptoms in rice such as membrane integrity, accumulation of osmoprotectants, and expression of marker genes. RoyChoudhury et al. (2007) and Kumar et al. (2014) reported that overexpression of two stress tolerance marker genes LEA and DHN provides abiotic stress tolerance in several crops as well as its expression is increased during the elevated expression of dehydration responsive element binding (DREB) in rice (Lata et al. 2011).

13.8.2 Salinity Tolerance by **Bacillus** *spp.*

Salinity is the commonest of all the abiotic stresses in modern agriculture. It affects the growth of plants and microorganisms through its osmotic effect and ion poisoning. Salinity has detrimental effects on seed germination and disrupts the physicochemical and ecological balance (Shrivastava and Kumar 2015). The role of PGPM and endophytic microorganisms in the mitigation of salt stress in various plants has been reported in detail. The formation of PGPB biofilms under salt pressure is the most effective strategy for mitigating adverse effects (Kasim et al. 2016). Under salt stress, in order to maintain cell turgor and plant growth, crop inoculated with PGPB accumulates osmolytes in their cytoplasm, which counteract osmotic stress. Damodaran et al. (2013) demonstrated that *B. pumilus* and *B. subtilis* isolated from rhizospheric part of saline soil displayed PGP properties, *viz.* IAA production, ammonia production, phosphate, zinc, and potassium solubilization, and hydrogen cyanide (HCN) and also salinity tolerance.

Inoculation of paddy with *B. pumilus* and *B. subtilis* showed PGPR characteristics (Damodaran et al. 2013). Many PGPB such as *Mycobacterium phlei, P. alcaligenes* PsA15, *B. polymyxa* BcP26, and MbP18 have also shown their potential to survive in saline soils like calcisol soil. Rice plants inoculated with *P. pseudoalcaligenes* and *B. pumilus* will increase salinity as they accumulate high concentrations of glycine betaine (Jha et al. 2011). Jha and Subramanian (2014) investigated that the PGPB inoculated crops reduce lipid peroxidation and superoxide dismutase activity in GJ-17 in salt-sensitive rice during salt stress. Inoculation of salt-tolerant *B. amyloliquefaciens* NBRISN13 (SN13) into rice plants increases growth and salt tolerance through upregulation and repression of 14 genes in rice plants (Tank and Saraf 2010; Bal et al. 2013; Nautiyal et al. 2013). Whereas the inoculation of rice with *B. amyloliquefaciens* (SN13) downregulates the expression of LEA and DHN the stress marker genes under salt and heat stress, this demonstrates the crucial role of (SN13) in rice seedling stress alleviation as well as its active participation in osmolyte biosynthesis and subsequent osmotic adjustment (Tiwari et al. 2017). Shahzad et al. (2017) reported that inoculation of rice with *B. amyloliquefaciens* RWL-1 induced salinity stress by the production of ABA and auxin. Hashem et al. (2015) proved that *B. subtilis* enhanced lipid synthesis, *viz.* oleic, linolenic, and linoleic acids as well as phospholipids, in plants grown in saline conditions.

Jha and Subramanian (2014) investigated that *B. pumilus* associated with the roots plants reduce the activity of caspase, which is a protease that belongs to the cysteine endopeptidase family and engages in programmed cell death in plants. Inoculation with *B. subtilis* RH-4 enhances seed germination and plant growth and improves the synthesis of carbohydrates, photosynthetic pigments, proteins, proline, betaine, glycine, and choline in the injured chickpea plants. Plant-associated bacteria increases secondary metabolites such as gallic acid, caffeic acid, syringic acid, vanillic acid, ferulic acid, cinnamic acid, and quercetin, which allow plants to tolerate salinity and stimulate the expression of NADP-Me2 (NADP-malic enzyme 2), EREBP (ethylene-responsive element-binding protein), SOSI (salt overly sensitive 1), BADH (betaine aldehyde dihydride), and SERK1 (somatic embryogenesis receptor-like kinase 1) (Nautiyal et al. 2013; Tiwari et al. 2011). They found that reducing caspase activity reduced ROS formation and programmed cell death and restored the activity of antioxidants to increase plant tolerance. During saline conditions, SAPK4 (serine-threonine protein kinase) and GIG (gigantea) genes in plants are downregulated.

13.8.3 *Heavy Metal Tolerance by* **Bacillus** *spp.*

The use of microbial diversity to aid the solution of heavy metals is a recent strategy to protect and preserve the environment from their toxic effects. Microbes are good indicators for mitigating stress exhibited by heavy metals (Chen et al. 2014, Broos et al. 2004;). Mycorrhizae, heavy metal tolerant organisms, and plant-related microorganisms, have the potential for plant growth promotion and development under metal stress. Glick (2010) reported that these microorganisms are involved in several mechanisms, including EPS sequestration, water flow, metal insufficiency, destabilization, enzymatic detoxification, and metal complexation. This PGPM promotes the growth of plants by producing growth regulators such as IAA, ACC deaminase, and disease suppression. In addition, nitrogen fixation ability, nutrient accumulation, siderophore production, and phosphate, zinc, and potassium solubilization increase plant growth and removal of heavy metals (Verma et al. 2013; Ahmad et al. 2011). Vigliotta et al. (2016) reported that the combined use of heavy metal resistant bacteria from *Bacillus, Lysinibacillus (*Shabanamol et al. 2018*),* and *P. aeruginosa* might improve phytoremediation of heavy metals. *Bacillus* spp. associated plants showed to mitigate this stress effect by increasing amylase and protease, thereby reducing lipid peroxidation and SOD activity (Pandey et al. 2013). Some *Bacillus* spp. increase P and Ca content and reduce Ni accumulation in plants growing in contaminated soil (Jamil et al. 2014). Also, the inoculation *B. licheniformis* enhances Cu, Zn, Cd, Cr, and Pb accumulation and distribution in plants grown in heavy metal-contaminated soil, which causes a decrease in soil toxicity (Brunetti et al. 2012).

Ma et al. (2016) stated that heavy metal remediation by PGP beneficial microorganisms goes through different mechanisms such as the removal of metals, biosorption, precipitation, extrusion, active removal, biotransformation, or bioaccumulation of metals in the external and intracellular spaces of plants. When wheat is inoculated with *Bacillus* spp., AMP2 in chromium salts ($CrCl_3$, K_2CrO_4, $K_2Cr_2O_7$) reduced the Cr acquisition by 10–20 μg mL^{-1}.

Bacillus subtilis inoculated rice (*Oryza sativa* L.) can help to reduce Cd accumulation in all parts of rice roots, shoots (45 days), and grains (120 days) (Treesubsuntorn et al. 2018). Here, the heavy metals were effectively absorbed by *B. subtilis* than *B. cereus*, which might be the chief mechanism to decrease the heavy metal transport Cd, Ca, and Mg+ in rice plants. Plants that were inoculated with *B. subtilis* had the highest levels of Ca and Mg compared to plants inoculated with *B. cereus.* In addition to other stress factors, *B. subtilis* could increase the dry weight of the rice plant and protect them from Cd stress due to the ability to produce IAA, ACC deaminase activity, and phosphate solublization. They proved that sub-stylus can increase the dry weight of paddy and protect it from CD stress. Survival and plant growth intoxicated soil is acquired by the synthesis of mineral nutrients and pigments and by the *Bacillus* spp. association.

13.8.4 Temperature Tolerance by **Bacillus** *spp.*

Temperature plays an important role in the physiochemical and metabolic regulation of microorganisms living in extreme temperature environments. Plant-associated microbial enzymes help to adapt to low and high temperatures. Microorganisms contain special mechanisms that protect their proteins, membranes, and nucleic acids. Yadav et al. (2015) have shown that *Brundimonas terrei, P. cedrina, and Arthrobacter nicotianae* are suitable for thriving at lower temperatures and showing the ability to multifunctional plant growth promotion. Psychrophiles and psychotrophs are two groups of microorganisms. Psychotrophic microorganisms thrive at temperatures above 15°C, while the maximum growth of psychrophilic organisms is below 15°C. In comparing with other stress conditions, plant-associated *Bacillus* spp. makes a sustainable growth in a wide range of temperatures ranging from 31°C to 76°C (Desriac et al. 2013).

Chang et al. (2007) reported that during heat stress (HS) condition causes the production of heat shock proteins and production of normal cellular proteins. This allows the paddy plants to develop HS tolerance. Heat stress is an induced expression of microbial survival. The DnaK gene expression in *Alicyclobacillus acidoterrestris* increases the HSP code during heat stress, which protects microorganisms from thermal stress.

Bacillus subtilis strain 330-2 contains many genes associated with heat tolerance (elongation factor Tu; aspartokinase II; and dihydroorotase, *pyrC*) and cold stress (sporulation cortex protein, and *coxA*), and they are differentially expressed in the *B. subtilis* strain 330-2. Trehalose stimulates synthesis during heat stress and protects plant-associated microorganisms from temperature shock injuries and oxidative stress. The accumulation of trehalose in bacteria and fungi increases the heat stress by several times. Li et al. (2009) showed that trehalose deposited in the microbial cell during hot and cold pressures protects them from thermal deformation. Trehalose is the most active against freezing and desiccation. Trehalose can protect proteins from heat-induced protein denaturation. The amount of metabolites varies according to the plant and the microorganism. Psychrophilic bacteria from Antarctica show antimicrobial activity. In agricultural sectors, inoculation of thermotolerant phosphate solubilizing microbes or stress-tolerant microorganisms acts as a multifunctional biofertilizer (Javani et al. 2015).

13.9 Conclusion

Different types of abiotic stresses such as drought, salinity, heavy metals, cold, and heat stress adversely affect the different growth and transformation stages of crop plants. Morpho-physiological changes commonly occurring in rice due to abiotic stress include impaired seed germination due to cessation of meristematic cell activity in young seedlings, decreased photosynthetic rate, decreased leaf/plant size, reduced leaf size, and reduced plant height. Crops and plants can change their physical and biological properties by exposure to cold, heat, drought, and saltwater, as well as alkali-tolerant proteins when subjected to stress. These barriers are major barriers to crop production, food quality, and global food security. Continuous effects include hormonal imbalances, nutrient instability, ion toxicity, and the risk of disease. The only long-term and environmentally friendly solution to this problem is to develop some microbial tools and technologies that will benefit from plant-microbial-soil interactions. Application of stress-tolerant microbial consortium of *Bacillus* genera PGPM strains and mycorrhizal fungus can be used to enhance plant growth. These microorganisms can lead to plant growth promotion by enhancing the antioxidant system by regulating plant hormones, nutrient enhancement, and siderophore production. Other systems include induced systemic resistance and systemic acquired resistance (SAR) in multiple mechanisms. Arbuscular mycorrhizae increase the supply of nutrients and water under stress conditions and increase the tolerance of plants to various stressful conditions. The use of microorganisms can solve future food security problems and maintain soil health. Therefore, this chapter aims to shed light on the important role that microorganisms play as environmental engineers in solving environmental stress problems. Based on these reviewed data, we recommend scientific societies and policy planners to use the versatile, *Bacillus* spp. in biofertilization formulations, which can be used to solve problems of biotic and biotic stress and to have a better plan for their impact globally. Therefore, considering the current situation, future research is needed to identify highly effective stress-tolerant PGPM. Of course, the diversity of microbial strains should be tested to form an effective microbial consortium to overcome the negative impact of the changing environment.

References

Adrees M, Ali S, Rizwan M, Zia-ur-Rehman M, Ibrahim M, Abbas F et al (2015) Mechanisms of silicon-mediated alleviation of heavy metal toxicity in plants: a review. Ecotoxicol Environ Saf 119:186–197

Ahmadi A, Baker DA (2001) The effect of water stress on the activities of key regulatory enzymes of the sucrose to starch pathway in wheat. Plant Growth Regul 35(1):81–91

Ahmad S, Ahmad R, Ashraf MY, Ashraf M, Waraich EA (2009) Sunflower (*Helianthus annuus* L.) response to drought stress at germination and seedling growth stages. Pak. J. Bot 41(2):647–654

Ahmad M, Zahir ZA, Asghar HN, Asghar M (2011) Inducing salt tolerance in mung bean through coinoculation with rhizobia and plant-growth-promoting rhizobacteria containing 1-aminocyclopropane-1-carboxylate deaminase. Can J Microbiol 57(7):578–589

Ahmad M, Zahir ZA, Asghar HN, Asghar M (2011) Inducing salt tolerance in mung bean through coinoculation with rhizobia and plant-growth-promoting rhizobacteria containing 1-aminocyc lopropane-1-carboxylate deaminase. Can J Microbiol 57(7):578–589

Ahmad P, Hashem A, Abd-Allah EF, Alqarawi AA, John R, Egamberdieva D, Gucel S (2015) Role of *Trichoderma harzianum* in mitigating NaCl stress in Indian mustard (*Brassica juncea* L) through antioxidative defense system. Front Plant Sci 6:868

Ahmad Z, Wu J, Chen L, Dong W (2017) Isolated *Bacillus subtilis* strain 330-2 and its antagonistic genes identified by the removing PCR. Sci Rep 7(1):1–13

Akram R, Fahad S, Masood N, Rasool A, Ijaz M, Ihsan MZ et al (2019) Plant growth and morphological changes in rice under abiotic stress. In: Advances in rice research for abiotic stress tolerance. Woodhead Publishing, pp 69–85

Alami Y, Achouak W, Marol C, Heulin T (2000) Rhizosphere soil aggregation and plant growth promotion of sunflowers by an exopolysaccharide-producing *Rhizobium* sp. strain isolated from sunflower roots. Appl Environ Microbiol 66(8):3393–3398

Ali SZ, Sandhya V, Grover M, Kishore N, Rao LV, Venkateswarlu B (2009) *Pseudomonas* sp. strain AKM-P6 enhances tolerance of sorghum seedlings to elevated temperatures. Biol Fertil Soils 46(1):45–55

Ali S, Charles TC, Glick BR (2014) Amelioration of high salinity stress damage by plant growth-promoting bacterial endophytes that contain ACC deaminase. Plant Physiol Biochem 80:160–167

Andrews JH (1992) Biological control in the phyllosphere. Annu Rev Phytopathol 30(1):603–635

Andreasson E, Ellis B (2010) Convergence and specificity in the Arabidopsis MAPK nexus. Trends Plant Sci 15(2):106–113

Anjum SA, Xie XY, Wang LC, Saleem MF, Man C, Lei W (2011) Morphological, physiological and biochemical responses of plants to drought stress. Afr J Agricult Res 6(9):2026–2032

Anjum SA, Tanveer M, Hussain S, Bao M, Wang L, Khan I et al (2015) Cadmium toxicity in Maize (*Zea mays* L.): consequences on antioxidative systems, reactive oxygen species and cadmium accumulation. Environ Sci Pollut Res 22(21):17022–17030

Anjum AS, Zada R, Tareen WH (2016a) Organic farming: Hope for the sustainable livelihoods of future generations in Pakistan. J Pure Appl Agriculture 1(1):20–29

Anjum SA, Tanveer M, Hussain S, Ullah E, Wang L, Khan I et al (2016b) Morpho-physiological growth and yield responses of two contrasting maize cultivars to cadmium exposure. CLEAN–Soil, Air, Water 44(1):29–36

Arkhipova TN, Veselov SU, Melentiev AI, Martynenko EV, Kudoyarova GR (2005) Ability of bacterium *Bacillus subtilis* to produce cytokinins and to influence the growth and endogenous hormone content of lettuce plants. Plant and Soil 272(1-2):201–209

Armada E, Roldán A, Azcon R (2014) Differential activity of autochthonous bacteria in controlling drought stress in native *Lavandula* and *Salvia* plants species under drought conditions in natural arid soil. Microb Ecol 67(2):410–420

Asgari HR, Cornelis W, Van Damme P (2012) Salt stress effect on wheat (*Triticum aestivum* L.) growth and leaf ion concentrations.

Ashraf U, Kanu AS, Mo Z, Hussain S, Anjum SA, Khan I et al (2015) Lead toxicity in rice: effects, mechanisms, and mitigation strategies—a mini review. Environ Sci Pollut Res 22(23):18318–18332

Atkinson NJ, Lilley CJ, Urwin PE (2013) Identification of genes involved in the response of arabidopsis to simultaneous biotic and abiotic stresses. Plant Physiol 162:2028–2041. https://doi.org/10.1104/pp.113

Awais M, Wajid A, Nasim W, Ahmad A, Saleem MF, Raza MAS et al (2017a) Modeling the water and nitrogen productivity of sunflower using OILCROP-SUN model in Pakistan. Field Crop Res 205:67–77

Awais M, Wajid A, Bashir MU, Habib-ur-Rahman M, Raza MAS, Ahmad A et al (2017b) Nitrogen and plant population change radiation capture and utilization capacity of sunflower in semi-arid environment. Environ Sci Pollut Res 24(21):17511–17525

Backman PA, Sikora RA (2008) Endophytes: an emerging tool for biological control. Biol Contr 46(1):1–3

Bacon CW, Hinton DM (2007) Bacterial endophytes: the endophytic niche, its occupants, and its utility. In plant-associated bacteria. Springer, Dordrecht, pp 155–194

Bahuguna RN, Jha J, Pal M, Shah D, Lawas LM, Khetarpal S, Jagadish KS (2015) Physiological and biochemical characterization of NERICA-L-44: a novel source of heat tolerance at the vegetative and reproductive stages in rice. Physiol Plant 154(4):543–559

Bakhat HF, Zia Z, Fahad S, Abbas S, Hammad HM, Shahzad AN et al (2017) Arsenic uptake, accumulation and toxicity in rice plants: possible remedies for its detoxification: a review. Environ Sci Pollut Res 24(10):9142–9158

Bal HB, Nayak L, Das S, Adhya TK (2013) Isolation of ACC deaminase producing PGPR from rice rhizosphere and evaluating their plant growth promoting activity under salt stress. Plant and Soil 366(1-2):93–105

Barnawal D, Maji D, Bharti N, Chanotiya CS, Kalra A (2013) ACC deaminase-containing *Bacillus subtilis* reduces stress ethylene-induced damage and improves mycorrhizal colonization and rhizobial nodulation in *Trigonella foenum-graecum* under drought stress. J Plant Growth Regulat 32(4):809–822

Bashan Y, De-Bashan LE (2005) Plant growth-promoting. Encyclopedia Soil Environ 1:103–115

Beneduzi A, Ambrosini A, Passaglia LM (2012) Plant growth-promoting rhizobacteria (PGPR): their potential as antagonists and biocontrol agents. Gen Mol Biol 35:1044–1051

Beauregard PB, Chai Y, Vlamakis H, Losick R, Kolter R (2013) *Bacillus subtilis* biofilm induction by plant polysaccharides. Proc Natl Acad Sci 110(17):E1621–E1630

Bhambure AB, Mahajan GR, Kerkar S (2018) Salt tolerant bacterial inoculants as promoters of rice growth and microbial activity in coastal saline soil. Proceedings of the National Academy of Sciences, India Section B: Biological Sciences 88(4):1531–1538

Brannen PM, Kenney DS (1997) Kodiak®—a successful biological-control product for suppression of soilborne plant pathogens of cotton. J Ind Microbiol Biotechnol 19(3):169–171

Biswas MS, Mano JI (2015) Lipid peroxide-derived short-chain carbonyls mediate hydrogen peroxide-induced and salt-induced programmed cell death in plants. Plant Physiol 168(3):885–898

Broos K, Uyttebroek M, Mertens J, Smolders E (2004) A survey of symbiotic nitrogen fixation by white clover grown on metal contaminated soils. Soil Biol Biochem 36(4):633–640

Brunetti G, Farrag K, Soler-Rovira P, Ferrara M, Nigro F, Senesi N (2012) The effect of compost and *Bacillus licheniformis* on the phytoextraction of Cr, Cu, Pb and Zn by three brassicaceae species from contaminated soils in the Apulia region, Southern Italy. Geoderma 170:322–330

Cawoy H, Bettiol W, Fickers P, Ongena M (2011) *Bacillus*-based biological control of plant diseases. Pesticides in the modern world-pesticides use and management, pp 273–302

Chatterjee C, Dube BK, Sinha P, Srivastava P (2004) Detrimental effects of lead phytotoxicity on growth, yield, and metabolism of rice. Commun Soil Sci Plant Anal 35(1-2):255–265

Chauhan H, Bagyaraj DJ, Selvakumar G, Sundaram SP (2015) Novel plant growth promoting rhizobacteria—prospects and potential. Appl Soil Ecol 95:38–53

Chang PFL, Jinn TL, Huang WK, Chen Y, Chang HM, Wang CW (2007) Induction of a cDNA clone from rice encoding a class II small heat shock protein by heat stress, mechanical injury, and salicylic acid. Plant Sci 172:64–75

Chen W, Yao X, Cai K, Chen J (2011) Silicon alleviates drought stress of rice plants by improving plant water status, photosynthesis and mineral nutrient absorption. Biol Trace Elem Res 142:67–76

Chen L, Luo S, Li X, Wan Y, Chen J, Liu C (2014) Interaction of Cd-hyperaccumulator *Solanum nigrum* L. and functional endophyte *Pseudomonas* sp. Lk9 on soil heavy metals uptake. Soil Biol Biochem 68:300–308

Chinnusamy V, Schumaker K, Zhu JK (2004) Molecular genetic perspectives on cross-talk and specificity in abiotic stress signalling in plants. J Exp Bot 55(395):225–236

Chitrasree AC, Udayasankar S, Reddy MS, Srinivas C (2011) Plant growth promoting rhizobacteria mediated induced systemic resistance in rice against bacterial leaf blight caused by *Xanthomonas oryzae* pv. *oryzae*. Biol Control 59(2):114–122

Chodak M, Gołębiewski M, Morawska-Płoskonka J, Kuduk K, Niklińska M (2015) Soil chemical properties affect the reaction of forest soil bacteria to drought and rewetting stress. Ann Microbiol 65(3):1627–1637

Choudhary DK (2011) Plant growth-promotion (PGP) activities and molecular characterization of rhizobacterial strains isolated from soybean (*Glycine max* L. Merril) plants against charcoal rot pathogen, Macrophomina phaseolina. Biotechnol Lett 33(11):2287–2295

Creus CM, Sueldo RJ, Barassi CA (2004) Water relations and yield in *Azospirillum*-inoculated wheat exposed to drought in the field. Can J Bot 82(2):273–281

Dahro B, Wang F, Peng T, Liu JH (2016) PtrA/NINV, an alkaline/neutral invertase gene of *Poncirus trifoliata*, confers enhanced tolerance to multiple abiotic stresses by modulating ROS levels and maintaining photosynthetic efficiency. BMC Plant Biol 16(1):76

Damodaran T, Sah V, Rai RB, Sharma DK, Mishra VK, Jha SK, Kannan R (2013) Isolation of salt tolerant endophytic and rhizospheric bacteria by natural selection and screening for promising plant growth-promoting rhizobacteria (PGPR) and growth vigour in tomato under sodic environment. Afr J Microbiol Res 7(44):5082–5089

Desriac N, Broussolle V, Postollec F, Mathot AG, Sohier D, Coroller L, Leguerinel I (2013) Bacillus cereus cell response upon exposure to acid environment: toward the identification of potential biomarkers. Front Microbiol 4:284

Ding Y, Wang J, Liu Y, Chen S (2005) Isolation and identification of nitrogen-fixing bacilli from plant rhizospheres in Beijing region. J Appl Microbiol 99(5):1271–1281

Du Y, Hu XF, Wu XH, Shu Y, Jiang Y, Yan XJ (2013) Affects of mining activities on Cd pollution to the paddy soils and rice grain in Hunan province, Central South China. Environ Monit Assess 185(12):9843–9856

Egamberdieva D, Wirth SJ, Shurigin VV, Hashem A, Abd Allah EF (2017) Endophytic bacteria improve plant growth, symbiotic performance of chickpea (*Cicer arietinum* L.) and induce suppression of root rot caused by *Fusarium solani* under salt stress. Front Microbiol 8:1887

Enerstvedt KS, Sydnes MO, Pampanin DM (2017) Study of the plasma proteome of Atlantic cod (*Gadus morhua*): Effect of exposure to two PAHs and their corresponding diols. Chemosphere 183:294–304

Fahad S, Hussain S, Bano A, Saud S, Hassan S, Shan D et al (2015a) Potential role of phytohormones and plant growth-promoting rhizobacteria in abiotic stresses: consequences for changing environment. Environ Sci Pollut Res 22(7):4907–4921

Fahad S, Hussain S, Matloob A, Khan FA, Khaliq A, Saud S et al (2015b) Phytohormones and plant responses to salinity stress: a review. Plant Growth Regul 75(2):391–404

Fahad S, Hussain S, Saud S, Tanveer M, Bajwa AA, Hassan S et al (2015c) A biochar application protects rice pollen from high-temperature stress. Plant Physiol Biochem 96:281–287

Fahad S, Nie L, Chen Y, Wu C, Xiong D, Saud S et al (2015d) Crop plant hormones and environmental stress. In: Sustainable agriculture reviews. Springer, Cham, pp 371–400

Fahad S, Hussain S, Khan F, Wu C, Saud S, Hassan S et al (2015e) Effects of tire rubber ash and zinc sulfate on crop productivity and cadmium accumulation in five rice cultivars under field conditions. Environ Sci Pollut Res 22(16):12424–12434

Fahad S, Hussain S, Saud S, Hassan S, Chauhan BS, Khan F et al (2016a) Responses of rapid viscoanalyzer profile and other rice grain qualities to exogenously applied plant growth regulators under high day and high night temperatures. PLoS One 11(7):e0159590

Fahad S, Hussain S, Saud S, Khan F, Hassan S, Nasim W et al (2016b) Exogenously applied plant growth regulators affect heat-stressed rice pollens. J Agronomy & Crop Sci 202(2):139–150

Fahad S, Hussain S, Saud S, Hassan S, Ihsan Z, Shah AN et al (2016c) Exogenously applied plant growth regulators enhance the morpho-physiological growth and yield of rice under high temperature. Front Plant Sci 7:1250

Fahad S, Ihsan MZ, Khaliq A, Daur I, Saud S, Alzamanan S et al (2018) Consequences of high temperature under changing climate optima for rice pollen characteristics-concepts and perspectives. Arch Agron Soil Sci 64(11):1473–1488

Fahad S, Rehman A, Shahzad B, Tanveer M, Saud S, Kamran M et al (2019) Rice responses and tolerance to metal/metalloid toxicity. In: Advances in rice research for abiotic stress tolerance. Woodhead Publishing, pp 299–312

FAO (2007) Food and Agriculture Organization (FAO) of the United Nations, Rome

Farooq M, Wahid A, Kobayashi N, Fujita DBSMA, Basra SMA (2009) Plant drought stress: effects, mechanisms and management. In: Sustainable agriculture. Springer, Dordrecht, pp 153–188
Fatnassi IC, Chiboub M, Saadani O, Jebara M, Jebara SH (2015) Impact of dual inoculation with rhizobium and PGPR on growth and antioxidant status of *Vicia faba* L. under copper stress. C R Biol 338(4):241–254
Feller U, Vaseva II (2014) Extreme climatic events: impacts of drought and high temperature on physiological processes in agronomically important plants. Front Environ Sci 2:39
Flexas J, Bota J, Loreto F, Cornic G, Sharkey TD (2004) Diffusive and metabolic limitations to photosynthesis under drought and salinity in C3 plants. Plant biology 6(03):269–279
Gaind S, Gaur AC (1991) Thermotolerant phosphate solubilizing microorganisms and their interaction with mung bean. Plant and Soil 133(1):141–149
Gao S, Yan R, Cao M, Yang W, Wang S, Chen F (2008) Effects of copper on growth, antioxidant enzymes and phenylalanine ammonia-lyase activities in *Jatropha curcas* L. seedling. Plant Soil Environ 54(3):117–122
García-Fraile P, Menéndez E, Rivas R (2015) Role of bacterial biofertilizers in agriculture and forestry. AIMS Bioeng 2(3):183–205
Ghosh P, Rathinasabapathi B, Ma LQ (2015) Phosphorus solubilization and plant growth enhancement by arsenic-resistant bacteria. Chemosphere 134:1–6
Ghosh B, Md NA, Gantait S (2016) Response of rice under salinity stress: a review update. Rice Res Open Access:1–8
Gill SS, Tuteja N (2010) Reactive oxygen species and antioxidant machinery in abiotic stress tolerance in crop plants. Plant Physiol Biochem 48(12):909–930
Glick BR (2010) Using soil bacteria to facilitate phytoremediation. Biotech Adv 28(3):367–374
Glick BR, Cheng Z, Czarny J, Duan J (2007) Promotion of plant growth by ACC deaminase-producing soil bacteria. In: New perspectives and approaches in plant growth-promoting rhizobacteria research. Springer, Dordrecht, pp 329–339
Golldack D, Li C, Mohan H, Probst N (2014) Tolerance to drought and salt stress in plants: unraveling the signaling networks. Front Plant Sci 5:151
Gond SK, Bergen MS, Torres MS, White JF Jr (2015) Endophytic *Bacillus* spp. produce antifungal lipopeptides and induce host defence gene expression in maize. Microbiol Res 172:79–87
Goyer RA (1997) Toxic and essential metal interactions. Annu Rev Nutr 17(1):37–50
Grichko VP, Glick BR (2001) Amelioration of flooding stress by ACC deaminase-containing plant growth-promoting bacteria. Plant Physiol Biochem 39(1):11–17
Gusain YS, Singh US, Sharma A (2014) Enzymatic amelioration of drought stress in rice through the application of plant growth promoting rhizobacteria (PGPR). Int J Curr Res 6(1):4487–4491
Gusain YS, Singh US, Sharma AK (2015) Bacterial mediated amelioration of drought stress in drought tolerant and susceptible cultivars of rice (*Oryza sativa* L.). Afr J Biotechnol 14(9):764–773
Haas D, Défago G (2005) Biological control of soil-borne pathogens by fluorescent pseudomonads. Nat Rev Microbiol 3(4):307–319
Hashem A, Abd-Allah EF, Alqarawi AA, Aldubise A, Egamberdieva D (2015) Arbuscular mycorrhizal fungi enhances salinity tolerance of *Panicum turgidum* Forssk by altering photosynthetic and antioxidant pathways. J Plant Interact 10(1):230–242
Hashem A, Abd Allah EF, Alqarawi AA, Al Huqail AA, Egamberdieva D, Wirth S (2016) Alleviation of cadmium stress in *Solanum lycopersicum* L. by arbuscular mycorrhizal fungi via induction of acquired systemic tolerance. Saudi J Biological Sci 23(2):272–281
Hassan SED (2017) Plant growth-promoting activities for bacterial and fungal endophytes isolated from medicinal plant of *Teucrium polium* L. J Adv Res 8(6):687–695
Hayat R, Ali S, Amara U, Khalid R, Ahmed I (2010) Soil beneficial bacteria and their role in plant growth promotion: a review. Ann Microbiol 60(4):579–598
Hu HJ, Chen YL, Wang YF, Tang YY, Chen SL, Yan SZ (2017) Endophytic *Bacillus cereus* effectively controls *Meloidogyne incognita* on tomato plants through rapid rhizosphere occupation and repellent action. Plant Dis 101(3):448–455

Huang D, Wu W, Abrams SR, Cutler AJ (2008) The relationship of drought-related gene expression in *Arabidopsis thaliana* to hormonal and environmental factors. J Exp Bot 59(11):2991–3007
Hughes MA, Dunn MA (1996) The molecular biology of plant acclimation to low temperature. J Exp Bot 47(3):291–305
Huner N, Dahal K, Hollis L, Bode R, Rosso D, Krol M, Ivanov AG (2012) Chloroplast redox imbalance governs phenotypic plasticity: the "grand design of photosynthesis" revisited. Front Plant Sci 3:255
Hussain AI, Anwar F, Sherazi STH, Przybylski R (2008) Chemical composition, antioxidant and antimicrobial activities of basil (*Ocimum basilicum*) essential oils depends on seasonal variations. Food Chem 108(3):986–995
Hussain MB, Zahir ZA, Asghar HN, Asgher M (2014) Can catalase and exopolysaccharides producing rhizobia ameliorate drought stress in wheat? Int J Agricult & Biol 16(1)
Islam F, Wang J, Farooq MA, Yang C, Jan M, Mwamba TM et al (2019) Rice responses and tolerance to salt stress: deciphering the physiological and molecular mechanisms of salinity adaptation. In: Advances in rice research for abiotic stress tolerance. Woodhead Publishing, pp 791–819
Jagadish SVK, Muthurajan R, Oane R, Wheeler TR, Heuer S, Bennett J, Craufurd PQ (2010) Physiological and proteomic approaches to address heat tolerance during anthesis in rice (*Oryza sativa* L.). J Exp Bot 61(1):143–156
Jallad KN (2015) Heavy metal exposure from ingesting rice and its related potential hazardous health risks to humans. Environ Sci Pollut Res 22(20):15449–15458
Jamil M, Zeb S, Anees M, Roohi A, Ahmed I, ur Rehman S, Rha ES (2014) Role of *Bacillus licheniformis* in phytoremediation of nickel contaminated soil cultivated with rice. Int J Phytoremediation 16(6):554–571
Janoušková M, Pavlíková D, Vosátka M (2006) Potential contribution of arbuscular mycorrhiza to cadmium immobilisation in soil. Chemosphere 65(11):1959–1965
Jasim B, Sreelakshmi KS, Mathew J, Radhakrishnan EK (2016a) Surfactin, iturin, and fengycin biosynthesis by endophytic *Bacillus* sp. from *Bacopa monnieri*. Microb Ecol 72(1):106–119
Jasim B, Mathew J, Radhakrishnan EK (2016b) Identification of a novel endophytic *Bacillus* sp. from *Capsicum annuum* with highly efficient and broad spectrum plant probiotic effect. J Appl Microbiol 121(4):1079–1094
Javani S, Marín I, Amils R, Abad JP (2015) Four psychrophilic bacteria from Antarctica extracellularly biosynthesize at low temperature highly stable silver nanoparticles with outstanding antimicrobial activity. Colloids Surf A Physicochem Eng Asp 483:60–69
Jesus JM, Danko AS, Fiúza A, Borges MT (2015) Phytoremediation of salt-affected soils: a review of processes, applicability, and the impact of climate change. Environ Sci Pollut Res 22(9):6511–6525
Jha Y, Subramanian RB (2014) PGPR regulate caspase-like activity, programmed cell death, and antioxidant enzyme activity in paddy under salinity. Physiol Mol Biol Plants 20(2):201–207
Jha Y, Subramanian RB, Patel S (2011) Combination of endophytic and rhizospheric plant growth promoting rhizobacteria in *Oryza sativa* shows higher accumulation of osmoprotectant against saline stress. Acta Physiologiae Plantarum 33(3):797–802
Kabata-Pendias A (2010) Trace elements in soils and plants. CRC 394 Press227
Kang SM, Radhakrishnan R, Lee KE, You YH, Ko JH, Kim JH, Lee IJ (2015a) Mechanism of plant growth promotion elicited by *Bacillus* sp. LKE15 in oriental melon. Acta Agriculturae Scandinavica, Section B—Soil & Plant Science 65(7):637–647
Kang SM, Radhakrishnan R, Lee IJ (2015b) *Bacillus amyloliquefaciens* subsp. *plantarum* GR53, a potent biocontrol agent resists Rhizoctonia disease on Chinese cabbage through hormonal and antioxidants regulation. World J Microbiol Biotechnol 31(10):1517–1527
Karthika S, Midhun SJ, Jisha MS (2020) A potential antifungal and growth-promoting bacterium *Bacillus* sp. KTMA4 from tomato rhizosphere. Microb Pathog 142:104049
Kasim WA, Gaafar RM, Abou-Ali RM, Omar MN, Hewait HM (2016) Effect of biofilm forming plant growth promoting rhizobacteria on salinity tolerance in barley. Annal Agricul Sci 61(2):217–227

Kaya MD, Okçu G, Atak M, Cıkılı Y, Kolsarıcı Ö (2006) Seed treatments to overcome salt and drought stress during germination in sunflower (*Helianthus annuus* L.). Eur J Agron 24(4):291–295

Kilian M, Steiner U, Krebs B, Junge H, Schmiedeknecht G, Hain R (2000) FZB24® *Bacillus subtilis*–mode of action of a microbial agent enhancing plant vitality. Pflanzenschutz-Nachrichten Bayer 1(00):1

Kim YH, Khan AL, Kim DH, Lee SY, Kim KM, Waqas M et al (2014a) Silicon mitigates heavy metal stress by regulating P-type heavy metal ATPases, *Oryza sativa* low silicon genes, and endogenous phytohormones. BMC Plant Biol 14(1):1–13

Kim K, Jang YJ, Lee SM, Oh BT, Chae JC, Lee KJ (2014b) Alleviation of salt stress by *Enterobacter* sp. EJ01 in tomato and *Arabidopsis* is accompanied by up-regulation of conserved salinity responsive factors in plants. Mol Cells 37(2):109

Kobata T, Uemuki N (2004) High temperatures during the grain-filling period do not reduce the potential grain dry matter increase of rice. Agron J 96(2):406–414

Kotak S, Larkindale J, Lee U et al (2007) Complexity of the heat stress response in plants. Curr Opin Plant Biol 10:310–316

Koussevitzky S, Suzuki N, Huntington S, Armijo L, Sha W, Cortes D et al (2008) Ascorbate peroxidase 1 plays a key role in the response of *Arabidopsis thaliana* to stress combination. J Biol Chem 283(49):34197–34203

Krishnan P, Ramakrishnan B, Reddy KR, Reddy VR (2011) High-temperature effects on rice growth, yield, and grain quality. In: Advances in agronomy, vol 111. Academic Press, pp 87–206

Kuan KB, Othman R, Rahim KA, Shamsuddin ZH (2016) Plant growth-promoting rhizobacteria inoculation to enhance vegetative growth, nitrogen fixation and nitrogen remobilisation of maize under greenhouse conditions. PLoS One 11(3)

Kumar M, Lee SC, Kim JY, Kim SJ, Kim SR (2014) Over-expression of dehydrin gene, OsDhn1, improves drought and salt stress tolerance through scavenging of reactive oxygen species in rice (*Oryza sativa* L.). J Plant Biol 57(6):383–393

Kumar A, Mosa KA, Ji L et al (2018) Metabolomics-assisted biotechnological interventions for developing plant-based functional foods and nutraceuticals. Crit Rev Food Sci Nutr 58(11):1791–1807

Kumar AS, Sridar R, Uthandi S (2017) Mitigation of drought in rice by a phyllosphere bacterium *Bacillus altitudinis* FD48. Afr J Microbiol Res 11(45):1614–1625

Kundan R, Pant G, Jadon N, Agrawal PK (2015) Plant growth promoting rhizobacteria: mechanism and current prospective. J FertilPestic 6(2):9

Lafitte HR, Price AH, Courtois B (2004) Yield response to water deficit in an upland rice mapping population: associations among traits and genetic markers. Theor Appl Genet 109(6):1237–1246

Lata C, Prasad M (2011) Role of DREBs in regulation of abiotic stress responses in plants. J Exp Bot 62(14):4731–4748

Lata C, Yadav A, Prasad M (2011) Role of plant transcription factors in abiotic stress tolerance. Abiotic Stress Response in Plants, INTECH Open Access Publishers 10:269–296

Lee HJ, Abdula SE, Jang DW, Park SH, Yoon UH, Jung YJ et al (2013) Overexpression of the glutamine synthetase gene modulates oxidative stress response in rice after exposure to cadmium stress. Plant Cell Rep 32(10):1521–1529

Li L, Ye Y, Pan L, Zhu Y, Zheng S, Lin Y (2009) The induction of trehalose and glycerol in *Saccharomyces cerevisiae* in response to various stresses. Biochem Biophys Res Commun 387(4):778–783

Lima JM, Nath M, Dokku P, Raman KV, Kulkarni KP, Vishwakarma C et al (2015) Physiological, anatomical and transcriptional alterations in a rice mutant leading to enhanced water stress tolerance. AoB Plant 7

Lin YC, Kao CH (2005) Nickel toxicity of rice seedlings: cell wall peroxidase, lignin, and NiSO4-inhibited root growth. Crop, Env & Bioinfor 2(2):131–136

Liu JX, Liao DQ, Oane R, Estenor L, Yang XE, Li ZC, Bennett J (2006) Genetic variation in the sensitivity of anther dehiscence to drought stress in rice. Field Crop Res 97(1):87–100

Liu Z, Zhang Q, Han T, Ding Y, Sun J, Wang F, Zhu C (2016) Heavy metal pollution in a soil-rice system in the Yangtze river region of China. Int J Environ Res Public Health 13(1):63
Lopes RBM, de Oliveira Costa LE, Vanetti MCD, de Araújo EF, de Queiroz MV (2015) Endophytic bacteria isolated from common bean (*Phaseolus vulgaris*) exhibiting high variability showed antimicrobial activity and quorum sensing inhibition. Curr Microbiol 71(4):509–516
Lopes R, Tsui S, Gonçalves PJ, de Queiroz MV (2018) A look into a multifunctional toolbox: endophytic *Bacillus* species provide broad and underexploited benefits for plants. World J Microbiol Biotechnol 34(7):94
Lyngwi NA, Joshi SR (2014) Economically important *Bacillus* and related genera: a mini review. Bio of Useful Plants and Microbes 3:33–43
Ma Y, Rajkumar M, Zhang C, Freitas H (2016) Beneficial role of bacterial endophytes in heavy metal phytoremediation. J Environ Manage 174:14–25
Madhaiyan M, Poonguzhali S, Sa T (2007) Metal tolerating methylotrophic bacteria reduces nickel and cadmium toxicity and promotes plant growth of tomato (*Lycopersicon esculentum* L.). Chemosphere 69(2):220–228
Maksimov IV, Veselova SV, Nuzhnaya TV, Sarvarova ER, Khairullin RM (2015) Plant growth-promoting bacteria in regulation of plant resistance to stress factors. Russian J Plant Physiol 62(6):715–726
Manara A (2012) Plant responses to heavy metal toxicity. In: Plants and heavy metals. Springer, Dordrecht, pp 27–53
Mariani L, Ferrante A (2017) Agronomic management for enhancing plant tolerance to abiotic stresses—drought, salinity, hypoxia, and lodging. Horticulturae 3(4):52
Martin M, Dragoš A, Hölscher T, Maróti G, Bálint B, Westermann M, Kovács ÁT (2017) De novo evolved interference competition promotes the spread of biofilm defectors. Nat Commun 8:15127
Martins SJ, Rocha GA, de Melo HC, de Castro Georg R, Ulhôa CJ, de Campos Dianese É et al (2018) Plant-associated bacteria mitigate drought stress in soybean. Environ Sci Pollut Res 25(14):13676–13686
Marulanda A, Barea JM, Azcón R (2009) Stimulation of plant growth and drought tolerance by native microorganisms (AM fungi and bacteria) from dry environments: mechanisms related to bacterial effectiveness. J Plant Growth Regulat 28(2):115–124
Mass EV, Hoffman GJ (1977) Crop salt tolerance-current assessment. J Irrig Drain Div 103(2):115–134
Meena KK, Kumar M, Kalyuzhnaya MG, Yandigeri MS, Singh DP, Saxena AK, Arora DK (2012) Epiphytic pink-pigmented methylotrophic bacteria enhance germination and seedling growth of wheat (*Triticum aestivum*) by producing phytohormone. Antonie Van Leeuwenhoek 101(4):777–786
Meena KK, Sorty AM, Bitla UM, Choudhary K, Gupta P, Pareek A et al (2017) Abiotic stress responses and microbe-mediated mitigation in plants: the omics strategies. Front Plant Sci 8:172
Mickelbart MV, Hasegawa PM, Bailey-Serres J (2015) Genetic mechanisms of abiotic stress tolerance that translate to crop yield stability. Nat Rev Genet 16(4):237–251
Miller GAD, Suzuki N, Ciftci-Yilmaz SULTAN, Mittler RON (2010) Reactive oxygen species homeostasis and signalling during drought and salinity stresses. Plant Cell Environ 33(4):453–467
Munns R (2002) Comparative physiology of salt and water stress. Plant Cell Environ 25(2):239–250
Munns R, Tester M (2008) Mechanisms of salinity tolerance. Annu Rev Plant Biol 59:651–681
Naeem MS, Jin ZL, Wan GL, Liu D, Liu HB, Yoneyama K, Zhou WJ (2010) 5-Aminolevulinic acid improves photosynthetic gas exchange capacity and ion uptake under salinity stress in oilseed rape (*Brassica napus* L.). Plant and Soil 332(1-2):405–415
Nadeem SM, Ahmad M, Zahir ZA, Javaid A, Ashraf M (2014) The role of mycorrhizae and plant growth promoting rhizobacteria (PGPR) in improving crop productivity under stressful environments. Biotechnol Adv 32(2):429–448

Nadeem SM, Zahir ZA, Naveed M, Arshad M (2007) Preliminary investigations on inducing salt tolerance in maize through inoculation with rhizobacteria containing ACC deaminase activity. Can J Microbiol 53(10):1141–1149

Ngugi RK, Nyariki DM (2005) Rural livelihoods in the arid and semi-arid environments of Kenya: sustainable alternatives and challenges. Agric Hum Values 22(1):65–71

Nair AS, Abraham TK, Jaya DS (2008) Studies on the changes in lipid peroxidation and antioxidants in drought stress induced cowpea (*Vigna unguiculata* L.) varieties. J Environ Biol 29(5):689–691

Naseem S, Yasin M, Faisal M, Ahmed A (2016) Comparative study of plant growth promoting bacteria in minimizing toxic effects of chromium on growth and metabolic activities in wheat (*Triticum aestivum*). J Chem Soc Pak 38(3):509–516

Nasim W, Belhouchette H, Ahmad A, Habib-ur-Rahman M, Jabran K, Ullah K et al (2016) Modelling climate change impacts and adaptation strategies for sunflower in Pakistan. Outlook Agricul 45(1):39–45

Nautiyal CS, Srivastava S, Chauhan PS, Seem K, Mishra A, Sopory SK (2013) Plant growth-promoting bacteria *Bacillus amyloliquefaciens* NBRISN13 modulates gene expression profile of leaf and rhizosphere community in rice during salt stress. Plant Physiol Biochem 66:1–9

Niu L, Yang F, Xu C, Yang H, Liu W (2013) Status of metal accumulation in farmland soils across China: from distribution to risk assessment. Environ Pollut 176:55–62

Noel M, Bathke DJ, Haigh T, Smith KH, Svoboda MD, Hayes MJ (2018) Developing a framework to link drought impacts and drought severity state by state. In: AGU fall meeting abstracts, vol 2018, pp PA14B-03B

Oh-e I, Saitoh K, Kuroda T (2007) Effects of high temperature on growth, yield and dry-matter production of rice grown in the paddy field. Plant Prod Sci 10(4):412–422

Omar MNA, Osman MEH, Kasim WA, El-Daim IA (2009) Improvement of salt tolerance mechanisms of barley cultivated under salt stress using *Azospirillum brasilense*. In: Salinity and water stress. Springer, Dordrecht, pp 133–147

Ortiz N, Armada E, Duque E, Roldán A, Azcón R (2015) Contribution of arbuscular mycorrhizal fungi and/or bacteria to enhancing plant drought tolerance under natural soil conditions: effectiveness of autochthonous or allochthonous strains. J Plant Physiol 174:87–96

Pandey PK, Yadav SK, Singh A, Sarma BK, Mishra A, Singh HB (2012) Cross-species alleviation of biotic and abiotic stresses by the endophyte *Pseudomonas aeruginosa* PW09. J Phytopathol 160(10):532–539

Pandey S, Ghosh PK, Ghosh S, De TK, Maiti TK (2013) Role of heavy metal resistant *Ochrobactrum* sp. and *Bacillus* spp. strains in bioremediation of a rice cultivar and their PGPR like activities. J Microbiol 51(1):11–17

Panta S, Flowers T, Lane P, Doyle R, Haros G, Shabala S (2014) Halophyte agriculture: success stories. Environ Exp Bot 107:71–83

Potters G, Pasternak TP, Guisez Y, Palme KJ, Jansen MA (2007) Stress-induced morphogenic responses: growing out of trouble? Trends Plant Sci 12(3):98–105

Prakash C, Mithra SA, Singh PK, Mohapatra T, Singh NK (2016) Unraveling the molecular basis of oxidative stress management in a drought tolerant rice genotype Nagina 22. BMC Genomics 17(1):774

Prasch CM, Sonnewald U (2013) Simultaneous application of heat, drought, and virus to *Arabidopsis* plants reveals significant shifts in signaling networks. Plant Physiol 162(4):1849–1866

Qu AL, Ding YF, Jiang Q, Zhu C (2013) Molecular mechanisms of the plant heat stress response. Biochem Biophys Res Commun 432(2):203–207

Radhakrishnan R, Lee IJ (2016) Gibberellins producing *Bacillus methylotrophicus* KE2 supports plant growth and enhances nutritional metabolites and food values of lettuce. Plant Physiol Biochem 109:181–189

Radhakrishnan R, Kang SM, Baek IY, Lee IJ (2014) Characterization of plant growth-promoting traits of *Penicillium* species against the effects of high soil salinity and root disease. J Plant Interact 9(1):754–762

Radhakrishnan R, Hashem A, Abd-Allah EF (2017) *Bacillus*: a biological tool for crop improvement through bio-molecular changes in adverse environments. Front Physiol 8:667
Rahdari P, Hoseini SM (2012) Drought stress: a review. Int J Agronomy & Plant Prod 3(10):443–446
Rahman MA, Thomson MJ, De Ocampo M, Egdane JA, Salam MA, Shah-E-Alam M, Ismail AM (2019) Assessing trait contribution and mapping novel QTL for salinity tolerance using the Bangladeshi rice landrace capsule. Rice 12(1):1–18
Rang ZW, Jagadish SVK, Zhou QM, Craufurd PQ, Heuer S (2011) Effect of high temperature and water stress on pollen germination and spikelet fertility in rice. Environ Exp Bot 70(1):58–65
Rath M, Mitchell TR, Gold SE (2018) Volatiles produced by *Bacillus mojavensis* RRC101 act as plant growth modulators and are strongly culture-dependent. Microbiol Res 208:76–84
Remans R, Ramaekers L, Schelkens S, Hernandez G, Garcia A, Reyes JL et al (2008) Effect of Rhizobium–*Azospirillumc* oinoculation on nitrogen fixation and yield of two contrasting *Phaseolus vulgaris* L. genotypes cultivated across different environments in Cuba. Plant and Soil 312(1-2):25–37
Ritchie JT (1993) Genetic specific data for crop modeling. In: Systems approaches for agricultural development. Springer, Dordrecht, pp 77–93
RoyChoudhury A, Roy C, Sengupta DN (2007) Transgenic tobacco plants overexpressing the heterologous lea gene Rab16A from rice during high salt and water deficit display enhanced tolerance to salinity stress. Plant Cell Rep 26(10):1839–1859
Ruan CJ, da Silva JAT, Mopper S, Qin P, Lutts S (2010) Halophyte improvement for a salinized world. Crit Rev Plant Sci 29(6):329–359
Ryan RP, Germaine K, Franks A, Ryan DJ, Dowling DN (2008) Bacterial endophytes: recent developments and applications. FEMS Microbiol Lett 278(1):1–9
Sandhya V, Ali SZ, Grover M, Reddy G, Bandi V (2011) Drought- tolerant plant growth promoting *Bacillus* spp.: effect on growth, osmolytes, and antioxidant status of maize under drought stress. J Plant Interact 6:1–14. https://doi.org/10.1080/17429145.2010.535178
Sahi C, Singh A, Kumar K, Blumwald E, Grover A (2006) Salt stress response in rice: genetics, molecular biology, and comparative genomics. Funct Integr Genomics 6(4):263–284
Sahoo RK, Ansari MW, Pradhan M, Dangar TK, Mohanty S, Tuteja N (2014a) Phenotypic and molecular characterization of native *Azospirillum* strains from rice fields to improve crop productivity. Protoplasma 251(4):943–953
Sahoo RK, Ansari MW, Pradhan M, Dangar TK, Mohanty S, Tuteja N (2014b) A novel *Azotobacter vinellandii* (SRI Az 3) functions in salinity stress tolerance in rice. Plant Signal & Behav 9(7):511–523
Santarius KA (1992) Freezing of isolated thylakoid membranes in complex media. VIII. Differential cryoprotection by sucrose, proline and glycerol. Physiol Plant 84(1):87–93
Sato Y, Masuta Y, Saito K, Murayama S, Ozawa K (2011) Enhanced chilling tolerance at the booting stage in rice by transgenic overexpression of the ascorbate peroxidase gene, OsAPXa. Plant Cell Rep 30(3):399–406
Schulz B, Boyle C, Draeger S, Römmert AK, Krohn K (2002) Endophytic fungi: a source of novel biologically active secondary metabolites. Mycol Res 106(9):996–1004
Sethy SK, Ghosh S (2013) Effect of heavy metals on germination of seeds. J Nat Sci, Biol & Med 4(2):272
Shannon MC, Rhoades JD, Draper JH, Scardaci SC, Spyres MD (1998) Assessment of salt tolerance in rice cultivars in response to salinity problems in California. Crop Sci 38(2):394–398
Shabanamol S, Divya K, George TK, Rishad KS, Sreekumar TS, Jisha MS (2018) Characterization and in planta nitrogen fixation of plant growth promoting endophytic diazotrophic *Lysinibacillus sphaericus* isolated from rice (*Oryza sativa*). Physiol Mol Plant Pathol 102:46–54
Shabanamol S, Varghese EM, Thampi M, Karthika S, Sreekumar J, Jisha MS (2020) Enhancement of growth and yield of rice (*Oryza sativa*) by plant probiotic endophyte, *Lysinibacillus sphaericus* under greenhouse conditions. Commun Soil Sci Plant Anal 51(9):1268–1282
Shahzad B, Tanveer M, Rehman A, Cheema SA, Fahad S, Rehman S, Sharma A (2018) Nickel; whether toxic or essential for plants and environment-A review. Plant Physiol Biochem 132:641–651

Shahzad R, Khan AL, Bilal S, Waqas M, Kang SM, Lee IJ (2017) Inoculation of abscisic acid-producing endophytic bacteria enhances salinity stress tolerance in *Oryza sativa*. Environ Exp Bot 136:68–77

Shahzad R, Waqas M, Khan AL, Asaf S, Khan MA, Kang SM et al (2016) Seed-borne endophytic *Bacillus amyloliquefaciens* RWL-1 produces gibberellins and regulates endogenous phytohormones of *Oryza sativa*. Plant Physiol Biochem 106:236–243

Shrivastava P, Kumar R (2015) Soil salinity: a serious environmental issue and plant growth promoting bacteria as one of the tools for its alleviation. Saudi J Bio Sci 22(2):123–131

Shukla N, Awasthi RP, Rawat L, Kumar J (2012) Biochemical and physiological responses of rice (*Oryza sativa* L.) as influenced by *Trichoderma harzianum* under drought stress. Plant Physiol Biochem 54:78–88

Simontacchi M, Galatro A, Ramos-Artuso F, Santa-María GE (2015) Plant survival in a changing environment: the role of nitric oxide in plant responses to abiotic stress. Front Plant Sci 6:977 Singh DP, Prabha R, Yandigeri MS, Arora DK (2011) Cyanobacteria-mediated phenylpropanoids and phytohormones in rice (*Oryza sativa*) enhance plant growth and stress tolerance. Antonie Van Leeuwenhoek 100(4):557–568

Singh R, Tripathi RD, Dwivedi S, Kumar A, Trivedi PK, Chakrabarty D (2010) Lead bioaccumulation potential of an aquatic macrophyte *Najas indica* are related to antioxidant system. Bioresour Technol 101(9):3025–3032

Song S-Y, Chen Y, Chen J (2011) Physiological mechanisms underlying OsNAC5-dependent tolerance of rice plants to abiotic stress. Planta 234:331–345. https://doi.org/10.1007/s00425-011-1403-2

Sorty AM, Meena KK, Choudhary K, Bitla UM, Minhas PS, Krishnani KK (2016) Effect of plant growth promoting bacteria associated with halophytic weed (*Psoralea corylifolia* L) on germination and seedling growth of wheat under saline conditions. Appl Biochem Biotechnol 180(5):872–882

Srivastava RK, Pandey P, Rajpoot R, Rani A, Dubey RS (2014) Cadmium and lead interactive effects on oxidative stress and antioxidative responses in rice seedlings. Protoplasma 251(5):1047–1065

Strobel G, Daisy B, Castillo U, Harper J (2004) Natural products from endophytic microorganisms. J Nat Prod 67(2):257–268

Swain P, Mall AK, Bose LK, Baig MJ, Singh DN (2010, November) Drought susceptibility index as a parameter to identify drought tolerant rice genotypes for rainfed uplands. In Abs. National Symposium on Sustainable Rice Production System Under Changed Climate. CRRI, Cuttack, Orissa, India (pp. 27–29).

Swaine EK, Swaine MD, Killham K (2007) Effects of drought on isolates of *Bradyrhizobium elkanii* cultured from *Albizia adianthifolia* seedlings of different provenances. Agr Syst 69(2):135–145

Tahir HA, Gu Q, Wu H, Raza W, Hanif A, Wu L et al (2017) Plant growth promotion by volatile organic compounds produced by *Bacillus subtilis* SYST2. Front Microbiol 8:171

Tank N, Saraf M (2010) Salinity-resistant plant growth promoting rhizobacteria ameliorates sodium chloride stress on tomato plants. J Plant Interact 5(1):51–58

Taiz L, Zeiger E (2002) Photosynthesis: physiological and ecological considerations. Plant Physiol 9:172–174

Todaka D, Shinozaki K, Yamaguchi-Shinozaki K (2015) Recent advances in the dissection of drought-stress regulatory networks and strategies for development of drought-tolerant transgenic rice plants. Front Plant Sci 6:84

Thakur P, Kumar S, Malik JA, Berger JD, Nayyar H (2010) Cold stress effects on reproductive development in grain crops: an overview. Environ Exp Bot 67(3):429–443

Timm CM, Campbell AG, Utturkar SM, Jun SR, Parales RE, Tan WA et al (2015) Metabolic functions of *Pseudomonas fluorescens* strains from *Populus deltoides* depend on rhizosphere or endosphere isolation compartment. Front Microbiol 6:1118

Tittabutr P, Piromyou P, Longtonglang A, Noisa-Ngiam R, Boonkerd N, Teaumroong N (2013) Alleviation of the effect of environmental stresses using co-inoculation of mungbean by

Bradyrhizobium and rhizobacteria containing stress-induced ACC deaminase enzyme. Soil Sci Plant Nutr 59(4):559–571

Tiwari S, Singh P, Tiwari R, Meena KK, Yandigeri M, Singh DP et al (2011) Salt-tolerant rhizobacteria-mediated induced tolerance in wheat (*Triticum aestivum*) and chemical diversity in rhizosphere enhance plant growth. Biol Fertil Soils 47:907–916. https://doi.org/10.1007/s00374-011-0598-5

Tiwari S, Lata C, Chauhan PS, Nautiyal CS (2016) *Pseudomonas putida* attunes morphophysiological, biochemical and molecular responses in *Cicer arietinum* L. during drought stress and recovery. Plant Physiol Biochem 99:108–117

Tiwari S, Prasad V, Chauhan PS, Lata C (2017) *Bacillus amyloliquefaciens* confers tolerance to various abiotic stresses and modulates plant response to phytohormones through osmoprotection and gene expression regulation in rice. Front Plant Sci 8:1510

Tonelli ML, Taurian T, Ibáñez F, Angelini J, Fabra A (2010) Selection and in vitro characterization of biocontrol agents with potential to protect peanut plants against fungal pathogens. J Plant Pathol:73–82

Treesubsuntorn C, Dhurakit P, Khaksar G, Thiravetyan P (2018) Effect of microorganisms on reducing cadmium uptake and toxicity in rice (*Oryza sativa* L.). Environ Sci Pollut Res 25(26):25690–25701

Turan M, Gulluce M, Şahin F (2012) Effects of plant-growth-promoting rhizobacteria on yield, growth, and some physiological characteristics of wheat and barley plants. Commun Soil SciPlant Anal 43(12):1658–1673

Ullah A, Heng S, Munis MFH, Fahad S, Yang X (2015) Phytoremediation of heavy metals assisted by plant growth promoting (PGP) bacteria: a review. Environ Exp Bot 117:28–40

Vardharajula S, Zulfikar Ali S, Grover M, Reddy G, Bandi V (2011) Drought-tolerant plant growth promoting *Bacillus* spp.: effect on growth, osmolytes, and antioxidant status of maize under drought stress. J Plant Interact 6(1):1–14

Verma JP, Yadav J, Tiwari KN, Kumar A (2013) Effect of indigenous *Mesorhizobium* spp. and plant growth promoting rhizobacteria on yields and nutrients uptake of chickpea (*Cicer arietinum* L.) under sustainable agriculture. Ecol Eng 51:282–286

Vigliotta G, Matrella S, Cicatelli A, Guarino F, Castiglione S (2016) Effects of heavy metals and chelants on phytoremediation capacity and on rhizobacterial communities of maize. J Environ Manag 179:93–102

Wani PA, Wahid S, Singh R, Kehinde AM (2018) Antioxidant and chromium reductase assisted chromium (VI) reduction and Cr (III) immobilization by the rhizospheric *Bacillus* helps in the remediation of Cr (VI) and growth promotion of soybean crop. Rhizosphere 6:23–30

Wenzel WW, Bunkowski M, Puschenreiter M, Horak O (2003) Rhizosphere characteristics of indigenously growing nickel hyperaccumulator and excluder plants on serpentine soil. Environ Pollut 123(1):131–138

Xie SS, Wu HJ, Zang HY, Wu LM, Zhu QQ, Gao XW (2014) Plant growth promotion by spermidine-producing *Bacillus subtilis* OKB105. Mol Plant-Microbe Interact 27(7):655–663

Yadav AN, Verma P, Kumar M, Pal KK, Dey R, Gupta A et al (2015) Diversity and phylogenetic profiling of niche-specific bacilli from extreme environments of India. Ann Microbiol 65(2):611–629

Yaish MW (2017) Draft genome sequence of the endophytic *Bacillus aryabhattai* strain SQU-R12, identified from *Phoenix dactylifera* L. roots. Genome Announc 5(32):e00718–e00717

Yaish MW, Antony I, Glick BR (2015) Isolation and characterization of endophytic plant growth-promoting bacteria from date palm tree (*Phoenix dactylifera* L.) and their potential role in salinity tolerance. Antonie Van Leeuwenhoek 107(6):1519–1532

Yang Y, Hu C, Abu-Omar MM (2012) Conversion of glucose into furans in the presence of AlCl3 in an ethanol–water solvent system. Bioresour Technol 116:190–194

Yousaf M, Rehman Y, Hasnain S (2017) High-yielding wheat varieties harbour superior plant growth promoting-bacterial endophytes. Appl Food Biotechnol 4(3):143–154

Yu X, Ai C, Xin L, Zhou G (2011) The siderophore-producing bacterium, *Bacillus subtilis* CAS15, has a biocontrol effect on Fusarium wilt and promotes the growth of pepper. Eur J Soil Biol 47(2):138–145

Yu F, Liu K, Li M, Zhou Z, Deng H, Chen B (2013) Effects of cadmium on enzymatic and non-enzymatic antioxidative defences of rice (*Oryza sativa* L.). Int J Phytoremediation 15(6):513–521

Yu HY, Ding X, Li F, Wang X, Zhang S, Yi J et al (2016) The availabilities of arsenic and cadmium in rice paddy fields from a mining area: the role of soil extractable and plant silicon. Environ Pollut 215:258–265

Zaidi S, Usmani S, Singh BR, Musarrat J (2006) Significance of *Bacillus subtilis* strain SJ-101 as a bioinoculant for concurrent plant growth promotion and nickel accumulation in *Brassica juncea*. Chemosphere 64(6):991–997

Zahir ZA, Munir A, Asghar HN, Shaharoona B, Arshad M (2008) Effectiveness of rhizobacteria containing ACC deaminase for growth promotion of peas (*Pisum sativum*) under drought conditions. J Microbiol Biotechnol 18(5):958–963

Zawadzka AM, Abergel RJ, Nichiporuk R, Andersen UN, Raymond KN (2009) Siderophore-mediated iron acquisition systems in *Bacillus cereus*: identification of receptors for anthrax virulence-associated petrobactin. Biochemistry 48(16):3645–3657

Zeid IM, Shedeed ZA (2006) Response of alfalfa to putrescine treatment under drought stress. Biologia Plantarum 50(4):635

Zeng L, Lesch SM, Grieve CM (2003) Rice growth and yield respond to changes in water depth and salinity stress. Agric Water Manag 59(1):67–75

Zgallaï H, Steppe K, Lemeur R (2005) Photosynthetic, physiological and biochemical responses of tomato plants to polyethylene glycol-induced water deficit. J Integr Plant Biol 47(12):1470–1478

Zhang Q, Chen Q, Wang S, Hong Y, Wang Z (2014) Rice and cold stress: methods for its evaluation and summary of cold tolerance-related quantitative trait loci. Rice 7:24

Zhang H, Li D, Zhou Z, Zahoor R, Chen B, Meng Y (2017) Soil water and salt affect cotton (*Gossypium hirsutum* L.) photosynthesis, yield and fiber quality in coastal saline soil. Agric Water Manag 187:112–121

Zhao L, Xu Y, Sun R, Deng Z, Yang W, Wei G (2011) Identification and characterization of the endophytic plant growth prompter *Bacillus cereus* strain MQ23 isolated from *Sophora alopecuroides* root nodules. Braz J Microbiol 42(2):567–575

Zhao L, Xu Y, Lai XH, Shan C, Deng Z, Ji Y (2015) Screening and characterization of endophytic *Bacillus* and *Paenibacillus* strains from medicinal plant *Lonicera japonica* for use as potential plant growth promoters. Braz J Microbiol 46(4):977–989

Zheng N, Liu J, Wang Q, Liang Z (2010) Health risk assessment of heavy metal exposure to street dust in the zinc smelting district, Northeast of China. Sci Total Environ 408(4):726–733

Chapter 14
Growth Enhancement and Bioremediation of Heavy Metal in Crop Plants Through *Bacillus* Species Application

Saidu Abdullahi, M. A. Baset Mia, and Amir Hamzah Ghazali

Abstract Mitigating the effects of heavy metals and their subsequent remediation on plants is one of the hot topics of environmental research studies. Application of *Bacillus* species in this area of research has received considerable attention probably due to the high rate of adaptability and survivability of the species under extreme environments. *Bacillus* spp. have shown great potential in plant growth enhancement and bioremediation of heavy metal–contaminated soils. More *Bacillus*–plant physiological studies are required for better understanding of the mitigation mechanisms of *Bacillus* spp. against heavy metal stress conditions and plant growth promotion. Our findings have successfully shown that Bacilli have multiple beneficial traits which assisted the crop plants either directly or indirectly through plant growth–promoting activities and heavy metal tolerance enhancements.

Keywords Bioremediation · Heavy metal toxicity · Growth enhancement · Crop plants · Environmental pollution

S. Abdullahi
School of Biological Sciences, Universiti Sains Malaysia, Pulau Pinang, Malaysia

Department of Botany, Ahmadu Bello University, Zaria, Nigeria

M. A. B. Mia (✉)
Department of Crop Botany, Bangabandhu Sheikh Mujibur Rahman Agricultural University, Gazipur, Bangladesh

A. H. Ghazali (✉)
School of Biological Sciences, Universiti Sains Malaysia, Pulau Pinang, Malaysia
e-mail: amirhg@usm.my

M. T. Islam et al. (eds.), *Bacilli in Agrobiotechnology*, Bacilli in Climate Resilient Agriculture and Bioprospecting,
https://doi.org/10.1007/978-3-030-85465-2_14

14.1 Introduction

Bacillus species represents one of the most important and dominant groups of microbes that exist in various environments with a high rate of survival in extreme and adverse environmental conditions (Islam et al. 2019). The gram-positive or facultative gram-negative spore-forming bacteria can survive for a long time under unfavourable environmental conditions (Radhakrishnan et al. 2017), and spore formation is one of the most important characteristics of *Bacillus* spp. The aerobic endospore-forming bacteria are ubiquitous in agricultural systems due to possession of certain physiological traits associated with their survival which include production of a multilayered cell wall, production of endospores that are stress-resistant, secretion of antibiotics, peptide signal molecules, and extracellular enzymes (Gardener 2004). The cell-wall-degrading substances (protease, cellulase, chitinase, glucanase, hydrogen cyanide, and lipopeptides) from *Bacillus* spp. damage the pathogenic organisms including bacteria, fungi, viruses, nematodes, and pests to control their populations (Radhakrishnan et al. 2017). *Bacillus* spp. also have great potential for applications in agriculture, industry, and medicine and are a good source of important metabolites and enzymes of various biotechnological interests. As an example, it was shown that *B. thuringiensis* H-14 is able to kill mosquito larvae within 24 h of treatment with the lowest concentration of spore-crystal suspension (0.05 mg/L) (Fun et al. 2016).

There is an increasing interest in *Bacillus* spp. on the aspects of agricultural and environmental biotechnology due to increasing demand for food production, recovery of degraded soils as well as the development of biofertilizers and biopesticides (Nascimento et al. 2020). The unfavourable biotic and abiotic stimuli affect normal plant metabolism, suppressing the growth and yield of plants, however, the stress factors on crops are mitigated by *Bacillus*-induced physiological changes (Radhakrishnan et al. 2017). Under unfavourable environmental conditions including heavy metal stress, *Bacillus* produces important substances such as siderophores and exopolysaccharides, which prevent the movement of toxic ions and adjust the ionic balance and water transport in plant tissues (Radhakrishnan et al. 2017). *Bacillus* can live both outside and within plant tissues, facilitating plant growth and development as well as protection from harsh environmental conditions through several mechanisms including N_2 fixation, production of indole acetic acid (IAA), siderophore, and 1-aminocyclopropane-1-carboxylate (ACC) deaminase. Additionally, phosphate and potassium solubilization and production of exopolysaccharides show antagonistic actions against pathogens and other pests. There are many bacterial species that are being applied as plant growth-promoting bacteria (PGPB) to crop plants; however, members of the *Bacillus* group are more favoured for commercialization as PGPB due to some features including their ability to produce the heat and desiccation-tolerant endospores, which help to maintain high cell viability and prolong the shelf life of the bacteria in the carrier formulations (Akinrinlola et al. 2018). Bacterial colonization of roots provides a nutrient source, and the plants on the other hand, receive bacterial metabolites and other stimuli that enhance growth and stress resistance (Hashem et al. 2019). In this relationship, the microsymbiont (e.g. *B. subtilis*) will form a thin biofilm layer on the root surfaces for long-term bacterial colonization of the host plants (Hashem et al. 2019). Additionally, the rhizosphere of the host plants is also heavily colonized by beneficial bacterial populations. Nowadays, plant growth-promoting *Bacillus* spp. are gaining prominence as biofertilizer, biopesticides, and bioremediator of certain mineral elements that are toxic to crop plants (Amir et al. 2005; Fun et al. 2016; Mia et al. 2010; Salwani et al. 2012; Tang et al. 2020). Salwani et al. (2012) revealed that the beneficial diazotrophic microsymbionts of leguminous cover crop *Mucuna bracteata* were not only from the conventional Alphaproteobacteria class (e.g. *Brevundimonas* sp.) but were also from the Betaproteobacteria class (e.g. *Achromobacter* sp. and *Burkholderia* sp.) and the Gammaproteobacteria class (e.g. *Stenotrophomonas* sp.). The authors also successfully obtained and identified *Bacillus* sp. from the root nodules of the host plants.

Heavy metal may be found naturally in the soils or added to the soils through various anthropogenic activities such as exploitation of mines and smelters, fossil fuel combustion, metallurgical and electronics industries, indiscriminate waste disposal, and military training. Additionally, the agricultural practices involving excessive application of inorganic fertilizers and chemicals for optimum crop improvement are also contributing towards the escalated accumulation of heavy metal in soils (Kapahi and Sachdeva 2019; Oves et al. 2016). Examples of heavy metals which are contaminating soils and water bodies include As, Cd, Pb, Hg, Cr, Co, Cu, Hg, Ni, Zn, Se, U, Mn, and Ni (Weissmannová et al. 2019). The continuous release of heavy metals into the environment severely affects soil and water quality. Despite the fact that waste water may serve as an important source of essential nutrients for plants, many risks can be attributed to the use of such waste water for crop irrigation due to presence of toxic contaminants including heavy metal (Khalid et al. 2018). Many heavy metal are toxic even at very low concentrations, and some (e.g. As, Cd, Cr, Hg, Ni, Cu, Pb, and Zn) are not only cytotoxic but also mutagenic and carcinogenic in their nature (Dixit et al. 2015).

Heavy metal contaminated soils are generally deficient in essential nutrients, and if the contaminated soils are used for crop production, there is a huge risk of metal being transferred into food chain at higher concentration. It can cause severe human health problems (Oves et al. 2016). Certain heavy metal is essential for plant growth and health at certain concentrations. For example, Co is required in nitrogenase enzyme as metal activator for N_2 fixation both in legume and non-legume crops, and Mo is needed as catalytic centre of many enzymes namely nitrate reductase. However, when the elements exceed the required quantities, they may cause serious injury or even lead to death of plants (Oves et al. 2016). Biological methods for remediation of heavy metal known as bioremediation are considered most eco-friendly, cost-effective, reliable, and have no adverse effects to the environment (Dixit et al. 2015). There is an increasing interest towards studying the role of microorganisms in biotransformation and detoxification of heavy metals and the production of plant growth promoting substances by them under stress conditions (El-Meihy et al. 2019; Nayak et al. 2018). Understanding various mechanisms of metal accumulation and plant growth promotion has numerous biotechnological implications for bioremediation of heavy metal and crop productions in heavy metal contaminated soils. Mitigating the effects of heavy metal and their subsequent remediation on plants is one of the hot topics of environmental research nowadays. Application of *Bacillus* species in this area of research has received considerable attention (El-Meihy et al. 2019; Li et al. 2019; Tang et al. 2020) probably due to the high rate of adaptability and survivability of the species under extreme environments (e.g. heavy metal stress tolerance and their multiple beneficial mechanisms of actions) (Radhakrishnan et al. 2017). However, very few study reports are available on the application of *Bacillus* in bioremediation of heavy metal in plants. Most of the relevant studies were carried out on the physiological mechanisms of plant growth promotion or heavy metal tolerance of the individual *Bacillus* spp.. Rhizobacteria viz. *Bacillus* spp. (*B. megaterium, B. cereus,* and *B. pumilus*), isolated from *Ludwigia octovalvis,* were capable of absorbing As and can be good candidates for bioremediation of toxic effects of that element on plants (Titah et al. 2018). More studies on heavy metal tolerant and plant growth promoting *Bacillus* and other bacteria to harness their potentials as bioremediation or phytoremediation agents are therefore still needed. In this chapter, potentials of *Bacillus* spp. on promoting growth of associated selected host plants and their ability to remediate heavy metal are discussed.

14.2 Production of Plant Growth Promoting (PGP) Substances by *Bacillus* Species

14.2.1 Production of PGP Substances by Individual Bacillus *Species Under Culture Conditions Supplemented with Heavy Metal*

Some of *Bacillus* spp. namely *B. sphaericus* strain UPMB10 are able to produce phytohormone like auxin and gibberellin which resulted in increased root growth in banana and oil palm under hydroponics and pot culture conditions (Mia et al. 2010; Amir et al. 2005). A large body of studies reflected that *Bacillus* species can produce plant growth promoting substances even under heavy metal stress conditions (Wu et al. 2019). Rizvi et al. (2019) also stated that heavy metal tolerant *B. subtilis* strain BM2 was able to synthesise variable concentrations of indole acetic acid (IAA) when cultured under different concentrations of Pb and Ni. However, the highest concentration of IAA was produced by the bacterial strain (BM2) when cultured in metal free medium (51.6 μg/mL), which kept declining with increasing concentrations of the added metal. At 400 μg/mL Pb and Ni, IAA secretion was reduced by 58 and 47%, respectively. Despite the toxicity and inhibitory effects of the metal, the strain maintained reasonable amount of IAA production potentiality. The strain was also able to produce substantial amount of ACC deaminase and siderophore when cultured in media supplemented with high concentrations of metal. However, the siderophore production by the strain was not detected on CAS agar with metal supplementation but detected under liquid medium with the metal supplementation. The cause of siderophore not detected in CAS agar but in liquid medium with metal supplementation remains unknown, but likely due to the presence of HDTMA (hexadecyltrimethyl-ammonium bromide) in the solid agar (Rizvi et al. 2019).

Bacterial IAA induce root elongation and development to overcome heavy metal toxicity and improve plant growth, while ACC deaminase, on the other hand, metabolizes the ethylene precursor ACC, thereby alleviating the effects of ethylene stress due to the metal and helps in plant survival under stress conditions (Mishra et al. 2017). Siderophore produced by *Bacillus* and other plant growth promoting bacteria is one the most important substances that can directly alleviate heavy metal toxicity by binding/chelating heavy metal ions and increasing the supply of iron to plants (Nayak et al. 2018). Therefore, maintenance in the production of plant growth promoting substances by *Bacillus* spp. under heavy metal stress could play an important role in improving growth of the host plants in addition to reduce the toxic effects of the metal on the plant.

14.2.2 *Production of PGP Substances by* **Bacillus** *Species in Association with Plants*

Nitrogen is an important essential nutrient which is required for growth and development of plants, taken up as NO_3^-, NH_4^+, and N_2 (*via* biological nitrogen fixation process) (Bellogín et al. 2014). Atmospheric nitrogen fixing bacteria form symbiotic association on the roots of plants where they convert nitrogen into form that can be used by plants, a process mainly restricted to leguminous plants. However, several non-symbiotic bacteria have been recognized including many *Bacillus* spp. as free living nitrogen fixers in many plants of agronomic importance. Various species of *Bacillus* were reported to fix atmospheric nitrogen which include *B. cereus, B. marisflavi, B. megaterium, Paenibacillus polymyxa,* and *P. massiliensis* (Ding et al. 2005; Halim et al. 2016; Nayak et al. 2018; Tang et al. 2020). Others were able to solubilize phosphates like *B. cereus, B. thuringiensis, B. megaterium, B. safensis, P. cineris,* and *B. subtilis* (Akinrinlola et al. 2018; Babu et al. 2013; Huang et al. 2020; Zaidi et al. 2006). Phosphorus, which mostly occurs in the soil in the form of non-soluble compounds and thus not always available for plant consumption (Bellogín et al. 2014), is converted to soluble form that can be absorbed and utilized by plants in the process of phosphate solubilization. Besides that, PGPB inoculation can also stimulate and increase uptake of other nutrients such as K, Ca, Fe, Cu, and Zn which may be achieved through acidification of the rhizosphere (e.g. by production of organic acids), and in general, decrease in soil pH increases solubilization of nutrients (Bellogín et al. 2014). Some *Bacilli* have iron-chelating properties through the production of siderophores that are known to solubilize iron in the rhizosphere from minerals and organic compounds (Radhakrishnan et al. 2017). The siderophores produced bind insoluble Fe^{3+} and reduce it to soluble Fe^{2+}, which can then be utilized by plants.

14.2.3 Role of Phytohormone (IAA) and Enzyme (ACC Deaminase) in Mitigating the Heavy Metal Stress in Plants

The IAA production by plant-associated bacteria plays an important role in plant-bacterial interactions. IAA synthesized by beneficial bacteria can increase the number of root hairs and lateral roots as well as the total root surface area, leading to an enhancement of root exudation and mineral uptake from the soil (Kong and Glick 2017). Number of studies have also suggested that PGPB producing IAA may play essential role in improving metal phytoremediation, however, most did not provide definitive proof of the direct involvement of the IAA (Kong and Glick 2017). In a previous study, two PGPB strains *B. paralicheniformis* YSP151 and *Brevibacterium frigoritolerans* YSP40 isolated from *Brassica juncea* in a Pb-contaminated mine soils were used to inoculate similar host plants grown in a metal-contaminated soil (Yahaghi et al. 2018). Results from the trial showed enhanced growth and Pb uptake by the inoculated host plants and both bacterial strains were reported to have high IAA producing ability. Multiple plant growth promoting traits may together be responsible for survival and enhanced metal tolerance of the host plants grown in heavy metal contaminated soils. Some bacteria can stimulate plant growth either by synthesizing more IAA or by degrading excess synthesized IAA when it is detrimentally higher than normal levels (Kong and Glick 2017).

Increase in ethylene synthesis is mostly associated with various environmental stresses such as heavy metal stress, extreme temperatures, drought, high salinity, organic pollution, insect damage, radiation, wounding, and various pathogens including viruses, bacteria, and fungi. Much of the plant growth inhibition that occurs due to environmental stresses is the result of increased levels of stress induced ethylene, which affects the plant's response to the stressor (Kong and Glick 2017). Some PGPB have great potential to decrease ethylene production that in turn enhance plant growth. This potential is due to the production of enzyme ACC (1-Aminocyclopropane-1-Carboxylate) deaminase. ACC is the precursor to ethylene synthesis and ACC deaminase enzyme hydrolyses ACC into a-ketobutyrate and ammonia (Ashraf et al. 2017). The ammonia generated add to the nitrogen source of the plants. ACC deaminase producing PGPB can facilitate plant growth and development by converting the ACC into α-ketobutyrate and ammonia, thereby reducing the levels of plant ethylene and providing some protection against growth inhibition caused by heavy metal and other stress factors (Kong and Glick 2017).

14.3 *Bacillus* Species and Heavy Metal Transformation, Detoxification, and Mobilization

Bacteria are known to enhance plant growth and survival under heavy metal stress conditions due to their ability of consuming and converting complex waste into simple non-toxic products/compounds (Tiwari and Lata 2018). Some heavy metal tolerant bacteria are also involved in enzymatic oxidation or reduction of toxic metal to a non-toxic or less toxic forms. Application of *Bacillus* spp. into heavy metal-contaminated soils can reduce the deleterious and toxic effects of heavy metal on growth of crop plants. For example, the Cr and As resistant *B. firmus* strain TE7 isolated from tannery effluent was able to reduce Cr(VI) to Cr(III) and oxidize As(III) to As(V) (Bachate et al. 2013). It reduced 100 mg/L Cr(VI) and oxidized 150 mg/L As(III) within 60 hours in nutrient broth and 10 hours in minimal medium. The hexavalent chromium [Cr(VI)] is the most toxic form of chromium, and arsenite [As(III)] the most toxic form of arsenic, and thus, their respective reduction and oxidation have great environmental importance as they affect their toxicity and mobility (Bachate et al. 2013). Furthermore, two *B. cereus* strains (*B. cereus* D and 332) were recently shown to have high Cr tolerance and reduction ability, and *B. cereus* D achieved 87.8% Cr(VI) removal in 24 h (Li et al. 2020). According to the authors, Cu (II) significantly increased the removal rate of the Cr (VI). Inoculation of *B. cereus* WSE01 was reported to increase the level of leaf enzymatic activity in *Myriophyllum verticillatum*, which, on the other hand, mitigate the effects of oxidative damage caused by reactive oxygen species resulting from heavy metal stress. Li et al. (2019) found that metal-resistant *B. thuringiensis* HC-2 was able to reduce the concentrations of Cd and Pb by extracellular adsorption and increasing the pH of the solution. Treatments of metal contaminated substrates with *B. thuringiensis* HC-2, biochar, and biochar combined with *B. thuringiensis* HC-2 significantly reduced water-soluble Cd and Pb concentrations by 34–56% and 31–54%, respectively and also increased the pH and NH_4^+ concentration in solution, in comparison to the values in a control.

14.4 Physiological and Biochemical Mechanisms of Heavy Metal Tolerance of *Bacillus*

The PGPB may contribute in reducing metal phytotoxicity through biosorption and bioaccumulation (Ma et al. 2016). Metal biosorption by bacteria comprised of two steps, which are passive and active biosorption (Ma et al. 2016). Passive biosorption of metal usually by living and dead or inactive cells take place in the cell wall due to number of metabolism-independent processes. Here, metal ions are adsorbed rapidly to the cell surface by reactions between metal and functional groups on the cell surface. Metal binding mechanisms like ion exchange, complexation, coordination, sorption, chelation, and precipitation may be involved. Active biosorption (bioaccumulation), on the other hand, is referred to the uptake of metal by living cells through a much slower active metabolism-dependent transport of metal into bacterial cells. In heavy metal detoxification, studies carried out using different species, *B. licheniformis* NSPA5, *B. cereus* NSPA8, and *B. subtilis* NSPA13, indicated a significant level of Pb biosorption, with *B. cereus* having the maximum of 87–90% (Syed and Chinthala 2015). The negatively charged functional groups such as hydroxyl groups, phosphate groups, carbonyl groups, etc. that are present in biomolecules of microbial cell wall surfaces bind readily to heavy metal ions (Ojuederie and Babalola 2017). One of the most essential constituent in bacterial cells having ion sequestration capability is exopolysaccharide (EPS) (Ojuederie and Babalola 2017) mainly composed of complex high molecular weight organic macromolecules. Exopolysaccharides are known to protect bacteria against environmental stresses such as heavy metal toxicity.

Other abiotic and biotic stress factors with adverse effects on growth and survival of crop plants are mitigated by *Bacillus*-induced physiological changes, which include the activation of antioxidant and defence systems (Huang et al. 2020), regulation of water transport, nutrient uptake, and enhancement of photosynthetic pigments (Babu et al. 2013), leading to increased crop tolerance and productivity.

14.5 Molecular and Genetic Basis for Plant Growth Promotion and Heavy Metal Tolerance Ability of *Bacillus* Species

Bacillus association triggers plant resistance and immunity against different stresses (including heavy metal stress) by altering stress-responsive genes, proteins, phytohormones, and related metabolites (Radhakrishnan et al. 2017). Studies on genomic analysis of species of *Bacillus* have revealed rich genetic elements involved in important plant growth promoting and heavy metal tolerance activities. For example, the genome of *B. megaterium* STB1 possess genes related to rhizosphere colonization, xenobiotic degradation, pathogen antagonistic activities and several other genes for multiple stress resistance (Nascimento et al. 2020). Genes associated with different plant growth promoting activities are found in many species of *Bacillus* under heavy metal stress. Ding et al. (2005) selected 29 isolates in a study aimed at identifying the possible nitrogen-fixing Bacilli from plant rhizospheres based on their ability to grow on nitrogen-free medium, out of which seven had *nifH* gene belonging to *Bacillus* and *Paenibacillus* genera. It was the first report of nitrogen fixation in *B. marisflavi* and *P. massiliensis* and the first of the *nifH* gene from *B. megaterium* and *B. cereus*. Also, the whole genome of *B. aryabhattai* AB211 was sequenced, with main focus on genomic elements related to plant microbe interaction (Bhattacharyya et al. 2017). Genome comparisons between the strain AB211 and other related strains of *B. aryabhattai* revealed about 3,558 genes which were conserved among all the genomes, with most genes involved in plant growth promotion activities found within core genes of all the genomes used for comparison. The findings showed possible common plant growth promoting traits shared among the strains of *B. aryabhattai*. Functional annotation of genes in the *B. aryabhattai* strain AB211 revealed the presence of many PGP genes which include those responsible for phosphate solubilization, siderophore production, exopolysaccharides production, and IAA production, most of which were experimentally verified in the study. Earlier findings by Halim et al. (2016) also proven that *Paenibacillus durus* ATCC 35681 is a free-living nitrogen-fixing bacterium, and the complete genome of the strain was successfully sequenced too. Interestingly, the isolate can also be found in a symbiotic relationship with plant roots.

Similarly, genes for the resistance, mobilization, detoxification, or transformation were detected in many *Bacillus* species. For example, genomic analysis of *B. megaterium* STB1 revealed abundance of genes for heavy metal resistance and transport, which include those encoding arsenate (*arsC*) and chromate (*chrR*) reductases and several others for Zn, Cd, Cu, Co, Mn, Ni, Cr, and As (Nascimento et al. 2020). One of the arsenate reductase genes in the genome occur in a cluster that contain other arsenate transport and resistance related genes. BLASTn analysis showed that the cluster is rare, and that close homologs were only found in *B. weihaiensis* Alg07 chromosome and *B. oceanisediminis* 2691 pB01. These bring in new insights into the capabilities and roles of *Bacillus* spp. as potential and important plant growth promoting bacteria with ability of not only improving plant growth but also enhancing the tolerance and survivability of the host plants under heavy metal stress conditions.

14.6 Enhancement of Plant Growth and Heavy Metal Tolerance of Plants in Association with Different Species of *Bacillus*

Representatives of plant-beneficial bacteria are widely spread among gram-negative and gram-positive bacteria, with *Pseudomonas* and *Bacillus* attracting main attention (Borriss 2014; Radhakrishnan et al. 2017). Despite the great progress in *Pseudomonas* research, its commercial use in agriculture is unfortunately limited due to difficulties in preparing stable and long-living bioformulations compared to *Bacillus* (Borriss 2014). Bacilli are increasingly and interestingly used commercially in agriculture to enhance yield of crops and reduce use of harmful agrochemicals (Ahmad and Saghir 2010; El-Meihy et al. 2019; Minaxi et al., 2012). When the plant growth promoting attributes of *Bacillus* sp. RM-2 were tested in both laboratory and field conditions, the isolate significantly increased growth ($P<0.05$) and yield parameters and nutrients uptake of cowpea plants (Minaxi et al., 2012). Metal tolerant *B. subtilis* BM2 was able to mediate the phytotoxic impact and significantly reduced Ni and Pb uptake in winter wheat and also improved plant growth (Rizvi et al. 2019). The BM2 strain was able to increase the grain yield significantly by

49% and 50% under 870 mg/kg Ni and 585 mg/kg Pb, respectively. The exceptional ability of the strain to produce plant growth promoting substances like siderophore, IAA, ammonia, and ACC deaminase under the metal stress condition might have played important role in achieving the overall performance and enabling continuous growth and survival of wheat even under the metal stress condition (Rizvi et al. 2019).

The *Bacillus* spp. are as well employed in the field of phytoremediation aimed at improving the uptake of heavy metal by hyperaccumulator plants (Babu et al. 2013; Ndeddy Aka and Babalola 2016; Tang et al. 2020). Recently, *B. cereus* strain WSE01, that can tolerate up to 1500 mg/L Mn, was shown to increase the growth and leaf enzymatic activities in *Myriophyllum verticillatum* under 400 mg/L Mn stress condition. The bacterium was also able to increase the Mn content in the stems and leaves of the inoculated plants by 36.4% and 54.7% , respectively, compared to non-inoculated plants (Tang et al. 2020). In a study to assess bacteria with potential to enhance growth and metal accumulation in hyperaccumulator *Alnus firma* (Babu et al. 2013), the bacterium identified as *Bacillus thuringiensis* GDB-1 had better capacity and increased significantly the biomass, chlorophyll content, and accumulation of As, Pb, Cu, Ni, and Zn in the *A. firma* seedlings. A summary of different species of *Bacillus* with plant growth promoting and heavy metal tolerant activity applied for growth enhancement and bioremediation of heavy metal in plant is presented in Tables 14.1 and 14.2.

In our recent study, one of the most promising and potential heavy metal tolerant plant growth-promoting isolate, identified as CCB-MBL5001 *Bacillus cereus* 2M1, was tested for its ability to alleviate metal stress and improve the growth of rice seedlings under Pb stress. Sterilized rice seeds (treated with or without 2M1 inoculation) were sown in sterile petri plates before uniformly germinated seeds were transferred aseptically to test tubes containing Yoshida medium supplemented with 0, 100 and 150 mg/L Pb. For both inoculated and non-inoculated seedlings, all growth parameters were noted to decline as affected by increasing concentrations of metal tested. However, better improvement in plant growth parameters were observed for seedlings inoculated with the bacteria compared to the control (non-inoculated, supplied with similar Pb concentrations) (Table 14.3). On the effects on photosynthetic pigments, the bacterial inoculated treatments significantly increased chlorophyll a, chlorophyll b, and total chlorophyll contents in the rice seedlings despite the occurrence of high metal stress (Fig. 14.1). Increase in the chlorophyll contents due to bacterial inoculation was not only found under Pb stress but also under non-stressed (control) conditions. Toxicity of heavy metal can affect photosynthesis by causing distortion in the ultrastructure of chloroplast, preventing the synthesis of photosynthetic pigment in chlorophyll content and enzymes of Calvin cycle. To assess the extent of oxidative stress and degree of damage caused by the heavy metal (Pb), the electrolyte leakage was estimated as presented in Fig. 14.2a. Stress resulting from the tested metal was shown to cause oxidative stress in the leaves of the host plants, which increased in response to increasing concentrations of the metal. However, there was significant decrease in the electrolyte leakage in 2M1 inoculated plants, which is a clear indication in the stability of the membrane. The % EL was decreased by 58% under the highest concentration of Pb, and the

Table 14.1 Plant growth enhancement and bioremediation of heavy metal in crop plants as influenced by plant growth promoting *Bacillus* species

Bacillus species	Plant species	Heavy metal	Function/activity	Condition	References
B. thuringiensis HC-2	Radish plant	Cd and Pb	Decreased Cd (28–94%) and Pb (22–63%) content in radish; increased dry weight of roots (18.4–22.8%) and leaves (37.8–39.9%)	Field	Li et al. (2019)
B. cereus MG257494.1	Sorghum	Cu, Cd, Zn and Pb	Decreased metal uptake; increased growth and biomass, antioxidant enzymes activity, and photosynthetic pigments	Pot	El-Meihy et al. (2019)
B. subtilis BM2	Wheat	Ni and Pb	Decreased metal uptake; improved growth parameters; relieved metal stress; decreased in proline, malondialdehyde content, and antioxidant enzymes activities	Pot	Rizvi et al. (2019)
P. mucilaginosus	Alfalfa	Cu	Increased nutrient content in plants; improved plant growth; increased antioxidant enzymes activities; improved growth and Cu uptake	Pot	Ju et al. (2019)
Bacillus sp. PSB10	Chickpea	Cr	Improved growth and biomass; increased nodulation and chlorophyll content; reduced Cr uptake by chickpea plants	Pot	Ahmad and Saghir (2010)

lower percentages of EL in the inoculated rice plants proved metal stress alleviation by the microbe to the plant. Plants are always being exposed to several stress factors in the field which can include heavy metal, that can affect their growth, development, and productivity. The adverse effects generally induce the accumulation of reactive oxygen species (ROS), which can cause severe oxidative damage including electrolyte leakage to plants. The balance between the production and detoxification of ROS is usually sustained by enzymatic and nonenzymatic antioxidants. The catalase (CAT) and superoxide dismutase (SOD) enzymes activities determined in the leaves of rice seedlings were higher in plants inoculated with 2M1 compared to their uninoculated counterparts (Fig. 14.2b and c). It was also observed that the inoculated controls without any heavy metal had the least activities. Increase in CAT and SOD activities by bacterial inoculation indicated strong response towards coping with oxidative stress generated by exposure to the heavy metal.

Table 14.2 Influence of *Bacillus* species on growth enhancement of hyperaccumulator plants and phytoremediation potentials of soils contaminated with heavy metal

Bacillus species	Plant species	Heavy metal	Function/activity	Condition	References
B. cereus HM5 *B. thuringiensis* HM7	*Broussonetia papyrifera*	Mn	Higher growth and biomass; increased accumulation of Mn by plant; reduced malondialdehyde and antioxidant activities in leaves	Pot	Huang et al. (2020)
B. thuringiensis GDB-1	*Alnus firma*	As, Ni, Cu, Pb and Zn	Increased growth and biomass, chlorophyll content, and heavy metal accumulation	Pot	Babu et al. (2013)
B. cereus WSE01	*Myriophyllum verticillatum*	Mn	Improved growth; increased leaf enzymatic activity; increased Mn accumulation in stems (36.4%) and leaves (54.7%)	Hydroponic culture	Tang et al. (2020)
B. subtilis SJ-101	*Brassica juncea*	Ni	Decreased Ni toxicity; enhanced Ni accumulation in B. juncea	Pot	Zaidi et al. (2006)
B. subtilis KP717559	*B. juncea*	Cd, Cr, and Ni	Increased growth and biomass; increased metal accumulation	Pot	Ndeddy Aka and Babalola (2016)
B. cereus T1B3	*Vetiveria zizanoides*	Cr and Fe	Enhanced growth and biomass; enhanced Cr and Fe accumulation	Pot	Nayak et al. (2018)

Table 14.3 Growth parameters of rice seedlings inoculated with locally isolated heavy metal tolerance *B. cereus* 2M1 under Pb stress condition

	Treatments		Plant growth parameter					
	Pb (mg/L)	Inoculation (2M1)	Root length (cm)	Shoot length (cm)	Root fresh weight (mg)	Shoot fresh weight (mg)	Root dry weight (mg)	Shoot dry weight (mg)
1	0	–	7.50 ± 0.29^{a}	25.67 ± 2.90^{a}	50.00 ± 0.00^{a}	103.33 ± 3.33^{a}	3.63 ± 0.58^{ns}	19.10 ± 0.55^{a}
2	0	+	7.83 ± 0.44^{a}	26.50 ± 3.28^{a}	50.00 ± 0.00^{a}	106.67 ± 3.33^{a}	3.80 ± 0.55^{ns}	17.87 ± 1.49^{a}
3	100	–	5.83 ± 0.44^{bc}	15.73 ± 0.37^{bc}	40.00 ± 5.77^{ab}	73.33 ± 3.33^{bc}	3.30 ± 0.44^{ns}	8.33 ± 0.20^{b}
4	100	+	5.33 ± 0.33^{c}	17.03 ± 0.26^{b}	43.33 ± 6.67^{a}	80.00 ± 5.77^{b}	3.50 ± 0.29^{ns}	10.26 ± 0.68^{b}
5	150	–	5.33 ± 0.60^{c}	10.83 ± 0.60^{c}	26.67 ± 3.33^{c}	56.67 ± 6.67^{c}	2.83 ± 0.44^{ns}	7.97 ± 0.55^{b}
6	150	+	7.07 ± 0.58^{ab}	13.83 ± 0.44^{bc}	40.00 ± 5.77^{ab}	73.33 ± 3.33^{bc}	3.57 ± 0.30^{ns}	9.83 ± 0.95^{b}

Values are presented as means ± standard error. Values with different superscripts along the columns are significantly different ($p < 0.05$)

Notes: – = uninoculated; + = inoculated; ns = not significant

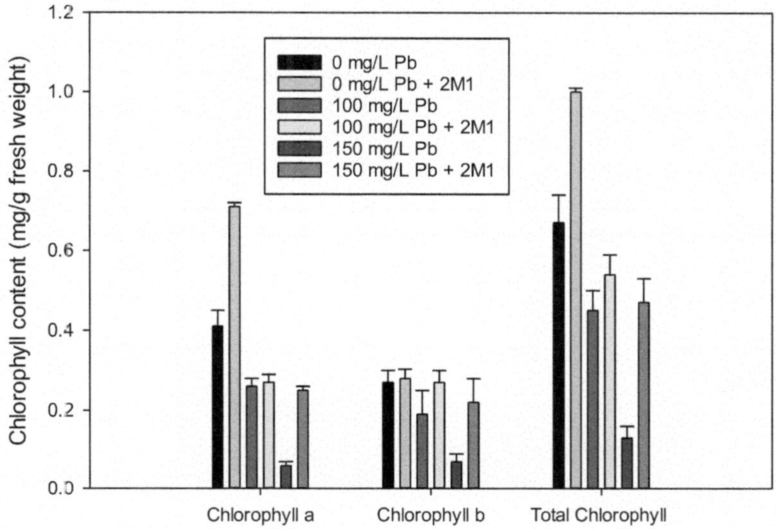

Fig. 14.1 Chlorophyll content in rice seedlings inoculated with locally isolated heavy metal tolerant *B. cereus* 2M1 after 15 days of growth under Pb stress and aseptic conditions

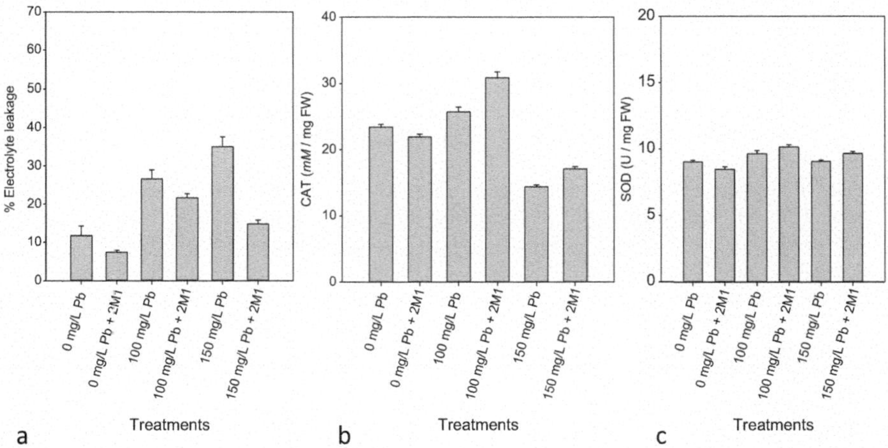

Fig. 14.2 Electrolyte leakage (**a**) CAT activity (**b**) and SOD activity (**c**) in rice seedlings inoculated with locally isolated heavy metal tolerant *B. cereus* 2M1 after 15 days of growth under Pb stress and aseptic conditions

14.7 Conclusions and Future Perspective

Bacilli have multiple beneficial traits, which mediate better growth and development of crop either directly or indirectly through their different plant growth-promoting activities and heavy metal tolerance enhancement. This has great application towards growth enhancement and bioremediation of heavy metal in crop plants. Information made available in this chapter are testament of great potential of

Bacillus spp. in growth enhancement and bioremediation of heavy metal in crop plants. However, most of the experiments were carried out under control conditions (lab or pot experiments), with paucity of information from field trials. Though the *Bacillus* group is one of the most commercially exploited bacteria in the agrobiotechnology industry, its potential has still not been realized sufficiently, and thus, the emphasis should be towards translating the relevant technologies from laboratory to the real world situation in the field (Saxena et al. 2019). More *Bacillus*-plant physiological studies are required to have better understanding of the *Bacillus* spp. mediated mitigation mechanisms against heavy metal stresses in plants.

References

Ahmad P, Saghir M (2010) Bacillus species enhance growth parameters of chickpea (Cicer arietinum L.) in chromium stressed soils. Food Chem Toxicol 48(11):3262–3267. https://doi.org/10.1016/j.fct.2010.08.035

Akinrinlola RJ, Yuen GY, Drijber RA, Adesemoye AO (2018) Evaluation of Bacillus strains for plant growth promotion and predictability of efficacy by in vitro physiological traits. Int J Microbiol

Amir HG, Shamsuddin ZH, Halimi MS, Marziah M, Ramlan MF (2005) Enhancement in nutrient accumulation and growth of oil palm seedlings caused by PGPR under field nursery conditions. Commun Soil Sci Plant Anal 36(15-16):2059–2066

Ashraf MA, Hussain I, Rasheed R, Iqbal M, Riaz M, Arif MS (2017) Advances in microbe-assisted reclamation of heavy metal contaminated soils over the last decade: A review. J Environ Manage 198:132–143. https://doi.org/10.1016/j.jenvman.2017.04.060

Babu AG, Kim JD, Oh BT (2013) Enhancement of heavy metal phytoremediation by Alnus firma with endophytic Bacillus thuringiensis GDB-1. J Hazard Mater 250–251:477–483. https://doi.org/10.1016/j.jhazmat.2013.02.014

Bachate SP, Nandre VS, Ghatpande NS, Kodam KM (2013) Chemosphere Simultaneous reduction of Cr (VI) and oxidation of As (III) by Bacillus firmus TE7 isolated from tannery effluent. Chemosphere 90(8):2273–2278. https://doi.org/10.1016/j.chemosphere.2012.10.081

Bellogín RA, Cerro P, Espuny MR, Pérez-monta F, Ollero FJ, Cubo T (2014) Plant growth promotion in cereal and leguminous agricultural important plants: From microorganism capacities to crop production. Microbiol Res 169:325–336. https://doi.org/10.1016/j.micres.2013.09.011

Bhattacharyya C, Bakshi U, Mallick I, Mukherji S, Bera B, Ghosh A (2017) Genome-guided insights into the plant growth promotion capabilities of the physiologically versatile Bacillus aryabhattai strain AB211. Front Microbiol 8(MAR):1–16. https://doi.org/10.3389/fmicb.2017.00411

Borriss, R. (2014). Use of plant-associated Bacillus strains as biofertilizers and biocontrol agents in agriculture (Issue June). https://doi.org/10.1007/978-3-642-20332-9

Ding Y, Wang J, Liu Y, Chen S (2005) Isolation and identification of nitrogen-fixing Bacilli from plant rhizospheres in Beijing region. J Appl Microbiol:1271–1281. https://doi.org/10.1111/j.1365-2672.2005.02738.x

Dixit R, Wasiullah Malaviya D, Pandiyan K, Singh UB, Sahu A, Shukla R, Singh BP, Rai JP, Sharma PK, Lade H, Paul D (2015) Bioremediation of heavy metal from soil and aquatic environment: An overview of principles and criteria of fundamental processes. Sustainability (Switzerland) 7(2):2189–2212. https://doi.org/10.3390/su7022189

El-Meihy RM, Abou-Aly HE, Youssef AM, Tewfike TA, El-Alkshar EA (2019) Efficiency of heavy metal-tolerant plant growth promoting bacteria for alleviating heavy metal toxicity on sorghum. Environ Exp Bot 162(February):295–301. https://doi.org/10.1016/j.envexpbot.2019.03.005

Fun CP, Rathinam X, Amir HG, Subramaniam S (2016) Characterization of a native Bacillus thuringiensis isolates from Malaysia that produces exosporium-enclosed parasporal inclusion. Emirate J Food & Agric 28(9):653–659

Gardener BBM (2004) Ecology of Bacillus and Paenibacillus spp. in Agricultural Systems. Phytopathology 94(December 2004):1252–1258. https://doi.org/10.1094/PHYTO.2004.94.11.1252

Halim MA, Rahman AY, Sim KS, Yam HC, Rahim AA, Amir HG, Najimudin N (2016) Genome sequence of a Gram-positive diazotroph, Paenibacillus durus type strain ATCC 35681. Genome Announc 4(1):e00005–e00016. https://doi.org/10.1128/genomeA.00005-16

Hashem A, Tabassum B, Fathi E, Allah A (2019) Bacillus subtilis : A plant-growth promoting rhizobacterium that also impacts biotic stress. Saudi J Biol Sci 26(6):1291–1297. https://doi.org/10.1016/j.sjbs.2019.05.004

Huang H, Zhao Y, Fan L, Jin Q, Yang G, Xu Z (2020) Improvement of manganese phytoremediation by Broussonetia papyrifera with two plant growth promoting (PGP) Bacillus species. Chemosphere 260:127614. https://doi.org/10.1016/j.chemosphere.2020.127614

Islam T, Rahman M, Pandey P, Boehme MH, Haesaert G (2019) Bacilli and agrobiotechnology: Phytostimulation and Biocontrol, vol 2. Springer Nature Switzerland AG

Ju W, Liu L, Fang L, Cui Y, Duan C, Wu H (2019) Impact of co-inoculation with plant-growth-promoting rhizobacteria and rhizobium on the biochemical responses of alfalfa-soil system in copper contaminated soil. Ecotoxicol Environ Saf 167(May 2018):218–226. https://doi.org/10.1016/j.ecoenv.2018.10.016

Kapahi M, Sachdeva S (2019) Bioremediation options for heavy metal pollution. J Health & Pollut 9(24). https://doi.org/10.5696/2156-9614-9.24.191203

Khalid S, Shahid M, Bibi I, Shah AH, Niazi NK (2018) A review of environmental contamination and health risk assessment of wastewater use for crop irrigation with a focus on low and high-income countries. Int J Environ Res Public Health 15(895):1–36. https://doi.org/10.3390/ijerph15050895

Kong Z, Glick BR (2017) The role of plant growth-promoting bacteria in metal phytoremediation. Adv Microb Physiol 71(July):97–132. https://doi.org/10.1016/bs.ampbs.2017.04.001

Li Z, Wang P, Yue X, Wang J, Ren B, Qu L (2019) Effects of Bacillus thuringiensis HC-2 combined with biochar on the growth and Cd and Pb accumulation of radish in a heavy metal-contaminated farmland under field conditions. Int J Environ Res Public Health 16

Li M, Gao X, Li C, Yang C, Fu C, Liu J, Wang R (2020) Isolation and identification of chromium reducing Bacillus Cereus species from chromium-contaminated soil for the biological detoxification of chromium. Int J Environ Res Public Health Vi

Ma Y, Rajkumar M, Zhang C, Freitas H (2016) Beneficial role of bacterial endophytes in heavy metal phytoremediation. J Environ Manage 174:14–25. https://doi.org/10.1016/j.jenvman.2016.02.047

Mia AB, Shamsuddin ZH, Wahab Z, Marziah M (2010) Effect of plant growth promoting rhizobacterial (PGPR) inoculation on growth and nitrogen incorporation of tissue-cultured Musa plantlets under nitrogen-free hydroponics condition. Aust J Crop Sci 4(2):85–90

Minaxi NL, Yadav RC, Saxena J (2012) Characterization of multifaceted Bacillus sp. RM-2 for its use as plant growth promoting bioinoculant for crops grown in semi arid deserts. Appl Soil Ecol 59:124–135. https://doi.org/10.1016/j.apsoil.2011.08.001

Mishra J, Singh R, Arora NK (2017) Alleviation of heavy metal stress in plants and remediation of soil by rhizosphere microorganisms. Front Microbiol 8(SEP):1–7. https://doi.org/10.3389/fmicb.2017.01706

Nascimento FX, Hernández AG, Glick BR, Rossi MJ (2020) Plant growth-promoting activities and genomic analysis of the stress-resistant Bacillus megaterium STB1, a bacterium of agricultural and biotechnological interest. Biotechnol Report 25:1–9. https://doi.org/10.1016/j.btre.2019.e00406

Nayak AK Panda SS, Basu A, Dhal NK (2018) Enhancement of toxic Cr (VI), Fe, and other heavy metal phytoremediation by the synergistic combination of native Bacillus cereus strain and Vetiveria zizanioides L. Int J Phytoremediation 20(7):682–691. https://doi.org/10.1080/15226514.2017.1413332

Ndeddy Aka RJ, Babalola OO (2016) Effect of bacterial inoculation of strains of Pseudomonas aeruginosa, Alcaligenes feacalis and Bacillus subtilis on germination, growth and heavy metal (Cd, Cr, and Ni) uptake of Brassica juncea. Int J Phytoremediation 18(2):200–209. https://doi.org/10.1080/15226514.2015.1073671

Ojuederie OB, Babalola OO (2017) Microbial and plant-assisted bioremediation of heavy metal polluted environments: a review. Int J Environ Res Public Health 14(12). https://doi.org/10.3390/ijerph14121504

Oves M, Saghir KM, Huda QA, Nadeen FM, Almeelbi T (2016) Heavy metal: Biological importance and detoxification strategies. J Bioremed Biodegr 7(2). https://doi.org/10.4172/2155-6199.1000334

Radhakrishnan R, Hashem A, Abd Allah EF (2017) Bacillus: A biological tool for crop improvement through bio-molecular changes in adverse environments. Front Physiol 8(SEP):1–14. https://doi.org/10.3389/fphys.2017.00667

Rizvi A, Ahmed B, Zaidi A, Khan MS (2019) Heavy metal mediated phytotoxic impact on winter wheat: oxidative stress and microbial management of toxicity by Bacillus subtilis BM2. RSC Adv:6125–6142. https://doi.org/10.1039/c9ra00333a

Salwani S, Amir HG, Najimudin N (2012) Evidence of diazotrophic symbionts in the leguminous cover crop mucuna bracteata. Pertanika J Tropic Agric Sci 35(3):537–552

Saxena AK, Kumar M, Chakdar H, Anuroopa N, Bagyaraj DJ (2019) Bacillus species in soil as a natural resource for plant health and nutrition. J Appl Microbiol:1583–1594. https://doi.org/10.1111/jam.14506

Syed S, Chinthala P (2015) Heavy metal detoxification by different Bacillus species isolated from solar salterns. Scientifica 2015:1–8. https://doi.org/10.1155/2015/319760

Tang Y, Kang H, Qin Z, Zhang K, Zhong Y, Li H, Mo L (2020) Significance of manganese resistant Bacillus cereus strain WSE01 as a bioinoculant for promotion of plant growth and manganese accumulation in Myriophyllum verticillatum. Sci Total Environ 707:135867. https://doi.org/10.1016/j.scitotenv.2019.135867

Titah HS, Abdullah SRS, Idris M, Anuar N, Basri H, Mukhlisin M, Tangahu BV, Purwanti IF, Kurniawan SB (2018) Arsenic Resistance and Biosorption by Isolated Rhizobacteria from the Roots of Ludwigia octovalvis. Int J Microbiol. https://doi.org/10.1155/2018/3101498

Tiwari S, Lata C (2018) Heavy metal stress, signaling, and tolerance due to plant-associated microbes: an overview. Front Plant Sci 9(April):1–12. https://doi.org/10.3389/fpls.2018.00452

Weissmannová HD, Miho˘cová S, Chovanec P, Pavlovský J (2019) Potential ecological risk and human health risk assessment of heavy metal pollution in industrial affected soils by coal mining and metallurgy in Ostrava, Czech Republic. Int J Environ Res Public Health 16(2019)

Wu J, Kamal N, Hao H, Qian C, Liu Z, Shao Y, Zhong X, Xu B (2019) Endophytic Bacillus megaterium BM18-2 mutated for cadmium accumulation and improving plant growth in Hybrid Pennisetum. Biotechnol Report 24. https://doi.org/10.1016/j.btre.2019.e00374

Yahaghi Z, Shirvani M, Nourbakhsh F, de la Peña TC, Pueyo JJ, Talebi M (2018) Isolation and characterization of Pb-solubilizing bacteria and their effects on pb uptake by Brassica juncea: Implications for microbe-assisted phytoremediation. J Microbiol Biotechnol 28(7):1156–1167. https://doi.org/10.4014/jmb.1712.12038

Zaidi S, Usmani S, Singh BR, Musarrat J (2006) Significance of Bacillus subtilis strain SJ-101 as a bioinoculant for concurrent plant growth promotion and nickel accumulation in Brassica juncea. Chemosphere 64:991–997. https://doi.org/10.1016/j.chemosphere.2005.12.057

Chapter 15
Bacillus Probiotics and Bioremediation: An Aquaculture Perspective

Dibyendu Kamilya and Wangkheimayum Malemnganbi Devi

Abstract Probiotics confer health benefits to the host, and their application to control infectious diseases in aquaculture is well recognized. Probiotic bacteria exert beneficial effects in fish/shellfish when applied through the feed. Besides the application of probiotics as a feed additive, an additional concept of "water probiotic" is also pertinent in aquaculture, owing to the intricate relationship of aquatic organisms with their surrounding environment. In fact, an extended definition of probiotics has been suggested in aquaculture, incorporating the idea of improving the quality of the ambient aquatic environment by probiotics. Among different bacteria tested as probiotics in aquaculture, *Bacillus* constitutes the dominant genera. Besides the use of *Bacillus* as a feed additive to obtain the probiotic effect, administration of this bacterial group in the rearing water has also been investigated to improve water quality. The majority of the studies indicated an effective bioremediation capability of *Bacillus* spp. in terms of improving water quality. However, in a few studies, *Bacillus* spp. did not show any effect on all or some of the water quality parameters. This chapter provides an overview of the concept, application, and beneficial effects of *Bacillus* as a bioremediating probiotic in aquaculture.

Keywords Probiotic bacteria · Feed additive · Water quality · Bioremediation · Aquaculture

15.1 Introduction

Owing to the increasing demand for fish and fishery products by the ever-increasing global population, there has been a remarkable development in the aquaculture industry across the globe. To satisfy the increasing demand, a shift in aquaculture practices, from traditional to intensive systems, has taken place in several parts of

D. Kamilya (✉) · W. M. Devi
Department of Aquatic Health and Environment, College of Fisheries, Central Agricultural University, West Tripura, Tripura, India

M. T. Islam et al. (eds.), *Bacilli in Agrobiotechnology*, Bacilli in Climate Resilient Agriculture and Bioprospecting,
https://doi.org/10.1007/978-3-030-85465-2_15

the world. The intensification and diversification of aquaculture have led to the frequent occurrence of infectious diseases and deterioration of environmental conditions (Bondad-Reantaso et al. 2005). To counter the disease problems, antibiotics and other chemotherapeutants have been widely used in aquaculture. However, the emergence of antibiotic and drug-resistant bacteria has become a serious problem, leading to the restriction on their rampant usage (Akinbowale et al. 2006; Balcázar et al. 2006). The use of probiotic bacteria to control infectious diseases in aquaculture is increasingly being considered as an eco-friendly alternative to drug and antibiotic treatments (Balcázar et al. 2006; Dawood et al. 2019; Doan et al. 2020).

Even though various definitions have been proposed for probiotics, the two most widely quoted definitions are – "a live microbial feed supplement which beneficially affects the host animal by improving its intestinal balance" (Fuller 1989) and "live microorganisms that when administered in adequate amounts confer health benefits on the host" (FAO/WHO 2001). In all the definitions, the concept of probiotic effects revolves around the realization that active modulation of the gastrointestinal tract could confer protection against pathogenic microorganisms, help stimulate the immune system, provide nutritional benefits, and assist the mucosal barrier of the intestine (Vaughan et al. 2002). Essentially, probiotics are considered as health-promoting "functional foods" for humans, as well as therapeutic, prophylactic, and growth supplements in animal production and human health (Kesarcodi-Watson et al. 2008). Like the human and animal production sectors, the concept and application of probiotics as health-promoting "functional foods" in aquaculture has also been extensive and well researched.

Besides the health-promoting attributes, probiotics have an extended application in aquaculture. The intricate relationship between an aquatic organism and its surrounding environment, as compared to that of terrestrial animals, legitimate the extension of the probiotic concept beyond the health-promoting feed supplements. Thus, bacterial treatments to improve water quality by directly applying the probiotics into the rearing water have been included in the broad concept of probiotics in aquaculture. In fact, this extended concept has been proposed to describe probiotics in aquaculture as microbial "water additives" (Moriarty 1998). It is expected that the bacteria which improve water quality may be beneficial to the health of the aquatic animals, and interestingly, several commercial products, referred to as "probiotics," have sought to exploit this idea (Gatesoupe 1999).

A large number of gram-positive and gram-negative bacteria have been investigated as probiotics in aquaculture. Among different probiotic bacteria, *Bacillus* constitutes the dominant genera, and the beneficial effects of their applications in the aquaculture field are well established (Gatesoupe 1999, 2008; Kuebutornye et al. 2019; Soltani et al. 2019). Apart from the health-promoting effects (as dietary additives), several species in the genus *Bacillus* have also been investigated for their bioremediation capability when applied directly into the rearing water. The gram-positive, endospore-forming, and rod-shaped *Bacillus* spp. are almost ubiquitous in nature, and the important physiological traits, such as their sporulation capacity and ability to produce a diverse range of antimicrobial compounds and extracellular enzymes, are responsible for contributing to the beneficial attributes (Soltani et al. 2019).

In view of the above, this chapter aims to present an overview of the published information on the concept, application, and beneficial effects of *Bacillus* as a bioremediator in aquaculture. The research gaps existing in the current state of knowledge about the bioremediation potential of *Bacillus* in aquaculture and suggestions for future studies are also delineated.

15.2 The Concept of "Water Probiotic" and Microbial Community Manipulation

Given the close relationship of aquatic organisms with their external environment, the classical probiotic concept was modified with respect to their aquatic usage. Accordingly, an extended definition was proposed by Verschuere et al. (2000) to define probiotic as "a live microbial adjunct which has a beneficial effect on the host by modifying the host-associated or ambient microbial community, by ensuring improved use of the feed or enhancing its nutritional value, by enhancing the host response towards disease, or by improving the quality of its ambient environment." Unlike the terrestrial animals, probiotic application to the aquatic animals can be accomplished either through feed ("feed probiotic") or by directly applying them into the aquaculture pond or tank ('water probiotic"). Probiotics applied through feed, directly reach the intestine and modify the host-associated microbial community, resulting in health benefits. On the other hand, water probiotics can modify the ambient microbial community by competitive exclusion principle, thereby, creating a better environment from the health perspective. An additional advantage of water probiotics is their ability to degrade organic wastes present in the pond water or sediment. The addition of beneficial microbes in the rearing water improves the water quality, leading to stress minimization and, hence, render the animals less susceptible to infectious diseases.

The beneficial effects of water probiotics in terms of exclusion of pathogenic bacteria and improvement of water quality parameters in the aquaculture system can be attributed to the interaction of the added bacteria with the resident microbial community and the factors facilitating those interactions. The species composition and development of a microbial community are determined by deterministic and stochastic factors. Deterministic factors include various environmental parameters that create a habitat in the aquaculture system in which a well-defined range of microbes can grow and proliferate more rapidly than others (Verschuere et al. 1997; Moriarty 1999). On the other hand, the species composition of a microbial community is also influenced by stochastic phenomena where chance favors those organisms which happen to be in the right place at the right time to respond and proliferate if the conditions are suitable (Moriarty and Body 1995; Verschuere et al. 1997). Thus, the species composition in an aquaculture environment can be manipulated by the addition of probiotic bacteria. If the probiotic cultures are well adapted to the prevailing environmental conditions, a single addition of a probiotic culture may suffice to achieve colonization and persistence in its ambient environment (Verschuere et al. 2000). However, regular addition of probiotic cultures is recommended to maintain a persistent numerical dominance when there is a stable and well-established microbial community in the environment (Verschuere et al. 2000).

Typically, a change in species composition in an aquaculture environment can occur via competitive exclusion principles (Smith 1993). The probiotics can compete with the resident microflora by secreting antimicrobials substances. For example, *B. subtilis* is known to secrete several antimicrobial substances, such as coagulin, amicoumacin, and subtilisin, to suppress the growth of competing microbes (Cutting 2011).

Change in species composition can also occur via the addition of water probiotics that can favor the growth of other groups of beneficial bacteria. The dominance of probiotics and other beneficial bacteria facilitates efficient utilization of the available substrates, leading to improvement in water quality. For example, the addition of *B. subtilis* SC02 at a dose of 1×10^9 CFU m^{-3} per seven days in fish rearing water significantly altered the microbial diversity and improved the water quality, compared to the control group (Zhang et al. 2013). The 454-pyrosequencing analysis revealed that the *B. subtilis* SC02 treated group was dominated by Proteobacteria, Bacteroidetes, and Actinobacteria, all of which are in one or other way beneficial and might have facilitated the improvement of chemical quality of the culture water. A commercial probiotic significantly improved the water quality and increased the population density of *Bacillus* sp., ammonifying bacteria, and protein mineralizing bacteria in shrimp ponds (Wang et al. 2005). In another study, the addition of *B. subtilis* FY99–01 at a dose of 5×10^4 CFU ml^{-1} once every week had a pronounced effect on bacterial community structure and composition of shrimp culture water (Wu et al. 2016). The addition also improved the water quality, promoted beneficial microalgal growth, and inhibited the growth of pathogenic bacteria belong to Vibrionaceae. Thus, the water probiotics can inhibit pathogenic microorganisms and simultaneously improve the water quality, ensuring better health of the resident fish population.

A lot of commercial products and technologies are now available to establish the dominance of a particular or a group of microbial species to treat an aquaculture environment. This practice of bioremediation is being applied in many areas with varying degrees of success. The success of a bacterium or bacterial consortium as bioremediator depends on the selection of bacteria capable of performing specific functions compliant to bioremediation, and the addition of bacteria into the environment at a high concentration, and under favorable environmental conditions (Moriarty 1999).

15.3 Application of *Bacillus* as Bioremediator in Aquaculture

The addition of feed, generation of fecal matters, and inclusion of other organic matters into the aquaculture system lead to the accumulation of organic pollutants in the pond. This, in turn, deteriorates the pond water quality, making the environment unsuitable for the reared animals. Bioremediation of such contaminated rearing environment can be accomplished by the water probiotics. It has been suggested that gram-positive bacteria like *Bacillus* spp. are more efficient in converting organic matter back to CO_2 than gram-negative bacteria (Verschuere et al. 2000). Thus, the buildup of dissolved and particulate organic carbon during a culture cycle can be minimized by maintaining a high level of these bacteria in the culture pond (Balcázar et al. 2006). Additionally, *Bacillus* spp. can participate in nitrification, denitrification, nitrogen fixation, iron precipitation as well as oxidation, and reduction of metals (Priest 1993; Slepecky and Hemphill 2006). These attributes make *Bacillus* spp. a potent bioremediator of the contaminants in the aquaculture environment. Several bacterial species in the genus *Bacillus* have been found to improve the water quality parameters of the aquaculture pond or tank. Modulation of ammonia, nitrite, nitrate, pH, dissolved oxygen, biochemical oxygen demand, alkalinity, chemical oxygen demand, total dissolved solids, etc. has been recorded in rearing water after treatment with the *Bacillus* spp. A summary of the impacts of *Bacillus* spp. on water quality parameters in aquaculture systems is presented in Table 15.1.

15.4 Mechanism of Bioremediation by *Bacillus* spp.

The ability of *Bacillus* spp. to act as heterotrophic nitrifiers, denitrifiers, nitrogen fixers, metal oxidizers and reducers, iron precipitators, and others reflects a considerable diversity in their physiological properties (Abriouel et al. 2011). All these properties exhibited by a great variety of *Bacillus* strains indicate that these bacteria can colonize a wide variety of substrates and can perform a range of functions pertinent to bioremediation. Their widespread ecological diversification is potentiated by the production of endospores, which are resistant to environmental stress and can remain dormant for a considerable period (Abriouel et al. 2011).

It has been suggested that *Bacillus* spp. compete with the resident bacterial community for available organic matter in the aquaculture environment. The outcome of the competitive microbial interactions is greatly impacted not only by the relevant enzyme-producing capacity of the strains but also by the environmental conditions (Soltani et al. 2019). In fact, temperature as well as pH are considered as critical factors in influencing the growth of bacteria and their ability to absorb and utilize substrates (as enzyme activities are highly temperature-dependent) (Verstraete and Focht 1977; Wilks et al. 2009). For example, among nine strains of bacteria of wastewater origin, *B. amyloliquefaciens* showed the highest ammonia nitrogen

Table 15.1 *Bacillus* probiotics used as bioremediator in aquaculture

Species	Reared animal/ media	Origin	Dose	Impacts on water quality	References
Bacillus subtilis, B. megaterium, B. polymyxa, B. licheniformis	*Ictalurus punctarus*	Commercial	2×10^9 CFU mL^{-1}	↓COD, ↑BOD, ↑TAN, ↑ NO_2^--N	Queiroz and Boyd (1998)
B. subtilis, B. megaterium, B. polymyxa, B. licheniformis	*Litopenaeus vannamei*	Commercial	Unknown	↓ammonia, ↓ NO_3^--N, ↓ NO_2^--N, ↓phosphorus, ↓COD, ↓BOD, ↓TSS	McIntosh et al. (2000)
B. subtilis, B. megaterium	*C. citrinellum* × *C. synspilum*	Commercial	10^{10} CFU per trail, twice a week	↓TAN, ↓COD, ↑Transparency	Chen and Chen (2001)
Bacillus sp.	*Penaeus monodon*	Commercial	7×10^{11} cells in 0.8 ha pond	↑transparency, ↓total organic carbon	Dalmin et al. (2001)
Bacillus spp.+ non-*Bacillus* spp.	*P. monodon*	Commercial	10^8 CFU mL^{-1}	↑TAN, ↑ NO_{2-}^{-}-N, ↑ammonia, ↑transparency, ↓hardness, ↔COD, ↔pH, ↔hydrogen sulfide, ↔ DO, ↔ BOD	Matias et al. (2002)
B. licheniformis, B. pumilus, B. subtilis	Shrimp hatchery	Brackish water	Unknown	↓TAN	Foon (2004)
Bacillus spp.+ non-*Bacillus* spp.	*L. vannamei*	Commercial	10^4–10^9 CFU mL^{-1}	↔pH, ↓ phosphate, ↑transparency, ↑DO, ↓total inorganic nitrogen, ↓COD	Wang et al. (2005)
B. subtilis, B. cereus, B. licheniformis	*Cyprinus carpio*	*C. carpio*	Each at 10^5 CFU mL^{-1}	↓ NO_2^--N, ↓NO_3^--N, ↓ammonia	Lalloo et al. (2007)
B. subtilis	*Xiphophorus helleri, X. maculates, Poecilia reticulata and P. sphenops*	*Cirrhinus mrigala*	5×10^5–5×10^8 cells mL^{-1}	↓dissolved organic matter, ↓TAN	Ghosh et al. (2008)
B. pumilus	*P. monodon*	Native	10^6 CFU mL^{-1}, every 3 days	↓TAN, ↓ NO_3^--N, ↔DO, ↔salinity,↔pH	Banerjee et al. (2010)

Bacillus sp.	*M. rosenbergii*	Native prawn, water, sediment	10^6 cells mL^{-1}	↓ammonia, ↓ NO_2^--N, ↓pH, ↔DO	Mujeeb-Rahiman et al. (2010)
B. subtilis, B. licheniformis, B. megaterium, B. laterosporous	*Thunnus albacares*	Commercial	1.5×10^6 CFU mL^{-1}	↓TAN, ↓un-ionized ammonia, ↑DO	Zink et al. (2011)
B. megaterium, B. licheniformis, B. thuringiensis, B. polymyxa, B. subtilis, B. circulans, B. pumilus	*L. vannamei*	Commercial	Each probiotic at 10^9 CFU mL^{-1}	↓pH, ↓ammonia, ↓ NO_2^--N, ↔ NO_3^--N, ↔salinity, ↔DO, ↔phosphate	Nimrat et al. (2012)
B. licheniformis, *B. natto, B. subtilis* + non-*Bacillus* spp.	Crucian carp	Unknown	Unknown	↑DO, ↓COD, ↓ NO_3^--N, ↓ NO_2^--N, ↓sulfide, ↔pH, ↓ammonia	Zhu et al. (2012)
B. subtilis, B. pumilus, B. licheniformis	*Paralichthys olivaceus*	Shrimp intestine	10^{10} CFU g^{-1}	↓ammonia	Cha et al. (2013)
B. amyloliquefaciens	Simulated aquaculture water	Activated sludge of a polluted river	Unknown	↓ NO_2^--N, ↓ammonia	Xie et al. (2013)
B. subtilis	*Ctenopharyngodon idellus*	Grass carp pond	10^9 CFU m^{-3}	↓NH_3, ↓ NO_3^--N, ↓total nitrogen	Zhang et al. (2013)
Bacillus sp. + non- *Bacillus* sp.	*L. vannamei*	Commercial	1.65×10^{10} CFU L^{-1}	↑pyrrophyta density, ↑water sediment quality, ↑water quality	de Paiva-Maia et al. (2013)
B. pumilus, B. licheniformis, B. subtilis	*P. monodon*	Native marine water and soil	10^6 CFU mL^{-1}	↓ NO_3^--N, ↔DO, ↔salinity, ↔pH	Devaraja et al. (2013)
B. subtilis	*L. vannamei*	Fermented pickle	10^5 and 10^8 CFU mL^{-1}	↓ammonia, ↓ NO_2^--N, ↓ NO_3^--N	Zokaeifar et al. (2014)
B. vietnamensis	Shrimp wastewater effluent	Shrimp wastewater	Unknown	↓ TAN, ↓ NO_2^--N	Muthukrishnan et al. (2015)
B. subtilis + non-*Bacillus* spp.	*C. carpio* var. koi	Commercial	Bacterial mixture @ 0.25, 0.50 and 1.0 g m^{-3}	↓phosphate-P, ↓BOD, ↓COD, ↓DO, ↓TAN	Sonia et al. (2015)

(continued)

Table 15.1 (continued)

Species	Reared animal/ media	Origin	Dose	Impacts on water quality	References
Bacillus sp. + non-*Bacillus* sp.	*Clarias gariepinus*	Sampled fish farm water	1 ml inoculum L^{-1} wastewater	↓phosphate, ↓sulphate, ↓NO_3^-, ↓ammonia$_3$, ↓BOD, ↓COD, ↓total suspended solids, ↑DO, ↑pH	Omitoyin et al. (2016)
Paenibacillus polymyxa	*C. carpio*	Lab culture	10^3–10^5 CFU mL^{-1}	↔pH, ↔ NO_2^--N, ↔ammonia	Gupta et al. (2016)
B. licheniformis, *B. laterosporus*, *B. polymyxa*, *B. subtilis*, and *B. circulans*	Common carp wastewater effluent	Commercial product	10^8 CFU L^{-1}	↓TAN, ↑NO_3^- –N, ↓Turbidity	Naderi-Samani et al. (2016)
B. subtilis	*Litopenaeus vannamei*	Shrimp pond soil	5×10^4 CFU mL^{-1}	↓pH, ↓ NO_2^--N, ↑transparency, ↑COD, ↓soluble reactive phosphorus, ↑Chlorophyll a density	Wu et al. (2016)
B. pumilus + non-*Bacillus* sp.	*C. carpio*	Lab culture	62.5×10^8 cells mL^{-1}	↓TAN, ↔pH, ↔DO, ↑ total suspended solids, ↑ total dissolved solids	Dash et al. (2018)
B. subtilis, *B. pumilus*, *B. licheniformis*	*Oreochromis niloticus*	Commercial probiotic	10^{10} CFU g^{-1}	↓ammonia, ↑electrical conductivity, ↑ total dissolved solids, ↑pH	Elsabagh et al. (2018)
B. cereus	*P. monodon*	Native pond water and soil	Unknown	↑DO, ↓pH, ↓NO_2^--N	Barman et al. (2018)
B. velezensis	*I. punctatus*	Soil and channel catfish	10^{9} CFU mL^{-1}	↓eutrophication, ↓TAN, ↓NO_3^- –N, ↓ total nitrogen, ↓total phosphorus	Thurlow et al. (2019)
B. cereus, *B. amyloliquefaciens*, + non-*Bacillus* sp.	*O. mossambicus*	Water, sediment, soil samples from discharge points	10^8 CFU mL^{-1}	↓ammonia, ↓ NO_3^--N ↓ NO_2^--N	John et al. (2020)

'↑' increase; '↓' Decrease; '↔' No change; COD – chemical oxygen demand; TAN – total ammonia nitrogen; BOD- biochemical oxygen demand; TAN- total ammonia nitrogen; NO_3^- –N – nitrate nitrogen; NO_2^--N - nitrate nitrogen; DO – dissolved oxygen degrading activity at 35 °C and pH 7.0, and a gradual reduction in this activity was observed at higher temperatures and pH (Yu et al. 2012). Additionally, bacterial cell surface hydrophobicity is considered to play a crucial role in the adhesion of microorganisms to bioremediation surfaces (Krasowska and Sigler 2014). With the increasing cell surface hydrophobicity, bacteria tend to aggregate and form cell-to-cell adhesion, and degradation of phosphorus, nitrogenous and other wastes occurs via various metabolic pathways (Adav et al. 2008; Liu et al. 2009). In a study, the role of hydrophobicity of two *Bacillus* probiotics was investigated in bioremediation of aquaculture, and the results showed a significantly higher bioremediation ability of the strain with greater cell hydrophobicity (Wang and Han 2007).

The feed and excreta of the reared animals in aquaculture are the major source of nitrogenous wastes. *Bacillus* spp. can effectively utilize these nitrogenous substances, and under aerobic, facultative aerobic, and anaerobic conditions, they can greatly impact nitrogen metabolism. Among different nitrogen species, nitrate or nitrite is used by *Bacillus* spp. as sole sources of nitrogen and also as electron acceptors for anaerobic respiration (Lalloo et al. 2007). *Bacillus* spp. have also been reported to carry out heterotrophic nitrification in which nitrate is produced by the oxidation of inorganic and organic reduced forms of nitrogen (Mevel and Prieur 2000). Heterotrophic nitrification is particularly more pronounced in acid soils or where heterotrophic bacterial biomass and carbon:nitrogen ratios are high (Prosser 2005). Simultaneous heterotrophic nitrification and aerobic denitrification is a recently discovered alternative nitrogen removal process where NH_4-N can be aerobically converted to nitrogenous gas (Khanichaidecha et al. 2019). *Bacillus* spp. isolated from soils and wastewater treatment systems have been found to have the capability of performing heterotrophic nitrification and aerobic denitrification simultaneously (Takenaka et al. 2007). Another interesting mechanism of direct uptake and assimilation of ammonia occurs in *Bacillus* spp. Different species in this bacterial group can employ either glutamate dehydrogenase or glutamine synthetase-glutamate synthase pathway to assimilate ammonia into glutamic acid (Kanamori et al. 1987). Thus, *Bacillus* spp. can effectively employ a wide range of mechanisms to act as a potent bioremediator.

15.5 Conclusions

Except for a few, the majority of the published literature indicates the potential of *Bacillus* spp. as a water probiotic in reducing the environmental impacts in the aquaculture system. Several commercial probiotic products are also available where different *Bacillus* spp. have been used to function as bioremediator to improve the pond water quality. These imply the ability and potential of *Bacillus* spp. in controlling the toxic contaminants of pond rearing water, particularly the nitrogenous wastes. Improvement of water quality parameters is translated into the reduction of stress, which ensures a better environment for growing a healthy population of fish and shellfish. Moreover, the addition of *Bacillus* spp. into the rearing water promote the growth of other groups of beneficial bacteria and reduce the load of pathogenic bacteria. Therefore, the use of *Bacillus* as a water additive not only facilitates bioremediation of the aquaculture environment but improves the health status of the animals. However, it is important to have a thorough knowledge of the mechanisms by which *Bacillus* spp. control water quality and improve the health of the reared animals. A comprehensive understanding of the interaction of *Bacillus* spp. with the natural microflora, the degree and frequency of their addition into the rearing water to achieve dominance and persistence in the ambient environment, the influence of the prevailing environmental conditions on their growth dynamics, and identification of suitable strains capable for functioning optimally are some of the important considerations which warrant further research.

References

Abriouel H, Franz CM, Omar NB, Gálvez A (2011) Diversity and applications of *Bacillus* bacteriocins. FEMS Microbiol Rev 35(1):201–232

Adav SS, Lee DJ, Show KY, Tay JH (2008) Aerobic granular sludge: recent advances. Biotechnol Adv 26(5):411–423

Akinbowale OL, Peng H, Barton MD (2006) Antimicrobial resistance in bacteria isolated from aquaculture sources in Australia. J Appl Microbiol 100(5):1103–1113

Balcázar JL, De Blas I, Ruiz-Zarzuela I, Cunningham D, Vendrell D, Múzquiz JL (2006) The role of probiotics in aquaculture. Vet Microbiol 114(3–4):173–186

Banerjee S, Khatoon H, Shariff M, Yusoff FM (2010) Enhancement of *Penaeus monodon* shrimp postlarvae growth and survival without water exchange using marine *Bacillus pumilus* and periphytic microalgae. Fish Sci 76(3):481–487

Barman P, Bandyopadhyay P, Kati A, Paul T, Mandal AK, Mondal KC, Mohapatra PK (2018) Characterization and strain improvement of aerobic denitrifying EPS producing bacterium *Bacillus cereus* PB88 for shrimp water quality management. Waste Biomass Valori 9(8):1319–1330

Bondad-Reantaso MG, Subasinghe RP, Arthur JR, Ogawa K, Chinabut S, Adlard R, Tan Z, Shariff M (2005) Disease and health management in Asian aquaculture. Vet Parasitol 132(3–4):249–272

Cha JH, Rahimnejad S, Yang SY, Kim KW, Lee KJ (2013) Evaluations of *Bacillus* spp. as dietary additives on growth performance, innate immunity and disease resistance of olive flounder (*Paralichthys olivaceus)* against *Streptococcus iniae* and as water additives. Aquaculture 402:50–57

Chen C-C, Chen S-N (2001) Water quality management with *Bacillus* spp. in the high-density culture of red-parrot fish *Cichlasoma citrinellum* × *C. synspilum*. N Am J Aquacult 63(1):66–73

Cutting SM (2011) *Bacillus* probiotics. Food Microbiol 28(2):214–220

Dalmin G, Kathiresan K, Purushothaman A (2001) Effect of probiotics on bacterial population and health status of shrimp in culture pond ecosystem. Indian J Exp Biol 39:939–942

Dash P, Tandel RS, Bhat RA, Mallik S, Pandey NN, Singh AK, Sarma D (2018) The addition of probiotic bacteria to microbial floc: water quality, growth, non-specific immune response and disease resistance of *Cyprinus carpio* in mid-Himalayan altitude. Aquaculture 495:961–969

Dawood MA, Koshio S, Abdel-Daim MM, Doan HV (2019) Probiotic application for sustainable aquaculture. Rev Aquacult 11(3):907–924

de Paiva-Maia E, Alves-Modesto G, Otavio-Brito L, Vasconcelos-Gesteira TC, Olivera A (2013) Effect of a commercial probiotic on bacterial and phytoplankton concentration in intensive shrimp farming (*Litopenaeus vannamei*) recirculation systems. Lat Am J Aquat Res 41(1):126–137

Devaraja T, Banerjee S, Yusoff F, Shariff M, Khatoon H (2013) A holistic approach for selection of *Bacillus* spp. as a bioremediator for shrimp postlarvae culture. Turk J Biol 37(1):92–100

Doan HV, Hoseinifar SH, Ringo E, Esteban MA, Dadar M, Dawood MAO, Faggio C (2020) Host-associated probiotics: a key factor in sustainable aquaculture. Rev Fish Sci Aquac 28:16–42

Elsabagh M, Mohamed R, Moustafa EM, Hamza A, Farrag F, Decamp O, Dawood MAO, Eltholth M (2018) Assessing the impact of *Bacillus* strains mixture probiotic on water quality, growth performance, blood profile and intestinal morphology of Nile tilapia, *Oreochromis niloticus*. Aquac Nutr 24(6):1613–1622

FAO/WHO (2001) Report of a joint FAO/WHO expert consultation on evaluation of health and nutritional properties of probiotics in food including powder milk with live lactic acid bacteria, Cordoba, Argentina

Foon NG (2004) Effectiveness of *Bacillus* spp. on ammonia reduction and improvement of water quality in shrimp hatchery. PhD Thesis, Serdang (MY), Universiti Putra Malaysia

Fuller R (1989) Probiotics in man and animals. J Appl Bacteriol 66(5):365–378

Gatesoupe FJ (1999) The use of probiotics in aquaculture. Aquaculture 180(1–2):147–165

Gatesoupe FJ (2008) Updating the importance of lactic acid bacteria in fish farming: natural occurrence and probiotic treatments. J Mol Microbiol Biotechnol 14(1–3):107–114

Ghosh S, Sinha A, Sahu C (2008) Bioaugmentation in the growth and water quality of live bearing ornamental fishes. Aquacult Int 16(5):393–403

Gupta A, Gupta P, Dhawan A (2016) *Paenibacillus polymyxa* as a water additive improved immune response of *Cyprinus carpio* and disease resistance against *Aeromonas hydrophila*. Aquacult Rep 4:86–92

John EM, Krishnapriya K, Sankar TV (2020) Treatment of ammonia and nitrite in aquaculture wastewater by an assembled bacterial consortium. Aquaculture 20:735390

Kanamori K, Weiss RL, Roberts JD (1987) Role of glutamate dehydrogenase in ammonia assimilation in nitrogen-fixing *Bacillus macerans*. J Bacteriol 169(10):4692–4695

Kesarcodi-Watson A, Kaspar H, Lategan MJ, Gibson L (2008) Probiotics in aquaculture: the need, principles and mechanisms of action and screening processes. Aquaculture 274(1):1–14

Khanichaidecha W, Nakaruk A, Ratananikom K, Eamrat R, Kazama F (2019) Heterotrophic nitrification and aerobic denitrification using pure-culture bacteria for wastewater treatment. J Water Reuse Desal 9(1):10–17

Krasowska A, Sigler K (2014) How microorganisms use hydrophobicity and what does this mean for human needs? Front Cell Infect Microbiol 4:112

Kuebutornye FK, Lu Y, Abarike ED, Wang Z, Li Y, Sakyi ME (2019) *In vitro* assessment of the probiotic characteristics of three *Bacillus* species from the gut of Nile tilapia, *Oreochromis niloticus*. Prob Antimicrob Prot 17:1–13

Lalloo R, Ramchuran S, Ramduth D, Georgens J, Gardiner N (2007) Isolation and selection of *Bacillus* spp. as potential biological agents for enhancement of water quality in culture of ornamental fish. J Appl Microbiol 103:1471–1479

Liu XW, Sheng GP, Yu HQ (2009) Physicochemical characteristics of microbial granules. Biotechnol Adv 27(6):1061–1070

Matias HB, Yusoff FM, Shariff M, Azhar O (2002) Effects of commercial microbial products on water quality in tropical shrimp culture ponds. Asian Fish Sci 15(3):239–248

McIntosh D, Samocha TM, Jones ER, Lawrence AL, McKee DA, Horowitz S, Horowitz A (2000) The effect of a commercial bacterial supplement on the high-density culturing of *Litopenaeus vannamei* with a low-protein diet in an outdoor tank system and no water exchange. Aquac Eng 21(3):215–227

Mevel G, Prieur D (2000) Heterotrophic nitrification by a thermophilic *Bacillus* species as influenced by different culture conditions. Can J Microbiol 46:465–473

Moriarty DJW (1998) Control of luminous Vibrio species in penaeid aquaculture ponds. Aquaculture 164:351–358

Moriarty DJW (1999) Disease control in shrimp aquaculture with probiotic bacteria. In: Bell CR, Brylinsky M, Johnson-Green P (eds) Proceedings of the 8th international symposium on microbial ecology. Atlantic Canada Society for Microbial Ecology Halifax, Canada, pp 237–243

Moriarty DJ, Body AG (1995) Modifying microbial ecology in ponds: the key to sustainable aquaculture. In: Proceedings of fish Asia'95 conference: 2nd Asian aquaculture and fisheries exhibition and conference. RAI Exhibitions, Singapore, pp 1–10

Mujeeb Rahiman KM, Jesmi Y, Thomas AP, Mohamed Hatha AA (2010) Probiotic effect of *Bacillus* NL110 and Vibrio NE17 on the survival, growth performance and immune response of *Macrobrachium rosenbergii* (de Man). Aquac Res 41(9):120–134

Muthukrishnan S, Sabaratnam V, Tan GY, Chong VC (2015) Identification of indigenous bacteria isolated from shrimp aquaculture wastewater with bioremediation application: total ammoniacal nitrogen (TAN) and nitrite removal. Sains Malays 44(8):1103–1110

Naderi-Samani M, Jafaryan H, Gholipour H, Harsij M, Farhangi M (2016) Effect of different concentration of profitable *Bacillus* on bioremediation of common carp (*Cyprinus carpio*) pond discharge. Iran J Aquat Anim Health 2(2):44–54

Nimrat S, Suksawat S, Boonthai T, Vuthiphandchai V (2012) Potential *Bacillus* probiotics enhance bacterial numbers, water quality and growth during early development of white shrimp (*Litopenaeus vannamei*). Vet Microbiol 159(3–4):443–450

Omitoyin BO, Ajani EK, Okeleye OI, Akpoilih AO, Demola EA (2016) Efficiency of toxic substance removal from aquaculture wastewater by duckweed (*Lemna minor*) and bacteria (*Bacillus* sp.). Afr J Fish Aquat Resour Manage 1(1):31–40

Priest FG (1993) Systematics and ecology of *Bacillus*. In: Sonenshein AL, Hoch JA, Losick R (eds) *Bacillus subtilis* and other gram-positive bacteria, biochemistry, physiology, and molecular genetics. American Society for Microbiology, Washington DC, pp 3–16

Prosser JI (2005) Nitrification. In: Hillel D, Hatfield JL (eds) The encyclopedia of soils in the environment, Vol, vol 3. Elsevier, Oxford, UK, pp 31–39

Queiroz JF, Boyd CE (1998) Effects of a bacterial inoculum in channel catfish ponds. J World Aquacult Soc 29(1):67–73

Slepecky R, Hemphill E (2006) The genus *Bacillus*. In: Dworkin M, Falkow S, Rosenberg E, Schleifer KH, Stackebrandt E (eds) Nonmedical. The Prokaryotes, vol 4. Springer, New York, pp 530–562

Smith VH (1993) Implications of resource-ratio theory for microbial ecology. In: Jones JG (ed) Advances in microbial ecology. Springer, Boston, pp 1–37

Soltani M, Ghosh K, Hoseinifar SH, Kumar V, Lymbery AJ, Roy S, Ringo E (2019) Genus *Bacillus*, promising probiotics in aquaculture: aquatic animal origin, bio-active components, bioremediation and efficacy in fish and shellfish. Rev Fish Sci Aquac 27:331–379

Sonia V, Rajagopalsamy CBT, Ahilan B, Francis T (2015) Influence of bioremediation on the growth and survival of *Cyprinus carpio* var *Koi* using aquaculture waste water. J Ind Pollut Control 31(2):243–248

Takenaka S, Zhou Q, Kuntiya A, Seesuriyachan P, Murakami S, Aoki K (2007) Isolation and characterization of thermotolerant bacterium utilizing ammonium and nitrate ions under aerobic conditions. Biotechnol Lett 29(3):385–390

Thurlow CM, Williams MA, Carrias A, Ran C, Newman M, Tweedie J, Allison E, Jescovitch LN, Wilson AE, Terhune JS, Liles MR (2019) *Bacillus velezensis* AP193 exerts probiotic effects in channel catfish *(Ictalurus punctatus)* and reduces aquaculture pond eutrophication. Aquaculture 503:347–356

Vaughan EE, de Vries MC, Zoetendal EG, Ben-Amor K, Akkermans AD, de Vos WM (2002) The intestinal LABs. In: Siezen RJ, Kok J, Abee T, Schaafsma G (eds) Lactic acid bacteria: genetics, metabolism and applications. Springer, Dordrecht, pp 341–352

Verschuere L, Dhont J, Sorgeloos P, Verstraete W (1997) Monitoring biolog patterns and r/K--strategists in the intensive culture of Artemia juveniles. J Appl Microbiol 83(5):603–612

Verschuere L, Rombaut G, Sorgeloos P, Verstraete W (2000) Probiotic bacteria as biological control agents in aquaculture. Microbiol Mol Biol Rev 64(4):655–671

Verstraete W, Focht DD (1977) Biochemical ecology of nitrification and denitrification. Adv Microb Ecol 1:135–214

Wang YB, Han JZ (2007) The role of probiotic cell wall hydrophobicity in bioremediation of aquaculture. Aquaculture 269(1–4):349–354

Wang YB, Xu ZR, Xia MS (2005) The effectiveness of commercial probiotics in northern white shrimp *Penaeus vannamei* ponds. Fish Sci 71(5):1036–1041

Wilks JC, Kitko RD, Cleeton SH, Lee GE, Ugwu CS, Jones BD, BonDurant SS, Slonczewski JL (2009) Acid and base stress and transcriptomic responses in *Bacillus subtilis*. Evolut Genomic Microbiol 75:981–990

Wu DX, Zhao SM, Peng N, Xu CP, Wang J, Liang YX (2016) Effects of a probiotic (*Bacillus subtilis* FY99-01) on the bacterial community structure and composition of shrimp (*Litopenaeus vannamei*, Boone) culture water assessed by denaturing gradient gel electrophoresis and high-throughput sequencing. Aquac Res 47(3):857–869

Xie F, Zhu T, Zhang F, Zhou K, Zhao Y, Li Z (2013) Using *bacillus amyloliquefaciens* for remediation of aquaculture water. Springerplus 2:119

Yu C-H, Wang Y, Guo T, Shen W-X, Gu M-X (2012) Isolation and identification of ammonia nitrogen degradation strains from industrial waste water. Eng Res 4:790–793

Zhang X, Fu L, Deng B, Liang Q, Zheng J, Sun J, Zhu H, Peng L, Wang Y, Wenying S, Li W (2013) *Bacillus subtilis* SC02 supplementation causes alterations of the microbial diversity in grass carp water. World J Microbiol Biotechnol 29(9):1645–1653

Zhu J, Qiu J, Yang J, Fang W, Liu H (2012) Study on a microbial preparation for improving water quality in aquaculture. Fresenius Environ Bull 21(10):2942–2947

Zink IC, Benetti DD, Douillet PA, Margulies D, Scholey VP (2011) Improvement of water chemistry with *Bacillus* probiotics inclusion during simulated transport of yellowfin tuna yolk sac larvae. N Am J Aquacult 73(1):42–48

Zokaeifar H, Babaei N, Saad CR, Kamarudin MS, Sijam K, Balcázar JL (2014) Administration of *Bacillus subtilis* strains in the rearing water enhances the water quality, growth performance, immune response, and resistance against *Vibrio harveyi* infection in juvenile white shrimp, *Litopenaeus vannamei*. Fish Shellfish Immunol 36(1):68–74

Chapter 16
Enhanced Nutrient Accumulation in Non-leguminous Crop Plants by the Application of Endophytic Bacteria *Bacillus* Species

Gazi Md. Akram Hossain, Amir Hamzah Ghazali, Tofazzal Islam, and M. A. Baset Mia

Abstract Endophytic bacteria exert beneficial effects on various crop plants especially the non-legumes, and the effects are carried out in multidimensional mechanisms unlike the biological N_2 fixation (BNF) process. They create a conducive environment in the apoplastic area of root tissues as well as in shoot for providing benefits to the host plant. Among the endophytic bacteria, recently, *Bacillus* spp. are gaining prominence as a biofertilizer and bioenhancer for crop production. A large number of species and strains of this genera have been isolated and identified from the diversified crop plants such as rice, wheat, maize, alfalfa, banana, black pepper, canola, cucumber, clover, oil palm, and apple. Inoculation of plants with these endophytic Bacilli resulted in various beneficial effects on the colonization including better nutrition, improvement of growth, yield, and quality of crop plants. A significant amount of atmospheric N_2 is fixed and incorporated into a good number of non-legumes like rice, wheat, maize, banana, oil palm etc. that are confirmed by the ^{15}N isotopic dilution technique. The *Bacillus* spp. are found in the apoplastic area, produce phytohormone especially auxin, and excrete their fixed N_2 as ammonium to

---

G. M. A. Hossain
Soils and Nutrition Division, Bangladesh Sugarcrop Research Institute, Ishurdi, Pabna, Bangladesh

A. H. Ghazali
School of Biological Sciences, Universiti Sains Malaysia, Pulau Pinang, Malaysia

T. Islam
Institute of Biotechnology and Genetic Engineering (IBGE), Bangabandhu Sheikh Mujibur Rahman Agricultural University, Gazipur, Bangladesh

M. A. B. Mia (✉)
Department of Crop Botany, Bangabandhu Sheikh Mujibur Rahman Agricultural University, Gazipur, Bangladesh
e-mail: miabaset@bsmrau.edu.bd

M. T. Islam et al. (eds.), *Bacilli in Agrobiotechnology*, Bacilli in Climate Resilient Agriculture and Bioprospecting,
https://doi.org/10.1007/978-3-030-85465-2_16

the host plant cells. Additionally, these endophytes are also able to enhance the accumulation of P, K, and Ca through the stimulation of cell membrane-ATP-ase activity. The *Bacillus* spp. are also capable to solubilize complex rock phosphate to a simpler form of phosphate, i.e. dihydrogen orthophosphate and monohydrogen orthophosphate, and help plant phosphorus nutrition. Besides, they produce siderophores, which are very much effective in iron uptake in diversified crop plants. Therefore, the endophytic Bacilli are considered as microbial agents for enhancing uptake and better utilisation of nutrients in different crop plants under normal and harsh environmental conditions. This chapter updates our understanding of nutrient accumulation in non-legumes crop plants by endophytic *Bacillus* spp.

Keywords ^{15}N isotopic dilution · Phytohormones · Non-legumes · Endophytes · ATP-ase activity; · Cell membrane

16.1 Introduction

Endophytic bacteria thrive and colonize in the internal tissue of the plant showing no external sign of infection or adverse effect on the host plant (Afzal et al. 2019; Holliday 1989; Schulz and Boyle 2006). Generally, they complete their life cycle partially or wholly inside the plant and inhabited in the apoplastic area, i.e. intercellular space or middle lamella of epiblema, cortical tissue of roots, or in the xylem vessel of stem and even in the leaf as well (Xia et al. 2015). They can be regarded either as obligate or facultative endophytes where obligates are not culturable and require a more specific environment for their growth. On the contrary, the facultative endophytes are able to survive in soil, rhizosphere, artificial medium, and inside the plants. The facultative endophytes are widely distributed throughout the plant biota for their advantage of growth potentiality, higher adaptability under adverse conditions. The most common endophytes were isolated from wild or cultivated crops of monocotyledonous and dicotyledonous plants. They may be classified as actinomyces, bacteria, and fungi depending on the microorganism, with bacterial and fungal endophytes being the most studied organisms. The interactions between endophytes and the host plants are complex which involve mutualism and antagonism, and the association might be obligate or facultative. Nearly 300,000 plant species are present on earth, and each individual plant is host to one or more endophytes (Strobel et al. 2004). Only a few of these plants have ever been completely studied in relation to their endophytic biology. Hence, the potentiality to explore this arena is crucial, in order to boost up the biofertilizer development in a sustainable approach. Plant growth-promoting bacterial endophytes (PGPBE) have also been identified which promote plant growth and development in versatile approaches (Mia et al. 2016). Their identification as PGPEBE is attributed to their role in enhancing plant growth using various traits namely production of ammonia, indole-3-acetic acid (IAA), siderophores, N_2 fixation, and accumulation of essential nutrient elements viz. P, K, Ca, Mg, and Fe (Mia et al. 2009). Several rhizospheric or endophytic bacteria belonging to the genera *Azospirillum, Azotobacter, Burkholderia, Bacillus, Enterobacter, Pseudomonas, Rhizobium*, and *Agrobacterium* have been noted as plant growth-promoting (PGP) microorganisms (Vessey 2003). Among them, *Bacillus* is frequently reported as a potential biofertilizer due to its multifunctional PGP traits namely, phosphate solubilisation, IAA production accumulation of nutrients, siderophore (iron chelator) production, and biopesticidal activity (Bahadir et al. 2018; Bjelić et al. 2018; Mohite 2013; Wahyudi et al. 2011, 2020). In addition, biofertilizers containing *Bacillus* strains are considered important because of their spore-forming capacity, allowing their adaptation to extreme abiotic conditions, like extreme temperatures, pH, or pesticide exposure. These species have been shown to have positive effects on soybean seed germination by enhancing the root and shoot length or the number of lateral roots of the seedling, which is related to the production of phytohormone and siderophore, and the capability of these bacteria to solubilize P complex. It has also been reported that *Bacillus* promoted seed germination and growth of tomato, pepper, eggplant, and rice (Mia and Shamsuddin 2013; Beneduzi et al. 2008). Other studies also revealed that *Bacillus* have improved plant growth under drought stress and produced a variety of compounds that can be used for the management of a wide array of plant pathogens. Bacteria viz. *B. megaterium, B. cereus, B. pumilus, B. circulans, B. licheniformis, B. subtilis, B. brevis,* and *B. firmus* have been recognised as N_2 fixers based on their nitrogenase activity (Radhakrishnan et al. 2017). It is the most abundant genus in the rhizosphere, and the PGP activity of some of these strains has been known for many years. There are several metabolites that are released by these strains, which strongly affected the environment by increasing nutrient availability in the plants (Barriuso and Solano 2008).

Currently, one of the major challenges of agriculture is to boost up crop productivity under adverse environmental conditions, especially poor fertile soils having less availability of nutrients. It is interesting to note that microbes that are beneficial to plants are used to enhance crop yield and are alternatives to chemical fertilizers. Additionally, several endophytic *Bacillus* species have emerged as a complementary, efficient, and safe alternative to current crop management practices. It is reported that endophytic *Bacillus* is not host-specific, can colonize a wide array of crop plants, and this gives it a great potential tool for increasing crop productivity in a sustainable manner. This chapter reviews current knowledge on the roles of endophytic *Bacilli* in nutrient accumulation in non-legume crop plants.

16.2 Beneficial Effects of Endophytes on Accumulation of Nutrients in Non-legumes

The major contributions of beneficial and biofertilizer endophytes are to supply nutrients like fixed N_2, enhanced uptake of other essential nutrients like P, K, Zn, and Fe. However, limited information are available on the direct contribution of endophytes to host plant like legume-*Rhizobium* symbiosis process (Hardoim et al. 2008). Nevertheless, endophytes mediate plant growth promotion through direct and indirect mechanisms. Since they start their journey as rhizosphere bacteria, it is assumable that they may retain their attributes inside the plant. Their mechanisms of beneficial effects seem related to rhizosphere bacteria because most endophytes can be cultured and can survive outside of the host in the rhizosphere. A list of endophytic *Bacilli* performs beneficial effects has been presented in Table 16.1.

Table 16.1 Performance of endophytic *Bacillus* spp. inoculation on growth and development of different crop plants

Host plant	Name of endophytic *Bacillus*	Mode of beneficial effect	References
Rice (*Oryza sativa* L.)	*Bacillus aryabhattai* strain E7, *B. aryabhattai* MN1, *B. fortis* strain T9. *B. aryabhattai* strain HS-S05; *B. megaterium* strain KW7-R08; *B. subtilis* strain CB-R05	Increase rice plant growth Synthesis of indoleacetic acid (IAA)	Shen et al. (2019) Ji et al. (2014)
Sweet sorghum (*Sorghum bicolor*)	*B.* spp.	Increase plant growth	Mareque et al. (2015)
Banana (*Musa* spp.)	*B. sphaericus* strain UPMB10	Increase growth and nutrient uptake	Mia et al. (2007)
Beet (*Beta vulgaris* L.)	*B. pumilus*	Increased concentration of carbohydrates	Shi et al. (2010)
Sugar cane (*Saccharum officinarum* L.)	*Bacillus* sp. strain H15	Increase plant growth	Chauhan et al. (2012); Wang et al. (2020)
Cocco (*Theobroma cacao* L.)	*B. subtilis*	Promote plant growth	Leite et al. (2013)
Sunflower (*Helianthus annus* L.)	*B. pumilus*	Improve plant growth	Forchetti et al. (2007)
Maize (*Zea mays* L.)	*B. subtilis, B. lentimorbus*	Nitrogen fixation; IAA synthesis; growth promotion	Wang et al. (2010); Szilagyi-Zecchin et al. (2014)
Soybean (*Glycin max* L.)	*B. amyloliquefaciens, B. japonicum*	Production of siderophores; IAA synthesis; ACC-deaminase; antifungal activity; phytases; N_2 fixation	Sharma et al. (2013) and Hungria et al. (2013)
Wheat (*Triticum aestivum* L.)	*B. subtilis, Bacillus* spp.	IAA synthesis; phosphate solubilisation; growth promotion; increase in grain yield	Wang et al. (2010); Hungria (2011) and Upadhyay et al. (2012) Zhao et al. (2015)

16.2.1 *Mechanism of Beneficial Effects of* Bacillus *Spp. on Plants*

Unlike nutrient accumulation, they can directly benefit plants by providing nitrogen supply via N_2 fixing abilities iron chelators and siderophore production (Long et al. 2008), P complex solubilizing compounds and (Knoth 2014). In addition, they influence on plant growth through the production of phytohormones especially auxin which increased root growth through hair formation, increased lateral roots, volume, and surface area. Also, several S oxidizing endophytes are known which can oxidize elemental S into SO_4^{2+} to be used by plants (Banerjee and Yesmin 2009). Moreover, endophytes are prolific sources of phytochemicals (Nisa 2015) which impede plant hygiene (Chen 2011; Benhamou et al. 1998). They are also excellent sources of biologically active secondary metabolites and contributing to the production of metabolites (Brader 2014; Schulz and Boyle 2002).

16.2.2 *Phytohormone Production in Relation to More Root Growth for Higher Nutrient Uptake*

Plant growth promotion by endophytic *Bacillus* spp., through phytohormone production, is perhaps the well-agreed method which causes morphological and anatomical changes in roots of the plant. The mechanism of phytohormones production by endophytic *Bacillus* in host plants is similar to plant growth-promoting rhizobacteria. But here, the roots get extra benefits for proper utilisation of synthesized auxin, since the endophytes remain in the apoplastic areas. There are several pathways to produce auxin where endophytes mainly follow the tryptophan precursor–mediated pathways. Whatever the pathways followed for auxin synthesis, the products are released in the apoplastic area of root cortical zone transported to endodermis through apoplastic and symplastic pathways via downhill process. The endodermis releases auxin to the pericycle, located just beneath it; the auxin stimulates the pericycle to convert into meristem consequently form the lateral roots primordia and finally, produced the lateral roots as well as the root hair (Péret et al. 2009; Waidmann et al. 2020). The increased root growth facilitated the plants to absorb more nutrients through root interception, which is one of the mechanisms of nutrient uptake (Marschner 1995).

16.2.3 *Atmospheric Nitrogen (N_2) Fixation*

Nitrogen is one of the most important nutrient elements which limits crop yield under deficient conditions. The main source of N is the atmospheric N_2, which is unavailable to crop plants due to strong triple bond ($N \equiv N$), which must be reduced by the fixation process. Biological N_2 fixation is the most significant where a large amount of N_2 is fixed by various types of symbiotic associative free-living and endophytic bacteria in association with plants, both of dicotyledon and monocotyledon. The process is mediated by the activity of enzyme nitrogenase using ATP and the overall reaction is shown as follows:

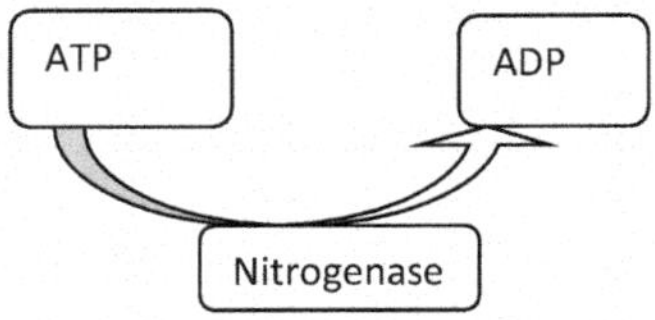

$$N_2 + 8e^- + 8H^+ 16\ ATP \longrightarrow 2NH_3 + H_2 + 16\ ADP + 16Pi$$

Diazotrophic endophytes can fix a handsome amount of N_2, which have been recorded in rice, wheat, maize, sugarcane, oil palm, and bananas, and have been documented by various researchers (Mia et al. 2007, 2010a; Amir et al. 2001; Bashan and Holguin 1997). The mechanism of fixing N_2 by nitrogenase enzyme present in the endophytes are very much similar to *Rhizobium* species symbiosis with legumes, and their genomes possess an *nifHDK* operon, encoding both nitrogenase components: the nitrogenase protein (MoFe protein, *NifDK*), which contains a molybdenum–iron cofactor in the prosthetic group (Carvalho et al. 2014). The nitrogenase enzyme is encoded by ~20 genes which are termed as *nif* genes (N-fixation genes) that are organised in seven operons (*nif* cluster) spanning over 24 kb. The *nifH* gene encodes the Fe protein and *nif D* and *nifK* genes encode the Mo-Fe protein of the nitrogenase enzyme (Rubio and Ludden 2008; Seefedt et al. 2009). The fixed N_2 could be confirmed and measured by several methods like acetylene reduction assay (ARA), ^{15}N isotopic both enrichment and dilution techniques. The ARA technique is a qualitative measure for the activity of enzyme nitrogenase and can be used under certain cases and conditions (Mia and Shamsuddin 2010; Danso 1985). The ^{15}N isotopic technique is used to provide a direct method for detecting BNF (Danso 1995), which comprises enrichment and dilution techniques. The ^{15}N isotope dilution technique was used to quantify the contribution of BNF to clover legume by McAuliffe et al. (1958) which has the capacity to separate out any plant-associated contribution of BNF to plants and is recognised as a more accurate method (Boddey et al. 1996; Roger and Ladha 1992).

The microbial bioassay where isolated endophytes could be cultured specific N-free cultural media and specific conditions and can be quantified by the degree of growth of endophytic bacteria (Das and De 2018). Cell sap analysis is another technique of estimating N_2-fixation where analysis of N solute in xylem exudates and plant parts is based on the determination of the composition of N compounds in plant tissues or N flowing through the xylem sap to the shoot, although the method has some limitation as only a small proportion of known N_2 fixing plants are ureide exporters (Mia and Shamsuddin 2010). A list of N_2-fixing *Bacilli* spp. has been presented in Table 16.2.

Table 16.2 List of N_2-fixing endophytic *Bacillus* spp. beneficial to different non-legumes

Endophytic *Bacillus* sp.	Isolated from crop plant	Colonized into crop	Method used to confirm N_2-fixing ability	References
B. aryabhattai strain HS-S05 *B. megaterium* *B. subtilis* strain CB-R05	Rice (*Oryza sativa* sub sp. *japonica*)	Rice (*Oryza sativa* sub sp. *japonica*)	Amplification of *nifH* genes	Ji et al. (2014)
B. subtilis strain EB-04; *B. pumilus* strain EB-64, EB-169; *Paenibacillus* sp. strain EB-144	Banana cultivar 'Prata Anã' (*Musa acuminata* × *balbisiana*)	–	Amplification of *nifH* genes; acetylene reduction assay (ARA)	Andrade et al. (2013)
B. sp. strain CNPSo 2476, CNPSo 2477, CNPSo 2478;	Corn (*Zea mays* L.)	Corn (*Zea mays* L.)	Amplification of *nifH* genes; ARA	Szilagyi-Zecchin et al. (2014)
B. cereus strain KU097330	Sugarcane (*Saccharum officinarum* L.)	Sugarcane	ARA	Hossain et al. (2020)
B. sphaericus strain UPMB10	Oilpalm roots (*Elaeis guineensis*)	Oil palm, banana, soybean, sweet potato	ARA, ^{15}N isotopic dilution technique. ARA	Amir et al. (2001) Mia et al. (2007, 2010a, 2010b), Saad et al. (1999)

A good number of endophytic *Bacillus* spp. have been identified based on the nitrogenase activity like *B. megaterium, B. cereus, B. pumilus, B. circulans, B. licheniformis, B. subtilis, B. brevis,* and *B. firmus* (Xie et al. 1998). One of the plant-associated members, viz. *B. sphaericus* strain UPMB10, fixes a substantial amount of atmospheric N_2 in association with various non-legumes, including, banana, oil palms, and sweet potato (Mia et al. 2007, 2010a; Mia and Shamsuddin 2010; Mia et al. 2013). The strain UPMB10 has been evaluated for their N_2-fixing capacities in association with oil palm seedlings, banana, rice, and vegetable soybean through ARA and ^{15}N isotopic dilution technique (Mia et al. 2007). Inoculation with *B. spharecus* strain UPMB10 could fix 37–39%N_2 in banana tissue cultured plantlets enumerated by ^{15}N isotopic dilution technique. The highest % Ndfa (nitrogen derived from atmosphere) was recorded at the lowest fertilizer-N applied plant with strain UPMB10. Generally, % Ndfa declined with an increase in fertilizer N application, making a lower contribution of N_2 fixed to the total plant N. The confirmation of N_2 fixation was supported by higher ARA values (129 ηmole plant^{-1} hour^{-1}) in roots of inoculated plants (Mia et al. 2007). In field conditions, rice plants could obtain 20% of their total N requirement and 25–50% of N for oil palm in nursery condition by rhizobacterial inoculation (Shrestha and Ladha 1996; Amir et al. 2001). In a greenhouse study on sugarbeet, three different *Bacillus* isolates fixed N_2 and increased growth (Çakmakçi et al. 2006). Enhanced accumulation of N in sweet sorghum by inoculation of *Bacillus* spp. were recorded by Ribeiro et al. (2018). In another study, Hafeez et al. (2006) noted that selected *Bacillus* sp. used as bioinoculants on wheat resulted in increased root length, plant biomass, and higher accumulation of N and P. Generally, they are widely used in crop production system and are important P-solubilizing inocula, resulting in improved growth and yield of crops (Prakash and Arora 2019).

The production of plant growth regulators such as auxin, cytokinin, and gibberellin by these bacteria may also give additional support to the growth and development of host plant species (Joo et al. 2009; Kang et al. 2009). Recent literature indicates that *Bacillus* sp. have a positive role in plant growth enhancement and biologically active metabolites production (Indiragandhi et al. 2008; Kang et al. 2009). Shen et al. (2019) isolated *B. aryabhattai* E7, *B. aryabhattai* MN1, and *B. fortis* T9 from rice seedling roots and found that these strains have both N_2 fixation potential and IAA production abilities. Sequence analyses of endophytic *Bacilli* in banana cv. 'Prata Anã' roots revealed that PCR amplification of the *nifH* gene was detected in 24 of the 102 bacterial isolates. Seven species of *Bacillus* were *nifH*-positive, including: *B. amyloliquefaciens*, *B. cereus*, *B. flexus*, *B. licheniformis*, *B. pumilus*, *B. subtilis*, and *B. tequilenses* (Suzane et al. 2013). Çakmakçi et al. (2006) observed that three different *Bacilli* isolates fixed N_2 and increased growth in sugar beet in a greenhouse study. A good number of endophytic and associative *Bacillus* spp. have been commercialized as bioinoculants for the growth of crop plants (Islam and Hossain 2013).

16.2.4 Transfer of Fixed N_2 to Host Cell

Whatever the amount of N_2 fixed by endophytes, the release and transportation of fixed N_2 to the host tissues is of the greatest importance. The release or excretion of NH_4^+ and subsequent transportation to the host cell is not clearly identified in the case of associative bacteria since very little amount of N_2 is fixed, which is most probably being utilized by the bacteria itself. Nevertheless, several strains of endophytic bacteria have a unique N regulation system where the NH_4^+ is excreted out of the cell through simple diffusion despite accumulation in its own cell (Brewin et al. 1999; Castorph and Kleiner 1984; Day et al. 2001; Kleiner 1982). The excreted NH_4^+ are transported to cytoplasm via ion channel or plasmodesmata either by downhill or uphill process thereafter being utilized via GS-GOGATT pathway for the synthesis of amino acid and subsequently, other amino acids through transamination process (Mia 2015).

16.2.5 Solubilisation of Soil Insoluble Phosphates

Apart from N_2 fixation, several endophytes release organic acids into the soil which solubilize the phosphate complexes and convert them into ortho-phosphate for plant uptake and utilisation. Although phosphorous exist abundantly in soils, most of it remains unavailable as an insoluble form (Miller 2010). Endophytic *Bacillus* is capable to solubilize the complex unavailable rock phosphate to an available form of P where inoculation of *Bacillus* spp. could solubilize the P-Fe minerals and enhanced uptake of P and Fe in pearl millet (Ribeiro et al. 2018). A good number of *Bacillus* spp. have been isolated from bananas and aerobic rice by Matos et al. (2017) and Panhwar et al. (2009), which are very much effective in the solubilisation of rock phosphate. Rhizobacteria *B. methylotrophicus* CKAM isolated from apple roots showed higher P- solubilisation (Mehta et al. 2014). Similarly, the application of *B. megatorium* in combination with fish bone produced the highest amount of available P for crop plants (Saeid et al. 2018). One of the most important mechanisms is the production of organic acid and release to the soil to solubilize monocalcium, bicalcium, and tricalcium complex phosphate to simpler phosphate as well as increase the activity of phosphatase. The concentration of organic acid is released through root exudates is decreased with increasing the presence of complex phosphate in the rhizosphere.

16.2.6 Enhancement of K Uptake by Plants

Potassium is an essential macronutrient element that contributes to the growth and development of crop plants by influencing the activities of various enzymes system. This element has non-specific functions not directly involved in any reaction system in the cell and do physiological functions in the various arena through osmoregulatory, enzymetic regulation, charge balance in the cytosole (Mia 2015). The greatest amount of K is found in the plant body which is highly mobile in the plant system. Although higher quantities of K present in the soil plant cannot take this element as those are fixed and unavailable for the plants. Recently, the use of potassium solubilizing bacteria (KSB) viz. *Acidothiobacillus ferrooxidans, Paenibacillus* spp., *B. mucilaginosus, B. edaphicus,* and *B. circulans* have shown to solubilize K- bearing minerals, like biotite, feldspar, illite, muscovite, orthoclase, and mica. This type of bacteria can dissolve silicate minerals and release K through the production of organic and inorganic acids, acidolysis, polysaccharides, and complexolysis, chelation, and exchange reactions. The application of *B. sphaericus* strain UPMB10 could not increase the K concentration but its total accumulation was greatly increased due to higher (36%) dry matter production in tissue-cultured banana plantlets growth under a hydroponic condition where dry matter production is directly related to K accumulation. The higher accumulation of K might be due to enhanced root proliferation from the inoculation process (Mia et al. 2010a). Anyway, the KSB have a greater impact on the release of K by various mechanisms. Inoculation of KSB *Bacillus* strain B2084 and B2088 increased the higher accumulation K in a shoot of pearl millet (Ribeiro et al. 2018). The endophytes may also have functions in providing K for a plant by increasing root exudates through supplying different chemicals which can lower the pH by proton efflux, enhancing chelation of the cation bonds to K, and acidolysis of the fixed K area of the soil (Etesami et al. 2017; Meena et al. 2014).

16.2.7 Improvement of Ca Uptake and Absorption by Plants

The endophytic PGPB inoculation greatly increased Ca concentration (nearly 14%) in root especially in plants inoculated along with supplemental 33% fertilizer-N in banana. However, inoculation without fertilizer-N, which could not show any increment, indicated the requirement of starter inorganic N. As Ca is a non-mobile element in the plant system, a higher concentration is found in the root and no influence has been observed in the pesudostem and leaf. Similarly, inoculated plants provided with 33% fertilizer-N also showed lower concentrations when compared to inoculated plants without fertilizer-N. However, in leaf, application of 100% fertilizer-N showed lower Ca concentration (Mia et al. 2009).

The higher Ca uptake by the inoculated plants might not only be due to higher plant growth but also higher uptake capacity, which is induced by bacterial interaction through the acceleration of proton efflux, which results in the acidification of the rhizosphere. It is one of the important mechanisms in cation uptake by the roots, which is the consequence of stimulation of root membrane ATP-ase activity. The endophyte could increase this enzyme activity through bacteria-root interaction as both the strain successfully colonized banana roots. Higher activity of this enzyme resulted in higher Ca concentration in roots (Marschner 1995). Similarly, Bashan et al. (1989) concluded that *Azospirillum* could increase the proton efflux by stimulating ATP-ase activity in the root of wheat seedlings. Increased accumulation of Ca in the cytosole may be the stimulation of the Ca^{2+} ion channel consequently greater influx of Ca^{2+} from the apoplastic area.

16.2.8 Influence of Endophytic Bacilli on Uptake and Absorption of Mg by Plants

Magnesium is an essential macronutrient element which is mobile in the plant system. It plays an important role in the synthesis of chlorophyll molecule as it is the central atom of chlorophyll. Application of endophytic bacteria *B. sphericus* strain UPMB10 could not increase the concentration of Mg rather enhanced the total accumulation which is due to increased root growth not to general uptake rate as in Ca.

16.2.9 *Influence of* Bacillus *Spp. on Siderophores Production and Iron Uptake*

Iron is an important micronutrient element that performs many metabolic activities in the plant through the activation of many enzymes taking part as a prosthetic group. However, it is a problematic element in case of uptake and absorption by the plants because the ferric form (Fe^{3+}) readily becomes insoluble in the soil. Therefore, crop plants suffer deficiency despite having sufficiency in the soil. Recently, the microbial siderophore technique in the uptake of iron is gaining prominence for boosting up crop productivity. It is a smaller molecule organic compound synthesize by microorganisms and is capable of absorbing ferric even under Fe-limited soil conditions thereby solving iron nutrition problems in crop plants. The Fe^{3+} siderophore complex can easily enter into cytosole and release the Fe^{3+} where it reduces as $Fe^{2+}$for further utilisation (Saha et al. 2016). Similarly, endophytes also synthesize siderophore which can absorb and utilize Fe^{3+}as like as rhizosphere bacteria. Endophytes produce small molecular compounds called siderophores, which are iron-chelating compounds that can make available iron to plants and deprive pathogens of iron (Compant et al. 2005). Out of the range of siderophore produced by endophytes, one with biocontrol properties are catecholate, hydroxymate, and/or phenolate types (Rajkumar 2010). Also, siderophores specifically help iron-deficient plants in fixing N_2, since diazotrophs require Fe^{2+} and Mo factors for nitrogenase synthesis and functioning (Kraepiel 2009). The interaction between plants and beneficial bacteria can have a profound effect on plant health, growth, development, production, and soil quality.

Ribeiro et al. (2018) reported the capacity of four endophytic *Bacillus* strains to solubilize iron phosphate (Fe-P), produce siderophores and indole-acetic acid (IAA) in vitro, and evaluate their plant growth promotion ability in greenhouse conditions by inoculation into pearl millet cultivated in a P-deficient soil. All strains solubilized Fe-P, and three of them produced carboxylate-type siderophores and high levels of IAA in the presence of tryptophan.

16.3 Conclusion and Future Perspectives

Endophytic *Bacillus* spp. promote better nutrition and increase the yield of crop plants. These biologicals are highly potential tools for boosting crop productivity in a sustainable agricultural system. They exert beneficial effects on plants through multifaceted mechanisms. Recently, *Bacillus* spp. gaining prominence as potential candidates for endophytic biofertilisation. This chapter reviews the effect of these bacteria on the nutrient accumulation of non-legumes and discusses their mode of action. Our knowledge of molecular crosstalks between endophytic *Bacillus* spp. and host plants are limited (Alvin et al. 2014). Recent progress in genomics and post-genomics analytical methods enhances our understanding about the underlying molecular mechanisms of the beneficial effects of endophytic Bacilli on plants. Exploration of elite strains of endophytic Bacilli and their genes involved in exerting beneficial effects on plants would help us to apply them in sustainable and climate-smart agricultural systems. Furthermore, the application of the recently developed genome editing by the CRISPR-Cas toolkit would be helpful to design industrially powerful biofertilizers and growth promoters from the genus of plant endophytic *Bacillus*.

References

Afzal I, Shinwari ZK, Sikandar S, Shahzad S (2019) Plant beneficial endophytic bacteria: mechanisms, diversity, host range and genetic determinants. Microbiol Res 221:36–49

Alvin A, Miller KI, Neilan BA (2014) Exploring the potential of endophytes from medicinal plants as sources of antimycobacterial compounds. Microbiol Res 169(2014):483–495

Amir HG, Shamsuddin ZH, Halimi MS, Ramlan MF, Marziah M (2001) Effects of *Azospirillum* inoculation on N_2 fixation and growth of oil palm plantlets at nursery stage. J Oil Palm Res 13:42–49

Andrade LF, De Souza GL, Nietsche S, Xavier AA, Costa MR, Cardoso AM, Pereira MC, Pereira DF (2013) Analysis of the abilities of endophytic bacteria associated with banana tree roots to promote plant growth. J Microbiol 52(1):27–34

Bahadir PS, Liaqat F, Eltem R (2018) Plant growth promoting properties of phosphate solubilizing *Bacillus* species isolated from the Aegean region of Turkey. Turk J Bot 42:1–14

Banerjee MR, Yesmin L (2009) Sulfur–oxidizing plant growth promoting rhizobacteria for enhanced canola performance. Google Patents

Barriuso J, Solano BR (2008) Ecology, genetic diversity and screening strategies of plant growth promoting Rhizobacteria (PGPR). J Plant Nutr:1–17

Bashan Y, Holguin G (1997) *Azospireillum*/plant relationship: environmental and physiological advances (1990–1996). Can J Microbiol 43:103–121

Bashan Y, Levanony H, Mitiku G (1989) Changes in proton efflux in intact wheat roots inoculated by *A. brasilense* Cd. Can J Microbiol 35:691–697

Beneduzi A, Peres D, Vargas LK, Bodanese-Zanettini MH, Passaglia LMP (2008) Evaluation of genetic diversity and plant growth promoting activities of nitrogen-fixing *Bacilli* isolated from rice fields in South Brazil. Appl Soil Ecol 39:311–320

Benhamou N, Kloepper JW, Tuzun S (1998) Induction of resistance against fusarium wilt of tomato by combination of chitosan with an endophytic bacterial strain:ultrastructure and cytochemistry of the host response. Planta 204(2):153–168

Bjelić D, Ignjatov M, Marinković J, Milošević D, Nikolić Z, Gvozdanović-Varga J, Karaman M (2018) *Bacillus* isolates as potential biocontrol agents of *Fusarium* clove rot of garlic. Zemdirbyste-Agriculture 105(4):369–376

Boddey RM, Alves BJR, Urquiaga S (1996) Nitrogen cycling and sustainability of improved pastures in the Brazilian cerrados. In: Pereira RC, Nasser LCB (eds) Proceedings of VIII Symp. on cerrados: Biodiv. sustain. prod. food fibres in cerrados, Proc. 1st int. symp. trop. savannas, Brasilia, D. F., Brazil. 24–29 March 1996

Brader G (2014) Metabolic potential of endophytic bacteria. Curr Opin Biotechnol 27:30–37
Brewin B, Woodley P, Drummond M (1999) The basis of ammonium release in nifL mutants of *Azotobacter vinelandii*. J Bacteriol 181:7356–7362
Çakmakçi R, Donmez F, Aydın A, Sahin F (2006) Growth promotion of plants by plant growth promoting rhizobacteria under greenhouse and two different field soil conditions. Soil Biol Biochem 38:1482–1487
Carvalho TLGE, Balsemão-Pires RM, Saraiva PCG, Ferreira ASH (2014) Nitrogen signalling in plant interactions with associative and endophytic diazotrophic bacteria. J Exp Bot 65(19):5631–5642
Castorph H, Kleiner D (1984) Some properties of a *Klebsiella pneumoniae* ammonium transport negative mutant (Amt$^-$). Arch Microbiol 139:245–247
Chauhan H, Bagyaraj DJ, Sharma A (2012) Plant growth-promoting bacterial endophytes from sugarcane and their potential in promoting growth of the host under field conditions. Exp Agric 49(1):43–52
Chen B (2011) A bacterial endophyte from banana:its isolation, identification, activity to Fusarium Wilt and PGPR effect to banana seedlings. Microbiol/Weishengwuxue Tongbao 38(2):199–205
Compant S, Duffy B, Nowak J, Clément C, Barka EA (2005) Use of plant growth-promoting bacteria for biocontrol of plant diseases: principles, mechanisms of action, and future prospects. Appl Environ Microbiol 71(9):4951–4959
Danso SKA (1985) Methods for estimating biological nitrogen fixation. In: Ssali H, Keya SO (eds) Biological nitrogen fixation in Africa. Nairobi MIRCENS, Nairobi, pp 213–244
Danso SKA (1995) Assessment of biological nitrogen fixation. Fertilizer Res 42:33–41
Das S, De TK (2018) Microbial assay of N_2 fixation rate, a simple alternate for acetylene reduction assay. Methods X 5:909–914
Day DA, Poole PS, Tyerman SD, Rosendahl L (2001) Ammonia and amino acid transport across symbiotic membranes in nitrogen-fixing legume nodules. Cell Mol Life Sci 58:61–71
Etesami H, Emami S, Alikhani HA (2017) Potassium solubilizing bacteria (KSB): mechanisms, promotion of plant growth, and future prospects, A review. J Soil Sci Plant Nutr 17(4)
Forchetti G, Masciarell O, Alemano S, Abdala G (2007) Endophytic bacteria in sunflower (*Helianthus annuus* L.): isolation, characterization, and production of jasmonates and abscisic acid in culture medium. Appl Microbiol Biotechnol 76(5):1145–1152
Hafeez FY, Yasmin S, Ariani D, Zafar Y, Malik KA (2006) Plant growth promoting bacteria as biofertilizer. Agron Sustain Dev 26:143–150
Hardoim PR, van Overbeek LS, Elsas JD (2008) Properties of bacterial endophytes and their proposed role in plant growth. Trends Microbiol 16:463–471
Holliday P (1989) A dictionary of plant pathology. Cambridge University Press, Cambridge
Hossain GMA, Solaiman ARM, Karim AJMS, Rahman GKMM, Mia MAB (2020) Influence of Diazotrophic bacteria on growth and biomass production of sugarcane *in vitro*. Int J Curr Microbiol Appl Sci 9(3):3077–3088
Hungria M (2011) Inoculação com *Azospirillum brasilense*: inovação em rendimento a baixo custo. Embrapa Soja Documentos 325
Hungria M, Nogueira MA, Araujo RS (2013) Co-inoculation of soybeans and common beans with rhizobia and azospirilla: strategies to improve sustainability. Bioll Fertil Soils 49(7):791–801
Indiragandhi, P., Anandham, R. Madhaiyan, M. & Sa, T.M. (2008). Characterization of plant growth-promoting traits of bacteria isolated from larval guts of diamondback moth *Plutella xylostella* (Lepidoptera: Plutellidae). Curr Microbiol, 56, 327–333
Islam MT, Hossain MM (2013) Biological control of *Peronosporomycete phytopathogens* by antagonistic bacteria. In: Maheshwari DK (ed) Bacteria in agrobiology: plant disease management. Springer, Berlin, pp 167–218
Ji SH, Gururani MA, Chun SC (2014) Isolation and characterization of plant growth pro- moting endophytic diazotrophic bacteria from Korean rice cultivars. Microbiol Res 169(1):83–98

Joo, G.J., Kang, S.M., Hamayun, M., Kim,S.K., Na,C.I., Shin, D.H. & Lee, I.J. (2009). *Burkhoderia* sp. KCTC 11096BPas a newly isolated gibberellins producing bacterium. J Microbiol, 47, 167–171

Kang, S.M., Joo, G.J., Hamayun,M. Na,C.I.,Shin, D.H., Kim,Y.K., Hong,J.K. & Lee,I.J. (2009). Gibberellin production and phosphate solubilization by newly isolated strain of *Acinetobacter calcoaceticus* and its effect on plant growth. Biotechnol Lett, 31, 277–281

Kleiner D (1982) Amonium (methylamonium) transport by Klebsiella pneumonia. Biochim Biophys Acta 688:702–708

Knoth JL (2014) Biological nitrogen fixation and biomass accumulation within poplar clones as a result of inoculations with diazotrophic endophyte consortia. New Phytol 201(2):599–609

Kraepiel A (2009) Multiple roles of siderophores in free–living nitrogen–fixing bacteria. Biometals 22(4):573–581

Leite HA, Silva AB, Gomes FP, et al. (2013) *Bacillus subtilis* and *Enterobacter cloacae* endophytes from healthy *Theobroma cacao* L. trees can systemically colonize seedlings and promote growth. Appl Microbiol Biotechnol, 97,2639–2651

Long H, Schmidt D, Baldwin I (2008) Native bacterial endophytes promote host growth in a species–specific manner; phytohormone manipulations do not result in common growth responses. PLoS One 3(7):e2702

Mareque C, Taulé C, Beracochea M et al (2015) Isolation, characterization and plant growth promotion effects of putative bacterial endophytes associated with sweet sorghum (*Sorghum bicolor* (L) Moench). Ann Microbiol 65:1057–1067

Marschner, H. (1995). *Mineral nutrition of high plants* (2nd Edn), London aca-demic press, London, 889

Matos ADM, Izabela CP, Nietsche S, Xavier AA, Gomes WS, Neto JDS, Pereira MCT (2017) Phosphate solubilization by endophytic bacteria isolated from banana trees. An Acad Bras Cienc 89(4):2945–2954

McAuliffe C, Chamblee DS, Uribe-Arango H, Woodhouse WW (1958) Influence of inorganic nitrogen on nitrogen fixation by legumes ilS revealed by N-15. Agron J 50:334–347

Meena VS, Maurya BR, Verma JP, Meena RS (2014) Potassium solubilizing microorganisms for sustainable agriculture. Springer

Mehta P, Walia A, Kakkar N, Shirkot CK (2014) Tricalcium phosphate solubilisation by new endophyte Bacillus methylotrophicus CKAM isolated from apple root endosphere and its plant growth-promoting activities. Acta Physiol Plant 36:2033–2045

Mia MAB (2015) Nutrition of crop. Plants. Nova Science Publisher, New York

Mia MAB, Shamsuddin ZH (2010) Nitrogen fixation and transportation by rhizobacteria: A scenario of rice and bananas. Int J Bot 6(3):235–242

Mia MAB, Shamsuddin ZH (2013) Biofertilizer for banana production. Lambert Academic Publisher, Germany

Mia MAB, Shamsuddin ZH, Marziah M (2007) Associative nitrogen fixation by *Azospirillum* and *Bacillus* spp. in bananas. Infomusa 16(1&2):11–15

Mia MAB, Shamsuddin ZH, Wahab Z, Maziah M (2009) The effect of rhizobacterial inoculation on growth and nutrient accumulation of tissue-cultured banana plantlets under low N-fertilizer regime. Afr J Biotechnol 8(21):5855–5866

Mia MAB, Shamsuddin ZH, Zakaria W, Marziah M (2010a) Rhizobacteria as bioenhancer for growth and yield of banana (*Musa* spp. cv. "Berangan"). Sci Hortic 126(2):80–87

Mia MAB, Shamsuddin ZH, Marziah M (2010b) Rhizobacterial inoculation on growth and nitrogen incorporation in tissue-cultured *Musa* plantlets under nitrogen-free hydroponics condition. Austr J Crop Sci 4(2):85–90

Mia MAB, Hossain MM, Shamsuddin ZH, Islam MT (2013) Plant-associated bacteria in nitrogen nutrition in crops, with special reference to rice and banana. In: Maheshwari DK (ed) Bacteria in agrobiology: crop productivity. Springer, Berlin/Heidelberg

Mia MAB, Naher UA, Panhwar QA, Islam MT (2016) Growth promotion of non-legumes by the inoculation of *bacillus* species. In: Islam et al (eds) Bacillus and agrobiotechnology. Springer

Miller SH (2010) Biochemical and genomic comparison of inorganic phosphate solubilization in pseudomonas species. Environ Microbiol Rep 2(3):403–411
Mohite B (2013) Isolation and characterization of indole acetic acid (IAA) producing bacteria from rhizospheric soil and its effect on plant growth. J Soil Sci Plant Nutr 13(3):638–649
Nisa H (2015) Fungal endophytes as prolific source of phytochemicals and other bioactive natural products:A review. Microb Pathog 82:50–59
Panhwar QA, Radziah O, Sariah M, MohdRazi I (2009) Solubilization of phosphate forms by phosphate solubilizing bacteria isolated from aerobic rice. Int J Agric Biol 11:667–673
Péret B, BertDe R, Casimiro I, Benková E, Swarup R· Laplaze L, Beeckman T, Bennett MJ (2009) *Arabidopsis* lateral root development: an emerging story. Trends Plant Sci, 14 (7):399–408
Prakash J, Arora NK (2019) Phosphate-solubilizing Bacillus sp. enhances growth, phosphorus uptake and oil yield of *Mentha arvensis* L. 3 Biotech 9:126
Radhakrishnan R, Hashem A, AbdAllah EFA (2017) *Bacillus*: a biological tool for crop improvement through bio-molecular changes in adverse environments. Front Physiol, 8:667
Rajkumar M (2010) Potential of siderophore–producing bacteria for improving heavy metal phytoextraction. Trends Biotechnol 28(3):142–149
Ribeiro VP, Marriel IE, Evódio I, de Sousa SM, Lana Gomes U, Mattos BB, de Oliveira CA, Gomes EA (2018) Endophytic *Bacillus* strains enhance pearl millet growth and nutrient uptake under low-P. Braz J Microbiol 49:SuI
Roger PA, Ladha JK (1992) Biological N_2 fixation in wetland rice fields: estimation and contribution to nitrogen balance. Plant Soil 141:41–55
Rubio LM, Ludden PW (2008) Biosynthesis of the iron-molybdenum cofactor nitrogenase. Annu Rev Microbiol 62:93–111
Saad MS, Shabuddin ASA, Yunus AG, Shamsuddin ZH (1999) Effects of *Azospirillum* inoculation on sweet potato grown on sandy tin-tailing soil. Commun Soil Sci Plant Anal 30(11&12):1583–1592
Saeid A, Prochownik E, Dobrowolska-Iwanek J (2018) Phosphorus solubilization by *Bacillus* species. Molecules 23(11):2897
Saha M, Sarkar S, Sarkar B, Kumar B, Bhattacharjee S, Tribedi P (2016) Microbial siderophores and their potential applications: a review. Environ Sci Pollut Res 23:3984–3999
Schulz B, Boyle C (2002) Siegfried Draeger, et al. Endophytic fungi: a source of novel biologically active secondary metabolites. Mycol Res 106(9):996–1004
Schulz B, Boyle C (2006) What are endophytes? In: Schulz BJE, Boyle CJC, Sieber TN (eds) Microbial root endophytes, Soil biology, 9. Springer, Berlin/Heidelberg
Seefedt LC, Hoffman BM, Dean DR (2009) Mechanism of Mo-dependent nitrogenase. Annu Rev Biochem 78:701–722
Sharma SB, Sayyed RZ, Trivedi MH, Gobi TA (2013) Phosphate solubilizing microbes: sustainable approach for managing phosphorus deficiency in agricultural soils. Springer Plus 2(587):1–14
Shen FT, Yen JH, Liao CS, Chen WC, Chao YT (2019) Screening of rice endophytic biofertilizers with fungicide tolerance and plant growth-promoting characteristics. Sustainability 11:1133
Shi Y, Lou K, Li C (2010) Growth and photosynthetic efficiency promotion of sugar beet (*Beta vulgaris* L.) by endophytic bacteria. Photosynth Res 105:5–13
Shrestha RK, Ladha JK (1996) Genotypic variation in promotion of rice dinitrogen fixation as determined by nitrogen-15 dilution. Soil Sci Soc Am J 60:1815–1821
Strobel G, Daisy B, Castillo U, Harper J (2004) Natural products from endophytic microorganisms. J Nat Prod 67(2):257–268
Suzane AS, Adelica AX, Márcia RC, Acleide MSC (2013) Endophytic bacterial diversity in banana 'Prata Anã' (Musa spp.) roots. Genet Mol Biol 36(2):252–264
Szilagyi-Zecchin VJ, Ikeda AC, Hungria M, Adamoski D, Kava-Cordeiro V, Glienke C, Galli-Terasawa LV (2014) Identification and characterization of endophytic bacteria from corn (Zea mays L) roots with biotechnological potential in agriculture. AMB Express 4:26
Upadhyay SK, Singh JS, Saxena AK, Singh DP (2012) Impact of PGPR inoculation on growth and antioxidant status of wheat under saline conditions. Plant Biol 4(14):605–611

Vessey JK (2003) Plant growth promoting rhizobacteria as biofertilizers. Plant Soil 255:571–586

Wahyudi AT, Astuti RP, Widyawati A, Meryandini A, Nawangsih AA (2011) Characterization of *Bacillus sp.* strains isolated from rhizosphere of soybean plants for their use as potential plant growth for promoting Rhizobacteria. J Microbiol Antimicrob 3:34–40

Wahyudi AT, Astuti RP, Widyawati A, Meryandini A, Nawangsih AA (2020) Characterization of bacillus sp. strains isolated from rhizosphere of soybean plants for their use as potential plant growth for promoting Rhizobacteria. J Microbiol Antimicrob 3(2):34–40

Waidmann S, Sarkel E, Kleine-Vehn J (2020) Same same, but different: growth responses of primary and lateral roots. J Exp Bot 23 71(8):2397–2411

Wang Y, Li H, Zhao W, He X, Chen J, Geng X, Xiao M (2010) Induction of toluene degradation and growth promotion in corn and wheat by horizontal gene transfer within endophytic bacteria. Soil Biol Biochem 42(7):1051–1057

Wang Z, Yu ZX, Solanki MK, Yang L-T, Xing Y-X, Dong D-F, Li Y-R (2020) Diversity of sugarcane root-associated endophytic *Bacillus* and their activities in enhancing plant growth. J Appl Microbiol 128, 3:814–827

Xia Y, De Bolt S, Dreyer J, Scott D, Williams MA (2015) Characterization of culturable bacterial endophytes and their capacity to promote plant growth from plants grown using organic or conventional practices. Front Plant Sci 6:490

Xie G, Su B, Cui Z (1998) Isolation and identification of N_2-fixing strains of *bacillus* in rice rhizosphere of the Yangtze River valley. Wei Sheng Wu XueBao (Article in Chinese) 38(6):480–483

Zhao L, Xu Y, Lai XH, Shan C, Deng Z, Ji Y (2015) Screening and characterization of endophytic *Bacillus* and *Paenibacillus* strains from medicinal plant Lonicera japonica for use as potential plant growth promoters. Braz J Microbiol 46(4):977–989

Chapter 17
Role of *Bacillus* Species in Alleviating Biotic Stress in Crops

Neha Chausali and Jyoti Saxena

Abstract Feeding the growing world population has become a crucial issue with each passing year. At present, the prime focus of farmers and scientists is on maximizing yield and minimizing the damages to food crops by diseases and harsh environmental conditions. Synthetic pesticides and fertilizers are being used abundantly in agricultural fields to increase productivity but the indiscriminate use of synthetic chemicals has resulted in severe pollution of soil and water. Consequently, practices as the use of biopesticides and biofertilizers have become an eco-friendly alternative for harmful agrochemicals, thus encouraging sustainable agriculture. A group of bacteria characterized as plant growth-promoting rhizobacteria (PGPR) has been known to reinforce plant growth and development and also mitigating abiotic and biotic stresses. Many weeds and phytopathogens such as bacteria, fungi, viruses, and nematodes may induce biotic stress in their plant hosts resulting in reduced biomass, crop quality, and yield. Various species of *Bacillus* are well-known PGPR and are also considered as potential biocontrol agents for many plant diseases. These are used to combat biotic stresses by inducing physiological changes in plants and secreting several metabolites in response. The present chapter focuses on the biotic stress management by *Bacillus* spp. and the various mechanisms involved in it.

Keywords Biopesticides · Biofertilizer · Plant growth-promoting bacteria · Phytopathogens · Biocontrol · Biotic stress

17.1 Introduction

Human beings depend on agriculture to a large extent for their food necessities. India is an agriculture-based economy with 18% of its GDP contributed by the agriculture sector. Also, 70% of its rural households and 58% of the total population

N. Chausali · J. Saxena (✉)
Biochemical Engineering Department, Bipin Tripathi Kumaon Institute of Technology, Almora, India

M. T. Islam et al. (eds.), *Bacilli in Agrobiotechnology*, Bacilli in Climate Resilient Agriculture and Bioprospecting,
https://doi.org/10.1007/978-3-030-85465-2_17

depend predominantly on agriculture for their livelihood (FAO 2020; Tripathi et al. 2020). The world population at present is about 7.7 billion (https://www.worldometers.info/worldpopulation) and is projected to increase by 10 billion in the next 50 years (Etesami and Maheshwari 2018; Glick 2014), hence, requiring 70% more food production (FAO 2009). To achieve this, the expansion of agricultural land and a significant increase in production will be the major target. Inadequate food supply may create an alarming situation worldwide in the future. So the agriculture sector requires much more attention from the research community and the government.

For many years, various synthetic chemicals have been used to enhance food production which has caused serious threats to the environment and human health. Moreover, the environmental stresses including abiotic and biotic stresses have also been a big hurdle and limiting factor for agricultural production (Etesami et al. 2020). Therefore, the use of biological environment-friendly alternatives to agrochemicals came into a trend to overcome problems associated with chemical-based products. Microorganisms play a significant role in enhancing plant growth and mitigating biotic and abiotic stresses posed by harsh environmental conditions and phytopathogens. PGPR is a well-known group of bacteria used extensively in plant growth and health promotion of various crops. Several microorganisms belonging to genera *Acetobacter*, *Azospirillum*, *Azotobacter*, *Bacillus*, *Burkholderia*, *Klebsiella*, *Pseudomonas*, and *Serratia* are the most studied plant growth promoting bacteria and also reported for abiotic and biotic stress mitigation (Jha et al. 2013; Mishra et al. 2017; Verma et al. 2019). *Bacillus* and *Pseudomonas* are dominantly used PGPR for agricultural applications due to their beneficial role in plant growth and development, however *Bacillus* based biofertilizers are more effective than *Pseudomonas* due to their spore-forming nature and more efficient metabolite production capability which increases their commercial applicability (Haas and Défago 2005; Ongena and Jacques 2008). *Bacillus* species possess some unique characteristics that make it a potential candidate for biological control as it replicates rapidly and has large genetic biodiversity. Also, due to spore-forming ability, it can survive in extreme environmental conditions like high or low temperatures, unsuitable pH, and insufficiency of nutrients or water (Albayrak 2019). *Bacillus* is a ubiquitously found genera in nature, some species are free-living, while others are endophytic and can colonize the rhizospheric zone of plant root and internal tissues. It is a gram-positive spore-forming bacterium having immense applications in industry, agriculture, and medicinal fields (Lyngwi and Joshi 2014). Further, *Bacillus* spp. have been identified for their presence in stressed environments and also reported for alleviating biotic and abiotic stress (Yadav et al. 2016; Mishra et al. 2017; Ahmad et al. 2018). It has shown a good response in tolerating abiotic stresses like salinity, water deficit, heavy metal toxicity, flooding, extreme temperatures, and nutrient deficiency (Etesami et al. 2020). On the other hand, biotic stresses like weeds, nematodes, and phytopathogens (bacteria, fungi, and viruses) affect crop quality, biomass, and yield negatively and the species of *Bacillus* have been found very effective against them. Therefore, various species of *Bacillus* have been reported to act as a biocontrol agent for various phytopathogens and pests (García-Fraile et al. 2015; Kang et al. 2015). The phytopathogens can be controlled by the action of several cell wall degrading enzymes produced by

Bacillus such as cellulase, chitosanase, glucanase, protease, and other compounds viz. hydrogen cyanide and lipopeptides (Radhakrishnan et al. 2017). Also, a range of metabolites produced by *Bacillus* spp. including antibiotics, lipopolysaccharides (LPS), salicylic acid (SA), siderophores, and hydrolytic enzymes (Hassan et al. 2010, 2015; Qin et al. 2011) were found to be responsible for suppressing the growth of pathogens and boosting up the plant defense mechanisms (Rais et al. 2017).

As shown in Fig. 17.1, *Bacillus* can control plant diseases through various mechanisms such as competition for nutrients and ecological niche in the rhizosphere, production of inhibitory chemicals and metabolites, and induced systemic resistance (ISR) in plants (Cawoy et al. 2011; Rais et al. 2017). *Bacillus* spp. are also responsible for enhancing plant immunity by altering stress-responsive genes, phytohormones, proteins, and allied metabolites and also induce physiological changes including nutrient uptake, regulation of water transport, etc. (Radhakrishnan et al. 2017). In addition, *Bacillus* species significantly stimulate the production of antioxidant defense enzymes like superoxide dismutase, peroxidase, and other enzymes, which are known to suppress diseases in plants by diminishing the reactive oxygen species (ROS) causing oxidative stress (Liu et al. 2011; Shi et al. 2006; Yasmin et al. 2016). The association of *Bacillus* spp. with plant roots promoted plant growth by the formation of biofilm (Beauregard et al. 2013) and enhanced the availability of nutrients such as phosphate (by P solubilization) for plant uptake (Jha et al. 2012; Minaxi et al. 2012). Various species of *Bacillus* genera can produce an enzyme 1-aminocyclopropane-1-carboxylate (ACC) deaminase, which in turn can alleviate environmental stresses by reducing the ethylene level in the host plant (Minaxi et al. 2012; Misra and Chauhan 2020). This enzyme cleaves ACC (a precondition of ethylene production) to α-ketobutyrate and ammonia and thus reduces ethylene levels in plants (Etesami et al. 2020). In addition to ACC deaminase, the genera also produced indole-3-acetic acid and gibberellic acid that regulated intracellular

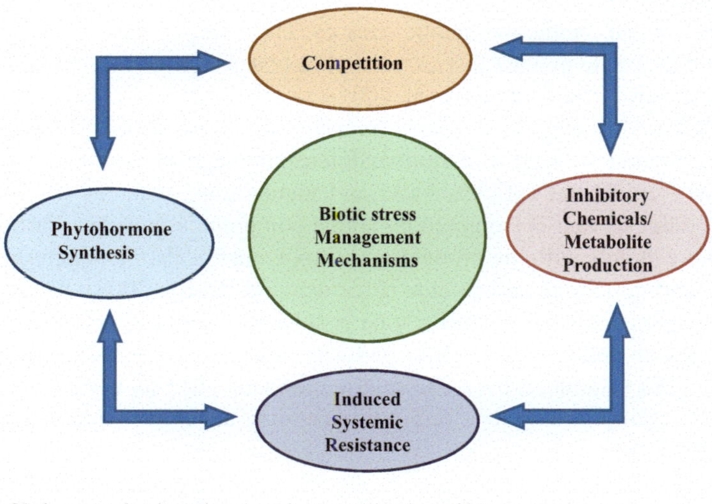

Fig. 17.1 Various mechanisms involved in the mitigation of biotic stresses

phytohormone metabolism, which consequently increased plant stress tolerance considerably (Minaxi et al. 2012; Radhakrishnan et al. 2017).

The global biopesticide market was estimated to grow about 3.0 billion USD in 2018 and is expected to grow about 6.4 billion USD by 2023, at a CAGR of 15.99%. Furthermore, the major driving force for the growth of the biopesticide market is the rise in the organic industry, increase in the cost of synthetic pesticides, growing insect resistance to these chemicals, and awareness towards hazards caused by chemical pesticides to the environment (https://www.marketsandmarkets.com). Among all bacterial biocontrol agents, approximately 70% of the total sale is contributed by *Bacillus thuringiensis* (Cawoy et al. 2011). This bacterium is the source of the Bt gene used in "Bt GMO crops" and about half of the commercial bacterial biocontrol agents belong to this species (Cawoy et al. 2011). There are numerous advantages of using biopesticides over chemical products. Microbial pesticides do not cause pollution as they decompose quickly and are not toxic for nontarget species (Cawoy et al. 2011). Also, they do not have any bad impact on health and the environment.

A limited number of studies are available on physiological changes induced by *Bacillus* species that occur in plants in stressed conditions. The present chapter deals with different biological stresses in crops and the beneficial effects of *Bacillus* species in alleviating biotic stresses through different mechanisms.

17.2 Alleviation of Biotic Stress in Plants by the *Bacillus* Species

Plants may encounter biotic stress due to the presence of weeds, phytopathogens, nematodes, etc. in agricultural fields which affect crop productivity inversely. *Bacillus* species and other PGPR have the capacity to promote the growth of plants as well as mitigate biotic and abiotic stresses. The effect of these PGPR on plant growth and their role in plant disease control has been well demonstrated (Etesami and Maheshwari 2018; Compant et al. 2005).

As illustrated in Fig. 17.2, plants under biotic stress generally employ two defense mechanisms. First, constitutive defense includes performed barriers like walls, waxy epidermal cuticles, bark, and metabolites, whereas the second is inducible defense, that is triggered by signal compounds, invaders, or herbivore attack and responds with the production of toxic chemicals, pathogen-degrading enzymes, and deliberate cell suicide (Freeman and Beattie 2008). Again, inducible mechanism has two categories, one is systemic acquired resistance (SAR), which relies on salicylic acid (SA) pathway, and another is induced systemic resistance (ISR), induced by some microorganisms such as mycorrhizal fungi and PGPR relying on ethylene (ET) and jasmonic acid (JA) signaling pathway (Boubakri 2020).

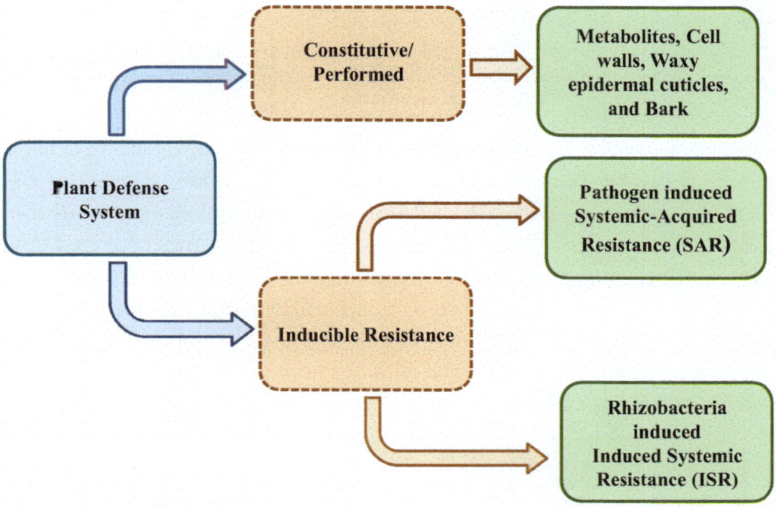

Fig. 17.2 Plant defense mechanism under biotic stress

Furthermore, Fig. 17.3 describes the direct and indirect mechanisms employed by *Bacillus* species to ameliorate biotic stress or plant diseases. Several PGPR including *Bacillus* species adopts one of these two basic mechanisms to combat biotic stress. The compound released in response to stress that stimulated plant growth and ameliorates stress comes under a direct mechanism (Goswami et al. 2016). This encompassed a number of compounds like secondary metabolites including antioxidants (superoxide dismutase and peroxidase) and antibiotics. Various hydrolytic enzymes (cellulose, chitosanase, glucanase, hydrogen cyanide, lipopeptides, and protease), siderophores, hormones (IAA, gibberellic acid, etc.), and other metabolites such as LPS and SA were found to be produced in response to biotic stresses directly by species of *Bacillus* (Hashem et al. 2019). Also, nitrogen fixation, mineralization of organic phosphates, and solubilization of insoluble inorganic phosphates are also part of this mechanism, through which plants get nutrition for their growth and are able to survive in stressed conditions (Etesami and Beattie 2017; Etesami and Maheshwari 2018; Glick 2012; Hayat et al. 2012). Further, induction of systemic resistance and competitive omission support plant growth through an indirect mechanism in stressed conditions (Tripathi et al. 2012).

17.2.1 Molecular Mechanisms Behind Inducible Resistance (SAR and ISR)

The ISR is a systemic resistance developed by some non-pathogenic rhizobacteria that are able to suppress disease in plants (Van Loon et al. 1998). In contrast, SAR is a type of induced resistance that is developed in plants by prior exposure to a

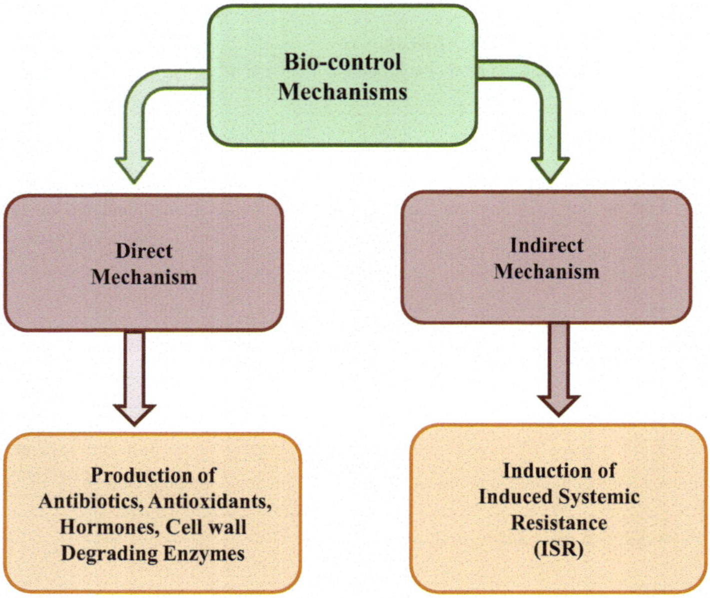

Fig. 17.3 Biocontrol mechanisms of *Bacillus* species

pathogen (Nie et al. 2017). A redox-sensitive transcription factor NIM1/NPR1 (nonexpressor of PR1) that regulates the expression of pathogenesis-related (PR) genes is a key player in both SAR and ISR mechanisms, as illustrated in Fig. 17.4 (Pieterse et al. 1998, 2014; Conrath et al. 2015). NPR1 induces the expression of pathogenesis-related (PR) genes in response to SAR signal molecule, salicylic acid (Hermann et al. 2013). The NPR1 translocates to the nucleus after getting activated by SA and functioning as a coactivator of PR genes providing SAR, whereas during the development of ISR, NPR1 was found to act in the cytosol, though its exact role is unidentified (Asari et al. 2017). Elicitation of ISR by plant-associated bacteria was first demonstrated in *Pseudomonas* spp. and other gram-negative bacteria. Besides *Pseudomonas*, various *Bacillus* spp. specifically *B. amyloliquifaciens*, *B. cereus*, *B. mycoides*, *B. pasteurii*, *B. pumilus*, *B. sphaericus*, and *B. subtilis* are also reported as elicitors of ISR (Kloepper et al. 2004). In most cases, these ISR eliciting species of *Bacillus* genera have also been found to elicit plant growth promotion (Kloepper et al. 2004). Several species of *Bacillus* were found independent of the salicylic acid pathway but dependent on jasmonic acid, ethylene, and the regulatory gene NPR1 in elicitation of ISR. However, some ISR eliciting species of *Bacillus* are independent of jasmonic acid and NPR1 and dependent on salicylic acid (Choudhary et al. 2007; Kloepper et al. 2004). Moreover, in some cases, ISR mediated by the rhizobacterium *Bacillus* species such as *B. cereus* strain AR156 employed both the JA/ET and SA signaling pathways, and NPR1 (Niu et al. 2011). Numerous *Arabidopsis* mutants and reporter lines revealed that the activation of JA-dependent genes VSP2 and PDF1.2 signifying the participation of MYC/ABA

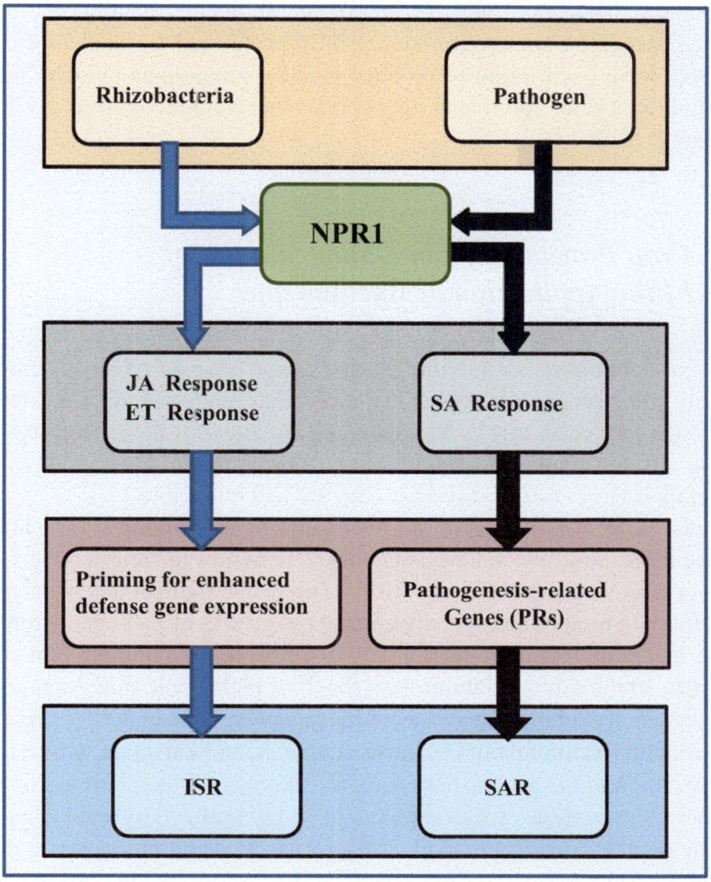

Fig. 17.4 Molecular mechanism of inducible resistance in plants

and ERF/ethylene, respectively (Pieterse et al. 2012). Further, SAR and ISR are well-characterized on the basis of key regulators such as NPR3 and NPR4 or COI1 (SAR) and MYB72 and MYC2 (ISR) with activation of defense genes, such as pathogenesis-related (PR) genes (SAR) or gene encoding plant defensin 1.2 (PDF1.2) and VSP2 (ISR) (Pieterse et al. 2014). However, the molecular mechanism for priming of ISR is not well acknowledged (Asari et al. 2017).

ISR is a sequential process involving three steps: (i) plant cells encounter elicitors produced by the inducing agents, (ii) initiation of signal transduction that propagates the induced state, and (iii) expression of defense mechanisms to inhibit the entry of the pathogen into the host tissues (Van Loon 2007). Salicylic acid and jasmonic acid pathways produce characteristic molecules like pathogenesis-related (PR) proteins (chitinases, β-1, 3-glucanases, proteinase inhibitors, etc.), phytoalexins (antimicrobial compounds), oxidative enzymes (peroxidases, polyphenol oxidases, and lipoxygenases) to diminish ROS and lignin for reinforcement of cell walls (Boubakri 2018; Van Loon 2007). ISR-based biocontrol strategies have been investigated and some trials were successfully performed under field conditions. *Bacillus* spp. have been found to produce volatile compounds (VOCs) such as 2, 3-butanediol (Ryu et al. 2005), and lipopeptides that were recognized as elicitors of ISR (Cawoy et al. 2011).

17.2.2 Crop Protection from Pathogenic Fungi by the Application of **Bacillus** *spp.*

Crops are susceptible to various fungal diseases. They can adversely affect crop productivity and their growth leading to major losses in food production and storage worldwide (Savary et al. 2012). Various trends of *Bacillus* have been reported for controlling a wide range of plant diseases. Different *Bacillus*-based biocontrol agents and their target fungal diseases/fungi are listed in Table 17.1.

Members of the *Bacillus* genus are distinguished as the good source of biologically active molecules, which have antagonistic activities towards a wide variety of phytopathogens (Meena and Kanwar 2015). Direct and indirect mechanisms as discussed before are used to biologically control the growth of pathogenic fungi in the host plant. Under direct mechanism, *Bacillus* spp. produce a number of metabolites and enzymes which directly inhibit the growth of pathogenic microorganisms and are effective against a broad spectrum of fungal species (Stein 2005). Lipopeptides such as surfactin (bacillomycin D), iturin, fengycin, and kurstakin, which are commonly found in *Bacillus* genera, have been well-known for their antimicrobial properties. These lipopeptides are composed of a lipophilic fatty acid chain and a hydrophilic peptide ring (Toure et al. 2004). Surfactins and iturins are amphiphilic cyclic peptides composed of 7 α-amino acids and fengycins by 10 α-amino acids. Moreover, iturins are linked to a single β-amino fatty acid, while surfactins and fengycins linked to a β-hydroxy fatty acid (Dimkić et al. 2017).

On the other hand, lytic enzymes like β-1, 3-glucanase, protease, and chitinase play a key role in controlling the growth of fungi through their cell wall degrading activity. Other than that, the volatile organic compounds (VOC) recognized by their antifungal activity are 2, 3-butanediol, benzene acetic acid, benzaldehyde, 1-decene, phenylethyl alcohol, and tetradecane and have also been studied for their role in biocontrol activity against a variety of fungal pathogens by Ryu et al. (2005) and Dhouib et al. (2019). Studies on the indirect mechanism of biocontrol found in several *Bacilli* reveal that it has a significant role in enhancing and boosting up the plant defense system through inducible resistance, namely SAR and ISR. Characteristic molecules of inducible resistant such as pathogenesis-related (PR) proteins (chitinases, β-1, 3-glucanases, proteinase inhibitors, etc.), phytoalexins (antimicrobial compound), oxidative enzymes (peroxidases, polyphenol oxidases, and lipoxygenases), and VOCs have been studied and well-demonstrated in the findings of García-Gutiérrez et al. (2013), Jangir et al. (2018), Pingping et al. (2017), Myo et al. (2019), Rais et al. (2017), and Waewthongrak et al. (2014).

Table 17.1 *Bacillus* based biocontrol of fungal diseases

Biocontrol agent	Crops	Mechanism of control	Fungal diseases/fungi	References
B. subtilis	Tomato	Direct inhibition (lytic enzymes)	*Fusarium oxysporum*	Chebotar et al. (2009)
		Indirect inhibition (ISR)	*Fusarium semitectum*	Nihorimbere et al. (2010)
		Direct inhibition (antibiosis)	*F. oxysporum f.* sp. *lycopersici*	Abd-Allah et al. (2007) and Baysal et al. (2008)
	Banana	Direct inhibition (antibiosis)	*Pseudocercospora musae*, *Colletotrichum musae*	Fu et al. (2010)
	Cotton	Direct inhibition (hydrolytic enzymes)	*F. oxysporum*	Gajbhiye et al. (2010)
	Lettuce	Competition/direct inhibition (antibiosis)	*Pythium aphanidermatum*	Correa et al. (2010)
	Peach, nectarine	Direct inhibition (antibiosis/enzymes)	*Monilinialaxa*(Brown rot)	Casals et al. (2010)
			Moniliniafructicola	Fan et al. (2000)
	Pea	Direct inhibition (antibiosis/enzymes) and indirect inhibition (ISR)	*Fusarium* spp. (Fusarium wilt)	Khan et al. (2011)
	Chilli	Direct inhibition (mycolyticenzymes)	*Colletotrichum gloeosporioides*OGC1	Ashwini and Srividya (2014)
	Cucumber	Competition, direct inhibition (inhibitory metabolites), and indirect inhibition (ISR)	*Podosphaer axanthii* (powdery mildew), *Didymella bryoniae* (gummy stem blight) *F. oxysporum* f. sp. *radicis-cucumerinum* (Fusarium root), *Pythium* spp. (Pythium crown and root rot)	Ni and Punja (2019)
	Tomato	Direct inhibition (antibiosis and inhibitorymetabolites)	*P. aphanidermatum*	Kipngeno et al. (2015) and Shankar (2016)
		Direct inhibition (antifungal compounds) and indirect inhibition (ISR)	*F. oxysporum* f. sp. *lycopersici* (*Fusarium wilt*)	Abd-Allah et al. (2007), Akram and Anjum (2011) and Shafi et al. (2017)
		Direct inhibition (antibiosis)	*Penicillium* spp. (Blue mold rot)	Punja et al. (2016) and Soleyman et al. (2014)
*B. subtilis*BCB3-19	Tomato	Indirect inhibition (ISR)	*Botrytis cinerea* (Grey mold)	Siripornvisal (2010)

(continued)

Table 17.1 (continued)

Biocontrol agent	Crops	Mechanism of control	Fungal diseases/fungi	References
*B. subtilis*WXCDD105	Tomato	Direct inhibition (antibiosis)	*B. cinerea* (Grey mold)	Wang et al. (2018)
B. subtilis UMAF6639	Melon	Indirect inhibition (inducible resistance)	*P. xanthii* (Powdery mildew)	García-Gutiérrez et al. (2013)
B. subtilis CCTCC M207209	Table grape	Direct inhibition (inhibitory metabolites/ antibiosis)	*Aspergillus carbonarius* CCTCC AF2011004	Jiang et al. (2014)
B. subtilis 30VD-1	Pea	Direct inhibition (lytic enzymes/VOCs/ inhibitory metabolites)	*Fusarium* spp.	Khan et al. (2018)
B. subtilis CICC 10034	Apples	Direct inhibition (cell wall degrading enzymes/antibiosis)	*Penicillium expansum*	Wang et al. (2016b)
B. subtilis ABS-S14	Citrus fruit	Indirect inhibition (ISR)	*Penicillium digitatum*	Waewthongrak et al. (2014)
B. subtilis HZ-72	Flax	Direct inhibition (cell wall degrading enzymes/antibiosis)	*Rhizoctonia solani* (Flax seedling blight)	Tan et al. (2019)
B. subtilis, *B. megaterium*	Peanut	Direct inhibition (enzymatic lysis)	*Aspergillus niger* (Root rot disease)	Yuttavanichakul et al. (2012)
B. amyloliquefaciens 9001	Apple	Direct inhibition (lytic enzymes) and indirect inhibition (ISR)	*Botryosphaeria dothidea* (Apple ring rot)	Li et al. (2013)
B. amyloliquefaciens	Wheat	Direct inhibition (metabolites/ antifungal compounds)	*Fusarium graminearum* (Fusarium head blight)	Crane and Bergstrom (2014)
B. amyloliquefaciens Q-426	Spinach	Direct inhibition (antibiosis)	*F. oxysporum* f. sp. *Spinaciae*	Zhao et al. (2014)
B. Amyloliquefaciens subsp. *Plantarum*	Ginseng	Direct inhibition (antifungal activity/lytic enzymes) or antagonistic	*F.* cf. *incarnatum* (Ginseng root rot)	Song et al. (2014)
B. amyloliquefaciens W19	Banana	Direct inhibition (antibiosis, iturin and bacillomycin D)	*F. oxysporum* f. sp. *Cubense* (FOC)	Wang et al. (2016a)

B. amyloliquefaciens strain BLB369, *B. subtilis*BLB277, *Paenibacilluspolymyxa* BLB267	Durum wheat	Direct inhibition (antibiosis)	*F. graminearum*	Zalila-Kolsi et al. (2016)
B. amyloliquefaciens L-1	Pear	Direct inhibition (antioxidant enzymes)	*Botryosphaeria berengeriana* (Pear ring rot)	Pingping et al. (2017)
B. amyloliquefaciens, *B. megaterium* (B5)	Wheat plants	Direct inhibition (antagonistic)	*Cochliobolus sativus*, *Alternaria alternata*, and *F. graminearum*	El-Gremi et al. (2017)
B. amyloliquefaciens, *B. subtilis*	Pistachio	Direct inhibition (antibiosis)	*Aspergillus parasiticus*	Siahmoshteh et al. (2017)
B. pumilus	Tomato, melon	Direct inhibition (antibiosis/enzymatic lysis)	*Xanthomonas campestris* and *F. oxysporum* f. sp. *melonis*	Suárez-Estrella et al. (2013)
	Tomato	Direct inhibition (antibiosis/enzymatic digestion) and indirect inhibition (ISR)	*F. oxysporum* f. sp. *lycopersici* (Fusarium wilt)	Heidarzadeh and Baghaee-Ravari (2015)
B. cereus AR156	Sweet cherry fruit blue rot	Direct inhibition (cell wall degrading enzymes)	*P. expansum*	Wang et al. (2015)
B. toyonensis, *B. cereus*, *B. aryabhattai*, *B. megaterium*, *B. aerius*, *B. stratosphericus*, *Paenibacillus*, *Barcinonensis*	Tomato	Direct inhibition (antagonistic/enzymatic digestion)	*F. oxysporum*f. sp. *lycopersici* (Fusarium wilt)	Rocha et al. (2017)
B. stratosphericus (FW3)	Ginseng	Direct inhibition (antagonistic metabolites)	*Ilyonectria* sp., *Neurospora* sp., *Cladosporium* sp., *Eutypella* sp., *Aschersonia* sp. and *Fusarium* sp. (Ginseng root rot disease)	Durairaj et al. (2018)

(continued)

Table 17.1 (continued)

Biocontrol agent	Crops	Mechanism of control	Fungal diseases/fungi	References
B. atrophaeus B5	Anthracnose soursop and avocado	Direct inhibition (antibiosis)	*Colletotrichum, Gloeosporioides*	Guardado-Valdivia et al. (2018)
B. velezensis C2	Tomato	Direct inhibition (antibiosis, enzymatic lysis) and ISR (VOCs)	*Verticillium dahlia* (Verticillium wilt disease)	Dhouib et al. (2019)
B. velezensis NKG-2	Tomato	Direct inhibition (enzymatic lysis) and ISR (VOCs)	*F. oxysporum* (Wilt disease)	Myo et al. (2019)
Bacillus spp.	Rice	Indirect inhibition(ISR)	*Pyriculariaoryzae*	Rais et al. (2017)
	Tomato	Direct inhibition (enzymatic lysis, metabolite production) and indirect inhibition by ISR (VOCs)	*F. oxysporum* f. sp. *Lycopersici*	Jangir et al. (2018)
Bacillus sp. P12	Bean	Direct inhibition (metabolite production/lipopeptides)	*Macrophomina phaseolina*	Sabaté et al. (2019)

Pathogenic fungi cause diseases in plants and some of them also produce mycotoxins, which contaminate the food and feed. Mycotoxins are toxic secondary metabolites produced by toxigenic fungi (Albayrak 2019). In the literature survey, *Bacillus* spp. were also found active against a number of mycotoxin producing fungi and destroyed them by antibiosis. *B. subtilis* SQR9 synthesized fengycin and bacillomycin antibiotics which inhibit mycelial growth and conidial germination of *F. oxysporum* f. sp. *couperin* (Cao et al. 2012). Also, *B. subtilis* fmbJ produced bacillomycin D which was active against *Aspergillus flavus* and was liable for injury to cell wall and membrane (Gong et al. 2014). Ayed et al. (2014) reported that antibiotic fengycin, surfactin, and pumilacidin produced by *B. mojavensis* acted against gram (+ve), gram (−ve), and many fungal pathogens. Further, antibiotic bacillomycin D from *B. subtilis* fmbJ caused the distortion of mycelia and disruption of spores, induction of more ROS, and apoptosis of *Aspergillus ochraceus* through cell and DNA damage (Qian et al. 2016). Ochratoxin A (OTA), a mycotoxin mainly produced by species of *Aspergillus* and *Penicillium* was very efficiently removed by *Bacillus megaterium* through adsorption as reported by Shang et al. (2019).

17.2.3 **Bacillus** *spp. in Prevention of Bacterial Diseases*

A number of bacterial diseases that are biologically controlled by the various species of *Bacillus* are listed below in Table 17.2.

As it can be clearly seen from Table 17.2, the *B. subtilis* and *B. amyloliquefaciens* have emerged as the most potential biocontrol agent for bacterial diseases. Different strains of *B. subtilis* produced a good range of hydrolytic enzymes, including i.e., cellulases, beta-glucanases, and proteases. This bacterial species also produced several metabolites and antibiotics that could limit the growth of invading pathogens and microorganisms. It has been reported in the literature that the indirect mechanism like ISR played a significant role in suppressing bacterial diseases in plants. Remarkably, *B. subtilis* strains are well-recognized for synthesizing antibiotic lipopeptides, including fengycin, surfactin, and iturin (Hashem et al. 2019). Surfactants are antimicrobial compounds and can also have an important role behind inhibiting phytopathogens.

17.2.4 **Bacillus** *in Pest/Insect/Nematode Control and* **Bacillus**-*Based Commercial Products*

Some important species of *Bacillus* efficient in controlling pest/insects and nematodes are listed in Table 17.3.

*Bacillus thuringiensis*is, a well-known species of *Bacillus*, used as biopesticide worldwide since biopesticides came into existence. Approximately 95% of

Table 17.2 *Bacillus* species as a biocontrol agent for bacterial diseases

Biocontrol agent	Plant/crop	Mode of action	Target disease/bacteria	References
B. subtilis	Arabidopsis	Antibiosis (lipopeptide surfactin) and biofilm formation	*Pseudomonas syringae* pv. *tomato* DC3000 (root infection)	Bais et al. (2004)
	Mulberry	Biofilm formation and indirect inhibition (ISR)	*Ralstonia solanacearum* (bacterial wilt)	Ji et al. (2008)
	Tomato	Indirect inhibition (ISR and SAR)	*Xanthomonas euvesicatoria* *Xanthomonas perforans* (bacterial spot)	Roberts et al. (2008)
B. subtilis, *B. amyloliquefaciens*	Potato	Competition, fertilization, induction of antagonist (microbial) population	*R. solanacearum* (bacterial wilt)	Chen et al. (2013)
B. subtilis AP-01 (Larminar™), *Trichoderma harzianum* AP-001 (Trisan™)	Tobacco	Direct inhibition (antibiosis, metabolite secretion)	*R. solanacearum* (bacterial wilt)	Maketon et al. (2008)
B. subtilis 9407	Melon	Direct inhibition (antibiosis by surfactin)	*Acidovoraxcitrulli* (bacterial fruit blotch)	Fan et al. (2017)
Endophytic *B. subtilis* SR63	Grapes	Direct inhibition (antibiosis, metabolite secretion)	*Agrobacterium tumefaciens* (crown gall)	Ferrigo et al. (2017)
Endophytic *B. amyloliquefaciens*	Rice	Direct inhibition (metabolite secretion-siderophores, IAA etc. and root colonization)	*Xanthomonas oryzaepv. oryzae* (bacterial leaf blight)	El-shakh et al. (2017)
B. amyloliquefaciens BL10	Tomato	Direct inhibition (antibiosis)	*R. solanacearum* (bacterial wilt)	Nawangsih et al. (2011)
B. subtilis	Chili	Direct inhibition (growth promotion, metabolite secretion)	*R. solanacearum* (bacterial wilt)	Istifadah et al. (2017)
B. pseudomycoides NBRC 101232 *B. thuringiensis* ATCC 10792 *B. mycoides strain* 273		Direct inhibition (antibiosis etc.) and indirect inhibition (ISR)		Yanti et al. (2018)

Bacillus sp.	Eggplant	Direct inhibition (inhibitory compound production) and indirect inhibition (ISR)	*R. solanacearum* (bacterial wilt)	Achari and Ramesh (2014)
B. amyloliquefaciens strain S1	Tomato	Direct inhibition (antibacterial metabolite, siderophores and lytic enzymes production)	*Clavibactermichiganensis* ssp. *michiganensis* (bacterial canker)	Gautam et al. (2019)
B. subtilis, *B. amyloliquefaciens* (FZB 24), EPB 9, EPB10, EPCO 29 and EPCO 78	Rice	Indirect inhibition (ISR)	Bacterial leaf blight, sheath blight	Krishnan et al. (2013)
Endophyte B. velezensis	Citrus species	Direct inhibition (bioactive secondary metabolites production)	*Xanthomonas citri* subsp. *citri* (citrus bacterial canker)	Rabbee et al. (2019)
B. amyloliquefaciens and *Trichoderma asperellum*	Tomato	ISR and growth promotion by enhanced nutrients (P, K, Mg) availability	*Xanthomonas perforans* (bacterial spot)	Chien and Huang (2020)
B. velezensis	Potato	Bacteriostatic activity and antibacterial mechanisms	*Streptomyces* (potatp scab)	Cui et al. (2020)

Table 17.3 *Bacillus* species in controlling pests/nematodes/insects/weeds

Biocontrol agent	Crops	Pest	Mode of action	References
B. subtilis	Soybean	*Heterodera glycines*	Direct inhibition (antibiosis/metabolite production)	Araújo et al. (2002)
	Tomato	*Meloidogyne incognita* (root-knot nematode)	Direct inhibition (inhibitory metabolites)	Araújo and Marchesi (2009) and Siddiqui and Futai (2009)
	Pulses	*M. incognita* (root-krot nematode)	Direct inhibition (antibiosis) and indirect inhibition (ISR)	Khan et al. (2011)
Bacillus strains (EPCO 102 and EPCO 16)	Cotton	Cotton bollworm	Indirect inhibition (ISR)	Rajendran et al. (2007)
B. Thuringiensis	Potato	Coleopteran insects, boll weevil, Colorado potato beetle	Bt toxin	Herrnstadt et al. (1986)
	Soybean	Caterpillars, stink bugs	Membrane pore formation and cell lysis	Schünemann et al. (2014)
B. thuringiensis var. tenebrionis Xd3 (Btt-Xd3)	Alder	*Agelasticaalni* (Alder leaf beetle)	Bt toxin	Eski et al. (2017)
B. flexus JIM24	–	*Lathyrus aphaca* weed	Aminolevulinic acid production	Phour and Sindhu (2019)

biological control products for agricultural pests belong to this species (Lambert et al. 1992). It produces the toxic proteins Cry, Cyt, and vegetative (secretable) insecticidal proteins (Vip) known as Bt toxins, which are highly lethal against a wide range of insects but nontoxic for mammals (Schnep et al. 1998). Bt toxins present in spore get activated on cleaving with proteases in the alkaline environment of the insect gut. That is why they act in a very specific manner and do not have any toxic effect on nontarget species (Bravo et al. 2007). Coleoptera, Lepidoptera, and Diptera are the major insect families against which Bt toxins (Cry/Cyt) are very effective.

Due to the promising results in controlling a range of pathogenic diseases, the commercial applicability of *Bacillus*-based biocontrol agents has increased. Hence, a good range of *Bacillus*-based biocontrol agents is available in the market. Some commercial *Bacillus*-based biocontrol agents are given in Table 17.4.

Table 17.4 Commercial *Bacillus* based biocontrol agents

Biocontrol agent	Product	Manufacturer	Target disease/organism	Crops
B. subtilis	Avogreen®	Ocean Agriculture South Africa	*Colletotrichum gloeosporioides* and *Cercospora* spot	Avocado
	Biosubtilin	Biotech International Ltd. India	*Fusarium*, *Verticillium*, *Pythium*, *Cercospora*, *Colletotrichum*, *Alternaria*, *Ascochyta*, *Macrophomina*, *Myrothecium*, *Ramularia*, *Xanthomonas*, and *Erysiphe polygoni*	Cotton, cereals, ornamental plants, and vegetable crops
	Stanes Sting®	Stanes Company, India	*M. incognita*	Tomato
B. subtilis strain GB34	GB 34	Gustafon, USA	*Rhizoctonia*, *Fusarium*	Soyabean
B. subtilis strain GB03	Kodiac companion	Growth Products, USA	*Rhizoctonia*, *Aspergillus*	Wheat, barley, pea
B. megaterium	Bioarc®	Sphere Bio-Arc Pvt Ltd.	*Tylenchulus semipenetrans* (nematode)	Cooton, beans, orange
B. firmus	BioNem®	Agro-green Minrav group of Israel	*M. incognita*	Tomato
B. amyloliquefaciens	RhizoVital® 42 liand RhizoVital 42 TB	ABiTEP GmbH, Germany	Soilborne pathogens	Potato, corn, strawberry, tomato, cucumber, ornamental plants
B. thuringiensis aizawai	Agree-WP	Certis USA L.L.C., USA	Armyworms, diamondback moth	Fruits, nuts, vegetables
	Florbac	Valent Biosciences Libertyville, USA		
	XenTari WG	Nufarm, Canada		
	Xantari®	Valent Biosciences Libertyville, USA		

(continued)

Table 17.4 (continued)

Biocontrol agent	Product	Manufacturer	Target disease/organism	Crops
B. thuringiensis kurstaki	Biobit®	GroChem, New Zealand	Lepidoptera	Apple, avocado, citrus, flowers, grapes etc.
	Cordalen®	Agrichem Bio, Madrid, Spain		
	Costar-WG	SKL Biosynthesis, Italy		
B. thuringiensis israelensis	Teknar® SC, VectoBac®	Valent Biosciences, Libertyville, USA	Mosquitoes and black flies	–
	Vectobar™	AgriLife, AP, India		
B. thuringiensis tenebrionis	Novodor®	Valent Biosciences, Libertyville, Illinois, USA	Colorado potato beetle	Potato
	Trident®	Certis USA L.L.C., USA		
B. thuringiensis sphaericus	VectoLex®, VectoMax®	Valent Biosciences, Libertyville, USA	Mosquitoes	–
B. pumilus	Yield Shield®	Bayer Crop Science, USA	*R. solani* and *Fusarium* (root rot)	Soybean
B. thuringiensis var. kurstaki	DiPel 2x®	Nufarm, Canada	*M. incognita* , Lepidoptera pests	Several vegetables

17.3 A Comparison of Biopesticides and Synthetic Pesticides

There are several advantages of using biopesticides. *Bacillus* species are recognized as safe bacteria that produce substances that are beneficial for crops and the production of industrial compounds (Stein 2005). As a biocontrol agent, *Bacillus* has the advantage of long-term storage and reduced complexity of formulation process due to its ability to form spores that help it to survive in adverse environmental conditions (Collins and Jacobsen 2003). In addition, biopesticides are nontoxic and easily degradable, which makes them more beneficial than any chemical pesticide. Although biopesticides offer a lot of advantages but have not replaced conventional pesticides completely as they are not so popular and common in use, and have specific requirements. Since they are highly specific, farmers will need different biopesticides for different pathogens or insects. Furthermore, maintaining the viability of these biocontrol agents is extremely important (inside.battelle.org).

17.4 Future Perspective

Developing new biopesticides itself is a very tedious process due to several challenges like cost, efficacy, and commercialization process. Delay in the authorization process is common due to the lack of enough expertise and regulatory model for biopesticides in India (Tripathi et al. 2020). Besides the investigation of new biomolecules, recombinant DNA technology is also being used for improving the efficiency of biopesticides. Novel fusion proteins, made up of toxins combined with a carrier protein, have been developed as next-generation biopesticides, and this technology makes this fusion protein toxic to target insects or pests after it is consumed orally (Fitches et al. 2004). More research is required in order to have an effective pest management in production systems. Funding agencies and government policies are influencing factors in biopesticide research and promotion. Government can control the use of hazardous pesticides by enforcing laws and encouraging the biopesticide industry for organic agriculture (Moosavi and Zare 2016).

Also, a strict regulatory mechanism is equally important for the desired quality and reasonable cost of biopesticides (Kumar and Singh 2015). Other than that, biological control agents (BCAs) may behave differently in different environmental and climatic conditions, hence, every country needs to develop indigenous BCAs (Keswani 2020). Moreover, limitations like slow in killing pests, cost, production, and formulation problems are the major drawbacks associated with biological pesticides. Therefore, working on these limitations to improve the performance may help in the global acceptance of biopesticides. Nanoformulations may play a significant role in improving the residual action and stability of biopesticides (Damalas and Koutroubas 2018; Tripathi et al. 2020). Recombinant DNA technology, molecular biology, and biotechnology can help to enhance the performance of biopesticides in their field use.

17.5 Conclusion

Biological stress is considered as one of the major restrictions to crop production in agricultural fields, which also exacerbates with climate change (Etesami et al. 2020). The use of biological agents to control plant diseases has become a very good alternative to conventional pesticides as they are nonhazardous for living beings and the environment. Several species of *Bacillus* are able to suppress plant diseases through various mechanisms, categorized into direct and indirect mechanisms. These mechanisms are responsible for the production of a broad range of antibiotic compounds (lipopeptides), lytic enzymes, antioxidants, siderophores, formation of biofilms, and various other metabolites which inhibit the growth of pathogens by their action. Moreover, through indirect mechanisms such as ISR, *Bacillus*-based biocontrol agents induce the plant immune/defense system and help them to grow in harsh conditions of stress. Also, *Bacillus* has a prominent role in alleviating induced ethylene levels under biological and nonbiological stresses, which suppress plant growth. Biopesticides and *Bacillus*-based products are gaining much attention that is why huge numbers of commercial products are available in the market belonging to *Bacillus* species. It has been well demonstrated that *Bacillus* species have immense potential to mitigate biotic stresses and encourage the growth and development of plants. However, these agents are not able to provide full protection against diseases but biopesticides combination with synthetic pesticides, fertilizers, and different types of tillage, incorporated into integrated pest management systems can fulfill the purpose to some extent. Apart from this, extensive research in new active ingredients, biopesticide formulation, and efficacy will give a new insight into biopesticide application in agriculture.

References

Abd-Allah EF, Ezzat SM, Tohamy MR (2007) *Bacillus subtilis* as an alternative biologically based strategy for controlling *Fusarium* wilt disease in tomato: a histological study. Phytoparasitica 35:474–478

Achari GA, Ramesh R (2014) Diversity, biocontrol and plant growth promoting abilities of xylem residing bacteria from Solanaceous crops. Int J Microbiol 296521:1–14

Ahmad M, Ahmad I, Hilger TH, Nadeem SM, Akhtar MF, Jamil M et al (2018) Preliminary study on phosphate solubilizing *Bacillus subtilis* strain Q3 and *Paenibacillus* sp. strain Q6 for improving cotton growth under alkaline conditions. Peer J 4(6):e5122

Akram W, Anjum T (2011) Quantitative changes in defense system of tomato induced by two strains of *Bacillus* against *Fusarium* wilt. Ind J Fund Appl Life Sci 1(3):7–13

Albayrak ÇB (2019) *Bacillus* species as biocontrol agents for fungal plant pathogens. In: Islam M, Rahman M, Pandey P, Boehme M, Haesaert G (eds) Bacilli and agrobiotechnology: phytostimulation and biocontrol, Bacilli in climate resilient agriculture and bioprospecting, vol 2. Springer, Cham, pp 239–265

Araújo FF, Marchesi GVP (2009) Use of *Bacillus subtilis* in the control of root-knot nematode and the growth promotion in tomato. Cienc Rural 39(5):1558–1561

Araújo FF, Silva JFV, Araújo ASF (2002) Influence of *Bacillus subtilis* on the *Heterodera* glycine-seclosion, orientation and infection in soybean. Cienc Rural 32(2):197–203

Asari S, Ongega M, Debois D, Pawn ED, Chen K, Bejai S, Meijer J (2017) Insights into the molecular basis of biocontrol of *Brassica* pathogens by *Bacillus amyloliquefaciens* UCMB5113 lipopeptides. Ann Bot 120(4):551–562

Ashwini N, Srividya S (2014) Potentiality of *Bacillus subtilis* as biocontrol agent for management ofanthracnose disease of chilli caused by *Colletotrichum gloeosporioides* OGC1. 3 Biotech 4(2):127–136

Ayed HB, Hmidet N, Bechet M, Chollet M, Chataigné G, Leclère V et al (2014) Identification and biochemical characteristics of lipopeptides from *Bacillus mojavensis* A21. Process Biochem 49(10):1699–1707

Bais HP, Fall R, Vivanco JM (2004) Biocontrol of *Bacillus subtilis* against infection of Arabidopsis roots by *Pseudomonas syringae* is facilitated by biofilm formation and surfactin production. Plant Physiol 134(1):307–319

Baysal O, Caliskan M, Yesilova O (2008) An inhibitory effect of a new *Bacillus subtilis* strain (EU07) against *Fusarium oxysporum*f. sp*radicis-lycopersici*. Physiol Mol Plant Pathol 73(1–3):25–32

Beauregard PB, Chai Y, Vlamakis H, Losick R, Kolter R (2013) *Bacillus subtilis* biofilm induction by plant polysaccharides. PNAS 110(17):E1621–E1630

Boubakri H (2018) The role of ascorbic acid in plant–pathogen interactions. In: Hossain M, Munne-Bosch S, Burritt D, Diaz-Vivancos P, Fujita M, Lorence A (eds) Ascorbic acid in plant growth, development and stress tolerance. Springer, Cham, pp 255–271

Boubakri H (2020) Induced resistance to biotic stress in plants by natural compounds: possible mechanisms. In: Hossain MA, Liu F, Huang B (eds) Mediated stress and cross-stress tolerance in crop plants. Elsevier, London, pp 79–99

Bravo A, Gill SS, Soberón M (2007) Mode of action of bacillus thuringiensis cry and Cyt toxins and their potential for insect control. Toxicon 49(4):423–435

Cao Y, Xu Z, Ling N (2012) Isolation and identification of lipopeptides produced by *B. subtilis* SQR 9 for suppressing Fusarium wilt of cucumber. Sci Hortic 135:32–39

Casals C, Teixidó N, Viñas I, Silvera E, Lamarca N, Usall J (2010) Combination of hot water, *Bacillus subtilis* CPA-8 and sodium bicarbonate treatments to control postharvest brown rot on peaches and nectarines. Eur J Plant Pathol 128:51–63

Cawoy H, Bettiol W, Fickers P, Ongena M (2011) *Bacillus*-based biological control of plant diseases. In: Stoytcheva M (ed) Pesticides in the modern world – pesticides use and management. InTech Open, Rijeka, pp 274–302

Chebotar VK, Makarova NM, Shaposhnikov AI, Kravchenko LV (2009) Antifungal and phytostimulating characteristics of *Bacillus subtilis* Ch-13 rhizospheric strain, producer of biopreparations. Appl Biochem Microbiol 45(4):419–423

Chen W, Ding C, Shen Q, Zhang R (2013) Evaluation of rhizosphere bacteria and derived bio-organic fertilizers as potential biocontrol agents against bacterial wilt (*Ralstonia solanacearum*) of potato. Plant Soil 366:453–466

Chien Y, Huang C (2020) Biocontrol of bacterial spot on tomato by foliar spray and growth medium application of *Bacillus amyloliquefaciens* and *Trichoderma asperellum*. Eur J Plant Pathol 156:995–1003

Choudhary DK, Prakash A, Johri BN (2007) Induced systemic resistance (ISR) in plants: mechanism of action. Indian J Microbiol 47:289–297

Collins DP, Jacobsen BJ (2003) Optimizing a *Bacillus subtilis* isolate for biological control of sugar beet *Cercospora* leaf spot. Biol Control 26(2):153–161

Compant S, Duffy B, Nowak J, Clément C, Essaïd Barka A (2005) Use of plant growth-promoting bacteria for biocontrol of plant diseases: principles, mechanisms of action and future prospects. Appl Environ Microbiol 71(9):4951–4959

Conrath U, Beckers GJ, Langenbach CJ, Jaskiewicz MR (2015) Priming for enhanced defence. Annu Rev Phytopathol 53:97–119

Correa EB, Bettiol W, Sutton JC (2010) Biocontrol of root rot (*Pythium aphanidermatum*) and growth promotion with *Pseudomonas chlororaphis* 63-28 and *Bacillus subtilis* GB03 in hydroponic lettuce. Summa Phytopathol 36(4):275–281

Crane JM, Bergstrom GC (2014) Spatial distribution and antifungal interactions of a *bacillus* biological control agent on wheat surfaces. Biol Control 78:23–32

Cui L, Yang C, Wei L, Li T, Chen X (2020) Isolation and identification of an endophytic bacteria *Bacillus velezensis* 8-4 exhibiting biocontrol activity against potato scab. Biol Control 141:104156

Damalas CA, Koutroubas SD (2018) Current status and recent developments in biopesticide use. Agriculture 8(1):13

Dhouib H, Zouari I, Abdallah DB, Belbahri L, Taktak W, Triki MA, Tounsi S (2019) Potential of a novel endophytic *Bacillus velezensis* in tomato growth promotion and protection against *Verticillium* wilt disease. Biol Control 139:104092

Dimkić I, Stanković S, Nišavić M, Petković M, Ristivojević P, Fira D, Berić T (2017) The profile and antimicrobial activity of *Bacillus* lipopeptide extracts of five potential biocontrol strains. Front Microbiol 8:925

Durairaj K, Velmurugan P, Park JH, Chang WS, Park YJ, Senthilkumar P et al (2018) An investigation of biocontrol activity *Pseudomonas* and *Bacillus* strains against *Panax ginseng* root rot fungal phytopathogens. Biol Control 125:138–146

El-Gremi SM, Draz IS, Youssef WAE (2017) Biological control of pathogens associated with kernel black point disease of wheat. Crop Prot 91:13–19

El-shakh ASA, Kakar KU, Wang X, Almoneafy AA, Ojaghian MR et al (2017) New biopesticide from a local *Bacillus thuringiensis var. tenebrionis* (Xd3) against alder leaf beetle (Coleoptera: Chrysomelidae). World J Microbiol Biotechnol 33:95

Etesami H, Beattie GA (2017) Plant-microbe interactions in adaptation of agricultural crops to abiotic stress conditions. In: Kumar V, Kumar M, Sharma S, Prasad R (eds) Probiotics and plant health. Springer, Singapore, pp 163–200

Eski A, Demir İ, Sezen K, Demirbag Z (2017) A new biopesticide from a local bacillus thuringiensis var. tenebrionis (Xd3) against alder leaf beetle (Coleoptera: Chrysomelidae). World J Microbiol Biotechnol 33(5):95

Etesami H, Beattie GA (2017) Plant-microbe interactions in adaptation of agricultural crops to abiotic stress conditions. In: Kumar V, Kumar M, Sharma S, Prasad R (eds) Probiotics and plant health. Springer, Singapore, pp 163–200

Etesami H, Maheshwari DK (2018) Use of plant growth promoting rhizobacteria (PGPRs) with multiple plant growth promoting traits in stress agriculture: action mechanisms and future prospects. Ecotoxicol Environ Saf 156:225–246

Etesami H, Noori F, Ebadi A, Reiahi Samani N (2020) Alleviation of stress-induced ethylene-mediated negative impact on crop plants by bacterial ACC deaminase: perspectives and applications in stressed agriculture management. In: Yadav A, Singh J, Rastegari A, Yadav N (eds) Plant microbiomes for sustainable agriculture. Sustainable development and biodiversity, vol 25. Springer, Cham, pp 287–315

Fan Q, Tian SP, Li YX, Xu Y, Wang Y (2000) Biological control of postharvest brown rot in peach and nectarine fruits by *Bacillus subtilis* (B-912). Acta Bot Sin 42(11):1137–1143

Fan H, Zhang Z, Li Y, Zhang X, Duan Y, Wang Q (2017) Biocontrol of bacterial fruit blotch by *Bacillus subtilis* 9407 via surfactin-mediated antibacterial activity and colonization. Front Microbiol 8:1973

FAO (2009) Available from: http://www.fao.org/fileadmin/templates/wsfs/docs/expert_paper/How_to_Feed_the_World_in_2050.pdf. Accessed on 10 May 2020

FAO (2020) Available from: http://www.fao.org/india/fao-in-india/india-at-a-glance/en/. Accessed on 2 May 2020

Ferrigo D, Causin R, Raiola A (2017) Effect of potential biocontrol agents selected among grapevine endophytes and commercial products on crown gall disease. Biol Control 62(6):821–833

Fitches E, Edwards MG, Mee C, Grishin E, Gatehouse AMR (2004) Fusion proteins containing insect-specific toxins as pest control agents: snowdrop lectin delivers fused insecticidal spider venom toxin to insect haemolymph following oral ingestion. J Insect Physiol 50:61–71

Freeman BC, Beattie GA (2008) An overview of plant defenses against pathogens and herbivores. Plant Pathol Microbiol 94. Iowa State University. Available from: https://doi.org/10.1094/PHI-I-2008-0226-01

Fu G, Huang SL, Ye YF, Wu YG, Cen ZL, Lin SH (2010) Characterization of a bacterial biocontrol strain B106 and its efficacy in controlling banana leaf spot and post-harvest anthracnose diseases. Biol Control 55(1):1–10

Gajbhiye A, Rai AR, Meshram SU, Dongre AB (2010) Isolation, evaluation and characterization of *Bacillus subtilis* from cotton rhizospheric soil with biocontrol activity against *Fusarium oxysporum*. World J Microbiol Biotechnol 26(7):1187–1194

García-Fraile P, Menéndez E, Rivas R (2015) Role of bacterial biofertilizers in agriculture and forestry. AIMS Bioeng 2(3):183–205

García-Gutiérrez L, Zeriouh H, Romero D et al (2013) The antagonistic strain *Bacillus subtilis* UMAF6639 also confers protection to melon plants against cucurbit powdery mildew by activation of jasmonate- and salicylic acid-dependent defence responses. Microb Biotechnol 6:264–274

Gautam S, Sehgal R, Shirkot CK, Chauhan A, Sharma R (2019) Potential of *Bacillus amyloliquefaciens* for biocontrol of bacterial canker of tomato incited by *Clavibactermichiganensis ssp. michiganensis*. Microb Pathog 130:196–203

Glick BR (2012) Plant growth-promoting bacteria: mechanisms and applications. Scientifica 2012:963401. Hindawi

Glick BR (2014) Bacteria with ACC deaminase can promote plant growth and help to feed the world. Microbiol Res 169:30–39

Gong Q, Zhang C, Lu F, Zhao H, Bie X, Lu Z (2014) Identification of bacillomycin D from *Bacillus subtilis* fmbJ and its inhibition effects against *Aspergillus flavus*. Food Control 36(1):8–14

Goswami D, Thakker JN, Dhandhukia PC (2016) Portraying mechanics of plant growth promoting rhizobacteria (PGPR): a review. Cogent Food Agric 2(1):1–19

Guardado-Valdivia L, Tovar-Pérez E, Chacón-López A, et al (2018) Identification and characterization of a new bacillus atrophaeus strain B5 as biocontrol agent of postharvest anthracnose disease in soursop (Annona muricata) and avocado (Persea Americana). Microbiol Res 210:26–32

Haas D, Défago G (2005) Biological control of soil-borne pathogens by fluorescent pseudomonads. Nat Rev Microbiol 3:307–319

Hashem A, Tabassum B, Abd-Allah EF (2019) *Bacillus subtilis*: a plant-growth promoting rhizobacterium that also impacts biotic stress. Saudi J Biol Sci 26(6):1291–1297

Hassan MN, Afghan S, Hafeez FY (2010) Suppression of red rot caused by *Colletotrichum falcatum*on sugarcane plants using plant growth-promoting rhizobacteria. Biol Control 55:531–542

Hassan MN, Shah SZ-U-H, Afghan S, Hafeez FY (2015) Suppression of red rot disease by *Bacillus* sp. based biopesticide formulated in non-sterilized sugarcane filter cake. Biol Control 60:691–702

Hayat S, Hayat Q, Alyemeni MN, Wani AS, Pichtel J, Ahmad A (2012) Role of proline under changing environments. Plant Signal Behav 7(11):1456–1466

Heidarzadeh N, Baghaee-Ravari S (2015) Application of *Bacillus pumilus*as a potential biocontrol agent of *Fusarium* wilt of tomato. Arch Phytopathol Plant Protect 48(13–16):841–849

Hermann M, Maier F, Masroor A, Hirth S, Pfitzner AJP, Pfitzner UM (2013) The Arabidopsis NIMIN proteins affect NPR1 differentially. Front Plant Sci 4:88

Herrnstadt C Soares G, Wilcox ER, Edward DL (1986) A new strain of *Bacillus thuringiensis* with activity against coleopteran insects. Nat Biotechnol 4:305–308

Istifadah N, Ningtyas DNY, Suryatmana P, Fitriatin BN (2017) The abilities of endophytic and biofertilizing bacteria and their combinations to suppress bacterial wilt disease (*Ralstonia solanacearum*) of Chili. In: 2nd international conference on sustainable agriculture and food security: a comprehensive approach (ICSAFS), KnE Life Sciences, pp 296–304

Jangir M, Pathak R, Sharma S, Sharma S (2018) Biocontrol mechanisms of *Bacillus* sp., isolated from tomato rhizosphere, against *Fusarium oxysporum f.* sp. *Lycopersici*. Biol Control 123:60–70

Jha A, Sharma D, Saxena J (2012) Effect of single and dual phosphate-solubilizing bacterial strain inoculations on overall growth of mung bean plants. Arch Agron Soil Sci 58(9):967–981

Jha A, Saxena J, Sharma V (2013) Investigation on phosphate solubilization potential of agricultural soil bacteria as affected by different phosphorus sources, temperature, salt, and pH. Commun Soil Sci Plan 44(16):2443–2458

Ji X, Lu G, Gai Y, Zheng C (2008) Biological control against bacterial wilt and colonization of mulberry by an endophytic *Bacillus subtilis* strain. FEMS Microbiol Ecol 65(3):565–573

Jiang C, Shi, J, Liu Y, Zhu C (2014) Inhibition of aspergillus carbonarius and fungal contamination in table grapes using *Bacillus subtilis*. Food Control 35(1):41–48

Kang SM, Radhakrishnan R, Lee IJ (2015) *Bacillus amyloliquefaciens* subsp. plantarum GR53, a potent biocontrol agent resists *Rhizoctonia* disease on Chinese cabbage through hormonal and antioxidants regulation. World J Microbiol Biotechnol 31:1517–1527

Keswani C (ed) (2020) Bioeconomy for sustainable development. Available from: https://doi.org/10.1007/978-981-13-9431-7

Khan MR, Majid S, Mohidin FA, Khan N (2011) A new bioprocess to produce low cost powder formulations of biocontrol bacteria and fungi to control fusarial wilt and root-knot nematode of pulses. Biol Control 59(2):130–140

Khan N, Martinez-Hidalgo P, Ice TA, Maymon M, Humm EA, Nejat N, Sanders ER, Hirsch AM (2018) Antifungal activity of *Bacillus* species against *Fusarium* and analysis of the potential mechanisms used in biocontrol. Front Microbiol 9:2363

Kipngeno P, Losenge T, Maina N, Kahangi E, Juma P (2015) Efficacy of *Bacillus subtilis* and *Trichoderma asperellum* against *Pythium aphanidermatum* in tomatoes. Biol Control 90:92–95

Kloepper, JW, Ryu CM, Zhang S (2004) Induced systemic resistance and promotion of plant growth by *Bacillus* spp. Phytopathology 94(11):1259–1266

Krishnan N, Gandhi K, Faisal PM, Muthurajan R, Kuppusamy P, Thiruvengadam R (2013) Management of bacterial leaf blight disease in rice with endophytic bacteria. World Appl Sci J 28(12):2229–2241

Kumar S, Singh A (2015) Biopesticides: present status and the future prospects. J Fertil Pestic 6:e129

Lambert B, Höfte H, Annys K, Jansens S, Soetaert P, Peferoen M (1992) Novel *Bacillus thuringiensis* insecticidal crystal protein with a silent activity against coleopteran larvae. Appl Environ Microbiol 58(8):2536–2542

Li Y, Han LR, Zhang Y, Fu X, Chen X, Zhang L (2013) Biological control of apple ring rot on fruit by *Bacillus amyloliquefaciens* 9001. Plant Pathol J 29(2):168–173

Liu TT, Wu P, Wang LH, Zhou Q (2011) Response of soybean seed germination to cadmium and acid rain. Biol Trace Elem Res 144:1186–1196

Lyngwi NA, Joshi SR (2014) Economically important *Bacillus* and related genera: a mini review. In: Sen A (ed) Biology of useful plants and microbes. Narosa Publishing House, New Delhi, pp 33–43

Maketon M, Apisitsantikul J, Siriraweekul C (2008) Greenhouse evaluation of *Bacillus subtilis* AP-01 and *Trichoderma harzianum* AP-001 in controlling tobacco diseases. Braz J Microbiol 39(2):296–300

Meena KR, Kanwar SS (2015) Lipopeptides as the antifungal and antibacterial agents: applications in food safety and therapeutics. Biomed Res Int 2015:473050

Minaxi, Nain L, Yadav RC, Saxena J (2012) Characterization of multifaceted *Bacillus* sp. RM-2 for its use as plant growth promoting bioinoculant for crops grown in semi arid deserts. Appl Soil Ecol 59:124–135

Mishra J, Singh R, Arora NK (2017) Plant growth-promoting microbes: diverse roles in agriculture and environmental sustainability. In: Kumar V, Kumar M, Sharma S, Prasad R (eds) Probiotics and plant health. Springer, Singapore, pp 71–111

Misra S, Chauhan PS (2020) ACC deaminase-producing rhizosphere competent *Bacillus* spp mitigate salt stress and promote *Zea mays* growth by modulating ethylene metabolism. 3 Biotech 10:19

Moosavi MR, Zare R (2016) Present status and the future prospects of microbial biopesticides in Iran. In: Singh H, Sarma B, Keswani C (eds) Agriculturally important microorganisms. Springer, Singapore, pp 293–305

Myo EM, Liu B, Ma J, Shi L, Jiang M, Zhang K, Ge B (2019) Evaluation of *Bacillus velezensis* NKG-2 for bio-control activities against fungal diseases and potential plant growth. Biol Control 134:23–31

Nawangsih AA, Damayanti I, Wiyono S, Kartika JG (2011) Selection and characterization of endophytic bacteria as biocontrol agents of tomato bacterial wilt disease. Hayati J Biosci 18(2):66–70

Ni L, Punja ZK (2019) Management of fungal diseases on cucumber (*Cucumis sativus* L.) and tomato (*Solanum lycopersicum* L.) crops in greenhouses using *Bacillus subtilis*. In: Islam M, Rahman M, Pandey P, Boehme M, Haesaert G (eds) Bacilli and agrobiotechnology: phytostimulation and biocontrol. Bacilli in climate resilient agriculture and bioprospecting, vol 2. Springer, Cham, pp 1–28

Nie P, Li X, Wang S, Guo J, Zhao H, Niu D (2017) Induced systemic resistance against *Botrytis cinerea* by *Bacillus cereus* AR156 through a JA/ET- and *NPR1*-dependent Signaling pathway and activates PAMP-triggered immunity in Arabidopsis. Front Plant Sci 8:238

Nihorimbere V, Ongena M, Cawoy H, Brostaux Y, Kakana P, Jourdan E, Thonart P (2010) Beneficial effects of *Bacillus subtilis* on field-grown tomato in Burundi: reduction of local *fusarium* disease and growth promotion. Afr J Microbiol Res 4(11):1135–1142

Niu DD, Liu HX, Jiang CH, Wang YP, Wang QY, Jin HL et al (2011) The plant growth-promoting rhizobacterium *Bacillus cereus* AR156 induces systemic resistance in *Arabidopsis thaliana* by simultaneously activating salicylate- and jasmonate/ethylene-dependent signaling pathways. Mol Plant-Microbe Interact 24:533–542

Ongena M, Jacques P (2008) *Bacillus* lipopeptides: versatile weapons for plant disease biocontrol. Trends Microbiol 16 (3):115–125

Phour M, Sindhu SS (2019) Bio-herbicidal effect of 5-aminoleveulinic acid producing rhizobacteria in sup pression of *Lathyrus aphaca* weed growth. Biol Control 64:221–232

Pieterse CM, Van Wees SC, Van Pelt JA, Knoester M, Laan R, Gerrits H et al (1998) Plant Cell 10(9):157–180

Pieterse CM, Van der Does D, Zamioudis C, Leon-Reyes A, Van Wees SC (2012) Hormonal modulation of plant immunity. Annu Rev Cell Dev Biol 28:489–521

Pieterse CM, Zamioudis C, Berendsen RL, Weller DM, Van Wees SC, Bakker PA (2014) Induced systemic resistance by beneficial microbes. Annu Rev Phytopathol 52:347–375

Pingping S, Jianchao C, Xiaohui J, Wenhui W (2017) Isolation and characterization of *Bacillus amyloliquefaciens* L-1 for biocontrol of pear ring rot. Hortic Plant J 3(5):183–189

Punja ZK, Rodriguez G, Tirajoh A (2016) Effects of *Bacillus subtilis* strain QST 713 and storage temperatures on post-harvest disease development on greenhouse tomatoes. Crop Prot 84:98–104

Qian S, Lu H, Sun J, Zhang C, Zhao H, Lu F, Bie X, Lu Z (2016) Antifungal activity mode of *Aspergillus ochraceus* by bacillomycin D and its inhibition of ochratoxin A (OTA) production in food samples. Food Control 60:281–288

Qin S, Xing K, Jiang J-H, Xu LH, Li W-J (2011) Biodiversity, bioactive natural products and biotechnological potential of plant-associated endophytic actinobacteria. Appl Microbiol Biotechnol 89:457–473

Rabbee MF, Ali MS, Baek KH (2019) Endophyte *Bacillus velezensis* isolated from citrus spp. controls streptomycin-resistant *Xanthomonas citri* subsp. *citri* that causes citrus bacterial canker. Agronomy 9(8):470

Radhakrishnan R, Hashem A, Abd-Allah EF (2017) *Bacillus*: a biological tool for crop improvement through bio-molecular changes in adverse environments. Front Physiol 8:667

Rais A, Jabeen Z, Shair F, Hafeez FY, Hassan MN (2017) *Bacillus* spp., a bio-control agent enhances the activity of antioxidant defense enzymes in rice against *Pyriculariaoryzae*. PLoS One 12(11):e0187412

Rajendran L, Samiyappan R, Raguchander T, Saravanakumar D (2007) Endophytic bacteria mediate plant resistance against cotton bollworm. J Plant Interact 2(1):1–10

Roberts PD, Momol MT, Ritchie L, Olson SM, Jones JB, Balogh B (2008) Evaluation of spray programs containing famoxadone plus cymoxanil, acibenzolar-S-methyl, and *Bacillus subtilis* compared to copper sprays for management of bacterial spot on tomato. Crop Prot 27(12):1519–1526

Rocha FYO, de Oliveira CM, da Silva PRA, de Melo LHV, Do Carmo MGF, Baldani JI (2017) Taxonomical and functional characterization of *Bacillus* strains isolated from tomato plants and their biocontrol activity against races 1, 2 and 3 of *Fusarium oxysporum f. sp. Lycopersici*. Appl Soil Ecol 120:8–19

Ryu CM, Farag MA, Pare P, Kloepper JW (2005) Invisible signals from the underground: bacterial volatiles elicit plant growth promotion and induce systemic resistance. Plnt Pthol J 21(1):7–12

Sabaté DC, Petroselli G, Erra-Balsells R, Carina Audisio M, Brandan CP (2019) Beneficial effect of *Bacillus* sp. P12 on soil biological activities and pathogen control in common bean. Biol Control 141:104131

Savary S, Ficke A, Aubertot JN, Hollier C (2012) Crop losses due to diseases and their implications for global food production losses and food security. Food Sec 4(4):519–537

Schnep E, Crickmore N, Van Rie J, Lereclus D, Baum J, Feitelson J, Zeigler DR, Dean DH (1998) *Bacillus thuringiensis* and its pesticidal crystal proteins. Microbiol Mol Biol Rev 62:775–806

Schünemann R, Knaak N, Fiuza LM (2014) Mode of action and specificity of *Bacillus thuringiensis* toxins in the control of caterpillars and stink bugs in soybean culture. ISRN Microbiol 2014:135675

Shafi J, Tian H, Ji M (2017) *Bacillus* species as versatile weapons for plant pathogens: a review. Biotechnol Biotechnol Equip 31(3):446–459

Shang L, Bai X, Chen C, Liu L, Li M, Xia X et al (2019) Isolation and identification of a *Bacillus megaterium* strain with ochratoxin a removal ability and antifungal activity. Food Control 106(8):106743

Shankar SM (2016) Epidemiology and management of damping-off of tomato (*Solanum esculentum* L.). Plant Pathology Department of N.M. College of Agriculture, Navsari Agricultural University. Available from: http://krishikosh.egranth.ac.in/handle/1/5810036276

Shi QH, Zhu ZJ, Juan L, Qian QQ (2006) Combined effects of excess Mn and low pH on oxidative stress and antioxidant enzymes in cucumber roots. Agric Sci China 5(10):767–772

Siahmoshteh F, Siciliano I, Banani H, Hamidi-Esfahani Z, Razzaghi-Abyaneh M, Gullino ML, Spadaro D (2017) Efficacy of *Bacillus subtilis* and *Bacillusamyloliquefaciens* in the control of *Aspergillus parasiticus* growth and aflatoxins production on pistachio. Int J Food Microbiol 254:47–53

Siddiqui ZA, Futai K (2009) Biocontrol of *Meloidogyne incognita* on tomato using antagonistic fungi, plant-growth-promoting rhizobacteria and cattle manure. Pest Manag Sci 65(9):943–948

Siripornvisal S (2010) Biocontrol efficacy of *Bacillus subtilis* BCB3-19 against tomato gray mold. Sci Technol J 10:37–44

Soleyman G, Masoud A, Siavash T (2014) Biological control of *Alternaria* rot of tomato by two bacterial strains, *Pseudomonas fluorescens* UTPF68, and *Bacillus subtilis* UTB96. Iran J Plant Prot Sci 44:299–305

Song M, Yun HY, Kim YH (2014) Antagonistic *Bacillus* species as a biological control of ginseng root rot caused by *Fusarium cf. incarnatum*. J Ginseng Res 38(2):136–145

Stein T (2005) *Bacillus subtilis* antibiotics: structures, syntheses and specific functions. Mol Microbiol 56:845–857

Suárez-Estrella F, Arcos-Nievas MA, López MJ, Vargas-García MC, Moreno J (2013) Biological control of plant pathogens by microorganisms isolated from agro-industrial composts. Biol Control 67(3):509–515

Tan T, Zhu J, Shen A et al (2019) Isolation and identification of a *Bacillus subtilis* HZ-72 exhibiting biocontrol activity against flax seedling blight. Eur J Plant Pathol 153:825–836

Toure Y, Ongena M, Jacques P, Guiro A, Thonart P (2004) Role of lipopeptides produced by *Bacillus subtilis* GA1 in the reduction of grey mould disease caused by *Botrytis cinerea* on apple. J Appl Microbiol 96:1151–1160

Tripathi DK, Singh VP, Kumar D, Chauhan DK (2012) Impact of exogenous silicon addition on chromium uptake, growth, mineral elements, oxidative stress, antioxidant capacity, and leaf and root structures in rice seedlings exposed to hexavalent chromium. Acta Physiol Plant 34(1):279–289

Tripathi YN et al (2020) Biopesticides: current status and future prospects in India. In: Keswani C (ed) Bioeconomy for sustainable development. Springer, Singapore, pp 79–109

Van Loon LC (2007) Plant responses to plant growth-promoting rhizobacteria. Eur J Plant Pathol 119:243–254

Van Loon LC, Bakker PA, Pieterse CMJ (1998) Systemic resistance induced by rhizosphere bacteria. Annu Rev Phytopathol 36:453–483

Verma M, Mishra J, Arora NK (2019) Plant growth-promoting rhizobacteria: diversity and applications. In: Sobti RC, Arora NK, Kothari R (eds) Environmental biotechnology: for sustainable future. Springer, Singapore, pp 129–173

Waewthongrak W, Leelasuphakul W, McCollum G (2014) Cyclic lipopeptides from *Bacillus subtilis* ABS-S14 elicit defense-related gene expression in citrus fruit. PLoS One 9(10):e109386

Wang L, Jin P, Wang J, Jiang L, Zhang S, Gong H et al (2015) In vitro inhibition and in vivo induction of defense response against *Penicillium expansum* in sweet cherry fruit by postharvest applications of *Bacillus cereus* AR156. Postharvest Biol Technol 101:15–17

Wang B, Shen Z, Zhang F, Raza W, Yuan J, Huang R, Ruan Y, Li R, Shen Q (2016a) *Bacillus amyloliquefaciens* strain W19 can promote growth and yield and suppress *Fusarium* wilt in banana under greenhouse and field conditions. Pedosphere 26(5):733–744

Wang Y, Yuan Y, Liu B, Zhang Z, Yue T (2016b) Biocontrol activity and patulin removal effects of *Bacillus subtilis*, *Rhodobactersphaeroides* and *agrobacterium tumefaciens* against *Penicilliumexpansum*. J Appl Microbiol 121(5):1384–1393

Wang H, Shi Y, Wang D, Yao Z, Wang Y, Liu J, Zhang S, Wang A (2018) A biocontrol strain of *Bacillus subtilis* WXCDD105 used to control tomato *Botrytis cinerea* and *Cladosporium fulvum* and promote the growth of seedlings. Int J Mol Sci 19:1371

Yadav AN, Sachan SG, Verma P, Saxena AK (2016) Bio-prospecting of plant growth promoting psychrotrophic Bacilli from the cold desert of north western Indian Himalayas. Indian J Exp Biol 54:142–150

Yanti Y, Warnita R, Busniah M (2018) Indigenous endophyte bacteria ability to control *Ralstonia* and *Fusarium* wilt disease on chili pepper. Biodiversitas 19(4):1532–1538

Yasmin S, Zaka A, Imran A, Zahid MA, Yousaf S, Rasul G (2016) Plant growth promotion and suppression of bacterial leaf blight in rice by inoculated bacteria. PLoS One 11:e0160688

Yuttavanichakul W, Lawongsa P, Wongkaew S, Teaumroong N, Boonkerd N, Nomura N, Tittabutr P (2012) Improvement of peanut rhizobial inoculant by incorporation of plant growth promoting rhizobacteria (PGPR) as biocontrol against the seed borne fungus, *Aspergillus niger*. Biol Control 63(2):87–97

Zalila-Kolsi I, Mahmoud AB, Ali H, Sellami S, Nasfi Z, Tounsi S, Jamoussi K (2016) Antagonist effects of *Bacillus* spp. strains against *Fusarium graminearum* for protection of durum wheat (*Triticum turgidum L. subsp. durum*). Microbiol Res 192:148–158

Zhao P, Quan C, Wang Y, Wang J, Fan S (2014) *Bacillus amyloliquefaciens* Q-426 as a potential biocontrol agent against *Fusarium oxysporum*f. sp. spinaciae. J Basic Microbiol 54(5):448–456

Chapter 18
Bacilli and Polyhydroxyalkanoates: An Intracellular Granule Having Promising Feature as a Resource for Production of Bioplastics

Priya Patel and Nasreen S. Munshi

Abstract Polyhydroxyalkanoate (PHA) is biodegradable biopolymer produced by microorganisms as lipid inclusion body under the stressful environmental conditions. They possess the properties analogous to petrochemically derived synthetic plastics and can serve as novel resource for production of bioplastics. Varieties of prokaryotes in diverse niches have been reported to accumulate PHA when there is excess carbon and/or limited nitrogen or phosphorous. *Bacillus* spp. are prominent source for industrial production of PHA as they are predominant in nature. Different *Bacillus* spp. are reported to utilize a wide range of substrates such as sucrose, glucose, fructose, starch and others for production of PHA. On the other hand, few *Bacilli* accumulate PHA while using inexpensive biowastes such as pea-shell slurry, fish solid waste, activated sludge, sugar industry wastewater and others as substrates. This allows sustainable management of waste along with generating a valuable by-product. They are known to synthesize PHA homopolymer as well as copolymers. The Food and Drug Administration (FDA) has considered *Bacillus* as Generally Regarded As Safe organisms (GRAS), allowing its application for large-scale bioplastic production. Further, the absence of immunogenic lipopolysaccharide layer in *Bacillus* spp. allows biomedical applications of produced PHA. The main emphasis of this article is to summarize the generalized, metabolic and genetic features of *Bacillus* spp. associated with PHA production and providing substantial information for exploiting capabilities of *Bacillus* spp. for industrial PHA production.

Keywords Polyhydroxyalkanoate · *Bacillus* spp. · Biodegradable · Biopolymer · Industrial PHA production · Bioplastic · Biowastes · PHA granule

---

P. Patel · N. S. Munshi (✉)
Institute of Science, Nirma University, Ahmedabad, Gujarat, India
e-mail: nasreen.munshi@nirmauni.ac.in

M. T. Islam et al. (eds.), *Bacilli in Agrobiotechnology*, Bacilli in Climate Resilient Agriculture and Bioprospecting,
https://doi.org/10.1007/978-3-030-85465-2_18

18.1 Introduction

Plastics and their products are basic necessity of individual's life. Low cost, light weight, water resistance, rust free and robustness are promising features of plastic, permitting its wide applications for societal benefits (Thompson et al. 2009). Concurrent to its enormous applications, plastic pollution has become threat for biosphere. Approximately more than 300 Mt. plastic is produced annually. Plastic manufacturing and utilization rate in India, China and Brazil are increasing exponentially (Koller 2017). These non-biodegradable petrochemically derived synthetic plastics accumulate as such in environment causing harm to wildlife, marine animals, humans and environment (Sathya et al. 2018).

Biodegradable polymers having properties similar to synthetic plastics are considered to be potential substitute of petrochemically derived plastics (Koller 2017). When discarded in surroundings, the polymers which get entirely converted into CO_2 and H_2O within fixed duration are designated as "biodegradable polymers" (Sathya et al. 2018). Nature-based and chemical-based polymers are the two main categories of biodegradable polymers. Further, as depicted in Fig. 18.1, animal-based, agro-based and microbe-based polymers are subcategorized under naturally derived polymers (Mulchandani and Katiyar 2020). Nature-based polymers are also called as "biodegradable biopolymers" as they are obtained from biological materials (Pittmann and Steinmetz 2017). Chemical-based polymers are initially extracted from biological sources in the monomeric form and then polymerized by using various chemicals. For example, polylactic acid (PLA) and polybutylene succinate

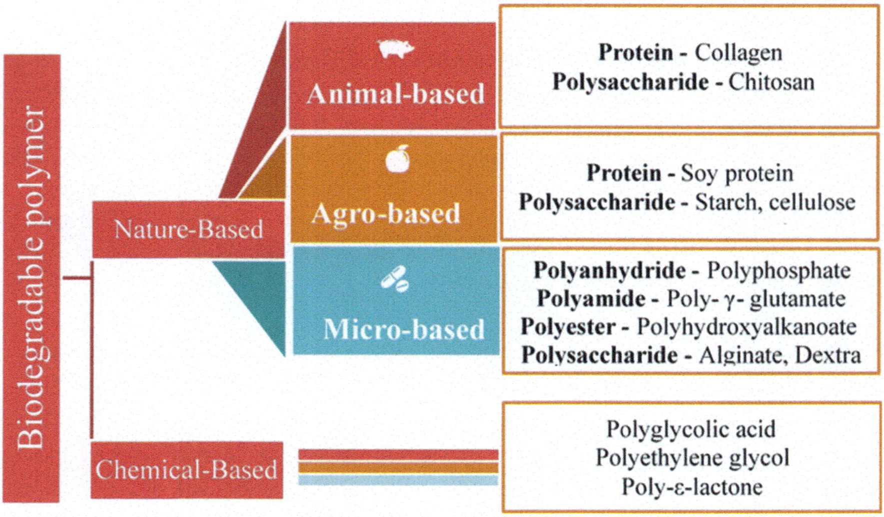

Fig. 18.1 Categories of biodegradable biopolymers: there are two main types of biodegradable polymers: nature-based and chemical-based. Further, nature-based polymer is divided into three subtypes, i.e. animal-based, agro-based and microbe-based. Examples are illustrated in different categories of polymers

(PBS) get polymerized in the presence of lactic acid and succinic acid, respectively. In contrast to this, nature-based biopolymers are advantageous as they do not require polymerization after extraction (Kourmentza et al. 2017). Amongst all the nature-based biopolymers, starch is widely used as packaging material due to its biodegradable nature other than being cost-effective and non-hazardous. But the major drawback is that it lacks thermoplastic properties (Sanyang et al. 2015). Chemical treatment of starch with plasticizer can impart thermoplastic properties to it (Coats et al. 2016). Microbe-based biopolymer – polyhydroxyalkanoate (PHA) – is an intracellular granule synthesized by bacteria in the presence of excess substrate and/or nutrient limiting condition. They possess features similar to polypropylene including thermoplasticity and is also biodegradable in nature (Pagliano et al. 2017; Hassan et al. 2016; Mokhtarani et al. 2012; Gamba et al. 2017). Also, biodegradability of PHA is higher than PLA and starch (Coats et al. 2016). Different bacterial strains have been reported to accumulate PHA by utilizing several carbon sources as well as agricultural and industrial waste as feedstock (Chua et al. 2003). The stated characteristics of PHA allow its wide acceptance as a source of bioplastics suitable for applications in packaging, medical, pharmaceutical, agriculture and food industries (Coats et al. 2016; Goudarztalejerdi et al. 2015).

Prokaryotes such as *Ralstonia eutropha*, *P. putida* CA-3, *P. putida* mt-2, *P. putida* F1, *Sphingobacterium* sp. ATM, *Bacillus odyssey* SUK3, *P. desmolyticum* NCIM 2112, etc. are capable of PHA production (Nikodinovic et al. 2008; Tamboli et al. 2010; Sato et al. 2008). Although many types of PHA producers are reported till date, *Bacillus* spp. are considered as valuable bioresource for industrial PHA production as they are predominant in nature (Mohapatra et al. 2017). They are capable of accumulating PHA homopolymer as well as copolymers (Singh et al. 2009). Moreover, they are also stated as "Generally Regarded As Safe organisms" (GRAS) by the Food and Drug Administration (FDA) (Singh et al. 2009; Mohapatra et al. 2017). Hence, *Bacillus* spp. are appropriate bio-resource for industrial PHA production. In this chapter, the generalized characteristics of PHA are described along with an account on how the abilities of *Bacillus* spp. can be exploited for industrial PHA production.

18.2 Biodegradable Biopolymer: Polyhydroxyalkanoates

Polyhydroxyalkanoates (PHA) are water insoluble polyester granules situated inside bacterial cytoplasm (Colombo et al. 2017; Goudarztalejerdi et al. 2015; Kumar et al. 2004). They are synthesized by bacteria in the presence of excess carbon and/or limited phosphorus, sulphur, oxygen or nitrogen (Goudarztalejerdi et al. 2015; Sathya et al. 2018).

18.2.1 Monomeric PHA and Its Derivatives

PHA biopolymer comprises of hydroxyalkanoic acid monomers (Goudarztalejerdi et al. 2015; Sathya et al. 2018). The general structural formula of hydroxyalkanoic acid or PHA monomer is shown in Fig. 18.2, where R represents an alkyl group and n ranges from 1 to 3. A total number of carbon atom in PHA monomer ranges from 3 to 14 on the basis of type of R group incorporated (Pradhan et al. 2020). The PHA biopolymer consists of long chain of monomeric PHA ranging from ~100 to 30,000 units (Pradhan et al. 2020; Jacquel et al. 2008; Jiang et al. 2016; Basnett and Roy 2010). Table 18.1 denotes the type of PHA monomer and a number of C atoms present in it according to incorporated alkyl group.

PHAs are classified into four categories on the basis of biopolymer chain length and the type of monomeric unit incorporated in biopolymer chain. On the basis of chain length, they are classified as short-chain length PHA (*scl*-PHA) and medium-chain length PHA (*mcl*-PHA). The *scl*-PHA comprises of 3–5 carbon atoms, whereas *mcl*-PHA contains 6–14 carbon atoms (Kourmentza et al. 2017; Ciesielska and Kiewisz 2016). On the basis of monomeric unit, two types of PHAs are there, one is homopolymer PHA and the other is copolymer PHA. In homopolymer PHA, biopolymer chain is consisted of identical type of PHA monomer, whereas biopolymer chain of copolymer PHA contains different types of PHA monomers (Pradhan et al. 2020). Classification of PHA is defined in Fig. 18.3.

The *mcl*-PHAs have properties similar to elastomers and are semi-crystalline in nature. They are used for preparation of drug delivery matrix, surgical sutures and implants. On the other hand *scl*-PHAs have tensile strength similar to polypropylene, they are used for formulating food packaging material and disposable items (Kourmentza et al. 2017; Pradhan et al. 2020). The *scl*-PHAs are predominantly accumulated in bacteria as compared to *mcl*-PHAs (Pradhan et al. 2020). More than 150 different kinds of PHAs have been reported so far (Pittmann and Steinmetz 2017; Sathya et al. 2018; Pradhan et al. 2020; Kourmentza et al. 2017). Table 18.2 illustrates the types of PHAs produced by different species of bacteria (Singh et al. 2015).

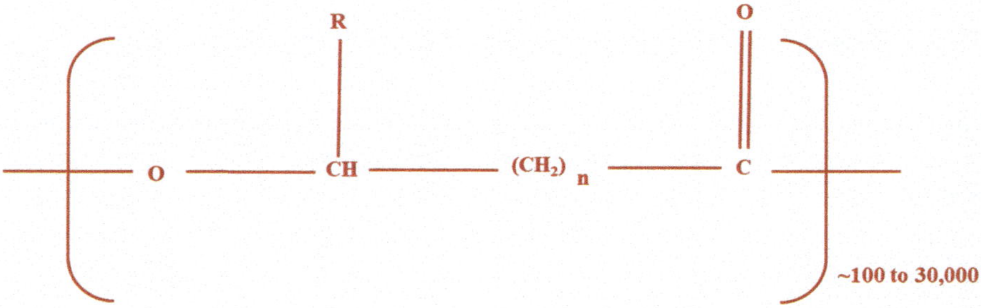

Fig. 18.2 General formula of monomeric PHA: the formula shown is of hydroxyalkanoic acid, the monomeric subunit of PHA biopolymer where R indicates an alkyl group and n ranges from 1 to 3. Long chain of hydroxyalkanoic acid in PHA biopolymer contains multiple monomers

Table 18.1 Type of alkyl group and number of C atoms present in monomeric PHA

Alkyl group (–R)				
Type	Molecular formula	Total number of C atoms in PHA monomer	n	Type of PHA monomer
Hydrogen	–H	C_3	1	Poly(3-hydroxypropionate)
		C_4	2	Poly(3-hydroxybutyrate)
		C_5	3	Poly(3-hydroxyvalerate)
Methyl	$-CH_2$	C_4	1	Poly(3-hydroxybutyrate)
		C_5	2	Poly(3-hydroxyvalerate)
		C_6	3	Poly(3-hydroxyhexanoate)
Ethyl	$-C_2H_6$	C_5	1	Poly(3-hydroxyvalerate)
		C_6	2	Poly(3-hydroxyhexanoate)
		C_7	3	Poly(3-hydroxyheptanoate)
Propyl	$-C_3H_8$	C_6	1	Poly(3-hydroxyhexanoate)
		C_7	2	Poly(3-hydroxyheptanoate)
		C_8	3	Poly(3-hydroxyoctanoate)
Butyl	$-C_4H_{10}$	C_7	1	Poly(3-hydroxyheptanoate)
		C_8	2	Poly(3-hydroxyoctanoate)
		C_9	3	Poly(3-hydroxynonanoate)
Pentyl	$-C_5H_{12}$	C_8	1	Poly(3-hydroxyoctanoate)
		C_9	2	Poly(3-hydroxynonanoate)
		C_{10}	3	Poly(3-hydroxydecanoate)
Hexyl	$-C_6H_{14}$	C_9	1	Poly(3-hydroxynonanoate)
		C_{10}	2	Poly(3-hydroxydecanoate)
		C_{11}	3	Poly(3-hydroxyundecanoate)
Heptyl	$-C_7H_{16}$	C_{10}	1	Poly(3-hydroxydecanoate)
		C_{11}	2	Poly(3-hydroxyundecanoate)
		C_{12}	3	Poly(3-hydroxydodecanoate)
Octyl	$-C_8H_{18}$	C_{11}	1	Poly(3-hydroxyundecanoate)
		C_{12}	2	Poly(3-hydroxydodecanoate)
		C_{13}	3	Poly(3-hydroxytridecanoate)
Nonyl	$-C_9\_H_{20}$	C_{12}	1	Poly(3-hydroxydodecanoate)
		C_{13}	2	Poly(3-hydroxytridecanoate)
		C_{14}	3	Poly(3-hydroxytetradecanoate)
Decyl	$-C_{10}\_H_{22}$	C_{13}	1	Poly(3-hydroxytridecanoate)
		C_{14}	2	Poly(3-hydroxytetradecanoate)
Undecyl	$-C_{11}\_H_{24}$	C_{14}	1	Poly(3-hydroxytetradecanoate)

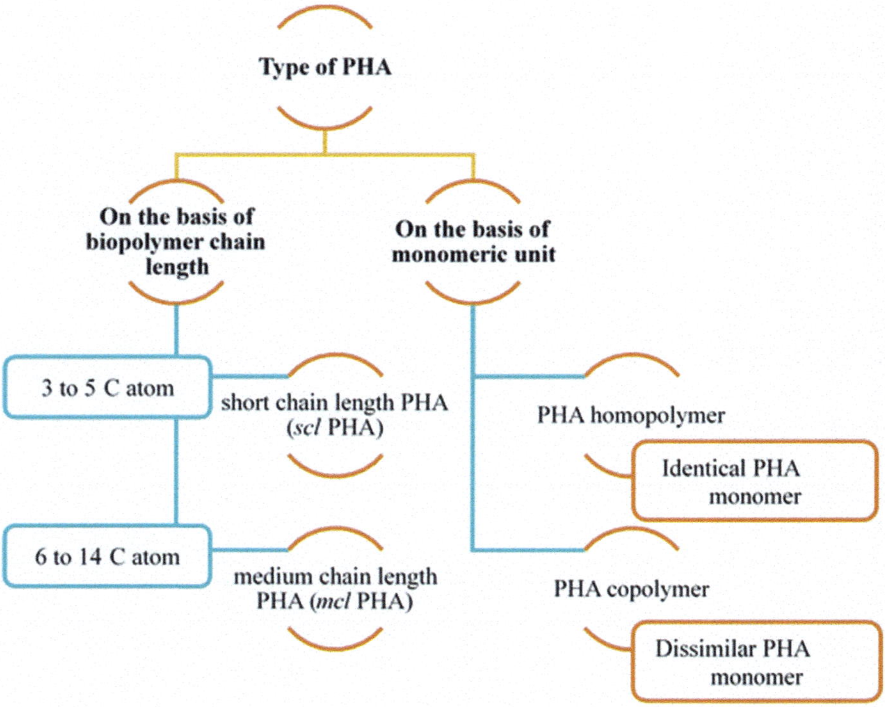

Fig. 18.3 Classification of PHA: there are four types of PHA classified on the basis of biopolymer chain length and on the basis of monomeric unit. Short-chain length PHA (*scl*-PHA) contains 3–5 carbon atoms, whereas medium chain length PHA (*mcl*-PHA) contains 6–14 carbon atoms. PHA biopolymer chain having identical monomeric unit is termed as PHA homopolymer while that with different monomeric units is termed as PHA copolymer

18.2.2 PHA as Carbon and Energy Reserves for Prokaryotes

PHAs are the inclusion bodies acting as intracellular carbon reservoir in bacterial strain (Sowinski et al. 2010). PHA granules constitute about 90% of the cell dry weight (CDW) (Bhuwal et al. 2013; Strong et al. 2016). *Ralstonia eutropha* was capable of accumulating 80% PHA of cell dry weight (Mokhtarani et al. 2012). When availability of carbon sinks in surrounding, PHA granules are used as carbon and energy source. Hence, they augment the survival of bacterial strain under stressful conditions existing in water and soil environments. UV irradiation, salinity, desiccation, osmotic shock, thermal stress and oxidative stress are the few stressful conditions occurring in water and soil environments where PHAs are used as carbon and energy source. Taxonomically diverse species of bacteria are inhabiting in soil niches. They have to deal with fluctuating conditions existing there. It has been reported that PHA-producing strain is protected during starvation as compared to mutant strain deficient in PHA production. But the co-relation

Table 18.2 Types of PHAs produced by bacteria

Name of PHA	Type of PHA		Name of PHA-producing bacteria	Substrate	Yield	References
	Chain length	Monomeric unit				
P(3HB)	NA	Homopolymer	*Enterococcus* sp. NAP11	Cardboard industry wastewater	79.27%	Bhuwal et al. (2013)
P(3HB)	NA	Homopolymer	*Brevundimonas* sp. NAC1	Cardboard industry wastewater	77.63%	Bhuwal et al. (2013)
Biopolymer having P(3HO), P(3HD), P(2HDDE), P(3HTD), P(3HHDE)	NA	Copolymer	*P. aeruginosa* strain SDS3(HQ 230975)	Crude oil	23.13%	Goudarztalejerdi et al. (2015)
Biopolymer having P(3HO), P(3HD), P(2HDDE), P(3HTD), P(3HHDE)	NA	Copolymer	*P. aeruginosa* strain XB7(KF 44738)	Crude oil	21.87%	Goudarztalejerdi et al. (2015)
Biopolymer having P(3HO), P(3HD), P(2HDDE), P(3HTD), P(3HHDE)	NA	Copolymer	*P. stutzeri* strain PS-SRU-ICU (JF 264901)	Crude oil	23.26%	Goudarztalejerdi et al. (2015)
Biopolymer having P(3HO), P(3HD), P(2HDDE), P(3HTD), P(3HHDE)	NA	Copolymer	*P. aeruginosa* strain H1 (JX 100389)	Crude oil	20%	Goudarztalejerdi et al. (2015)
Biopolymer having P(3HO), P(3HD), P(3HDDE), P(3HHDE).	*mcl*-PHA	Copolymer	*P. putida* F1, *P. putida* mt-2 and *P. putida* CA-3	BTEX mixture	0.25 ± 0.04 g/l	Nikodinovic et al. (2008)

NA not available, *P* poly, *(3HB)* 3-hydroxybutyrate, *(3HO)* 3-hydroxyoctanoate, *(3HD)* 3-hydroxydecanoate, *(2HDDE)* 2-hydroxydodecanoate, *(3HTD)* 3-hydroxytetradecanoate, *(3HHDE)* 3-hydroxydecahexanoate, *(3HHDE)* 3-hydroxydodecanoate between PHA accumulation and survival strategy is strain specific depending on the suboptimal growth preceding to starvation (Sowinski et al. 2010). PHA-producing bacteria are reported in microbial mats pursuing an essential role in stress tolerance and biofilm formation (Campisano et al. 2008; Sowinski et al. 2010). Reducing equivalents produced during PHA degradation plays a vital role in energizing chemotaxis process in the surrounding environments having lower quantity of reducing power (Kadouri et al. 2003; Sowinski et al. 2010). Moreover, EPS production is observed in PHA accumulating strains (Aneja et al. 2004). NADH generated during PHA degradation gets accumulated into EPS which serves as energy reservoir, useful for bacteria under stressed conditions (Sowinski et al. 2010).

Apart from this, biodegradable nature of PHA permits its utilization as carbon and energy source by microbial communities existing in the environment. PHA gets depolymerized into oligomer by intracellular PHA depolymerase (i-PhaZ) and extracellular depolymerase (e-PhaZ) (Grage et al. 2009; Philip et al. 2007). Former acts on native PHA granules and later converts the partially degraded PHA granules into oligomers. These oligomers are transformed into monomers by hydrolases. The monomers generated can be utilized as source of carbon and energy by surrounding microbial community (Sowinski et al. 2010). Figure 18.4 displays the possible role of PHA in stressful environment.

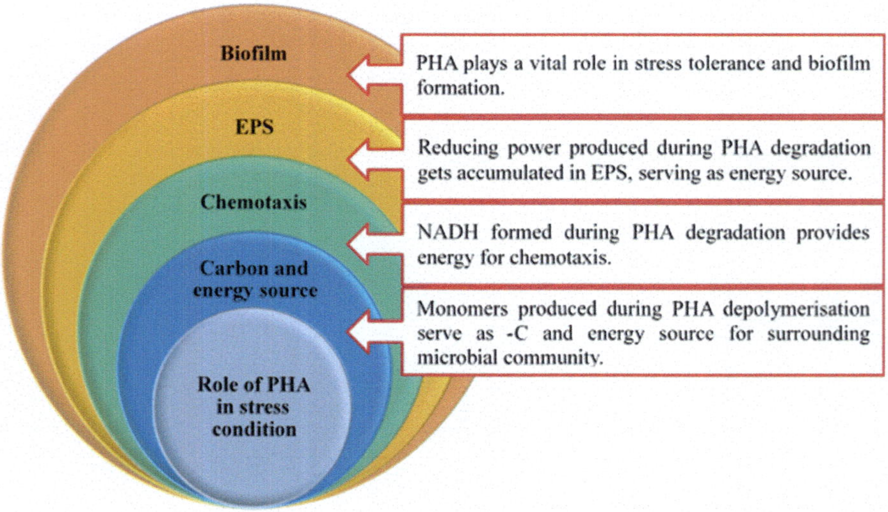

Fig. 18.4 Possible roles of PHA in stressful environment: monomeric PHA synthesized during biodegradation of polymeric PHA by depolymerase enzyme is used as carbon and energy source by diverse microbial communities existing in the surrounding environment. During its breakdown, reducing equivalent formed is used as energy source for chemotaxis. These reducing powers also get accumulated in EPS serving as energy source. Even PHA is essential in stress tolerance and for formation of microbial mats

18.2.3 Structure of PHA

Carbonosomes PHA granules are composed of proteins and phospholipid layer which is resistant to physical and chemical agents (Jendrossek 2009; Jendrossek and Handrick 2002). The innermost part of granule is made up of polyesters (Grage et al. 2009). PHA synthase (PhaC), PHA depolymerase (PhaZ), regulatory protein (PhaR) and phasins (PhaP) are four proteins present in PHA granule (Grage et al. 2009; Potter and Steinbuchel 2005). Structure of PHA granule is shown in Fig. 18.5.

PHA synthase plays a crucial role in converting hydroxyalkanoic acid into PHA polyester. They are classified into four classes – I, II, II and IV – as described in Fig. 18.5b. PHA synthase of class I and II is composed of 61–70 kDa PhaC subunit. Class III PHA synthase comprises of PhaC similar to that of class I PhaC and PhaE, both having 40 kDa molecular weight, whereas class IV PHA is composed of 40 kDa PhaC and 20 kDa PhaR. Class I, III and IV PhaC are reported to synthesize *scl*-PHA in contrast to class II PhaC which forms *mcl*-PHA (Grage et al. 2009; Potter and Steinbuchel 2005, 2006). As mentioned earlier, PhaZ are of two types, i-PhaZ present on the surface of PHA inclusion bodies and e-PhaZ secreted by

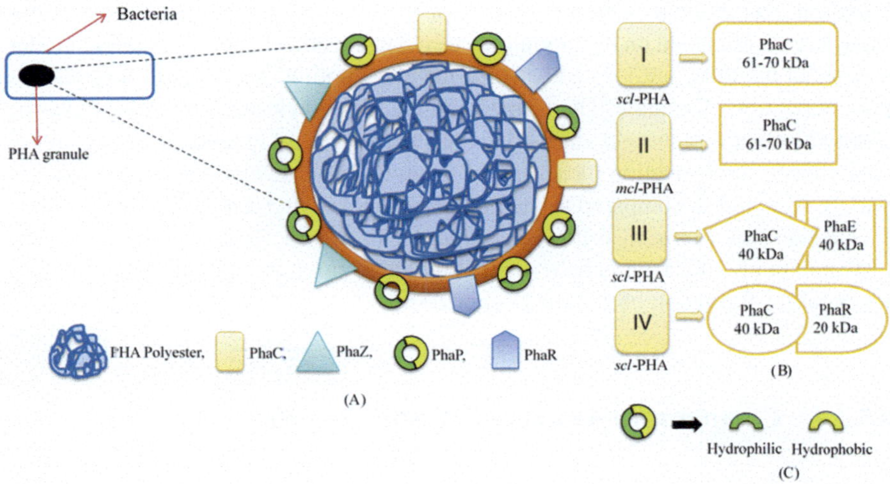

Fig. 18.5 Structure of PHA granule and its components: (**a**) intracellular PHA carbonosomes, (**b**) four classes of PhaC, (**c**) PhaP protein components. PHA carbonosomes are composed of hydroxyalkanoic acid polyester core surrounded by PhaC, PhaZ, PhaP and PhaR proteins. PhaZ are PHA depolymerase, important for degradation of PHA, while PhaR is a regulatory protein, essential for regulating the process of granule formation. PhaC is PHA synthase and PhaP is phasin proteins. There are four classes of PhaC which are essential for synthesizing biopolymer PHA. Class I and II of PhaC have single subunit of 61–70 kDa and are essential for synthesizing *scl*-PHA and *mcl*-PHA, respectively. Class III PhaC leads to formation of *scl*-PHA and is a combination of 40 kDa PhaC and 40 kDa PhaE. Class IV PhaC is made up of 40 kDa PhaC and 20 kDa PhaR, playing vital role in formation of *scl*-PHA. PhaP protein which forms a major part of the granule is amphibolic in nature having outer hydrophilic layer facing towards cytoplasm and inner hydrophobic layer facing granular inner side

many bacteria. Both are very essential for biodegradation of PHA granules (Grage et al. 2009; Sowinski et al. 2010). PhaR is the transcriptional regulatory protein required for regulation of PHA synthesis and PhaP production (Grage et al. 2009). Phasins are non-catalytic surface proteins having low molecular weight ranging between 11 and 25 kDa. They are produced in huge amount, comprising ~5% of cellular proteins (Grage et al. 2009). Their amphipathic layer as shown in Fig. 18.5c is composed of hydrophobic domain facing inside of granules and hydrophilic domain facing towards cytoplasm, thus, creating an interface between granule and cytoplasm (Grage et al. 2009; Mezzina and Pettinari 2016). They are essential for expression and activation of PhaC and PhaZ and, hence, are essential for PHA synthesis as well as degradation. They are also important for sorting of PHA granules and even affect size and number of granules. Few phasins act as chaperone proteins. Further, they belong to four families – PF09361, PF09602, PF09650 and PF05597 (Mezzina and Pettinari 2016).

Two models as presented in Fig. 18.6 have been reported till date for formation of PHA granules – micelle formation model and budding model (Potter and Steinbuchel 2006; Rehm 2006). According to the first model, initially PhaC proteins are randomly distributed in the cytoplasm, and as polymerization event initiates, it gets arranged in the form of micelle. Later on with increase in the biopolymer chain length, it gets distributed on the surface of PHA granules. After this, phasins and PhaR get accumulated on the granule surface. On the other hand, the budding model states that the PhaC are located between the phospholipid bilayer and carry out the polymerization of granule. The formed granule is then released into the cytoplasm. Then, the other proteins get attached to the outer surface of granule. Budding model is similar to the formation of eukaryotic neutral lipid. Micelle formation model is widely accepted as compared to the budding model (Potter and Steinbuchel 2006). *B. megaterium* seems to follow the budding model for PHA granule formation (Valappil et al. 2007a). It is believed that the PhaC possesses all the characteristics essential for formation of granule (Rehm 2006).

18.2.4 Comparative Aspects of Plastics and PHA

PHA have properties similar to petrochemically derived synthetic plastics like polyethylene and polypropylene (Numata et al. 2009; Bhuwal et al. 2013). Table 18.3 describes the general properties of PHA (Bugnicourt et al. 2014). Average molecular mass of PHA is 4.0×10^6 Da (Verlinden et al. 2007).

Further, the mentioned general property of PHA varies according to the type of biopolymer polymerized. Poly(3-hydroxybutyrate) denoted as P(3HB) is the most common type of PHA biopolymer accumulated by bacteria (Numata et al. 2009). It was the first biopolymer to be isolated from *B. megaterium* in the year 1920 by Maurice Lemoigne at the Pasteur Institute (Philip et al. 2007; Verlinden et al. 2007; Numata et al. 2009; Potter and Steinbuchel 2005). It has 162–181 °C melting temperature, −4 to 18 glass transition temperature, 19–44 MPa tensile strength, 1.2–4

Gpa Young's module, 0.8–4.5% elongation to break and 50–80% degree of crystallinity (Pradhan et al. 2020). Table 4 elucidates the properties of diverse types of PHA and polypropylene.

18.3 Bacilli and PHA

18.3.1 Diversity of PHA-Producing Bacillus *Species*

Members belonging to the genus *Bacillus* are able to accumulate varieties of PHA by utilizing ample of carbon sources. Table 18.5 displays the list of PHA-producing *Bacilli*. For PHA production, they were able to feed on simplest nutrient source such as glucose to complex hydrocarbons such as dyes and industrial effluents. Cost associated with PHA production decreases when any waste is used as substrates (Wen et al. 2010). *Bacillus* spp. were able to accumulate PHA while feeding on low-cost C source (Mohapatra et al. 2017).

18.3.2 Nutrients Essential for PHA Production by Bacilli

As described earlier, microorganisms accumulate PHA when there is excess substrate and/or growth limiting conditions. Feast and famine are the two terminologies describing the substrate availability. Former describes excess substrate and later indicates limited substrate availability. In famine conditions, the microorganisms will limit their cellular activity to minimum level essential for cell viability. They limit their activity by lowering the level of RNA transcription and/or enzyme activity. Further, when famine is followed by feast, initially the available substrate is utilized for PHA production instead of cellular growth. This happens because of the absence of essential enzymes required for cell growth. Thus, PHA accumulation is the physiological adaptation of microorganisms depending on substrate concentration (Albuquerque et al. 2010; Beun et al. 2002). In other words, such type of microorganisms does not require nutrient limiting conditions for PHA accumulation, and hence, they show growth-associated biopolymer production (Shi et al. 2007). Thus, type of microorganisms and microbial growth rate are the two internal factors affecting PHA accumulation. Moreover, the growth limiting conditions also arise by external factors such as limiting nutrients like N and P or in the presence of electron acceptors, viz. O_2, nitrate and phosphorous (Albuquerque et al. 2010; Shahid et al. 2013; Saharan et al. 2014). Growth-limiting nutrients will minimize the cellular growth as described earlier, and excess substrate is directed towards PHA production. Such microorganisms require limited nutrients and excess substrate for PHA production. In few organisms, it has been reported that C:N ratio also affects PHA

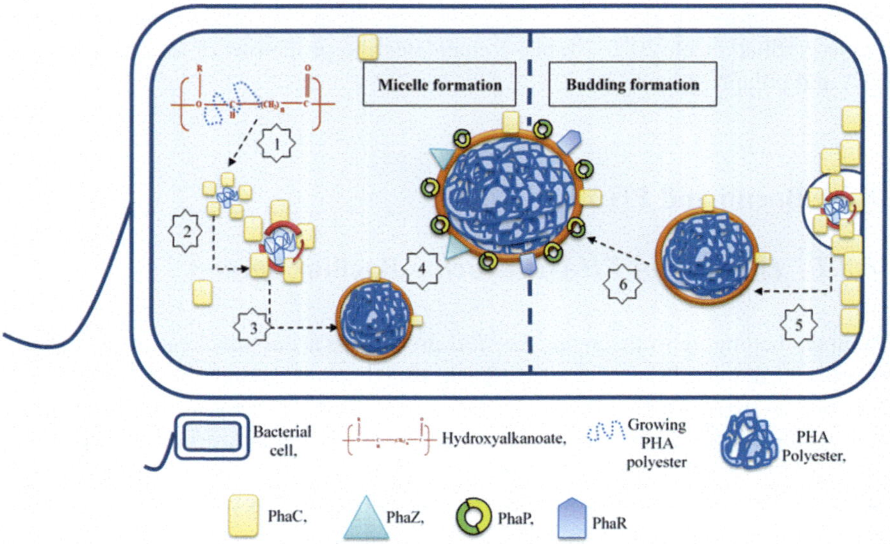

Fig. 18.6 Mechanism of PHA granule formation: two mechanisms for PHA granule formation are known – micelle formation (shown on left) and budding formation (shown on right). According to the most widely accepted micelle formation model, randomly distributed PhaC in cytoplasm initiates formation of granular body from hydroxyalkanoic acid (1) and (2). The granular body then enlarges into native granule acquiring PhaP, PhaZ and PhaR (3) and (4), while budding formation model depicts that the PhaC are located in the intracellular membrane and buds off into the granule (5) and (6)

accumulation (Shi et al. 2007; Wen et al. 2010; Saharan et al. 2014). Figure 18.7 indicates the pictorial outline of PHA production in varied nutrient conditions.

B. megaterium PNCM 1890 prefer urea as nitrogen source over sodium nitrate, ammonium chloride and ammonium sulphate, showing high PHB production. Researchers believe that uptake of low molecular weight, polar and uncharged urea for PHB formulation is higher as compared to remaining inorganic and ionic sources. The strain was capable to accumulate 3.91 g/L of PHB when C:N and C:P ratios were 14.3 g/g and 21.4 g/g, respectively, within 24 h (Danez et al. 2020). *B. megaterium* BA-019 produced 42% PHB of CDW in the presence of sugarcane molasses and urea as carbon and nitrogen source, respectively. The C:N ratio was 10 mol/mol, and incubation period was 24 h (Kulpreecha et al. 2009) (Table 18.4).

PHA production by *B. megaterium* DSM 509 was observed when grown in MM medium with different carbon sources such as glucose, glycerol, succinic acid, citric acid, acetic acid, pentanoic acid and octanoic acid. Then from this, medium cells were transferred into MM medium without N (MM-N), and PHA was extracted. Monomeric composition of PHA extracted from MM medium indicated *scl*-PHA, whereas that from MM-N indicated *mcl*-PHA. Authors suggested that the *scl*-PHA may get degraded when transferred from MM to MM-N in order to provide energy for synthesizing *mcl*-PHA (Shahid et al. 2013). *Bacillus flexus* is reported to

Table 18.3 General properties of PHA

Property	Unit	PHA
Glass transition temperature (T_g)	°C	2
Melting temperature (T_m)	°C	160–175
Degree of crystallinity (X_{cr})	%	40–60
Young's modulus (E)	GPa	1–2
Tensile strength (σ)	MPa	15–14
Elongation to break (ε)	%	1–15

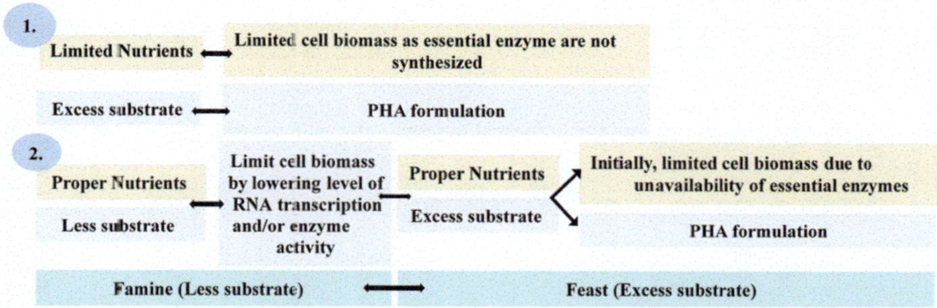

Fig. 18.7 Generalized mechanism for microbial PHA production under different nutrient conditions: as indicated in mechanism 1, limited nutrients hinder synthesizing of enzymes essential for cell biomass and hence limit growth. Excess substrate is directed towards PHA formulation, whereas as seen in mechanism 2, instead of nutrients, substrate is limited (famine) which again limits cell biomass development by lowering the level of RNA transcription and/or enzyme activity. When famine is followed by feast (excess substrate), initially cell biomass is limited as essential enzymes are unavailable, and excess substrate is directed for PHA formulation

produce high PHA in nitrogen-limiting conditions (Somashekara et al. 2009). In contrast to this, *Bacillus thuringiensis* EGU45 showed higher P(3HB-co-3 HV) (1.5–3.5 g/L) copolymer production in the presence of excess nitrogen or low C:N ratio and crude glycerol as carbon source (Kumar et al. 2015). *Bacillus mycoides* RLJ B-017 was unable to accumulate PHB at high oxygen transfer rate (OTR). It accumulated 69.4 ± 0.4% PHB of CDW in the presence of sucrose as carbon source and di-ammonium sulphate (Borah et al. 2002). The non-photosynthetic microorganisms, *Bacillus thuringiensis* EGU45 and *Bacillus cereus* EGU44 were capable of producing 11.3% PHB of CDW in the same medium where they were initially subjected for H_2 production (Patel et al. 2011) (Table 18.5).

Table 18.4 Properties of different types of PHA and polypropylene

Type of PHA and polypropylene	T_g (°C)	T_m (°C)	X_{cr} (%)	E (GPa)	σ (MPa)	ε (%)	References
Polypropylene	−14 to −6	160–169	50	1.1–2	28–40	20–75	Pradhan et al. (2020)
Polypropylene	−10	176	NA	1.7	38	400	Strong et al. (2016)
Polypropylene	−10	176	50–70	NA	38	400	Verlinden et al. (2007)
LDPE	−30	130	NA	0.2	10	620	Strong et al. (2016)
LDPE	−125 to −90	105–125	43	0.14–0.3	7–17	200–900	Pradhan et al. (2020)
HDPE	−125 to −90	130–137	79.8–81	0.7–1.4	20–40	100–1000	Pradhan et al. (2020)
P(3HB)	−4 to 18	162–181	50–80	1.2–4	19–44	50–80	Pradhan et al. (2020)
P(3HB)	−1	180	NA	3.5	40	5	Strong et al. (2016)
P(3HB)	2	177	60	NA	43	5	Verlinden et al. (2007)
P(3HO)	−35	60	30		10	300	Basnett and Roy (2010)
P(3HB-co-3 HV)	10 to −6	137–170	NA	0.7–2.9	Up to 690	30–38	Ciesielska and Kiewisz (2016)
P(3HB-co-3 HV)	−13 to 10	64–171	53–56	0.14 to 8.7	1.8 to 51	1–970	Pradhan et al. (2020)
P(3HB-co-3 HV)	−1	145	56	NA	20	50	Verlinden et al. (2007)
P(3HB-co-20%3HD)	−8	130	NA	NA	17	680	Ciesielska et al. (2016)
mcl-PHA	−48.46	43.95	NA	NA	NA	NA	Nikodinovic et al. (2008)
mcl-PHA	−40	80	NA	NA	20	300	Ciesielska and Kiewisz (2016)
mcl-PHA	~ − 40	~60	40	NA	Higher	Higher	Pradhan et al. (2020)
scl-PHA	179	4	NA	3.5	5	40	Ciesielska and Kiewisz (2016)
scl-PHA	~ − 180	~ − 0	70	NA	Lower	Lower	Pradhan et al. (2020)
P(3HB-co-20 mol% 3 HV)	−1	145	NA	0.8	20	50	Strong et al. (2016)
P(3HB-co-6 mol% 3 HV)	−8	133	NA	0.2	17	680	Strong et al. (2016)
P(3HB-co-mol16% 4B)	−7	150	45	NA	26	444	Verlinden et al. (2007)
P(3HB-co-mol10% Hx)	−10	176	34	NA	21	400	Verlinden et al. (2007)
P(3HB)	−11	161	NA	NA	NA	NA	Contreras et al. (2013)
P(3HB)	−16	136.8	NA	NA	NA	NA	Contreras et al. (2013)

NA not available, *LDPE* low-density polyethylene, *HDPE* high-density polyethylene, *P* poly, *(3HB)* 3-hydroxybutyrate, *(3HO)* 3-hydroxyoctanoate, *3(HV)* 3-hydroxyvalerate, *3(HD)* 3-hydroxydecanoate, *(4B)* butyrate, *(Hx)* hexanoate, T_g glass transition temperature, T_m, melting temperature, X_{cr} degree of crystallinity, *E* Young's module, ε elongation to break, σ tensile strength

Table 18.5 Diversity of PHA-producing *Bacilli*

Type of *Bacilli*	Type of PHA	PHA yield	Nutrient source	Incubation period	Study level	References
B. megaterium	P(3HB)	186.8 mg/g	Glycerol reagent grade (GRG)	48 h	Flask level	Cardozo et al. (2016)
B. megaterium uyuni S29	P(3HB) with 161Tm	30%	Glucose	NA	Bioreactor level	Contreras et al. (2013)
	P(3HB) with 136.8 T_m*	70%	Glucose	NA	Bioreactor level	
Isolate AWW belonging to genus *Bacillus*	P(3HB)	41.66%	Glucose	48 h	Flask level	Getachew and Woldesenbet (2016)
	P(3HB)	54.16%	Fructose	48 h	Flask level	
	P(3HB)	48.83%	Sucrose	48 h	Flask level	
	P(3HB)	51.61%	Corn cob	48 h	Flask level	
	P(3HB)	38.55%	Teff straw	48 h	Flask level	
	P(3HB)	26.92%	Banana peel	48 h	Flask level	
	P(3HB)	63.41%	Peptone	48 h	Flask level	
	P(3HB)	51.25%	Ammonium nitrate	48 h	Flask level	
Isolate ASS belonging to genus *Bacillus*	P(3HB)	35.45%	Glucose	48 h	Flask level	
Isolate LAW belonging to genus *Bacillus*	P(3HB)	28.88%	Glucose	48 h	Flask level	
Isolate FPS belonging to genus *Bacillus*	P(3HB)	23.59%	Glucose	48 h	Flask level	
Isolate KAS belonging to genus *Bacillus*	P(3HB)	16.66%	Glucose	48 h	Flask level	
Isolate KIS belonging to genus *Bacillus*	P(3HB)	28.57%	Glucose	48 h	Flask level	
Isolate WW belonging to genus *Bacillus*	PHA	46.28%	Glucose	48 h	Flask level	Getachew and Berhanu (2016)
Isolate RS belonging to genus *Bacillus*	PHA	35.45%	Glucose	48 h	Flask level	
Isolate SS belonging to genus *Bacillus*	PHA	34.04%	Glucose	48 h	Flask level	
Bacillus sp. N-2	P(3HB)	20%	Glucose	5 days	Flask level	Hassan et al. (2016)

(continued)

Table 18.5 (continued)

Type of *Bacilli*	Type of PHA	PHA yield	Nutrient source	Incubation period	Study level	References
6 BC1 *Bacillus* co-culture	P(3HB)	150 mg/L	Glucose	48 h	Flask level	Kumar et al. (2014)
	P(3HB)	855 mg/L	Pea-shell slurry	48 h	Flask level	
	P(3HV)	20 mg/L	Pea-shell slurry	48 h	Flask level	
5 BC1 *Bacillus* co-culture	P(3HB)	230 mg/L	Glucose	48 h	Flask level	
	P(3HB)	1620 mg/L	Pea-shell slurry	48 h	Flask level	
	P(3 HV)	25 mg/L	Pea-shell slurry	48 h	Flask level	
5 BC2 *Bacillus* co-culture	P(3HB)	230 mg/L	Glucose	48 h	Flask level	
	P(3HB)	1595 mg/L	Pea-shell slurry	48 h	Flask level	
	P(3HV)	15 mg/L	Pea-shell slurry	48 h	Flask level	
4 BC1 *Bacillus* co-culture	P(3HB)	250 mg/L	Glucose	48 h	Flask level	
	P(3HB)	430 mg/L	Pea-shell slurry	48 h	Flask level	
	P(3HV)	30 mg/L	Pea-shell slurry	48 h	Flask level	
4 BC2 *Bacillus* co-culture	P(3HB)	205 mg/L	Glucose	48 h	Flask level	
	P(3HB)	960 mg/L	Pea-shell slurry	48 h	Flask level	
	P(3HV)	20 mg/L	Pea-shell slurry	48 h	Flask level	
4 BC3 *Bacillus* co-culture	P(3HB)	190 mg/L	Glucose	48 h	Flask level	
	P(3HB)	570 mg/L	Pea-shell slurry	48 h	Flask level	
	P(3HV)	30 mg/L	Pea-shell slurry	48 h	Flask level	
3 BC1 *Bacillus* co-culture	P(3HB)	185 mg/L	Glucose	48 h	Flask level	
	P(3HB)	875 mg/L	Pea-shell slurry	48 h	Flask level	
	P(3HV)	25 mg/L	Pea-shell slurry	48 h	Flask level	
2 BC1 *Bacillus* co-culture	P(3HB)	230 mg/L	Glucose	48 h	Flask level	
	P(3HB)	780 mg/L	Pea-shell slurry	48 h	Flask level	
	P(3HV)	40 mg/L	Pea-shell slurry	48 h	Flask level	
2 BC2 *Bacillus* co-culture	P(3HB)	220 mg/L	Glucose	48 h	Flask level	
	P(3HB)	835 mg/L	Pea-shell slurry	48 h	Flask level	
	P(3HV)	35 mg/L	Pea-shell slurry	48 h	Flask level	

Bacillus sp.	PHA	NA	Marine sample	NA	NA	Wecker et al. (2015)
Bacillus odysseyi SUK3	PHA	58%	Mixture of red HE8B, red M5B, remazol red, orange 3R, rubine, golden yellow HER and direct blue GLL	48 h	Flask level	Tamboli et al. (2010)
Bacillus megaterium BA-019	P(3HB)	42%	Molasses and sucrose	24 h	Fed-batch	Pagliano et al. (2017)
Bacillus cereus SPV	P(3HB)	61.07%	Sugarcane molasses	50 h	Shaken flask	Akaraonye et al. (2012)
	P(3HB)	51.37%	Sugarcane molasses	50 h	Fermenter level	
B. megaterium DSM 509	PHA	48%	Octanoic acid	24 h	Flask level	Shahid et al. (2013)
Bacillus megaterium	PHB	5.61 g/L	Glucose and ammonium sulphate	64 h	Bioreactor	Mohanrasu et al. (2020)
Bacillus megaterium SRKP-3	PHB	11.32 g/L	Dairy waste, Rice bran and Sa water	36 h	Feed-batch	Pandian et al. (2010)
Bacillus drentensis BP17	PHB	5.55 g/L	Pineapple peel solution	36 h	Flask level	Penkhrue et al. (2020)
Bacillus sp. *CFR 67*	PHA	524 mg/L	Wheat bran hydrolysate	72 h	Flask level	Sreekanth et al. (2013)
Bacillus tequilensis	P(3HB-co-3 HV)	59%	Synthetic acids	48 h	Flask level	Reddy et al. (2014)
		36%	Acidogenic fermented food waste (AFW)			

NA not available, T_m melting temperature, *P* Poly, *3(HB)* 3-hydroxybutyrate, *3(HV)* 3-Hydroxyvalerate

18.3.3 Metabolic Overview for PHA Production in Bacilli

The monomeric PHA/hydroxyalkanoic acid is synthesized by different metabolic pathways. The type of –R group incorporated in monomeric PHA depends on the type of substrate and the capability of microorganisms to metabolize the available substrate (Sudesh et al. 2000). The common intermediate in the pathway is hydroxyacyl-Co-A, as shown in Fig. 18.8a (Chen 2010). The type of acyl group present in it reveals the formation of a particular type of PHA. For example, 3-hydroxybutyryl-Co-A contains butyryl group which gets polymerized into 3-hydroxybutyric acid/polyhydroxybutyrate [3(PHB)] (Chen 2010; Pradhan et al. 2020). Polymerization event is catalysed by PHA synthase (PhaC) (Grage et al. 2009; Chen 2010; Pradhan et al. 2020). It is the most common enzyme involved in all PHA production pathways. The metabolic pathway of PHB, the most common kind of PHA, is well established. As shown in Fig. 18.8b, it involves two enzymes, viz. β-ketothiolase (PhaA) and acetoacetyl-CoA reductase (PhaB), along with PhaC. The microorganisms act upon available substrate and convert it into acetyl CoA. Further, PhaA transforms it into acetoacetyl-CoA which gets converted into 3-hydroxybutyryl-CoA by PhaB (Pradhan et al. 2020; Chen 2010; Beun et al. 2002). Like all other biochemical pathways, PHA synthesis also involves wide varieties of enzymes and a few of them are listed in Table 18.6.

Apart from this, *B. thuringiensis* strain YBT-1520 was reported to show PHB production from acetyl Co-A via two dissimilar pathways. One involved three traditional enzymes, viz. PhaA, PhaB and PhaC, while other involved enzymes designated as AtoB, FadB and Crt. The enzyme AtoB converts acetyl CoA into acetoacetyl-CoA. Then it gets transformed into Crotonoyl CoA via

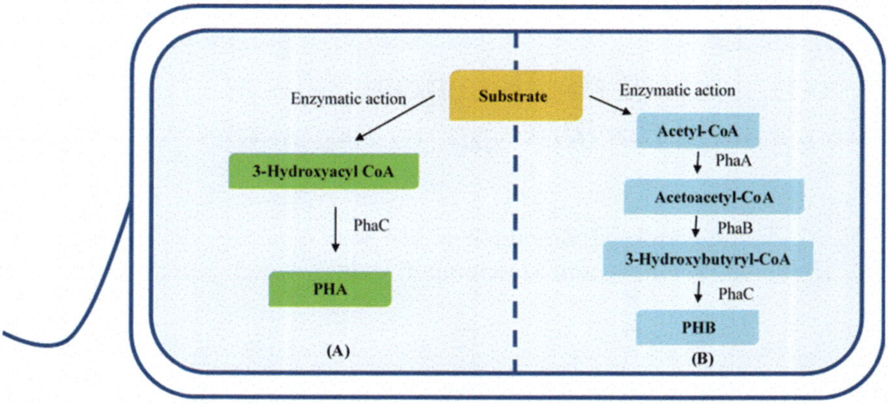

Fig. 18.8 Metabolic pathways for PHA production in bacterial cells. (**a**) PHA synthesis via 3-hydroxyacyl CoA intermediate and (**b**) PHB synthesis via acetyl-CoA: Formation of particular type of PHA depends on the type of acyl group. For example, butyryl is the acyl group leading to formation of PHB

Table 18.6 Enzymes involved in biosynthesis of PHA and gene encoding them

Enzyme	Gene	Type of PHA	Organism	References
β-ketothiolase	*pha*A	P(3HB-*co*-3HHx)-	*Aeromonas caviae*	Fukui and Doi (1997) and Fukui et al. (1998)
NADH-acetoacetyl-CoA dehydrogenase	*pha*B			
(*S*)-specific enoyl-CoA hydratase (crotonase)	*pha*J_{Ac}			
PHA synthase	*pha*C			
β-ketothiolase	NA	P(3HB)	*Rhizobium* (*Cicer*) sp. strain CC 1192	Chohan and Copeland (1998)
NADPH-dependent acetoacetyl-CoA reductase	NA			
PHA synthase	NA			
3-ketoacyl-ACP synthase III (FabH)	*fab*H	PHA	*Aeromonas caviae*	Taguchi et al. (1999)
Malonyl-CoA-ACP transacylase (FabD)	*fab*D			
PHA synthase (PhaC*Ac*)	*pha*C_{Ac}			
3-ketothiolase	NA	Poly(3HB-co-4HB)	*Clostridium kluyveri*	Valentin and Dennis (1997)
Acetoacetyl-CoA reductase	NA			
PHA synthase	NA			
Succinic semialdehyde dehydrogenase	*suc*D			
4-hydroxybutyrate dehydrogenase	4*hb*D			
4-hydroxybutyrate-CoA: CoA transferase	*orf*Z			
β -ketothiolase	NA	Poly(3HB-co-3HV-co-4 HV	*Alcaligenes eutrophus*	Valentin and Steinbuchel (1995)
Acetoacetyl-CoA reductase	NA			
3-hydroxyacyl CoA dehydrogenase	NA			
Acyl CoA dehydrogenase	NA			
Acyl CoA synthase	NA			
Acyl CoA transferase	NA			
β-hydroxyacyl-CoA hydrolase	NA			
PHA synthase	NA			
PHA synthase	*pha*C	PHA	*Bacillus megaterium*	Mccool and cannon (2001)
Class IV PHA synthase	*phaRC*$_{Bm}$	P(3HB)	*Bacillus megaterium* NBRC15308T	Tomizawa et al. (2011)
Class IV PHA synthase	*phaRC*$_{By}$	P(3HB)	*B. cereus* YB-4	Tomizawa et al. (2011)
PHA synthase	*pha*C and *pha*R	PHA	*Bacillus* sp. INT005	Satoh et al. (2002)

NA not available

(S)-3-hydroxybutynyl Co-A by FadB. Lastly, Crt acts upon crotonoyl CoA and forms (R)-3-hydroxybutynyl CoA, the common intermediate of both pathways which is polymerized by PhaC (Gong et al. 2012). *B. cereus* has been reported to formulate PHA via acetyl CoA metabolism or via fatty acid oxidation. It contains enzymes such as PhaA, PhaB, PhaJ, PhaRC and FadB for PHA production. PhaJ converts crotonoyl CoA to R-3-hydroxybutyryl CoA. PhaRC functions like PhaC while the remaining enzymes function as described before (Tsugeet et al. 2015).

Generally, PHA biosynthesis competes with TCA cycle for assimilating acetyl-CoA. Under stable growth condition, acetyl-CoA gets oxidized via TCA cycle and NADH generated is utilized for ATP production. In nutrient-deprived state, cell growth is limited, and the NADH accumulates (Faccin et al. 2013). This NADH will inhibit citrate synthase indicating the availability of sufficient amount of ATP and precursor for biosynthesis. Thus, rate of TCA cycle declines (Kim and Gadd 2008). Further, acetyl CoA generated due to the presence of excess substrate is directed towards PHA biosynthesis (Faccin et al. 2013). Decline in TCA cycle is even related with the absence of oxygen. Due to unavailability of O_2, NADH does not get oxidized via electron transport chain (ETC) and gets accumulated there. This inhibits citrate synthase, and rate of TCA cycle reduces (Kim and Gadd 2008; Faccin et al. 2013). Though unavailability of O_2 causes failure of TCA cycle, PHA accumulation is also limited. The reason behind this is oxidation of enzymes essential for PHA production. The inhibitory effect of O_2 over PHA production is solely dependent on the individual organisms (Borah et al. 2002). For appropriate PHA production, optimum aeration is required. Availability of O_2 above and below optimum level reduces PHA accumulation (Faccin et al. 2013).

18.3.4 Molecular Evidences for PHA Production by Bacilli

Biosynthesis of PHA includes numerous classes of enzymes. Three genes termed as *pha*A, *pha*B and *pha*C encode three key enzymes β-ketothiolase, acetoacetyl-CoA reductase and PHA synthase of PHB production pathway. Other genes involved in PHA production are listed in Table 18.6.

B. thuringiensis R1 possesses *pha*P, *pha*Q, *pha*R, *pha*B and *pha*C genes, as depicted in Fig. 18.9. Gene designated as *pha*P encodes for phasins protein and *pha*Q encodes for transcriptional regulatory protein essential which regulates *pha*P activity. Apart from this, *pha*R encodes for unknown product essential for activity of *pha*C. Gene length of these genes along with their promoter regions and

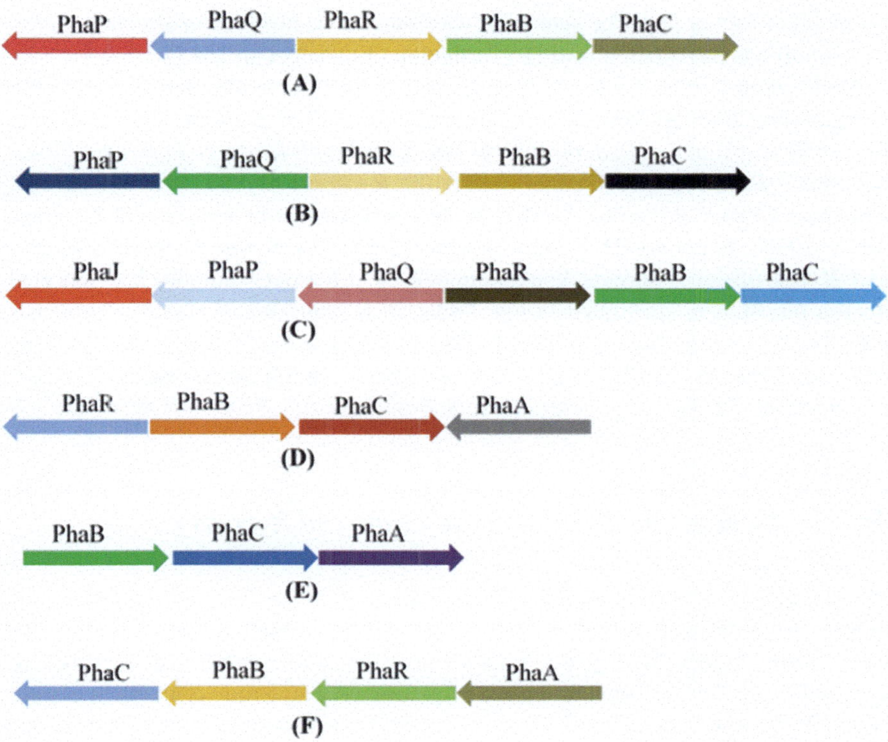

Fig. 18.9 Arrangement of PHA synthesis operon in different *Bacillus* spp. (**a**) *B. thuringiensis* R1, (**b**) *B. megaterium*, (**c**) *B. cereus* subgroups, (**d**) numerous *Bacillus* spp., (**e**) *B. thuringiensis* serovar chinensis CT-43 and *B. megaterium* QM B1551 and (**f**) *B. anthracis* CDC 684 and *B. megaterium* WSH-002

Table 18.7 Nucleotide sequence of R1 promoter and RBS of genes essential for PHA formulation in *Bacillus thuringiensis* (Desetty et al. 2008)

Gene	Size (bp)	Promoter sequence	RBS sequence
*pha*P	522	−10 (^{1099}CACATTTAA^{1091})	^{556}TTGGGG^{551}
*pha*Q	450	−35 (^{1119}AACTGA^{1114})	^{1053}GAGGTG^{1048}
*pha*R	528	−10 (^{1143}AAATAAAAT^{1130})	^{1171}CAGAAT^{1176}
*pha*B	741	−35 (^{1765}TTTCTA^{1770})	^{2682}AAGGAG^{2687}
*pha*C	1083	−10 (^{2606}ATATGTAAT^{2614})	^{2682}AAGGAG^{2687}

ribosomal binding sites (RBS) is listed in Table 18.7 (Desetty et al. 2008). *B. megaterium* have *phaP* (513 bp), *phaQ* (441 bp), *phaR* (609 bp), *phaB* (744 bp) and *phaC* (1089 bp) as shown in Fig. 18.9b (Valappil et al. 2007a). Few *B. cereus* subgroups as shown in Fig. 18.9c have *pha*R-*pha*B-*pha*C operon located in same direction and *pha*P-*pha*Q-*pha*J operon in its contradictory direction. Later operon is found upstream of former one. The gene *pha*J was found to be involved in directing monomer supply via β-oxidation for PHA formation (Tsugeet et al. 2015).

Wide variations in orientation of genes crucial for PHA production were reported in numerous *Bacillus* spp. It was reported that operon *pha*RBCA as shown in Fig. 18.9d was found in *B. anthracis* A0248, *B. anthracis* Ames, *B. anthracis* "Ames Ancestor", *B. cereus* Q1, *B. cereus* 03BB102, *B. cereus* AH820, *B. cereus* B4264, *B. cereus* G9842, *B. cereus* AH187, *B. cereus* biovar anthracis str. CI, *B. cereus* ATCC 10987, *B. cereus* E33L, *B. megaterium* DSM 319, *B. thuringiensis* BMB171 and *B. thuringiensis* serovar konkukian, 97–27. Moreover, authors had reported operon *pha*BCA as illustrated in Fig. 18.9e, in *B. thuringiensis* serovar chinensis CT-43 and *B. megaterium* QM B1551. They had even shown that *B. anthracis* CDC 684 and *B. megaterium* WSH-002 have *pha*CBRA operon, as shown in Fig. 18.9f (Kumar et al. 2013).

18.3.5 Strategy for PHA Accumulation and Recovery

Microorganisms show higher PHA accumulation in favourable environmental conditions. As described earlier in nutrient-depleted environments, few microorganisms ensure higher PHA yield. Growth conditions favourable for microbes to accumulate PHA vary from organism to organism (Albuquerque et al. 2010; Shahid et al. 2013; Saharan et al. 2014; Beun et al. 2002). Also, yield of PHA obtained varies with the time at which PHA is extracted (Pandian et al. 2010; Kumar et al. 2015; Pagliano et al. 2017; Rathika et al. 2018). Moreover, inoculum size and pH also affect the yield of extracted PHA. Higher PHA yield extracted from *B. subtilis* RS1 was obtained with 10% inoculum size and at pH 7 (Rathika et al. 2018). *Bacillus* sp. BPPI-14 and *Bacillus* sp. BPPI-19 showed higher PHA yield at pH 7 and 37 °C using glucose as sole source of carbon (Mohammed et al. 2019). Hence, optimization of PHA accumulation media is required for enhancing PHA yield. Utilization of cheap nutrient source will reduce the cost of industrial PHA production (Verlinden et al. 2011; Mohapatra et al. 2017; Kourmentza et al. 2017).

Another factor which affects industrial level PHA production is the cost associated with PHA recovery. Effective PHA recovery scheme plays an important role in gaining higher PHA yield. As demonstrated in Fig. 18.10, approaches for PHA recovery can be divided into six steps, viz. cellular biomass harvesting, pretreatment, non-PHA cellular biomass disruption, PHA extraction, drying and purification (Kourmentza et al. 2017; Sathya et al. 2018; Jacquel et al. 2008). As PHA is intracellular product, concentration of cellular biomass is carried out. It is harvested using centrifugation or filtration (Kourmentza et al. 2017). The harvested biomass is subjected for pretreatment prior to cell lysis. The main aim of pretreatment is to weaken the microbial cell wall, which involves physical techniques such as lyophilization, ultrasonication and high temperature. The cell lysis of pretreated biomass is carried out using chemical, enzymatic or biological method. The agents used for cell lysis should not deteriorate PHA. Chemical method involves usage of sodium

PHA recovery scheme

Cell biomass harvesting
Filtration, precipitation

Pre-treatment
Lyophilisation, ultrasonication and heating

Non PHA cellular biomass disruption
Chemical - SDS, NaCl, NaClO
Enzymatic - protease, lysozyme
Biological - virus, predator bacteria, rat, mealworm

PHA extraction
Non halogenated organic solvents - ethanol, acetone, hexane, propanol

Drying

Purification
H_2O_2, ozonation

Fig. 18.10 PHA recovery scheme: it includes six steps, viz. cell biomass harvesting, pretreatment, non-PHA cellular biomass disruption, PHA extraction, drying and purification

chloride (NaCl), sodium hypochlorite (NaClO) or sodium dodecyl sulphate (SDS). Enzymatic method includes application of enzymes such as proteases and lysozyme (Kourmentza et al. 2017; Jacquel et al. 2008). For biological cell lysis, researchers have used virus and predatory bacteria. Some researchers have fed the rats and mealworms with biomass, and PHA was recovered from faeces (Kourmentza et al. 2017). Amongst all, chemical methods are widely accepted as they are eco-friendly and do not involve the use of halogenated solvents. Moreover, enzymatic methods are costly, and biological methods are time-consuming (Kourmentza et al. 2017; Jacquel et al. 2008). Traditionally, PHA was extracted using hazardous chlorinated halogenated solvents, such as chloroform, 1, 2-dichloroethane and methylene chloride (Jacquel et al. 2008). Nowadays, PHA extraction is carried out using non-halogenated organic solvents in which PHA is soluble. It includes solvents such as acetone, n-hexane, propanol and ethanol (Kourmentza et al. 2017; Jacquel et al. 2008). Extracted PHA is dried and obtained in the form of powder (Kourmentza et al. 2017; Jacquel et al. 2008). For purification of biopolymer, H_2O_2 or ozonation has been employed (Horowitz et al. 2001; Madkour et al. 2013). Application of cheap and environment-friendly PHA recovery technique is beneficial which needs to be optimized (Kourmentza et al. 2017).

18.3.6 Techniques Involved in Characterization of PHA

A wide range of sophisticated techniques are employed for characterization of extracted PHA powder. Fourier-transform infrared spectroscopy (FTIR), nuclear magnetic resonance (NMR) and gas chromatography-mass spectroscopy (GC-MS) are the most common techniques used for determining the functional group incorporated in PHA. Apart from this, X-ray diffraction (XRD), gel permeation chromatography (GPC), differential scanning calorimetry (DSC) and thermogravimetric analysis (TGA) were used for determining mechanical and thermal properties of PHA (Pradhan et al. 2020; Sathya et al. 2018; Mohapatra et al. 2017; Gumel et al. 2012; Godbole 2016; Johnston et al. 2018). Details of all these techniques are described in Table 18.8.

PHA extracted from *B. licheniformis* AS3-2 was characterized using FTIR (Shah 2012). PHA extracted from *Bacillus cereus* was characterized using scanning electron microscopy (SEM), FTIR, XRD, NMR, DSC and thermal gravimetric analysis (TGA) and confirmed as PHB (Babruwad et al. 2015). PHB-co-PHV extracted from *Bacillus megaterium* OUAT 016 was characterized using FTIR, NMR, XRD and TGA (Mohapatra et al. 2020).

18.3.7 Challenges for Bacilli *to Produce PHA*

Bacillus spp. are capable of producing PHA using cheap substrates including waste materials. Moreover, they are capable of tolerating high pH and high osmotic pressure. Despite of such facts, the major drawback associated with its application for industrial scale PHA production is its sporulating nature (Wu et al. 2001; Mohapatra et al. 2017). Spore formation may utilize energy generated via PHB degradation (Wu et al. 2001). Normally, it was supposed that PHB degradation diverts energy and carbon source for sporulation (Kominek and Halvorson 1965). *B. cereus* has been reported to accumulate PHB prior to sporulation which subsequently gets degraded during spore formation event (Navarro et al. 2006; Kominek and Halvorson 1965). PHB accumulated by *B. thuringiensis* using glucose is utilized for spore formation (Benoit et al. 1990). *Bacillus* sp. JMa5 showed spore formation in nutrient-limiting conditions with low PHB yield. Besides this, it showed growth-associated PHB production. Hence, authors conclude that low levels of nutrients induce sporulation which may limit PHB accumulation (Wu et al. 2001). Further, acidic pH and low level of potassium show decline in spore formation (Mohapatra et al. 2017). *Bacillus* spp. SPV was unable to sporulate in acidic pH and showed PHB accumulation (Valappil et al. 2007b). The antisporogenic agent α-picolinic acid prevents conversion of vegetative *Bacillus cereus* T cells to sporulating cells. TCA cycle enzymes essential for sporulation seem to be synthesized during shift from vegetative phase to sporulation phase. α-picolinic acid is inhibitory to aconitase synthesis and prevents TCA cycle essential for spore formation and not for

Table 18.8 Techniques used for characterization of PHA

Name	Role	Information obtained			References
FTIR	Used for identification of functional group of PHA	–R group	Wave frequency (cm^{-1})		Pradhan et al. (2020)
		–CH	2962–2853		
		–C=O	1742–1709		
		–C–O or –C–C	1300–1709		
		–OH	3460–3407		
NMR	Used to recognize functional group and polymeric content of PHA biopolymer	–R group	^{1}H NMR (ppm)	^{13}C NMR (ppm)	Pradhan et al. (2020)
		–CH	5.2–5.26	67.8–68.5	
		$–CH_2$	2.17–2.7	31.09–41.3	
		$–CH_3$	1.25–1.6	19.95–21.4	
		–C=O	NA	169.1–169.5	
GC-MS	Used for analysing monomeric composition of PHA	Methyl esters of PHA are subjected to GC-MS analysis and functional groups are identified on the basis of retention time			Lee and Choi (1997)
XRD	Useful for studying crystalline nature of PHA	% Xc = At − Aa/At × 100 At = area of crystalline peak Aa = area of amorphous peak			Pradhan et al. (2020)
GPC	Used for determining mw, Mn and polydispersity index	Polydispersity index indicates molecular mass distribution and was calculated by determining Mw/Mn ratio			Pradhan et al. (2020)
DSC	Used for understanding thermal properties such as T_m and X_c of PHA	$\% X_c = \Delta H_m/\Delta H_f \times 100$ ΔH_m = measured enthalpy of polymer ΔH_f = enthalpy of 100% pure PHB (146 J/g)			Pradhan et al. (2020), Gunaratne and Shanks (2005), and Kulkarni et al. (2010)
TGA	Used for analysing thermal stability and degradation as well as resistance temperature of PHA	TGA graph indicates the reduction in weight of PHA with rise in temperature. PHA degradation takes place as the temperature increases. Degradation of PHA correlates with decrease in weight of PHA			Pradhan et al. (2020)

NA not available, *Mw* average molecular weight, *Mn* number average molecular weight, T_m, melting temperature, X_c crystallinity

biomass development. Thus, this antisporogenic agent prevents spore formation (Hanson et al. 1963). Such sort of manipulation of environmental conditions may be supportive in regulating the sporulating nature of *Bacillus* spp. and thus permitting its usage for industrial scale PHA synthesis. Moreover, study by Wang et al. (2016) shows that sporulation in *B. thuringiensis* is not associated with PHB degradation.

They even observed that many *Bacillus* spp. lack *pha*C and *pha*Z genes but still sporulate, indicating individuality of spore formation over PHB degradation. There arises a need of strategies for controlling sporulation for application of such strains in industrial PHA production.

Apart from this, the *Bacillus* spp. possess thick cell wall which makes PHA extraction difficult (Wu et al. 2001; Mohapatra et al. 2017). *B. flexus*, grown in inorganic rich medium contains less diaminopimelic acid and amino acids in cell wall. This allows easier cell lysis, and so higher PHA recovery was found (Divyashree and Shamala 2010). Similar techniques can be used for efficient recovery of PHA from *Bacillus* spp.

18.3.8 Approaches for Improving Properties of PHA for Industrial Application

Most of the biopolymers possess poor mechanical properties. Brittle and fragile nature of biopolymers limits their industrial scale application (Vieira et al. 2011). In order to improve their mechanical properties, they are either blended with plasticizer and/or cross-linking agent (Jantrawut et al. 2017) or blended with other polymers (Mohapatra et al. 2017; Narancic et al. 2018). Plasticizers have low molecular weight and are non-volatile compounds. They are widely used in polymer industry for enhancing the properties of polymers. They are known to reduce Tg of polymers and supplement their biodegradation (Vieira et al. 2011). Usage of eco-friendly nature-based biodegradable plasticizer is advantageous over conventional plasticizer such as phthalates (Vieira et al. 2011). Thermal and mechanical properties of PHBV films improve after blending with biodegradable plasticizers such as soybean oil (SO), epoxidized soybean oil (ESO), dibutyl phthalate (DBP) and triethyl citrate (TEC) (Vieira et al. 2011). Cross-linking agent forms intermolecular cross linkages with biopolymer, permitting suitable biopolymer film formation (Jantrawut et al. 2017). Further, application of some additives along with plasticizer enhances enzymatic degradation of PHB (Vieira et al. 2011).

Blending of P(3HB) extracted from *B. megaterium* Ti3 with polyethylene glycol enhances biocompatibility of P(3HB) film (Israni et al. 2020). The blends of PHB extracted from *B. cereus* strain VIT-SSR1 isolated from industrial effluents were prepared with chitosan. Biocompatibility of these blends was investigated on L929 mouse fibroblast cell line with MTT assay. They proved to be biocompatible and hence can be used for drug delivery (Evangeline and Sridharan 2019). PHBV extracted from *B. aryabhattai* PHB10 was blended with polyethylene glycol, and cytotoxicity was analysed on human keratinocytes (HaCat cells). Approximately 99% cells were viable, and hence this blend can be employed for skin graft application (Pillai et al. 2020).

18.3.9 *Commercial Applications of PHA Obtained from* Bacilli

Microbially originated biodegradable PHA have plenty of applications. They are used for manufacturing of packaging material and biomedical products (Mohapatra et al. 2017; Chen 2010; Sathya et al. 2018). Moreover, they are used as drug delivery carriers, as pharmaceutical products/drugs and as biofuels (Chen 2010). PHA are non-toxic in nature and hence are biocompatible, allowing its biomedical applications (Pradhan et al. 2020). They are even used for agricultural purposes (Pradhan et al. 2020; Sowinski et al. 2010).

Commercially, manufacturer entitled as PHB Industrial S.A., Brazil, have employed *Bacillus* spp. under Biocycle trademark. They are exploited for P(3HB) production from sugarcane (Ciesielska and Kiewisz 2016). PHB extracted from pigmented *Bacillus* sp. C1 (2013) (KF626477) was biocompatible in nature and hence can be used as drug delivery carrier (Pati et al. 2020). PHB of *B. thuringiensis* is non-toxic and is suitable for biomedical purpose (El-Abd et al. 2017). PHB-co-PHV extracted from *B. megaterium* OUAT 016 was found to be biocompatible and can be used as drug delivery carrier (Mohapatra et al. 2020). P(3HB-co-HV) extracted from *B. thermoamylovorans* was esterified using methanol and H_2SO_4. This, P(3HB-co-HV) methyl ester can be used as biofuel (Sangkharak et al. 2020). PHA levofloxacin nanoparticles were prepared using PHA extracted from *B. subtilis* NCDC0671, and its levofloxacin releasing efficacy was proved to be efficient. Hence, they can be used for delivering levofloxacin drug (Umesh et al. 2017).

18.3.10 *PHA Depolymerase of* Bacilli *and Biodegradation*

The most attractive feature of PHA is its biodegradability. It is composed of 100% natural biobased resources (Pradhan et al. 2020). In aerobic conditions, they get transformed into CO_2 and H_2O, whereas in anaerobic conditions, CH_4 is obtained additionally. It undergoes thermal degradation as well as enzymatic and non-enzymatic degradation (Pradhan et al. 2020; Nesic et al. 2020). It gets degraded when exposed to soil or compost and even in marine sediments (Nesic et al. 2020). It is prone to get degraded by microbial PHA depolymerase (i-PhaZ and e-PhaZ) or non-enzymatically inside animal tissues. It takes about few months to a year to get degraded in anaerobic conditions. Degradation rate boosts up in UV light. PHA polymers with high T_m take longer duration for degradation. Apart from this, PHA with low molecular weights gets degraded faster. Factors affecting rate of PHA degradation are temperature, pressure, moisture, surface area, pH and type of microorganism (Pradhan et al. 2020).

The role of i-PhaZ and e-PhaZ is described in earlier section, i.e. in functions of PHA and structure of PHA granule. *Bacillus megaterium* has i-PhaZ designated as PhaZ1 and degrades PHB into 3-hydroxybutyric acid monomers (Chen et al. 2009). *B. thuringiensis* possess P(3HB) depolymerase designated as PhaZ (Huang et al.

2012). *Bacillus* sp. strain NRRL B-14911, *B. megaterium*, *B. pseudofirmus* and *Bacillus* sp. strain SG-1 have e-PhaZ (Ma et al. 2011). *B. megaterium* N-18-25-9 contains e-PhaZ gene designated as $phaZ_{Bm}$ (Takaku et al. 2006). *B. thuringiensis* subsp. *israelensis* ATCC 35646 contains gene designated as *phaZ* have function similar to intracellular P(3HB) depolymerase (Tseng et al. 2006).

18.4 Future Prospects

Biodegradable biopolymers are recognized as potential substitute for conventional petrochemical-based plastics. *Bacillus* spp. are considered as promising agents for PHA production. A wide range of lab level investigations on *Bacillus* and PHA have been carried out till date. But their application in actual biopolymer industry needs attention. Moreover, varied spectrum of microbial PHA is known to exist, but most of the studies are constricted to P(3HB). Finding out additional forms of PHA having plastics like properties might be advantageous. Genetically engineered *Bacillus* strain ensuring higher PHA-producing capabilities can be the targeted aim of future research. Methyl esters of PHA have properties similar to biofuel so exploring the application of PHA is obtained using waste substrates, as biofuel will allow sustainable management of waste along with formation of valuable by-product. The PHA polymeric proteins such as PhaP, PhaZ and PhaC may be applicable as potential drug delivering tools.

18.5 Conclusions

Plastic pollution is one of the major concerns in the world. Biodegradable biopolymers being solutions to such issue have gained much attention amongst scientific community. Primarily, microbial PHA are significantly valuable as they are solely biobased and totally biodegradable. They are carbon-rich inclusion bodies synthesized by microorganisms in response to stress conditions. The polyester PHA granules are comprised of hydroxyalkanoic acid, PhaP, PhaC, PhaZ and PhaR. Micelle formation and budding are the two strategies known for PHA formulation in microorganisms. *Scl*-PHA have properties identical to polystyrene, whereas *mcl*-PHA are widely accepted in medical industry.

First and foremost concern for using PHA in bio-industry depends on the choice of microbial strain and cost-effective nutrient source. *Bacillus* spp. are known to serve as appropriate bacteria for industrial application. They are also recognized to produce PHA from varieties of inexpensive waste. They are known to produce PHA in the presence of excess substrate with either nutrient-deprived or non-nutrient-deprived conditions. Metabolic pathway of PHA production and operons encoding enzymes of pathway in *Bacilli* are well known. The media conditions known to enhance PHA accumulation in bacilli need to be optimized in order to increase the yield. Accomplishing appropriate cost-effective eco-friendly PHA recovery strategy can enable us to meet the need of industries. Advances in modern science have led to development of many sophisticated techniques for characterization of PHA. Features such as sporulation and thick cell wall associated with *Bacillus* spp. hinder its applicability for industrial scale PHA manufacturing. Abundant research had been done to solve such issues. Moreover, blending of PHA extracted from *Bacilli* with plasticizer or cross-linking agent has increased its biocompatibility, making them suitable for biomedical applications. Blending also decreases the time required for PHA degradation. The enzymes i-PhaZ as well as e-PhaZ from *Bacilli* origin play a vital role in biodegradation of PHA. Thus, *Bacillus* spp. are promising resources for bioplastic industry.

References

Akaraonye E, Moreno C, Knowles JC, Keshavarz T, Roy I (2012) Poly(3-hydroxybutyrate) production by *Bacillus cereus* SPV using sugarcane molasses as the main carbon source. Biotechnol J 7(2):293–303. https://doi.org/10.1002/biot.201100122

Albuquerque MGE, Torres CAV, Reis MAM (2010) Polyhydroxyalkanoate (PHA) production by a mixed microbial culture using sugar molasses: effect of the influent substrate concentration on culture selection. Water Res 44(11):3419–3433. https://doi.org/10.1016/j.watres.2010.03.021

Aneja P, Dai M, Lacorre DA, Pillon B, Charles TC (2004) Heterologous complementation of the exopolysaccharide synthesis and carbon utilization phenotypes of *Sinorhizobium meliloti* Rm1021 polyhydroxyalkanoate synthesis mutants. FEMS Microbiol Lett 239(2):277–283. https://doi.org/10.1016/j.femsle.2004.08.045

Babruwad PR, Shruti UP, Kishor PU, Hungund BS (2015) Production and characterization of a thermostable bioplastic (poly-s-hydroxybutyrate) from *Bacillus cereus* NRRL-b-3711. J Biochem Technol 6(3):990–995

Basnett P, Roy I (2010) Microbial production of biodegradable polymers and their role in cardiac stent development. Appl Microbiol Biotechnol 2:1405–1415

Benoit TG, Wilson GR, Baugh CL (1990) Fermentation during growth and sporulation of *Bacillus thuringiensis* HD-1. Lett Appl Microbiol 10(1):15–18. https://doi.org/10.1111/j.1472-765X.1990.tb00084.x

Beun JJ, Dircks K, Van Loosdrecht MCM, Heijnen JJ (2002) Poly-β-hydroxybutyrate metabolism in dynamically fed mixed microbial cultures. Water Res 36(5):1167–1180. https://doi.org/10.1016/S0043-1354(01)00317-7

Bhuwal AK, Singh G, Aggarwal NK, Goyal V, Yadav A (2013) Isolation and screening of polyhydroxyalkanoates producing bacteria from pulp, paper, and cardboard industry wastes. Int J Biomater 2013:1–10. https://doi.org/10.1155/2013/752821

Borah B, Thakur PS, Nigam JN (2002) The influence of nutritional and environmental conditions on the accumulation of poly-β-hydroxybutyrate in *Bacillus mycoides* RLJ B-017. J Appl Microbiol 92(4):776–783. https://doi.org/10.1046/j.1365-2672.2002.01590

Bugnicourt E, Cinelli P, Lazzeri A, Alvarez V (2014) Polyhydroxyalkanoate (PHA): review of synthesis, characteristics, processing and potential applications in packaging. Express Polym Lett 8(11):791–808. https://doi.org/10.3144/expresspolymlett.2014.82

Campisano A, Overhage J, Rehm BHA (2008) The polyhydroxyalkanoate biosynthesis genes are differentially regulated in planktonic- and biofilm-grown *Pseudomonas aeruginosa*. J Biotechnol 133(4):442–452. https://doi.org/10.1016/j.jbiotec.2007.11.007

Cardozo GJR, Martinez MAL, Perez YM, Londono CGA (2016) Production and characterization of polyhydroxyalkanoates and native microorganisms synthesized from fatty waste. Int J Polym Sci 2016:1–12. https://doi.org/10.1155/2016/6541718

Chen GU (2010) Plastics completely synthesized by bacteria: polyhydroxyalkanoates. Plastics from bacteria: natural functions and applications. Microbiol Monogr 14:17–37. https://doi.org/10.1007/978-3-642-03287_5_2

Chen HJ, Pan SC, Shaw GC (2009) Identification and characterization of a novel intracellular poly(3-hydroxybutyrate) depolymerase from *Bacillus megaterium*. Appl Environ Microbiol 75(16):5290–5299. https://doi.org/10.1128/AEM.00621-09

Chohan SN, Copeland L (1998) Acetoacetyl coenzyme a reductase and polyhydroxybutyrate synthesis in *Rhizobium* (Cicer) sp. strain CC 1192. Appl Environ Microbiol 64(8):2859–2863. https://doi.org/10.1128/aem.64.8.2859-2863.1998

Chua ASM, Takabatak H, Satoh H, Mino T (2003) Production of polyhydroxyalkanoates (PHA) by activated sludge treating municipal wastewater: effect of pH, sludge retention time (SRT), and acetate concentration in influent. Water Res 37(15):3602–3611. https://doi.org/10.1016/S0043-1354(03)00252-5

Ciesielska MJ, Kiewisz R (2016) Bacterial polyhydroxyalkanoates: still fabulous? Microbiol Res 192:271–282. https://doi.org/10.1016/j.micres.2016.07.010

Coats ER, Watson BS, Brinkman CK (2016) Polyhydroxyalkanoate synthesis by mixed microbial consortia cultured on fermented dairy manure: effect of aeration on process rates/yields and the associated microbial ecology. Water Res 106(208):26–40. https://doi.org/10.1016/j.watres.2016.09.039

Colombo B, Favini F, Scaglia B, Sciarria TP, D'Imporzano G, Pognani M, Alekseeva A, Eisele G, Cosentino C, Adani F (2017) Enhanced polyhydroxyalkanoate (PHA) production from the organic fraction of municipal solid waste by using mixed microbial culture. Biotechnol Biofuels 10(1):1–15. https://doi.org/10.1186/s13068-017-0888-8

Contreras RA, Koller M, Miranda-de Sousa Dias M, Calafell-Monfort M, Braunegg G, Marques-Calvo MS (2013) High production of poly(3-hydroxybutyrate) from a wild *Bacillus megaterium* Bolivian strain. J Appl Microbiol 114(5):1378–1387. https://doi.org/10.1111/jam.12151

Danez JCA, Requiso PJ, Alfafara CG, Nayve Jr. FRP, Ventura JRS (2020) Optimization of fermentation factors for polyhydroxybutyrate (PHB) production using *Bacillus megaterium* PNCM 1890 in simulated glucose-xylose hydrolysates from agricultural residues. Philip J Sci 149(1):163–175

Desetty RD, Mahajan VS, Khan BM, Rawal SK (2008) Isolation and heterologous expression of PHA synthesising genes from *Bacillus thuringiensis* R1. World J Microbiol Biotechnol 24(9):1769–1774. https://doi.org/10.1007/s11274-008-9669-7

Divyashree MS, Shamala TR (2010) Extractability of polyhydroxyalkanoate synthesized by *Bacillus flexus* cultivated in organic and inorganic nutrient media. Indian J Microbiol 50(1):63–69. https://doi.org/10.1007/s12088-010-0013-1

El-Abd MAEH, Sheikh HH, Desouky S, Shehab A (2017) Identification, biodegradation and bio-evaluation of biopolymer produced from *Bacillus thuringenesis*. J Appl Pharm Sci 7(4):103–110. https://doi.org/10.7324/JAPS.2017.70414

Evangeline S, Sridharan TB (2019) Biosynthesis and statistical optimization of polyhydroxyalkanoate (PHA) produced by *Bacillus cereus* VIT-SSR1 and fabrication of biopolymer films for sustained drug release. Int J Biol Macromol 135:945–958. https://doi.org/10.1016/j.ijbiomac.2019.05.163

Faccin DJL, Rech R, Secchi AR, Cardozo NSM, Ayub MAZ (2013) Influence of oxygen transfer rate on the accumulation of poly(3-hydroxybutyrate) by *Bacillus megaterium*. Process Biochem 48(3):420–425. https://doi.org/10.1016/j.procbio.2013.02.004

Fukui T, Doi Y (1997) Cloning and analysis of the poly(3-hydroxybutyrate-co-3-hydroxyhexanoate) biosynthesis genes of *Aeromonas caviae*. J Bacteriol 179(15):4821–4830. https://doi.org/10.1128/jb.179.15.4821-4830.1997

Fukui T, Shiomi N, Doi Y (1998) Expression and characterization of (R)-specific enoyl coenzyme a hydratase involved in polyhydroxyalkanoate biosynthesis by *Aeromonas caviae*. J Bacteriol 180(3):667–673. https://doi.org/10.1128/jb.180.3.667-673.1998

Gamba AM, Fonseca JS, Méndez DA, Viloria A, Fajardo D, Moreno NC, Cabeza IO (2017) Assessment of different plasticizer – polyhydroxyalkanoate mixtures to obtain biodegradable polymeric films. Chem Eng Trans 57:1363–1368. https://doi.org/10.3303/CET1757228

Getachew A, Berhanu A (2016) Production of sterilized medium chain length polyhydroxyalkanoates (Smcl-Pha) as a biofilm to tissue engineering application. J Tissue Sci Eng 7(2):1–9

Getachew A, Woldesenbet F (2016) Production of biodegradable plastic by polyhydroxybutyrate (PHB) accumulating bacteria using low cost agricultural waste material. BMC Res Notes 9(1):1–9. https://doi.org/10.1186/s13104-016-2321-y

Godbole S (2016) Methods for identification, quantification and characterization of polyhydroxyalkanoates. Int J Bioassays 5(4):4977–4983

Gong Y, Li M, Xu D, Wang H, He J, Wu D, Chen D, Qiu N, Bao Q, Sun M, Yu Z (2012) Comparative proteomic analysis revealed metabolic changes and the translational regulation of cry protein synthesis in *Bacillus thuringiensis*. J Proteome 75(4):1235–1246. https://doi.org/10.1016/j.jprot.2011.10.037

Goudarztalejerdi A, Tabatabaei M, Eskandari MH, Mowla D, Iraji A (2015) Evaluation of bioremediation potential and biopolymer production of *Pseudomonas* isolated from petroleum hydrocarbon-contaminated areas. Int J Environ Sci Technol 12(9):2801–2808. https://doi.org/10.1007/s13762-015-0779-0

Grage K, Jahns CA, Parlane N, Palanisamy R, Rasiah IA, Jane A, Rehm BHA (2009) Biomacromolecules 10:660–669. https://doi.org/10.1021/bm801394s

Gumel AM, Annuar MSM, Heidelberg T (2012) Biosynthesis and characterization of polyhydroxyalkanoates copolymers produced by *Pseudomonas putida* Bet001 isolated from palm oil mill effluent. PLoS One 7(9):1–8. https://doi.org/10.1371/journal.pone.0045214

Gunaratne LMWK, Shanks RA (2005) Melting and thermal history of poly(hydroxybutyrate-co-hydroxyvalerate) using step-scan DSC. Thermochim Acta 430(1–2):183–190. https://doi.org/10.1016/j.tca.2005.01.060

Hanson RS, Srinivasan VR, Halvorson HO (1963) Enzymatic changes during sporulation of *Bacillus cereus*. J Bacteriol 86:45–50. https://doi.org/10.1128/jb.86.1.45-50.1963

Hassan MA, Bakhiet EK, Ali SG, Hussien HR (2016) Production and characterization of polyhydroxybutyrate (PHB) produced by *Bacillus* sp. isolated from Egypt. J Appl Pharm Sci 6(4):46–51. https://doi.org/10.7324/JAPS.2016.60406

Horowitz DM, Somerville MA, Elaine MB, Somerville. Patent application publication. US 2001/0006802 A1, United States, July 5 2001

Huang YL, Chung TW, Chang CM, Chen CH, Liao CC, Tsay YG, Shaw GC, Liaw SH, Sun CM, Lin CH (2012) Qualitative analysis of the fluorophosphonate-based chemical probes using the serine hydrolases from mouse liver and poly-3-hydroxybutyrate depolymerase (PhaZ) from *Bacillus thuringiensis*. Anal Bioanal Chem 404(8):2387–2396. https://doi.org/10.1007/s00216-012-6349-0

Israni N, Venkatachalam P, Gajaraj B, Varalakshmi KN, Shivakumar S (2020) Whey valorization for sustainable polyhydroxyalkanoate production by *Bacillus megaterium*: production, characterization and in vitro biocompatibility evaluation. J Environ Manag 255(2020):109884. https://doi.org/10.1016/j.jenvman.2019.109884

Jacquel N, Lo CW, Wei YH, Wu HS, Wang SS (2008) Isolation and purification of bacterial poly(3-hydroxyalkanoates). Biochem Eng J 39(1):15–27. https://doi.org/10.1016/j.bej.2007.11.029

Jantrawut P, Chaiwarit T, Jantanasakulwong K, Brachais CH, Chambin O (2017) Effect of plasticizer type on tensile property and in vitro indomethacin release of thin films based on low-methoxyl pectin. Polymers 9(7):1–14. https://doi.org/10.3390/polym9070289

Jendrossek D (2009) Polyhydroxyalkanoate granules are complex subcellular organelles (carbonosomes). J Bacteriol 191(10):3195–3202. https://doi.org/10.1128/JB.01723-08

Jendrossek D, Handrick R (2002) Microbial degradation of polyhydroxyalkanoates. Annu Rev Microbiol 56:403–432. https://doi.org/10.1146/annurev.micro.56.012302.160838

Jiang G, Hill DJ, Kowalczuk M, Johnston B, Adamus G, Irorere V, Radecka I (2016) Carbon sources for polyhydroxyalkanoates and an integrated biorefinery. Int J Mol Sci 17(7):1157, 1–21. https://doi.org/10.3390/ijms17071157

Johnston B, Radecka I, Hill D, Chiellini E, Ilieva VI, Sikorska W, Musioł M, Zięba M, Marek AA, Keddie D, Mendrek B, Darbar S, Adamus G, Kowalczuk M (2018) The microbial production of polyhydroxyalkanoates from waste polystyrene fragments attained using oxidative degradation. Polymers 10(9):957, 1–22. https://doi.org/10.3390/polym10090957

Kadouri D, Jurkevitch E, Okon Y (2003) Involvement of the reserve material poly-β-hydroxybutyrate in *Azospirillum brasilense* stress endurance and root colonization. Appl Environ Microbiol 69(6):3244–3250. https://doi.org/10.1128/AEM.69.6.3244-3250.2003

Kim BH, Gadd GM (2008) Bacterial physiology and metabolism. Cambridge University Press, Cambridge, pp 60–80. www.cambridge.org/9780521846363

Koller M (2017) Advances in polyhydroxyalkanoate (PHA) production. Bioengineering 4:88. https://doi.org/10.3390/bioengineering4040088

Kominek LA, Halvorson HO (1965) Metabolism of poly-beta-hydroxybutyrate and acetoin in *Bacillus cereus*. J Bacteriol 90(5):1251–1259. https://doi.org/10.1128/jb.90.5.1251-1259.1965

Kourmentza C, Placido J, Venetsaneas N, Burniol-Figols A, Varrone C, Gavala HN, Rei MAM (2017) Recent advances and challenges towards sustainable polyhydroxyalkanoate (PHA) production. Bioengineering 4(2):1–43. https://doi.org/10.3390/bioengineering4020055

Kulkarni SO, Kanekar PP, Nilegaonkar SS, Sarnaik SS, Jog JP (2010) Production and characterization of a biodegradable poly (hydroxybutyrate-co-hydroxyvalerate) (PHB-co-PHV) copolymer by moderately haloalkalitolerant *Halomonas campisalis* MCM B-1027 isolated from Lonar Lake. India Bioresour Technol 101(24):9765–9771. https://doi.org/10.1016/j.biortech.2010.07.089

Kulpreecha S, Boonruangthavorn A, Meksiriporn B, Thongchul N (2009) Inexpensive fed-batch cultivation for high poly(3-hydroxybutyrate) production by a new isolate of *Bacillus megaterium*. J Biosci Bioeng 107(3):240–245. https://doi.org/10.1016/j.jbiosc.2008.10.006

Kumar SM, Mudliar SN, Reddy KMK, Chakrabarti T (2004) Production of biodegradable plastics from activated sludge generated from a food processing industrial wastewater treatment plant. Bioresour Technol 95:327–330. https://doi.org/10.1016/j.biortech

Kumar P, Patel SKS, Lee JK, Kalia VC (2013) Extending the limits of *Bacillus* for novel biotechnological applications. Biotechnol Adv 31(8):1543–1561. https://doi.org/10.1016/j.biotechadv.2013.08.007

Kumar P, Singh M, Mehariya S, Patel SKS, Lee JK, Kalia VC (2014) Ecobiotechnological approach for exploiting the abilities of *Bacillus* to produce co-polymer of polyhydroxyalkanoate. Indian J Microbiol 54(2):151–157. https://doi.org/10.1007/s12088-014-0457-9

Kumar P, Ray S, Patel SKS, Lee JK, Kalia VC (2015) Bioconversion of crude glycerol to polyhydroxyalkanoate by *bacillus thuringiensis* under non-limiting nitrogen conditions. Int J Biol Macromol 78:9–16. https://doi.org/10.1016/j.ijbiomac.2015.03.046

Lee EY, Choi CY (1997) Structural identification of polyhydroxyalkanoic acid (PHA) containing 4-hydroxyalkanoic acids by gas chromatography-mass spectrometry (GC-MS) and its application to bacteria screening. Biotechnol Tech 11(3):167–171. https://doi.org/10.1023/A:1018401513621

Ma WT, Lin JH, Chen HJ, Chen SY, Shaw GC (2011) Identification and characterization of a novel class of extracellular poly(3-hydroxybutyrate) depolymerase from *Bacillus* sp. strain NRRL B-14911. Appl Environ Microbiol 77(22):7924–7932. https://doi.org/10.1128/AEM.06069-11

Madkour MH, Heinrich D, Alghamdi MA, Shabbaj II, Steinbuchel A (2013) PHA recovery from biomass. Biomacromolecules 14(9):2963–2972. https://doi.org/10.1021/bm4010244

Mccool GJ, Cannon MC (2001) PhaC and PhaR are required for polyhydroxyalkanoic acid synthase activity in *Bacillus megaterium*. J Bacteriol 183(14):4235–4243. https://doi.org/10.1128/JB.183.14.4235-4243.2001

Mezzina MP, Pettinari MJ (2016) Phasins, multifaceted polyhydroxyalkanoate granule-associated proteins. Appl Environ Microbiol 82(17):5060–5067. https://doi.org/10.1128/AEM.01161-16
Mohammed S, Panda AN, Ray L (2019) An investigation for recovery of polyhydroxyalkanoates (PHA) from *Bacillus* sp. BPPI-14 and *Bacillus* sp. BPPI-19 isolated from plastic waste landfill. Int J Biol Macromol 134:1085–1096. https://doi.org/10.1016/j.ijbiomac.2019.05.155
Mohanrasu K, Rao RGR, Dinesh GH, Zhang K, Prakash GS, Song DP, Muniyasamy S, Pugazhendhi A, Jeyakanthan J, Arun A (2020) Optimization of media components and culture conditions for polyhydroxyalkanoates production by *Bacillus megaterium*. Fuel 271(March):1–9. https://doi.org/10.1016/j.fuel.2020.117522
Mohapatra S, Maity S, Dash HR, Das S, Pattnaik S, Rath C, Samantaray D (2017) *Bacillus* and biopolymer: prospects and challenges. Biochem Biophys Rep 12:206–213. https://doi.org/10.1016/j.bbrep.2017.10.001
Mohapatra S, Pattnaik S, Maity S, Sharma S, Akhtar J, Pati S, Samantaray DP, Varma A (2020) Comparative analysis of PHAs production by *Bacillus megaterium* OUAT 016 under submerged and solid-state fermentation. Saudi J Biol Sci 27(5):1242–1250. https://doi.org/10.1016/j.sjbs.2020.02.001
Mokhtarani N, Ganjidoust H, Vasheghani Farahani E (2012) Effect of process variables on the production of polyhydroxyalkanoates by activated sludge. Iran J Environ Health Sci Eng 9(6):1–7. https://doi.org/10.1186/1735-2746-9-6
Mulchandani N, Katiyar V (2020) Synthesis strategies for biomedical grade polymers. In: Katiyar V, Kumar A, Mulchandani N (eds) Advances in sustainable polymers, materials horizons: from nature to nanomaterials. Springer, Singapore, pp 1–20. https://doi.org/10.1007/978-981-15-1251-3_1
Narancic T, Verstichel S, Reddy Chaganti S, Morales-Gamez L, Kenny ST, De Wilde B, Babu Padamati R, O'Connor KE (2018) Biodegradable plastic blends create new possibilities for end-of-life management of plastics but they are not a panacea for plastic pollution. Environ Sci Technol 52(18):10441–10452. https://doi.org/10.1021/acs.est.8b02963
Navarro AK, Farrera RR, Lopez R, Perez-Guevara F (2006) Relationship between poly-β-hydroxybutyrate production and δ-endotoxin for *Bacillus thuringiensis* var. kurstaki. Biotechnol Lett 28(9):641–644. https://doi.org/10.1007/s10529-006-0029-0
Nesic A, Castillo C, Castan P (2020) Bio-based packaging materials. In: Galanakis CM (ed) Biobased products and industries. Elsevier, Amsterdam, pp 279–309. https://doi.org/10.1016/B978-0-12-818493-6.00008-7
Nikodinovic J, Kenny ST, Babu RP, Woods T, Blau WJ, O'Connor KE (2008) The conversion of BTEX compounds by single and defined mixed cultures to medium-chain-length polyhydroxyalkanoate. Appl Microbiol Biotechnol 80(4):665–673. https://doi.org/10.1007/s00253-008-1593-0
Numata K, Abe H, Iwata T (2009) Biodegradability of poly(hydroxyalkanoate) materials. Materials 2(3):1104–1126. https://doi.org/10.3390/ma2031104
Pagliano G, Ventorino V, Panico A, Pepe O (2017) Integrated systems for biopolymers and bioenergy production from organic waste and by-products: a review of microbial processes. Biotechnol Biofuels 10(1):1–24. https://doi.org/10.1186/s13068-017-0802-4
Pandian RS, Deepak V, Kalishwaralal K, Rameshkumar N, Jeyaraj M, Gurunathan S (2010) Optimization and fed-batch production of PHB utilizing dairy waste and sea water as nutrient sources by *Bacillus megaterium* SRKP-3. Bioresour Technol 101(2):705–711. https://doi.org/10.1016/j.biortech.2009.08.040
Patel SKS, Singh M, Kalia VC (2011) Hydrogen and polyhydroxybutyrate producing abilities of *Bacillus* spp. from glucose in two stage system. Indian J Microbiol 51(4):418–423. https://doi.org/10.1007/s12088-011-0236-9
Pati S, Maity S, Dash A, Jema S, Mohapatra S, Das S, Samantaray DP (2020) Biocompatible PHB production from *Bacillus* species under submerged and solid-state fermentation and extraction through different downstream processing. Curr Microbiol 77:1–7. https://doi.org/10.1007/s00284-020-01922-7

Penkhrue W, Jendrossek D, Khanongnuch C, Pathomareeid W, Aizawa T, Behrens RL, Lumyongid S (2020) Response surface method for polyhydroxybutyrate (PHB) bioplastic accumulation in *Bacillus* drentensis BP17 using pineapple peel. PLoS One 15(3):1–21. https://doi.org/10.1371/journal.pone.0230443

Philip S, Keshavarz T, Roy I (2007) Polyhydroxyalkanoates: biodegradable polymers with a range of applications. J Chem Technol Biotechnol 82:233–247. https://doi.org/10.1002/jctb.1667

Pillai BA, Kumar JA, Kumarapillai H (2020) Biosynthesis of poly(3-hydroxybutyrate-co-3-hydroxyvalerate) (PHBV) in *Bacillus aryabhattai* and cytotoxicity evaluation of PHBV/poly(ethylene glycol) blends. 3 Biotech 10(2):1–10. https://doi.org/10.1007/s13205-019-2017-9

Pittmann T, Steinmetz H (2017) Polyhydroxyalkanoate production on waste water treatment plants: process scheme, operating conditions and potential analysis for German and European municipal waste water treatment plants. Bioengineering 4(2):1–24. https://doi.org/10.3390/bioengineering4020054

Potter M, Steinbuchel A (2005) Poly(3-hydroxybutyrate) granule-associated proteins: impacts on poly(3-hydroxybutyrate) synthesis and degradation. Biomacromolecules 6:552–560. https://doi.org/10.1021/bm049401n

Potter M, Steinbuchel A (2006) Biogenesis and structure of polyhydroxyalkanoate granules in Shively JM (ed) Inclusions in prokaryotes. Springer, pp 110–129. https://doi.org/10.1007/7171_005/. Published online: 8 Apr 2006

Pradhan SP, Kumar D, Moholkar S (2020) Production, characterization, and application of biodegradable polymer: polyhydroxyalkanoates. In: Katiyar V, Kumar A, Mulchandani N (eds) Advances in sustainable polymers, materials horizons: from nature to nanomaterials. Springer, Singapore, pp 1–20. https://doi.org/10.1007/978-981-15-1251-3_1

Rathika R, Janaki V, Shanthi K, Kamala-Kannan S (2018) Bioconversion of agro-industrial effluents for polyhydroxyalkanoates production using *Bacillus subtilis* RS1. Int J Environ Sci Technol 16(10):5725–5734. https://doi.org/10.1007/s13762-018-2155-3

Reddy VM, Amulya K, Rohit MV, Sarma PN, Venkata MS (2014) Valorization of fatty acid waste for bioplastics production using Bacillus tequilensis: integration with dark-fermentative hydrogen production process. Int J Hydrog Energy 39(14):7616–7626. https://doi.org/10.1016/j.ijhydene.2013.09.157

Rehm BHA (2006) Genetics and biochemistry of polyhydroxyalkanoate granule self-assembly: the key role of polyester synthases. Biotechnol Lett 28(4):207–213. https://doi.org/10.1007/s10529-005-5521

Saharan SB, Grewal A, Kumar P (2014) Biotechnological production of polyhydroxyalkanoates: a review on trends and latest developments. Chin J Biol 2014:1–18. https://doi.org/10.1155/2014/802984

Sangkharak K, Khaithongkaeo P, Chuaikhunupakarn T, Choonut A, Prasertsan P (2020) The production of polyhydroxyalkanoate from waste cooking oil and its application in biofuel production. Biomass Convers Biorefin:1–14. https://doi.org/10.1007/s13399-020-00657-6

Sanyang ML, Sapuan SM, Jawaid M, Ishak MR, Sahari J (2015) Effect of plasticizer type and concentration on tensile, thermal and barrier properties of biodegradable films based on sugar palm (Arenga pinnata) starch. Polymers 7(6):1106–1124. https://doi.org/10.3390/polym7061106

Sathya AB, Sivasubramanian V, Santhiagu A, Sebastian C, Sivashankar R (2018) Production of polyhydroxyalkanoates from renewable sources using bacteria. J Polym Environ 26(9):3995–4012. https://doi.org/10.1007/s10924-018-1259-7

Sato S, Ono Y, Mochiyama Y, Sivaniah E, Kikkawa Y, Sudesh K, Hiraishi T, Doi Y, Abe H, Tsuge T (2008) Polyhydroxyalkanoate film formation and synthase activity during in vitro and in situ polymerization on hydrophobic surfaces. Biomacromolecules 9(10):2811–2818. https://doi.org/10.1021/bm800566s

Satoh Y, Minamoto N, Tajima K, Munekata M (2002) Polyhydroxyalkanoate synthase from *Bacillus* sp. INTO05 is composed of PhaC and PhaR. J Biosci Bioeng 94(4):343–350

Shah K (2012) FTIR analysis of polyhydroxyalkanoates by a locally isolated novel *Bacillus* sp. AS 3-2 from soil of Kadi region, North Gujarat, India. J Biochem Technol 3(4):380–383

Shahid S, Mosrati R, Ledauphin JO, Amiel C, Fontaine P, Gaillard JL, Corroler D (2013) Impact of carbon source and variable nitrogen conditions on bacterial biosynthesis of polyhydroxyalkanoates: evidence of an atypical metabolism in *Bacillus megaterium* DSM 509. J Biosci Bioeng 116(3):302–308. https://doi.org/10.1016/j.jbiosc.2013.02.017

Shi HP, Lee CM, Ma WH (2007) Influence of electron acceptor, carbon, nitrogen, and phosphorus on polyhydroxyalkanoate (PHA) production by *Brachymonas* sp. P12. World J Microbiol Biotechnol 23(5):625–632. https://doi.org/10.1007/s11274-006-9271-9

Singh M, Patel SK, Kalia VC (2009) *Bacillus subtilis* as potential producer for polyhydroxyalkanoates. Microb Cell Factories 8(1):38, 1–11. https://doi.org/10.1186/1475-2859-8-38

Singh M, Kumar P, Ray S, Kalia VC (2015) Challenges and opportunities for customizing polyhydroxyalkanoates. Indian J Microbiol 55(3):235–249. https://doi.org/10.1007/s12088-015-0528-6

Somashekara DM, Rastogib NK, Ramachandriah ST (2009) A simple kinetic model for growth and biosynthesis of polyhydroxyalkanoate in *Bacillus flexus*. New Biotechnol 26:92–98. https://doi.org/10.1016/j.nbt.2009.04.004

Sowinski CS, Burdman S, Matan O, Okon Y (2010) Natural functions of bacterial Polyhydroxyalkanoates. In: Chen G-Q (ed) Plastics from bacteria: natural functions and applications, vol 14. Springer, Berlin/Heidelberg, pp 39–61. https://doi.org/10.1007/978-3-642-03287_5_3

Sreekanth MS, Vijayendra SVN, Joshi GJ, Shamala TR (2013) Effect of carbon and nitrogen sources on simultaneous production of α-amylase and green food packaging polymer by *Bacillus* sp. CFR 67. J Food Sci Technol 50(2):404–408. https://doi.org/10.1007/s13197-012-0639-6

Strong P, Laycock B, Mahamud S, Jensen P, Lant P, Tyson G, Pratt S (2016) The opportunity for high-performance biomaterials from methane. Microorganisms 4(1):11, 2–10. https://doi.org/10.3390/microorganisms4010011

Sudesh K, Abe H, Doi Y (2000) Synthesis, structure and properties of polyhydroxyalkanoates: biological polyesters. Prog Polym Sci 25(10):1503–1555. https://doi.org/10.1016/S0079-6700(00)00035-6

Taguchi K, Aoyagi Y, Matsusaki H, Fukui T, Doi Y (1999) Over-expression of 3-ketoacyl-ACP synthase III or malonyl-CoA-ACP transacylase gene induces monomer supply for polyhydroxybutyrate production in *Escherichia coli* HB101. Biotechnol Lett 21(7):579–584. https://doi.org/10.1023/A:1005572526080

Takaku H, Kimoto A, Kodaira S, Nashimoto M, Takagi M (2006) Isolation of a Gram-positive poly(3-hydroxybutyrate) (PHB)-degrading bacterium from compost, and cloning and characterization of a gene encoding PHB depolymerase of *Bacillus megaterium* N-18-25-9. FEMS Microbiol Lett 264:152–159. https://doi.org/10.1111/j.1574-6968.2006.00448.x

Tamboli DP, Gomare SS, Kalme SS, Jadhav UU, Govindwar SP (2010) Degradation of orange 3R, mixture of dyes and textile effluent and production of polyhydroxyalkanoates from biomass obtained after degradation. Int Biodeterior Biodegradation 64(8):755–763. https://doi.org/10.1016/j.ibiod.2010.09.003

Thompson RC, Moore CJ, Saal FSV, Swan SH (2009) Plastics, the environment and human health: current consensus and future trends. Phil Trans R Soc B 364(1526):2153–2166. https://doi.org/10.1098/rstb.2009.0053

Tomizawa S, Hyakutake M, Saito Y, Agus J, Mizuno K, Abe H, Tsuge T (2011) Molecular weight change of polyhydroxyalkanoate (PHA) caused by the PhaC subunit of PHA synthase from *Bacillus cereus* YB-4 in recombinant *Escherichia coli*. Biomacromolecules 12:2660–2666. https://doi.org/10.1021/bm2004687

Tseng CL, Chen HJ, Shaw GC (2006) Identification and characterization of the *Bacillus thuringiensis* phaZ gene, encoding new intracellular poly-3-hydroxybutyrate depolymerase. J Bacteriol 188(21):7592–7599. https://doi.org/10.1128/JB.00729-06

Tsugeet T, Hyakutake M, Mizuno K (2015) Class IV polyhydroxyalkanoate (PHA) synthases and PHA-producing *Bacillus*. Appl Microbiol Biotechnol 99(15):6231–6240. https://doi.org/10.1007/s00253-015-6777-9

Umesh M, Priyanka K, Thazeem B, Preethi K (2017) Biogenic PHA nanoparticle synthesis and characterization from *Bacillus subtilis* NCDC0671 using orange peel medium. Int J Polym Mater 67(17):996–1004. https://doi.org/10.1080/00914037.2017.1417284

Valappil SP, Boccaccini AR, Bucke C, Roy I (2007a) Polyhydroxyalkanoates in Gram-positive bacteria: insights from the genera *Bacillus* and *Streptomyces*. Anton Leeuw Int J G 91(1):1–17. https://doi.org/10.1007/s10482-006-9095-5

Valappil SP, Misra SK, Boccaccini AR, Keshavarz T, Bucke C, Roy I (2007b) Large-scale production and efficient recovery of PHB with desirable material properties, from the newly characterised *Bacillus cereus* SPV. J Biotechnol 132(3):251–258. https://doi.org/10.1016/j.jbiotec.2007.03.013

Valentin HE, Dennis D (1997) Production of poly(3-hydroxybutyrate-co-4-hydroxybutyrate) in recombinant *Escherichia coli* grown on glucose. J Biotechnol 58(1):33–38. https://doi.org/10.1016/S0168-1656(97)00127-2

Valentin HE, Steinbuchel A (1995) Accumulation of poly(3-hydroxybutyric acid-co-3-hydroxyvaleric acid-co-4-hydroxyvaleric acid) by mutants and recombinant strains of *Alcaligenes eutrophus*. J Environ Polym Degrad 3(3):169–175. https://doi.org/10.1007/BF02068468

Verlinden RAJ, Hill DJ, Kenward MA, Williams CD, Radecka I (2007) Bacterial synthesis of biodegradable polyhydroxyalkanoates. J Appl Microbiol 102(6):1437–1449. https://doi.org/10.1111/j.1365-2672.2007.03335.x

Verlinden RAJ, Hill DJ, Kenward MA, Williams CD, Piotrowska-Seget Z, Radecka IK (2011) Production of polyhydroxyalkanoates from waste frying oil by cupriavidus necator. AMB Express 1(1):1–8. https://doi.org/10.1186/2191-0855-1-11

Vieira MGA, Da Silva MA, Dos Santos LO, Beppu MM (2011) Natural-based plasticizers and biopolymer films: a review. Eur Polym J 47(3):254–263. https://doi.org/10.1016/j.eurpolymj.2010.12.011

Wang X, Li Z, Li X, Qian H, Cai X, Li X, He J (2016) Poly-β-hydroxybutyrate metabolism is unrelated to the sporulation and parasporal crystal protein formation in *Bacillus thuringiensis*. Front Microbiol 7:1–9. https://doi.org/10.3389/fmicb.2016.00836

Wecker P, Moppert X, Simon-Colin C, Costa B, Berteaux-Lecellier V (2015) Discovery of a mcl-PHA with unexpected biotechnical properties: the marine environment of French Polynesia as a source for PHA-producing bacteria. AMB Express 5(1):1–9. https://doi.org/10.1186/s13568-015-0163-y

Wen Q, Chen Z, Tian T, Chen W (2010) Effects of phosphorus and nitrogen limitation on PHA production in activated sludge. J Enviorn Sci 22(10):1602–1607. https://doi.org/10.1016/S1001-0742(09)60295-3

Wu Q, Huang H, Hu G, Chen J, Ho KP, Chen GQ (2001) Production of poly-3-hydroxybutrate by *Bacillus* sp. JMa5 cultivated in molasses media. Anton Leeuw Int J G 80(2):111–118. https://doi.org/10.1023/A:1012222625201

Chapter 19
Bacillus as a Versatile Tool for Crop Improvement and Agro-Industry

Sudisha Mukherjee, Vivek Pandey, Amreesh Parvez, Xianghui Qi, and Touseef Hussain

Abstract Agriculture and the agro-industry form the backbone of any economy, whether traditional or modern sector. Though it contributes less in terms of GDP percentage but employs a significantly large amount of human power. Agriculture depends upon numerous biotic and abiotic factors which are essential for productivity. Alteration in the fine balance of these factors has the potential to result in the development of stress that may affect the plants. The chemical fertilizer and pesticide routinely tackle nutrient nonavailability and competition from weeds and pathogenic infection. However, the impact of these chemicals on the environment and health is a serious concern with addition to the increasing cost of production. Metabolites produced by various microbes have been considered as an effective biological agent to enhance productivity, prevent stress conditions, and provide chemical-free food products. *Bacillus* spp. have been a striking biological agent owing to its ability to produce a wide range of metabolites that enhance crop productivity by availing plant various micronutrients, volatile compounds, and antimicrobials targeting pathogens. Many species of *Bacillus* are known to produce antibiotics, chitinase, lipopeptides, glucanase, and hydrogen cyanide which kill bacteria, fungi, and pests. Some other *Bacillus* species promote plant growth indirectly by eliciting systematic resistance against pathogens or by producing growth hormones like cytokinins, IAA, gibberellin, and spermidines, which results in root and

Sudisha Mukherjee and Vivek Pandey contributed equally with all other contributors.

S. Mukherjee · V. Pandey
Faculty of Life Sciences and Biotechnology, South Asian University, New Delhi, India

A. Parvez · X. Qi
School of Food and Biological Engineering, Jiangsu University, Zhenjiang, Jiangsu, China

T. Hussain (✉)
Department of Botany, Aligarh Muslim University, Aligarh, Uttar Pradesh, India

M. T. Islam et al. (eds.), *Bacilli in Agrobiotechnology*, Bacilli in Climate Resilient Agriculture and Bioprospecting,
https://doi.org/10.1007/978-3-030-85465-2_19

shoot growth. Introduction of *Bacillus* in soil has also shown to provide tolerance towards harsh environmental stimuli like heavy metal, droughts, and salinity to the crops. In this chapter, we try to elaborate on these essential roles played by the *Bacilli* species in soil and its future directions.

Keywords Agriculture · *Bacillus* · Crop protection · Metabolites · Biological agents

19.1 Introduction

Agriculture plays a vital role in the world economy. However, crop cultivation is affected by various biotic and abiotic factors. Agriculture is affected by biotic stresses that include plant infection with viruses, bacteria, and fungi, as well as infestation by the pest and arachnids (Singh et al. 2018). Several weed species are competing for the nutrients with the plant also a major cause of biotic stresses. At the same time, the abiotic stress includes an imbalance in soil salinity, temperature, water, and nutrients deprivation. This abiotic and biotic stress upon manifestation creates a negative outcome on plant growth, development, and yield (Bacon et al. 2012; Shafi et al. 2017a). To overcome such stresses and enhance productivity, many chemical-based management strategies are being used since industrialization. Long-term and excessive utilization of these chemical and synthetic products on crop production has resulted in severe environmental pollution, degradation in the quality of soil, and adverse effect on human health. Due to these adverse effects of chemicals on human health and environment, the scientific community is trying to develop a sustainable and green solution for the improvement of plant productivity (Shafi et al. 2017a; Hussain and Khan 2020a; Dunlap et al. 2013; Torres et al. 2019).

Biological methods provide a new ray of hope to enhance plant productivity and simultaneously curtailing the role of chemical fertilizer and pesticide (Dhankar et al. 2020). These biological agents in the form of live cells or their by-products are a positive, sustainable source for agricultural development by increasing the productivity and reducing the health risk caused by the use of chemicals (Berg et al. 2005; Bacon et al. 2015; Hussain and Dhanker 2021).

Microbes associate with the plant as a symbiotic relationship at multiple surfaces such as leaves and the rhizome. Microbial endophytes dwelling within the plant tissues consist of the beneficial symbiont. Such plant-microbe interaction is a mutually beneficial relationship, unlike the pathogen, which negatively regulates plant growth. Of all the beneficial microbes in the entophytic system, bacteria are quite varied and are known to provide the plant with several essential products (Ahamd et al. 2019; Haris et al. 2020; Hussain et al. 2020e, f). These endophytic bacteria have the potential for phytoremediation, biological control of pest/pathogens, and agronomic enhancements of plants' maturity and growth (Berg et al. 2005; Bacon et al. 2015; Hussain and Khan, 2020d; Hussain et al. 2021). The numerous bacteria *Bacillus* spp. are considerable interest for its role as plant growth-promoting rhizobacteria (PGPR) and as the producer of beneficial secondary metabolites (Kang et al. 2015). *Bacillus* consists of a diverse group of rod-shaped, gram-positive motile bacteria owing to peritrichous flagella. Their presence across a variety of environment can be attributed to their ability to form hard resistance endospore which helps to survive during the stress conditions and ability to live both as aerobe and facultative anaerobe (Radhakrishnan et al. 2017). These *Bacilli* form the active part of the (PGPR) that could be utilized in agricultural practice to alleviate stresses and to produce environmental-friendly management tool (Grover et al. 2011; Vejan et al. 2016).

The gram-positive bacteria are ubiquitous and are acquired from various environmental niches. Many of the *Bacillus* spp. are used commercially, as a fertilizer, resulting in a 40% increase in crop production (FZB24® Bacillus subtilis – mode of action of a microbial agent enhancing plant vitality n.d.). Some other *Bacillus*-based products such as Kodiak (*Bacillus subtilis* GB03), rhizovital (*B. amyloliquefaciens* FZB42) quantum-400 (*B. subtilis* GB03), serenade (*B. subtilis* QST 13), and YIB (*Bacillus* spp.) have also been commercialized for agricultural industry (Brannen and Kenney 1997; Ngugi et al. 2005; Cawoy et al. 2011). *Bacillus*-based products are more active compared to other biofertilizers due to their more effective metabolite production and spore-forming character, which enhances the viability of the cells for a long time (Haas and Défago 2005). Here we describe the different metabolites produced by the *Bacillus* and their role in increasing the agricultural productivity (Fig. 19.1).

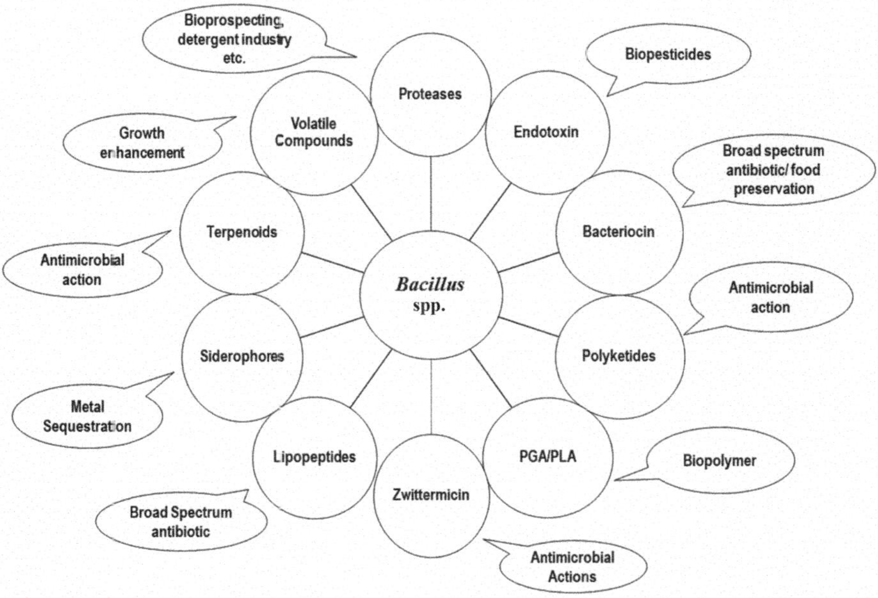

Fig. 19.1 Flow chart depicting the classification of various metabolites produced from different *Bacillus* spp. involving different synthesis pathways and their importance

Bacterial metabolites can be broadly classified into two categories, i.e. primary and secondary metabolites based upon the stage they are produced and their role. Primary metabolites are produced during the growth phase (log phase) of the bacterial life cycle and are essential for growth and survival. At the same time, the secondary metabolites are produced in the stationary phase and play subsidiary roles such as defence and pigmentation, etc. Owing to its presence across a broad diversity of environment, *Bacillus* produces several metabolites, especially the secondary metabolites.

19.2 Functions of *Bacillus* Various Metabolites

Bacillus as a PGPR has been described to stimulate plant growth through various direct and indirect mechanisms, or by a combination of both mechanisms. Direct interaction comprises of fixing of nitrogen, providing phytohormones to plants, solubilization of phosphates, iron uptake via bacterial siderophores, and production of ACC deaminases that can reduce the levels of stress-induced ethylene in plants (Glick 2004). Extracellular hydrolytic enzymes, antimicrobial agents, and siderophores produced by these microorganisms eliminate the pathogenic microbes present in the rhizosphere and the surrounding. This type of interaction between plant and microbes is termed as indirect interaction (Yuwono et al. 2005; Mandal et al. 2021).

19.2.1 Enzymes

Several *Bacillus* spp. are commercially important due to their ability to secrete industrially essential enzymes. These enzymes are essential in industries like food, detergent, textiles, feed, and beverages (Siddiqua et al. 2021). Alkaline serine proteases secreted from *Bacillus* are of considerable importance in several industries. Different species of *Bacillus* such as *B. halodurans*, *B. clausii*, *B. amyloliquefaciens*, and *B. subtilis* (Schallmey et al. 2004; Saeki et al. 2007) produce several proteases of commercial importance that includes amylases, cellulases, ligininlases, and lipases. α-galactosidase produced from *B. megaterium* and *B. stearothermophilus* has shown wide application in the pulp, beer sugar, paper, soyfood, and pharma industries.

Apart from the above-mentioned enzymes, the *Bacillus* species also produce peptidoglycan hydrolase enzymes which de-polymerizes and hydrolyse specific bonds in the cell wall. This type of hydrolase has great perspectives in different biotechnological fields such as in bioenergy production and as biocontrol against other microbes (Raddadi et al. 2004; Raddadi et al. 2005). Adding to the antimicrobial property of *Bacillus* are also the quorum quenching enzymes like AHL- lactonase (Dong et al. 2002).

19.2.2 Biopolymer

Biopolymers have gained extensive interest because of their eco-friendly production and vast areas of application. Various types of polymers have been reported to be produced by *Bacillus* spp. Poly (γ-glutamic acid) PGA, which is an anionic water-soluble polymer, has been used in pharmaceutical, cosmetics, food, and wastewater management as a heavy metal chelator or flocculant (Bajaj and Singhal 2011). *B. licheniformis*, *B. subtilis*, and *B. amyloliquefaciens* are known to be producing such biopolymers (Yamashiro et al. 2011; Cao et al. 2011).

19.2.3 Protein Crystal

Bacillus thuringiensis during the sporulation phase produce parasporal crystals. These Cry protoxins or δ endotoxin, when ingested by insects, convert into its active form which leads to pore formation (Lüthy and Wolfersberger 2000; Schwartz and Laprade 2000), and ultimately membrane permeability disrupts resulting in the death of the insect larvae (Porcar and Juárez-Pérez 2003). Cry proteins are active against dipteran, lepidopteran, coleopteran, and nematodes like *Nippostrongylus brasiliensis* (Schnepf et al. 1998; Wei et al. 2003; Kotze et al. 2005). Besides the endotoxins, *B. thuringiensis* also produced δ-exotoxins, a non-proteinaceous toxin which is secreted into the medium. This is active against lepidopterans, dipteran, and coleopteran but has also shown an adverse effect against beneficial organisms like *Apis mellifera* (honeybee) (Espinasse et al. 2002, 2004).

19.2.4 Antimicrobial Agents

19.2.4.1 Bacteriocin

Bacteriocin is 20–60 amino acid long ribosomally synthesized, cationic, and hydrophobic antimicrobial peptides (AMP) (Riley and Wertz 2002a, b). These peptides inhibit both gram-positive and gram-negative bacteria (Walsh et al. 2015; Suganthi and Mohanasrinivasan 2015; Nes and Holo 2000). *Bacillus* produces immunity proteins to protect themselves from the toxic effect of their bacteriocin (Oppegård et al. 2007). Bacteriocin synthesis is regulated by structural genes involved in the synthesis of bacteriocin peptide with an N terminal leader sequence and several accessory genes involved in the processing and transporting the peptide molecule outside the cell. The bacteriocin production from the bacterial cell is often regulated by cell density or quorum sensing (Lee and Kim n.d.).

Bacteriocin targets the bacterial membrane using membrane surface charge (Rashid et al. 2016; Moll et al. n.d.). Bacterial surfaces are negatively charged due to the presence of phosphatidylglycerol (PG), lipopolysaccharide (LPS), phatidylethanolamine (PE), cardiolipin (CL), and lipoteichoic acid (LTA). These serve as targets for the cationic bacteriocins (Rashid et al. 2016; López-Lara and Geiger 2017). The positively charged part of the bacteriocin interacts with the negative bacterial surface via electrostatic interactions. Simultaneously, the hydrophobic part of the bacteriocin penetrates the lipid membrane. Upon piercing the bilayer, the peptides self-polymerize and form channel complexes (Shahnawaz and Soto 2012), resulting in permeabilization of the cell membrane by ion-selective pores which result in proton motive force and exhaustion of intercellular ATP (Christensen and Hutkins 1992). This, in turn, results in intercellular substrate leakage and eventually death of the organism (Minahk et al. 2000).

Many *Bacillus* species are known to produce a range of bacteriocin, and these include lichenin, megacin, coagulin, and polyfermenticin from *B. licheniformis*, *B. megaterium*, *B. coagulans*, and *B. polyfermecticus*, respectively. Similarly, *B. thuringiensis* subsp. *kurstaki* produces a bacteriocin known as Bacthuricin F4. Bacthuricin F4 is a 3160.05 Da molecular weight bacteriocin and has demonstrated antibacterial and highest bacteriocin activity. Another bacteriocin produced by *B. amyloliquefaciens* from mangrove has shown to inhibit *Ralstonia solanacearum (*Hu et al. 2010*)*. Subtilin, a bacteriocin produced by *B. subtilis*, is prominently used in the dairy industry owing to its broad-spectrum antimicrobial activity. These bacteriocins not just protect the phytopathogen present in the soil they are living, but also their antimicrobial activity is being exploited for their use as food preservatives in packed food and beverages.

19.2.4.2 Lipopeptides

Lipopeptides are non-ribosomal peptides synthesized by large multienzyme peptide synthetase complex (Bacon et al. 2015). Lipopeptides are powerful biosurfactant having anti-mycoplasm, antibacterial, and antiviral activity. These lipopeptides are stable over an extensive range of pH 3–11 and temperature from 15 to 100 ℃ (Makkar and Cameotra 1998; Cameotra and Makkar 1998; Joshi et al. 2008). Many of these properties are commercially explored to provide sustainable solutions for the environment and industrial applications. Cyclic lipopeptides extracted from *Bacillus subtilis* ABS-S14 have shown to activate defence mechanism at translation and transcriptional levels by inducing defensive proteins and genes (Tunsagool et al. 2019). Surfactins are a class of cyclic lipopeptides synthesized by surfactin synthetase in *B. subtilis* synthetase (Das et al. 2008; Nicolas 2003). It has a broad-spectrum antimicrobial, anti-coagulation, anti-tumour activity. Apart from that, Bacillus also possesses bioremediation property (Franzetti et al. 2010) as well as exhibits antimicrobial property against phytopathogens (Mulligan 2005). Surfactins owing to their bioremediation activity are routinely used to reclaim the degraded land and making them fit for cultivation by degrading the organic pollutants (Franzetti et al. 2010).

Bacillus spp. producing a combination of these lipopeptides form prominent biocontrolling agents (Bais et al. 2004; Kearns et al. 2004; Ongena and Jacques 2008; Hussain et al. 2019a; Hussain and Khan 2018a). Globally about 33 *Bacillus* spp. strains have been isolated, which produce surfactins with varying carbon length ranging from 11 to 17. Most of these strains produce surfactins with 15 carbon lengths, which are amongst the most active kind of surfactins (Bacon et al. 2012; Snook et al. 2009); *B. mojavensis* produces varieties of surfactins that have shown to be plant-friendly and symptomlessly infect the tissues except for the xylem (Bacon and Hinton 2002; Bacon and Hinton n.d.). *B. mojavensis* is osomophile tolerant and anaerobic in its character (Folmsbee et al. 2004). It has been studied for its protective role in maize from *Fusarium (*Bacon and Hinton 2002*;* Bacon and Hinton 2011*)* and for their capability to decrease the building up of Fumonisin mycotoxins (Bacon et al. 2001). The surfactins are inhibitory in feature to bacteria, fungi, virus and insects; however, the in vitro fermentation capacity of *B. mojavensis* of these bio-surfactins is rather poor and limits its use to technological exploitation (Bacon and Hinton 2014). *B. subtilis* produces three different lipopeptides (mycosubtilin, fengycin, and surfactin) which show potent inhibitory activity (Bacon and Hinton 2014; Hussain and Khan 2018b, 2020b; Hussain et al. 2019b). Ornamental plant cultivation shares a vital place in agro-industry. One of the significant phytopathogenic threats to ornamental plants has been the soil-borne fungus *F. oxysporum*. Mycosubtilin-like fungicide will protect these ornamental plants from the fungus and help in the agro-industry development and growths (Mihalache et al. 2018).

19.2.4.3 Polyketides

Polyketides are a member of a large family of metabolites which consist of bioactive compounds with immunosuppressive, antibacterial, and other bioactivities. The biosynthesis of these compounds is accomplished in a stepwise decarboxylative condensation (claisen condensation) of malonyl-CoA units. The PKS (polyketide synthase) multienzyme system utilizes acyl carrier proteins to channel the extending chain in the course of elongation processes (Cane 1997). The condensation process takes place between the growing polyketide chain and the extender unit with a simultaneous generation of β-ketoacyl as intermediates.

Bacterial type I PKSs is modularly arranged synthase, with each module containing an acyltransferase (AT), a β-ketoacyl synthase (KS), the ACP, and some other domains. The order of these modules decides the sequence of the polyketide biosynthesis events (Cheng et al. 2003). Genome analysis revealed the presence of several gene clusters responsible for coding PKS in *B. amyloliquefaciens* FZB42 (Koumoutsi et al. 2004) and in *B. mojavensis* strain RRC101 (Gold et al. 2014).

B. amyloliquefaciens FZB42 has three operons for the synthesis of polyketides. psk1 and psk2 have been ascribed with bacillaene and difficidin/oxydifficidin synthesis, whereas the psk2 has been associated with the synthesis of four members from the macrolactin family, namely, macrolactin A, macrolactin D 7-O-malonyl, and 7-O succinyl macrolactin (Schneider et al. 2007). This macrolactin has also been related to an antimicrobial activity like other polyketides. Macrolactin A, 7-O-succinyl macrolactin A, 7-O malonyl macrolactin A, and two homologous bacillomycin D isolated from *B. amyloliquefaciens* NJN-6 showed antagonistic effects against *Ralstonia solanacearum* and *Fusarium oxysporum*, respectively (Yuan et al. 2012).

19.2.4.4 Zwittermicin A

Zwittermicin A, a linear aminopolyol, possesses antibiotic properties (Silo-Suh et al. 1998). Chemically, zwittermicin consists of D amino acids, gylcosyl moieties, ethanolamine, and an unusual terminal amide modifications (Sansinenea and Ortiz 2011). High polarity, sensitivity to alkaline condition, and charged state at physiological pH make its large-scale production difficult (Rogers and Molinski 2007). Biosynthesis of zwittermicin A follows mixed pathway involving both the NRPS and PKS systems.

B. cereus UW85 produces a novel antibiotic that showed activity against oomycetes, algal protist, bacteria, and plant fungi. When combined with kanosamine another antibiotic by *B. cereus*, they act synergistically against *Phytophthora medicaginis*, which is an oomycetes (Silo-Suh et al. 1998).

19.2.4.5 Terpenoids

Terpenoids are amongst the most diverse class of natural compounds. They occur in various chemical structures and have been reported to have antifungal, antiviral, anti-parasitic, antimicrobial, anti-allergic, immunomodulatory, and chemotherapeutic properties (Abdallah and Quax 2017). Isoprene(2-methyl-1-,3-butadiene), also known as isopentenyl diphosphate, is a precursor for all terpenoids. In archaea and eukaryotes, the precursor is formed via the mevalonate pathway (Kuzuyama 2002), whereas in plant plastids and some eubacteria, this is formed through the methylerythritol phosphate pathway (MEP) (Rohmer et al. 1993; Rohmer 1999). In *B. subtilis*, only homologous forms of the genes involved in MEP pathway have been reported (Wagner et al. 2000). *B. subtilis* for MEP pathways consists of five genes, namely, *yluB*, *yacM*, *yabH*, *yacN*, and *yqfP* to be essential for the isoprene production (Julsing et al. 2007). When a synthetic operon with up to eight regulated gene expression was introduced in *B. subtilis*, it resulted in a stable cell factory with high carotenoid production (Abdallah et al. 2020). Comparing the lipid profile of wild-type *B. subtilis* and SqhC-deficient strain also led to the detection of three novel tetracyclic terpenoids, the sporulenes from *B. subtilis* (Kontnik et al. 2008). These sporulenes are produced by the cyclization of regular polyprenes by

B. subtilis cyclase. The presence of these tetracyclic isoprenoids in the lipids helps the bacteria against oxygen species (Bosak et al. 2008).

19.2.5 Insecticides and Pesticides

An inappropriate use of agrochemicals can lead to decrease in the microbial metabolic activities (Dhankar et al. 2021). Insecticidal activity of the *Bacillus* proteins is utilized to produce transgenic plant lines resistant to various Lepidopteron, thus eliminating the need for insecticide. An elite rice line KMD1 formed by introducing cry1Ab gene into Japonica Chinese rice variety shows insect-resistant attributes, with no damage being caused by Lepidopteran pests like *Chilo suppressalis*, *Scirpophaga incertulas*, and *Cnaphalocrocis medinalis* (Shu et al. 2000). A similar introduction of a novel synthetic cry2A gene in Indica rice via *Agrobacterium*-mediated transformation has shown to develop transgenic rice lines resistant to Lepidopteran rice pest (Chen et al. 2005). Transgenic castor oilseed crop with cry1Aa gene has also been genetically developed to resist pests like *Achaea janata* and *Spodoptera litura* (Muddanuru et al. 2019). Apart from the Cry protein, another protein from *B. thuringiensis* has also proved to exhibit pest-controlling attributes. This protein Vip3A (vegetative insecticidal protein 3A) can be isolated from the culture medium and has a wide spectrum of targets. Vip3A protein has shown to have a toxic effect on larva growth, leading to operative mortality (Chakrabarty et al. 2020).

19.2.6 Nematocides

Plant-parasitic nematodes lead to massive loss in crop production sector every year worldwide. Proteins produced by various *Bacillus* sp. have gained a role as nematocide. A 28 kDa serine protease enzyme isolated from *Bacillus* sp. strain B16 has shown to hydrolyse substrates like collagen and nematode cuticle. The protease reportedly kills 80% of nematodes, *Panagrellus redivivus* within 24 hours (Qiuhong et al. 2006). Another instance of biocontrol of nematodes was reported by the dipeptide cyclo(D-Pro-I-Leu) from *B. amyloliquefaciens* Y1, *B. subtilis HussainT-AMU* (Hussain et al. 2020), and *B. firmus HussainT:Lab.66* (Hussain and Khan 2020c). This has shown to reduce gall per root of the tomato plant as well as a decrease in egg counts and stage 2 juvenile population of nematode, *Meloidogyne incognita*. The increased root, shoot length and fresh and dry weight have also been observed after the introduction of strain Y1 (Jamal et al. 2017). Few more examples of the functions are given in Table 19.1.

Table 19.1 Functions of various metabolites of *Bacillus* spp.

Metabolites	Compounds	Organisms	Functions	References
1. Enzymes	AHL lactonase	*B. thuringiensis*	Antipathogenic against *E. carotovora*	Dong et al. (2002)
2. Biopolymer	Poly γ-glutamic acid	*B. amyloliquefaciens*LL3	Hydrogels, flocculants, thickeners, feed additives, cosmetics, drug deliveries	Cao et al. (2011)
3. Protein crystal	Cry protoxin or δ endotoxin	*B. thuringiensis*	This microbial protein has been exploited to produced bioinsecticides and transgenic plants	Roh et al. (2007)
	Bac14B (bacteriocin)	*B. subtilis*14B	Antimicrobial effect against *agrobacterium tumefaciens*	Hammami et al. (2009)
	Tochicin (bacteriocin)	*B. thuringiensis* subsp. *toichigiensis*HD868	A narrow spectrum of bactericidal activity against *B. Thuringiensis* and *B. subtilis*	Paik et al. (1997)
4. Antimicrobial agents	Mycosubtilin (lipopetides)	*B. subtilis*	Fungicidal	Mihalache et al. (2018)
	Difficidins and bacilysin (polyketides)	*B. amyloliquefaciens*	Biocontrol against fire blight of orchard trees by *Erwinia amylovora*	Chen et al. (2009)
	Zwittermicin A	*B. cereus*	Suppress fungal diseases	Silo-Suh et al. (1998)
	Terpenoids	*B. amyloliquefaciens*S5Ia	Antibacterial and antifungal	Shafi et al. (2017b)
5. Insecticides	Cry9Cb1 (MP1489)	*B. thuringiensis strain* SP663	Insecticidal activity towards *Plutella xylostella*, *Chilo suppressalis*, and *Ostrinia furnacalis*	Shan et al. (2019)
6. Nematocides	Endotoxins (AI, AII, AIII, AIV)	*B. thuringiensis*	Inhibits *M. incognita*	Abd El-Moneim (2014)

19.3 Stress Tolerance

19.3.1 Water Stress

Drought refers to the non-availability of soil moisture over a prolonged period. Drought is one of the common stresses that plants suffer across the globe and serves as the deciding factors in the production and growth of agro-industry. Most of the

PGPRs have been reported to initiate drought tolerance and help the crops in arid and semi-arid regions to cover this problem. *B. licheniformis* K11 upon inoculation in the pepper plant led to the up-regulation of four differentially up-regulated proteins. These proteins were, namely, CaPR protein 10, sHSP, VA, and Cadhn (Lim and Kim 2013). CaPR protein 10 is related to plant defence against abiotic and biotic stress (Lee et al. 2008). sHSP (heat shock protein) generally works as molecular chaperon and plays a vital role during stress by preventing irreversible aggregation of denatured proteins (Sarkar et al. 2009). VA (V ATPase) energizes sodium sequestration in the central vacuoles (Golldack and Dietz 2001), and Cadhn, which functions through stabilization of large-scale hydrophobic interactions (Borovskii et al. 2002). Thus, the over-expression of all these four proteins may perhaps result in the induction of drought stress tolerance in pepper (Lim and Kim 2013).

Another example of an endophytic relationship with *Bacillus* mitigating drought stress was seen with *Phleum pratense* L (Timothy). The plant with *B. subtilis* strain B26 inoculation under stressed condition showed an increase in photosynthesis and an increase in shoot and root biomass by 26.6 and 63.6%, respectively. It also leads to an increase in stomatal conductance by 214.9% compared to non-inoculated plants. A higher amount of sugars like fructans, sucrose, and amino acids like glutamic acids, glutamine, and asparagine were also recorded in the shoot and roots of the colonized plants (Gagné-Bourque et al. 2016). Inoculation of PGPR also results in an increase in the production of secondary metabolites and antioxidants in plants under drought stress. *Bacillus pumilus*, when inoculated with Glycyrrhiza, had shown to increase its biomass by 34.9% along with the increase in oxygen species, antioxidants, and few important enzymes expression like that of SQS, HMGR, and beta-AS which plays a vital role in the synthesis of glycyrrhizic acid. Its association has also marked a decrease in lipid peroxidation level in the plant (Xie et al. 2019).

19.3.2 Salt Stress

Another type of stress which plants face is salt stress. Microbial volatile compounds (mVOCs) released from *B. amyloliquefaciens* GBo3 upon administration to *Mentha piperita* without any physical contact with the bacteria showed a generation of tolerance towards salt stress. VOCs-exposed plants showed a better morphological and chlorophyll content compared to the control. It also showed an increase in salicylic acid content and a decrease in abscisic acid, which aid in stress tolerance (del Rosario Cappellari and Banchio 2019). In a similar example, halotolerant bacteria *Bacillus subtilis* HG-15 has also shown to effectively improve the growth and systematic tolerance level of heat when associated with it (Ji et al. 2020). PGPR strains *B. siamensis* (PM13), *B. methylotrophicus* (PM19), and *Bacillus* sp. (PM15) have also exhibited induced NaCl tolerance. Wheat seeds showed an increase in germination and seedling growth under salt stress when grown along with bacteria. The emission of exopolysaccharides and ACC deaminase by these strains played an important role in providing tolerance towards salt (Amna et al. 2019).

19.3.3 Heavy Metal Stress

Contamination of soil with heavy metals poses great adverse stress scenario towards the growth of the crops in soil. Phytoremediation of heavy metal contaminated soil is an important issue for better crop production (Kumar et al. 2021; Haris et al. 2021; Hussain et al. 2021). Arsenic remediation is seen in *Valliseriadense serrulata–Bacillus* spp XZM union. There was an increased uptake of the arsenic when the plant was colonized with the bacteria. This arsenic is then detoxified and metabolized by the *V. denseserrulata* leaves (Irshad et al. 2020). Similarly, *Bacillus thuringiensis* and biochar both individually and when taken together showed their ability to immobilize Cd and Pb in culture solution. These treatments increase the dry weight of radish leaves and roots and also resulted in a decrease in the content of Pb and Cd in the roots. Simultaneously, this also leads to an increase in ammonium concentration, pH, organic content, and the $NH^{4+}/NO^{3-}$the ratio of the rhizosphere soil (Li et al. 2019). A mutant of *Bacillus megaterium* strain BM18, BM18–2, has shown to enhance the phytoremediation potential of the hybrid *Pennisetum* by improving its growth, and Cd absorption property, when grown on Cd-stressed soil (Wu et al. 2019). A wastewater bacterium *Bacillus* spp. KUJM2 has been reported to be efficient in removing potential toxic elements like As, Cd, Cu, and Ni under both mono- and co-contaminated conditions (Mondal et al. 2019). All these can be very useful in turning the land around the mining area to bearable again.

19.3.4 Temperature Stress

Temperature is another significant abiotic factor in plant growth and development. As the global temperature is increasing and negatively affecting the crop yield and quality, some sustainable solution needs to be worked out. Soybean, when grown with *B. tequilensis* SSB07, has shown promising effects countering to heat stress. It results in plant growth and increases in jasmonic acid and salicylic acid, whereas a considerable decrease in ABA as a stress response mechanism was noted (Kang et al. 2019a). Under heat stress, soybean plants when inoculated with *B. aryabhattai* SRB02 showed tolerance by containing steady levels of ABA and exhibiting ABA-induced stomatal closure (Park et al. 2017).

19.3.5 Radiation Stress

Bacillus spp. have shown to provide tolerance to plants against ozone toxicity. Ozone being a phytotoxic compound is capable of reducing crop productivity. Addition of *Bacillus* sp. N7 with *Oryza sativa* L. var. *indica* cv. *Pathum Thani1* (PTT1) and *Oryza sativa* L. var. *indica* cv. RD41(RD41) resulted in increased O_3

tolerance along with high rice quality. An evident decrease in malondialdehyde (MDA) due to increased levels of ascorbic acid was noticed. This colonization also resulted in high photosynthesis, stomatal conductance, and chlorophyll content in plants (Anusaraporn et al. 2020).

Another way by which the PGPR enhances ozone tolerance in plants was seen through the production of different kinds of volatile organic components, which in return triggers the induced systemic tolerance of plants. Such an example is seen when the Tobacco plant is colonized with PGPR *Bacillus velezensis* Gj11, which produces volatile organic compound consisting of acetoin and 2,3-butanediol. These compounds trigger the IST (induced systemic tolerance) and lead to stomatal closure and prevent plant against O_3 injury (Liu et al. 2020).

19.4 Plant Growth Promotion

19.4.1 Plant Growth Promotion Hormone

Volatile organic compounds (VOCs), apart from playing a significant role in stress tolerance and defence, also promote plant growth. These volatile compounds work similarly to plant growth hormones and enhance plant growth; some other stimulates the secretion of the phytohormone itself. VOCs composed of acetoin from *B. methylotrophicus* M4–96, when applied to strawberry plant without physical contact with bacteria, resulting in improved growth and development, and also it increases callus deposition by fivefold in leaves and increased petiole length and root number. Increase in root length and petiole number in strawberry plants were seen after subjecting them to *B. methylotrophicus* M4–96 inoculum (Vicente-Hernández et al. 2019). VOCs emitted by *Bacillus* sp. JC03 included tetrahydrofuran-3ol, 2-heptanone, and 2–ethyl-1-hexanol promoted the growth of *Arabidopsis* seedlings via the action of strigolactone and auxin. This also increases biomass accumulation in *Arabidopsis* (Jiang et al. 2019).

Many PGPRs stimulate the growth and development of plants by modulating the necessary phytohormones. Root architecture variability was also observed in rice plants upon association with *B. altitudinis* (strain FD48). This change of architecture in rice and FD48 partnership was attributed to increase in IAA accumulation in the root of the plant compared to non-inoculated control (Ambreetha et al. 2018). Similarly, *B. aryabhattai* SRB02, which produces IAA along with cytokinin, jasmonic acid, ABA, and different gibberellic acids (GA1, GA7, GA4) in culture, showed longer root and shoot for soybean plant (Park et al. 2017). Another bacterium, *B. aequilensis* (SSB07), which produced IAA with several gibberellins and abscisic acid also showed improvement in Chinese cabbage seedling and soybean cultivation (Kang et al. 2019a). PGPR like *Bacillus subtilis* JW1 was seen to produce high concentration of gibberellins, GA1, GA4, and GA7, along with other organic acids, fatty acids, and tricalcium phosphates. These chemicals showed to result in a substantial rise in plant height, chlorophyll contents, biomass, and nutrient uptake when Chinese cabbage was inoculated with it (Kang et al. 2019b).

Another critical aspect of plant growth is nutrient uptake from its surrounding. PGPR promotes growth by enhancing the accumulation of essential nutrients in plants. Endophytic *Bacillus* strains B1923, B2084, and B2088 when inoculated with pearl millet resulted in better growth and enhanced nutrient uptake. B1923 strain produced IAA and solubilized Fe-P, which also resulted in the increase of root and shoot dry weight and N, P content of root. At the same time, the B2084 and B2088 strain showed enhanced biomass production and N, K, and P accumulations in pearl millet shoot (Ribeiro et al. 2018).

Similarly, *B. subtilis* QST713 has shown to improve the P uptake in wheat grain. It has also reported improving Zn concentration in edible parts of the plants like grains by 24% relative to control, thus stimulating Zn biofortification effect (Moreno-Lora et al. 2019). *Bacillus subtilis* strain QST713 also promotes root growth in cucumber by enhancing the overall uptake of P by 40% (García-López and Delgado 2016). *Bacillus subtilis* STU6 has also shown to ameliorate iron deficiency in tomato plant (Zhou et al. 2019).

19.4.2 Siderophore

Iron is an essential micronutrient required by the plant for its growth, and iron deficiency leads to a significant loss in crop production, especially in the region of calcareous soils (Kobayashi et al. n.d.). Siderophores are low molecular iron chelators produced by microbes that bind with iron with high affinity and shuttle them into the cells (Zawadzka et al. 2009). Several types of siderophores have been found in various *Bacillus* spp. like schizokinen from *B. licheniformis*, *B. megaterium*, *B. cereus*, and *B. subtilis*, petrobactin and bacillibactin from *B. cereus (*Park et al. 2005*)*, *B. thuringiensis (*Wilson et al. 2006*)*, and *B. anthracis (*Cendrowski et al. 2004*)*. These siderophores also help in biocontrol of phytopathogens like in the case of *B. thuringiensis* against fungi (Raddadi et al. 2007).

19.5 Conclusion

Bacillus, a group of spore-forming bacteria, is known to be found in the varied habitats owing to its ability to produce a diverse array of metabolites upon interaction with different environmental condition. These bacteria synthesize various primary and secondary metabolites that help in their survival in the given environmental condition. *Bacillus* spp. are of considerable importance owing to its ability to produce new and essential metabolites. Metabolites secreted by them are up of considerable importance in agro-industry, pharmaceuticals, beverage, and other industries. Many of the metabolites secreted by *Bacilli* are significant to plant as they act as antimicrobial agents against the phytopathogens and many times as the nutrient quenchers. All these functions make these bacteria referred to as the "good bacteria" for the plant/soil, as they enrich the soil by various primary and secondary metabolites upon their presence there.

Metabolites such as bacteriocin and lipopeptides secreted by *Bacillus* are known for their antimicrobial activity against pathogenic bacteria, virus, and fungi and thereby acting as the phytoprotectant against these pathogens. Further, similar antimicrobial activity is also shown by numerous polyketides secreted by these bacteria to protect themselves and help the symbiont. Zwittermicin A, an aminopolyol known to be synthesized by the mixed pathway (NRPS/PKS), also shows antimicrobial activity against a broad range of microbes. Apart from providing protection against microbe, some other groups of secondary metabolites work by increasing the plant productivity. These include the molecules such as siderophores, which quench out micronutrient from the soil and provide the associated symbiont plant with a much-needed advantage over the others and enhances the plant productivity. Some *Bacillus* secret volatile compounds are similar to the IAA, gibberellin, and jasmonic acid that enhance the plant growth. Apart from having the antimicrobial and growth enhancement activity, they also help in combating various abiotic stresses. Heat shock protein V, ATPase secreted by these *Bacilli*, helps in water stress tolerance. Further, many other *Bacillus* provide stress tolerance to temperature and salinity by hormonally regulating the stomatal opening and closing. In addition to growth enhancement, microbial protection, and stress tolerance to various abiotic stresses, they are also essential agents of bioremediation. These microbes help in bioremediation by removing heavy metal pollutants from the soil and making soil suitable for other crops.

Several secondary metabolites produced by *Bacillus* have been commercialized, and many more are in the pipeline. These include various proteases that are used in the detergent industry, several antimicrobial used in the dairy industry. These products being natural are gaining popularity over synthetic fertilizer and pesticides for increasing the soil productivity and soil health, as they also eliminate the entry of toxic chemicals in the human food chain. Many of these bacteria are used as inoculum along with seeds which provides protection and increases productivity. Several protease-producing species of *Bacillus* are used in biofuel production from the lignocellulosic waste. In the current decade, when there is gaining popularity for organic food and greener and cleaner development, these microbes provide cost-effective, affordable, and easy benefits to fight with various biotic and abiotic stresses. Further, many of these secondary metabolites are being explored for their industrial-scale production through various fermentation processes. These processes suffer at the scale of per unit yield, downstream processing, and cost issues. However, they also provide a ray of hope to commercialize many of these bacteria and their metabolites to counter the agricultural distress developing due to varied reasons and add to the agricultural productivity.

References

Abd El-Moneim TS (2014) The effect of endo toxin produced by Bacillus thuringiensis (Bt.) against meloidogyne incognita. https://www.researchgate.net/publication/233961901. Accessed 18 May 2020

Abdallah II, Quax WJ (2017) A glimpse into the biosynthesis of terpenoids, KnE. Life Sci 3, 81. https://doi.org/10.18502/kls.v3i5.981

Abdallah II, Xue D, Pramastya H, van Merkerk R, Setroikromo R, Quax WJ (2020) A regulated synthetic operon facilitates stable overexpression of multigene terpenoid pathway in Bacillus subtilis. J Ind Microbiol Biotechnol 47:243–249. https://doi.org/10.1007/s10295-019-02257-4

Ahamd G, Nishat Y, Haris M, Danish M, Hussain T (2019) Efficiency of soil, plant and microbes for the healthy plant immunity and sustainable agricultural system. In: Varma A, Tripathi S, Prasad R (eds) Plant-Microbe Interface. Springer, Cham. https://doi.org/10.1007/978-3-030-19831-2_15

Ambreetha S, Chinnadurai C, Marimuthu P, Balachandar D (2018) Plant-associated Bacillus modulates the expression of auxin-responsive genes of rice and modifies the root architecture. Rhizosphere 5:57–66. https://doi.org/10.1016/j.rhisph.2017.12.001

Amna BUD, Sarfraz S, Xia Y, Kamran MA, Javed MT, Sultan T, Hussain MF, Munis HJ (2019) Chaudhary, Mechanistic elucidation of germination potential and growth of wheat inoculated with exopolysaccharide and ACC- deaminase producing Bacillus strains under induced salinity stress. Ecotoxicol Environ Saf 183:109466. https://doi.org/10.1016/j.ecoenv.2019.109466

Anusaraporn S, Autarmat S, Treesubsuntorn C, Thiravetyan P (2020) Application of Bacillus sp. N7 to enhance ozone tolerance of various Oryza sativa in vegetative phase: possible mechanism and rice productivity, Biocatal. Agric. Biotechnol 25:101591. https://doi.org/10.1016/j.bcab.2020.101591

Bacon CW, Hinton DM (2002) Endophytic and biological control potential of Bacillus mojavensis and related species. Biol Control 23:274–284. https://doi.org/10.1006/bcon.2001.1016

Bacon CW, Hinton DM (2011) In planta reduction of maize seedling stalk lesions by the bacterial endophyte *Bacillus mojavensis*. Can J Microbiol 57:485–492. https://doi.org/10.1139/w11-031

Bacon CW, Hinton DM (2014) Microbial endophytes: future challenges. In: Adv. Endophytic Res. Springer, pp 441–451. https://doi.org/10.1007/978-81-322-1575-2_22

Bacon CW, Hinton DM (n.d.) Bacillus mojavensis: its endophytic nature, the Surfactins, and their role in the plant response to infection by fusarium verticillioides. Bact Agrobiol Plant Growth Responses:21–39

Bacon CW, Yates IE, Hinton DM, Meredith F (2001) Biological control of fusarium moniliforme in maize. Environ Health Perspect 109:325–332. https://doi.org/10.1289/ehp.01109s2325

Bacon CW, Hinton DM, Mitchell TR, Snook ME, Olubajo B (2012) Characterization of endophytic strains of Bacillus mojavensis and their production of surfactin isomers. Biol Control 62:1–9. https://doi.org/10.1016/j.biocontrol.2012.03.006

Bacon CW, Palencia ER, Hinton DM (2015) Abiotic and biotic plant stress-tolerant and beneficial secondary metabolites produced by endophytic bacillus species, in: plant microbes Symbiosis Appl. Facet. Springer, pp 163–177. https://doi.org/10.1007/978-81-322-2068-8_8

Bais HP, Fall R, Vivanco JM (2004) Biocontrol of Bacillus subtilis against infection of Arabidopsis roots by pseudomonas syringae is facilitated by biofilm formation and Surfactin production. Plant Physiol 134:307–319. https://doi.org/10.1104/pp.103.028712

Bajaj I, Singhal R (2011) Poly (glutamic acid) – an emerging biopolymer of commercial interest. Bioresour Technol 102:5551–5561. https://doi.org/10.1016/j.biortech.2011.02.047

Berg G, Krechel A, Ditz M, Sikora RA, Ulrich A, Hallmann J (2005) Endophytic and ectophytic potato-associated bacterial communities differ in structure and antagonistic function against plant pathogenic fungi. FEMS Microbiol Ecol 51:215–229. https://doi.org/10.1016/j.femsec.2004.08.006

Borovskii GB, Stupnikova IV, Antipina AI, Vladimirova SV, Voinikov VK (2002) Accumulation of dehydrin-like proteins in the mitochondria of cereals in response to cold, freezing, drought and ABA treatment. BMC Plant Biol 2. https://doi.org/10.1186/1471-2229-2-5

Bosak T, Losick RM, Pearson A (2008) A polycyclic terpenoid that alleviates oxidative stress. Proc Natl Acad Sci U S A 105:6725–6729. https://doi.org/10.1073/pnas.0800199105

Brannen PM, Kenney DS (1997) Kodiak® – A successful biological-control product for suppression of soilborne plant pathogens of cotton. J Ind Microbiol Biotechnol 19:169–171. https://doi.org/10.1038/sj.jim.2900439

Cameotra SS, Makkar RS (1998) Synthesis of biosurfactants in extreme conditions. Appl Microbiol Biotechnol 50:520–529. https://doi.org/10.1007/s002530051329

Cane DE (1997) Introduction: polyketide and nonribosomal polypeptide biosynthesis. From Collie to Coli. Chem Rev 97:2463–2464. https://doi.org/10.1021/cr970097g

Cao M, Geng W, Liu L, Song C, Xie H, Guo W, Jin Y, Wang S (2011) Glutamic acid independent production of poly-γ-glutamic acid by bacillus amyloliquefaciens LL3 and cloning of pgsBCA genes. Bioresour Technol 102:4251–4257. https://doi.org/10.1016/j.biortech.2010.12.065

Cawoy H, Bettiol W, Fickers P, Onge M (2011) Bacillus-based biological control of plant diseases. In: Pestic. Mod. World – Pestic. Use Manag. InTech. https://doi.org/10.5772/17184

Cendrowski S, MacArthur W, Hanna P (2004) Bacillus anthracis requires siderophore biosynthesis for growth in macrophages and mouse virulence. Mol Microbiol 51:407–417. https://doi.org/10.1046/j.1365-2958.2003.03861.x

Chakrabarty S, Jin M, Wu C, Chakraborty P, Xiao Y (2020) *Bacillus thuringiensis* vegetative insecticidal protein family Vip3A and mode of action against pest Lepidoptera. Pest Manag Sci 76:1612–1617. https://doi.org/10.1002/ps.5804

Chen H, Tang W, Xu C, Li X, Lin Y, Zhang Q (2005) Transgenic indica rice plants harboring a synthetic cry2A* gene of bacillus thuringiensis exhibit enhanced resistance against lepidopteran rice pests. Theor Appl Genet 111:1330–1337. https://doi.org/10.1007/s00122-005-0062-8

Chen XH, Scholz R, Borriss M, Junge H, Mögel G, Kunz S, Borriss R (2009) Difficidin and bacilysin produced by plant-associated Bacillus amyloliquefaciens are efficient in controlling fire blight disease. J Biotechnol 140:38–44. https://doi.org/10.1016/j.jbiotec.2008.10.015

Cheng YQ, Tang GL, Shen B (2003) Type I polyketide synthase requiring a discrete acyltransferase for polyketide biosynthesis. Proc Natl Acad Sci U S A 100:3149–3154. https://doi.org/10.1073/pnas.0537286100

Christensen DP, Hutkins RW (1992) Collapse of the proton motive force in listeria monocytogenes caused by a bacteriocin produced by Pediococcus acidilactici. Appl Environ Microbiol 58:3312–3315. http://www.ncbi.nlm.nih.gov/pubmed/1444365. Accessed 3 June 2020

Das P, Mukherjee S, Sen R (2008) Genetic regulations of the biosynthesis of microbial surfactants: an overview. Biotechnol Genet Eng Rev 25:165–186. https://doi.org/10.5661/bger-25-165

del Rosario Cappellari L, Banchio E (2019) Microbial volatile organic compounds produced by Bacillus amyloliquefaciens GB03 ameliorate the effects of salt stress in mentha piperita principally through acetoin emission. J Plant Growth Regul:1–12. https://doi.org/10.1007/s00344-019-10020-3

Dhankar R, Goyal S, Kumar K, Hussain T (2021) Bacterial community response to pesticides polluted soil. In: Mandal S et al (eds) Recent Advancement in Microbial Biotechnology-Agricultural and Industrial Approach. Elsevier, pp 339–355. https://doi.org/10.1016/B978-0-12-822098-6.00010-0

Dhankar R, Tyagi P, Kamble SS, Gupta D, Hussain T (2020) Advances in fungi: Rejuvenation of polluted sites. In: Sharma VK, Shah MP, Parmar S, Kumar A (eds) Fungi Bio-Prospects in Sustainable Agriculture, Environment and Nano-Technology, 2. Elsevier AP, U.K, pp 251–275. https://doi.org/10.1016/B978-0-12-821925-6.00012-5

Dong YH, Gusti AR, Zhang Q, Xu JL, Zhang LH (2002) Identification of quorum-quenching N-acyl homoserine lactonases from bacillus species. Appl Environ Microbiol 68:1754–1759. https://doi.org/10.1128/AEM.68.4.1754-1759.2002

Dunlap CA, Bowman MJ, Schisler DA (2013) Genomic analysis and secondary metabolite production in Bacillus amyloliquefaciens AS 43.3: A biocontrol antagonist of Fusarium head blight. Biol Control 64:166–175. https://doi.org/10.1016/j.biocontrol.2012.11.002

Espinasse S, Gohar M, Lereclus D, Sanchis V (2002) An ABC transporter from bacillus thuringiensis is essential for β-exotoxin I production. J Bacteriol 184:5848–5854. https://doi.org/10.1128/JB.184.21.5848-5854.2002

Espinasse S, Gohar M, Lereclus D, Sanchis V (2004) An extracytoplasmic-function sigma factor is involved in a pathway controlling β-exotoxin I production in bacillus thuringiensis subsp. thuringiensis Strain 407-1. J Bacteriol 186:3108–3116. https://doi.org/10.1128/JB.186.10.3108-3116.2004

Folmsbee MJ, McInerney MJ, Nagle DP (2004) Anaerobic growth of Bacillus mojavensis and Bacillus subtilis requires deoxyribonucleosides or DNA. Appl Environ Microbiol 70:5252–5257. https://doi.org/10.1128/AEM.70.9.5252-5257.2004

Franzetti A, Tamburini E, Banat IM (2010) Applications of biological surface active compounds in remediation technologies. Adv Exp Med Biol 672:121–134. https://doi.org/10.1007/978-1-4419-5979-9_9

FZB24® Bacillus subtilis – mode of action of a microbial agent enhancing plant vitality (n.d.). https://www.cabdirect.org/cabdirect/abstract/20013014098. Accessed 18 May 2020

Gagné-Bourque F, Bertrand A, Claessens A, Aliferis KA, Jabaji S (2016) Alleviation of drought stress and metabolic changes in timothy (Phleum pratense L.) colonized with Bacillus subtilis B26. Front Plant Sci 7:584. https://doi.org/10.3389/fpls.2016.00584

García-López AM, Delgado A (2016) Effect of Bacillus subtilis on phosphorus uptake by cucumber as affected by iron oxides and the solubility of the phosphorus source. Agric Food Sci 25:216–224. https://doi.org/10.23986/afsci.56862

Glick BR (2004) Bacterial ACC deaminase and the alleviation of plant stress. Adv Appl Microbiol 56:291–312. https://doi.org/10.1016/S0065-2164(04)56009-4

Gold SE, Blacutt AA, Meinersmann RJ, Bacon CW (2014) Whole-genome shotgun sequence of Bacillus mojavensis strain RRC101, an endophytic bacterium antagonistic to the mycotoxigenic endophytic fungus Fusarium verticillioides. Genome Announc 2. https://doi.org/10.1128/genomeA.01090-14

Golldack D, Dietz KJ (2001) Salt-induced expression of the vacuolar H+-ATPase in the common ice plant is developmentally controlled and tissue specific. Plant Physiol 125:1643–1654. https://doi.org/10.1104/pp.125.4.1643

Grover M, Ali SZ, Sandhya V, Rasul A, Venkateswarlu B (2011) Role of microorganisms in adaptation of agriculture crops to abiotic stresses. World J Microbiol Biotechnol 27:1231–1240. https://doi.org/10.1007/s11274-010-0572-7

Haas D, Défago G (2005) Biological control of soilborne pathogens by fluorescent pseudomonads. Nat Rev Microbiol 3:307–319. https://doi.org/10.1038/nrmicro1129

Hammami I, Rhouma A, Jaouadi B, Rebai A, Nesme X (2009) Optimization and biochemical characterization of a bacteriocin from a newly isolated Bacillus subtilis strain 14B for biocontrol of Agrobacterium spp. strains. Lett Appl Microbiol 48:253–260. https://doi.org/10.1111/j.1472-765X.2008.02524.x

Haris M, Shakeel A, Hussain T, Ahmad G, Khan AA (2021) New Trends in Removing Heavy Metals from Industrial Wastewater Through Microbes. In: Shah MP (ed) Removal of Emerging Contaminants Through Microbial Processes. Springer, Singapore. https://doi.org/10.1007/978-981-15-5901-3_9

Haris M, Shakeel A, Ansari MA, Hussain T, Khan AA, Dhankar R (2020) Sustainable crop production and improvement through bio-prospecting of fungi. In: Sharma VK, Shah MP, Parmar S, Kumar A (eds) Fungi Bio-Prospects in Sustainable Agriculture, Environment and Nano-Technology, 1, vol 4. Elsevier AP, U.K, pp 7–428. https://doi.org/10.1016/B978-0-12-821394-0.00016-0

Hu HQ, Li XS, He H (2010) Characterization of an antimicrobial material from a newly isolated Bacillus amyloliquefaciens from mangrove for biocontrol of Capsicum bacterial wilt. Biol Control 54:359–365. https://doi.org/10.1016/j.biocontrol.2010.06.015

Hussain T, Dhanker R (2021) Science of Microorganisms for the Restoration of Polluted sites for Safe and Healthy Environment. In: Shah M, Rodriguez-Couto S (eds) Microbial Ecology of Wastewater Treatment Plants. Elsevier, U.K, pp 127–144. https://doi.org/10.1016/C2019-0-04695-X

Hussain T, Khan AA (2018a) *Bacillus firmusHussainT: Lab. 66:* A new Biosurfactant producing bacteria for the Biocontrol of Late Blight of potato caused by *Phytophthorainfestans* (Mont.) de Bary. National Seminar on New Paradigms of Plant Health Management:Sustainable Food Security under Climatic Scenario from 17th Nov. to 19th Nov. 2018 at Bihar Agricultural University, Sabour (Bhagalpur), Bihar, India under Indian Phytopathological Society, Eastern Zone region Annual meeting. OP, p 60

Hussain T, Khan AA (2018b) Biological control of black scurf of potato caused by *Rhizoctoniasolani* using lipopeptides produced by novel biosurfactant bacteria: *Bacillus subtilis HussainT-AMU* isolated from Paradwip port, Odisha, India. National Conference on Biointensive Approaches in Plant Protection Strategies and their Socio-economic impacts held on 29th–30th Oct. 2018 at Dept. of Plant Protection, Aligarh Muslim University, Aligarh, India under Indian Phytopathological society, Mid-Eastern Zone region Annual meeting. OP, p 31

Hussain T, Khan AA (2020a) Determining the antifungal activity and characterization of *Bacillus siamensis AMU03* against *Macrophominaphaseolina*(Tassi) Goid. Indian Phytopathol. https://doi.org/10.1007/s42360-020-00239-6

Hussain T, Khan AA (2020b) *Bacillus subtilis HussainT-AMU* and its antifungal activity against potato black scurf caused by*Rhizoctoniasolani*. Biocatal Agric Biotechnol 23:101433

Hussain T, Khan AA (2020c) *Bacillus firmus HussainT:Lab.66,* a new biosurfactant bioagent having potential bio-nematicidal activity against root knot nematode *Meloidogyne incognita*. National Conference on Recent Advances in Biological Science (NCRABS-2020) held on 05th March, 2020,organized by Dept. of Biosciences, Faculty of Natural Sciences, Jamia Millia Islamia University, New Delhi, OP-25

Hussain T, Khan AA (2020d) Isolation of potential biosurfactant bacteria from railway diesel engine yard soil, its PCR detection and antifungal activity *Rhizoctonia solani* of potato tuber. *J Agrica* 9:39–52

Hussain T, Khan A, Khan M (2021) Biocontrol of Soil Borne Pathogen of Potato Tuber Caused by Rhizoctonia solani through Biosurfactant based Bacillus strain. J Nepal Agric Research Council 7(1):54–66. https://doi.org/10.3126/jnarc.v7i1.36921

Hussain T, Khan AA, Tomar S (2019a) Biocontrol of dry rot of potato seed tubers caused by *Fusarium* sp. by using novel biosurfactant bacteria *Bacillus firmus HussainT-AMU:Lab.66* producing cyclic lipopeptides. Indian Plant science Congress 2019, from 23rd to 25th Jan.2019 at SRM Institute of Science and Technology, Kattankulathur, TN, p 101

Hussain T, Haris M, Shakeel A, Khan AA, Khan MA (2020) Bio-nematicidal activities by culture filtrate of *Bacillus subtilis HussainT-AMU*: new promising biosurfactant bioagent for the management of root Galling caused by *Meloidogyne incognita*. Vegetos 33:229–238. https://doi.org/10.1007/s42535-020-00099-5

Hussain T, Akhtar N, Aminedi R, Danish M, Nishat Y, Patel S (2020e) Role of the potent microbial based bio agent and their emerging strategies for the eco-friendly management of Agricultural Phytopathogens. In: Singh J, Yadav AN (eds) Natural Bioactive Products in Sustainable Agriculture. Springer, Singapore, pp 45–66. https://doi.org/10.1007/978-981-15-3024-1_14

Hussain T, Khan AA, Khan MA, Ahmad G, Haris M, Shakeel A (2019b) Cyclic Lipopeptides produced by a Noval Green Biosurfactant bacteria *Bacillus subtilis Hussain T-AMU* against Deadly Fungal Pathogens of Potato Seed Tubers at 2nd International Conference on Chemistry, Industry and Environment held on 18th & 19th Dept. Of Applied Chemistry, ZHCET, AMU, Aligarh. OP 68, p 95

Hussain T, Singh S, Danish M, Pervez R, Hussain K, Husain R (2020f) Natural Metabolites an eco-friendly approach to manage plant diseases and for better agricultural farming. In: Singh J, Yadav AN (eds) Natural Bioactive Products in Sustainable Agriculture. Springer Singapore, pp 1–13. https://doi.org/10.1007/978-981-15-3024-1_1

Hussain K, Haris M, Qamar H, Hussain T, Ahmad G, Ansari MS, Khan AA (2021) Bioremediation of Waste Gases and Polluted Soils. In: Panpatte DG, Jhala YK (eds) Microbial Rejuvenation of Polluted Environment. Microorganisms for Sustainability, 26. Springer, Singapore. https://doi.org/10.1007/978-981-15-7455-9_5

Irshad S, Xie Z, Wang J, Nawaz A, Luo Y, Wang Y, Mehmood S (2020) Faheem, indigenous strain bacillus XZM assisted phytoremediation and detoxification of arsenic in Vallisneria denseserrulata. J Hazard Mater 381:120903. https://doi.org/10.1016/j.jhazmat.2019.120903

Jamal Q, Cho J-Y, Moon J-H, Munir S, Anees M, Kim KY (2017) Identification for the first time of Cyclo(d-Pro-l-Leu) produced by bacillus amyloliquefaciens Y1 as a Nematocide for control of Meloidogyne incognita. Molecules 22:1839. https://doi.org/10.3390/molecules22111839

Ji C, Tian H, Wang X, Hao L, Wang C, Zhou Y, Xu R, Song X, Liu Y, Du J, Liu X (2020) Bacillus subtilis HG-15, a Halotolerant Rhizoplane Bacterium, Promotes Growth and Salinity Tolerance in Wheat (Triticum aestivum). https://doi.org/10.21203/RS.3.RS-19695/V1

Jiang C-H, Xie Y-S, Zhu K, Wang N, Li Z-J, Yu G-J, Guo J-H (2019) Volatile organic compounds emitted by Bacillus sp. JC03 promote plant growth through the action of auxin and strigolactone Indole acetic acid ABA Abscisic acid GA Gibberellins ETH Ethylene SLs Strigolactone MRM Multiple reaction monitoring mode SPME Solid phase microextraction LSD Least significant difference. 87:317–328. https://doi.org/10.1007/s10725-018-00473-z

Joshi S, Bharucha C, Jha S, Yadav S, Nerurkar A, Desai AJ (2008) Biosurfactant production using molasses and whey under thermophilic conditions. Bioresour Technol 99:195–199. https://doi.org/10.1016/j.biortech.2006.12.010

Julsing MK, Rijpkema M, Woerdenbag HJ, Quax WJ, Kayser O (2007) Functional analysis of genes involved in the biosynthesis of isoprene in Bacillus subtilis. Appl Microbiol Biotechnol 75:1377–1384. https://doi.org/10.1007/s00253-007-0953-5

Kang SM, Radhakrishnan R, Lee KE, You YH, Ko JH, Kim JH, Lee IJ (2015) Mechanism of plant growth promotion elicited by Bacillus sp. LKE15 in oriental melon. Acta Agric Scand Sect B Soil Plant Sci 65:637–647. https://doi.org/10.1080/09064710.2015.1040830

Kang SM, Khan AL, Waqas M, Asaf S, Lee KE, Park YG, Kim AY, Khan MA, You YH, Lee IJ (2019a) Integrated phytohormone production by the plant growth-promoting rhizobacterium Bacillus tequilensis SSB07 induced thermotolerance in soybean. J Plant Interact 14:416–423. https://doi.org/10.1080/17429145.2019.1640294

Kang SM, Hamayun M, Khan MA, Iqbal A, Lee IJ (2019b) Bacillus subtilis JW1 enhances plant growth and nutrient uptake of Chinese cabbage through gibberellins secretion. J Appl Bot Food Qual 92:172–178. https://doi.org/10.5073/JABFQ.2019.092.023

Kearns DB, Chu F, Branda SS, Kolter R, Losick R (2004) A master regulator for biofilm formation by Bacillus subtilis. Mol Microbiol 55:739–749. https://doi.org/10.1111/j.1365-2958.2004.04440.x

Kobayashi T, Suzuki M, Inoue H, Itai RN, Takahashi M, Nakanishi H, Mori S, Nishizawa NK (n.d.) Expression of iron-acquisition-related genes in iron-deficient rice is co-ordinately induced by partially conserved iron-deficiency-responsive elements. J Exp Bot. https://doi.org/10.1093/jxb/eri131

Kontnik R, Bosak T, Butcher R, Brocks J, Losick R, Pearson A, Clardy J (2008) Sporulenes, three new polycyclic terpenoids from Bacillus subtilis spores. Planta Med 74:PB98. https://doi.org/10.1055/s-0028-1084443

Kotze AC, O'Grady J, Gough JM, Pearson R, Bagnall NH, Kemp DH, Akhurst RJ (2005) Toxicity of Bacillus thuringiensis to parasitic and free-living life-stages of nematode parasites of livestock. Int J Parasitol 35:1013–1022. https://doi.org/10.1016/j.ijpara.2005.03.010

Koumoutsi A, Chen XH, Henne A, Liesegang H, Hitzeroth G, Franke P, Vater J, Borriss R (2004) Structural and functional characterization of gene clusters directing nonribosomal synthesis of bioactive cyclic lipopeptides in bacillus amyloliquefaciens strain FZB42. J Bacteriol Am Soc Microbiol J:1084–1096. https://doi.org/10.1128/JB.186.4.1084-1096.2004

Kuzuyama T (2002) Mevalonate and nonmevalonate pathways for the biosynthesis of isoprene units. Biosci Biotechnol Biochem 66:1619–1627. https://doi.org/10.1271/bbb.66.1619

Kumar A, Hussain T, Susmita C, Maurya DK, Danish M, Farooqui SA (2021) Microbial remediation and detoxification of heavy metals by plants and microbes. In: Shah M et al (eds) The Future of Effluent Treatment Plants-Biological Treatment Systems. Elsevier, U.K, pp 589–614. https://doi.org/10.1016/B978-0-12-822956-9.00030-1

Lee H, Kim H (n.d.) Lantibiotics, Class I Bacteriocins From the Genus Bacillus – PubMed. https://pubmed.ncbi.nlm.nih.gov/21464591/. Accessed 27 June 2020

Lee BR, Jung WJ, Lee BH, Avice JC, Ourry A, Kim TH (2008) Kinetics of drought-induced pathogenesis-related proteins and its physiological significance in white clover leaves. Physiol Plant 132:329–337. https://doi.org/10.1111/j.1399-3054.2007.01014.x

Li Z, Wang P, Yue X, Wang J, Ren B, Qu L, Han H (2019) Effects of bacillus thuringiensis HC-2 combined with biochar on the growth and Cd and Pb accumulation of radish in a heavy metal-contaminated farmland under field conditions. Int J Environ Res Public Health 16:3676. https://doi.org/10.3390/ijerph16193676

Lim JH, Kim SD (2013) Induction of drought stress resistance by multi-functional PGPR Bacillus licheniformis K11 in pepper. Plant Pathol J 29:201–208. https://doi.org/10.5423/PPJ.SI.02.2013.0021

L. Liu, X. Zhao, Y. Huang, L. Ke, R. Wang, G. Qi, Protecting tobacco plants from O $_3$ injury by *Bacillus velezensis* with production of acetoin, Physiol Plant. (2020) ppl.13120. doi:https://doi.org/10.1111/ppl.13120

López-Lara IM, Geiger O (2017) Bacterial lipid diversity. Biochim Biophys Acta Mol Cell Biol Lipids 1862:1287–1299. https://doi.org/10.1016/j.bbalip.2016.10.007

Lüthy P, Wolfersberger MG (2000) Pathogenesis of bacillus thuringiensis toxins, in: Entomopathog. Bact. from Lab. to F. Appl. Springer, pp 167–180. https://doi.org/10.1007/978-94-017-1429-7_9

Makkar RS, Cameotra SS (1998) Production of biosurfactant at mesophilic and thermophilic conditions by a strain of Bacillus subtilis. J Ind Microbiol Biotechnol 20:48–52. https://doi.org/10.1038/sj.jim.2900474

Mandal SD, Sonali, Singh S, Hussain K, Hussain T (2021) Plant Microbe Association for the mutual benefits for plant growth and soil health. In: Yadav AN et al (eds) Current Trends in Microbial Biotechnology for Sustainable Agriculture, Environmental and Microbial Biotechnology. Springer Nature, Singapore, pp 95–121. https://doi.org/10.1007/978-981-15-6949-4_5

Mihalache G, Balaes T, Gostin I, Stefan M, Coutte F, Krier F (2018) Lipopeptides produced by Bacillus subtilis as new biocontrol products against fusariosis in ornamental plants. Environ Sci Pollut Res 25:29784–29793. https://doi.org/10.1007/s11356-017-9162-7

Minahk CJ, FarÃas ME, Sesma F, Morero RD (2000) Effect of Enterocin CRL35 on Listeria monocytogenes cell membrane. FEMS Microbiol. Lett 192:79–83. https://doi.org/10.1111/j.1574-6968.2000.tb09362.x

Moll A.J, Konings GN, Driessen WN (n.d.) Bacteriocins: mechanism of membrane insertion and pore formation – PubMed. https://pubmed.ncbi.nlm.nih.gov/10532378/. Accessed 27 June 2020

Mondal M, Biswas JK, Tsang YF, Sarkar B, Sarkar D, Rai M, Sarkar SK, Hooda PS (2019) A wastewater bacterium Bacillus sp. KUJM2 acts as an agent for remediation of potentially toxic elements and promoter of plant (Lens culinaris) growth. Chemosphere 232:439–452. https://doi.org/10.1016/j.chemosphere.2019.05.156

Moreno-Lora A, Recena R, Delgado A (2019) Bacillus subtilis QST713 and cellulose amendment enhance phosphorus uptake while improving zinc biofortification in wheat. Appl Soil Ecol 142:81–89. https://doi.org/10.1016/j.apsoil.2019.04.013

Muddanuru T, Polumetla AK, Maddukuri L, Mulpuri S (2019) Development and evaluation of transgenic castor (Ricinus communis L.) expressing the insecticidal protein Cry1Aa of Bacillus thuringiensis against lepidopteran insect pests. Crop Prot 119:113–125. https://doi.org/10.1016/j.cropro.2019.01.016

Mulligan CN (2005) Environmental applications for biosurfactants. Environ Pollut 133:183–198. https://doi.org/10.1016/j.envpol.2004.06.009

Nes IF, Holo H (2000) Class II antimicrobial peptides from lactic acid bacteria. Pept Sci 55:50–61. https://doi.org/10.1002/1097-0282(2000)55:1<50::AID-BIP50>3.0.CO;2-3

Ngugi HK, Dedej S, Delaplane KS, Savelle AT, Scherm H (2005) Effect of flower-applied Serenade biofungicide (Bacillus subtilis) on pollination-related variables in rabbiteye blueberry. Biol Control 33:32–38. https://doi.org/10.1016/j.biocontrol.2005.01.002

Nicolas JP (2003) Molecular dynamics simulation of surfactin molecules at the water-hexane interface. Biophys J 85:1377–1391. https://doi.org/10.1016/S0006-3495(03)74571-8
Ongena M, Jacques P (2008) Bacillus lipopeptides: versatile weapons for plant disease biocontrol. Trends Microbiol 16:115–125. https://doi.org/10.1016/j.tim.2007.12.009
Oppegård C, Rogne P, Emanuelsen L, Kristiansen PE, Fimland G, Nissen-Meyer J (2007) The two-peptide class II Bacteriocins: structure, production, and mode of action. J Mol Microbiol Biotechnol 13:210–219. https://doi.org/10.1159/000104750
Paik HD, Bae SS, Park SH, Pan JG (1997) Identification and partial characterization of tochicin, a bacteriocin produced by Bacillus thuringiensis subsp tochigiensis. J Ind Microbiol Biotechnol 19:294–298. https://doi.org/10.1038/sj.jim.2900462
Park R-Y, Choi M-H, Sun H-Y, Shin S-H (2005) Production of catechol-Siderophore and utilization of transferrin-bound iron in Bacillus cereus. Biol Pharm Bull 28:1132–1135. https://doi.org/10.1248/bpb.28.1132
Park YG, Mun BG, Kang SM, Hussain A, Shahzad R, Seo CW, Kim AY, Lee SU, Oh KY, Lee DY, Lee IJ, Yun BW (2017) Bacillus aryabhattai SRB02 tolerates oxidative and nitrosative stress and promotes the growth of soybean by modulating the production of phytohormones. PLoS One 12. https://doi.org/10.1371/journal.pone.0173203
Porcar M, Juárez-Pérez V (2003) PCR-based identification of bacillus thuringiensis pesticidal crystal genes. FEMS Microbiol Rev 26:419–432. https://doi.org/10.1111/j.1574-6976.2003.tb00624.x
Qiuhong N, Xiaowei H, Baoyu T, Jinkui Y, Jiang L, Lin Z, Keqin Z (2006) Bacillus sp. B16 kills nematodes with a serine protease identified as a pathogenic factor. Appl Microbiol Biotechnol 69:722–730. https://doi.org/10.1007/s00253-005-0019-5
Raddadi N, Cherif A, Mora D, Ouzari H, Boudabous A, Molinari F, Daffonchio D (2004) The autolytic phenotype of Bacillus thuringiensis. J Appl Microbiol 97:158–168. https://doi.org/10.1111/j.1365-2672.2004.02287.x
Raddadi N, Cherif A, Mora D, Brusetti L, Borin S, Boudabous A, Daffonchio D (2005) The autolytic phenotype of the Bacillus cereus group. J Appl Microbiol 99:1070–1081. https://doi.org/10.1111/j.1365-2672.2005.02713.x
Raddadi N, Cherif A, Ouzari H, Marzorati M, Brusetti L, Boudabous A, Daffonchio D (2007) Bacillus thuringiensis beyond insect biocontrol: plant growth promotion and biosafety of polyvalent strains. Ann Microbiol 57:481–494. https://doi.org/10.1007/BF03175344
Radhakrishnan R, Hashem A, Abd Allah EF (2017) Bacillus: A biological tool for crop improvement through bio-molecular changes in adverse environments. Front Physiol 8:667. https://doi.org/10.3389/fphys.2017.00667
Rashid R, Veleba M, Kline KA (2016) Focal targeting of the bacterial envelope by antimicrobial peptides. Front Cell Dev Biol 4:55. https://doi.org/10.3389/fcell.2016.00055
Ribeiro VP, Marriel IE, de Sousa SM, de Lana UGP, Mattos BB, de Oliveira CA, Gomes EA (2018) Endophytic Bacillus strains enhance pearl millet growth and nutrient uptake under low-P. Brazilian J Microbiol 49:40–46. https://doi.org/10.1016/j.bjm.2018.06.005
Riley MA, Wertz JE (2002a) Bacteriocin diversity: ecological and evolutionary perspectives. Biochimie 84:357–364. https://doi.org/10.1016/S0300-9084(02)01421-9
Riley MA, Wertz JE (2002b) Bacteriocins: evolution, ecology, and application. Annu Rev Microbiol 56:117–137. https://doi.org/10.1146/annurev.micro.56.012302.161024
Rogers EW, Molinski TF (2007) Asymmetric synthesis of diastereomeric diaminoheptanetetraols. A proposal for the configuration of (+)-zwittermicin A. Org Lett 9:437–440. https://doi.org/10.1021/ol062804a
Roh JY, Choi JY, Li MS, Jin BR, Je YH (2007) Bacillus thuringiensis as a specific, safe, and effective tool for insect pest control. J Microbiol Biotechnol 17:547–559. http://europepmc.org/article/med/18051264. Accessed 27 June 2020
Rohmer M (1999) The discovery of a mevalonate-independent pathway for isoprenoid biosynthesis in bacteria, algae and higher plants. Nat Prod Rep 16:565–574. https://doi.org/10.1039/a709175c

Rohmer M, Knani M, Simonin P, Sutter B, Sahm H (1993) Isoprenoid biosynthesis in bacteria: A novel pathway for the early steps leading to isopentenyl diphosphate. Biochem J 295:517–524. https://doi.org/10.1042/bj2950517

Saeki K, Ozaki K, Kobayashi T, Ito S (2007) Detergent alkaline proteases: enzymatic properties, genes, and crystal structures. J Biosci Bioeng 103:501–508. https://doi.org/10.1263/jbb.103.501

Sansinenea E, Ortiz A (2011) Secondary metabolites of soil bacillus spp. Biotechnol Lett 33:1523–1538. https://doi.org/10.1007/s10529-011-0617-5

Sarkar NK, Kim YK, Grover A (2009) Rice sHsp genes: genomic organization and expression profiling under stress and development. BMC Genomics 10:393. https://doi.org/10.1186/1471-2164-10-393

Schallmey M, Singh A, Ward OP (2004) Developments in the use of Bacillus species for industrial production. Can J Microbiol 50:1–17. https://doi.org/10.1139/w03-076

Schneider K, Chen XH, Vater J, Franke P, Nicholson G, Borriss R, Süssmuth RD (2007) Macrolactin is the polyketide biosynthesis product of the pks2 cluster of Bacillus amyloliquefaciens FZB42. J Nat Prod 70:1417–1423. https://doi.org/10.1021/np070070k

Schnepf E, Crickmore N, Van Rie J, Lereclus D, Baum J, Feitelson J, Zeigler DR, Dean DH (1998) Bacillus thuringiensis and its Pesticidal crystal proteins. Microbiol Mol Biol Rev 62:775–806. https://doi.org/10.1128/mmbr.62.3.775-806.1998

Schwartz J-L, Laprade R (2000) Membrane permeabilization by Bacillus thuringiensis toxins: protein insertion and pore formation, in: Entomopathog. Bact. from Lab. to F. Appl. Springer, pp 199–217. https://doi.org/10.1007/978-94-017-1429-7_11

Shafi J, Tian H, Ji M (2017a) *Bacillus* species as versatile weapons for plant pathogens: a review. Biotechnol Biotechnol Equip 31:446–459. https://doi.org/10.1080/13102818.2017.1286950

Shafi S, Reda FM, Ismail M (2017b) Production of Terpenoids, terpene alcohol, fatty acids and N 2 compounds by bacillus amyloliquefaciens S5i4 isolated from archaeological Egyptian soil. http://www.imedpub.com/advanced-techniques-in-clinical-microbiology/. Accessed 27 June 2020

Shahnawaz M, Soto C (2012) Microcin amyloid fibrils A are reservoir of toxic oligomeric species. J Biol Chem 287:11665–11676. https://doi.org/10.1074/jbc.M111.282533

Shan Y, Shu C, He K, Cheng X, Geng L, Xiang W, Zhang J (2019) Characterization of a novel insecticidal protein Cry9Cb1 from bacillus thuringiensis. J Agric Food Chem 67:3781–3788. https://doi.org/10.1021/acs.jafc.9b00385

Shu Q, Ye G, Cui H, Cheng X, Xiang Y, Wu D, Gao M, Xia Y, Hu C, Sardana R, Altosaar I (2000) Transgenic rice plants with a synthetic cry1Ab gene from bacillus thuringiensis were highly resistant to eight lepidopteran rice pest species. Mol Breed 6:433–439. https://doi.org/10.1023/A:1009658024114

Siddiqua KS, Farooqui SA, Hussain T, Mohamed HI (2021) Microbial Enzymes and Their Role in Phytoremediation. In: Mohamed HI, El-Beltagi HEDS, Abd-Elsalam KA (eds) Plant Growth-Promoting Microbes for Sustainable Biotic and Abiotic Stress Management. Springer, Cham. https://doi.org/10.1007/978-3-030-66587-6_22

Silo-Suh LA, Stabb EV, Raffel SJ, Handelsman J (1998) Target range of Zwittermicin A, an aminopolyol antibiotic from Bacillus cereus. Curr Microbiol 37:6–11. https://doi.org/10.1007/978-3-319-75910-4_16

Singh P, Hussain T, Patel S, Akhtar N (2018) Impact of Climate Change on Root–Pathogen Interactions. In: Giri B, Prasad R, Varma A (eds) Root Biology. Soil Biology. Springer, Cham, p 52. https://doi.org/10.1007/s002849900328

Snook ME, Mitchell T, Hinton DM, Bacon CW (2009) Isolation and characterization of Leu 7-surfactin from the endophytic bacterium bacillus mojavensis RRC 101, a biocontrol agent for fusarium Verticillioides. J Agric Food Chem 57:4287–4292. https://doi.org/10.1021/jf900164h

Suganthi V, Mohanasrinivasan V (2015) Optimization studies for enhanced bacteriocin production by Pediococcus pentosaceus KC692718 using response surface methodology. J Food Sci Technol 52:3773–3783. https://doi.org/10.1007/s13197-014-1440-5

Torres M, Llamas I, Torres B, Toral L, Sampedro I, Béjar V (2019) Growth promotion on horticultural crops and antifungal activity of Bacillus velezensis XT1. Appl Soil Ecol 103453. https://doi.org/10.1016/j.apsoil.2019.103453

Tunsagool P, Jutidamrongphan W, Phaonakrop N, Jaresitthikunchai J, Roytrakul S, Leelasuphakul W (2019) Insights into stress responses in mandarins triggered by Bacillus subtilis cyclic lipopeptides and exogenous plant hormones upon Penicillium digitatum infection. Plant Cell Rep 38:559–575. https://doi.org/10.1007/s00299-019-02386-1

Vejan P, Abdullah R, Khadiran T, Ismail S, Nasrulhaq Boyce A (2016) Role of plant growth promoting Rhizobacteria in agricultural sustainability—a review. Molecules 21:573. https://doi.org/10.3390/molecules21050573

Vicente-Hernández A, Salgado-Garciglia R, Valencia-Cantero E, Ramírez-Ordorica A, Hernández-García A, García-Juárez P, Macías-Rodríguez L (2019) Bacillus methylotrophicus M4-96 stimulates the growth of strawberry (Fragaria × ananassa 'aromas') plants in vitro and slows Botrytis cinerea infection by two different methods of interaction. J Plant Growth Regul 38:765–777. https://doi.org/10.1007/s00344-018-9888-6

Wagner WP, Helmig D, Fall R (2000) Isoprene biosynthesis in Bacillus subtilis via the methylerythritol phosphate pathway. J Nat Prod 63:37–40. https://doi.org/10.1021/np990286p

Walsh CJ, Guinane CM, Hill C, Ross RP, O'Toole PW, Cotter PD (2015) In silico identification of bacteriocin gene clusters in the gastrointestinal tract, based on the human microbiome Project's reference genome database. BMC Microbiol 15:183. https://doi.org/10.1186/s12866-015-0515-4

Wei JZ, Hale K, Carta L, Platzer E, Wong C, Fang SC, Aroian RV (2003) Bacillus thuringiensis crystal proteins that target nematodes. Proc Natl Acad Sci U S A 100:2760–2765. https://doi.org/10.1073/pnas.0538072100

Wilson MK, Abergel RJ, Raymond KN, Arceneaux JEL, Byers BR (2006) Siderophores of bacillus anthracis, Bacillus cereus, and bacillus thuringiensis. Biochem Biophys Res Commun 348:320–325. https://doi.org/10.1016/j.bbrc.2006.07.055

Wu J, Kamal N, Hao H, Qian C, Liu Z, Shao Y, Zhong X, Xu B (2019) Endophytic bacillus megaterium BM18-2 mutated for cadmium accumulation and improving plant growth in hybrid Pennisetum. Biotechnol Rep 24:e00374. https://doi.org/10.1016/j.btre.2019.e00374

Xie Z, Chu Y, Zhang W, Lang D, Zhang X (2019) Bacillus pumilus alleviates drought stress and increases metabolite accumulation in Glycyrrhiza uralensis Fisch. Environ Exp Bot 158:99–106. https://doi.org/10.1016/j.envexpbot.2018.11.021

Yamashiro D, Yoshioka M, Ashiuchi M (2011) Bacillus subtilis pgsE (formerly ywtC) stimulates poly-γ-glutamate production in the presence of zinc. Biotechnol Bioeng 108:226–230. https://doi.org/10.1002/bit.22913

Yuan J, Li B, Zhang N, Waseem R, Shen Q, Huang Q (2012) Production of bacillomycin- and macrolactin-type antibiotics by bacillus amyloliquefaciens NJN-6 for suppressing soilborne plant pathogens. J Agric Food Chem 60:2976–2981. https://doi.org/10.1021/jf204868z

Yuwono T, Handayani D, Soedarsono J (2005) The role of osmotolerant rhizobacteria in rice growth under different drought conditions. Aust J Agric Res 56:715. https://doi.org/10.1071/AR04082

Zawadzka AM, Abergel RJ, Nichiporuk R, Andersen UN, Raymond KN (2009) Siderophore-mediated iron acquisition systems in Bacillus cereus: identification of receptors for anthrax virulence-associated petrobactin. Biochemistry 48:3645–3657. https://doi.org/10.1021/bi8018674

Zhou C, Zhu L, Guo J, Xiao X, Ma Z, Wang J (2019) Bacillus subtilis STU6 ameliorates iron deficiency in tomato by enhancement of polyamine-mediated iron remobilization. J Agric Food Chem 67:320–330. https://doi.org/10.1021/acs.jafc.8b05851

Chapter 20
Mechanisms of the Beneficial Effects of Probiotic *Bacillus* spp. in Aquaculture

A. Q. M. Robiul Kawser, Tofazzal Islam, Md Shah Alam, Md. Mahbubur Rahman, and Mohammad Abdus Salam

Abstract The usage of probiotic bacteria in aquaculture is profoundly ranging from growth promotion of cultured fishes to improvement of the rearing environment. Among the different groups of probiotic bacteria, the *Bacillus* spp. stand out due to their capability to withstand the harsh environment in nature and in fish gut. Although the probiotic *Bacillus* spp. in sustainable aquaculture is widely applicable, the mechanisms of their beneficial effects are not comprehensively reviewed. This chapter aimed to update our knowledge of the mode of actions of *Bacillus* spp. as probiotic in the promotion of aquaculture. The probiotic *Bacillus* spp. secrete diverse enzymes that help fishes in the digestion of complex feedstuffs thus decrease the feed conversion ratio and increase the specific growth rate of the host fishes. They can colonize the fish gut and produce diverse antimicrobial substances like bacteriocins (antibiotics) and lytic enzymes and promote immunity in fishes to protect from pathogens. They also compete for nutrient, energy, and binding sites and thus exclude pathogens from the fish guts. The probiotic *Bacillus* spp. also boost the host's genes expression associated with the growth and development of the fishes. Conversion of toxic nitrogenous wastes by *Bacillus* spp. detoxifies the rearing water

A. Q. M. Robiul Kawser
Department of Aquaculture, Bangabandhu Sheikh Mujibur Rahman Agricultural University, Gazipur, Bangladesh

School of Veterinary Medicine and Science, University of Nottingham, Nottingham, UK

T. Islam (✉) · M. M. Rahman
Institute of Biotechnology and Genetic Engineering (IBGE), Bangabandhu Sheikh Mujibur Rahman Agricultural University, Gazipur, Bangladesh
e-mail: tofazzalislam@bsmrau.edu.bd

M. S. Alam
Aquaculture Program, Center for Marine Science, University of North Carolina Wilmington, Wilmington, USA

M. A. Salam
Department of Genetics and Fish Breeding, Bangabandhu Sheikh Mujibur Rahman Agricultural University, Gazipur, Bangladesh

M. T. Islam et al. (eds.), *Bacilli in Agrobiotechnology*, Bacilli in Climate Resilient Agriculture and Bioprospecting,
https://doi.org/10.1007/978-3-030-85465-2_20

and environment. This chapter is compendium of existing knowledge on the mechanism of beneficial actions of *Bacillus* spp. in aquaculture.

Keywords Green aquaculture · Biocontrol of fish diseases · Quorum quenching · Gene expression · ISR · Fish probiotics

20.1 Introduction

The attention of researchers to the field of beneficial bacteria or microorganisms has arisen many folds since the first hypothesis of beneficial microorganisms was given by Elie Metchnikoff in 1907. The term "probiotic" has gained popularity over the years, and its application has expanded in many areas of life today. "Probiotics are living microorganisms, which when deliver in adequate amount provides benefits to the health of the host" (FAO/WHO 2002). There are various formulations of probiotic microorganisms that are being established and standardized for both human and animal intake. To prevent gastrointestinal infections, probiotics are used extensively in the aquaculture and poultry feed industries (Hong et al. 2005). They are also applied as organic fertilizers and biopesticides in crop production (Yadav et al. 2017). Due to consumer awareness, the exploration of safe and cost-effective treatment/feed and the pathogens antibiotic resistance capability have forced researchers to find an alternative to current therapeutic practice, which mainly rely on antibiotics. Moreover, it is expensive to use antibiotics in large water volumes (Harikrishnan et al. 2011). Probiotic therapy appears to be the most effective one, with a long history of guaranteed safety, among loads of recommended options.

Bacillus is the globally recognized genus of bacteria that are being used for their potential benefits to plants (Rahman et al. 2018; Surovy et al. 2019, Dutta et al. 2018), animals (Xu et al. 2017; Sun et al. 2013), and even in humans (Piewngam et al. 2018), and the significance of *Bacillus* sp. for sustainable aquaculture is systematically reviewed (Rahman et al. 2019). Some *Bacillus* strains have been selected for their possible probiotic application, in several in vitro and in vivo representations. Beside maintaining probiotic standard, *Bacillus* spp. can tolerate higher acidity and stability in heat processing and low storage temperature (Bader et al. 2012). They can exclude pathogens and possess antioxidant, antimicrobial, and immunomodulatory activities (Lefevre et al. 2015; Shobharani et al. 2015; Ripert et al. 2016). Species in the genus *Bacillus* are capable of endospore production. They are Gram-positive bacteria with a rod-shaped structure and can be either aerobic or facultative anaerobes, sized between 2.5 and 10 mm. *Bacillus* spp. belong to the large phylum Firmicutes and classified as *Bacilli* in the order *Bacillales*. About 200 species are included in the genus, which are almost omnipresent in the environment. *Bacillus* spp. can be found in compost (*B. composti* and *B. thermophiles*) (Yang et al. 2013), in high temperature (*B. thermophiles*) (Yang et al. 2013), in high pH conditions (*B. firmus* OF4) (Sturr et al. 1994), and high salt (*B. halodurans*) (Annamalai et al. 2013), in all types of the waterbody (Ichimatsu et al. 2000) and the gut of aquatic animals (Ray et al. 2012). Yi et al. (2018) reported that endospore formation capability enables *Bacillus* spp. to withstand extreme stresses. The capacity to produce tannase, chitinase, protease, phytase, lipase, cellulase, and xylanase are some of their varied biological characteristics (Ray et al. 2012, Ghosh et al. 2019). Bacteriocins and antimicrobial substances are among the other important features of *Bacillus* spp. (Abriouel et al. 2011). Therefore, they earn the ability to take possession of different habitats and patronize host nutrition, and they also serve as environmental and feed probiotics (Moriarty 1998; Hong et al. 2005) in fish farming. It is proven that the host health heavily relies on the optimal and effective functionality of the gastrointestinal (GI) tract and its gut microbiota (Clemente et al. 2012; Xiong et al. 2017), and numerous intricate mechanisms are involved. General immunity and function are weakened, reduced protection against infections, and decreasing gut health is reported in the absence of intestinal microbiota (Ringø et al. 2018; Rawls et al. 2004, 2006; Gómez and Balcázar 2008; Ray et al. 2012; Wang et al. 2018a, b; Li et al. 2019).

Bacteria, virus, and parasites are the attributes of aquaculture diseases (Carbone and Faggio 2016; Bastos Gomes et al. 2017) that linked with the genus *Streptococcus*, *Aeromonas*, *Vibrio*, *Yersinia*, *Acinetobacter*, *Lactococcus*, and *Pseudomonas* (Santos et al. 2018). *Bacillus* spp. as probiotics have already verified experimentally to fight against the diseases causing pathogens (Balcázar et al. 2006; Kavitha et al. 2018; Ramesh and Souissi 2018; Yi et al. 2018) and improve feed utilization, which increases growth (Aly et al. 2008; Goda et al. 2018), to boost the immunity of farmed fish species (Yu et al. 2019; Buruiană et al. 2014) and ameliorate the culture water quality (Camargo and Alonso 2006), as well as to reduce environmental stresses (Eissa et al. 2018).

Although the significance of *Bacillus* spp. and their commercial application in aquaculture as fish probiotics have been reported (Rahman et al. 2019; Acosta et al. 2016; Adorian et al. 2019), the underlying molecular mechanisms of their beneficial effects on fishes are poorly understood (Kuebutornye et al. 2020). The probiotic *Bacillus* spp. exert beneficial effects to the host fishes through various mechanisms that include the antibiotics and lytic enzymes production, promotion digestion of feeds, trigger immunity against disease pathogens, and gene expressions associated with growth and tolerance to the abiotic stresses (Lin et al. 2019; Liu et al. 2012). Some of the probiotic *Bacilli* also improve the water quality in the cultured environment (Zokaeifar et al. 2014). However, no comprehensive review has so far been published on the mechanisms of the beneficial effects of probiotic *Bacilli* in aquaculture. This chapter summarizes the mechanisms of probiotic *Bacillus* spp. to improve the productivity and sustainability in aquaculture.

20.2 *Bacillus* in Aquaculture

In previous decade, the evaluation of different types of probiotic bacteria has tremendously been increased particularly in aquaculture (Hoseinifar et al. 2018). The beneficial effects of probiotic *Bacillus* spp. on growth performance and disease resistance are well documented for many fish species like carps and salmonids (Zhang et al. 2014; Dawood and Koshio 2016; Hoseinifar et al. 2016; Dawood et al. 2018; Fečkaninová et al. 2017). Not only in fish but also in white shrimp, *Litopenaeus vannamei* farming, the probiotic *Bacillus* spp. have been proved promising (Zokaeifar et al. 2012a, b) with a different mode of actions that include the secretion of antimicrobial peptides and extracellular substances. These findings encourage Pacific white shrimp (*L. vannamei*), tiger shrimp (*Penaeus monodon*), and giant freshwater prawn (*Macrobrachium rosenbergii*) health and immunity to combat pathogens thus increases the chance of survival as well as promote growth by improved digestion (Shen et al. 2010; Chai et al. 2016; Laranja et al. 2017). The use of prebiotics as a growth promoter in aquaculture (Gatlin III et al. 2006; Li and Gatlin 2004; Ng and Koh 2016) stimulates the production of exogenous enzymes from probiotic *Bacillus* spp. that improves the intestinal physiology (Gatlin and Peredo 2012; Zhou et al. 2010a, b) with the dietary administration in different fish species such as sturgeon (*Acipenser transmontanus*) (Hoseinifar et al. 2017), tilapia (*Oreochromis niloticus*), and totoaba (*Totoaba macdonaldi*) (Van Hai 2015; González-Félix et al. 2018). Application of probiotic bacteria is vital to increase the feed efficiency as feed constitutes around half of the production cost of rainbow trout (*Oncorhynchus mykiss*) in aquaculture (Amiri et al. 2018). *B. subtilis* significantly improved growth parameter and feed efficiency at the rate of 3.0×10^9 CFU/kg in grass carp (Wu et al. 2012), and in Nile tilapia with *B. subtilis* at either 5×10^6 CFU/g (Telli et al. 2014), 1×10^7CFU/g for 2 months (Aly et al. 2008), or 1×10^8 CFU/g for 2 months (Liu et al. 2017), and in olive flounder (*B. subtilis* at 1×10^{10} CFU/g). A higher dosage of *B. subtilis* ($0.42–1.35 \times 10^7$ CFU/g) administered with feed juvenile large yellow croaker (*Larimichthys crocea*) (Ai et al. 2011). In contrast to this, *B. licheniformis* Dahb1 was more productive at a lower dosage (1×10^5 CFU/mL) in the growth of catfish (*Pangasius hypophthalmus*) than a higher dosage (1×10^7 CFU/mL) (Gobi et al. 2016). Besides oral administration, *Bacillus* is also effective in bath application as reported by Zhou et al. (2010a, b) with *B. coagulans* (1×10^7 CFU/mL) and *B. subtilis* (1×10^7 CFU/mL) in Nile tilapia which had significantly greater final weight and SGR compared to control.

The intestinal microbiota of fish can colonize the host's epithelial surface, allied with the microvilli, or in the lumen (Ringø 1999). Gastrointestinal (GI) tract is among the key infection routes for some pathogens in fish (Groff and LaPatra 2000; Birkbeck and Ringø 2005; Ringø et al. 2007, 2010; Bøgwald and Dalmo 2014); therefore, evaluation of the intestinal microbiota is of immense importance. Figure 20.1 represents the different fish species from where the *Bacillus* is reported with especial emphasis on the Indian major carps where the presence of *Bacillus* reported most.

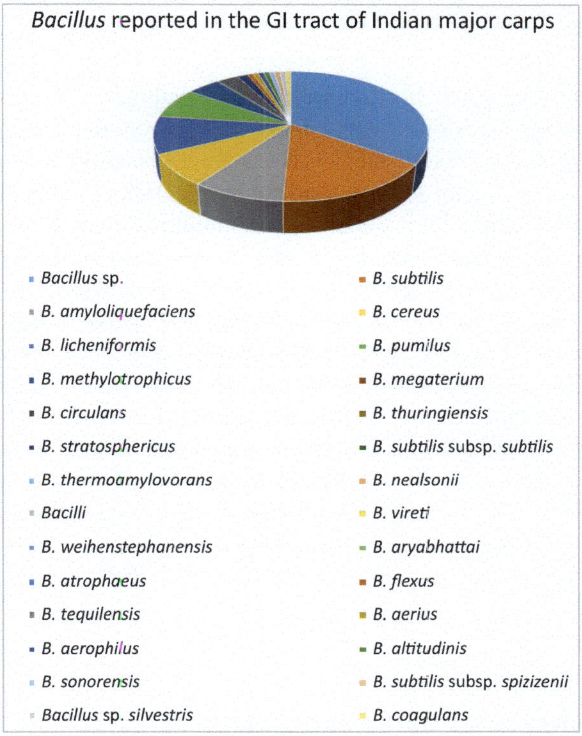

Fig. 20.1 *Bacillus* reported in different fish species and in the GI tract of Indian major carp Sources: Trust (1975); Ramirez-Torrez et al. (2018); Kim et al. (2007); Trust et al. (1979); Sugita et al. (1985); Ramachandran et al. (2005); Sugita et al. (1988b); Yang et al. (2017); Nedoluha and Westhoff (1995); Nedoluha and Westhoff (1997); Sugita et al. (1987); Sugita et al. (1998); Al-Harbi and Uddin (2005); Ramirez and Romero (2017); Rasheeda et al. (2017); Bindiya et al. (2015); Sanchez et al. (2012); Sirtori et al. (2006); Kavitha et al. (2018); Ran et al. (2012); Giri et al. (2011); He et al. (2013); Nyman et al. (2017); Bairagi et al. (2002); Miao et al. (2018); An et al. (2015); Chen et al. (2016a); Chen et al. (2016b); Ghosh et al. (2002); Esakkiraj et al. (2009); Austin and Al-Zahrani (1988); Ray et al. (2007); Saha et al. (2006); Yilmaz et al. (2018); Ma et al. (2010); Kumar et al. (2018); Banerjee et al. (2017); Zhu et al. (2016); Jasmin et al. (2016); Gao et al. (2019); Tzeng et al. (2015); Vargas-Albores et al. (2017); Hindu et al. (2018a); Al-Hisnawi et al. (2015); Ray et al. (2010); Nandi et al. (2017b); Dey et al. (2016); Khan and Ghosh (2012); Mukherjee et al. (2016); Ghosh et al. (2010); Mondal et al. (2010); Ghosh et al. (2017); Banerjee et al. (2016); Mukherjee and Ghosh (2016); Khan and Ghosh (2013); Ringø et al. (2006); Zhou et al. (2013); Dutta et al. (2015); Dutta and Ghosh (2015); Kim et al. (2013); Ringø et al. (2016); Askarian et al. (2012); Banerjee et al. (2013); Mukherjee et al. (2017); Talukdar et al. (2016); Green et al. (2013); Mente et al. (2018); Sugita et al. (1988a); Sugita et al. (1989); Das et al. (2014); Sarkar and Ghosh (2014); Das and Ghosh (2013); Das and Ghosh (2015); Hovda et al. (2007); Hovda et al. (2012); Wong et al. (2013); Peixoto et al. (2011); Ramachandran and Ray (2007); Yang et al. (2018); Strøm and Olafsen (1990); Ricaud et al. (2018); Ringø et al. (2014); Li et al. (2015a,b); Guo et al. (2016); Salam et al. (2021).

20.3 Competition for Nutrients and Energy

Competition between probiotic bacteria and fish pathogenic microorganisms for nutrients and energy leads to the competitive exclusion of the loser (Fig. 20.2). As pathogens and probiotic bacteria hunt energy and nutrients from the same resources, this competition has a great influence on the domination of probiotic over pathogenic microbes (Verschuere et al. 2000a; Hassanein and Soliman 2010). Heterotrophs largely rely on carbon and other energy sources especially in the aquatic environment for their survival (Mohapatra et al. 2013). *Bacillus* species make starve the pathogens by higher organic carbon consumption and uptake of iron by synthesizing siderophores (Verschuere et al. 2000b; Winkelmann 2002; Kesarcodi-Watson et al. 2008; Lalloo et al. 2010) that result in reduced growth of the pathogens since most microbes depend heavily on iron and carbon for their growth (Braun and Killmann 1999). Lalloo et al. (2010) demonstrated that because of the siderophore production, *B. cereus* had significantly better growth with inadequate iron or glucose than pathogenic *A. hydrophila*. Similarly, winning over pathogens in the competition of carbon sources for energy makes *Bacillus* species as a strong probiotic as shown by many scientists in in vitro studies (Lee et al. 2017; Meidong et al. 2017; Kavitha et al. 2018).

20.4 Exo-enzymes Produced by *Bacillus*

Autochthonous gut-adherent bacteria reduce the chances of negative effects by the microorganisms or their metabolites as the microorganisms obtained from tropical and subtropical finfishes can safeguard their colonization and enzyme supplementation within the intestine which is ideal for aquaculture (Ghosh et al. 2019). *Bacillus*

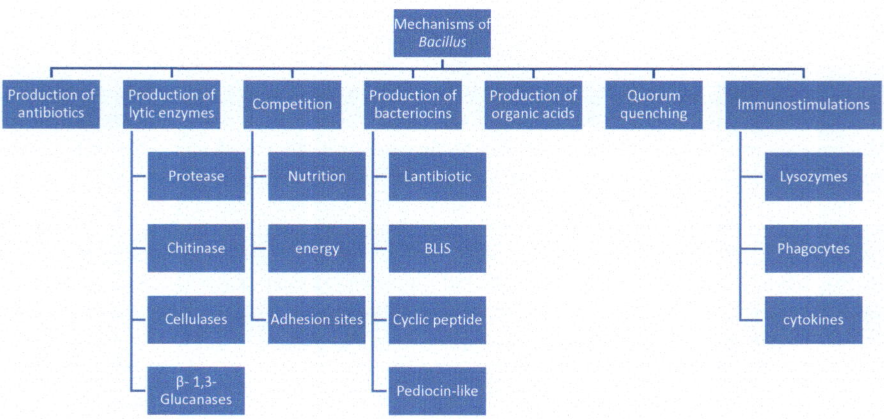

Fig. 20.2 Mechanisms of beneficial actions of probiotic Bacilli in aquaculture species
BLIS bacteriocin-like inhibitory substances

spp. produce a variety of beneficial enzymes, antimicrobial compounds (Prieto et al. 2012), and proteinaceous substances (Zokaeifar et al. 2012a) that include enzymes and bacteriocins or bacteriocin-like inhibitory substances (BLIS) with diverse biological functions. Numerous studies on teleosts like Atlantic salmon (*Salmo salar*), Arctic charr (*Salvelinus alpinus* L.), and Indian major carps revealed that autochthonous *Bacillus* spp. can produce exogenous enzymes such as carbohydrases, phosphatases, esterases, lipases, and peptidases which contribute to the better digestion process and nutrition in fish (Cahil 1990; Ringø et al. 1995, Ringø and Song 2016). Ringø et al. (1995) reported that either by enzymatic breakdown of triglyceride or by shifting pancreatic lipase action with bacterial proteases, gut bacteria might prompt lipolysis. Lipase-producing *Bacillus* spp. were identified in the guts of Atlantic salmon (Askarian et al. 2012), Indian major carps (Dutta and Ghosh 2015), wishers catfish (Das et al. 2014), catfishes (Dey et al. 2016), and Nile tilapia (Ghosh et al. 2017). Apart from these digestive enzymes, the non-starch polysaccharide (NSP) – degrading enzymes, phytase, tannase, and chitinase produced by gut bacteria also attributes significantly to fish nutrition and well-being where autochthonous gut *Bacilli* play a major part (Ray et al. 2012). Among the NSPs, cellulose and hemicelluloses (e.g., xylans) are abundant in natural food like algae, phytoplankton, detritus, aquatic macrophytes, and in different oil cakes which are mostly indigestible for fishes (monogastric animals) because beta-glucanases and beta-xylanases (essential to digest NSPs) are either rare or not present in fish (Kuzmina 1996). However, the degradation of cellulose and hemicelluloses may be accomplished by the fermentative symbiotic gut microorganisms for the host fish (Clements 1997). Recently, studies on the omnivorous fish such as rohu, catla, and mrigal (Ray et al. 2010); grass carp and tilapia (Saha et al. 2006); bata (Mondal et al. 2010); walking catfish (Dey et al. 2016); and striped dwarf catfish (Nandi et al. 2017a) revealed a huge number of cellulose-degrading *Bacilli*. Tannins, which are prevalent in plant feedstuffs, can be degraded by *Bacillus* spp. (Deschamps et al. 1980), and tannase in the alimentary canal of ruminants is well known (Goel et al. 2005). But reports on tannase-producing bacteria from the fish gut are rare. Talukdar et al. (2016) documented that *B. subtilis* KP765736 and *Brevibacillus agri* KP765734 isolated from Nile tilapia have tannase activity. Like tannins, the chitin is abundant in voluminous fish food organisms, viz., crustaceans, mollusks, protozoans, fungi, and green algae (Ray et al. 2012). *B. subtilis*, *B. thuringiensis*, *B. cereus*, and *Bacillus* sp. obtained from the gut of Atlantic salmon were reported to possess chitinase enzyme (Askarian et al. 2012), although the ability of *Bacilli* to degrade chitin was first recorded in *B. chitinovorus* in the early twentieth century from an aquatic source.

Plant feedstuffs for aquaculture feed can be processed by solid-state fermentation (SSF) using exo-enzymes produced by *Bacilli*. The *Bacillus* spp. minimize the level of anti-nutritional factors present in the plant ingredients (Bairagi et al. 2004; Ramachandran et al. 2005; Ramachandran and Ray 2007; Khan and Ghosh 2013) which is responsible for the bio-unavailability of different nutrients in feed. Banerjee and Ghosh (2016) predicted that during the bioprocessing of plant feedstuffs through SSF, essential biomolecules such as amino acids, fatty acids, and vitamins might increase in nutrient levels through microbial synthesis. This improves the bioavailability and digestibility of nutrients which can be ensured from plant feedstuffs. To avoid the detrimental effects of harmful metabolites, microbial symbionts isolated from the fish gut were used for the application of SSF-processed substrate as fish feed ingredients (Khan and Ghosh 2013). *Bacilli* can supplement cellulase (Bairagi et al. 2002) and phytase (Roy et al. 2014) in the processing of plant feedstuffs or in a complete diet. They are not only found potential in reducing phytic acid, cellulose, hemicelluloses, crude fiber, tannins, and trypsin inhibitor but also in increasing the levels of free amino acids (especially arginine, cysteine, valine, histidine, isoleucine, tryptophan, methionine, phenylalanine and threonine), fatty acids, and different minerals (Das and Ghosh 2015). Ray et al. (2007) reported enhanced cellulase-producing ability under SSF of *B. circulans* TP3 and *B. subtilis* CY5 from the gut of Mozambique tilapia and common carp, respectively, while maximum residual cellulase activity of *B. subtilis* P6 and *Bacillus velesensis* P11 was found at pH 7.0–9.0 (Peixoto et al. 2011). Phytase supplementation hydrolyzes the phytate compounds of minerals bound to phytic acid which might increase the availability of phosphorus and other minerals of plant feedstuffs (Oatway et al. 2001). Phytase-producing *Bacilli* first found within fish gut was *B. licheniformis* from rohu (Roy et al. 2009) and, subsequently, autochthonous *B. subtilis*, *B. thuringiensis*, and *B. cereus* from the gut of Atlantic salmon (Askarian et al. 2012). Overall, the autochthonous exo-enzyme-producing *Bacilli* are effective in the fermentation of plant feedstuffs to improve the digestibility and quality of aquafeed. Therefore, the application of these bacteria has a great prospect ahead.

20.5 Production of Organic Acids

Lindgren and Dobrogosz (1990) revealed that lactic acid bacteria (LAB) such as *Lactobacillus* sp. can produce organic acids during lactic fermentation subject to strain and type of LAB. Antimicrobial effects through the reduction of pH with organic acids of LAB result in inhibition of pathogenic microbes which is well documented (González et al. 2007; Maeda et al. 2014). The acidic pH causes an interruption of substrate transport systems across cell cytoplasm, and the undissociated acid collapses the electrochemical proton gradient when diffused passively across the membrane or by modifying the permeability of the cell membrane (Ammor et al. 2006; Musikasang et al. 2009). Organic acids play a robust inhibitory role against deleterious bacteria using this mechanism of action (Musikasang et al. 2009). *B. mojavensis* inhibits the fish pathogen *Edwardsiella piscicida* by secreting low-molecular-weight organic acids (Etyemez and Balcazar 2016). The organic acids secreted by *Bacillus* spp. also help to solubilize insoluble nutrients in the sediments that promote the growth of phytoplankton and zooplanktons in the aquaculture system.

20.6 Production of Antioxidant Enzymes by the *Bacillus* spp.

Various kinds of reactive oxygen species and superoxides are produced in the eukaryotic cells through the metabolic process and due to environmental stresses. Overproduction of ROS damages cell membranes and many macromolecules in cells including DNA. Neutralization of these ROS by antioxidants is therefore needed for maintaining the sound health of fishes. When the antioxidant capacity of cells or tissues fails to surpass the superoxides produced through phagocytosis and/or cellular metabolism (Gobi et al. 2018), it leads to oxidative stresses (Mouthuy et al. 2016; Bermejo-Nogales et al. 2016) in the cells. Neutralization of the oxidative stresses is directly linked to the production of antioxidant enzymes in the cells (Messaoudi et al. 2009; Hindu et al. 2018b). According to the findings of Li et al. (2012), probiotic *Bacillus* species can produce some common antioxidant enzymes that are also found in fishes (Di Giulio et al. 1993). Castex et al. (2010) and Li et al. (2015a, b) found that SOD decomposes O^{2-} to H_2O_2, CAT catalyzes the disproportionation of H_2O_2 into H_2O and O_2 (Wang et al. 2017), and glutathione reduces the free radicals effectively (Hindu et al. 2018a). *B. licheniformis* Dahb1 is capable of modulating antioxidant responses either in serum or in mucus as mentioned by Gobi et al. (2016, 2018) in the study of Asian catfish (*Pangasius hypophthalmus*) and Mozambique tilapia (*O. mossambicus*). Similar findings were reported by Esteban et al. (2014), in gilthead seabream (*Sparus aurata*) where expressions of antioxidant enzyme GPx and SOD were enhanced in mucus after dietary supplementation of *Shewanella putrefaciens* and *Bacillus* species. *B. coagulans* and *B. subtilis* supplementation improved the antioxidant potentiality of gibel carp (*Carassius auratus gibelio*), while a similar effect was observed with *B. licheniformis* supplementation in Nile tilapia (Abarike et al. 2018). These antioxidant enzymes produced by the colonized probiotic *Bacillus* spp. may also improve the abiotic stress (high salinity, pH, temperature fluctuation, etc.) tolerance capacities in the treated fishes.

20.7 Mechanisms of Suppression of Fish Pathogens by the Probiotic *Bacillus* spp.

As the antimicrobial action of bacterial strains are reliant on their ability to modulate varied substances in a very definitive spectrum (Urdaci and Pinchuk 2004), understanding the mode of actions such as the production of bacteriocins, bacteriolytic enzymes, and antibiotics by *Bacillus* spp. is needed for efficient application of probiotic *Bacillus* spp. in aquaculture. It is reported by many scientists that *Bacillus* sp. can provide immunostimulation to host that enable them to survive against many deadly fish pathogens. For instance, *B. subtilis* through immunostimulation can prevent *A. hydrophila*, *A. salmonicida*, *V. anguillarum*, *Y. ruckeri*, *E. ictaluri*, and *F. columnare* in Nile tilapia (*O. niloticus*) (Addo et al. 2017; Mohamed and Refat 2011), grass carp (*C. idella*) (Tang et al. 2019), rainbow trout (*O. mykiss*)

(Newaj-Fyzul et al. 2007; Raida et al. 2003), giant freshwater prawn (*Macrobrachium rosenbergii*) (Keysami and Mohammadpour 2013), and Pacific white shrimp (*L. vannamei*) (Cheng et al. 2017).

20.7.1 Production of Bacteriocins

A way out from the antibiotics can be the bacteriocins (Bierbaum and Sahl 2009) and discussed elaborately regarding their differences with conventional antibiotics by several investigators (Zou et al. 2018). Bacteriocins (form in the ribosome) are antimicrobial peptides that are bioactive and released extracellularly to kill or inhibit the prolification of pathogenic prokaryotes including bacteria. These natural products have the potential antimicrobial action against various pathogenic bacteria including antibiotic-resistant strains (Riley and Wertz 2002). Bacteriocinogenic *Bacillus* spp. have been identified in different aquatic habitats like in Amazon basin (Cladera-Olivera et al. 2004), in seaweeds (Prieto et al. 2012), and in marine sediment (Rajesh et al. 2012, Smitha and Bhat 2013, Chopra et al. 2014).

Subtilin, ericin S, ericin A, sublancin 168, mersacidin, and paenibacillin are the bacteriocins produced by *B. subtilis* (Parisot et al. 2008; Dubois et al. 2009). Some other important bacteriocins such as coagulin, thurincin H, cerein 7A, subtilosin A, and megacin A-216 are produced by *B. coagulans*, *B. thuringiensis*, *B. cereus*, *B. amyloliquefaciens*, and *B. megaterium*, respectively (Le Marrec et al. 2000; Gray et al. 2006; Oscariz and Pisabarro 2000; Kiss et al. 2008; Sutyak et al. 2008).

Bacteriocins that produce antimicrobial compounds (such as by *Bacillus*) are confirmed by genome sequencing (Grubbs et al. 2017; Rahman et al. 2020). They are antibiotics and nonribosomally produced compounds, i.e., amino sugars, polyketides, and phospholipids (Stein 2005). Bacteriocins and similar substances produced by genus *Bacillus* (Abriouel et al. 2011; Al-Thubiani et al. 2018) are effective against various fish pathogens (Urdaci and Pinchuk 2004). Yi et al. (2018) identified four bacteriocins gene clusters, five non-ribosomal peptide synthetase gene clusters, and three polyketide synthases which are bacteriocins-associated gene clusters from *B. velezensis* isolated from carp. The *B. velezensis* inhibits various fish pathogenic bacteria including *Aeromonas hydrophila*, *Lactococcus garvieae*, *Vibrio parahaemolyticus*, and *Streptococcus agalactiae*. Rahman et al. (2020) identified several peptide genes coded for bacillaene, subtilin, bacillibactin, surfactin, fengycin, bacilycin, and subtilosin A in the genome of *B. subtilis* strain WS1A isolated from a marine sponge from the Saint Martin's island of the Bay of Bengal. The *B. subtilis* strain WS1A exhibited antimicrobial activity against fish pathogenic *A. veronii*. Dietary supplementation of the strain and its extracellular products significantly strengthened the resistance against motile *Aeromonas septicaemia* in rohu fish (*L. rohita*) (Rahman et al. 2020). Bacteriocins from *Bacillus* species such as coagulin isolated from *Bacillus coagulans* acts as food preservatives (Gálvez et al. 2007; Fu et al. 2018). *B. licheniformis* can increase the shelf-life of food and feed by inhibiting spoilage bacteria through bacteriocins (Teixeira et al. 2009).

20.7.2 Quorum Quenching

Quorum quenching (QQ) can be defined as the interruption of quorum sensing (Roy et al. 2011). Quorum sensing (QS) is the bacterial communication through signaling molecules that regulates antibiotic resistance, biofilm formation, bioluminescence, and bacterial virulence. QS is bacterial coordinated gene expression performed by releasing and distinguishing tiny signaling molecules to regulate their density and harmonize target gene expression (Zhang and Dong 2004). QS can control phenotypes by virulence factors, bioluminescence, and biofilm development (Miller and Bassler 2001; Menit et al. 2003; Mellbye and Schuster 2011).

Many microbes including *Bacillus* spp. can produce enzymes that interrupt N-acyl homoserine lactone (Tang et al. 2013) signal to disrupt QS (Wee et al. 2018). The mechanism of QQ can be used effectively in aquaculture against fish pathogens (Defoirdt et al. 2004) since QS regulates the pathogenicity of bacteria. Musthafa et al. (2011) discovered that pathogenicity and biofilm production capability of *Chromobacterium violaceum* and *Pseudomonas aeruginosa* were reduced by *Bacillus* sp. SS4 through interfering the AHL signaling of the pathogenic bacteria. Decreased mortality of common carp with *A. hydrophila* was observed in a challenge test due to QQ attributed to AHL lactonase produced by *Bacillus* species (Chen et al. 2010). Similarly, regarding *Bacillus*, it is reported by Zhang et al. (2011) that a gene (AiiO-AIO6) can inhibit the virulence factors by blocking the signaling molecules of *A. hydrophila*. An N-acyl homoserine lactone degradation gene from *Bacillus* has been reported by Reimmann et al. (2002) that was effective against pathogenic *P. aeruginosa*. In zebrafish, AiiAAI96 gene of *Bacillus* species reduces *A. hydrophila* infection through oral administration (Cao et al. 2012, 2014). These results suggest that *Bacillus* species degrade the QS AHL of pathogenic bacteria by producing degradation enzymes, and genes thus protect the fishes in the aquaculture system (Fig. 20.2).

20.7.3 Suppression of Fish Pathogens by the Production of Lytic Enzymes by Probiotic Bacillus spp.

Bacillus spp. are reported to secrete variety of lytic enzymes with antimicrobial (Urdaci and Pinchuk 2004) and antifungal properties (Kim et al. 1999). However, how these enzymes alone act upon pathogens of fishes is not clear and needs further investigations. The cell wall of pathogenic microbes is made up of proteins, cellulose, chitins, etc. that can be broken by proteases, cellulases, and chitinases enzymes, respectively (Jadhav et al. 2017). *Bacillus* is reported to produce those cell wall lytic enzymes which make them important for aquaculture industry and in the digestion of fishes (Thankappan et al. 2015; Zaineldin et al. 2018).

20.7.4 Production of Antibiotics

The antibiotic production capability of *Bacillus* spp. is one of their key strengths against fish pathogens. The chances of the development of antibiotic-resistant in pathogens are less as the mode of actions of antibiotics produced by *Bacillus* species is diverse and wide. Genus *Bacillus* was reported to produce 167 antibiotics, 40% of which are from *B. subtilis* (Béahdy 1974; Mondol et al. 2013). Roughly 4–5% of the genome of *B. subtilis* dedicates to antibiotic production (Stein 2005), and to combat *Helicobacter pylori*, amicoumacin antibiotic synthesized by *B. subtilis* 2335 was found effective (Pinchuk et al. 2001). Antibiotics from *Bacillus* species include bacitracin, laterosporin, gramicidin, tyrocidin, polymyxin, mycobacillin, and zwittermicin (Suva et al. 2016).

20.7.5 Enhancement of Immune Response in Fish

Probiotic bacteria enhance the host's innate immunity when supplemented with the diet (Nayak 2010) by humoral immune responses modulation and expression of host genes related to immunity (Verschuere et al. 2000a). This section highlights the significance of *Bacillus* probiotics in mediating humoral immunity and host gene expression against fish pathogens.

20.7.5.1 Nonspecific Immunity

The nonspecific and the acquired immune system perform decisive roles against pathogens in the host's defense mechanism (Munir et al. 2016). Superoxide dismutase, lysozyme, peroxidase, protease and antiprotease, catalase, and myeloperoxidase are some of the immune parameters of serum of fishes (Nayak 2010). Higher activities of these parameters indicate the higher immune responses (Alexander et al. 2010; Rauta et al. 2012) against infection (Fig. 20.3). Modulation of these parameters and immune system of fishes by the supplementation of probiotic *Bacillus* species in the fish diet is well documented (Wilson 2017). Increased disease resistance attributed to enhanced humoral and cellular immune responses was observed in fish fed with *Bacillus* supplementation (Yi et al. 2018). As the frontliner against pathogens, mucus has vital function in the fish immune system (McNeilly et al. 2008). The mucus contains antimicrobial peptides, immunoglobulin, lysozyme, and proteolytic enzymes, (Nigam et al. 2012; Jung et al. 2012). The research on mucosal immunity enhancement with probiotic bacteria is scarce. However, there are a few reports that suggest that *Bacillus* species can increase the mucosal immunity in fishes (Sheikhzadeh et al. 2012; Sangma and Kamilya 2015). Thy et al. (2017) reported that dietary administration of a mixture of *B. amyloliquefaciens* and *B. pumilus* increases the phagocytic activity in striped catfish (*P.*

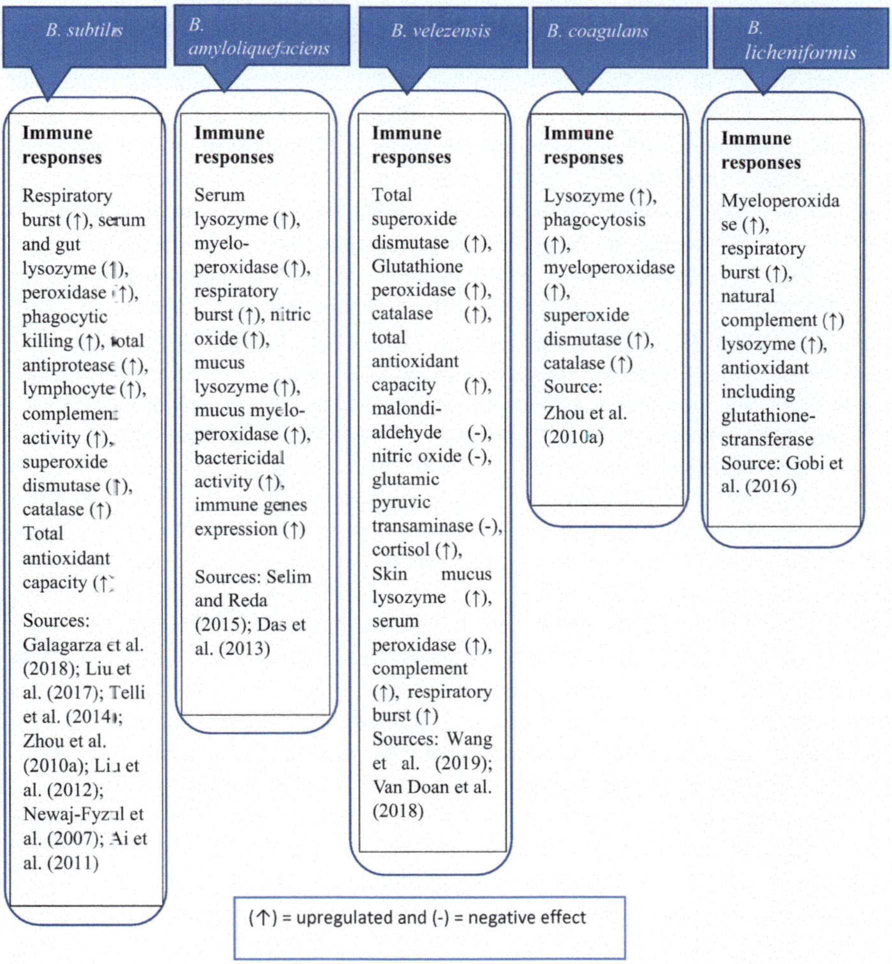

Fig. 20.3 Immune responses of different species of *Bacillus* in fish

hypophthalmus). Similarly, *B. subtilis* E20 and *B. licheniformis* proved phagocytic interaction with different fish species was observed after dietary supplementation (Gao et al 2018). Nandi et al. (2017a) observed enhanced activities of lysozymes in *L. rohita* after supplementation of *Bacillus* probiotics. *Bacillus* is also reported to have a positive effect on pro-inflammatory cytokines (Yan et al. 2016), IgM (Nandi et al. 2017a), and respiratory burst (Thy et al. 2017) of fishes.

20.7.5.2 Immune Gene Expression in Host

The importance of *Bacillus* spp. in improving fish growth and metabolism is well established. Expression of genes linked with antioxidants, digestion processes, growth, metabolism, inflammation, cytoskeleton, and transport proteins is enhanced by the application of *Bacillus* spp. (Esteban et al. 2014; Rajanbabu and Chen 2011; Peng et al. 2012; Avella et al. 2010). Diet supplemented with suitable strains of *Bacillus* spp. augments the expression of genes associated with the immunity of fishes (Fig. 20.4) (Midhun et al. 2019). Avella et al. (2010) demonstrated that the adjustment to the rearing environment of the seabream larvae was improved due to lessening cellular stress by suppressing the functionality of HSP70 gene with a mix of *Bacillus* species in rearing water. Immune-related genes involved in biosynthesis of peroxinectin, prophenoloxidase, lipopolysaccharide, β-1,3-glucanbinding protein, and serine protein are highly expressed in fishes by the application of *B. subtilis* L10 and G1. These gene expressions help white shrimp, *L. vannamei*, to combat pathogen *V. harveyi* (Zokaeifar et al. 2012b). Supplementation of probiotic *Bacillus* species modulates the expression of the mucosal gene, immune-related genes in the head kidney and pro-inflammatory cytokines in head kidney and intestine of gilthead seabream (*S. aurata*) (Esteban et al. 2014), goldfish *(C. auratus)*, orange-spotted grouper (*E. coioides*) (Wang et al. 2018b), and Nile tilapia (*O. niloticus*) (Galagarza et al. 2018), respectively. Yi et al. (2018) reported that *B. velezensis* JW can boost the expression of the interferon- gamma gene (IFN- γ) and tumor necrosis factor-α (TNF-α) plus the gene expression of interleukin-1 (IL-1) and interleukin-4 (IL-4) all of which has significance in the defense mechanism of goldfish (*C. auratus*). Pro-inflammatory genes, interleukin-8 (IL-8), and the expression of caspase-1 (CASP-1) were increased by *B. subtilis* plus microalgae (*Tetraselmis* chuii) fed a diet in gilthead seabream (*S. aurata* L.) (Cerezuela et al. 2013). Similarly, as feed additives in tilapia (*O. niloticus*), it was evident that probiotic *B. licheniformis* (HGA8B) can enhance growth-related genes like growth hormone receptor genes (GHR-1 and GHR-2), insulin-like growth factor genes (IGF-1 and IGF-2) expression in the tissue of liver, and anti-inflammatory cytokine such as IL-10 (Midhun et al. 2019). Significantly higher heat-shock protein (HSP70), interleukin (IL-1 β), tumor necrosis factor (TNF-α), and interferon-gamma (INF-γ) were also obtained in Nile tilapia (*O. niloticus*) fed with *B. subtilis* (BS7 and BS8) (Won et al. 2020).

20.8 Enhancement of Abiotic Stress Tolerance

Stress in commercial aquaculture is unavoidable. It is caused by handling, feeding, vaccination, water exchange, fish transportation, high stocking densities, and ammonia levels in water all of which distress the health and physiology of fishes (Abdollahi-Arpanahi et al. 2018; Fuchs et al. 2017; Hoseinifar et al. 2015; Segner et al. 2012; Thy et al. 2017; Liu et al. 2010). Higher levels of stress reduce the survivability and increase the incidence of diseases in aquaculture (Eissa and Wang

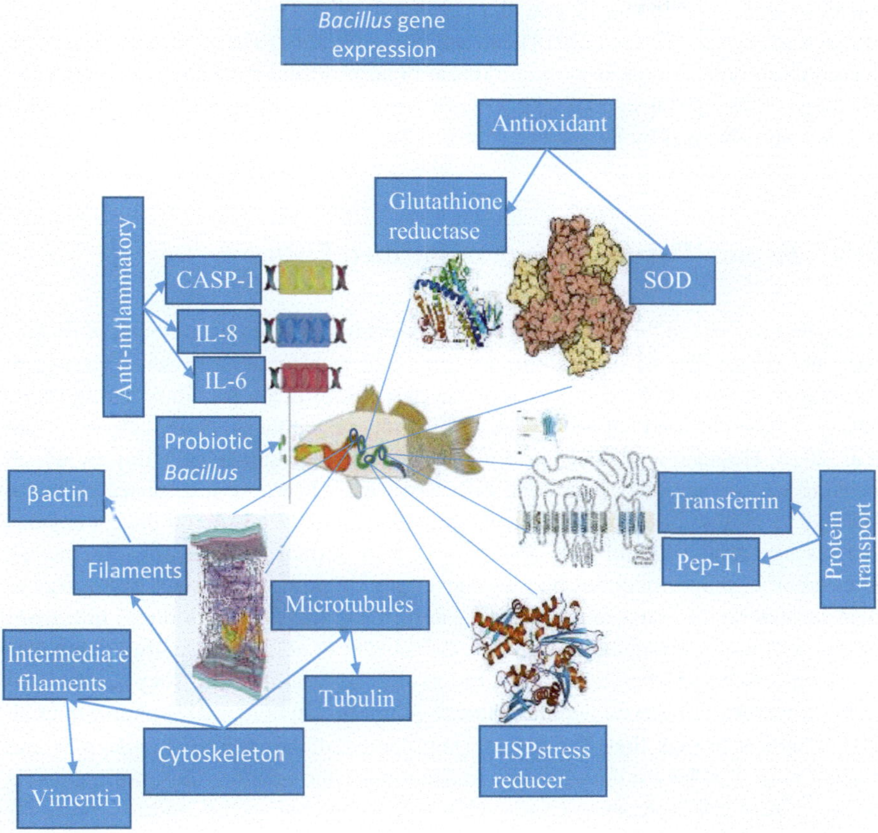

Fig. 20.4 Molecular mode of actions of probiotic Bacillus spp. on aquaculture species
HSP heat-shock proteins, *CASP-1* caspase-1, *Pep-T$_1$* peptide transporter 1, *SOD* superoxide dismutase, *IL-6* interleukin 6, *IL-8* interleukin 8

2016). The global climate change also exerts various stresses such as high temperature to the aquaculture system. In fishes, the retaliation of stress occurs at the cellular level indicator of which includes glucose and cortisol (Abdollahi-Arpanahi et al. 2018), and heat-shock proteins that have a self-protective role in preserving the hemostasis (Eissa and Wang 2016) by reducing the stressors that produce healthy fishes (Yamashita et al. 2010). The probiotic bacteria make the fish resilient to stresses through various mechanisms including the expression of host-related genes in the host fishes (Forsatkar et al. 2017; Shaheen et al. 2014). Stress resistance to asphyxiation and air exposure of yellow perch observed to be improved with a mixture of *B. subtilis*, *B. pumilus*, *B. amyloliquefaciens*, and *B. licheniformis* (Eissa et al. 2018) which is evident to the potentiality of *Bacillus* against different stressors. Similarly, Abdollahi-Arpanahi et al. (2018) reported that whiteleg shrimps, *Litopenaeus vannamei*, treated with *B. subtilis* and *B. licheniformis* were more resistant to environmental and chemicals stressors than the untreated control. Dietary administration of *B. subtilis* in Nile tilapia and *B. amyloliquefaciens* 54A and *B. pumilus* 47B in striped catfish (*P. hypophthalmus*) reduces the stress of high stocking density (Telli et al. 2014) and ammonia toxicity (Thy et al. 2017), respectively. As climate change is posing a threat to aquaculture productivity worldwide, advanced research is needed to explicate the precise mechanisms of stress tolerance in fishes and shrimps by means of probiotic *Bacillus* spp.

20.9 Repair Tissue Damage in Farmed Fishes

Tissue damage of the farmed fishes is often associated with environmental or dietary toxicants which lead to various diseases like gall syndrome and liver enlargement (Kunjiappan et al. 2015). *Bacillus* spp. can play a significant role in reducing tissue damage of the cultured fish as suggested by many researchers (Abdollahi-Arpanahi et al. 2018; Hassaan et al. 2018). To understand the mechanism of action by which *Bacillus* species exert its beneficial effect, we have to know the causes of tissue damage in fishes. Some biochemical reactions of metabolism produce enzymes such as aspartate transaminase and alanine transaminase that in turn synthesize other metabolic intermediates by interconversion of amino acids with an upsurge in their amount leads to tissue damage (in heart, liver, muscle, and kidney) in farmed fishes (Abdollahi-Arpanahi et al. 2018; Babazadeh et al. 2011). As these enzymes are cytoplasmic and liberated into the blood after cellular damage, they can be used as an indicator to diagnose hepatic damage (Kunjiappan et al. 2015). Adorian et al. (2019) demonstrated that the dietary supplementation of *B. licheniformis* and *B. subtilis* in feed lessen the aspartate transaminase and alanine transaminase levels in fishes. Similar findings were reported in Nile tilapia fed with *B. subtilis*, *B. megaterium*, and *B. licheniformis* (Sutthi et al. 2018; Hassaan et al. 2018), in Rohu (*L. rohita*) treated with *B. amyloliquefaciens* CCF7 (Nandi et al. 2017a, b), and in shrimp with *B. subtilis* and *B. licheniformis* (Abdollahi-Arpanahi et al. 2018). These results demonstrate that *Bacillus* species play a role in reducing the toxic effects of enzymes and ensuring the better liver functions of the treated fishes.

20.10 Water Quality Enhancement

Better water quality for rearing fish with optimal physicochemical properties is crucial for successful aquaculture since diseases and infections are largely attributed to poor water quality (Zokaeifar et al. 2014). It is well recognized that probiotic *Bacillus* spp. enhance the water quality by detoxifying the harmful substances and provide fish a better environment to live in (Eissa et al. 2010; Hura et al. 2018; Tuan et al. 2013). Extracellular enzymes with antimicrobial peptides produced by the *Bacillus* species help to develop the culture water quality apart from killing harmful microbes (NavinChandran et al. 2014; Nimrat et al. 2012; Xu et al. 2013). Administration of *Bacillus* spp. directly in shrimp aquaculture can maintain the ammonia level in a

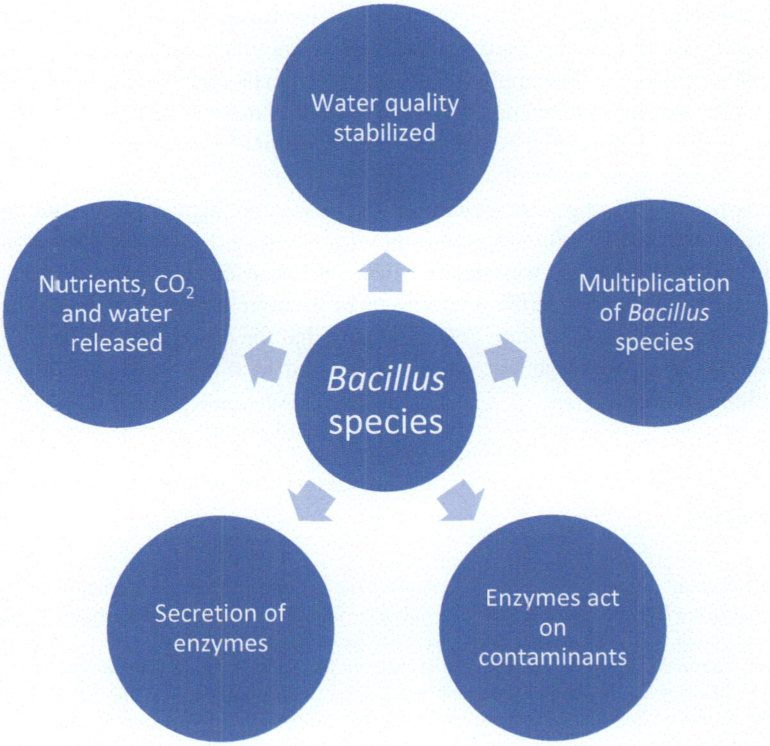

Fig. 20.5 Mode of actions of probiotic *Bacillus* spp. in improving water quality of aquaculture system

suitable range (Zokaeifar et al. 2014; Nimrat et al. 2012) through nitrification and/or denitrification of nitrogenous wastes and thus enhance water quality (Zink et al. 2011; Mujeeb Rahiman et al. 2010; Song et al. 2011; Lakshmanan and Soundarapandian 2008; Nimrat et al. 2011; Wang et al. 2005; Xie et al. 2013). Removing organic matter from rearing water is another important feature of *Bacillus* spp. that also improves water quality (Luis-Villaseñor et al. 2011). Figure 20.5 illustrated the mode of action of *Bacillus* species in improving water quality. The role of *Bacillus* spp. in modulating the microbiome in the aquaculture system and in the fishes is poorly understood. A better comprehension of the interactions of applied probiotics with fishes and environmental organisms would facilitate their effective use in the aquaculture industry.

20.11 Conclusions

Probiotic *Bacillus* spp. exert beneficial effects on treated fishes and aquaculture systems. Molecular systems involved in their beneficial effects are partially being understood. The mode of actions of *Bacillus* spp. to protect treated fishes includes the production of antibiotics and lytic enzymes, the formation of biofilm, induction of host immune system, and battling for scope and nourishment with the fish pathogenic microorganisms. The applied *Bacillus* spp. also promotes the growth of fishes and shrimps through enhancing digestion of feeds, metabolism of nutrients, and expression of growth-related genes. However, additional experiments are needed to reveal the underlying molecular cross-talks between probiotic *Bacillus* spp. with treated fishes and shrimps using genomic and post-genomic analyses. Although the beneficial effects of *Bacillus* spp. on improving the water quality and environment of aquaculture systems are well documented, further research is needed on how the applied probiotics modulate the microbiome of the aquaculture system and the guts of the treated fishes and shrimps. We anticipate that the proper utilization of probiotic *Bacillus* spp. would enhance productivity and facilitates sustainability in the aquaculture industry.

References

Abarike ED, Jian J, Tang J, Cai J, Yu H, Lihua C, Jun L (2018) Influence of traditional Chinese medicine and *Bacillus* species (TCMBS) on growth, immune response and disease resistance in Nile tilapia, *Oreochromis niloticus*. Aquacult Res 49(7):2366–2375

Abdollahi-Arpanahi D, Soltani E, Jafaryan H, Soltani M, Naderi-Samani M, Campa-Córdova AI (2018) Efficacy of two commercial and indigenous probiotics, *Bacillus subtilis* and *Bacillus licheniformis* on growth performance, immuno-physiology and resistance response of juvenile white shrimp (*Litopenaeus vannamei*). Aquaculture 496(1):43–49

Abriouel H, Franz CM, Omar NB, Gálvez A (2011) Diversity and applications of *Bacillus* bacteriocins. FEMS Microbiol Rev 35(1):201–232

Acosta F, Ramos-Vivas JOSE, Lazaro-Diez MARIA, Gonzalez-De Canales ANA, Vega B, Bravo J, Román L (2016) Effect of dietary supplementation with *Bacillus amyloliquefaciens* in the innate immunity in the European sea bass (*Dicentrarchus labrax*). Fish Shellfish Immunol 53:70

Addo S, Carrias AA, Williams MA, Liles MR, Terhune JS, Davis DA (2017) Effects of *Bacillus subtilis* strains and the prebiotic Previda® on growth, immune parameters and susceptibility to *Aeromonas hydrophila* infection in Nile tilapia, *Oreochromis niloticus*. Aquacult Res 48(9):4798–4810

Adorian TJ, Jamali H, Farsani HG, Darvishi P, Hasanpour S, Bagheri T, Roozbehfar R (2019) Effects of probiotic bacteria *Bacillus* on growth performance, digestive enzyme activity, and hematological parameters of Asian sea bass, *Lates calcarifer* (Bloch). Probiotic & Antimicrobial Protein 11(1):248–255

Ai Q, Xu H, Mai K, Xu W, Wang J, Zhang W (2011) Effects of dietary supplementation of *Bacillus subtilis* and fructooligosaccharide on growth performance, survival, non-specific immune response and disease resistance of juvenile large yellow croaker, *Larimichthys crocea*. Aquaculture 317(1-4):155–161

Alexander CP, Kirubakaran CJW, Michael RD (2010) Water soluble fraction of *Tinospora cordifolia* leaves enhanced the non-specific immune mechanisms and disease resistance in *Oreochromis mossambicus*. Fish Shellfish Immunol 29(5):765–772

Al-Harbi AH, Uddin MN (2005) Microbiological quality changes in the intestine of hybrid tilapia (*Oreochromis niloticus*× *Oreochromis aureus*) in fresh and frozen storage condition. Lett Appl Microbiol 40(6):486–490

Al-Hisnawi A, Ringø E, Davies SJ, Waines P, Bradley G, Merrifield DL (2015) First report on the autochthonous gut microbiota of brown trout (*Salmo trutta* Linnaeus). Aquacult Res 46(12):2962–2971

Al-Thubiani AS, Maher YA, Fathi A, Abourehab MA, Alarjah M, Khan MS, Al-Ghamdi SB (2018) Identification and characterization of a novel antimicrobial peptide compound produced by *Bacillus megaterium* strain isolated from oral microflora. Saudi Pharma J 26(8):1089–1097

Aly SM, Ahmed YAG, Ghareeb AAA, Mohamed MF (2008) Studies on *Bacillus subtilis* and *Lactobacillus acidophilus*, as potential probiotics, on the immune response and resistance of Tilapia nilotica (*Oreochromis niloticus*) to challenge infections. Fish Shellfish Immunol 25(1-2):128–136

Amiri O, Miandare HK, Hoseinifar SH, Shabni A, Safari R (2018) Skin mucus protein profile, immune parameters, immune-related gene expression, and growth performance of rainbow trout (*Oncorhynchus mykiss*) fed white button mushroom (*Agaricus bisporus*) powder. Int J Med Mushroom 20(4)

Ammor S, Tauveron G, Dufour E, Chevallier I (2006) Antibacterial activity of lactic acid bacteria against spoilage and pathogenic bacteria isolated from the same meat small-scale facility: 1-Screening and characterization of the antibacterial compounds. Food Control 17(6):454–461

An J, Zhu W, Liu Y, Zhang X, Sun L, Hong P et al (2015) Purification and characterization of a novel bacteriocin CAMT2 produced by *Bacillus amyloliquefaciens* isolated from marine fish Epinephelus areolatus. Food Control 51:278–282

Annamalai N, Rajeswari MV, Elayaraja S, Balasubramanian T (2013) Thermostable, haloalkaline cellulase from *Bacillus halodurans* CAS 1 by conversion of lignocellulosic wastes. Carbohydr Polym 94(1):409–415

Askarian F, Zhou Z, Olsen RE, Sperstad S, Ringø E (2012) Culturable autochthonous gut bacteria in Atlantic salmon (Salmo salar L.) fed diets with or without chitin. Characterization by 16S rRNA gene sequencing, ability to produce enzymes and in vitro growth inhibition of four fish pathogens. Aquaculture 326:1–8

Austin B, Al-Zahrani AMJ (1988) The effect of antimicrobial compounds on the gastrointestinal microflora of rainbow trout, Salmo gairdneri Richardson. J Fish Biol 33(1):1–14

Avella MA, Gioacchini G, Decamp O, Makridis P, Bracciatelli C, Carnevali O (2010) Application of multi-species of *Bacillus* in sea bream larviculture. Aquaculture 305(1-4):12–19

Babazadeh D, Vahdatpour T, Nikpiran H, Jafargholipour MA, Vahdatpour S (2011) Effects of probiotic, prebiotic and synbiotic intake on blood enzymes and performance of Japanese quails (Coturnix japonica). Indian J Anim Sci 81(8):870

Bader J, Albin A, Stahl U (2012) Spore-forming bacteria and their utilization as probiotics. Beneficial Microbes 3(1):67–75

Bairagi A, Ghosh KS, Sen SK, Ray AK (2002) Duckweed (*Lemna polyrhiza*) leaf meal as a source of feedstuff in formulated diets for rohu (*Labeo rohita* Ham.) fingerlings after fermentation with a fish intestinal bacterium. Bioresour Technol 85(1):17–24

Bairagi A, Sarkar Ghosh K, Sen SK, Ray AK (2004) Evaluation of the nutritive value of *Leucaena leucocephala* leaf meal, inoculated with fish intestinal bacteria *Bacillus subtilis* and *Bacillus circulans* in formulated diets for rohu, *Labeo rohita* (Hamilton) fingerlings. Aquacult Res 35(5):436–446

Balcázar JL, De Blas I, Ruiz-Zarzuela I, Cunningham D, Vendrell D, Múzquiz JL (2006) The role of probiotics in aquaculture. Vet Microbiol 114(3-4):173–186

Banerjee G, Ray AK, Askarian F, Ringø E (2013) Characterisation and identification of enzyme-producing autochthonous bacteria from the gastrointestinal tract of two Indian air-breathing fish. Beneficial Microbes 4(3):277–284

Banerjee G, Dan SK, Nandi A, Ghosh P, Ray AK (2015) Autochthonous gut bacteria in two Indian air-breathing fish, climbing perch (*Anabas testudineus*) and walking catfish (*Clarias batrachus*): mode of association, identification, and enzyme producing ability. Pol J Microbiol 64(4):361–368

Banerjee S, Mukherjee A, Dutta D, Ghosh K (2016) Non-starch polysaccharide degrading gut bacteria in Indian major carps and exotic carps. Jordan J Biol Sci 147(3384):1–10
Banerjee G, Nandi A, Ray AK (2017) Assessment of hemolytic activity, enzyme production and bacteriocin characterization of *Bacillus subtilis* LR1 isolated from the gastrointestinal tract of fish. Arch Microbiol 199(1):115–124
Bastos Gomes G, Jerry DR, Miller TL, Hutson KS (2017) Current status of parasitic ciliates *Chilodonella* spp. (Phyllopharyngea: Chilodonellidae) in freshwater fish aquaculture. J Fish Dis 40(5):703–715
Béahdy J (1974) Recent developments of antibiotic research and classification of antibiotics according to chemical structure. In: Advances in applied microbiology, vol 18. Academic Press, pp 309–406
Bermejo-Nogales A, Fernández M, Fernández-Cruz ML, Navas JM (2016) Effects of a silver nanomaterial on cellular organelles and time course of oxidative stress in a fish cell line (PLHC-1). Comp Biochem Physiol Part C: Toxicol & Pharmacol 190:54–65
Bierbaum G, Sahl HG (2009) Lantibiotics: mode of action, biosynthesis and bioengineering. Curr Pharm Biotechnol 10(1):2–18
Bindiya ES, Tina KJ, Raghul SS, Bhat SG (2015) Characterization of Deep Sea Fish Gut Bacteria with Antagonistic Potential, from *Centroscyllium fabricii* (Deep Sea Shark). Probiotic & Antimicrobial Protein 7(2):157–163
Birkbeck TH, Ringø E (2005) Pathogenesis and the gastrointestinal tract of growing fish. In: Biology of growing animals, vol 2. Elsevier, pp 208–234
Bøgwald J, Dalmo RA (2014) Gastrointestinal pathogenesis in aquatic animals. In: Aquaculture nutrition: gut health, probiotics and prebiotics. Wiley-Blackwell, Hoboken, NJ, pp 53–74
Braun V, Killmann H (1999) Bacterial solutions to the iron-supply problem. Trends Biochem Sci 24(3):104–109
Buruiană CT, Profir AG, Vizireanu C (2014) Effects of probiotic *Bacillus* species in aquaculture–an overview. Annal Univ Dunarea de Jos of Galati Fascicle VI-Food Technol 38(2):9–17
Cahill MM (1990) Bacterial flora of fishes: a review. Microb Ecol 19(1):21–41
Camargo JA, Alonso Á (2006) Ecological and toxicological effects of inorganic nitrogen pollution in aquatic ecosystems: a global assessment. Environ Int 32(6):831–849
Cao Y, He S, Zhou Z, Zhang M, Mao W, Zhang H, Yao B (2012) Orally administered thermostable N-acyl homoserine lactonase from *Bacillus* sp. strain AI96 attenuates *Aeromonas hydrophila* infection in zebrafish. Appl Environ Microbiol 78(6):1899–1908
Cao Y, Liu Y, Mao W, Chen R, He S, Gao X et al (2014) Effect of dietary N-acyl homoserin lactonase on the immune response and the gut microbiota of zebrafish, Danio rerio, infected with Aeromonas hydrophila. J World Aquac Soc 45(2):149–162
Carbone D, Faggio C (2016) Importance of prebiotics in aquaculture as immunostimulants. Effects on immune system of *Sparus aurata* and *Dicentrarchus labrax*. Fish Shellfish Immunol 54:172–178
Castex M, Lemaire P, Wabete N, Chim L (2010) Effect of probiotic *Pediococcus acidilactici* on antioxidant defences and oxidative stress of *Litopenaeus stylirostris* under *Vibrio nigripulchritudo* challenge. Fish Shellfish Immunol 28(4):622–631
Cerezuela R, Meseguer J, Esteban MÁ (2013) Effects of dietary inulin, *Bacillus subtilis* and microalgae on intestinal gene expression in gilthead seabream (*Sparus aurata* L.). Fish Shellfish Immunol 34(3):843–848
Chai PC, Song XL, Chen GF, Xu H, Huang J (2016) Dietary supplementation of probiotic Bacillus PC465 isolated from the gut of *Fenneropenaeus chinensis* improves the health status and resistance of *Litopenaeus vannamei* against white spot syndrome virus. Fish Shellfish Immunol 54:602–611
Chen R, Zhou Z, Cao Y, Bai Y, Yao B (2010) High yield expression of an AHL-lactonase from *Bacillus* sp. B546 in *Pichia pastoris* and its application to reduce *Aeromonas hydrophila* mortality in aquaculture. Microb Cell Fact 9(1):39

Chen Y, Li J, Xiao P, Li GY, Yue S, Huang J et al (2016a) Isolation and characterization of *Bacillus* spp. M 001 for potential application in turbot (*Scophthalmus maximus* L.) against *Vibrio anguillarum*. Aquacult Nutr 22(2):374–381
Chen Y, Li J, Xiao P, Zhu W, Mo Z (2016b) The ability of marine *Bacillus* spp. isolated from fish gastrointestinal tract and culture pond sediment to inhibit growth of aquatic pathogenic bacteria. Iranian J Fish Sci 15(2):701–714
Cheng AC, Lin HL, Shiu YL, Tyan YC, Liu CH (2017) Isolation and characterization of antimicrobial peptides derived from *Bacillus subtilis* E20-fermented soybean meal and its use for preventing *Vibrio* infection in shrimp aquaculture. Fish Shellfish Immunol 67:270–279
Chopra L, Singh G, Choudhary V, Sahoo DK (2014) Sonorensin: an antimicrobial peptide, belonging to the heterocycloanthracin subfamily of bacteriocins, from a new marine isolate, *Bacillus sonorensis* MT93. Appl Environ Microbiol 80(10):2981–2990
Cladera-Olivera F, Caron GR, Brandelli A (2004) Bacteriocin-like substance production by *Bacillus licheniformis* strain P40. Lett Appl Microbiol 38(4):251–256
Clemente JC, Ursell LK, Parfrey LW, Knight R (2012) The impact of the gut microbiota on human health: an integrative view. Cell 148(6):1258–1270
Clements KD (1997) Fermentation and gastrointestinal microorganisms in fishes. In: Gastrointestinal microbiology. Springer, Boston, MA, pp 156–198
Das P, Ghosh K (2013) Evaluation of phytase-producing ability by a fish gut bacterium, *Bacillus subtilis* subsp. *subtilis*. J Biol Sci 13(8):691
Das P, Ghosh K (2015) Improvement of nutritive value of sesame oil cake in formulated diets for rohu, *Labeo rohita* (Hamilton) after bio-processing through solid state fermentation by a phytase-producing fish gut bacterium. Int J Aquatic Biol 3(2):89–101
Das A, Nakhro K, Chowdhury S, Kamilya D (2013) Effects of potential probiotic Bacillus amyloliquefaciens FPTB16 on systemic and cutaneous mucosal immune responses and disease resistance of catla (Catla catla). Fish Shellfish Immunol 35(5):1547–1553
Das P, Mandal S, Khan A, Manna SK, Ghosh K (2014) Distribution of extracellular enzyme-producing bacteria in the digestive tracts of 4 brackish water fish species. Turkish J Zool 38(1):79–88
Dawood MA, Koshio S (2016) Recent advances in the role of probiotics and prebiotics in carp aquaculture: a review. Aquaculture 454:243–251
Dawood MA, Koshio S, Esteban MÁ (2018) Beneficial roles of feed additives as immunostimulants in aquaculture: a review. Rev Aquacult 10(4):950–974
Defoirdt T, Boon N, Bossier P, Verstraete W (2004) Disruption of bacterial quorum sensing: an unexplored strategy to fight infections in aquaculture. Aquaculture 240(1-4):69–88
Deschamps AM, Mahoudeau G, Lebeault JM (1980) Fast degradation of kraft lignin by bacteria. Europ J Appl Microbiol & Biotechnol 9(1):45–51
Dey A, Ghosh K, Hazra N (2016) Evaluation of extracellular enzyme-producing autochthonous gut bacteria in walking catfish, *Clarias batrachus* (L.). J Fish 4(1):345–352
Di Giulio RT, Habig C, Gallagher EP (1993) Effects of Black Rock Harbor sediments on indices of biotransformation, oxidative stress, and DNA integrity in channel catfish. Aquat Toxicol 26(1-2):1–22
Dubois JYF, Kouwen TR, Schurich AK, Reis CR, Ensing HT, Trip EN et al (2009) Immunity to the bacteriocin sublancin 168 is determined by the SunI (YolF) protein of *Bacillus subtilis*. Antimicrob Agents Chemother 53(2):651–661
Dutta D, Ghosh K (2015) Screening of extracellular enzyme-producing and pathogen inhibitory gut bacteria as putative probiotics in mrigal, *Cirrhinus mrigala* (Hamilton, 1822). Int J Fish & Aquatic Stud 2(4):310–318
Dutta D, Banerjee S, Mukherjee A, Ghosh K (2015) Selection and probiotic characterization of exoenzyme-producing bacteria isolated from the gut of Catla catla (Actinopterygii: Cypriniformes: Cyprinidae). Acta Ichthyologica et Piscatoria 45(4)

Dutta S, Surovy MZ, Gupta DR, Mahmud NU, Chanclud E, Win J et al (2018) Genomic analyses reveal that biocontrol of wheat blast by *Bacillus* spp. may be linked with production of antimicrobial compounds and induced systemic resistance in host plants. Figshare 17:48

Eissa N, Wang HP (2016) Transcriptional stress responses to environmental and husbandry stressors in aquaculture species. Rev Aquacult 8(1):61–88

Eissa AE, Zaki MM, Aziz AA (2010) Flavobacterium columnare/Myxobolus tilapiae concurrent infection in the earthen pond reared Nile tilapia (*Oreochromis niloticus*) during the early summer. Interdiscip Bio Cent 2(2):1–10

Eissa N, Wang HP, Yao H, Abou-ElGheit E (2018) Mixed *Bacillus* species enhance the innate immune response and stress tolerance in yellow perch subjected to hypoxia and air-exposure stress. Sci Rep 8(1):1–10

Esakkiraj P, Immanuel G, Sowmya SM, Iyapparaj P, Palavesam A (2009) Evaluation of protease-producing ability of fish gut isolate *Bacillus cereus* for aqua feed. Food Bioproc Tech 2(4):383

Esteban MA, Cordero H, Martínez-Tomé M, Jiménez-Monreal AM, Bakhrouf A, Mahdhi A (2014) Effect of dietary supplementation of probiotics and palm fruits extracts on the antioxidant enzyme gene expression in the mucosae of gilthead seabream (*Sparus aurata* L.). Fish Shellfish Immunol 39(2):532–540

Etyemez M, Balcazar JL (2016) Isolation and characterization of bacteria with antibacterial properties from Nile tilapia (*Oreochromis niloticus*). Res Vet Sci 105:62–64

Fečkaninová A, Koščová J, Mudroňová D, Popelka P, Toropilova J (2017) The use of probiotic bacteria against Aeromonas infections in salmonid aquaculture. Aquaculture 469:1–8

Forsatkar MN, Nematollahi MA, Rafiee G, Farahmand H, Lawrence C (2017) Effects of the prebiotic mannan-oligosaccharide on the stress response of feed deprived zebrafish (*Danio rerio*). Physiol Behav 180:70–77

Fu L, Wang C, Ruan X, Li G, Zhao Y, Wang Y (2018) Preservation of large yellow croaker (*Pseudosciaena crocea*) by Coagulin L1208, a novel bacteriocin produced by *Bacillus coagulans* L1208. Int J Food Microbiol 266:60–68

Fuchs VI, Schmidt J, Slater MJ, Buck BH, Steinhagen D (2017) Influence of immunostimulant polysaccharides, nucleic acids, and *Bacillus* strains on the innate immune and acute stress response in turbots (*Scophthalmus maximus*) fed soy bean-and wheat-based diets. Fish Physiol Biochem 43(6):1501–1515

Galagarza OA, Smith SA, Drahos DJ, Eifert JD, Williams RC, Kuhn DD (2018) Modulation of innate immunity in Nile tilapia (*Oreochromis niloticus*) by dietary supplementation of *Bacillus subtilis* endospores. Fish Shellfish Immunol 83:171–179

Gálvez A, Abriouel H, López RL, Omar NB (2007) Bacteriocin-based strategies for food biopreservation. Int J Food Microbiol 120(1-2):51–70

Gao X, Zhang M, Li X, Han Y, Wu F, Liu Y (2018) Effects of a probiotic (*Bacillus licheniformis*) on the growth, immunity, and disease resistance of *Haliotis discus* hannai Ino. Fish Shellfish Immunol 76:143–152

Gao S, Pan L, Huang F, Song M, Tian C, Zhang M (2019) Metagenomic insights into the structure and function of intestinal microbiota of the farmed Pacific white shrimp (*Litopenaeus vannamei*). Aquaculture 499:109–118

Gatlin DM III, Peredo AM (2012) Prebiotics and probiotics: definitions and applications. Prebiotics and Probiotics: Definitions and Applications 4711

Gatlin DM III, Li P, Wang X, Burr GS, Castille F, Lawrence AL (2006) Potential application of prebiotics in aquaculture. Avances en Nutrición Acuicola

Ghosh K, Sen SK, Ray AK (2002) Characterization of Bacilli isolated from the gut of rohu, *Labeo rohita*, fingerlings and its significance in digestion. J Appl Aquacul 12(3):33–42

Ghosh K, Roy M, Kar N, Ringø E (2010) Gastrointestinal bacteria in Rohu, *Labeo rohita* (Actinopterygii: Cypriniformes: Cyprinidae): scanning electron microscopy and bacteriological study. Acta Ichthyologica et Piscatoria 40(2)

Ghosh K, Banerjee S, Moon UM, Khan HA, Dutta D (2017) Evaluation of gut associated extracellular enzyme-producing and pathogen inhibitory microbial community as potential probiotics in Nile tilapia, *Oreochromis niloticus*. Int J Aquacul 7(23):143–158

Ghosh K, Ray AK, Ringø E (2019) Applications of plant ingredients for tropical and subtropical freshwater finfish: possibilities and challenges. Rev Aquacult 11(3):793–815

Giri SS, Sukumaran V, Sen SS, Vinumonia J, Banu BN, Jena PK (2011) Antagonistic activity of cellular components of potential probiotic bacteria, isolated from the gut of *Labeo rohita*, against *Aeromonas hydrophila*. Probiotic & Antimicrobial Protein 3(3-4):214–222

Gobi N, Malaikozhundan B, Sekar V, Shanthi S, Vaseeharan B, Jayakumar R, Nazar AK (2016) GFP tagged *Vibrio parahaemolyticus* Dahv2 infection and the protective effects of the probiotic *Bacillus licheniformis* Dahb1 on the growth, immune and antioxidant responses in *Pangasius hypophthalmus*. Fish Shellfish Immunol 52:230–238

Gobi N, Vaseeharan B, Chen JC, Rekha R, Vijayakumar S, Anjugam M, Iswarya A (2018) Dietary supplementation of probiotic *Bacillus licheniformis* Dahb1 improves growth performance, mucus and serum immune parameters, antioxidant enzyme activity as well as resistance against *Aeromonas hydrophila* in tilapia *Oreochromis mossambicus*. Fish Shellfish Immunol 74:501–508

Goda AM, Omar EA, Srour TM, Kotiet AM, El-Haroun E, Davies SJ (2018) Effect of diets supplemented with feed additives on growth, feed utilization, survival, body composition and intestinal bacterial load of early weaning European seabass, *Dicentrarchus labrax* post-larvae. Aquac Int 26(1):169–183

Goel G, Puniya AK, Aguilar CN, Singh K (2005) Interaction of gut microflora with tannins in feeds. Naturwissenschaften 92(11):497–503

Gómez GD, Balcázar JL (2008) A review on the interactions between gut microbiota and innate immunity of fish. FEMS Immunol & Med Microbiol 52(2):145–154

González L, Sandoval H, Sacristán N, Castro JM, Fresno JM, Tornadijo ME (2007) Identification of lactic acid bacteria isolated from Genestoso cheese throughout ripening and study of their antimicrobial activity. Food Control 18(6):716–722

González-Félix ML, Gatlin DM III, Urquidez-Bejarano P, de la Reé-Rodríguez C, Duarte-Rodríguez L, Sánchez F et al (2018) Effects of commercial dietary prebiotic and probiotic supplements on growth, innate immune responses, and intestinal microbiota and histology of *Totoaba macdonaldi*. Aquaculture 491:239–251

Gray EJ, Di Falco M, Souleimanov A, Smith DL (2006) Proteomic analysis of the bacteriocin thuricin 17 produced by *Bacillus thuringiensis* NEB17. FEMS Microbiol Lett 255(1):27–32

Green TJ, Smullen R, Barnes AC (2013) Dietary soybean protein concentrate-induced intestinal disorder in marine farmed Atlantic salmon, Salmo salar is associated with alterations in gut microbiota. Vet Microbiol 166(1-2):286–292

Groff JM, Lapatra SE (2000) Infectious diseases impacting the commercial culture of salmonids. J Appl Aquacul 10(4):17–90

Grubbs KJ, Bleich RM, Santa Maria KC, Allen SE, Farag S, Team A et al (2017) Large-scale bioinformatics analysis of *Bacillus* genomes uncovers conserved roles of natural products in bacterial physiology. MSystems 2(6):e00040–e00017

Guo X, Chen DD, Peng KS, Cui ZW, Zhang XJ, Li S, Zhang YA (2016) Identification and characterization of *Bacillus subtilis* from grass carp (*Ctenopharyngodon idellus*) for use as probiotic additives in aquatic feed. Fish Shellfish Immunol 52:74–84

Harikrishnan R, Kim MC, Kim JS, Balasundaram C, Heo MS (2011) Probiotics and herbal mixtures enhance the growth, blood constituents, and nonspecific immune response in *Paralichthys olivaceus* against *Streptococcus parauberis*. Fish Shellfish Immunol 31(2):310–317

Hassaan MS, Soltan MA, Jarmołowicz S, Abdo HS (2018) Combined effects of dietary malic acid and *Bacillus subtilis* on growth, gut microbiota and blood parameters of Nile tilapia (*Oreochromis niloticus*). Aquacult Nutr 24(1):83–93

Hassanein SM, Soliman NK (2010) Effect of probiotic (*Saccharomyces cerevisiae*) adding to diets on intestinal microflora and performance of Hy-Line layers hens. J Am Sci 6(11):159–169

He S, Wu Z, Liu Y, Wu N, Tao Y, Xu L et al (2013) Effects of dietary 60 gkg^{-1} dried distiller's grains in least-cost practical diets on production and gut allochthonous bacterial composition of cage-cultured fish: comparison among fish species with different natural food habits. Aquacult Nutr 19(5):765–772

Hindu SV, Chandrasekaran N, Mukherjee A, Thomas J (2018a) Effect of dietary supplementation of novel probiotic bacteria *Bacillus vireti* 01 on antioxidant defence system of freshwater prawn challenged with *Pseudomonas aeruginosa*. Probiotic & Antimicrobial Protein 10(2):356–366

Hindu SV, Thanigaivel S, Vijayakumar S, Chandrasekaran N, Mukherjee A, Thomas J (2018b) Effect of microencapsulated probiotic *Bacillus vireti* 01-polysaccharide extract of *Gracilaria folifera* with alginate-chitosan on immunity, antioxidant activity and disease resistance of *Macrobrachium rosenbergii* against *Aeromonas hydrophila* infection. Fish Shellfish Immunol 73:112–120

Hong HA, Duc LH, Cutting SM (2005) The use of bacterial spore formers as probiotics. FEMS Microbiol Rev 29(4):813–835

Hoseinifar SH, Roosta Z, Hajimoradloo A, Vakili F (2015) The effects of *Lactobacillus acidophilus* as feed supplement on skin mucosal immune parameters, intestinal microbiota, stress resistance and growth performance of black swordtail (*Xiphophorus helleri*). Fish Shellfish Immunol 42(2):533–538

Hoseinifar SH, Ringø E, Shenavar Masouleh A, Esteban MÁ (2016) Probiotic, prebiotic and synbiotic supplements in sturgeon aquaculture: a review. Rev Aquacult 8(1):89–102

Hoseinifar SH, Dadar M, Ringø E (2017) Modulation of nutrient digestibility and digestive enzyme activities in aquatic animals: the functional feed additives scenario. Aquacult Res 48(8):3987–4000

Hoseinifar SH, Sun YZ, Wang A, Zhou Z (2018) Probiotics as means of diseases control in aquaculture, a review of current knowledge and future perspectives. Front Microbiol 9:2429

Hovda MB, Lunestad BT, Fontanillas R, Rosnes JT (2007) Molecular characterisation of the intestinal microbiota of farmed Atlantic salmon (*Salmo salar* L.). Aquaculture 272(1-4):581–588

Hovda MB, Fontanillas R, McGurk C, Obach A, Rosnes JT (2012) Seasonal variations in the intestinal microbiota of farmed Atlantic salmon (*Salmo salar* L.). Aquacult Res 43(1):154–159

Hura MUD, Zafar T, Borana K, Prasad JR, Iqbal J (2018) Effect of commercial probiotic *Bacillus megaterium* on water quality in composite culture of major carps. Int J Curr Agricult Sci 8(1):268–273

Ichimatsu T, Mizuki E, Nishimura K, Akao T, Saitoh H, Higuchi K, Ohba M (2000) Occurrence of *Bacillus thuringiensis* in fresh waters of Japan. Curr Microbiol 40(4):217–220

Jadhav HP, Shaikh SS, Sayyed RZ (2017) Role of hydrolytic enzymes of rhizoflora in biocontrol of fungal phytopathogens: an overview. In: Rhizotrophs: Plant growth promotion to bioremediation. Springer, Singapore, pp 183–203

Jasmin MY, Wagaman H, Yin TA, Ina-Salwany MY, Daud HM, Karim M (2016) Screening and evaluation of local bacteria isolated from shellfish as potential probiotics against pathogenic Vibrios. J Environ Biol 37(4 Spec):801–809

Joint FAO/WHO Working Group (2002) Guidelines for the evaluation of probiotics in food: report of a joint FAO/WHO working group on drafting guidelines for the evaluation of probiotics in food. World Health Organization and Food and Agriculture organization of the United Nations, Rome, Italy.

Jung TS, del Castillo CS, Javaregowda PK, Dalvi RS, Nho SW, Park SB et al (2012) Seasonal variation and comparative analysis of non-specific humoral immune substances in the skin mucus of olive flounder (*Paralichthys olivaceus*). Dev Comp Immunol 38(2):295–301

Kavitha M, Raja M, Perumal P (2018) Evaluation of probiotic potential of *Bacillus* spp. isolated from the digestive tract of freshwater fish *Labeo calbasu* (Hamilton, 1822). Aquac Report 11:59–69

Kesarcodi-Watson A, Kaspar H, Lategan MJ, Gibson L (2008) Probiotics in aquaculture: the need, principles and mechanisms of action and screening processes. Aquaculture 274(1):1–14

Keysami MA, Mohammadpour M (2013) Effect of *Bacillus subtilis* on *Aeromonas hydrophila* infection resistance in juvenile freshwater prawn, *Macrobrachium rosenbergii* (de Man). Aquac Int 21(3):553–562
Khan A, Ghosh K (2012) Characterization and identification of gut-associated phytase-producing bacteria in some freshwater fish cultured in ponds. Acta Ichthyologica et Piscatoria 42(1)
Khan A, Ghosh K (2013) Evaluation of phytase production by fish gut bacterium, *Bacillus subtilis*, for processing of *Ipomea aquatica* leaves as probable aquafeed ingredient. J Aquatic Food Product Technol 22(5):508–519
Kim SY, OHK SH, BAI DH, YU, J. H. (1999) Purification and properties of bacteriolytic enzymes from *Bacillus licheniformis* YS-1005 against *Streptococcus mutans*. Biosci Biotechnol Biochem 63(1):73–77
Kim DH, Brunt J, Austin B (2007) Microbial diversity of intestinal contents and mucus in rainbow trout (*Oncorhynchus mykiss*). J Appl Microbiol 102(6):1654–1664
Kim YR, Kim EY, Lee JM, Kim JK, Kong IS (2013) Characterisation of a novel *Bacillus* sp. SJ-10 β-1, 3–1, 4-glucanase isolated from jeotgal, a traditional Korean fermented fish. Bioprocess Biosyst Eng 36(6):721–727
Kiss A, Balikó G, Csorba A, Chuluunbaatar T, Medzihradszky KF, Alföldi L (2008) Cloning and characterization of the DNA region responsible for Megacin A-216 production in *Bacillus megaterium* 216. J Bacteriol 190(19):6448–6457
Kuebutornye FK, Abarike ED, Lu Y, Hlordzi V, Sakyi ME, Afriyie G et al (2020) Mechanisms and the role of probiotic *Bacillus* in mitigating fish pathogens in aquaculture. Fish Physiol Biochem:1–23
Kumar P, Jain KK, Sardar P (2018) Effects of dietary synbiotic on innate immunity, antioxidant activity and disease resistance of *Cirrhinus mrigala* juveniles. Fish Shellfish Immunol 80:124–132
Kunjiappan S, Bhattacharjee C, Chowdhury R (2015) In vitro antioxidant and hepatoprotective potential of *Azolla microphylla* phytochemically synthesized gold nanoparticles on acetaminophen–induced hepatocyte damage in *Cyprinus carpio* L. In Vitro Cell & Develop Biol-Anim 51(6):630–643
Kuz'Mina VV (1996) Influence of age on digestive enzyme activity in some freshwater teleosts. Aquaculture 148(1):25–37
Lakshmanan R, Soundarapandian P (2008) Effect of commercial probiotics on large scale culture of black tiger shrimp *Penaeus monodon* (Fabricius). Res J Microbiol 3(3):198–203
Lalloo R, Moonsamy G, Ramchuran S, Görgens J, Gardiner N (2010) Competitive exclusion as a mode of action of a novel *Bacillus cereus* aquaculture biological agent. Lett Appl Microbiol 50(6):563–570
Laranja JLQ, Amar EC, Ludevese-Pascual GL, Niu Y, Geaga MJ, De Schryver P, Bossier P (2017) A probiotic *Bacillus* strain containing amorphous poly-beta-hydroxybutyrate (PHB) stimulates the innate immune response of *Penaeus monodon* postlarvae. Fish Shellfish Immunol 68:202–210
Le Marrec C, Hyronimus B, Bressollier P, Verneuil B, Urdaci MC (2000) Biochemical and genetic characterization of coagulin, a new antilisterial bacteriocin in the pediocin family of bacteriocins, produced by *Bacillus coagulans* I4. Appl Environ Microbiol 66(12):5213–5220
Lee S, Lee J, Jin Y, Jeong J, Chang YH, Lee Y et al (2017) Probiotic characteristics of *Bacillus* strains isolated from Korean traditional soy sauce. LWT- Food Sci Technol 79:518–524
Lefevre M, Racedo SM, Ripert G, Housez B, Cazaubiel M, Maudet C et al (2015) Probiotic strain *Bacillus subtilis* CU1 stimulates immune system of elderly during common infectious disease period: a randomized, double-blind placebo-controlled study. Immun & Age 12(1):1–11
Li P, Gatlin DM III (2004) Dietary brewers yeast and the prebiotic Grobiotic™ AE influence growth performance, immune responses and resistance of hybrid striped bass (*Morone chrysops*× *M saxatilis*) to *Streptococcus iniae* infection. Aquaculture 231(1-4):445–456

Li WF, Deng B, Cui ZW, Fu LQ, Chen NN, Zhou XX et al (2012) Several indicators of immunity and antioxidant activities improved in grass carp given a diet containing *Bacillus* additive. J Anim Vet Adv 11(14):2392–2397

Li J, Xu Y, Jin L, Li X (2015a) Effects of a probiotic mixture (*Bacillus subtilis* YB-1 and *Bacillus cereus* YB-2) on disease resistance and non-specific immunity of sea cucumber, *Apostichopus japonicus* (Selenka). Aquacult Res 46(12):3008–3019

Li T, Long M, Gatesoupe FJ, Zhang Q, Li A, Gong X (2015b) Comparative analysis of the intestinal bacterial communities in different species of carp by pyrosequencing. Microb Ecol 69(1):25–36

Li X, Ringø E, Hoseinifar SH, Lauzon HL, Birkbeck H, Yang D (2019) The adherence and colonization of microorganisms in fish gastrointestinal tract. Rev Aquacult 11(3):603–618

Lin YS, Saputra F, Chen YC, Hu SY (2019) Dietary administration of *Bacillus amyloliquefaciens* R8 reduces hepatic oxidative stress and enhances nutrient metabolism and immunity against *Aeromonas hydrophila* and *Streptococcus agalactiae* in zebrafish (*Danio rerio*). Fish Shellfish Immunol 86:410–419

Lindgren SE, Dobrogosz WJ (1990) Antagonistic activities of lactic acid bacteria in food and feed fermentations. FEMS Microbiol Rev 7(1-2):149–163

Liu KF, Chiu CH, Shiu YL, Cheng W, Liu CH (2010) Effects of the probiotic, *Bacillus subtilis* E20, on the survival, development, stress tolerance, and immune status of white shrimp, *Litopenaeus vannamei* larvae. Fish Shellfish Immunol 28(5-6):837–844

Liu CH, Chiu CH, Wang SW, Cheng W (2012) Dietary administration of the probiotic, *Bacillus subtilis* E20, enhances the growth, innate immune responses, and disease resistance of the grouper, *Epinephelus coioides*. Fish Shellfish Immunol 33(4):699–706

Liu H, Wang S, Cai Y, Guo X, Cao Z, Zhang Y et al (2017) Dietary administration of *Bacillus subtilis* HAINUP40 enhances growth, digestive enzyme activities, innate immune responses and disease resistance of tilapia, *Oreochromis niloticus*. Fish Shellfish Immunol 60:326–333

Luis-Villaseñor IE, Macías-Rodríguez ME, Gómez-Gil B, Ascencio-Valle F, Campa-Córdova ÁI (2011) Beneficial effects of four Bacillus strains on the larval cultivation of *Litopenaeus vannamei*. Aquaculture 321(1-2):136–144

Ma R, Yang H, Sun Y, Ye J, Zhang C, Lin W (2010) Biological characteristics of two *Bacillus* strains isolated from gut of grouper, *Epinephelus coioides*. Fish Sci (Dalian) 29(9):505–509

Maeda M, Shibata A, Biswas G, Korenaga H, Kono T, Itami T, Sakai M (2014) Isolation of lactic acid bacteria from kuruma shrimp (*Marsupenaeus japonicus*) intestine and assessment of immunomodulatory role of a selected strain as probiotic. Marine Biotechnol 16(2):181–192

McNeilly TN, Naylor SW, Mahajan A, Mitchell MC, McAteer S, Deane D et al (2008) *Escherichia coli* O157: H7 colonization in cattle following systemic and mucosal immunization with purified H7 flagellin. Infect Immun 76(6):2594–2602

Meidong R, Doolgindachbaporn S, Jamjan W, Sakai K, Tashiro Y, Okugawa Y, Tongpim S (2017) A novel probiotic *Bacillus siamensis* B44v isolated from Thai pickled vegetables (Phak-dong) for potential use as a feed supplement in aquaculture. J Gen Appl Microbiol 63(4):246–253

Mellbye B, Schuster M (2011) The sociomicrobiology of antivirulence drug resistance: a proof of concept. MBio 2(5):e00131-11

Menit J, Qi F, Goodman SD (2003) Mutation of luxS affects biofilm formation in *Streptococcus mutan*. Infect Immun 71(4):1972–1979

Mente E, Nikouli E, Antonopoulou E, Martin SA, Kormas KA (2018) Core versus diet-associated and postprandial bacterial communities of the rainbow trout (*Oncorhynchus mykiss*) midgut and faeces. Biology Bpen 7(6):bio034397

Messaoudi I, Barhoumi S, Saïd K, Kerken A (2009) Study on the sensitivity to cadmium of marine fish *Salaria basilisca* (Pisces: Blennidae). J Environ Sci 21(11):1620–1624

Miao S, Zhao C, Zhu J, Hu J, Dong X, Sun L (2018) Dietary soybean meal affects intestinal homoeostasis by altering the microbiota, morphology and inflammatory cytokine gene expression in northern snakehead. Sci Rep 8(1):1–10

Midhun SJ, Neethu S, Arun D, Vysakh A, Divya L, Radhakrishnan EK, Jyothis M (2019) Dietary supplementation of *Bacillus licheniformis* HGA8B improves growth parameters, enzymatic profile and gene expression of *Oreochromis niloticus*. Aquaculture 505:289–296

Miller MB, Bassler BL (2001) Quorum sensing in bacteria. Annu Rev Microbiol 55(1):165–199

Mohamed MH, Refat NAA (2011) Pathological evaluation of probiotic, *Bacillus subtilis*, against *Flavobacterium columnare* in Tilapia nilotica (*Oreochromis niloticus*) fish in Sharkia Governorate, Egypt. J Am Sci 7(2):244–256

Mohapatra S, Chakraborty T, Kumar V, DeBoeck G, Mohanta KN (2013) Aquaculture and stress management: a review of probiotic intervention. J Anim Physiol Anim Nutr 97(3):405–430

Mondal S, Roy T, Ray AK (2010) Characterization and identification of enzyme-producing bacteria isolated from the digestive tract of bata, *Labeo bata*. J World Aquac Soc 41(3):369–377

Mondol MAM, Shin HJ, Islam MT (2013) Diversity of secondary metabolites from marine *Bacillus* species: chemistry and biological activity. Mar Drugs 11(8):2846–2872

Moriarty DJW (1998) Control of luminous Vibrio species in penaeid aquaculture ponds. Aquaculture 164(1–4):351–358

Mouthuy PA, Snelling SJB, Dakin SG, Milković L, Gašparović AČ, Carr AJ, Žarković N (2016) Biocompatibility of implantable materials: an oxidative stress viewpoint. Biomaterials 109:55–68

Mujeeb Rahiman KM, Jesmi Y, Thomas AP, Mohamed Hatha AA (2010) Probiotic effect of *Bacillus* NL110 and Vibrio NE17 on the survival, growth performance and immune response of *Macrobrachium rosenbergii* (de Man). Aquacult Res 41(9):e120–e134

Mukherjee A, Ghosh K (2016) Antagonism against fish pathogens by cellular components and verification of probiotic properties in autochthonous bacteria isolated from the gut of an Indian major carp, *Catla catla* (Hamilton). Aquacult Res 47(7):2243–2255

Mukherjee A, Dutta D, Banerjee S, Ringø E, Breines EM, Hareide E et al (2016) Potential probiotics from Indian major carp, *Cirrhinus mrigala*. Characterization, pathogen inhibitory activity, partial characterization of bacteriocin and production of exoenzymes. Res Vet Sci 108:76–84

Mukherjee A, Dutta D, Banerjee S, Ringø E, Breines EM, Hareide E et al (2017) Culturable autochthonous gut bacteria in rohu, *Labeo rohita*. In vitro growth inhibition against pathogenic *Aeromonas* spp., stability in gut, bio-safety and identification by 16S rRNA gene sequencing. Symbiosis 73(3):165–177

Munir MB, Hashim R, Chai YH, Marsh TL, Nor SAM (2016) Dietary prebiotics and probiotics influence growth performance, nutrient digestibility and the expression of immune regulatory genes in snakehead (*Channa striata*) fingerlings. Aquaculture 460:59–68

Musikasang H, Tani A, H-kittikun A, Maneerat S (2009) Probiotic potential of lactic acid bacteria isolated from chicken gastrointestinal digestive tract. World J Microbiol Biotechnol 25(8):1337–1345

Musthafa KS, Saroja V, Pandian SK, Ravi AV (2011) Antipathogenic potential of marine *Bacillus* sp. SS4 on N-acyl-homoserine-lactone-mediated virulence factors production in *Pseudomonas aeruginosa* (PAO1). J Biosci 36(1):55–67

Nandi A, Banerjee G, Dan SK, Ghosh P, Ghosh K, Ray AK (2017a) Screening of autochthonous intestinal microbiota as candidate probiotics isolated from four freshwater teleosts. Curr Sci 113(4):767

Nandi A, Dan SK, Banerjee G, Ghosh P, Ghosh K, Ringø E, Ray AK (2017b) Probiotic potential of autochthonous bacteria isolated from the gastrointestinal tract of four freshwater teleosts. Probiotic & Antimicrobial Protein 9(1):12–21

NavinChandran M, Iyapparaj P, Moovendhan S, Ramasubburayan R, Prakash S, Immanuel G, Palavesam A (2014) Influence of probiotic bacterium *Bacillus cereus* isolated from the gut of wild shrimp *Penaeus monodon* in turn as a potent growth promoter and immune enhancer in *P. monodon*. Fish Shellfish Immunol 36(1):38–45

Nayak SK (2010) Probiotics and immunity: a fish perspective. Fish Shellfish Immunol 29(1):2–14

Nedoluha PC, Westhoff D (1995) Microbiological analysis of striped bass (*Morone saxatilis*) grown in flow-through tanks. J Food Prot 58(12):1363–1368

Nedoluha PC, Westhoff D (1997) Microbiological analysis of striped bass (*Morone saxatilis*) grown in a recirculating system. J Food Prot 60(8):948–953

Newaj-Fyzul A, Adesiyun AA, Mutani A, Ramsubhag A, Brunt J, Austin B (2007) Bacillus subtilis AB1 controls *Aeromonas* infection in rainbow trout (*Oncorhynchus mykiss*, Walbaum). J Appl Microbiol 103(5):1699–1706

Ng W-K, Koh C-B (2016) The utilization and mode of action of organic acids in the feeds of cultured aquatic animals. Rev Aquacult 9:342–368

Nigam AK, Kumari U, Mittal S, Mittal AK (2012) Comparative analysis of innate immune parameters of the skin mucous secretions from certain freshwater teleosts, inhabiting different ecological niches. Fish Physiol Biochem 38(5):1245–1256

Nimrat S, Boonthai T, Vuthiphandchai V (2011) Effects of probiotic forms, compositions of and mode of probiotic administration on rearing of Pacific white shrimp (*Litopenaeus vannamei*) larvae and postlarvae. Anim Feed Sci Technol 169(3-4):244–258

Nimrat S, Suksawat S, Boonthai T, Vuthiphandchai V (2012) Potential Bacillus probiotics enhance bacterial numbers, water quality and growth during early development of white shrimp (*Litopenaeus vannamei*). Vet Microbiol 159(3-4):443–450

Nyman A, Huyben D, Lundh T, Dicksved J (2017) Effects of microbe-and mussel-based diets on the gut microbiota in Arctic charr (*Salvelinus alpinus*). Aquac Report 5:34–40

Oatway L, Vasanthan T, Helm JH (2001) Phytic acid. Food Rev Intl 17(4):419–431

Oscariz JC, Pisabarro AG (2000) Characterization and mechanism of action of cerein 7, a bacteriocin produced by *Bacillus cereus* Bc7. J Appl Microbiol 89(2):361–369

Parisot J, Carey S, Breukink E, Chan WC, Narbad A, Bonev B (2008) Molecular mechanism of target recognition by subtilin, a class I lanthionine antibiotic. Antimicrob Agents Chemother 52(2):612–618

Peixoto SB, Cladera-Olivera F, Daroit DJ, Brandelli A (2011) Cellulase-producing *Bacillus* strains isolated from the intestine of Amazon basin fish. Aquacult Res 42(6):887–891

Peng KC, Lee SH, Hour AL, Pan CY, Lee LH, Chen JY (2012) Five different piscidins from Nile tilapia, Oreochromis niloticus: analysis of their expressions and biological functions. PLoS One 7(11)

Piewngam P, Zheng Y, Nguyen TH, Dickey SW, Joo HS, Villaruz AE et al (2018) Pathogen elimination by probiotic *Bacillus* via signalling interference. Nature 562(7728):532–537

Pinchuk IV, Bressollier P, Verneuil B, Fenet B, Sorokulova IB, Mégraud F, Urdaci MC (2001) In vitro anti-helicobacter pyloriactivity of the probiotic strain *Bacillus subtilis* 3 is due to secretion of antibiotics. Antimicrob Agents Chemother 45(11):3156–3161

Prieto ML, O'Sullivan L, Tan SP, McLoughlin P, Hughes H, O'Connor PM, Cotter PD, Lawlor PG, Gardiner GE (2012) Assessment of the bacteriocinogenic potential of marine bacteria reveals lichenicidin production by seaweed derived *Bacillus* spp. Mar Drugs 10:2280–2299

Rahman MM (2020) Final report on development of native probiotics, herbal extracts and antimicrobial agents for sustainable management of major fish diseases of Bangladesh. Research Management Wing, Bangabandhu Sheikh Mujibur Rahman Agricultural University, Gazipur, Bangladesh. pp. 83.

Rahman M, Sabir AA, Mukta JA, Khan MMA, Mohi-Ud-Din M, Miah MG et al (2018) Plant probiotic bacteria *Bacillus* and *Paraburkholderia* improve growth, yield and content of antioxidants in strawberry fruit. Sci Rep 8(1):1–11

Rahman MM, Kawser AQMR, Islam MT (2019) Probiotic Bacilli in sustainable aquaculture. In: Bacilli and agrobiotechnology: phytostimulation and biocontrol. Springer, Cham, pp 305–335

Rahman MM, Paul SI, Akter T, Tay ACY, Foysal MJ, Islam MT (2020) Whole-Genome sequence of *Bacillus subtilis* WS1A, a promising fish probiotic strain isolated from Marine Sponge of the Bay of Bengal. Microbiol Res Announcement 9(39):e00641-20

Raida MK, Larsen JL, Nielsen ME, Buchmann K (2003) Enhanced resistance of rainbow trout, *Oncorhynchus mykiss* (Walbaum), against *Yersinia ruckeri* challenge following oral administration of *Bacillus subtilis* and *B. licheniformis* (BioPlus2B). J Fish Dis 26(8):495–498

Rajanbabu V, Chen JY (2011) Antiviral function of tilapia hepcidin 1–5 and its modulation of immune-related gene expressions against infectious pancreatic necrosis virus (IPNV) in Chinook salmon embryo (CHSE)-214 cells. Fish Shellfish Immunol 30(1):39–44

Rajesh D, Karthikeyan S, Jayaraman G (2012) Isolation and partial characterization of a new bacteriocin from *Bacillus pumilus* DR2 isolated from sea water. CJM 1(2-3):33–41

Ramachandran S, Ray AK (2007) Nutritional evaluation of fermented black gram (*Phaseolus mungo*) seed meal in compound diets for rohu, *Labeo rohita* (Hamilton), fingerlings. J Appl Ichthyol 23(1):74–79

Ramachandran S, Bairagi A, Ray AK (2005) Improvement of nutritive value of grass pea (*Lathyrus sativus*) seed meal in the formulated diets for rohu, *Labeo rohita* (Hamilton) fingerlings after fermentation with a fish gut bacterium. Bioresour Technol 96(13):1465–1472

Ramesh D, Souissi S (2018) Effects of potential probiotic *Bacillus subtilis* KADR1 and its subcellular components on immune responses and disease resistance in *Labeo rohita*. Aquacult Res 49(1):367–377

Ramírez C, Romero J (2017) Fine flounder (Paralichthys adspersus) microbiome showed important differences between wild and reared specimens. Front Microbiol 8:271

Ramírez-Torrez JA, Monroy-Dosta MDC, Hernández-Hernández LH, Castro-Mejía J, Bustos-Martínez JA, Hamdan-Partida A (2018) Presumptive probiotic isolated from Oncorhynchus mykiss (Walbaum, 1792), cultivated in Mexico. Int J Aquatic Sci 9(1):3–12

Ran C, Carrias A, Williams MA, Capps N, Dan BC, Newton JC et al (2012) Identification of *Bacillus* strains for biological control of catfish pathogens. PLoS One 7(9)

Rasheeda MK, Rangamaran VR, Srinivasan S, Ramaiah SK, Gunasekaran R, Jaypal S et al (2017) Comparative profiling of microbial community of three economically important fishes reared in sea cages under tropical offshore environment. Marine Genomics 34:57–65

Rauta PR, Nayak B, Das S (2012) Immune system and immune responses in fish and their role in comparative immunity study: a model for higher organisms. Immunol Lett 148(1):23–33

Rawls JF, Samuel BS, Gordon JI (2004) Gnotobiotic zebrafish reveal evolutionarily conserved responses to the gut microbiota. Proc Natl Acad Sci 101(13):4596–4601

Rawls JF, Mahowald MA, Ley RE, Gordon JI (2006) Reciprocal gut microbiota transplants from zebrafish and mice to germ-free recipients reveal host habitat selection. Cell 127(2):423–433

Ray AK, Bairagi A, Ghosh KS, Sen SK (2007) Optimization of fermentation conditions for cellulase production by *Bacillus subtilis* CY5 and *Bacillus circulans* TP3 isolated from fish gut. Acta Ichthyologica et Piscatoria 1(37):47–53

Ray AK, Roy T, Mondal S, Ringø E (2010) Identification of gut-associated amylase, cellulase and protease-producing bacteria in three species of Indian major carps. Aquacult Res 41(10):1462–1469

Ray AK, Ghosh K, Ringø E (2012) Enzyme-producing bacteria isolated from fish gut: a review. Aquacult Nutr 18(5):465–492

Reimmann C, Ginet N, Michel L, Keel C, Michaux P, Krishnapillai V et al (2002) Genetically programmed autoinducer destruction reduces virulence gene expression and swarming motility in *Pseudomonas aeruginosa* PAO1The GenBank accession number for the aiiA nucleotide sequence is AF397400. The GenBank accession numbers for the nucleotide sequences of the 16S rRNA genes of strains A23 and A24 are AF397398 and AF397399. Microbiology 148(4):923–932

Ricaud K, Rey M, Plagnes-Juan E, Larroquet L, Even M, Quillet E et al (2018) Composition of intestinal microbiota in two lines of rainbow trout (*Oncorhynchus Mykiss*) divergently selected for muscle fat content. Open Microbiol J 12:308

Riley MA, Wertz JE (2002) Bacteriocin diversity: ecological and evolutionary perspectives. Biochimie 84(5-6):357–364

Ringø E (1999) Intestinal microflora of fish larvae and fry. Aquacult Res 30:73–93

Ringø E, Song SK (2016) Application of dietary supplements (synbiotics and probiotics in combination with plant products and β-glucans) in aquaculture. Aquacult Nutr 22(1):4–24

Ringø E, Strøm E, Tabachek JA (1995) Intestinal microflora of salmonids: a review. Aquacult Res 26(10):773–789

Ringø E, Birkbeck TH, Munro PO, Vadstein O, Hjelmeland K (1996) The effect of early exposure to *Vibrio pelagius* on the aerobic bacterial flora of turbot, *Scophthalmus maximus* (L.) larvae. J Appl Bacteriol 81(2):207–211

Ringø E, Sperstad S, Myklebust R, Mayhew TM, Olsen RE (2006) The effect of dietary inulin on aerobic bacteria associated with hindgut of Arctic charr (*Salvelinus alpinus* L.). Aquacult Res 37(9):891–897

Ringø E, Myklebust R, Mayhew TM, Olsen RE (2007) Bacterial translocation and pathogenesis in the digestive tract of larvae and fry. Aquaculture 268(1-4):251–264

Ringø E, Løvmo L, Kristiansen M, Bakken Y, Salinas I, Myklebust R et al (2010) Lactic acid bacteria vs. pathogens in the gastrointestinal tract of fish: a review. Aquacult Res 41(4):451–467

Ringø E, Zhou Z, He S, Olsen RE (2014) Effect of stress on intestinal microbiota of Arctic charr, Atlantic salmon, rainbow trout and Atlantic cod: a review. Afr J Microbiol Res 8(7):609–618

Ringø E, Zhou Z, Vecino JG, Wadsworth S, Romero J, Krogdahl Å et al (2016) Effect of dietary components on the gut microbiota of aquatic animals. A never-ending story? Aquacult Nutr 22(2):219–282

Ringø E, Hoseinifar SH, Ghosh K, Doan HV, Beck BR, Song SK (2018) Lactic acid bacteria in finfish—An update. Front Microbiol 9:1818

Ripert G, Racedo SM, Elie AM, Jacquot C, Bressollier P, Urdaci MC (2016) Secreted compounds of the probiotic *Bacillus clausii* strain O/C inhibit the cytotoxic effects induced by *Clostridium difficile* and *Bacillus cereus* toxins. Antimicrob Agents Chemother 60(6):3445–3454

Roy T, Mondal S, Ray AK (2009) Phytase-producing bacteria in the digestive tracts of some freshwater fish. Aquacult Res 40(3):344–353

Roy V, Adams BL, Bentley WE (2011) Developing next generation antimicrobials by intercepting AI-2 mediated quorum sensing. Enzyme Microb Technol 49(2):113–123

Roy T, Banerjee G, Dan SK, Ghosh P, Ray AK (2014) Improvement of nutritive value of sesame oilseed meal in formulated diets for rohu, Labeo rohita (Hamilton), fingerlings after fermentation with two phytase-producing bacterial strains isolated from fish gut. Aquac Int 22(2):633–652

Saha S, Roy RN, Sen SK, Ray AK (2006) Characterization of cellulase-producing bacteria from the digestive tract of tilapia, *Oreochromis mossambica* (Peters) and grass carp, *Ctenopharyngodon idella* (Valenciennes). Aquacult Res 37(4):380–388

Salam MA, Islam, MA, Paul SI, Rahman MM, Rahman ML, Islam F, Rahman A, Shaha DC, Alam MS, Islam T (2021) Gut probiotic bacteria of Barbonymus gonionotus improve growth, hematological parameters and reproductive performances of the host. Sci Rep 11, 10692. https://doi.org/10.1038/s41598-021-90158-x

Sanchez LM, Wong WR, Riener RM, Schulze CJ, Linington RG (2012) Examining the fish microbiome: vertebrate-derived bacteria as an environmental niche for the discovery of unique marine natural products. PLoS One 7(5)

Sangma T, Kamilya D (2015) Dietary *Bacillus subtilis* FPTB13 and chitin, single or combined, modulate systemic and cutaneous mucosal immunity and resistance of catla, *Catla catla* (Hamilton) against edwardsiellosis. Comp Immunol Microbiol Infect Dis 43:8–15

Santos RA, Oliva-Teles A, Saavedra MJ, Enes P, Serra CR (2018) *Bacillus* spp. as source of Natural Antimicrobial Compounds to control aquaculture bacterial fish pathogens. Front Mar Sci:5

Sarkar B, Ghosh K (2014) Gastrointestinal microbiota in *Oreochromis mossambicus* (Peters) and *Oreochromis niloticus* (Linnaeus): scanning electron microscopy and microbiological study. Int J Fish & Aquatic Stud 2(2):78–88

Segner H, Sundh H, Buchmann K, Douxfils J, Sundell KS, Mathieu C et al (2012) Health of farmed fish: its relation to fish welfare and its utility as welfare indicator. Fish Physiol Biochem 38(1):85–105

Selim KM, Reda RM (2015) Improvement of immunity and disease resistance in the Nile tilapia, Oreochromis niloticus, by dietary supplementation with Bacillus amyloliquefaciens. Fish Shellfish Immunol 44(2):496–503

Shaheen AA, Eissa N, Abou-ElGheit EN, Yao H, Wang HP (2014) Probiotic effect on molecular antioxidant profiles in yellow perch, *Perca flavescens*. Global J Fish & Aquacult Res 1(2):16–29
Sheikhzadeh N, Pashaki AK, Nofouzi K, Heidarieh M, Tayefi-Nasrabadi H (2012) Effects of dietary Ergosan on cutaneous mucosal immune response in rainbow trout (*Oncorhynchus mykiss*). Fish Shellfish Immunol 32(3):407–410
Shen WY, Fu LL, Li WF, Zhu YR (2010) Effect of dietary supplementation with Bacillus subtilis on the growth, performance, immune response and antioxidant activities of the shrimp (*Litopenaeus vannamei*). Aquacult Res 41(11):1691–1698
Shobharani P, Padmaja RJ, Halami PM (2015) Diversity in the antibacterial potential of probiotic cultures *Bacillus licheniformis* MCC2514 and *Bacillus licheniformis* MCC2512. Res Microbiol 166(6):546–554
Sirtori LR, Cladera-Olivera F, Lorenzini DM, Tsai SM, Brandelli A (2006) Purification and partial characterization of an antimicrobial peptide produced by *Bacillus* sp. strain P45, a bacterium from the Amazon basin fish *Piaractus mesopotamicus*. J Gen Appl Microbiol 52(6):357–363
Smitha S, Bhat SG (2013) Thermostable Bacteriocin BL 8 from *Bacillus licheniformis* isolated from marine sediment. J Appl Microbiol 114(3):688–694
Song ZF, An J, Fu GH, Yang XL (2011) Isolation and characterization of an aerobic denitrifying Bacillus sp. YX-6 from shrimp culture ponds. Aquaculture 319(1-2):188–193
Stein T (2005) *Bacillus subtilis* antibiotics: structures, syntheses and specific functions. Mol Microbiol 56(4):845–857
Strøm E, Olafsen JA (1990) The indigenous microflora of wild-captured juvenile cod in net-pen rearing. Microbiol Poecilotherm:181–185
Sturr MG, Guffanti AA, Krulwich TA (1994) Growth and bioenergetics of alkaliphilic *Bacillus firmus* OF4 in continuous culture at high pH. J Bacteriol 176(11):3111–3116
Sugita H, Tokuyama K, Deguchi Y (1985) The intestinal microflora of carp *Cyprinus carpio*, grass carp *Ctenopharyngodon idella* and tilapia *Sarotherodon niloticus*. Bull Japan Soc Sci Fish 51(8):1325–1329
Sugita H, Tsunohara M, Fukumoto M, Deguchi Y (1987) Comparison of microflora between intestinal contents and fecal pellets of freshwater fishes. Nippon Suisan Gakkaishi 53(2):287–290
Sugita H, Fukumoto M, Koyama H, Deguchi Y (1988a) Changes in the fecal microflora of goldfish *Carassius auratus* with the oral administration of oxytetracycline. Nippon Suisan Gakkaishi 54(12):2181–2187
Sugita H, Miyajima C, Iwata J-I, Arai S, Kubo T, Igarashi S, Deguchi Y (1988b) Intestinal microflora of Japanese coastal fishes. Nippon Suisan Gakkaishi 54:875–882
Sugita H, Miyajima C, Fukumoto M, Koyama H, Deguchi Y (1989) Effect of oxolinic acid on fecal microflora of goldfish (Carassius auratus). Aquaculture 80(1-2):163–174
Sugita H, Hirose Y, Matsuo N, Deguchi Y (1998) Production of the antibacterial substance by *Bacillus* sp. strain NM 12, an intestinal bacterium of Japanese coastal fish. Aquaculture 165(3-4):269–280
Sun P, Wang JQ, Deng LF (2013) Effects of *Bacillus subtilis* natto on milk production, rumen fermentation and ruminal microbiome of dairy cows. Anim Int J Anim Biosci 7(2):216
Surovy MZ, Gupta DR, Mahmud NU, Dame ZT, Roy PK, Islam MT (2019) Genomics and post-genomics approaches for elucidating molecular mechanisms of plant growth-promoting Bacilli. In: Bacilli and Agrobiotechnology: Phytostimulation and Biocontrol. Springer, Cham, pp 161–200
Sutthi N, Thaimuangphol W, Rodmongkoldee M, Leelapatra W, Panase P (2018) Growth performances, survival rate, and biochemical parameters of Nile tilapia (*Oreochromis niloticus*) reared in water treated with probiotic. Comp Clin Pathol 27(3):597–603
Sutyak KE, Wirawan RE, Aroutcheva AA, Chikindas ML (2008) Isolation of the *Bacillus subtilis* antimicrobial peptide subtilosin from the dairy product-derived *Bacillus amyloliquefaciens*. J Appl Microbiol 104(4):1067–1074
Suva MA, Sureja VP, Kheni DB (2016) Novel insight on probiotic *Bacillus subtilis*: mechanism of action and clinical applications. J Curr Res Scientific Med 2(2):65

Talukdar S, Ringø E, Ghosh K (2016) Extracellular tannase-producing bacteria detected in the digestive tracts of freshwater fishes (Actinopterygii: Cyprinidae and Cichlidae). Acta Ichthyologica Et Piscatoria 46(3):201–210

Tang K, Zhang Y, Yu M, Shi X, Coenye T, Bossier P, Zhang XH (2013) Evaluation of a new high-throughput method for identifying quorum quenching bacteria. Sci Rep 3:2935

Tang Y, Han L, Chen X, Xie M, Kong W, Wu Z (2019) Dietary supplementation of probiotic *Bacillus subtilis* affects antioxidant defenses and immune response in grass carp under *Aeromonas hydrophila* challenge. Probiotic & Antimicrobial Protein 11(2):545–558

Teixeira ML, Cladera-Olivera F, dos Santos J, Brandelli A (2009) Purification and characterization of a peptide from *Bacillus licheniformis* showing dual antimicrobial and emulsifying activities. Food Res Int 42(1):63–68

Telli GS, Ranzani-Paiva MJT, de Carla Dias D, Sussel FR, Ishikawa CM, Tachibana L (2014) Dietary administration of *Bacillus subtilis* on hematology and non-specific immunity of Nile tilapia *Oreochromis niloticus* raised at different stocking densities. Fish Shellfish Immunol 39(2):305–311

Thankappan B, Ramesh D, Ramkumar S, Natarajaseenivasan K, Anbarasu K (2015) Characterization of *Bacillus* spp. from the gastrointestinal tract of *Labeo rohita*—towards to identify novel probiotics against fish pathogens. Appl Biochem Biotechnol 175(1):340–353

Thy HTT, Tri NN, Quy OM, Fotedar R, Kannika K, Unajak S, Areechon N (2017) Effects of the dietary supplementation of mixed probiotic spores of *Bacillus amyloliquefaciens* 54A, and *Bacillus pumilus* 47B on growth, innate immunity and stress responses of striped catfish (*Pangasianodon hypophthalmus*). Fish Shellfish Immunol 60:391–399

Trust TJ (1975) Facultative anaerobic bacteria in the digestive tract of chum salmon (*Oncorhynchus keta*) maintained in fresh water under defined culture conditions. Appl Microbiol 29(5):663

Trust TJ, Bull LM, Currie BR, Buckley JT (1979) Obligate anaerobic bacteria in the gastrointestinal microflora of the grass carp (*Ctenopharyngodon idella*), goldfish (*Carassius auratus*), and rainbow trout (*Salmo gairdneri*). J Fish Board Canada 36(10):1174–1179

Tuan TN, Duc PM, Hatai K (2013) Overview of the use of probiotics in aquaculture. Int J Res Fish & Aquacul 3(3):89–97

Tzeng TD, Pao YY, Chen PC, Weng FCH, Jean WD, Wang D (2015) Effects of host phylogeny and habitats on gut microbiomes of oriental river prawn (*Macrobrachium nipponense*). PLoS One 10(7)

Urdaci MC, Pinchuk I (2004) Antimicrobial activity of *Bacillus* probiotics. In: Bacterial spore formers–Probiotics and emerging applications. Horizon Bioscience, Norfolk, UK, pp 171–182

Van Doan H, Hoseinifar SH, Khanongnuch C, Kanpiengjai A, Unban K, Srichaiyo S (2018) Host-associated probiotics boosted mucosal and serum immunity, disease resistance and growth performance of Nile tilapia (Oreochromis niloticus). Aquaculture 491:94–100

Van Hai N (2015) Research findings from the use of probiotics in tilapia aquaculture: a review. Fish Shellfish Immunol 45(2):592–597

Vargas-Albores F, Porchas-Cornejo MA, Martínez-Porchas M, Villalpando-Canchola E, Gollas-Galván T, Martínez-Córdova LR (2017) Bacterial biota of shrimp intestine is significantly modified by the use of a probiotic mixture: a high throughput sequencing approach. Helgoland Marine Res 71(1):5

Verschuere L, Rombaut G, Sorgeloos P, Verstraete W (2000a) Probiotic bacteria as biological control agents in aquaculture. Microbiol Mol Biol Rev 64(4):655–671

Verschuere L, Heang H, Criel G, Sorgeloos P, Verstraete W (2000b) Selected bacterial strains protect *Artemia* spp. from the pathogenic effects of *Vibrio proteolyticus* CW8T2. Appl Environ Microbiol 66(3):1139–1146

Wang YB, Xu ZR, Xia MS (2005) The effectiveness of commercial probiotics in northern white shrimp *Penaeus vannamei* ponds. Fish Sci 71(5):1036–1041

Wang L, Ge C, Wang J, Dai J, Zhang P, Li Y (2017) Effects of different combinations of *Bacillus* on immunity and antioxidant activities in common carp. Aquac Int 25(6):2091–2099

Wang AR, Ran C, Ringø E, Zhou ZG (2018a) Progress in fish gastrointestinal microbiota research. Rev Aquacult 10(3):626–640

Wang J, Yang HL, Xia HQ, Ye JD, Lu KL, Hu X et al (2018b) Supplementation of heat-inactivated *Bacillus clausii* DE 5 in diets for grouper, *Epinephelus coioides*, improves feed utilization, intestinal and systemic immune responses and not growth performance. Aquacult Nutr 24(2):821–831

Wang C, Liu Y, Sun G, Li X, Liu Z (2019) Growth, immune response, antioxidant capability, and disease resistance of juvenile Atlantic salmon (Salmo salar L.) fed Bacillus velezensis V4 and Rhodotorula mucilaginosa compound. Aquaculture 500:65–74

Wee WC, Mok CH, Romano N, Ebrahimi M, Natrah I (2018) Dietary supplementation use of *Bacillus cereus* as quorum sensing degrader and their effects on growth performance and response of Malaysian giant river prawn *Macrobrachium rosenbergii* juvenile towards *Aeromonas hydrophila*. Aquacult Nutr 24(6):1804–1812

Wilson AB (2017) MHC and adaptive immunity in teleost fishes. Immunogenetics 69(8-9):521–528

Winkelmann G (2002) Microbial siderophore-mediated transport. Biochem Soc Trans 30:691–696

Won S, Hamidoghli A, Choi W, Park Y, Jang WJ, Kong IS, Bai SC (2020) Effects of *Bacillus subtilis* WB60 and *Lactococcus lactis* on growth, immune responses, histology and gene expression in Nile Tilapia, *Oreochromis niloticus*. Microorganisms 8(1):67

Wong S, Waldrop T, Summerfelt S, Davidson J, Barrows F, Kenney PB et al (2013) Aquacultured rainbow trout (*Oncorhynchus mykiss*) possess a large core intestinal microbiota that is resistant to variation in diet and rearing density. Appl Environ Microbiol 79(16):4974–4984

Wu ZX, Feng X, Xie LL, Peng XY, Yuan J, Chen XX (2012) Effect of probiotic *Bacillus subtilis* Ch9 for grass carp, *Ctenopharyngodon idella* (Valenciennes, 1844), on growth performance, digestive enzyme activities and intestinal microflora. J Appl Ichthyol 28(5):721–727

Xie F, Zhu T, Zhang F, Zhou K, Zhao Y, Li Z (2013) Using *Bacillus amyloliquefaciens* for remediation of aquaculture water. SpringerPlus 2(1):119

Xiong J, Zhu J, Dai W, Dong C, Qiu Q, Li C (2017) Integrating gut microbiota immaturity and disease-discriminatory taxa to diagnose the initiation and severity of shrimp disease. Environ Microbiol 19(4):1490–1501

Xu D, Wang Y, Sun L, Liu H, Li J (2013) Inhibitory activity of a novel antibacterial peptide AMPNT-6 from *Bacillus subtilis* against *Vibrio parahaemolyticus* in shrimp. Food Control 30(1):58–61

Xu H, Huang W, Hou Q, Kwok LY, Sun Z, Ma H et al (2017) The effects of probiotics administration on the milk production, milk components and fecal bacteria microbiota of dairy cows. Sci Bull 62(11):767–774

Yadav G, Vishwakarma K, Sharma S, Kumar V, Upadhyay N, Kumar N et al (2017) Emerging significance of rhizospheric probiotics and its impact on plant health: current perspective towards sustainable agriculture. In: Probiotics and plant health. Springer, Singapore, pp 233–251

Yamashita M, Yabu T, Ojima N (2010) Stress protein HSP70 in fish. Aqua BioSci Monogr 3(4):111–141

Yan YY, Xia HQ, Yang HL, Hoseinifar SH, Sun YZ (2016) Effects of dietary live or heat-inactivated autochthonous *Bacillus pumilus* SE 5 on growth performance, immune responses and immune gene expression in grouper *Epinephelus coioides*. Aquacult Nutr 22(3):698–707

Yang G, Chen M, Yu Z, Lu Q, Zhou S (2013) *Bacillus composti* sp. nov. and *Bacillus thermophilus* sp. nov., two thermophilic, Fe (III)-reducing bacteria isolated from compost. Int J Syst Evol Microbiol 63(8):3030–3036

Yang HT, Zou SS, Zhai LJ, Wang Y, Zhang FM, An LG, Yang GW (2017) Pathogen invasion changes the intestinal microbiota composition and induces innate immune responses in the zebrafish intestine. Fish Shellfish Immunol 71:35–42

Yang P, Hu H, Liu Y, Li Y, Ai Q, Xu W et al (2018) Dietary stachyose altered the intestinal microbiota profile and improved the intestinal mucosal barrier function of juvenile turbot, *Scophthalmus maximus* L. Aquaculture 486:98–106

Yi Y, Zhang Z, Zhao F, Liu H, Yu L, Zha J, Wang G (2018) Probiotic potential of *Bacillus velezensis* JW: antimicrobial activity against fish pathogenic bacteria and immune enhancement effects on *Carassius auratus*. Fish Shellfish Immunol 78:322–330

Yilmaz S, Sova M, Ergün S (2018) Antimicrobial activity of *trans*-cinnamic acid and commonly used antibiotics against important fish pathogens and nonpathogenic isolates. J Appl Microbiol 125(6):1714–1727

Yu HH, Liang XF, Chen P, Wu XF, Zheng YH, Luo L et al (2019) Dietary supplementation of Grobiotic®-A increases short-term inflammatory responses and improves long-term growth performance and liver health in largemouth bass (*Micropterus salmoides*). Aquaculture 500:327–337

Zaineldin AI, Hegazi S, Koshio S, Ishikawa M, Bakr A, El-Keredy AM et al (2018) *Bacillus subtilis* as probiotic candidate for red sea bream: Growth performance, oxidative status, and immune response traits. Fish Shellfish Immunol 79:303–312

Zhang LH, Dong YH (2004) Quorum sensing and signal interference: diverse implications. Mol Microbiol 53(6):1563–1571

Zhang M, Cao Y, Yao B, Bai D, Zhou Z (2011) Characteristics of quenching enzyme AiiO-AIO6 and its effect on *Aeromonas hydrophila* virulence factors expression. J Fish China 35(11):1720–1728

Zhang Q, Yu H, Tong T, Tong W, Dong L, Xu M, Wang Z (2014) Dietary supplementation of *Bacillus subtilis* and fructooligosaccharide enhance the growth, non-specific immunity of juvenile ovate pompano, *Trachinotus ovatus* and its disease resistance against *Vibrio vulnificus*. Fish Shellfish Immunol 38(1):7–14

Zhou QC, Buentello JA, Gatlin DM III (2010a) Effects of dietary prebiotics on growth performance, immune response and intestinal morphology of red drum (*Sciaenops ocellatus*). Aquaculture 309(1-4):253–257

Zhou X, Tian Z, Wang Y, Li W (2010b) Effect of treatment with probiotics as water additives on tilapia (*Oreochromis niloticus*) growth performance and immune response. Fish Physiol Biochem 36(3):501–509

Zhou Z, Karlsen Ø, He S, Olsen RE, Yao B, Ringø E (2013) The effect of dietary chitin on the autochthonous gut bacteria of Atlantic cod (*Gadus morhua* L.). Aquacult Res 44(12):1889–1900

Zhu J, Dai W, Qiu Q, Dong C, Zhang J, Xiong J (2016) Contrasting ecological processes and functional compositions between intestinal bacterial community in healthy and diseased shrimp. Microb Ecol 72(4):975–985

Zink IC, Benetti DD, Douillet PA, Margulies D, Scholey VP (2011) Improvement of water chemistry with *Bacillus* probiotics inclusion during simulated transport of yellowfin tuna yolk sac larvae. North Am J Aquacul 73(1):42–48

Zokaeifar H, Balcázar JL, Kamarudin MS, Sijam K, Arshad A, Saad CR (2012a) Selection and identification of non-pathogenic bacteria isolated from fermented pickles with antagonistic properties against two shrimp pathogens. J Antibiot 65(6):289–294

Zokaeifar H, Balcázar JL, Saad CR, Kamarudin MS, Sijam K, Arshad A, Nejat N (2012b) Effects of *Bacillus subtilis* on the growth performance, digestive enzymes, immune gene expression and disease resistance of white shrimp, *Litopenaeus vannamei*. Fish Shellfish Immunol 33(4):683–689

Zokaeifar H, Babaei N, Saad CR, Kamarudin MS, Sijam K, Balcazar JL (2014) Administration of *Bacillus subtilis* strains in the rearing water enhances the water quality, growth performance, immune response, and resistance against *Vibrio harveyi* infection in juvenile white shrimp, *Litopenaeus vannamei*. Fish Shellfish Immunol 36(1):68–74

Zou J, Jiang H, Cheng H, Fang J, Huang G (2018) Strategies for screening, purification and characterization of bacteriocins. Int J Biol Macromol 117:781–789

Chapter 21
Bacillus spp.-Mediated Drought Stress Tolerance in Plants: Current and Future Prospects

Rachid Lahlali, Said Ezrari, Nabil Radouane, Zineb Belabess, Yunfei Jiang, Fouad Mokrini, Abdessalem Tahiri, and Gary Peng

Abstract Drought is a common environmental stress that threatens the sustainability of agricultural productivity in many parts of the world. Numerous morphological, physiological, and molecular changes may occur in plants as a response to this stress, thus affecting their growth and development. Some rhizosphere microorganisms, including *Bacillus* species, have proved efficiency in increasing drought tolerance to plants growing in regions with water scarcity. These endophytic bacteria can survive in the soil under a wide range of conditions and, when colonizing rhizosphere, may enhance drought stress tolerance of plants by promoting their growth.

R. Lahlali (✉) · A. Tahiri
Phytopathology Unit, Department of Plant Protection, Ecole Nationale d'Agriculture de Meknes, Meknes, Morocco
e-mail: rlahlali@enameknes.ac.ma

S. Ezrari · N. Radouane
Phytopathology Unit, Department of Plant Protection, Ecole Nationale d'Agriculture de Meknes, Meknes, Morocco

Department of Biology, Laboratory of Functional Ecology and Environmental Engineering, FST-Fez, Sidi Mohamed Ben Abdellah University, Fez, Morocco

Z. Belabess
Plant Protection Laboratory. INRA, Centre Régional de la Recherche Agronomique (CRRA), Oujda, Morocco

Y. Jiang
Aquatic and Crop Resource Development, National Research Council Canada, Saskatoon, SK, Canada

F. Mokrini
Plant Protection Laboratory, INRA, Centre Régional de la Recherche Agronomique (CRRA) Maroc, Rabat, Morocco

G. Peng
Saskatoon Research and Development Centre, Agriculture and Agri-Food Canada, Saskatoon, SK, Canada

M. T. Islam et al. (eds.), *Bacilli in Agrobiotechnology*, Bacilli in Climate Resilient Agriculture and Bioprospecting,
https://doi.org/10.1007/978-3-030-85465-2_21

This can be ensured directly by improving the uptake of water and essential nutrients; the production of phytohormones, siderophores, and enzymes; and indirectly by suppressing phytopathogens and/or activating induced systemic resistance (ISR). To mitigate the impact of drought stress, *Bacillus* spp. employ various mechanisms including the inhibition of the formation of reactive oxygen species (ROS); the modification of phytohormonal activities; the accumulation of osmolytes, soluble sugars, and other molecules; and the induction of stress-response genes expression. This paper provides an overview of the most remarkable achievements in *Bacillus* spp. research done in agricultural crop plants. These researches could improve the understanding of the physiological and molecular mechanisms deployed by these bacteria to alleviate drought stress and lead to cope with the impacts of climate change and satisfy food demands, especially when there is scarcity of water or drought.

Keywords PGPR · *Bacillus* spp. · Drought stress · Drought stress tolerance · Proline · Antioxidant

21.1 Introduction

Crop production is constantly confronted with a range of biotic and abiotic stresses that negatively affect its yield and economic returns. General signaling pathways often are activated to help plants withstand dominant abiotic stresses (Lastochkina 2019), including pathways that target cell regulation, as well as morphological, physiological, biochemical, molecular, and genetic changes. Abiotic stresses, including drought, are among the main limiting factors for agricultural production all over the world (Abd El-Daim et al. 2019). Drought is considered as a severe environmental constraint to crop productivity (Vurukonda et al. 2016) and is one of the most known environmental stresses that has brought much attention from researchers in agriculture. Several aspects of crop growth can be affected by drought stress, including genes responses, transpiration, plant cell size, and physiology (premature senescence), which all contribute to decreased crop productivity (Disante et al. 2011; Kumar and Verma 2018; Mishra and Singh 2010).

The most known approach to deal with this challenge is the development of drought-tolerant varieties via genetic engineering and/or conventional breeding methods. However, this approach was shown to have limited effectiveness in improving plant tolerance to drought (Singh et al. 2015). Recent research indicates that microorganisms, in particular plant growth-promoting rhizobacteria (PGPR), can help crops mitigate abiotic stress. The role of these PGPR in promoting plant growth, improving crop nutrition, and suppressing diseases has been well documented (Grover et al. 2014). In other words, the application of PGPR may provide tolerance to drought, thus allowing plant survival even under a lower water potential. *Bacillus* species, gram-positive bacteria, are a major group of rhizobacteria that has been investigated broadly. This is directly linked with the population stability of these bacteria which facilitates their storage, encapsulation, and subsequent applications in the field, and their ability to survive in the soil for a long period even under harsh conditions (Hashem et al. 2019).

Bacillus spp. can enhance plant tolerance to drought by regulating water transport and nutrient uptake, activating antioxidants, and other defense systems (e.g., the production of phytohormones and related metabolites), and regulating stress-responsive genes (e.g., the activity of 1-aminocyclopropane-1-carboxylate (ACC) deaminase) (Radhakrishnan et al. 2017).

The impact of drought stress on crop productivity, the ability of *Bacillus* spp. to colonize root system, and its role in plant growth improvement as well as in mitigating drought stress are discussed in this review. Furthermore, the mechanisms deployed by *Bacillus* spp. to mediate drought stress tolerance, molecular basis involved in alleviating drought stress, and the role of *Bacillus* spp.-based commercial products in agricultural applications will be highlighted.

21.2 Impact of Drought on Crop Productivity

Climate change and the rapid worldwide population growth affect the global food security (Lesk et al. 2016). Indeed, the climatic conditions that prevail in different parts of the world result in anomalies in precipitation, thus generating severe periods of drought. This abiotic stress significantly affects agriculture due to the low availability of water resources and its management issues (Guoju et al. 2016).

The negative impact of drought stress on the yield has been investigated on many crops (Table 21.1). It has been shown that its impact depends largely on two factors: (*i*) the duration and the severity of the stress period and (*ii*) the crop stage. Several studies highlighted the occurrence of various physiological, biochemical, and

Table 21.1 Examples of yield losses caused by drought stress in field crops

Crop	Yield loss (%)	References
Maize (*Zea mays* L.)	21–40	NeSmith and Ritchie (1992)
	79–81	Monneveux et al. (2006)
	25–92	Atteya (2003)
	15	Edmeades (2008)
Pigeon pea (*Cajanus cajan* L.)	50	Nam et al. (2001)
Soybean (*Glycine max* L.)	46–71	Samarah et al. (2006)
Wheat (*Triticum aestivum* L.)	57	Balla et al. (2011)
Chickpea (*Cicer arietinum* L.)	45–69	Nayyar et al. (2006)
Rice (*Oryza saliva*)	24–84	Venuprasad et al. (2007)
	53–92	Lafitte et al. (2007)
Sunflower (*Helianthus annuus* L.)	60	Mazahery-Laghab et al. (2003)
Barley (*Hordeum vulgare*)	49–57	Samarah (2005)
Common bean (*Phaseolus vulgaris*)	58–87	Martínez et al. (2007)

molecular changes during plant growth, under drought stress (Kunert and Vorster 2020; Melandri et al. 2020; Ye et al. 2020). In cereals, for example, the negative effects of drought on pollen development, kernel number (Ji et al. 2010), size, and filling (Yang and Zhang 2006) have been widely investigated. Estrada-Campuzano et al. (2008) showed that drought occurring after anthesis shortened the period of grain filling while that occurring at pre-anthesis stages reduced the time to anthesis in cereals. The process of grain filling in cereals is attributed to four enzymes, viz., sucrose synthase, adenosine diphosphate-glucose-pyrophosphorylase, starch-branching enzyme, and starch synthase (Ahmadi and Baker 2001). The same author reported that the reduction in grain yield of cereals could be the result of the reduction in the activity of these enzymes under drought stress.

21.3 *Bacillus* spp. Root Colonization and its Role in Plant Growth Improvement

Plant root colonization by *Bacillus* spp. can be mutually beneficial to plants and bacteria. This is a remarkable strategy adopted by members of this genus to use carbon-rich root exudates, thus facilitating rhizobacterial colonization and mutualistic associations with the host plant (Allard-massicotte et al. 2016). In particular, up to 30% of the carbon fixed by plants are secreted via its root exudates, thereby feeding the bacteria (Allard-massicotte et al. 2016). In exchange, the bacteria provide plants with several growth-promoting elements (Allard-massicotte et al. 2016). The role of intact chemotaxis machinery in the localization and early root colonization by *Bacillus* spp. has been highlighted by several studies, especially with *Bacillus subtilis* for efficient colonization of plant rhizosphere in natural environments (Allard-massicotte et al. 2016). Furthermore, it has been shown that *B. subtilis* is attracted particularly by some components of root exudates, such as amino acids, as environmental cues to locate the source of nutrients and initiate symbiosis or even pathogenesis (Yang et al. 2015).

The long-term root colonization by the bacteria is determined by the formation of a biofilm which is a multicellular bacterial community covered by an extracellular matrix (Allard-massicotte et al. 2016). This biofilm may be formed hours after motile bacterial cells first settle, but for *B. subtilis*, the root colonization and biofilm formation often require 24 h (Beauregard et al. 2013). It is important to note that the root colonization is robustly stimulated by plants (Beauregard et al. 2013); the biofilm formation depends on the production of certain plant-derived molecules. Plant polysaccharides, the major component of plant cell wall, are among these essential compounds. The polysaccharides can act as an environmental cue for the biofilm formation and can also be used by bacteria as a carbon source to build the extracellular matrix (Beauregard et al. 2013). It has been demonstrated that another plant-derived molecule, malic acid (MA), can be involved in biofilm formation; elevated levels of MA promote the binding and biofilm formation of *B. subtilis* FB17 on

Arabidopsis thaliana roots (Rudrappa et al. 2008). It was later shown that MA would stimulate biofilm formation in a histidine kinase-dependent manner, but only at high concentrations (Chen et al. 2012). As mentioned above, root exudates are known to recruit beneficial bacteria and trigger biofilm formation. The *B. subtilis* RR4 strain can stimulate MA biosynthesis and exudation, and it is highly attracted to rice roots (Rekha et al. 2018). Specific molecules secreted by *Bacillus* spp., especially exopolysaccharides, appear important to the root colonization; an exopolysaccharide-deficient mutant, constructed from *B. amyloliquefaciens* FZB42, was less able to support biofilm formation and, subsequently, the colonization of *A. thaliana* roots (Lu et al. 2018).

In some cases, rhizobacteria penetrate the roots and enter the plant tissues as a harmless endophyte. This entry of commensal microbes into plants may involve the breaking of plant defense system (Deng et al. 2019). However, the mechanism deployed by endophytes in overcoming plant defense in order to establish a steady commensal relationship with their host plants has rarely been reported (Hacquard et al. 2017). Most of the compromised plant defense mechanisms are due to the adaptation by pathogens. In an attempt to decrypt how an endophyte evades plant defense and achieve harmonious commensalism, Deng et al. (2019) used *B. subtilis* BSn5 and showed that this strain would reduce plant defense responses via the production of antibiotic subtilomycin. Bond to self-produced flagellin, the antibiotic producer demonstrates advantages in colonizing plant root systems on *A. thaliana* and *Amorphophallus konjac* (Deng et al. 2019).

Endophytic bacteria have shown notable benefits to their plant hosts directly or indirectly (Afzal et al. 2019), especially endophytic *Bacillus* spp. These endophytes, some considered as PGPR, can significantly increase plant growth in early stages, forming a mutualistic plant-microbe relationship. *Bacillus* spp. are one of the predominant PGPRs (Radhakrishnan et al. 2017) with spores that can survive in the soil for long periods even under harsh environmental conditions (Hashem et al. 2019). Moreover, the ability to form spores is a characteristic trait of *Bacillus* spp., which differentiates them from other PGPRs, such as *Pseudomonas* spp. (Radhakrishnan et al. 2017). It has been shown that endophytic bacteria ameliorate plant growth in two different ways: (*i*) directly through the production of plant growth beneficial substances, including solubilization/transformation of mineral nutrients such as phosphate, nitrogen, and potassium; phytohormones; siderophores; and enzymes, and (*ii*) indirectly by controlling plant pathogens and/or activating induced systemic resistance (ISR) of plants (Ma et al. 2016). These mechanisms can help plant's growth and development under normal and stressful conditions (Ma et al. 2016), especially due to the fact that *Bacillus* spp. can survive for a long time under unfavorable environmental conditions (Radhakrishnan et al. 2017). For instance, *Bacillus* spp. can produce exopolysaccharides and siderophores under the conditions such as lack of water, salinity, and heavy metal accumulation in soil, thus preventing the movement of toxic ions and adjusting the ionic balance and water transport in plant tissues while suppressing soil-borne pathogens (Radhakrishnan et al. 2017). The synthesis of indole acetic acid (IAA), gibberellic acid, and ACC deaminase by *Bacillus* spp. may help regulate intracellular

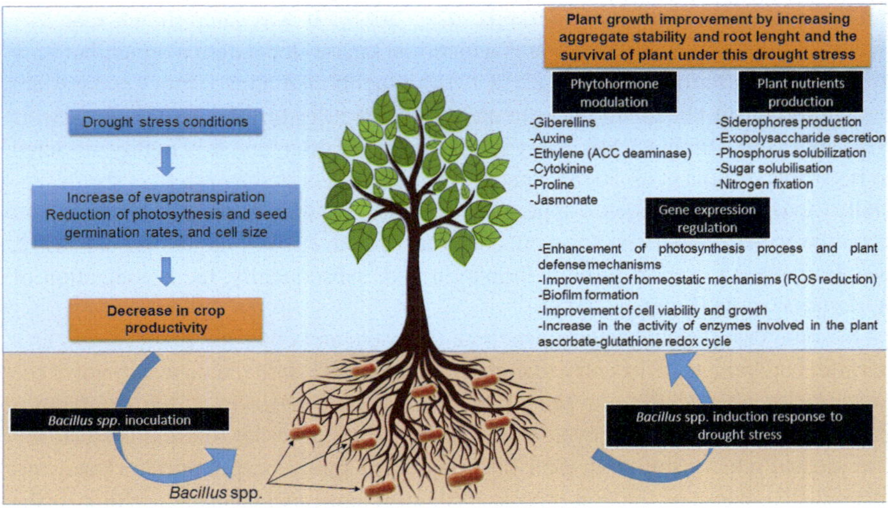

Fig. 21.1 Impact of drought stress on plant health and productivity and molecular and physiological mechanisms offered by *Bacillus* spp. to alleviate the adverse effect of drought stress

phytohormone metabolism and increase plant stress tolerance under certain conditions (Radhakrishnan et al. 2017). Cell-wall-degrading substances released by these bacteria, including chitosanase, protease, cellulose, glucanase, lipopeptides, and hydrogen cyanide, may be antagonistic to a range of pathogens and crop pests (Radhakrishnan et al. 2017). Taking together, it is noteworthy that biotic and abiotic stress factors to crops may be mitigated by *Bacillus*-induced changes, including the regulation of water transport, nutrient uptake, and the activation of the antioxidant and defense systems (Radhakrishnan et al. 2017). Below and in Fig. 21.1 are some of the examples on *Bacillus* spp. and their mechanisms involved in plant growth promotion.

21.3.1 Nutrient Acquisition and Phytohormones Production by Bacillus spp. in Plants

Both seed germination and plant growth are significantly influenced by nutrients available in the soil (Afzal et al. 2019; Fernández Bidondo et al. 2012). Plants absorb phosphorous and nitrogen from the soil via root transporters, but soils usually lack a sufficient quantity of the bioavailable forms of these nutrients for optimal plant growth (Afzal et al. 2019; Fernández Bidondo et al. 2012). Some *Bacillus* spp. can convert the complex form of essential nutrients, including phosphorous and nitrogen, to simpler forms for easy uptake by plant roots (Kang et al. 2015). *Bacillus megaterium* CDK25 showed a remarkable solubilizing ability for zinc which is indispensable for better growth and development of plants (Bhatt and Maheshwari

2020). This isolate can also help solubilize both inorganic and organic phosphate, thus leading to the amelioration of phosphate uptake by chili plants (*Capsicum annuum* L.), resulting in improved vegetative growth and other biological outputs. Furthermore, this *B. megaterium* isolate was able to produce siderophore, which positively influenced plant growth through the amelioration of soluble iron which is considered one of the requisite micronutrients for proper functions of several plant species. Thus, the availability of this element to plants led to the increase of their vegetative growth (Bhatt and Maheshwari 2020). The ability of four endophytic *Bacillus* spp., including *Bacillus cereus* EPP5, *B. cereus* EPP71, *B. amyloliquefaciens* EPP52, and *B. subtilis* subsp. *subtilis* EPP65, to transform insoluble nutrients to soluble forms has been reviewed by Kushwaha et al. (2020). *B. megaterium* HX-2 was also able to dissolve phosphate, providing more nutrients for improved plant growth (Li et al. 2019).

A broad range of genomic traits pertaining to plant growth promotion and potential agricultural applications has been raised following the study of *B. megaterium* RmBm31. This strain is proven to be an IAA-producing endophytic bacterium possessing a large set of genes that regulate plant growth-promoting traits. This bacterium enhanced the biomass and modified the root architecture of *A. thaliana* seedlings through a physical contact with roots and the production of volatile organic compounds (Dahmani et al. 2020). The *B. megaterium* CDK25 strain is endowed also with the IAA production (Bhatt and Maheshwari 2020). The *B. subtilis* RR4 strain has been shown to influence the biosynthesis of MA profoundly; the biosynthesis and accumulation of MA by this bacterium can improve primary metabolism, thus leading to plant growth promotion (Rekha et al. 2018). On pearl millet (*Pennisetum glaucum* (L.) R. Br.), rapid activation of several biochemical and molecular mechanisms was the key to the growth promotion occurred during plant-*Bacillus* interactions (Kushwaha et al. 2020), including the biosynthesis of IAA and siderophores; rapid solubilization and mobilization of nutrients such as zinc, potassium, and phosphate; and increased production of extracellular hydrolytic enzymes, etc. (Kushwaha et al. 2020). *Bacillus cereus*, for instance, has been shown to remarkably improve plant biomass increase by escalating nutrient uptake efficiency and releasing plant growth hormones (Kushwaha et al. 2020). *B. subtilis* can help solubilize phosphate in soil, improve nitrogen fixation, and produce siderophores that promote plant growth and suppress pathogens (Hashem et al. 2019). It has also been shown that *B. amyloliquefaciens* possesses immense potential to increase the growth of chili plants (*C. annuum*); a seed treatment with this bacterium resulted in substantial increases in germination (85%), seedling vigor, and vegetative growth (Gowtham et al. 2018). Another strain of *B. amyloliquefaciens*, Bc2, promoted the growth and development of strawberry seedlings, and its role in plant growth promotion has been demonstrated (Es-Soufi et al. 2020). The *B. megaterium* HX-2 strain, which also improved the growth of maize, was believed to confer the effect via IAA production and phosphate solubilization; seed soaked in a HX-2 suspension showed significantly improved plant growth, including the increase in plant height, root length, and plant dry weight (Li et al. 2019).

21.3.2 *Indirect Plant Growth Promotion by Suppressing Phytopathogens*

It has been well documented that the application of *Bacillus* spp. can activate ISR in several crop species as well as promote the growth of these plants (Hashem et al. 2019). The majority of endophytic strains of *Bacillus* spp. dwelling inside pearl millet showed antagonism against several important plant pathogens, including *Rhizoctonia solani*, *Sclerotium rolfsii*, and *Fusarium solani*. This is achieved via several mechanisms, including the secretion of antimicrobial peptides (AMP). An isolate of *B. cereus*, for instance, was able to increase plant biomass by disrupting the mycelium growth of fungal pathogens through the biosynthesis of siderophores and extracellular fungal cell wall degrading enzymes (Kushwaha et al. 2020). In another study, an isolate of *B. subtilis* was reported as one of the promising microorganisms for sustainable agriculture as it promoted plant growth and showed efficacy against a wide range of pathogens both in vitro and in greenhouse/field trials. Multiple mechanisms have been identified for disease suppression by *B. subtilis*, including direct plant growth promotion, antibiosis, competition for space and nutrients, lysis of pathogen hyphae, and ISR (Wang et al. 2018). Many *B. subtilis* strains are able to activate ISR in plants, thus increasing disease resistance as reflected by lower levels of infections. Indeed, the number and diversity of microorganisms used for the production of commercial biocontrol products have increased (Hashem et al. 2019). The antimicrobial nature of *B. subtilis* BSn5 against a bacterial pathogen, *Erwinia carotovora* subsp. *carotovora*, is a good example of such application, although the exact mechanism underlying the antagonism remains unknown (Deng et al. 2011). Antifungal activity against fungal pathogens has been reported for *Bacillus stratosphericus* LW-03. This strain targets several fungal pathogens known to cause serious diseases on several crops, with considerable inhibition of *Fusarium oxysporum*, *Botryosphaeria dothidea*, *F. fujikuroi*, and *Botrytis cinerea* (Khan et al. 2020). Significant disease reduction (~71%) against the anthracnose disease has been observed on chili plants treated with *B. amyloliquefaciens* Bc2 (Gowtham et al. 2018).

21.4 *Bacillus* spp. for Mitigating the Impact of Drought Stress

Several adaptation and mitigation strategies of physiological, biochemical, and cellular/molecular nature can be activated in plants exposed to drought to maintain cell homeostasis (Krasensky and Jonak 2012; Lastochkina 2019). Those strategies include (*i*) accumulation of abscisic acid (ABA); (*ii*) reduction in photosynthetic rate; (*iii*) expression of aquaporins and vacuolar (H+)-pyrophosphatases for maintaining cell turgor through osmotic adjustments; (*iv*) accumulation of compatible osmolytes like proline, sugars, betaines, polyamines, quaternary ammonium compounds, and amino acids; and (*v*) accumulation of drought stress-related proteins like dehydrins that maintain cell water status and help bacteria to maintain membrane permeability, proteins in their functional forms, and enzymes to maintain their integrity under stress conditions (Kumar and Verma 2018). In addition, drought stress induces reactive oxygen species (ROS) enzymes which are critical to cell integrity and functionality, as well as plant survival. ROS signaling is activated with the increase of ABA levels and calcium flows (Kaur and Asthir 2017). Therefore, a better understanding of the strategies used by plants, in response to drought stress, may facilitate the development of treatment that helps to alleviate the impact of drought and improve crop productivity.

21.4.1 Antioxidant Defenses and ROS

The lowered uptake of water and nutrients can affect plant physiological processes and generate ROS including superoxide radicals, hydrogen peroxide, hydroxyl radicals, and singlet oxygen. Under drought stress, these ROS can damage plant organization at various levels including lipid peroxidation initiation, electrolyte leakage, membrane integrity alteration, and degradation of proteins, lipids, and nucleic acids (Vurukonda et al. 2016). The plant's antioxidant defense system involves several mechanisms to reduce the biosynthesis of ROS, including increased activities of specialized enzymes such as superoxide dismutase (SOD), peroxidase (POD), catalase (CAT), ascorbate peroxidase (APX), and glutathione reductase (GR), as well as other synthesis (ascorbic acid, cysteine, and glutathione) that alleviate the impact of oxidative stresses (Kaushal and Wani 2016). Several studies identified that *Bacillus* spp. played the role of an elicitor to mediate tolerance to drought in plants via a variety of mechanisms (Lastochkina 2019). Plant inoculation with these bacteria was reported to improve the efficiency of water uptake that is considered as an important mechanism in mitigating the impact of drought stress (Marulanda et al. 2009). *Bacillus* spp. were also found to enhance the levels of antioxidant enzyme activities in wheat (Chakraborty et al. 2013), maize (Vardharajula et al. 2011), and cucumber (Hou et al. 2018). For example, an increase in the activity of antioxidant enzymes like APX, CAT, and glutathione peroxidase (GPX) has been reported in drought-stressed maize plants inoculated with drought-tolerant *Bacillus* spp. (Vardharajula et al. 2011). Treatment of drought-stressed cucumber plants with a *Bacillus* spp. strain, *Bacillus methylotrophicus* CSY-F1, was often accompanied by the improvement of plant growth and the enhancement of antioxidant enzyme activities, including those of SOD, catalase, guaiacol peroxidase, APX, dehydroascorbate reductase, monodehydroascorbate reductase, and GR. Those enzymes are generally considered as biochemical indicators of drought tolerance (Hou et al. 2018). It is important to emphasize that the use of *Bacillus* spp. under drought stress has led to the improvement of plant growth and nutrition, as well as the enhancement of physiological and metabolic plant activities which may contribute to drought stress mitigation. This includes, for example, the deployment of *B. megaterium*,

Bacillus thuringiensis, and *Bacillus* spp. in drought-stressed lavandula and salvia plants which was accompanied by the increase of the concentration of some plant nutrients such as calcium, magnesium, zinc, manganese, and copper (Armada et al. 2014). Calcium is known for its role as a stabilizer of membrane systems. As to magnesium, it is involved in regulating ion homeostasis in some organelles such as chloroplasts, vacuoles, and stomata (Radhakrishnan et al. 2017).

21.4.2 Modification of Phytohormonal Activities to Impart Drought Tolerance in Plants

Phytohormones, in particular IAA, gibberellins, ethylene (ET), ABA, salicylic acid (SA), and cytokinins, play an essential role in plant growth and development by regulating some of the critical biological processes. The latter include the induction of germination, cell division and enlargement, and root extension. These processes are often involved in supporting plants to cope with environmental stresses (Forni et al. 2016). The role of *Bacillus* spp. in enhancing drought stress tolerance by altering endogenous phytohormone levels has already been highlighted. Such is the case, for instance, of *B. subtilis* for which a role in mediating tolerance to wheat against drought stress has been emphasized. Indeed, this bacterial strain increases the IAA content of wheat exposed to drought stress. Furthermore, it counteracts the increase of ABA and ACC content under the same conditions (Barnawal et al. 2017). IAA plays a crucial role in regulating plant growth, including the differentiation of vascular tissues and roots, cell division, and shoot growth (Goswami et al. 2016). ABA accumulation in plants improves drought tolerance by inducing stomatal closure, which helps to limit the water loss, activate antioxidant enzymes and water-deficit-induced genes, and improve root hydraulic conductivity (Forni et al. 2016). The role of cytokinins under the stress may be due to the antioxidant activity related to the protection of purine breakdown (Javid et al. 2011) since a low level of this hormone would favor carbon reallocation to root growth and prompt stomatal closure (Arkhipova et al. 2007). Jasmonic acid (JA) is also involved in drought tolerance by enhancing the activities of antioxidant enzymes and pathogenesis-related proteins (Forni et al. 2016). Both salicylic acid (SA) and ethylene (ET) were reported as stress-response hormones; SA can induce the expression of drought-responsive genes involved in maintaining the cell membrane stability and in preventing ROS accumulation (Jumali et al. 2011).

21.4.3 Control of ET Levels by ACC Deaminase

Under drought stresses, plants generally respond by synthesizing ACC synthase, which is the precursor for ET, a gaseous phytohormone produced when plants are exposed to abiotic (flooding, drought, temperature, and various contaminants) or biotic (phytopathogens) stressors. The increase in ET levels in plants leads to senescence, chlorosis, and abscission (Afzal et al. 2019). The modulation of ACC deaminase expression is the key to ACC biosynthesis and consequently to ET production (Kumar and Verma 2018). PGPR that stimulate the activity of ACC deaminase can prevent a jump in plant ET levels via ACC hydrolysis to convert the immediate ET precursor into products which can be easily assimilated by plants (α-ketobutyrate and ammonia). This helps reduce the negative effect of drought stresses on plants (Afzal et al. 2019).

21.4.4 ROS-Scavenging Antioxidant Enzymes

Bacillus spp. are known for their capacity as "biofertilizers" which can help plants grow under drought stress by optimizing the uptake of water and essential nutrients due to suppressed ROS levels as well as reduced cellular disruption to plants (Lopes et al. 2018). Treated plants may show increased physiological activities, especially photosynthesis, accompanied by a reduction in ROS levels (Armada et al. 2014).

21.4.5 Drought Stress Tolerance Induced by the Accumulation of Osmolytes, Soluble Sugars, Exopolysaccharides (EPS), and Bacterium-Derived Volatiles

The accumulation of osmoprotectants—a group of small molecules with low molecular weight, such as osmolytes, soluble sugars, sugar alcohols, and amino acids—is another mechanism deployed by plants to maintain homeostasis under drought stress (Singh et al. 2015; Vardharajula et al. 2011). Exopolysaccharides are carbohydrates with high molecular weight and are produced by some bacteria known to be involved in drought stress tolerance (Kumar and Verma 2018). These molecules, reported to be produced by *Bacillus* spp., may facilitate water uptake by roots and maintain plant ionic balance under drought stress (Hashem et al. 2019).

Volatile compounds (VOCs) produced by some bacteria may play an important role in inducing drought tolerance by serving as a signal to plants for developing proper responses (Vurukonda et al. 2016). Although ROS-targeting enzymes, including GR, monodehydroascorbate reductase, superoxide dismutase, and catalase (Timmusk et al. 2014), are activated commonly, VOCs produced by *Bacillus* spp. may also induce plants to produce hydrogen peroxide and nitric oxide that are associated with plant drought tolerance (Liu and Zhang 2015).

21.5 Mechanisms of *Bacillus* spp.-Mediated Drought Tolerance

The rhizosphere nests a variety of microbe fauna, and their diversity has been characterized using metagenomics and high-throughput screening tools. A better understanding of this plant-microbe relationship helps lay the foundation for studying plant stress tolerance interacting with soil microorganisms (Kumar and Dubey 2020).

During the natural evolution, some microorganisms have developed mechanisms that could help them to survive harsh conditions, such as the lack of water supplies for an extended period of time. Some of these microbes have evolved with plants in the rhizosphere, becoming mutualistic to each other. *Bacillus* spp. are well known to promote plant growth through several direct and indirect mechanisms, including the production of phytohormones, bacterial exopolysaccharides, induction of systemic stress tolerance, and ACC deaminase (Kumar and Verma 2018). An intimate relationship between plants and their rhizosphere microbiota appears essential for improved plant tolerance to drought stresses (Goswami and Deka 2020). Application of biostimulants, which could induce physiological and molecular responses of plant for nutritional and functional improvements, has been reported to reduce the impact of drought stress (Basile et al. 2020). PGPR are among the microorganisms that produce a range of plant hormones to help reset some of the plant physiological process in response to drought stresses (Kumar and Verma 2018). Some of the hormones may act as a signal to trigger a cascade of further plant responses against drought stresses, including enzymatic activities and modification of root structure (Goswami and Deka 2020).

Drought tolerance involves several mechanisms in plants, including physiological, biochemical, and molecular responses to adapt to the drought stress (Goswami and Deka 2020). Under drought stress conditions, some rhizobacteria may accumulate solutes like, quaternary amines, sugar, and amino acids which can prevent degenerative processes and help cell growth under osmotic-deficient conditions (Vardharajula et al. 2011).

The knowledge of plant-microbe communication may be used to manipulate the abundance of microorganisms, altering plant development, and/or defense against pests or diseases (Arif et al. 2020). PGPR, especially *Bacillus* spp., have been shown repeatedly to promote plant growth and development and reduce the impact of stresses. Many *Bacillus* spp. are known to elicit tolerance against drought stress via the augmentation of antioxidants (AOX), followed by modification of root architecture, activation of early signaling, increase in osmoregulation, and finally the reduction of ET through ACC deaminase (Carlson et al. 2020). Some *Bacillus* spp. including *B. circulans*, *B. firmus*, *B. globisporus*, *B. licheniformis*, and

B. mojavensis have been reported to alleviate drought stress through ACC deaminase activities (Carlson et al. 2020).

B. sphaericus has been reported to absorb heavy metals and bind them extracellularly (Ramakrishna et al. 2020). Four *Bacillus* spp. isolates exhibited growth-promoting properties and drought stress tolerance in vitro under sterile and non-sterile soil conditions, via the production of ammonia and IAA, and solubilization of phosphorus (Grover et al. 2014). The treatment improved root growth and lateral root formation. *B. licheniformis* produces ACC deaminase under drought stress conditions, which limits the production of ET that impairs normal plants growth (Bleecker and Kende 2000; Lim and Kim 2013).

B. amyloliquefaciens, *B. licheniformis*, *B. thuringiensis*, and *B. subtilis* have been observed to stimulate root exudation and promote plant growth. These *Bacillus* spp. also were able to synthetize osmolytes under stress conditions, thus alleviating the negative impact on plants (Timmusk 2003) through the maintenance of the stability of plant cell membrane and proteins (Kogut and Russell 1987; Vardharajula et al. 2011).

Bacillus spp. strains also secrete exopolysaccharides under drought conditions that can form an organo-mineral sheath around the cells. The exopolysaccharide layer may lead to increased microaggregates as a secondary effect that increases the aggregate stability and root survival under drought stress (Alami et al. 2000). *Bacillus* spp. also increase the proline content in roots, which helps protect plant cell membranes, proteins, and water status (Yoshiba et al. 1997). Moreover, these bacteria are able to increase soluble sugars in root tissues, which support the biosynthesis against the effects of drought stress (Vardharajula et al. 2011). Armada et al. (2014) reported that *B. thuringiensis* mitigated the effect of drought stress depressing stomatal conductance to reduce water losses, increasing the production of potassium to support plant functionality under drought conditions, and controlling proline secretion and accumulation. Drought tolerance conferred by *B. thuringiensis* was linked to the decreased level of APX and GR which reduced the oxidative damage to plant cells (Armada et al. 2014). The rhizobacteria were also able to produce IAA under stress conditions, which is responsible for root growth enhancement. IAA also improves water use efficiency (Armada et al. 2014). *B. thuringiensis* produces a great amount of IAA and proline under drought stress conditions; this bacterial strain was able to survive and multiply under drought conditions. The use of *B. thuringiensis*, in combination with a consortium of arbuscular mycorrhizal (AM) in maize, decreased the oxidative damage to lipids, as well as the accumulation of proline in plant cells under drought stress conditions. The combined use enhanced the function of aquaporins for improved transport of water, carbon dioxide, glycerol, urea hydrogen peroxide, and boric acid (Armada et al. 2015).

Timmusk et al. (2014) reported that *B. thuringiensis AZP2* strain increased the growth of wheat seedlings, which was attributed to higher rates of photosynthesis and lower stress volatile emissions. *B. amyloliquefaciens* also increased the proline and total soluble sugar (TSS) levels in rice seedlings subjected to drought stress and positively modulates the expression of stress-responsive genes including glutathione S-transferases (GSTs) (Tiwari et al. 2017). GSTs are an enzyme with known antioxidant properties that contribute to the reduction of glutathione (Tiwari et al. 2017).

The reduction of stomatal conductance is often the first response to drought stresses. The stomatal closure, however, has the disadvantage of reduced rates of photosynthesis due to limited gas exchange, but *B. subtilis* (B26 strain) treatment seemed also help to reduce the negative impact on photosynthesis under drought stress due to optimized stomatal conductance and increased root biomass (Gagné-Bourque et al. 2016). The same study also reported notably higher sucrose and fructan concentrations in shoots and roots of treated plants under drought conditions. The increase of sugar biosynthesis may allow an osmotic adjustment, contributing to mitigating the impact of drought stress (Gagné-Bourque et al. 2016). Furthermore, the treatment also leads to the production of amino acids in greater quantities in plant shoots and roots. These amino acids mainly belong to glutamate and aspartate families.

Inoculation of *Platycladus orientalis* (L.) Franco (conifers) with *B. subtilis* induced an increased level of ABA and stomatal conductance in shoots, in connection with drought-tolerance responses (Liu et al. 2013). Inoculation of *Lavandula dentata* L. (fringed lavender) with the IAA-producing strains of *B. thuringiensis* induced the increase in both proline and potassium levels but reduced APX and GR activities (Armada et al. 2014). Cytokinin-producing *B. subtilis* strains also proved their role in alleviating drought stress (Liu et al. 2013). Under drought stress, *B. megaterium* (BOFC15 strain) was able to excrete spermidine, a type of polyamine that plays a major role in plant growth and development. Zhou et al. (2016) showed that the bacterium enhanced the plant tolerance against drought stress, and this was related to the decrease in ABA levels and the improvement of adaptive responses. The treatment also increased plant biomass, photosynthetic capacity, proline, and soluble sugars under drought stress conditions (Zhou et al. 2016). Moreover, *B. megaterium* var. *phosphaticum* and *B. mucilaginosus* can solubilize phosphorous for plant by releasing organic acids (Arif et al. 2020), which may further add to plant drought stress tolerance.

Bacillus altitudinis is known for its ability of IAA and ACC deaminase production in rice under drought stress via an increase in relative water content, chlorophyll and membrane stability, proline and phenolics content, catalase activity, and a reduction in malondialdehyde (MDA). In addition, a reduction of ET emission under drought stress was reported with inoculated plants (Kumar et al. 2017). Forty-four *Bacillus* spp. isolates have been tested by Chari et al. (2018) in India, with 28 of them showing plant growth-promoting activities, 4 exhibiting water-stress tolerance ranging from −0.05 to −0.73 Mpa, and 6 displaying high-temperature tolerance up to 50 °C (Damodara Chari et al. 2018). Some isolates showed high phosphate and potassium solubilization, and many produced high levels of ACC. Rhizobacterial strains were able to utilize ACC as a source of nitrogen to variable degrees, and all the strains had ACC-deaminase activity (Damodara Chari et al. 2018).

B. subtilis (RJ46 strain) alleviated the impact of drought stress on two different plant species, *Pisum sativum* L. and *Vigna mungo* L. via the activation of ACC deaminase (Saikia et al. 2018). The study also showed that several physiological changes occurred in treated plants during a period of drought including an increase in seed germination and shoot and root biomass. Furthermore, the treatment showed a higher-level osmolytes, which helps decrease the osmotic potential in plant cells. An upsurge in relative water potential and leaf chlorophyll contents was detected after the bacterial treatment compared to the non-inoculated plants under the same drought conditions (Saikia et al. 2018).

Hou et al. (2018) showed that *B. methylotrophicus* could alleviate drought stress even in the presence of a plant-growth inhibitor such as ferulic acid (FA). The study showed that the applied bacterial treatment decreased FA in the soil, with the capacity to increase soil water content, catalase, sucrose activities, and polysaccharide levels at each of the tested FA concentration. The bacterial treatment seemed to limit leaf wilting under drought stress. As in some of the other cases of *Bacillus* spp. treatment, proline, soluble sugars, ascorbate, and glutathione contents were all increased by the *B. methylotrophicus* (Hou et al. 2018). *Glycyrrhiza uralensis* Fisch. ex DC (Chinese liquorice) treated with *B. pumilus* under drought stress showed an increase in total biomass, flavonoids, polysaccharide, glycyrrhizic acid, and antioxidant enzymes, but a reduction in ROS and lipid peroxidation. These changes were believed to confer drought tolerance to the cultivated *G. uralensis* (Xie et al. 2019).

21.6 Molecular Mechanisms in Alleviating Drought Stress by *Bacillus* spp.

Despite the benefits of PGPR in alleviating drought stress to crop species (Saikia et al. 2018), key mechanisms that elicit the plant stress tolerance are yet to be well understood (Abd El-Daim et al. 2018, 2019; Lu et al. 2018; Wang et al. 2019). PGPR have been shown to play a role in plant resistance and adaptation to drought stress (Vurukonda et al. 2016), but changes at molecular levels in response to PGPR priming are not well characterized. Some of the possible molecular modes of action induced by *Bacillus* spp. in alleviating drought stresses are illustrated in Fig. 21.2. Transcriptomic analysis appears promising for revealing key pathways affected by plant-PGPR interactions (Malviya et al. 2020), and several studies have been carried out, with some candidate genes identified for potential drought stress tolerance (Abd El-Daim et al. 2018, 2019; Lu et al. 2018; Wang et al. 2019). Studies on the application of *Bacillus* spp. for drought tolerance are summarized in Table 21.2.

PGPR can trigger osmotic responses and activate genes involved in pathways that help plant survival under drought stresses (Vurukonda et al. 2016). Several *Bacillus* species that improve plant tolerance to drought stresses (Hou et al. 2018; Kasim et al. 2013; Li et al. 2019; Lim and Kim 2013; Vardharajula et al. 2011) have been studied for molecular modes of action (Table 21.3). Differential RNA and protein accumulation patterns were observed in pepper plants (*C. annuum*) inoculated with *B. licheniformis* K11 under drought stress conditions via two-dimensional

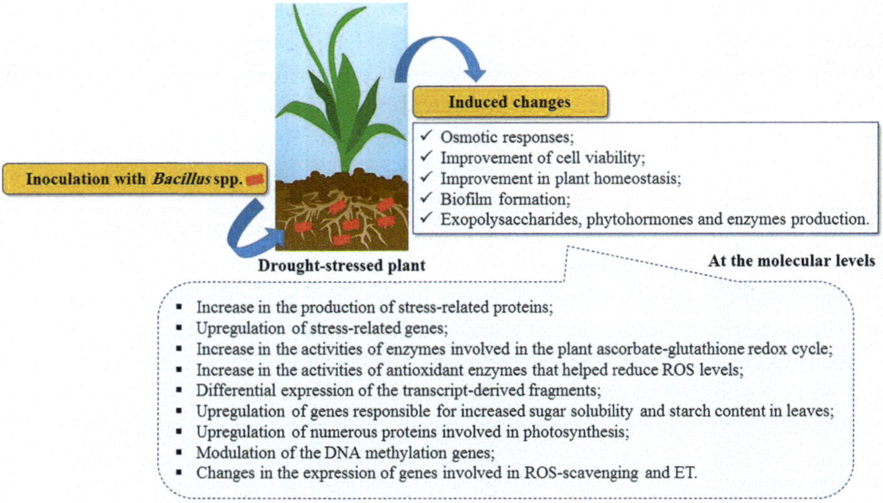

Fig. 21.2 Molecular mechanisms induced by *Bacillus* spp. to mediate drought stress tolerance in plants

Table 21.2 Examples of using *Bacillus* species for drought-stress tolerance in different plant species.

Bacillus species (strain)	Plant	Effects or putative mechanisms	References
Bacillus methylotrophicus (CSY-F1)	*Cucumber (Cucumis sativus* L.*)*	The increase of polysaccharide levels, plant relative water content and osmotic potential; Increased antioxidant enzyme activities (superoxide dismutase (SOD), catalase, guaiacol peroxidase, ascorbate peroxidase, dehydroascorbate reductase, monodehydroascorbate reductase, and glutathione reductase); Increased proline and soluble sugar ascorbate and glutathione contents levels in soil; Decreased levels of superoxide radical, hydrogen peroxide, and malonaldehyde; Increased gene expression of drought-responsive genes CsPYL1 and CsPYL2	Hou et al. (2018)
B. amyloliquefaciens SN13	Rice (*Oryza sativa* L.)	Phytohormone induction; Increased proline and total soluble sugar content	Tiwari et al. (2017)

(continued)

Table 21.2 (continued)

Bacillus species (strain)	Plant	Effects or putative mechanisms	References
B. megaterium (BOFC15)	*Arabidopsis* (*A. thaliana* (L.) Heynh.)	The production of spermidine (Spd) and a type of polyamine (PA); Increased the level of abscisic acid (ABA) and polyethylene glycol (PEG) in plants	Zhou et al. (2016)
B. subtilis (B26)	Timothy grass (*Phleum pratense* L.)	Metabolic changes; Increased root and shoot biomass; Increased the activity of photosynthetic and stomatic conductance, synthesis of chlorophylls a and b and carotenoids which increases photosynthesis, and decreasing concentration of carbohydrates and amino acids	Gagné-Bourque et al. (2016)
B. subtilis (B26)	Purple false brome *Brachypodium distachyon* (L.) P. Beauv.	Increased root and shoot weights, accelerated growth rate and seed yield, and total soluble sugars and starch contents; Upregulation of the drought- responsive gene expression (DREB2B-like, DHN3-like, and LEA-14-A-like) and modulation of the process of DNA methylation genes (MET1B-like, DRM2-like, and CMT3-like)	Gagné-Bourque et al. (2015)
B. licheniformis (Rt4M10)	Grapes (*Vitis vinifera* L. *cv. Malbec*)	Increased ABA synthesis, indole-3-acetic acid (IAA) and the gibberellins.	Salomon et al. (2014)
B. thuringiensis	Fringed lavender (*Lavandula dentate* L.)	increasing K content and IAA, by depressing stomatal conductance, and the control of shoot and proline accumulation, decreased the activity of glutathione reductase and ascorbate peroxidase which improve nutritional, physiological, and metabolic activities	Armada et al. (2014)
B. amyloliquefaciens (5113)	Wheat (*Triticum aestivum* L)	The increase of enzyme activity involved in the plant ascorbate–glutathione redox cycle; the improvement of homeostasis	Kasim et al. (2013)
B. licheniformis (K11)	Chili/pepper (*Capsicum annuum L.*	Modification of proteins and gene expression patterns and the increase of production of auxin and ACC deaminase	Lim and Kim (2013)

(continued)

Table 21.2 (continued)

Bacillus species (strain)	Plant	Effects or putative mechanisms	References
Bacillus spp.	Maize (*Zea mays* L.)	The increase of plant biomass, longer root and shoot length, relative water content, leaf water potential, root adhering soil/root tissue ratio, aggregate stability, decreasing leaf water loss; The increase of production and accumulation of osmoregulation proline, sugars, free amino acids and exopolysaccharides and decreased electrolyte leakage reduced the activity of antioxidants enzyme (catalase, glutathione peroxidase)	Vardharajula et al. (2011)
B. altitudinis	Rice	Production of aminocyclopropane-1-carboxylate (ACC) and indole acetic acid; Increased relative water content, chlorophyll stability index, and membrane stability index Increased proline content, phenolics content, catalase activity and reduced malondialdehyde (MDA)	Kumar et al. (2017)
B. subtilis RJ46	Pea and mungo bean (*Pisum sativum L.* and *Vigna mungo L.*)	Increased seed germination rate; Increased shoot and root length and biomass; Decreased in ACC accumulation and expression of the ACC oxidase gene regulating plant ethylene levels; Increased ROS-absorbing cellular enzymes and osmolytes, leaf chlorophyll content, relative water content (RWC) and intensified root recovery	Saikia et al. (2018)
B. thuringiensis	Maize	Decreased oxidative damage due to the accumulation of lipids and proline Regulate several aquaporins that are able to transport water, CO_2 and other compounds such as glycerol, urea hydrogen peroxide (H_2O_2) or boric acid	Armada et al. (2015)
B. pumilus	Chinese liquorice *Glycyrrhiza uralensis* Ledeb.)	Increased total biomass, total flavonoids, total polysaccharide, glycyrrhizic acid and antioxidant enzymes; Reduced ROS and the level of lipid peroxidation	Xie et al. (2019)
B. thuringiensis AZP2	wheat	Increased plant biomass; Reduction of stress volatile emission; Higher photosynthesis activity	Timmusk et al. (2014)

Table 21.3 Induction of drought-responsive genes in plants by *Bacillus* spp.

Bacillus species and strain	Plant	Drought responsive genes	References
B. cereus AR156, *B. subtilis* SM21*	Cucumber (*Cucumis sativa*)	Reduced expression of drought-triggered genes *cAPX, rbcL, rbcS*.	Wang et al. (2012)
B. methylotrophicus CSY-F1	Cucumber (*Cucumis sativa*)	Enhancement of transcript levels of the drought-responsive genes *CsPYL1* and *CsPYL2*.	Hou et al. (2018)
B. licheniformis K11	Pepper (*Capsicum annuum*)	Increased of stress protein genes *Cadhn, VA, sHSP* and *CaPR-10*.	Lim and Kim (2013)
B. pumilus DH-11 and *B. firmus* 40	Potato (*Solanum tuberosum*)	Putative changes in the expression of genes encoding the ROS-scavenging enzymes and ethylene biosynthesis.	Gururani et al. (2013)
B. amyloliquefaciens 5113*	Wheat (*Triticum aestivum*)	Upregulation of stress related genes *APX1, SAMS1*, and *HSP17.8*.	Kasim et al. (2013)
B. amyloliquefaciens 5113	Wheat (*Triticum aestivum*)	Upregulation of stress related genes *IA20* and *IA24* which are homologous to abscisic acid (ABA) responsive proteins.	Abd El-Daim et al. (2018)
B. amyloliquefaciens 54	Tomato (*Solanum lycopersicum*)	Increase in the expression levels of stress-responsive genes such as lea, tdi65, and ltpg2.	Wang et al. (2019)
B. amyloliquefaciens FZB42	Arabidopsis (*Arabidopsis thaliana*)	Increase in the relative expression levels of drought defense-related marker genes such as *RD29A, RD17, ERD1*, and *LEA14*.	Lu et al. (2018)
B. velezensis 5113	Wheat (*Triticum aestivum*)	Upregulation of proteins ID 194, 192 and 218 involved in photosynthesis process.	Abd El-Daim et al. (2019)
B. subtilis LDR2	Wheat (*Triticum aestivum*)	Upregulation of *TaCTR1* gene-a regulatory component of ethylene signaling pathway- and the increase in the expression of *TaDREB2* gene encoding a transcription factor which is crucial for improving the tolerance of plants to abiotic stress conditions, including drought.	Barnawal et al. (2017)
B. subtilis B26	Stiff brome (*Brachypodium distachyon* Bd21)	Upregulation of the drought-response genes, *DREB2B*-like, *DHN3*-like and *LEA-14-A*-like and the modulation of the DNA methylation genes, *MET1B*-like, *CMT3*-like and *DRM2*-like.	Gagné-Bourque et al. (2015)

**B. cereus* AR156, *B. subtilis* SM21 and *B. amyloliquefaciens* 5113 have been tested in mixture with *Serratia* sp. XY21 and *Azospirillum brasilense* NO40, respectively.

electrophoresis (2D-PAGE) and differential-display PCR (DD-PCR), respectively. Six differentially accumulated stress-related proteins were identified in the inoculated plants, among which Cadhn, VA, sHSP, and CaPR-10 showed more than 1.5-fold increases over those in non-inoculated plants. These changes were attributed to the increases in auxins and ACC deaminase in inoculated plants (Lim and Kim 2013). The use of quantitative reverse transcriptase-polymerase chain reaction (RT-qPCR) discovered the upregulation of stress-related genes APX1, SAMS1, and HSP17.8 in the leaves of wheat and increased activities of enzymes involved in the plant ascorbate-glutathione redox cycle in plants treated with *B. amyloliquefaciens* 5113 and *Azospirillum brasilense* NO40. The treatment showed an attenuation in transcript levels that indicates an improvement in plant homeostasis. In other words, the treatment resulted in increased activities of antioxidant enzymes that helped reduce ROS levels in drought-stressed plants (Kasim et al. 2013). Also on wheat, these molecular changes caused by the treatment of *B. amyloliquefaciens* 5113 helped extend the water holding in plants for 5 additional days, transcript profiling based on cDNA-amplified fragment length polymorphism (cDNA-AFLP) showed an upregulation of numerous genes in the leaves under drought stress condition, while the majority of these genes were downregulated in control plants. Differential expression of the transcript-derived fragments (TDFs) IA20 and IA24 in the leaves of treated wheat seedlings was also observed; functional studies by virus-induced gene silencing (VIGS) suggested that IA20 and IA24 silencing can compromise the effect of *B. amyloliquefaciens* 5113 treatment in mediating the drought tolerance (Abd El-Daim et al. 2018). Gagné-Bourque et al. (2015) have demonstrated that *B. subtilis* (B26 strain) confers drought stress resistance through the upregulation of genes responsible for increased sugar solubility and starch content in leaves, and the production of phytohormones, especially the solubilization of phosphorous and production of IAA, cytokinin, and zeatin riboside.

The mechanisms of another *B. amyloliquefaciens* strain (54) in drought stress relief were further elucidated by analyzing the expression of several stress-related genes in tomato plants (*Lycopersicon esculentum* Mill.); the relative transcript levels of *lea*, *tdi65*, and *ltpg2* genes increased under drought stress conditions, based on RT-qPCR (Wang et al. 2019). Biofilm formation also stimulated helpful to the overexpression of these genes and the drought stress tolerance. The *B. amyloliquefaciens* FZB42 strain induced systemic drought tolerance in *Arabidopsis*, with the involvement of RD29A, RD17, ERD1, and LEA14, genes although the underlying molecular processes associated with the improvement of cell viability under the stress still need to be elucidated (Lu et al. 2018). It seems that the bacterial treatment affects phytohormone-mediated pathways, especially those influenced by ET and jasmonic acid (JA) in the induction of drought tolerance. Exopolysaccharides produced by the bacteria also appeared beneficial in protecting the plants from drought stress, based on the root length and plant branching, biomass, and survival rate. It was hypothesized that exopolysaccharides may comprise a microbe-associated molecular pattern that, once recognized by plants, induces drought tolerance (Lu et al. 2018). Inoculation of wheat with the *B. velezensis* strain 5113 induced metabolic modulation and the abundance of several proteins in leaves, which resulted in increased drought stress tolerance (Abd El-Daim et al. 2019). Proteomic profiling of leaves from treated plants showed differential accumulation of numerous proteins involved in photosynthesis, including the ID 194, 192, and 218 proteins which were downregulated in control plants under drought. This suggests that *B. velezensis* 5113 provides the protection of photosynthesis in stressed plants through adjusting the expression of several proteins pertaining to stress defense, as well as energy supply (Abd El-Daim et al. 2019).

The molecular effect of *B. subtilis* B26 on the full life cycle of Purple false brome grass (*Brachypodium distachyon* L.), a well-established model species used for functional genomics in cereal crops and temperate grasses, has been studied, and both seedlings and mature plants inoculated with the bacterial inoculum showed reduced phenotypic effect under acute and chronic drought stress via the upregulation of drought response genes, including DHN3-like and LEA-14-A-like genes and the transcription factor that modulates the DREB2B-like gene, which binds dehydration-responsive elements (Gagné-Bourque et al. 2015). The modulation of the DNA methylation genes, i.e., MET1B-like, CMT3-like, and DRM2-like, involved in the process regulation is also involved in drought tolerance mediated by *B. subtilis* B26 (Gagné-Bourque et al. 2015). Another strain of *B. subtilis* (LDR2), when applied with other PGPR, enhanced the drought stress tolerance in wheat by modulating the regulatory component of ET signaling pathway, with increased expression of TaDREB2 gene. It is worth pointing out that this gene encodes a transcription factor crucial to improved plant tolerance to abiotic stresses (Barnawal et al. 2017). Treating cucumber plants (*Cucumis sativus* L) with a consortium of three PGPR strains, i.e., *B. cereus* AR156, *B. subtilis* SM21, and *Serratia* sp. XY21, produced a range of physiological indicators relevant to drought tolerance, without involving ACC deaminase (Wang et al. 2012). Drought tolerance was achieved via multiple mechanisms that help protect leaf cell membranes from being damaged by reduced water availability. The reduced expression of drought-triggered genes such as cAPX, rbcL, and rbcS that encode cytosolic ascorbate peroxidase and ribulose-1,5-bisphosphate carboxy/oxygenase (Rubisco) subunits was among those mechanisms (Wang et al. 2012). Changes in the expression of genes involved in ROS-scavenging and ET biosynthesis have been identified following the inoculation of potato plants with *B. pumilus* DH-11 and *B. firmus* 40 under drought stress conditions (Gururani et al. 2013).

21.7 Mass Production, Formulation, and Applications of *Bacillus* spp.

So far, not many PGPR have been commercialized, mainly due to their poor shelf lives and resilience under field conditions (Gotor-Vila et al. 2017; Radhakrishnan et al. 2017). *Bacillus* spp. can be a good candidate for commercial applications because this genus of bacterium forms endospores that resist adverse environmental conditions and are able to maintain the stability and efficacy throughout the commercial process (Mari et al. 2014; Stojanović et al. 2019). Efficient mass production and effective formulation are crucial factors for the viable development of bacterial inoculants (Lobo et al. 2019; Schisler et al. 2004). Optimized medium composition and fermentation conditions are often the first step for stable and efficacious bacterial products (Schisler et al. 2004).

Bioformulation begins with microbial fermentation to obtain cell biomass with desired density and sporulation efficiency (Stojanović et al. 2019). Standard medium such as nutrient broth or Luria Bertani (LB) is commonly used to start inoculum cultures before they are transferred to an industrial growth medium consisting of sources of carbon, nitrogen, inorganic salts, and other ingredients needed for bacterial growth. The onset of sporulation is often initiated or triggered by a reduction in glucose concentration after the exponential phase of growth (Stojanović et al. 2019). The optimal temperature for fermentation is between 30 °C and 37 °C, depending on the strain, with pH ranging from six to seven for most bacteria. Increasing the concentration of available oxygen can sometimes enhance cell numbers, sporulation, and the yield of metabolites (Suresh et al. 2009), as members of the genus *Bacillus* are aerobic or facultative anaerobic bacteria. Agitation during fermentation helps increase the amount of oxygen dissolved in liquid media but may affect the sporulation negatively, as higher agitation rate may result in shear stress, which may reduce the sporulation. For example, an agitation rate of 200 rpm was recommended for efficient sporulation of *B. amyloliquefaciens* in liquid fermentation (Tzeng et al. 2008). The scaled-up fermentation and formulation may use dry and liquid processes (Schisler et al. 2004), and the products can be wettable powders, dusts granules, cell suspensions in water, oil, or emulsions (Schisler et al. 2004). Additives and carriers (fillers, extenders) often are added during the mass fermentation and formulation, and these materials should be cost efficient and easy to handle. Carriers are used also to enhance the chemical, physical, and/or nutritional properties of formulated products; additives can act as stimulant to enhance the product efficacy (Schisler et al. 2004). Commonly used carriers include talc, coal, peat, clays, vermiculite, perlite, and plant waste materials (Tripti et al. 2017). Additional substances in carriers may include polymers, sugars, milk, honey, polyols, amino acids, albumin, vegetable oils, pero-dexin (a by-product of the starch industry), and coconut water (an industrial waste) (Gotor-Vila et al. 2017). Waste proteins have been used as additives in the solid-state fermentation of *B. amyloliquefaciens* as a bio-organic fertilizer (Huang et al. 2015). With a low water content, dry formulations generally allow PGPR to survive longer and tolerate higher temperatures better than in a liquid. Several methods may be used for drying, including air drying, desiccation, freeze drying (lyophilization), spray drying, and shade drying, with freeze drying being used most widely in industrial scales. Freeze drying and spray drying are more expensive than other drying methods due to the need for specialized equipment (Gotor-Vila et al. 2017; Lobo et al. 2019). In contrast, liquid formulations can be used directly without rehydration (Lobo et al. 2019). To meet the quality standards, the inoculum concentration often is required in the range of 10^7 to 10^9 colony-forming units (CFU) per mL (Malusá and Vassilev 2014). Table 21.4 lists several examples of liquid and dry formulations from recent publications.

Table 21.4 Selected examples of *Bacillus*-based formulations from recent publications

Microorganism	Carrier/ additive	Drying method	Number of viable cells	Shelf life	Assayed crop	Application	References
<u>Liquid formulation</u>							
Bacillus siamensis SCFB3-1	carrageenan	-	10^9 CFU/ml	-	Sweet pepper (*Capsicum annuum*)	biofertilizer	Pastor-Bueis et al. (2017)
B. polymyxa, *B. macerans*, *B. circulans*		-	10^8 CFU/ml	-	Potato (*Solanum tuberosum*)	biofertilizer, biocontrol	Abbas et al. (2014)
<u>dry formulation</u>							
Bacillus sp. A30	biochar, flyash	-	1.35×10^8 CFU/g (biochar); 10^7 CFU/g (flyash)	240 days at 27°C	Tomato (*Lycopersicon esculentum*)	biofertilizer	Tripti et al. (2017)
B. endophyticus 77-NS2, *B. sphaericus* 77-CS-S1, *B. safensis* PSB5, *B. megaterium* PSB12	biogas sludge, soil	-	10^6-10^8 CFU/g	-	Wheat (*Triticum aestivium*)	biofertilizer	Mukhtar et al. (2017)
B. amyloliquefaciens FZB42	alginate, gum arabic, talc	freeze-drying	1.3×10^7 CFU/g	-	Maize (*Zea mays*)	biocontrol	Berninger et al. (2016)
B. cereus B25	talc	55°C for 36 hours	10^9 CFU/g	360 days at room temperature	Maize	biocontrol	Martínez-Álvarez et al. (2016)
B. subtilis SB-MYP-1	soybean flour	freeze-drying	1.3×10^8 CFU/g	-	-	-	Mahidsanan et al. (2017)
Bacillus sp. CaB85	talc	shade air drying	9×10^8 CFU/ml in culture	45 days	Cowpea (*Vigna unguiculata*), lady's finger (*Abelmoschus esculentus*)	biofertilizer	Basheer et al. (2019)

CFU: colony forming units

Bacillus spp.-based products are among the most promising biofertilizers, biopesticides, and bioremediators for agro-industrial applications. These products often are positioned as environmentally friendly alternatives to conventional agrochemicals. Biofertilizers are promoted to improve plant nutritional status by either increasing soil nutrients or making them more accessible to plants (Malusá and Vassilev 2014). *Bacillus* spp.-based biofertilizers can increase the intake of nitrogen and phosphorus by plants; the first bacterial fertilizer Alinit was developed from *Bacillus* spp. in 1897 by Bayer AG, which enhanced crop yield by 40% (Kilian et al. 2000). Another biofertilizer based on the *Bacillus* sp. strain A30 promoted the growth and yield of tomato (*L. esculentum*) (Tripti et al. 2017). Biopesticides (also called bioprotectants) based on *Bacillus* spp. may promote plant growth by controlling diseases and pests (Malusá and Vassilev 2014). Be present in rhizosphere in a large quantity, *Bacillus* spp. may stimulate plants to release a wide range of metabolites, including plant hormones, siderophores, cyanides, and antibiotics (Stojanović et al. 2019), as well as enzymes that degrade fungal cell walls (Hashem et al. 2019). *B. amyloliquefaciens* (SF14 and SP10 strains) controlled brown rot disease caused by *Monilinia* spp. on apple by reducing the mycelial growth and inhibiting spore germination of pathogen (Lahlali et al. 2020). Furthermore, *Bacillus* spp.-mediated bioremediation of environmental pollutants has been shown to break down contaminants in soils (Stojanović et al. 2019). A mixture of bacterial suspension consisting of *B. pumilus* and *B. cereus* degraded up to 84% of total petroleum hydrocarbons in contaminated soils (Patowary et al. 2016).

21.8 Conclusion and Future Prospects

Drought is a destructive abiotic stress that significantly influences crop yield and life span by inducing negative morphological, physiological, biochemical, molecular, and genetic changes. These changes often coincide with the production of free radicals, decrease in chlorophyll, and increase in the biosynthesis of ET membrane lipid peroxidation. Plant rhizosphere has intense microbial activities with a wealth of exuded metabolites. *Bacillus* spp. are frequently attracted to the rhizosphere, resulting in the biofilm formation. The root colonization by *Bacillus* spp. is mutually beneficial, and the bacteria have often been observed to assist plant survival and growth under stress conditions by inducing a range of stress tolerance mechanisms in plants, including stress signaling, improved regulation of water transport, efficient uptake of nutrients from soil, and activation of antioxidants and plant defense systems. Many of these pathways are triggered by the production of phytohormones and volatile compounds (VOCs)-related metabolites, activity of ACC deaminase, accumulation of osmoprotectants and exopolysaccharides, and the regulation of stress-responsive genes in response to drought stresses. The research on *Bacillus* spp. for stress mitigation has only scratched the surface, and more can be done to develop a variety of viable products for novel options in addressing drought and other environmental stresses.

New studies to further decipher the mechanisms employed by *Bacillus* spp. to mitigate the impact of drought stress will be valuable to develop sustainable strategies to cope with climate changes. Nowadays, research on the involved mechanisms is in progress but remains largely speculative. In this context, a combination of proteomics and metagenomics (high-throughput DNA sequencing), metabolomics, and advancing computational data-mining approaches might be useful to understand bacteria belonging to *Bacillus* genus and their interactions with their host plants and to reveal more information on mechanisms underlying tolerance to drought stress. Similarly, it is important to conduct a deep analysis of both plant and *Bacillus* genes expressed during drought stress to increase our knowledge on how the inoculation with *Bacillus* spp. confers tolerance to stress in several plant species.

Although the research mentioned here highlights the effectiveness of *Bacillus* spp. to alleviate the impacts of drought stress in agricultural crop plants, more systemic studies seem to be needed. One weakness of the research described here is the use of sterilized soils to study the role of *Bacillus* spp. in drought stress alleviating. In other words, it is difficult to take results achieved through the use of sterilized soils and to generalize them for "real" field conditions. The applicability of these results to real conditions is not straightforward. Furthermore, as it has already been raised by Ngumbi and Kloepper (2016) for PGPR, the diversity of physical and chemical characteristics of soils, which may be impacted by both drought stress and the PGPR and which may also further impact beneficial microorganisms, need to be taken into account when studying the role of *Bacillus* spp. in alleviating drought stress. Since under natural conditions, drought often occurs with other biotic and abiotic stresses; further studies are needed to address the role of *Bacillus* spp. to simultaneously mitigate numerous stresses. Biotic stressors such as pathogens and insects are among the most important stresses damaging crops. Since numerous *Bacillus* spp. showed effectiveness to suppress phytopathogens (Deng et al. 2011; Gowtham et al. 2018; Khan et al. 2020; Kushwaha et al. 2020; Lahlali et al. 2020), the identification of *Bacillus* spp. strains with potential to confer tolerance to both drought and biotic stress would be of great interest. Furthermore, the comprehension of the basic mechanisms behind ISR and those behind the observed *Bacillus*-mediated drought tolerance overlap would be of great benefit for future applications.

References

Abbas MT, Hamza MA, Youssef HH, Youssef GH, Fayez M, Monib M, Hegazi NA (2014) Biopreparates support the productivity of potato plants grown under desert farming conditions of north Sinai: Five years of field trials. *Journal of Advanced Research, 5*(1), 41–48. https://doi.org/10.1016/j.jare.2012.11.004

Abd El-Daim IA, Bejai S, Fridborg I, Meijer J (2018) Identifying potential molecular factors involved in Bacillus amyloliquefaciens 5113 mediated abiotic stress tolerance in wheat. Plant Biol 20:271–279. https://doi.org/10.1111/plb.12680

Abd El-Daim IA, Bejai S, Meijer J (2019) Bacillus velezensis 5113 induced metabolic and molecular reprogramming during abiotic stress tolerance in wheat. Sci Rep 9:1–18. https://doi.org/10.1038/s41598-019-52567-x

Afzal I, Shinwari ZK, Sikandar S, Shahzad S (2019) Plant beneficial endophytic bacteria: mechanisms, diversity, host range and genetic determinants. Microbiol Res 221:36–49. https://doi.org/10.1016/j.micres.2019.02.001

Ahmadi A, Baker DA (2001) The effect of water stress on the activities of key regulatory enzymes of the sucrose to starch pathway in wheat. Plant Growth Regul 35:81–91

Alami Y, Achouak W, Marol C, Heulin T (2000) Rhizosphere soil aggregation and plant growth promotion of sunflowers by an exopolysaccharide-producing rhizobium sp. strain isolated from sunflower roots. Appl Environ Microbiol 66:3393–3398. https://doi.org/10.1128/AEM.66.8.3393-3398.2000

Allard-massicotte R, Tessier L, Lécuyer F, Lakshmanan V, Lucier JF, Garneau D, Caudwell L, Vlamakis H, Bais HP, Beauregard PB (2016) Bacillus subtilis early colonization of Arabidopsis thaliana roots involves multiple chemotaxis receptors. MBio 7:1–10. https://doi.org/10.1128/mBio.01664-16.Invited

Arif I, Batool M, Schenk PM (2020) Plant microbiome engineering: expected benefits for improved crop growth and resilience. Trends Biotechnol 1–12. https://doi.org/10.1016/j.tibtech.2020.04.015

Arkhipova TN, Prinsen E, Veselov SU, Martinenko EV, Melentiev AI, Kudoyarova GR (2007) Cytokinin producing bacteria enhance plant growth in drying soil. Plant Soil 292:305–315. https://doi.org/10.1007/s11104-007-9233-5

Armada E, Roldán A, Azcon R (2014) Differential activity of autochthonous bacteria in controlling drought stress in native Lavandula and Salvia plants species under drought conditions in natural arid soil. Microb Ecol 67:410–420. https://doi.org/10.1007/s00248-013-0326-9

Armada E, Azcon R, Lopez-Castillo OM, Calvo-Polanco M, Ruiz-Lozano JM (2015) Autochthonous arbuscular mycorrhizal fungi and Bacillus thuringiensis from a degraded Mediterranean area can be used to improve physiological traits and performance of a plant of agronomic interest under drought conditions. Plant Physiol Biochem J 90:64–74. https://doi.org/10.1016/j.plaphy.2015.03.004

Atteya A (2003) Alteration of water relations and yield of corn genotypes in response to drought stress. Bulg J Plant Physiol 29:63–76

Balla K, Rakszegi M, Li Z, Békés F, Bencze S, Veisz O (2011) Quality of winter wheat in relation to heat and drought shock after anthesis. Czech J Food Sci 29:117–128

Barnawal D, Bharti N, Pandey SS, Pandey A, Chanotiya CS, Kalra A (2017) Plant growth promoting rhizobacteria enhances wheat salt and drought stress tolerance by altering endogenous phytohormone levels and TaCTR1/TaDREB2 expression. Physiol Plant 161:502–514. https://doi.org/10.1111/ppl.12614

Basheer J, Ravi A, Mathew J, Krishnankutty RE (2019) Assessment of plant-probiotic performance of novel rndophytic Bacillus sp. in talc-based formulation. *Probiotics and Antimicrobial Proteins, 11*(1), 256–263. https://doi.org/10.1007/s12602-018-9386-y

Basile B, Rouphael Y, Colla G, Soppelsa S, Andreotti C (2020) Appraisal of emerging crop management opportunities in fruit trees, grapevines and berry crops facilitated by the application of biostimulants. Sci Hortic (Amsterdam) 267:109330. https://doi.org/10.1016/j.scienta.2020.109330

Beauregard PB, Chai Y, Vlamakis H, Losick R, Kolter R (2013) Bacillus subtilis biofilm induction by plant polysaccharides. Proc Natl Acad Sci U S A 110:1621–1630. https://doi.org/10.1073/pnas.1218984110

Berninger T, Mitter B, Preininger C (2016) The smaller, the better? The size effect of alginate beads carrying plant growth-promoting bacteria for seed coating. Journal of Microencapsulation, *33*(2), 127–136. https://doi.org/10.3109/02652048.2015.1134690

Bhatt K, Maheshwari DK (2020) Zinc solubilizing bacteria (Bacillus megaterium) with multifarious plant growth promoting activities alleviates growth in Capsicum annuum L. 3. Biotech 10:1–11. https://doi.org/10.1007/s13205-019-2033-9

Bleecker AB, Kende H (2000) Ethylene: a gaseous signal molecule in plants. Annu Rev Cell Dev Biol 16:1–18. https://doi.org/10.1146/annurev.cellbio.16.1.1

Carlson R, Tugizimana F, Steenkamp PA, Dubery IA, Hassen AI, Labuschagne N (2020) Rhizobacteria-induced systemic tolerance against drought stress in Sorghum bicolor (L.) Moench. Microbiol Res 232:126388. https://doi.org/10.1016/j.micres.2019.126388

Chakraborty U, Chakraborty BN, Chakraborty AP, Dey PL (2013) Water stress amelioration and plant growth promotion in wheat plants by osmotic stress tolerant bacteria. World J Microbiol Biotechnol 29:789–803. https://doi.org/10.1007/s11274-012-1234-8

Chari KD, Reddy RS, Triveni S, Trimurtulu N, Rani CVD, Sreedhar M (2018) Isolation and characterization of abiotic stress tolerant plant growth promoting Bacillus spp. from different rhizospheric soils of Telangana. Biosci Biotech Res Asia 2018:15(2)

Chen Y, Cao S, Chai Y, Clardy J, Kolter R, Guo JH, Losick R (2012) A Bacillus subtilis sensor kinase involved in triggering biofilm formation on the roots of tomato plants. Mol Microbiol 85:418–430. https://doi.org/10.1111/j.1365-2958.2012.08109.x

Dahmani MA, Desrut A, Moumen B, Verdon J, Mermouri L, Kacem M, Coutos-Thévenot P, Kaid-Harche M, Bergès T, Vriet C (2020) Unearthing the plant growth-promoting traits of bacillus megaterium RmBm31, an endophytic bacterium isolated from root nodules of Retama monosperma. Front Plant Sci 11:1–15. https://doi.org/10.3389/fpls.2020.00124

Damodara Chari K, Subhas Reddy R, Triveni S, Trimurtulu N, Durga Rani V, Sreedhar M (2018) Screening and characterization of stress tolerant rhizobia from different soils of Telangana. Int J Microbiol Res 10:1038. https://doi.org/10.9735/0975-5276.10.3.1038-1042

Deng Y, Zhu Y, Wang P, Zhu L, Zheng J, Li R, Ruan L, Peng D, Sun M (2011) Complete genome sequence of Bacillus subtilis BSn5, an endophytic bacterium of Amorphophallus konjac with antimicrobial activity for the plant pathogen Erwinia carotovora subsp. carotovora. J Bacteriol 193:2070–2071. https://doi.org/10.1128/JB.00129-11

Deng Y, Chen H, Li C, Xu J, Qi Q, Xu Y, Zhu Y, Zheng J, Peng D, Ruan L, Sun M (2019) Endophyte Bacillus subtilis evade plant defense by producing lantibiotic subtilomycin to mask self-produced flagellin. Commun Biol 2:1–12. https://doi.org/10.1038/s42003-019-0614-0

Disante KB, Fuentes D, Cortina J (2011) Response to drought of Zn-stressed Quercus suber L. seedlings. Environ Exp Bot 70:96–103. https://doi.org/10.1016/j.envexpbot.2010.08.008

Edmeade GO (2008) Drought tolerance in maize: An emerging reality. A feature. In: James, Clive. 2008. Global Status of Commercialized Biotech/GM Crops: 2008. ISAAA Brief No. 39. ISAAA: Ithaca, NY

Es-Soufi R, Tahiri H, El Oualkadi A, Azaroual L, Martin P, Badoc A, Lamarti A (2020) Evaluation of plant growth promoting ability of Bacillus amyloliquefaciens Bc2 and Trichoderma harzianum TR in vivo. Agric Sci 11:247–259. https://doi.org/10.4236/as.2020.113016

Estrada-Campuzano G, Miralles DJ, Slafer GA (2008) Genotypic variability and response to water stress of pre- and post-anthesis phases in triticale. Eur J Agron 28:171–177. https://doi.org/10.1016/j.eja.2007.07.005

Fernández Bidondo L, Bompadre J, Pergola M, Silvani V, Colombo R, Bracamonte F, Godeas A (2012) Differential interaction between two glomus intraradices strains and a phosphate solubilizing bacterium in maize rhizosphere. Pedobiologia (Jena) 55:227–232. https://doi.org/10.1016/j.pedobi.2012.04.001

Forni C, Duca D, Glick BR (2016) Mechanisms of plant response to salt and drought stress and their alteration by rhizobacteria. Plant Soil. https://doi.org/10.1007/s11104-016-3007-x

Gagné-Bourque F, Mayer BF, Charron JB, Vali H, Bertrand A, Jabaji S (2015) Accelerated growth rate and increased drought stress resilience of the model grass Brachypodium distachyon colonized by Bacillus subtilis B26. PLoS One 10:1–23. https://doi.org/10.1371/journal.pone.0130456

Gagné-Bourque F, Bertrand A, Claessens A, Aliferis KA, Jabaji S (2016) Alleviation of drought stress and metabolic changes in Timothy (Phleum pratense L.) colonized with Bacillus subtilis B26. Front Plant Sci 7:1–16. https://doi.org/10.3389/fpls.2016.00584

Goswami M, Deka S (2020) Plant growth-promoting rhizobacteria—alleviators of abiotic stresses in soil: a review. Pedosphere 30:40–61. https://doi.org/10.1016/S1002-0160(19)60839-8

Goswami D, Thakker JN, Dhandhukia PC (2016) Portraying mechanics of plant growth promoting rhizobacteria (PGPR): a review. Cogent Food Agric 2016(2):1–19. https://doi.org/10.1080/23311932.2015.1127500

Gotor-Vila A, Usall J, Torres R, Abadias M, Teixidó N (2017) Formulation of the biocontrol agent Bacillus amyloliquefaciens CPA-8 using different approaches: liquid, freeze-drying and fluid-bed spray-drying. BioControl 62:545–555. https://doi.org/10.1007/s10526-017-9802-3

Gowtham HG, Murali M, Singh SB, Lakshmeesha TR, Narasimha Murthy K, Amruthesh KN, Niranjana SR (2018) Plant growth promoting rhizobacteria Bacillus amyloliquefaciens improves plant growth and induces resistance in chilli against anthracnose disease. Biol Control 126:209–217. https://doi.org/10.1016/j.biocontrol.2018.05.022

Grover M, Madhubala R, Ali SZ, Yadav SK, Venkateswarlu B (2014) Influence of Bacillus spp. strains on seedling growth and physiological parameters of sorghum under moisture stress conditions. J Basic Microbiol 54:951–961. https://doi.org/10.1002/jobm.201300250

Guoju X, Fengju Z, Juying H, Chengke L, Jing W, Fei M, Yubi Y, Runyuan W, Zhengji Q (2016) Response of bean cultures' water use efficiency against climate warming in semiarid regions of China. Agric Water Manag 173:84–90. https://doi.org/10.1016/j.agwat.2016.05.010

Gururani MA, Upadhyaya CP, Baskar V, Venkatesh J, Nookaraju A, Park SW (2013) Plant growth-promoting Rhizobacteria enhance abiotic stress tolerance in Solanum tuberosum through inducing changes in the expression of ROS-scavenging enzymes and improved photosynthetic performance. J Plant Growth Regul 32:245–258. https://doi.org/10.1007/s00344-012-9292-6

Hacquard S, Spaepen S, Garrido-Oter R, Schulze-Lefert P (2017) Interplay between innate immunity and the plant microbiota. Annu Rev Phytopathol 55:565–589. https://doi.org/10.1146/annurev-phyto-080516-035623

Hashem A, Kumar A, Al-Dbass AM, Alqarawi AA, Al-Arjani A-BF, Singh G, Farooq M, Abd-Allah EF (2019) Arbuscular mycorrhizal fungi and biochar improves drought tolerance in chickpea. Saudi J Biol Sci 26:614–624

Hou X, Wu F, Wang X-J, Sun Z-T, Zhang Y, Yang M-T, Bai H, Li S, Bai J-G (2018) Bacillus methylotrophicus CSY-F1 alleviates drought stress in cucumber (Cucumis sativus) grown in soil with high ferulic acid levels. Plant Soil 431:89–105. https://doi.org/10.1007/s11104-018-3748-9

Huang Y, Sun L, Zhao J, Huang R, Li R, Shen Q (2015) Utilization of different waste proteins to create a novel PGPR-containing bio-organic fertilizer. Sci Rep 5:1–11. https://doi.org/10.1038/srep07766

Javid MG, Sorooshzadeh A, Moradi F, Modarres Sanavy SAM, Allahdadi I (2011) The role of phytohormones in alleviating salt stress in crop plants. Aust J Crop Sci 5:726

Ji X, Shiran B, Wan J, Lewis DC, Jenkins CLD, Condon AG, Richards RA, Dolferus R (2010) Importance of pre-anthesis anther sink strength for maintenance of grain number during reproductive stage water stress in wheat. Plant Cell Environ 33:926–942. https://doi.org/10.1111/j.1365-3040.2010.02130.x

Jumali SS, Said IM, Ismail I, Zainal Z (2011) Genes induced by high concentration of salicylic acid in Mitragyna speciosa. Aust J Crop Sci 5:296–303

Kang SM, Radhakrishnan R, Lee KE, You YH, Ko JH, Kim JH, Lee IJ (2015) Mechanism of plant growth promotion elicited by Bacillus sp. LKE15 in oriental melon. Acta Agric. Scand Sect B Soil Plant Sci 65:637–647. https://doi.org/10.1080/09064710.2015.1040830

Kasim WA, Osman ME, Omar MN, Abd El-Daim IA, Bejai S, Meijer J (2013) Control of drought stress in wheat using plant-growth-promoting bacteria. J Plant Growth Regul 32:122–130. https://doi.org/10.1007/s00344-012-9283-7

Kaur G, Asthir B (2017) Molecular responses to drought stress in plants. Biol Plant 61:201–209. https://doi.org/10.1007/s10535-016-0700-9

Kaushal M, Wani SP (2016) Plant-growth-promoting rhizobacteria: drought stress alleviators to ameliorate crop production in drylands. Ann Microbiol 66:35–42. https://doi.org/10.1007/s13213-015-1112-3

Khan MS, Gao J, Zhang M, Chen X, Moe TS, Du Y, Yang F, Xue J, Zhang X (2020) Isolation and characterization of plant growth-promoting endophytic bacteria Bacillus stratosphericus LW-03 from Lilium wardii. 3. Biotech 10:1–15. https://doi.org/10.1007/s13205-020-02294-2

Kilian M, Steiner U, Krebs B, Junge H, Schmiedeknecht G, Hain R (2000) Bacillus subtilis – mode of action of a microbial agent enhancing plant vitality. Pflanzenschutz-Nachrichten Bayer 1:72–93

Kogut M, Russell NJ (1987) Life at the limits: considerations on how bacteria can grow at extremes of temperature and pressure, or with high concentrations of ions and solutes. Sci Prog 71:381–399

Krasensky J, Jonak C (2012) Drought, salt, and temperature stress-induced metabolic rearrangements and regulatory networks. J Exp Bot 63:1593–1608. https://doi.org/10.1093/jxb/err460

Kumar A, Dubey A (2020) Rhizosphere microbiome: engineering bacterial competitiveness for enhancing crop production. J Adv Res 24:337–352. https://doi.org/10.1016/j.jare.2020.04.014

Kumar A, Verma JP (2018) Does plant—microbe interaction confer stress tolerance in plants: a review? Microbiol Res 207:41–52. https://doi.org/10.1016/j.micres.2017.11.004

Kumar AS, Sridar R, Uthandi S (2017) Mitigation of drought in rice by a phyllosphere bacterium Bacillus altitudinis FD48. African J Microbiol Res 11:1614–1625. https://doi.org/10.5897/AJMR2017.8610

Kunert K, Vorster BJ (2020) In search for drought-tolerant soybean: is the slow-wilting phenotype more than just a curiosity? J Exp Bot 71:457

Kushwaha P, Kashyap PL, Srivastava AK, Tiwari RK (2020) Plant growth promoting and antifungal activity in endophytic Bacillus strains from pearl millet (Pennisetum glaucum). Brazilian J Microbiol 51:229–241. https://doi.org/10.1007/s42770-019-00172-5

Lafitte HR, Yongsheng G, Yan S, Li Z-K (2007) Whole plant responses, key processes, and adaptation to drought stress: the case of rice. J Exp Bot 58(2):169–175

Lahlali R, Aksissou W, Lyousfi N, Ezrari S, Blenzar A, Tahiri A, Ennahli S, Hrustić J, MacLean D, Amiri S (2020) Biocontrol activity and putative mechanism of Bacillus amyloliquefaciens (SF14 and SP10), Alcaligenes faecalis ACBC1, and Pantoea agglomerans ACBP1 against brown rot disease of fruit. Microb Pathog 139:103914. https://doi.org/10.1016/j.micpath.2019.103914

Lastochkina O (2019) Bacillus subtilis- mediated abiotic stress tolerance in plants. In: Islam MT et al (eds) Bacilli and Agrobiotechnology: Phytostimulation and biocontrol, Bacilli in climate resilient agriculture and bioprospecting. Springer, Cham

Lesk C, Rowhani P, Ramankutty N (2016) Influence of extreme weather disasters on global crop production. Nature 529:84–87. https://doi.org/10.1038/nature16467

Li H, Zhao Y, Jiang X (2019) Seed soaking with Bacillus sp. strain HX-2 alleviates negative effects of drought stress on maize seedlings. Chil J Agric Res 79:396–404. https://doi.org/10.4067/S0718-58392019000300396

Lim J-H, Kim S-D (2013) Induction of drought stress resistance by multi-functional PGPR Bacillus licheniformis K11 in pepper. Plant Pathol J 29:201–208. https://doi.org/10.5423/PPJ.SI.02.2013.0021

Liu X, Zhang H (2015) The effects of bacterial volatile emissions on plant abiotic stress tolerance. Front Plant Sci 6:1–6. https://doi.org/10.3389/fpls.2015.00774

Liu F, Xing S, Ma H, Du Z, Ma B (2013) Cytokinin-producing, plant growth-promoting rhizobacteria that confer resistance to drought stress in Platycladus orientalis container seedlings. Appl Microbiol Biotechnol 97:9155–9164. https://doi.org/10.1007/s00253-013-5193-2

Lobo CB, Juárez Tomás MS, Viruel E, Ferrero MA, Lucca ME (2019) Development of low-cost formulations of plant growth-promoting bacteria to be used as inoculants in beneficial agricultural technologies. Microbiol Res 219:12–25. https://doi.org/10.1016/j.micres.2018.10.012

Lopes, R., Tsui, S., Gonçalves, P.J.R.O., de Queiroz, M.V., 2018. A look into a multifunctional toolbox: endophytic Bacillus species provide broad and underexploited benefits for plants. World J Microbiol Biotechnol 34, 0. doi:https://doi.org/10.1007/s11274-018-2479-7

Lu X, Liu SF, Yue L, Zhao X, Zhang YB, Xie ZK, Wang RY (2018) Epsc involved in the encoding of exopolysaccharides produced by Bacillus amyloliquefaciens FZB42 act to boost the drought tolerance of Arabidopsis thaliana. Int J Mol Sci 19:1–18. https://doi.org/10.3390/ijms19123795

Ma Y, Rajkumar M, Zhang C, Freitas H (2016) Beneficial role of bacterial endophytes in heavy metal phytoremediation. J Environ Manag 174:14–25. https://doi.org/10.1016/j.jenvman.2016.02.047

Mahidsanan T, Gasaluck P, Eumkeb G (2017) A novel soybean flour as a cryoprotectant in freeze-dried Bacillus subtilis SB-MYP-1. *LWT, 77*:152–159. https://doi.org/https://doi.org/10.1016/j.lwt.2016.11.015

Malusá E, Vassilev N (2014) A contribution to set a legal framework for biofertilisers. Appl Microbiol Biotechnol 98:6599–6607. https://doi.org/10.1007/s00253-014-5828-y

Malviya MK, Li CN, Solanki MK, Singh RK, Htun R, Singh P, Verma KK, Yang LT, Li YR (2020) Comparative analysis of sugarcane root transcriptome in response to the plant growth-promoting Burkholderia anthina MYSP113. PLoS One 15:1–20. https://doi.org/10.1371/journal.pone.0231206

Martínez JP, Silva H, Ledent JF, Pinto M (2007) Effect of drought stress on the osmotic adjustment, cell wall elasticity and cell volume of six cultivars of common beans (Phaseolus vulgaris L.). Eur J Agron 26:30–38

Mari M, Di Francesco A, Bertolini P (2014) Control of fruit postharvest diseases: old issues and innovative approaches. Stewart Postharvest Rev 10:1–4. https://doi.org/10.2212/spr.2014.1.1

Marulanda A, Barea JM, Azcón R (2009) Stimulation of plant growth and drought tolerance by native microorganisms (AM fungi and bacteria) from dry environments: mechanisms related to bacterial effectiveness. J Plant Growth Regul 28:115–124. https://doi.org/10.1007/s00344-009-9079-6

Martínez-Álvarez JC, Castro-Martínez C, Sánchez-Peña P, Gutiérrez-Dorado R, Maldonado-Mendoza IE (2016) Development of a powder formulation based on Bacillus cereus sensu lato strain B25 spores for biological control of Fusarium verticillioides in maize plants. *World Journal of Microbiology and Biotechnology, 32*(5), 75. https://doi.org/10.1007/s11274-015-2000-5

Melandri G, AbdElgawad H, Riewe D, Hageman JA, Asard H, Beemster GTS, Kadam N, Jagadish K, Altmann T, Ruyter-Spira C (2020) Biomarkers for grain yield stability in rice under drought stress. J Exp Bot 71:669–683

Monneveux P, Sánchez C, Beck D, Edmeades GO (2006) Drought tolerance improvement in tropical maize source populations: Evidence of progress. Crop Sci 46:180–191

Mishra AK, Singh VP (2010) A review of drought concepts. J Hydrol 391:202–216. https://doi.org/10.1016/j.jhydrol.2010.07.012

Mazahery-Laghab H, Nouri F, Abianeh HZ (2003) Effects of the reduction of drought stress using supplementary irrigation for sunflower (Helianthus annuus) in dry farming conditions. Pajouheshva Sazandegi Agron Hortic 59:81–86

Mukhtar S, Shahid I, Mehnaz S, Malik KA (2017) Assessment of two carrier materials for phosphate solubilizing biofertilizers and their effect on growth of wheat (Triticum aestivum L.). *Microbiological Research, 205*(May), 107–117. https://doi.org/10.1016/j.micres.2017.08.011

Nam NH, Chauhan YS, Johansen C (2001) Effect of timing of drought stress on growth and grain yield of extrashort-duration pigeonpea lines. J Agr Sci 136:179–189

NeSmith DS, Ritchie JT (1992) Maize (Zea mays L.) response to a severe soil water-deficit during grain-filling. Field Crops Res 29:23–35

Ngumbi E, Kloepper J (2016) Bacterial-mediated drought tolerance: current and future prospects. Appl Soil Ecol 105:109–125

Nayyar H, Kaur S, Singh S, Upadhyaya HD (2006) Differential sensitivityof Desi (small-seeded) and Kabuli (large-seeded) chickpeagenotypes to water stress during seed filling: effects on accumulationof seed reserves and yield. J Sci Food Agric 86:2076–2082

Pastor-Bueis, R., Mulas, R., Gómez, X., & González-Andrés, F. (2017). Innovative liquid formulation of digestates for producing a biofertilizer based on Bacillus siamensis: Field testing on sweet pepper. *Journal of Plant Nutrition and Soil Science, 180*(6), 748–758. https://doi.org/10.1002/jpln.201700200

Patowary K, Patowary R, Kalita MC, Deka S (2016) Development of an efficient bacterial consortium for the potential remediation of hydrocarbons from contaminated sites. Front Microbiol 7:1–14. https://doi.org/10.3389/fmicb.2016.01092

Radhakrishnan R, Hashem A, Abd-Allah EF (2017) Bacillus : a biological tool for crop improvement through bio-molecular changes in adverse environments. Front Physiol 8:1–14. https://doi.org/10.3389/fphys.2017.00667

Ramakrishna W, Rathore P, Kumari R, Yadav R (2020) Brown gold of marginal soil: plant growth promoting bacteria to overcome plant abiotic stress for agriculture, biofuels and carbon sequestration. Sci Total Environ 711:135062. https://doi.org/10.1016/j.scitotenv.2019.135062

Rekha K, Baskar B, Srinath S, Usha B (2018) Plant growth promoting rhizobacteria Bacillus subtilis RR4 isolated from rice rhizosphere induces malic acid biosynthesis in rice roots. Can J Microbiol 64:20–27. https://doi.org/10.1139/cjm-2017-0409

Rudrappa T, Czymmek KJ, Paré PW, Bais HP (2008) Root-secreted malic acid recruits beneficial soil bacteria. Plant Physiol 148:1547–1556. https://doi.org/10.1104/pp.108.127613

Saikia J, Sarma RK, Dhandia R, Yadav A, Bharali R, Gupta VK, Saikia R (2018) Alleviation of drought stress in pulse crops with ACC deaminase producing rhizobacteria isolated from acidic soil of Northeast India. Sci Rep 8:1–16. https://doi.org/10.1038/s41598-018-21921-w

Samarah NH (2005) Effects of drought stress on growth and yield of barley. Agron Sustain Dev 25:145–149

Samarah NH, Mullen RE, Cianzio SR, Scott P (2006) Dehydrin-like proteins in soybean seeds in response to drought stress during seed filling. Crop Sci 46:2141–2150

Schisler DA, Slininger PJ, Behle RW, Jackson MA (2004) Formulation of bacillus spp. for biological control of plant diseases. Phytopathology 94:1267–1271. https://doi.org/10.1094/PHYTO.2004.94.11.1267

Singh M, Kumar J, Singh S, Singh VP, Prasad SM (2015) Roles of osmoprotectants in improving salinity and drought tolerance in plants: a review. Rev Environ Sci Biotechnol 14:407–426. https://doi.org/10.1007/s11157-015-9372-8

Stojanović SS, Karabegović I, Beškoski V, Nikolić N, Lazić M (2019) Bacillus based microbial formulations: optimization of the production process. Hem Ind 73:169–182. https://doi.org/10.2298/HEMIND190214014S

Suresh S, Srivastava VC, Mishra IM (2009) Techniques for oxygen transfer measurement in bioreactors: a review. J Chem Technol Biotechnol 84:1091–1103. https://doi.org/10.1002/jctb.2154

Timmusk S (2003) Mechanism of action of the plant growth promoting bacterium Paenibacillus polymyxa. Compr. Summ. Uppsala Diss. From Fac. Sci. Technol. NV - 908. Acta Universitatis Upsaliensis, Department of Cell and Molecular Biology, Biology, Faculty of Science and Technology, Teknisk-naturvetenskapliga vetenskapsområdet, Uppsala University

Timmusk S, Abd El-Daim IA, Copolovici L, Tanilas T, Kännaste A, Behers L, Nevo E, Seisenbaeva G, Stenström E, Niinemets Ü (2014) Drought-tolerance of wheat improved by rhizosphere bacteria from harsh environments: enhanced biomass production and reduced emissions of stress volatiles. PLoS One 9. https://doi.org/10.1371/journal.pone.0096086

Tiwari S, Prasad V, Chauhan PS, Lata C (2017) Bacillus amyloliquefaciens confers tolerance to various abiotic stresses and modulates plant response to Phytohormones through Osmoprotection and gene expression regulation in Rice. Front Plant Sci 8:1–13. https://doi.org/10.3389/fpls.2017.01510

Tripti, Kumar A, Usmani Z, Kumar V, Anshumali (2017) Biochar and flyash inoculated with plant growth promoting rhizobacteria act as potential biofertilizer for luxuriant growth and yield of tomato plant. J Environ Manag 190:20–27. https://doi.org/10.1016/j.jenvman.2016.11.060

Tzeng YM, Rao YK, Tsay KJ, Wu WS (2008) Effect of cultivation conditions on spore production from Bacillus amyloliquefaciens B128 and its antagonism to Botrytis elliptica. J Appl Microbiol 104:1275–1282. https://doi.org/10.1111/j.1365-2672.2007.03683.x

Vardharajula S, Ali SZ, Grover M, Reddy G, Bandi V (2011) Drought-tolerant plant growth promoting bacillus spp.: Effect on growth, osmolytes, and antioxidant status of maize under drought stress. J Plant Interact 6:1–14. https://doi.org/10.1080/17429145.2010.535178

Venuprasad R, Lafitte HR, Atlin G N (2007) Response to direct selection for grain yield under drought stress in rice. Crop Sci 47:285–293

Vurukonda SSK, Vardharajula S, Shrivastava M, SkZ A (2016) Enhancement of drought stress tolerance in crops by plant growth promoting rhizobacteria. Microbiol Res 184:13–24. https://doi.org/10.1016/j.micres.2015.12.003

Wang CJ, Yang W, Wang C, Gu C, Niu DD, Liu HX, Wang YP, Guo JH (2012) Induction of drought tolerance in cucumber plants by a consortium of three plant growth-promoting Rhizobacterium strains. PLoS One 7:1–10. https://doi.org/10.1371/journal.pone.0052565

Wang XQ, Zhao DL, Shen LL, Jing CL, Zhang CS (2018) Application and mechanisms of Bacillus subtilis in biological control of plant disease. In: Meena VS (ed) Role of Rhizospheric microbes in soil, pp 225–250. https://doi.org/10.1007/978-981-10-8402-7

Wang DC, Jiang CH, Zhang LN, Chen L, Zhang XY, Guo JH (2019) Biofilms positively contribute to bacillus amyloliquefaciens 54-induced drought tolerance in tomato plants. Int J Mol Sci 20:1–16. https://doi.org/10.3390/ijms20246271

Xie Z, Chu Y, Zhang W, Lang D, Zhang X (2019) Bacillus pumilus alleviates drought stress and increases metabolite accumulation in Glycyrrhiza uralensis Fisch. Environ Exp Bot 158:99–106. https://doi.org/10.1016/j.envexpbot.2018.11.021

Yang J, Zhang J (2006) Grain filling of cereals under soil drying. New Phytol 169:223–236

Yang Y, Pollard AM, Höfler C, Poschet G, Wirtz M, Hell R, Sourjik V (2015) Relation between chemotaxis and consumption of amino acids in bacteria. Mol Microbiol 96:1272–1282. https://doi.org/10.1111/mmi.13006

Ye H, Song L, Schapaugh WT, Ali ML, Sinclair TR, Riar MK, Mutava RN, Li Y, Vuong T, Valliyodan B (2020) The importance of slow canopy wilting in drought tolerance in soybean. J Exp Bot 71:642–652

Yoshiba Y, Kiyosue T, Nakashima K, Yamaguchi-Shinozaki K, Shinozaki K (1997) Regulation of levels of proline as an Osmolyte in plants under water stress. Plant Cell Physiol 38:1095–1102. https://doi.org/10.1093/oxfordjournals.pcp.a029093

Zhou C, Ma Z, Zhu L, Xiao X, Xie Y, Zhu J, Wang J (2016) Rhizobacterial strain Bacillus megaterium BOFC15 induces cellular polyamine changes that improve plant growth and drought resistance. Int J Mol Sci 17. https://doi.org/10.3390/ijms17060976

Chapter 22
Unveiling the Potential of *Bacillus* sp. in Bioremediation and Biocontrol

Abdul Qadir, Muhammad Mahroz Hussain, Muhammad Saad Bin Zafar, Muhammad Ashir Hameed, and Zia Ur Rahman Farooqi

Abstract Bioremediation emerges as sustainable and environment-friendly approach in which role of microbes is most prominent after hyperaccumulator plants. Microbes including many *Bacillus* spp. and *Pseudomonas* spp. are highly efficient by producing various enzymes and secondary metabolites that enhance the bioremediation process. Many *Bacillus* spp. produce diverse secondary metabolites such as lipopeptides, polypeptides, macrolactones, fatty acids, polyketides, and isocoumarins. These metabolites have the ability to decontaminate the heavy metal(liods) and organic pollutants. Bioremediation is an efficient and desirable management tool that can be used to regain the environmental sustainability. On the other hand, biocontrol of plant diseases by the application of some elite strains of *Bacillus* spp. is an ecofriendly approach alternative to the hazardous synthetic pesticides. This chapter is focused on the use of *Bacillus* sp. for bioremediation. The present chapter describes the bioremediation processes by the application of *Bacillus* species and briefly discusses the mode of action of *Bacillus* spp. on these two processes.

Keywords *Bacillus* sp. · Bioremediation · PGPR · Secondary metabolites · Types of bioremediations

22.1 Introduction

Bioremediation is the process of removing contaminants from the environment using microbial agents (such as fungi, yeast, or bacteria). Out of many definitions, the most appropriate and self-explanatory one is "the use of biological processes to eliminate, mitigate, or transform contaminated substances" (Abatenh et al. 2017). In

A. Qadir · M. M. Hussain (✉) · M. S. B. Zafar · M. A. Hameed · Z. U. R. Farooqi
Institute of Soil and Environmental Sciences, University of Agriculture Faisalabad, Faisalabad, Pakistan

M. T. Islam et al. (eds.), *Bacilli in Agrobiotechnology*, Bacilli in Climate Resilient Agriculture and Bioprospecting,
https://doi.org/10.1007/978-3-030-85465-2_22

many countries, aquatic ecosystems are mostly affected by point and nonpoint sources of pollution. There are many sources of pollution in water which depend upon directly discharged of pollutants in the aquatic water bodies. Among the known point sources, municipal and industrial wastewaters are the most common. However, landfill leachate, industrial site runoff and drainage, and ship runoff are the nonselective water sources in farmland and orchard rivers and urban runoff in open areas (Divya et al. 2015).

It is known that microorganisms can remediate pollutants from contaminated soil and water systems either by transforming, sequestering, or immobilizing these. To survive in an environment contaminated by high concentrations of pollutants, many microorganisms have developed resistance mechanism either by producing certain enzymes or secondary metabolites, and it is known that most microorganisms have specific genes for tolerating such high concentrations of pollutants (Adams et al. 2015). Bacteria are the most frequently used microorganisms and make up a large part of the total terrestrial biomass of 1018 g (Jayashree et al. 2012). Bacteria are used as remediating agent due to its ubiquitous capacity to grow and nature of adoptability in adverse conditions (Venosa et al. 2002). Some microorganisms used in the pollutants remediation and sequestration processes are *Achromobacter*, *Alcaligenes*, *Arthrobacter*, *Bacillus*, *Acinetobacter*, *Corynebacterium*, *Flavobacterium*, *Micrococcus*, *Mycobacterium*, *Nocardia bacteria*, *Pseudomonas*, and *Vibrio*. These microbes have also been reported to remediate the highly polluted sites even with multiple kinds of pollutants either organic or inorganic (Abatenh et al. 2017).

Bacillus sp. are rod-shaped, Gram-positive bacterium that normally needs aerobic or facultative aerobic environments. Under stress conditions, these can produce oval endospores that can retain their efficacy for both plant growth promotion and remediation (Arkhipova et al. 2005). *Bacillus* sp. produces a variety of enzymes that allow them to degrade/decontaminate various complex substrates under dynamic environmental conditions (Santoyo et al. 2012). The biological remediation of anthropogenic pollutants depends on the nature of the microbes, the chemical species of pollutants, geo-environmental conditions of the polluted site (Ojuederie and Babalola 2017). Microbial degradation depends on the nature of microorganism and the production of certain enzymes which are required for the degradation/decontamination of the pollutants (Goyal et al. 2019).

Bacillus sp. has been used due to its potential to remediate under biological remediation of water pollution in septic tanks and fish-farming ponds. The use of *Bacillus brevis*, *B. sphaericus*, and *B. megaterium* leads to the decomposition of 3-hydroxybenzoate (Zokaeifar et al. 2012). Various bacteria can vigorously metabolize contaminants by using them as an energy source, but others produce the enzymes which break down pollutants into a less toxic substance (Igiri et al. 2018). *Bacillus* bacterium can selectively take in gold from solutions that include Cu, Zn, Fe, and Au (Boopathy 2000). Biodegradability of petroleum hydrocarbon contaminated soil by micrococcus, *Bacillus*, and coryneform bacteria. Regarding the biological decolorization and biodegradation of dye molecules, methods for the removal of dyes from wastewater were investigated (Mondol et al. 2013).

Bioremediation of heavy metals in polluted soil can be done by using *Bacillus*, *Pseudomonas*, and fungi under various environments and under influence of geochemical behavior of the polluted sites (Idris et al. 2004).

This chapter includes brief introduction of *Bacillus* sp. and its efficacy in removing pollutants, while its metabolites and enzymes have also been elucidated. Many studies and books had been published on these aspects of the *Bacillus* spp., but this chapter has the brief and better understandable review of the topic.

22.2 Role of *Bacillus* sp. in Bioremediation Process

Remediation technology using microbes can inexpensively remove pollutants to protect human health and the environment (Abatenh et al. 2017). In many facilities, exogenous or genetically modified microorganisms are used for biodegradation/decontamination of pollutants. An efficacious, cost-effective microbial bioremediation process depends on hydrogeological circumstances, pollutants concentration, microbial ecosystems, and other considerations that change over time which can either be favorable or harsh for the process. In bioremediation procedure, microbes use pollutants as nutritional supplements or energy (C) resources. Enhancing nutrients source (N and P), electron acceptors (O_2), and a substrate (CH_4, C_6H_6O, and C_7H_8) or by introducing microorganisms with the necessary catalytic capabilities can improve the bioremediation activities in soil, and this process can either be biostimulation or bioaugmentation later discussed in this chapter (Jayashree et al. 2012).

Heterotrophic microbes use electron acceptor nitrite or NO_3 rather than O_2 and use organic matter as a source of energy, carbon to reduce oxidized nitrogenous substances (nitrite to nitrate) to nitrogen gas. Denitrification is widespread in Gram-negative bacteria like the *Bacillus*, *Pseudomonas*, *Alcaligenes*, *Paracoccus*, and *Thiobacillus*. Denitrification is a slowdown process, especially for industrial wastewater with a high nitrate concentration. It was found that after adding *Bacillus thuringiensis* FQ1, the total amount of cadmium increased by 14.29% to 97.67%. When treated with bacterial and fungal consortium, 95.07% of the phenanthrene was removed, and the peak concentration was 500 mg/kg phenanthrene (Goyal et al. 2019; Pratush et al. 2018). Liu et al. (2015) reported that both lead and cadmium can affect the degradability of *Bacillus*. *Bacillus* sp. affects the level, composition, and activity of enzymes during the degradation of phenanthrene, while cadmium is more toxic to *Bacillus* sp. than lead, which also interferes with the *Bacillus* sp. efficiency to remediate the polluted sites.

22.2.1 *Mechanism of Microbial Bioremediation*

Microbes use different strategies to convert heavy metals. Microbes can alter the ionic nature of the different noxious elements which reduced their mobility, translocation, and bioavailability (Ayangbenro and Babalola 2017). The bioremediation of heavy metals can be achieved by the following methods, like biosorption, bioconcentration, immobilization, biomethylation, or organometallic complexation. The substance turns into a radionuclide converted. A brief description of the basic metalworking processes used by different microorganisms is as under:

22.2.1.1 Mobilization

Redox reaction technique enables to dissolve and transform hazardous metals and their radionuclides into inorganic-organic acids in this process. It can lower the pH of contaminants as well. Four subprocesses are further divided into complete mobilization processes, namely, enzymatic oxidation, enzymatic reduction, complexation, and siderophore. Soil Pb immobilization mechanisms were explored using *B. subtilis* DBM, a bacterial strain extracted from heavy-metal-contaminated soil. In the removal of Pb^{2+} from the liquid process, adsorption and desorption experiments with active bacterial cells and dead cells showed that both extracellular adsorption and intracellular accumulation were involved. Of the sequestered Pb(II), 8.5 percent was retained inside the cell wall by physical entrapment, 43.3 percent was held by ion exchange, 9.7 percent was complexed or precipitated on the cell surface with cell surface functional groups, and 38.5 percent was accumulated intracellularly. Fourier transform infrared spectroscopic study revealed the complexity of Pb^{2+} with carboxyl, hydroxyl, carbonyl, amino, and phosphate groups. X-ray diffraction study revealed that precipitates of $Pb_5(PO_4)_3OH$, $Pb_5(PO_4)_3Cl$, and $Pb_{10}(PO_4)_6(OH)_2$ formed on the cell surface during the biosorption process (Bai et al. 2014).

22.2.1.2 Enzymatic Oxidation

In general, in more than one oxidation state, certain inorganic compounds exist, and it is commonly found that the higher oxidation state is less soluble than the lower ionic phase. In this case, by oxidizing higher states to lower states, the enzymatic oxidation mechanism (catalyzed by enzymes released by microorganisms) plays a very important role and increases the solubility of compounds. The elimination of inorganic materials from the solution is one of the essential methods. Heavy metals lose electrons in this step and turn into a useful or less toxic state. The more frequent example of enzymatic oxidation is *Thiobacillus ferrooxidans*, and *T. thiooxidans* oxidizes uranium from ore (Cumberland et al. 2016).

22.2.1.3 Enzymatic Reduction

The mechanism is reversed in contrast to the enzymatic oxidation process. Inorganic compounds with several oxidation states in their reduced form remain insoluble. The enzymatic reduction mechanism was found to be useful in removing these elements from the solution. Enzymatic reduction reactions are carried out by facultative and obligatory anaerobic microorganisms, in an in situ bioremediation. Božič et al. (2010) carried a study in which A novel β-nicotinamide adenine dinucleotide (NADH) disodium salt-dependent reductase has been isolated from *B. subtilis*, which can convert the redox dye CI Acid Blue 74, CI Natural Orange 6, and CI Vat Blue 1 is reduced to a leuco form soluble in water. In reference to pH and temperature conditions, enzymatic reactions were enhanced. In the presence of NADH cofactor consumed stoichiometrically, the reductase can reduce acid blue 74 and natural orange 6. At the same time, Vat Blue 1 required the presence of the intermediate 1,8-dihydroxyanthraquinone (Božič et al. 2010).

22.2.1.4 Complexation

When complexing, inorganic metal complexes are produced by adding ligands. Due to the formation of metal complexes, hazardous inorganic substances are transferred and can be easily extracted from solid waste (Ayangbenro and Babalola 2017). There are two main types of microbial chelating agents: (i) organic acids with low molecular weight (citric acid, tricarboxylic acid, and alcohol) and (ii) high molecular weight ligands, siderophores, and toxic metal bond compounds. In addition, certain amino acids that are synthesized by certain bacteria can also be used as complexing agents. It has been observed that the complexation of heavy metals and radionuclides with microorganisms is strongly dependent on the pH value (Ayangbenro and Babalola 2017). Gorman-Lewis (2014) Gibbs energy, enthalpy, and entropy of Cd and Zn complexation between *B. licheniformis* and zinc on *B. subtilis* were measured by surface complexation model and isothermal titration calorimetry. Their results indicate that the Cd and Zn complexes of *B. licheniformis* are driven by entropy at low pH and enthalpy at ambient pH. Complexation of zinc on *B. subtilis* is driven by entropy, indicating that *B. licheniformis* has different donor ligands, which dominate the pH response of the surrounding medium.

22.2.1.5 Siderophores Production

When microorganisms grow in iron-deficient media, some microorganisms produce iron-specific chelating agents, also called siderophores. These siderophores have specific linking groups, such as catecholate, phenolate, or hydroxamate. Due to the presence of these specific groups, the complexing properties of these siderophores are improved, and they form various complexes with toxic metals, thus increasing their solubility (Khan et al. 2018). In recent years, from different biological systems, several siderophores or siderophore-like compounds have been identified. Siderophores are unique to iron (III), according to literature records, but recent studies have shown that they can also be complex with other metals and radionuclides (Ahmed and Holmström 2014). *Bacillus* sp. PZ-1's siderophore-producing characteristics and conditions were studied, and the improvement of siderophores on Pb uptake and translocation in *Brassica juncea* was determined. Single-factor experiment results showed that glucose, $Pb(NO_3)^2$, and pH could stimulate PZ-1 growth and development of siderophores. Optimized by the response surface method, under the conditions of glucose concentration of 21.84 g/l, pH of 6.18, and Pb (NO3) 2 concentration of 245.04 μmol/l, the maximum yield of siderophores obtained is 90.52%. The type of siderophore is hydroxamate, and its concentration in the fermentation broth is 32.24 μg/ml. The pot experiment results showed that the siderophore made the mustard rape absorb more Pb from the soil, with an absorption rate of 1.04–2.74, and absorbed more Pb from the underground soil to the soil with a TF value of 1.21–1.48. The results showed that the genus *Bacillus* PZ-1 can produce abundant siderophores, which can be used to increase the extraction of lead in the soil by plants (Yu et al. 2017).

22.2.1.6 Immobilization

To restore metal polluted soils, ex situ and in situ restriction procedures are used. Ex situ technology is used in heavily polluted areas. The soil in such areas is excavated from the original location and then stored in such a location for fixing metallic ions in that soil by treating it with various microbial systems. In terms of local technology, the soil contaminated with metals must be treated in its original location. During the fixation process, organic nitrogen is resulted from nitrate nitrogen conversion, which is the opposite to the process of mineralization. The fixation process is controlled by several soil bacteria due to which it is considered a biological process. The process is influenced by moisture and temperature of the soil. The fixation takes place through energy-dependent (active or passive) methods. According to the observations, passive fixation is not specific for metallic species, while active fixation depends mainly on metabolism of microbial cells and is relatively slow. In most cases, specific proteins such as metallothionein are used in active immobilization, which forms complexes with heavy metals. Immobilization of heavy metals occurs mainly through bioaccumulation, precipitation, and biosorption. These methods are commonly used to treat wastewater polluted by heavy metals (Ayangbenro and Babalola 2017). Effects of interaction between soil minerals and *B. subtilis* DBM for the adsorption of Pb(II) and Cu(II) were studied (Bai et al. 2019). The adsorption capacities of kaolinite and goethite for Pb(II) and Cu(II) improved after combination with DBM as compared to the independent mineral application. After combination with DBM, the site concentrations increased by 80% and 30% of kaolinite and goethite, respectively, based on the modeling outcomes by potentiometric titration. However, lower concentrations of observable sites in comparison to theoretical values were observed due to the presence of functional groups in DBM/

mineral combinations, and the combinations contributed to an improvement in NH_4NO_3 and EDTA-Na_2 desorption rates (Bai et al. 2019).

22.2.1.7 Precipitation or Solidification

Metal ions in solution or soil can be precipitated or solidified by a variety of approaches. Sulfate reduction is the most common method of precipitation. Reducing sulfate bacteria (SRB) are used to treat metal contaminants in technical natural systems such as built-up wetlands. SRB forms metal sulfide precipitates by removing toxic metals from the solution. Most toxic metal sulfides have very low solubility, and their toxicity is almost negligible to the environment after precipitation. SRB also creates solution conditions that are conducive to the chemical reduction of metals. The conversion of organophosphates to orthophosphates by microorganisms can also proceed to the precipitation of metals (intracellular phosphates can also fix metals) by forming metal phosphates (especially at a pH above 7) (Nguyen et al. 2020). The process of precipitation can be performed ex situ or in situ. The interactive effect of each approach has been discussed later in this chapter.

22.3 Factors Affecting Bioremediation Process

Natural soil microorganisms play a significant role in soil bioremediation as biogeochemical reagents and can convert complicated organic substances into simple inorganic substances and their derivatives; this process is known as mineralization (Adams et al. 2015). As a whole, soil particles are negatively charged, so by multivalent cations, soil and bacteria can be linked together (Bai et al. 2014). In bioremediation, microbes reduce, remove, and convert pollutants from soil, sediment, air, and water. Bioremediation explains how to use microbes to obliterate or remediate waste (Shanahan 2004). Such a process is dependent upon the optimal use of nutrients and other factors that support the biological functions (Divya et al. 2015; Jayashree et al. 2012). These are the following:

Pollutant concentration: This directly affects the activity of microorganisms if the concentration is too high. Conversely, small amounts of contaminants can prevent the degradation/decontamination by resisting the bacterial enzymes production.

Bioavailability of contaminants: This depends on the extent to which they are adsorbed on the solid or sequestered by molecules in the contaminated medium, the extent to which they diffuse into the macropores of the soil or sediment and on other factors like its form. If the bioavailability of the pollutant is low, then microbial reaction is lower because the pollutants are more strongly adsorbed in the solid, are contained in the molecular matrix in the contaminated medium, and continue to diffuse into the macropores of the aquatic bodies, soil, and sediments.

Geo-environmental conditions: This has a major impact on the efficacy of bioremediation strategies. For bioremediation applications, pH (6–8 optimal range), nutrient utilization, water content, temperature, and redox potential are important environmental conditions at the site.

Redox potential and oxygen content: These represent oxidation or reduction circumstances. Due to electron acceptors (like sulfate, manganese oxide, iron oxide, and nitrate), redox potential is influenced.

Nutrients: Microbial cell development and division are required. There are usually micronutrients that are suitable for bacterial growth, but these nutrients can be applied in available forms or by organic substrate modifiers (which can also be used as electron donors) to improve bioremediation.

Moisture content: The growth of microorganisms requires the optimal available of water in the ecological matrix. Microorganisms need 12%–25% water for optimal growth and proliferation.

Temperature: This directly disturbs the metabolic rate of microorganisms, which in turn affects the activity of microorganisms in the environment. To a certain extent, the rate of biodegradation increases or decreases with the increasing or decreasing temperatures (Goyal et al. 2019; Igiri et al. 2018; Ojuederie and Babalola 2017).

22.4 Role of *Bacillus* sp. in Remediating Different Contaminants

A few years ago, it is assumed that plants can only receive cytokinin from soils that have been synthesized by rhizosphere microorganisms (Arkhipova et al. 2007). Many researchers have explained the fact how plants evacuate their hormones, like cytokinins (Arkhipova et al. 2007). But *Bacillus* bacteria have demonstrated the capability to generate and secrete cytokinins in the rhizosphere, which has a positive impact on plant growth. Arkhipova et al. (2005) analyzed various *B. subtilis* strains to synthesize zeatin riboside-type cytokinin (ZR), dihydro-cellulose nucleoside (DHZR), and pentene to synthesize adenosine capacity (IPA). In an experiment inoculated with *Bacillus* strains in lettuce plants, that it had been detected later 2 weeks, root and shoot tissues contained more cytokinin than non-inoculated plants. The frequency of cytokinin is related to a 30% rise in plant weight. It is worth noting that extreme concentrations of plant hormones like indolyl-3-acetic acid (IAA) and abscisic acid (ABA) have also been noticed.

It indicates that *Bacillus* strains get double response on the growing plant. Enrichment of cytokinin and other ways to increase hormone synthesis (like IAA and ABA) promote plant growth and destroy the synthesis of other hormones such as erythromycin GA (Arkhipova et al. 2005). *Bacillus* bacteria synthesize various hormones that affect plant development, such as auxin (Valencia-Cantero et al. 2007). In the laboratory experiments, these were used to inoculate legumes with

B. megaterium UMCV1. Compared to non-inoculated plants, the tissues of legumes have higher iron (Santoyo et al. 2012).

22.4.1 Volatile Organic Compounds

It suggested that *B. megaterium* UMCV1 promotes development in an unknown way in *Arabidopsis* model plants. Ryu et al. (2004) provide a clear suggestion of PGPR's new PGP mechanism in *Arabidopsis*, including *B. subtilis* GB-3 and *B. amyloliquefaciens* IN-937a. By applying distinct compartments (separate petri dishes), one on the side of the petri dish is inoculated with different PGPR, while the seeds of *Arabidopsis* are placed on the far side to prevent physical contact. Attractively, the existence of *Bacillus* strains appreciably stimulated plant biomass. Since no plants showed an induction, *E. coli* DH5a bacteria were used as negative controls. The author believes that volatile chemicals are responsible for the chemical signal transmission among *Bacillus* strains and *Arabidopsis*. It was demonstrated that volatile substances, such as 2,3-butanediol and acetosyringone, are synthesized with bacteria and therefore play a crucial role in this interface.

In a supplement, individual compounds have similar effects on plants. Compared to the wild strain, this mutant strain showed a significant plant growth-promoting effect in the production of 2,3-butanediol and acetylase by *B. subtilis* BSIP1173 and BSIP1174. Several compounds identified involving aldehydes, ketones, and alcohols by gas chromatography may be involved in root development (Santoyo et al. 2012). Other volatile substances like 1-octen-3-ol and butyrolactone can also be involved in this collaboration between plants and bacteria. The effects of volatile compounds, particularly on plants, whether specific overall or for tissue, impact the growth of plant. This is also essential to explore the requirements for PGPR growth that can induce production of compounds, whether they have the beneficial effects, because of a particular compound or one synergies activity of numerous compounds (the combination shows better stimulatory activity) (Blom et al. 2011). To determine whether the discoveries are common, more detailed analysis of plant is needed as the results of using agricultural plants have significant interaction with microbes.

22.4.2 Heavy Metal Remediation by Bacillus sp.

Since bioremediation is the phenomenon of surface, the cell wall is of the genus *Bacillus*. The morphology and physiology change after the adsorption of cadmium. Research was carried out using FTIR and SEM-EDX (Nithya et al. 2011). After the adsorption of cadmium, the zeta potential analysis of *Bacillus cereus* RC1 confirmed that electrostatic interaction and surface complexation are of key importance for biosorption (Huang et al. 2014). This biosorbent capacity of *Bacillus* is due to several functional groups that are present on cell wall. Extracellular polymers (EPS) are also involved in biosorption in addition to these functional groups. The composition of EPS includes carbohydrates and proteins as well as their derivatives in the form of homopolymers or heteropolymers (Shameer 2016). Bacterial cells have many benefits from EPS production, but they are essentially benefitting from biofilm formation, which contributes to antibiotic and metal resistance (Shameer 2016). According to a study, *B. subtilis* NSPA13, *B. licheniformis* NSPA5, and *B. cereus* NSPA8 produce EPS, which makes them cadmium resistant and helps reduce cadmium levels (Shameer 2016). Another study confirmed these results (Chauhan et al. 2017). A TEM analysis of the EPS of the unidentified isolate showed the presence of cadmium entrainment in the same study. This may be due to the existence of active carboxyl groups as determined by FTIR of the *Bacillus* extracellular polymer substrate (Chauhan et al. 2017; Shameer 2016)).

Studies conducted by several scientists have observed that dead and living cells of *Bacillus* can remove nickel time to time which highlights the probability that biosorption is the main mechanism of nickel bioremediation. Aryal (2015) recommended that with the help of FTIR, it could be proven that the -NH_2 and -COOH groups of *B. sphaericus* are accountable for the binding of nickel. It was further proved by studying the removal of nickel under the influence of temperature. It was observed that with increasing temperature from 20 °C to 40 °C, the bioremediation capacity of organisms also increased (Aryal 2015). However, if the temperature continues to rise, the potential decreases because of the decrease in surface activity. The study was concluded with the report that the bioremediation of nickel by *B. sphaericus* occurred through exothermic processes and biosorption. One more published report showed related results (Sari et al. 2008) in which after getting exposed to nickel, the surface morphology of the dead *B. subtilis* MTCC 1628 examined by SEM-EDX, which further confirms that biosorption was responsible for the bioremediation of nickel. In *B. licheniformis*, similar changes were observed (Jain et al. 2017). The results show that it is a passive process, namely, the biosorption of surface phenomena.

Bacillus cell walls supply many functional groups with negative charge such as hydroxyl, amino, sulfate, and carboxylate groups (Vijayaraghavan and Yun 2008). After the initial formation of metal complexes and the neutralization of chemically active sites, these functional groups showed ionic interactions with Ni^{2+} ions (Shekharaiah et al. 2020; Abou-Aly et al. 2019). When the cells of *B. subtilis* 117S were pretreated with formaldehyde, mercury chloride, and sodium azide, it was observed that the Ni-removing ability of the cells was decreased. It occurs due to methylation of amino groups and esterification of carboxyl groups. It concludes that in nickel bioremediation, the functional groups present on the cell wall of *Bacillus* are important (Abdel-Monem et al. 2010).

De et al. (2008) used bacteria which were mercury-resistant, such as *Pseudomonas aeruginosa*, *Brevibacterium* iodide, and *B. pumilus*, to remove lead and cadmium. In this study, *Pseudomonas aeruginosa* and *Aspergillus faecium* removed 70% and 75% cadmium, and *Pseudomonas aeruginosa* reduced cadmium (by 1000 mg/L and A by 19.2 mg/L. The removal is approximately 72 hours. *Brevibacterium iodinum* and *B. pumilus* can remove more than 96% lead by more than 87% and 88% from

1000 mg/L to 1.8 mg/L. In another study by Singh et al. (2013), native facultative anaerobic *Bacillus cereus* was used to detoxify hexavalent chromium. *Bacillus cereus* has an ability to remove chromate of 72% at a chromate concentration of 1000 µg/ml. The bacteria can remediate chromate in a wide temperature range (25 to 40 °C) and pH value (6 to 10). The best temperature range is 37 °C, and the initial pH is 8.0 (Singh et al. 2013).

Various heavy metals were tested using the bacterial species (such as *Pseudomonas*, *Flavobacterium*, *Micrococcus*, *Enterobacter*, and *Bacillus*). Their high bioremediation capacity is based on their potential active chemisorption sites on the cell wall and high surface-to-volume ratio (Mosa et al. 2016). Abioye et al. (2018) used *B. subtilis*, *B. megaterium*, *Aspergillus niger*, and *Penicillium* to investigate the bioremediation of lead, chromium, and cadmium in tannery wastewater. *B. megaterium* recorded the highest reduction in lead, i.e., 2.13 to 0.03 mg/L, followed by *B. subtilis*, i.e., 2.13 to 0.04 mg/L. *Aspergillus niger* showed the highest ability to reduce chromium concentration (1.38–0.08 mg/L), followed by *Penicillium* (1.38–0.13 mg/L), while *B. subtilis* showed the highest ability to reduce the cadmium concentration (0.4–0.03 mg/L) after 20 days, followed by *B. megaterium* (0.04–0.06 mg/L).

22.5 Production of Bioactive Compounds as Tools for Biocontrol of Phytopathogens

22.5.1 Polypeptides

Non-ribosomal peptides (NRP) are synthesized by expanding the activated monomers of amino acid building blocks through non-ribosomal peptide synthesized (NRPS) modules. The NRPS is organized in three consecutive stages per module. These modules are liable for the integration of particular amino acids incorporated in the polypeptide chain: as well as for adenylation, thiolation, and condensation (Fickers 2012). *Bacillus* strains generally produce two types of polypeptides, linear peptides, and de-peptide peptides (Table 22.1). The polypeptide antibiotics are produced by *Bacillus* (Zokaeifar et al. 2012), and these antibiotics are efficient and effective against Gram-positive bacteria, though there is relatively little production of broad-spectrum antibiotics and antifungals effective against Gram-negative bacteria (Deng et al. 2011). *Bacillus* has a broad range of antibacterial activities because they are used as antifungal agents, antiviral agents, and cellular anti-mutant agents as well as anti-mycoplasma drugs (Aziz 2018).

Table 22.1 List of some important bioactive compounds isolated from *Bacillus* strains

Compounds	Producing strains	Inhibitory concentrations	Nature of bioactivities	References
Halobacillin	*Bacillus* sp. CND-914	0.98 μg/mL (IC50)	Anticancer	(Kim et al. 2010)
Mixirin	*Bacillus* sp.	0.68 μg/mL (IC50)	Anticancer	(Zhang et al. 2004)
Bogorol A	*Bacillus* sp.	2 μg/mL (MIC)	Antibacterial	(Blunt et al. 2017)
Loloatin B	*Bacillus* sp.	1–2 μg/mL (MIC)	Antibacterial	(Yang and Yousef 2018)
Bacillistatins 1	*B. silvestris*	10–4–10 – 5 μg/mL (GI50)	Anticancer	(Shukla 2016)
Bacillamide	*Bacillus* sp.	LC50 after 6 h: 3.2 μg/mL	Antialgal	(Jeong 2019)
Bacilosarcin A	*B. subtilis*	82% inhibition at 50 μM	Plant growth regulator	(Komaki et al. 2016)
Macrolactin S	*B. Amyloliquefaciens*	0.3 and 0.1 μg/mL (MIC)	Antibacterial	(Raimundo et al. 2018)
Macrolactin V	*B. Amyloliquefaciens*	0.1 μg/mL (MIC)	Antibacterial	(Raimundo et al. 2018)
Basiliskamides A	*B. Laterosporus*	1.0 and 3.1 μg/mL 2.5 and 5.0 μg/mL	Antifungal	(Baindara et al. 2017)
α-Amylase	*Bacillus subtilis S8–18*	*N/A*	*N/A*	(Karakus et al. 2018)
Lipase	*Bacillus smithii* BTMS11	*N/A*	*N/A*	(Zin et al. 2017)

22.5.2 *Lipopeptides*

Lipopeptide (LP) biosurfactants consist of lipid tails associated with the short linear as well as the circular oligopeptides. They are generated by fungi (involving *Aspergillus*) and different types of bacteria like *Streptomyces*, *Pseudomonas*, and *Bacillus*. Lipopeptide has got great attention due to its antimicrobial, cytotoxic, antitumor, immunosuppressive, and surfactant-containing attributes (Pirri et al. 2009). The main mechanism of LP is the formation of pores in the membrane, which leads to a discrepancy in transmembrane plasma flow and cell demise (Baltz 2009). Cyclic lipopeptide (CLP) is a universal metabolite delivered by a wide range of bacterial species. They consist of short loop oligopeptides that are associated with the fatty acid tail and have effective surfactant characteristics (Ongena and Jacques 2008). CLP is produced in the non-ribosome of the large NRPS hybridase synthase (non-ribosomal peptide synthase) -PKS (polyketide synthase). Formation of endospores *B. subtilis* produces various antibacterial peptides, which are synthesized by ribosomes and modified after translation (such as lantibiotic and lantibiotic-type peptides) or non-ribosomes (Caulier et al. 2019).

Wheat blast is a devastating disease caused a filamentous fungus *Magnaporthe oryzae Trticum* pathotype which poses a serious threat to food and nutritional security in South America, South Asia, and Africa. Several strains of *Bacillus* spp. significantly biocontrol this destructive fungus both in vitro and in vivo. Recently, some linear lipopeptides produced by a strain of marine *B. subtilis* have been suppressed devastating wheat blast fungus *M. oryzae Triticum*.

Due to its antibiotic action, CLP has received considerable attention from humans and phytopathogenic microbes, including mycoplasmas, viruses, trypanosomes, fungi, bacteria, and sporangia (Raaijmakers et al. 2006). Several strains of *Bacillus* can synthesize lipopeptide chemical compound, and the antagonistic activity of lipopeptide compounds against phytopathogenic agents has been studied (Leclere et al. 2005). *Bacillus* LPs are divided into three types of cyclic complexes: surfactants, iturin, and fengycin. Each family includes variants that have an identical peptide length then separate residues at certain positions. Besides, every variant may have several homologs with different lengths of fatty acid chains and different isomers, which leads to significant structural differences (Ongena et al. 2007), the heptapeptide variants of the esperin, lichenin, pumilacidin, and surfactin groups in surfactin family. The peptide part is combined with β-hydroxy fatty acids (C12-C16) with linear, hybrid, or pro-branched acids. In the iturin family, iturin A and C; bacitracin D, F, and L; and the subtilisin of mold are all six key variants. In all circumstances, the heptapeptide binds to chains of β-amino acid fatty acids of varying lengths (C14-C17) (Branquinho et al. 2014). The third family contains toyomycins A and B. If Tyr 9 is in configuration D, it is also called pravastatin. These decapeptides are combined with β-hydroxy fatty acid chains (C14-C18). The β-hydroxy fatty acid chains are also in linear, hetero- or trans-isomeric forms and can be saturated or unsaturated. In addition, the three *Bacillus* LP families showed a specific cyclization model (Ongena and Jacques 2008).

Surface actin and toymycin are large lactone rings, but for surfactants, the closed-loop between β-hydroxy fatty acids and the C-terminal peptide is catalyzed, while toymycin cyclase is in the peptide sequence. Like many *Pseudomonas* LP, the Tyr side chain in position 3 forms the ester bond together with the terminal C deposit, establishing an internal loop. Similar to surfactants, fatty acid chains are also implicated in the cyclization of iturins, but because of their β-amino nature, amide bonds are developed along with C-terminal groups that form large lactic acid structures (Ongena and Jacques 2008). Recently, two-dimensional NMR and the neutron reflectometer were used to carry out a detailed structural characterization of the representative members of these three families (Tsan and Gao 2007).

22.5.3 Fatty Acids

Bacillus sp. is well-known to use the conventional way of FAS (fatty acid synthase) to synthesize fatty acids with chain sizes between C12 and C19 (Fang and Kato 2007). The fatty acids in precursors of acetyl-CoA and malonyl-CoA are produced by an action of an enzyme called fatty acid synthase (FAS). In many organisms, there are fatty acids (methyl branched) together with methyl groups on linear carbon atoms (methyl substituents on carbons 2, 4, and 6), the source of which is selectively incorporated by the fatty acid synthase methylmalonyl-CoA (Mondol et al. 2013). The equivalent isocyanate branch in new methyl branched fatty acids is likely originating from leucine and isoleucine and then undergoes a sequence of extensions via malonyl-CoA. In the penultimate extension step, methylmalonyl-CoA appears to be a selective way to be incorporated by fatty acid synthase from bacteria that produce methyl branched-chain fatty acids (Mondol et al. 2013). So far, the question of whether it is a methyl branched fatty acid synthase known or unknown in bacteria remains a speculative question. The natural activities carried out of these new fatty acids may not have been described.

22.5.4 Macrolactins

Macrolactins are cyclic polyene macrolactones composed of 24-membered cyclic lactones, and their modified forms (such as adding of the glucose-β-glucopyranoside) might also appear as linear analogs (Mondol et al. 2013). The carbon skeleton of the macrolactin comprises three distinct diene structural elements in the 24-membered lactone cycle. Macrolactone compounds are mostly produced by terrestrial and marine strains (Hamdache et al. 2011). Through the genome of *B. amyloliquefaciens* FZB42, plants have environmental strains colonized by the root system, which can stimulate plant growth in the rhizosphere and inhibit soil-bound plant pathogens (Idris et al. 2004). There are 3 positions with a total of approximately 196,340 lead operon PKS. The origin of genome replication is equivalent to about 1.4 Mbit/s (PKS2), 1.7 Mbit/s (PKS1), and 2.3 Mbit/s (PKS3) clockwise, and the size is of 3916 kb (Chen et al. 2006). These three groups of genes show the typical modular structure of the PKS type I system, which indicates that the FZB42 strain has a biosynthetic mechanism and can produce at minimum three distinct types of polyketides. PKS1 and PKS3 are ascribed to the production of bacteriocins of which pks2 is engaged in the biosynthesis of large endocrine substances. The macrolide ring of the macrolide is formed by the cyclization of a polyketide chain composed of the PKS type I enzyme. These enzymes carry out a condensation of repeated decarboxylation of carboxylic acids by activated carboxylic acid initiating units (Schneider et al. 2007).

22.5.5 *Induced Systemic Resistance as a Mode of Biocontrol by* Bacillus

So far, the genus *Bacillus* contains species with different genetic traits, making it an ideal candidate for biological control agents. Its capability to induce systemic resistance (ISR) is another significant feature of *Bacillus* species. Induced systemic resistance constitutes a system through which bacteria (mainly PGPR) can cause plant defense systems to resist infection by pathogens (Kloepper et al. 2004). Several authors have compared ISR to acquired system resistance (ASR), and that is bacteria-free. Even if the induction mechanism is different from the plant response, the phenotypic response may be similar (Compant et al. 2005). For *Bacillus*, it's interesting to determine the biodiversity that can cause resistance in plant systems. Some examples are *B. subtilis*, *B. amyloliquefaciens*, *B. pasteurii*, *B. cereus*, *B. sphaericus*, and *B. pumilus* (Santoyo et al. 2012). *Bacillus* can induce this reaction in plants, its beneficial effect being to protect plants against different pathogens like viruses, bacteria, nematodes, and fungi (Kloepper et al. 2004).

Certain lipopeptides produced by *B. subtilis*, such as surfactants and antibiotics, may also be able to induce ISR in plants (Ongena et al. 2007). During the past few years, some experiments have tried to analyze the mechanism of induction of resistance in plants and discovered that the volatile chemicals contribute a substantial role in that induction. The GBO3 and FB17 strains of *B. subtilis*, *B. cereus* AR156, and *B. amyloliquefaciens* IN937a are all specimens of strains that have ISR system by volatile emissions (Ryu et al. 2004). Lately, it has been proposed that the FB17 strain of *B. subtilis* produces acetal (3-hydroxy-2-butanone) and causes ISR in *Arabidopsis* (Rudrappa et al. 2010).

To interrupt the signaling pathway by volatile substances, *B. subtilis* GB03, ISR is induced by an ethylene-dependent pathway independent of jasmonic acid and salicylic acid (Ryu et al. 2004). However, some writers believe that PGPR has a generic mechanism used for triggering ISR in plants by pathways dependent on jasmonate and ethylene (Santoyo et al. 2012). ISR is essential to prevent or reduce plant pathogens. When a pathogen is detected, the plant synthesizes several hormones involved in ASR (like the ISR response). Most recent hormones, like brassinosteroids (BR), are important for plant development and have just recently been endowed with systemic resistance to various biological and abiotic loads. They are tolerating the various environmental impacts of PGP bacteria that produce 1-amino cyclopropane-1-carboxylic acid deaminase (ACC) on the environment. PGP bacteria can stimulate plant growth by reducing their ethylene content (Santoyo et al. 2012). It will be fascinating to assess whether the PGP bacteria cannot only induce protection from the pathogens but similarly resist environmental stress beyond ISR or any other processes.

22.6 Conclusions and Prospects

Bacillus genera are the rich resources of different secondary metabolites, comprising polypeptides, lipopeptides, macrolides, polyketides, fatty acids, lipoamides, isocoumarin, and carotenoids. The structure of these marine isolates has multiple uses and can be extracted from complicated biosynthetic ways. Several biologically active ingredients have great potential for advancement of potent drugs and agrochemicals. Owing to the genetic ability to adjust at extreme environments and the *Bacillus* strains isolated from unique conditions (such as hydrothermal vents, salt lakes, and deep sea, pH > 9.0) can create beneficial biologically active compounds. Still in marine isolates, recessive biosynthetic gene clusters can be triggered to determine new organic products through cultivation under changing stress environment (such as nutrients, temperature, pH, or salinity stress). Additional primary characteristic of *Bacillus* is that it will cause reduction of heavy metals, which can be viewed as a candidate for the remediation of toxic heavy metals. Bioremediation is an ecofriendly and inexpensive approach to eliminate xenobiotics from contaminated environments. The next generation of sequencing technology provides important insights into the molecular and biological mechanisms that play a role in the biological remediation of ecological pollutants (such as heavy metal pollutants). These findings will enhance the strategy of bacterial bioremediation, monitor its progress, and determine its success. Biological pesticides based on *Bacillus* can progress plant health by a unique mode of action and therefore have great commercial potential. As the occurrence of *B. subtilis* is frequent in ecosystem, its production capacity of large quantities of antibiotics should be accounted when combating pests. Biological pesticides based on *Bacillus* have great potential for future sustainable agricultural practices.

Acknowledgments The authors want to thank the Institute of Soil and Environmental Sciences, University of Agriculture Faisalabad, Pakistan. Muhammad Mahroz Hussain* wants to thank the Higher Education Commission Pakistan for providing fellowship under International Research Support Initiative Program (IRSIP) program (IRSIP No: 1-8/HEC/HRD/2020/10595; PIN: IRSIP 45 BMS 60).

References

Abatenh E, Gizaw B, Tsegaye Z, Wassie M (2017) Application of microorganisms in bioremediation-review. J Environ Microbiol 1(1):2–9

Abdel-Monem M, Al-Zubeiry A, Al-Gheethi A (2010) Biosorption of nickel by pseudomonas cepacia 120S and Bacillus subtilis 117S. Water Sci Technol 61(12):2994–3007

Abioye O, Oyewole O, Oyeleke S, Adeyemi M, Orukotan A (2018) Biosorption of lead, chromium and cadmium in tannery effluent using indigenous microorganisms. Brazilian J Biol Sci 5(9):25–32

Abou-Aly HE, Youssef AM, El-Meihy RM, Tawfik TA, El-Akshar EA (2019) Evaluation of heavy metals tolerant bacterial strains as antioxidant agents and plant growth promoters. Biocatalysis Agric Biotechnol 19:101110

Adams GO, Fufeyin PT, Okoro SE, Ehinomen I (2015) Bioremediation, biostimulation and bioaugmention: a review. Intern J Environ Biorem Biodegrad 3(1):28–39

Ahmed E, Holmström SJ (2014) Siderophores in environmental research: roles and applications. Microb Biotechnol 7(3):196–208

Arkhipova T, Veselov S, Melentiev A, Martynenko E, Kudoyarova G (2005) Ability of bacterium Bacillus subtilis to produce cytokinins and to influence the growth and endogenous hormone content of lettuce plants. Plant Soil 272(1–2):201–209

Arkhipova T, Prinsen E, Veselov S, Martinenko E, Melentiev A, Kudoyarova G (2007) Cytokinin producing bacteria enhance plant growth in drying soil. Plant Soil 292(1–2):305–315

Aryal M (2015) Removal and recovery of nickel ions from aqueous solutions using Bacillus sphaericus biomass. Inter J Environ Res 9(4):1147–1156

Ayangbenro AS, Babalola OO (2017) A new strategy for heavy metal polluted environments: a review of microbial biosorbents. Inter J Environ Res Pub Heal 14(1):94

Aziz HMA (2018) Antimicrobial activities of pseudomonas putida strain isolated from soil. Sciences 8(02):543–553

Bai J, Yang X, Du R, Chen Y, Wang S, Qiu R (2014) Biosorption mechanisms involved in immobilization of soil Pb by Bacillus subtilis DBM in a multi-metal-contaminated soil. J Environ Sci 26(10):2056–2064

Bai J, Chao Y, Chen Y, Wang S, Qiu R (2019) The effect of interaction between Bacillus subtilis DBM and soil minerals on Cu (II) and Pb (II) adsorption. J Environ Sci 78:328–337

Baindara P, Gautam A, Raghava G, Korpole S (2017) Anticancer properties of a defensin like class IId bacteriocin Laterosporulin10. Sci Rep 7:46541

Baltz RH (2009) Daptomycin: mechanisms of action and resistance, and biosynthetic engineering. Curr Opinion Chem Biol 13(2):144–151

Blom D, Fabbri C, Connor E, Schiestl F, Klauser D, Boller T, Eberl L, Weisskopf L (2011) Production of plant growth modulating volatiles is widespread among rhizosphere bacteria and strongly depends on culture conditions. Environ Microbiol 13(11):3047–3058

Blunt JW, Copp BR, Keyzers RA, Munro MH, Prinsep MR (2017) Marine natural products. Nat Prod Rep 34(3):235–294

Boopathy R (2000) Factors limiting bioremediation technologies. Bioresour Technol 74(1):63–67. https://doi.org/10.1016/S0960-8524(99)00144-3

Božič M, Pricelius S, Guebitz GM, Kokol V (2010) Enzymatic reduction of complex redox dyes using NADH-dependent reductase from Bacillus subtilis coupled with cofactor regeneration. Appl Microbiol Biotechnol 85(3):563–571

Branquinho R, Meirinhos-Soares L, Carriço JA, Pintado M, Peixe LV (2014) Phylogenetic and clonality analysis of Bacillus pumilus isolates uncovered a highly heterogeneous population of different closely related species and clones. FEMS Microbiol Ecol 90(3):689–698

Caulier S, Nannan C, Gillis A, Licciardi F, Bragard C, Mahillon J (2019) Overview of the antimicrobial compounds produced by members of the Bacillus subtilis group. Frontiers Microbiol 10:302

Chauhan M, Solanki M, Nehra K (2017) Putative mechanism of cadmium bioremediation employed by resistant bacteria. Jordan J Biol Sci 10(2)

Chen X-H, Vater J, Piel J, Franke P, Scholz R, Schneider K, Koumoutsi A, Hitzeroth G, Grammel N, Strittmatter AW (2006) Structural and functional characterization of three polyketide synthase gene clusters in Bacillus amyloliquefaciens FZB 42. J Bacteriol 188(11):4024–4036

Compant S, Duffy B, Nowak J, Clément C, Barka EA (2005) Use of plant growth-promoting bacteria for biocontrol of plant diseases: principles, mechanisms of action, and future prospects. Appl Environ Microbiol 71(9):4951–4959

Cumberland SA, Douglas G, Grice K, Moreau JW (2016) Uranium mobility in organic matter-rich sediments: a review of geological and geochemical processes. Earth-Science Rev 159:160–185

De J, Ramaiah N, Vardanyan L (2008) Detoxification of toxic heavy metals by marine bacteria highly resistant to mercury. Mar Biotechnol 10(4):471–477

Deng Y, Lu Z, Lu F, Wang Y, Bie X (2011) Study on an antimicrobial protein produced by Paenibacillus polymyxa JSa-9 isolated from soil. World J Microbiol Biotechnol 27(8):1803–1807

Divya M, Aanand S, Srinivasan A, Ahilan B (2015) Bioremediation–an eco-friendly tool for effluent treatment: a review. Inter J Appl Res 1(12):530–537

Fang J, Kato C (2007) FAS or PKS, lipid biosynthesis and stable carbon isotope fractionation in deep-sea piezophilic bacteria. Commun Curr Res Educ Top Trends Appl Microbiol 1:190–200

Fickers P (2012) Antibiotic compounds from Bacillus: why are they so amazing. Americ J Biochem Biotechnol 8:38–43

Gorman-Lewis D (2014) Enthalpies and entropies of Cd and Zn adsorption onto Bacillus licheniformis and enthalpies and entropies of Zn adsorption onto Bacillus subtilis from isothermal titration calorimetry and surface complexation modeling. Geomicrobiol J 31(5):383–395

Goyal P, Belapurkar P, Kar A (2019) A review on in vitro and in vivo bioremediation potential of environmental and probiotic species of bacillus and other probiotic microorganisms for two heavy metals, cadmium and nickel. Biosci Biotechnol Res Asia 16(1):01–13

Hamdache A, Lamarti A, Aleu J, Collado IG (2011) Non-peptide metabolites from the genus Bacillus. J Nat Prod 74(4):893–899

Huang F, Guo C-L, Lu G-N, Yi X-Y, Zhu L-D, Dang Z (2014) Bioaccumulation characterization of cadmium by growing Bacillus cereus RC-1 and its mechanism. Chemosphere 109:134–142

Idris EE, Bochow H, Ross H, Borriss R (2004) Use of Bacillus subtilis as biocontrol agent. VI. Phytohormone like action of culture filtrates prepared from plant growth-promoting Bacillus amyloliquefaciens FZB24, FZB42, FZB45 and Bacillus subtilis FZB37/Nutzung von Bacillus subtilis als Mittel für den biologischen Pflanzenschutz. VI. Phytohormonartige Wirkung von Kulturfiltraten von pflanzenwachstumsfördernden Bacillus amyloliquefaciens FZB24, FZB42, FZB45 und Bacillus subtilis FZB37. Zeitschrift für Pflanzenkrankheiten und Pflanzenschutz/ J Plant Diseases and Protection. 583–597

Igiri BE, Okoduwa SI, Idoko GO, Akabuogu EP, Adeyi AO, Ejiogu IK (2018) Toxicity and bioremediation of heavy metals contaminated ecosystem from tannery wastewater: a review. J Toxicol 2018

Jain A, Udayashankara T, Lokesh K, Sudarshan B (2017) Bioremediation of lead, nickel and copper by metal resistant Bacillus licheniformis isolated from mining site: optimization of operating parameters under laboratory conditions. Int J Res Eng Technol 5:13–32

Jayashree R, Nithya SE, Rajesh P, Krishnaraju M (2012) Biodegradation capability of bacterial species isolated from oil contaminated soil. J Academia Indust Res 1(3):127–135

Jeong S-Y (2019) Algicidal activity of strain MS-51 against a HAB causing Alexandrium tamarense. Inter J Appl Environ Sci 14(6):655–663

Karakus BZ, Korkmaz İ, Demirci K, Sinan Arslan K, Unlu O, Catal T (2018) A combined treatment using ethylmethane sulfonate and ultraviolet light to compare amylase production by three Bacillus sp. isolates. Prepara Biochem Biotechnol 48(9):815–822

Khan A, Singh P, Srivastava A (2018) Synthesis, nature and utility of universal iron chelator–Siderophore: a review. Microbiol Res 212:103–111

Kim P-I, Ryu J-W, Kim Y-H, Chi Y-T (2010) Production of biosurfactant lipopeptides iturin A, fengycin, and surfactin A from Bacillus subtilis CMB32 for control of Colletotrichum gloeosporioides. J Microbiol Biotechnol 20(1):138–145

Kloepper JW, Ryu C-M, Zhang S (2004) Induced systemic resistance and promotion of plant growth by Bacillus spp. Phytopathology 94(11):1259–1266

Komaki H, Hosoyama A, Ichikawa N, Igarashi Y (2016) Draft genome sequence of marine-derived Bacillus subtilis TP-B0611, a producer of bacilosarcins and amicoumacins. Genome Announc 4(5)

Leclere V, Béchet M, Adam A, Guez J-S, Wathelet B, Ongena M, Thonart P, Gancel F, Chollet-Imbert M, Jacques P (2005) Mycosubtilin overproduction by Bacillus subtilis BBG100

enhances the organism's antagonistic and biocontrol activities. Appl Environ Microbiol 71(8):4577–4584
Liu L, Oza S, Hogan D, Perin J, Rudan I, Lawn JE, Cousens S, Mathers C, Black RE (2015) Global, regional, and national causes of child mortality in 2000–13, with projections to inform post-2015 priorities: an updated systematic analysis. Lancet 385(9966):430–440. https://doi.org/10.1016/S0140-6736(14)61698-6
Mondol MAM, Shin HJ, Islam MT (2013) Diversity of secondary metabolites from marine Bacillus species: chemistry and biological activity. Mar Drugs 11(8):2846–2872
Mosa KA, Saadoun I, Kumar K, Helmy M, Dhankher OP (2016) Potential biotechnological strategies for the cleanup of heavy metals and metalloids. Front Plant Sci 7:303
Nguyen PT, Nguyen XT, Nguyen TV, Nguyen TT, Vu TQ, Nguyen HT, Pham NT, Thi Dinh TM (2020) Treatment of Cd2+ and Cu2+ ions using modified apatite ore. J Chem 2020
Nithya C, Gnanalakshmi B, Pandian SK (2011) Assessment and characterization of heavy metal resistance in Palk Bay sediment bacteria. Marine Environ Res 71(4):283–294
Ojuederie OB, Babalola OO (2017) Microbial and plant-assisted bioremediation of heavy metal polluted environments: a review. Inter J Environ Res Pub Heal. 14(12):1504
Ongena M, Jacques P (2008) Bacillus lipopeptides: versatile weapons for plant disease biocontrol. Trends Microbiol 16(3):115–125
Ongena M, Jourdan E, Adam A, Paquot M, Brans A, Joris B, Arpigny JL, Thonart P (2007) Surfactin and fengycin lipopeptides of Bacillus subtilis as elicitors of induced systemic resistance in plants. Environ Microbiol 9(4):1084–1090
Pirri G, Giuliani A, Nicoletto SF, Pizzuto L, Rinaldi AC (2009) Lipopeptides as anti-infectives: a practical perspective. Central European J Biol 4(3):258–273
Pratush A, Kumar A, Hu Z (2018) Adverse effect of heavy metals (As, Pb, Hg, and Cr) on health and their bioremediation strategies: a review. Inter Microbiol 21(3):97–106
Raaijmakers JM, De Bruijn I, de Kock MJ (2006) Cyclic lipopeptide production by plant-associated pseudomonas spp.: diversity, activity, biosynthesis, and regulation. Mol Plant-Microbe Interact 19(7):699–710
Raimundo I, Silva SG, Costa R, Keller-Costa T (2018) Bioactive secondary metabolites from octocoral-associated microbes—new chances for blue growth. Mar Drugs 16(12):485
Rudrappa T, Biedrzycki ML, Kunjeti SG, Donofrio NM, Czymmek KJ, Paul WP, Bais HP (2010) The rhizobacterial elicitor acetoin induces systemic resistance in Arabidopsis thaliana. Commun Integrat Biol 3(2):130–138
Ryu CM, Murphy JF, Mysore KS, Kloepper JW (2004) Plant growth-promoting rhizobacteria systemically protect Arabidopsis thaliana against cucumber mosaic virus by a salicylic acid and NPR1-independent and jasmonic acid-dependent signaling pathway. Plant J 39(3):381–392
Santoyo G, Orozco-Mosqueda MC, Govindappa M (2012) Mechanisms of biocontrol and plant growth-promoting activity in soil bacterial species of Bacillus and Pseudomonas: a review. Biocont Sci Technol 22(8):855–872
Sari A, Mendil D, Tuzen M, Soylak M (2008) Biosorption of Cd (II) and Cr (III) from aqueous solution by moss (Hylocomium splendens) biomass: equilibrium, kinetic and thermodynamic studies. Chem Engin J 144(1):1–9
Schneider K, Chen X-H, Vater J, Franke P, Nicholson G, Borriss R, Süssmuth RD (2007) Macrolactin is the polyketide biosynthesis product of the pks2 cluster of Bacillus amyloliquefaciens FZB42. J Nat Prod 70(9):1417–1423
Shameer S (2016) Biosorption of lead, copper and cadmium using the extracellular polysaccharides (EPS) of Bacillus sp., from solar salterns. 3. Biotech 6(2):194
Shanahan P (2004) Bioremediation. Waste Containment and Remediation Technology, Spring 2004, Massachusetts Institute of Technology. MIT Open Course Ware,
Shekharaiah PC, Sanyal D, Dasgupta S, Sapre A, Banik A (2020) Heavy metal mitigation with special reference to bioremediation by Mixotrophic algae-bacterial Protocooperation. In: Cellular and molecular phytotoxicity of heavy metals. Springer, pp. 305–334

Shukla S (2016) Secondary metabolites from marine microorganisms and therapeutic efficacy: a mini review

Singh N, Verma T, Gaur R (2013) Detoxification of hexavalent chromium by an indigenous facultative anaerobic Bacillus cereus strain isolated from tannery effluent. Afr J Biotechnol 12(10)

Tsan M-F, Gao B (2007) Pathogen-associated molecular pattern contamination as putative endogenous ligands of toll-like receptors. J Endotoxin Res 13(1):6–14

Valencia-Cantero E, Hernández-Calderón E, Velázquez-Becerra C, López-Meza JE, Alfaro-Cuevas R, López-Bucio J (2007) Role of dissimilatory fermentative iron-reducing bacteria in Fe uptake by common bean (Phaseolus vulgaris L.) plants grown in alkaline soil. Plant Soil 291(1–2):263–273

Venosa AD, Lee K, Suidan MT, Garcia-Blanco S, Cobanli S, Moteleb M, Haines JR, Tremblay G, Hazelwood M (2002) Bioremediation and biorestoration of a crube oil-contaminated freshwater wetland on the St. Lawrence River. Biorem J 6(3):261–281

Vijayaraghavan K, Yun Y-S (2008) Bacterial biosorbents and biosorption. Biotechnol Adv 26(3):266–291

Yang X, Yousef AE (2018) Antimicrobial peptides produced by Brevibacillus spp.: structure, classification and bioactivity: a mini review. W J Microbiol Biotechnol 34(4):57

Yu S, Teng C, Bai X, Liang J, Song T, Dong L, Jin Y, Qu J (2017) Optimization of siderophore production by Bacillus sp. PZ-1 and its potential enhancement of phytoextration of Pb from soil. J Microbiol Biotechnol 27(8):1500–1512

Zhang HL, Hua HM, Pei YH, Yao XS (2004) Three new cytotoxic cyclic acylpeptides from marine Bacillus sp. Chem Pharmaceutical Bullet 52(8):1029–1030

Zin NBM, Yusof BM, Oslan SN, Wasoh H, Tan JS, Ariff AB, Halim M (2017) Utilization of acid pre-treated coconut dregs as a substrate for production of detergent compatible lipase by Bacillus stratosphericus. AMB Express 7(1):1–13

Zokaeifar H, Balcázar JL, Saad CR, Kamarudin MS, Sijam K, Arshad A, Nejat N (2012) Effects of Bacillus subtilis on the growth performance, digestive enzymes, immune gene expression and disease resistance of white shrimp, Litopenaeus vannamei. Fish Shellfish Immunol 33(4):683–689

Chapter 23
Impacts of *Bt* Brinjal on Economic Benefit of Farmers and Environmental Sustainability in Bangladesh

Sanjoy Kumar Paul, Nur Uddin Mahmud, and Tofazzal Islam

Abstract Brinjal or eggplant (*Solanum melongena* L.) is severely attacked by the brinjal shoot and fruit borer (BSFB) insect, which reduces the yield and causes massive loss in brinjal production. This crop plant belongs to the family of Solanaceae, which is one of the common cultivated vegetables throughout the Asia including Bangladesh. This chapter aimed to comprehensively review the knowledge and information of genetically engineered (GE) brinjal called *Bt* brinjal on economic benefit of growers and environmental sustainability in Bangladesh. As a potential solution to protect eggplant from BSFB, Bangladesh developed and released genetically modified (GM) four *Bt* brinjal varieties. This transgenic brinjal carries an additional *Cry1Ac* gene from *Bacillus thuringiensis* (*Bt*) that provides built-in protection against the devastating BSFB insect. The generated *Bt* protein is activated in the insect's alkaline gut and ruptures the gut membrane which is detrimental for the insects, and thus no insecticide spray is required. But, in case of non-*Bt* brinjal, farmers have to spray insecticides frequently to control that insect but without much success. The residue of applied insecticides is detrimental for the environment and human health. The recent data showed that approximately 17% of total brinjal growers of the country were adopted and directly benefited from this GM technology. By cultivating the *Bt* brinjal, farmers save approximately USD 115 per hectare as an input cost mainly due to no or less insecticide spray is required for *Bt* brinjal compared to the non-*Bt*. By the use of *Bt* technology, the cost for insecticide use could be reduced up to 61% compared to conventional varieties. Additionally, *Bt* brinjal farmers receive 30% price premium compared to non-*Bt*. As a result, *Bt* brinjal farmers get five times higher returns per hectare than the farmers cultivating non-*Bt* varieties. It is worth noting that reduced application of insecticide is remarkably beneficial to the environment. No significant negative impacts of *Bt* brinjal have so far been reported on soil arthropods and other nontarget beneficial organ-

S. K. Paul · N. U. Mahmud · T. Islam (✉)
Institute of Biotechnology and Genetic Engineering (IBGE), Bangabandhu Sheikh Mujibur Rahman Agricultural University, Gazipur, Bangladesh
e-mail: tofazzalislam@bsmrau.edu.bd

M. T. Islam et al. (eds.), *Bacilli in Agrobiotechnology*, Bacilli in Climate Resilient Agriculture and Bioprospecting,
https://doi.org/10.1007/978-3-030-85465-2_23

isms. However, the durability of insect resistance by a single *Cry1Ac* gene and its consequence to the environment are still unclear. Overall, *Bt* brinjal is a successful GM crop in Bangladesh, which is found profitable for the farmers and good for the environment and human health.

Keywords Genetically modified (GM) crops · *Bacillus thuringiensis* · *Bt* brinjal · *Cry1Ac* gene · Sustainable environment · Economic benefit

23.1 Introduction

Brinjal or eggplant (*Solanum melongena* L.) is one of the most popular vegetables cultivated all over Asia, including Bangladesh (Prodhan et al. 2018). It is an important year-round vegetable which is popularly known as begun in this country. The quantity of brinjal is increasing in terms of consumption, which ranks third after potato among the vegetables cultivated in the country (Rashid et al. 2018). Brinjal is grown year-round in this country, and approximately, 50,956 hectares of land are under cultivation by 150,000 farmers (BBS 2018). But, in the field level, the vegetable is seriously infested with brinjal shoot and fruit borer (BSFB) (*Leucinodes orbonalis* Guenee), one of the top destructive pests among others. In Bangladesh, up to 86% yield losses of brinjal have been described due to BSFB. To reduce brinjal shoot and fruit injury by BSFB, farmers depend on repeated application of insecticides (Prodhan et al. 2018). Practically, the larva of BSFB burrows into the petioles as well as tender shoots. As a result, leaves drooping and shedding of flower buds are common phenomena. In the developing fruits, severe damage is mainly occurring because the caterpillars make tunnel inside the fruits through feeding the fleshy part. However, Raza et al. (2018) reported that under different trade names, 16 insecticides were usually sprayed to tackle this devastating enemy of brinjal. Additionally, for controlling this insect, farmers usually apply broad-spectrum insecticides two to three sprays per week and in some cases twice a day (Del Prado-Lu 2015). Consequently, it is not surprising to spray over 100 times per growing season, resulting in high residual effects on consumable fruits. Such a hazardous insecticide-dependent strategy poses threats to the atmosphere as well as human health. In addition to health and environmental concerns, populations of BSFB have also developed resistance to various foliar insecticides (Shirale et al. 2017). Most significantly, about 68.75% of brinjal growers have awareness, while 31.25% brinjal growers did not have proper knowledge about the deleterious effects of using high amounts of insecticides. The brinjal growers are frequently using same active ingredients under different commercial names (Raza et al. 2018). However, inefficient use of insecticides poses a series of ecological problems, such as development of pesticide resistance, nonpoint pollution, bioaccumulation in food chain, and a huge loss of biodiversity (Zhao et al. 2017). Moreover, the loss and decomposition rate of insecticides on crop foliar is typically more than 99%, which are caused by runoff, spray drift, and rolling down at the time of field application (Nuruzzaman et al. 2016; Song et al. 2017; Zhao et al. 2017). As a result, the insect management based on insecticides is detrimental to humans, livestock, and environment (Javed et al. 2017).

A potential solution to control BSFB is the genetically modified *Bt* eggplant which carries an additional gene *Cry1Ac* from a soil bacterium, *Bacillus thuringiensis* (*Bt*). This gene provides brinjal built-in protection against the *Lepidopteran* insect pest (Rashid et al. 2018). As the first South Asian country, Bangladesh has developed and released four GM brinjal varieties in 2014 by introducing *Cry1Ac* gene (*Bt* brinjal) for commercial cultivation of brinjal as food crops (Ahmed et al. 2019; Prodhan et al. 2018). In the last two decades, utilization of *Bt* technology has revolutionized the agricultural practices in different countries by developing and commercializing the transgenic plants that have a built-in insecticidal toxin-expressing genes (Castagnola and Jurat-Fuentes 2012). The remarkable increase in acceptance of biotech crop technique can be recognized to its contribution such as higher food production which improves food availability for the native community and global levels. On the other hand, improvement of food quality through bio-fortification of important nutrients as well as social positive impact of the growers through economic development are also considered as important role of this technology (James 2014a, b). However, the approval issue of GM crops for commercial cultivation remains controversial in Bangladesh and globally. The common criticisms include GM crops, which are destructive to the ecology and environment and harmful to human health. On the other hand, small farmers will lose their accessibility due to higher cost and intellectual property rights of the seed producing company. Additionally, it is also claimed that GM crops, including *Bt* brinjal, deliver no yield paybacks (Ahmed et al. 2019). This book chapter aimed to provide an independent rigorous scientific information based on review of reputed research articles that could address some of these key criticisms. The specific objectives of this book chapter were as follows: (i) to describe the concept of *Bt* brinjal development; (ii) to illustrate the economic benefit of *Bt* brinjal cultivation over non-*Bt* varieties; and (iii) to demonstrate the impact of *Bt* brinjal cultivation on maintenance of environmental sustainability.

23.2 The Developmental Concept of *Bt* Brinjal in Bangladesh

23.2.1 History of Bt Brinjal

The development of transgenic plants and commercialization of their cultivation have revolutionized agricultural practices. Improvement of this ground-breaking insect pest control technology was backed by the identification and characterization of insecticidal *Bt* proteins. Moreover, the development was assisted by the advanced plant transformation technique and genetic engineering. In most of the cases, the developed transgenic crops are occupying the space for conventional crop varieties due to their insect resistance characters, lesser insecticide requirements, as well as greater yield potentials (Castagnola and Jurat-Fuentes 2012). In the year 2005, two BARI (Bangladesh Agricultural Research Institute) scientists visited the Mahyco Agricultural Research Station, Jalna, Maharashtra, India, with nine popular brinjal cultivars from Bangladesh. These cultivars were hybridized with the candidate brinjal plants (Fig. 23.1) containing the insecticidal toxin-expressing *Bt* gene from the bacterium *Bacillus thuringiensis* (*Bt*) as well as collected F1 seeds.

After that, backcrosses (BC1) were done there, and the resultant seeds were brought back to Bangladesh. Later on, successive backcrosses were produced at BARI from 2006 to 2011. While targeted plants were selected, multilocation confined field trials were conducted at seven different locations of BARI (APAARI 2018).

Over a 10-year period, public sector Bangladeshi agricultural researchers, with support from Maharashtra Hybrid Seeds Co. Pvt. Ltd. (Mahyco) and researchers based in the United States, have developed a series of GM varieties of *Bt* brinjal that are resistant to BSFB (APAARI 2018). Therefore, in 2013 after compiling all the results, BARI applied to National Committee of Biosafety (NCB), MoEF (Ministry of Environment and Forest) following Biosafety Guidelines/Rules of the country through MoA (Ministry of Agriculture) for the approval of four *Bt* brinjal genotypes. The applied *Bt* brinjal genotypes were BARI *Bt* begun-1, BARI *Bt* begun-2, BARI *Bt* begun-3, and BARI *Bt* begun-4.

Finally, on October 28, 2013, Bangladesh's National Committee on Biosafety (NCB) approved for commercial cultivation of four indigenous varieties of *Bt* brinjal, which were resistant to attacks by the BSFB, a common pest in South and Southeast Asia (Shelton et al. 2018).

Fig. 23.1 Cultivar Uttara (brinjal variety) showing injury by the brinjal shoot and fruit borer to non-*Bt* brinjal (**a**) and lack of injury in *Bt* brinjal (**b**) (Shelton et al. 2019)

23.3 Structure and Mode of Action of *Cry1Ac* Protein

Different insecticidal proteins such as Vip, Cyt, or Cry toxins were produced against different insect specificities by a gram-positive bacterium, *Bacillus thuringiensis* (*Bt*). Among these toxins, the three-domain *Cry* toxins (3d-*Cry*), the biggest *Cry* toxin family, have been shown to be highly effective for agricultural pest control (Pardo-López et al. 2013). Some of these 3d-Cry toxins like *Cry1Ab*, *Cry1Fa*, and *Cry1Ac* have already been introduced and expressed successfully to make transgenic crops such as maize, cotton, soybean, or brinjal reducing the use of chemical insecticides and in some cases increasing yields (Sanahuja et al. 2011; APAARI 2018). Herein, some of the *Bt* crops which are commercially available for cultivation are presented in Table 23.1. Generally, the *Cry1Ac* structure comprises seven distinct domains. The domain (D) number starts from D-I to D-VII. Among them, three canonical toxin core domains are D-I, D-II, and D-III, while four protoxin domains are D-IV, D-V, D-VI, and D-VII. *Cry1Ac*-FL is a sickle-shaped structure (Fig. 23.2) where the toxic core looks like handle and the protoxin domains as the blade (Evdokimov et al. 2014).

However, around thousands of individual toxin proteins have already been discovered. Among them, the majority belong to the "three-domain" toxin family. The insecticidal *Bt* proteins contain three structural domains within the toxic core. This is the most populous fraternity with thousands of known members, of which more than 50% are produced by their parent *Bt* strains as -1200-residue protoxins. These protoxins are consisting of a proteolytically labile C-terminal segment that is sometimes mentioned as the protoxin domain. On the other hand, another part is N-terminal -600-residue segment that encodes the three-domain toxic core (Schnepf et al. 1998; Crickmore et al. 1998; de Maagd et al. 2001). In fact, the isolation of *Cry1Ac* gene is done from the soil-living bacterium *Bacillus thuringiensis*. After the ingestion of *Cry1Ac* gene introduced plant parts by the BSFB larvae, the *Bt* protein is activated in the insect's alkaline gut and binds to the gut wall. After that, the gut wall breaks down and allowing the *Bt* spores to invade the insect's body cavity. However, current data suggested that the mode of action of *Cry1Ac* protoxins may not similar with that of activated toxins. Because the pore formation resulted from the binding of protoxins to insect receptor is different from that resulting from activated toxin. Based on these data, Gómez et al. (2014) proposed a dual mode of action of *Cry* proteins. They suggested that the two pre-pore structures may have differential roles in toxicity against selected targets by increasing their range of action. Moreover, insect may harbor different receptors, but different midgut proteases would influence the rate of protoxin/toxin activation. In support to the dual mode of action of *Cry* proteins, it was reported that several insect populations have developed resistance to *Cry* proteins and showed significantly lower resistance ratios to *Cry1Ac* protoxins than the activated toxins (Tabashnik et al. 2015). Therefore, by activating with the insect gut proteases, *Bt* 3d-*Cry* toxins are produced as protoxins which yield an active three-domain toxin. Herein, either *Cry1Ac* activated toxin or protoxin will create binding with the cadherin receptor forming

Table 23.1 *Bt* crops which are commercially available for cultivation (Adapted from ISAAA's GM Approval Database, 2020 http://www.isaaa.org/gmapprovaldatabase/; Castagnola and Jurat-Fuentes 2012)

Crops	Trade/commercial name	Targeted insects[a]	Involved genes	Method of transgene insertion[b]	GM developers
Maize (*Zea mays* L.)	Agrisure™ CB/LL	ECB	*Cry1Ab*	Direct DNA transfer	Syngenta
	Agrisure™ GT/CB/LL				
	Agrisure™ 3000GT	ECB, WCR	*Cry1Ab*, *Cry3Aa*	Direct DNA transfer, Agrobacterium-mediated transformation	
	Agrisure™ CB/LL/RW				
	Agrisure® Viptera™ 3110	ECB, FAW, CEW, BCW, WBC	*Cry1Ab,Vip3Aa20*	Direct DNA transfer, Agrobacterium-mediated transformation	
	Agrisure® Viptera™ 3111	ECB, WCR, FAW, CEW, BCW, WBC	*Cry1Ab*, *Vip3Aa20*		
	Agrisure® Viptera™ 4				
	YieldGard™ CB + RR	ECB	*Cry1Ab*	Biolistics	Monsanto
	YieldGard™ VT Triple	CRW	*Cry1Ab*		
	YieldGard™ Plus	ECB	*Cry3Bb1*		
	Genuity® VT Double Pro™	CEW, ECB, FAW	*Cry1A.105/Cry2Ab2*	Agrobacterium-mediated transformation	

Maize (*Zea mays* L.)	Genuity® VT Triple Pro™	CEW, CRW, ECB, FAW	*Cry1A.105/Cry2ab2*, *Cry3Bb1*	Agrobacterium-mediated transformation Agrobacterium-mediated transformation	Monsanto
	Genuity® SmartStax™	BCW, CEW, CRW, ECB, FAW, WBC, SCB, SWCB, SCSB, CEW, SCB, WBC, WCR	*Cry1A.105/Cry2Ab2,Cry1Fa2* *Cry3Bb1*, *Cry34/35Ab1*	Agrobacterium-mediated transformation Biolistics Biolistics Agrobacterium-mediated transformation	
Cotton (*Gossypium hirsutum L.*)	Bollgard™ Cotton, Ingard™	PBW, TBW	*Cry1Ac*	Agrobacterium-mediated transformation	Monsanto
	Bollgard II™ Cotton	CBW, PBW, TBW	*Cry1Ac/Cry2Ab2*	Agrobacterium-mediated transformation/biolistics	
	Optimum™ Intrasect	ECB	*Cry1Ab*	Biolistics	DuPont
		ECB, WBC, BCW, FAW	*Cry1F, Cry1Ab*	Biolistics, biolistics	
Cotton (*Gossypium hirsutum L.*)	WideStrike™ Cotton	*Lepidopteran* insect resistance	*Cry1AcCry1F*	NA	Dow AgroSciences LLC
Eggplant (*Solanum melongena*)	BARI Bt Begun-1, 2, 3 and 4	BSFB	*Cry1Ac*	Agrobacterium-mediated transformation	MAHYCO
Soybean (*Glycine max L.*)	Conkesta Enlist E3™ Soybean	*Lepidopteran* insect resistance	*Cry1AcCry1F*	NA	Dow AgroSciences LLC
	Intacta™ Roundup Ready™ 2 Pro		*Cry1Ac*	NA	Monsanto

(continued)

Table 23.1 (continued)

Crops	Trade/commercial name	Targeted insects[a]	Involved genes	Method of transgene insertion[b]	GM developers
Rice (*Oryza sativa L.*)	BT Shanyou 63	*Lepidopteran* insect resistance	*Cry1AbCry1Ac*	NA	Huazhong Agricultural University (China)
	Huahui-1				
Potato (*Solanum tuberosum L.*)	Atlantic NewLeaf™ potato	*Coleopteran* insect resistance	*Cry3A*	Agrobacterium-mediated transformation	Monsanto
	New Leaf™ Russet Burbank potato		*Cry3A*		
	New Leaf™ Plus Russet Burbank potato		*Cry3A*		
	Superior NewLeaf™ potato		*Cry3A*		

[a]Common insect name abbreviations are *ECB* European corn borer, *WCR* western corn rootworm, *FAW* fall armyworm, *CEW* corn earworm, *BCW* black cutworm, *WBC* western bean cutworm, *CBW* cotton bollworm, *TBW* tobacco budworm, *PBW* pink bollworm, *SCB* sugar cane borer, *SWCB* southwestern corn borer, *SCSB* southern cornstalk borer, *SPB* spotted bollworm, *BSFB* brinjal shoot and fruit borer

[b]NA, not available

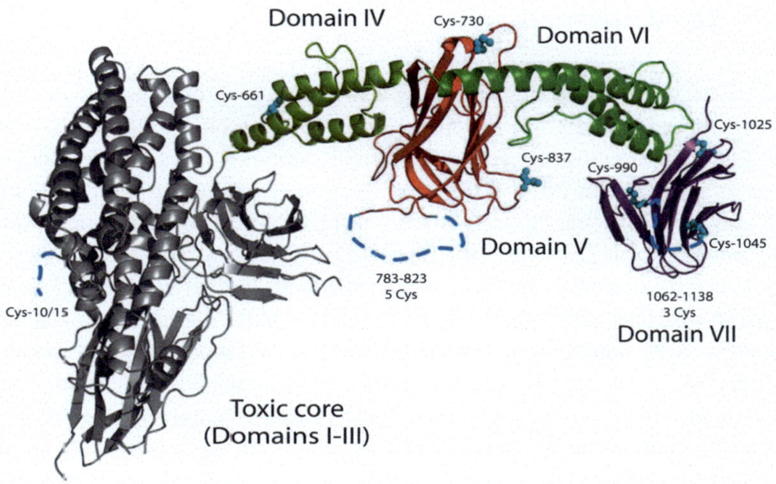

Fig. 23.2 Overall structure of *Cry1Ac* monomer that is domain-wise colored. The toxic core (domains I to III) are light gray, domain IV is dark green, domain V is red, domain VI is green, and domain VII is violet in color (Evdokimov et al. 2014)

distinct oligomers that insert into the membrane forming lytic pores. This consequences the reduction of potential evolution of resistance and possibly to broaden the range of insect targets. However, for the sustainable use of *Bt* crops, it is important to achieve a stable expression of *Cry* full-length proteins which will pose the same consequences. These phenomena are delaying resistance and giving protection from a wider number of insect pests (Soberón et al. 2016).

23.4 The Economic Benefit Incurred by the *Bt* Brinjal Growers

BSFB is a destructive pest of brinjal, which adversely affects quality and yield of brinjal fruit throughout the world (Javed et al. 2017). In modern world, widespread adoption of productivity-enhancing technologies has shifted production, with economic and environmental effects. Agricultural technologies like *Bt* brinjal offer new opportunities that must be evaluated in an increasingly complex world. A number of factors influence the effect of new or improved agricultural technologies on production and consumption. These include the characteristics of the existing agricultural and market systems, the agroecological conditions, socioeconomic status, and sources of information about these technologies, as well as beliefs, norms, and cultural practices. Adoption of agricultural technologies has proven to be effective in improving food availability and food quality, increasing farmers benefit from agricultural production system, as well as responsive to environmental risks and uncertainties (Ahmed et al. 2019).

23.4.1 Rapid Adoption

In Bangladesh, the first *Bt* brinjal seedling distribution program was held on January 22, 2014, where 20 farmers received seedlings to grow on a trial basis (APAARI 2018), and in the following years, *Bt* brinjal adoption increased tremendously (Shelton et al. 2018). In 2014–2015, BARI conducted demonstration trials in 108 farmer's fields of different climatic conditions (19 districts) of Bangladesh through its On-farm Research Division (OFRD). In 2015–2016 and 2016–2017, they conducted demonstration trials in 250 farmer's fields of 25 districts and 512 farmer's fields of 36 districts, respectively. In 2017–2018, BARI distributed seeds among 569 growers in 40 districts of Bangladesh. Besides, the Department of Agricultural Extension (DAE) also distributed the *Bt* brinjal seeds to the farmers. In this connection, for the production of *Bt* brinjal on commercial basis at the field level, 6000 and 7001 farmers got seeds in 2016–2017 and 2017–2018, respectively. In 2018, Bangladesh Agricultural Development Corporation (BADC) took the initiative to sell the seeds to an additional 17,950 farmers (Fig. 23.3).

In the year 2018, an estimate proved that approximately 150,000 farmers grown brinjal crops in Bangladesh. Among them, an estimated ~17% of brinjal farmers enjoyed the benefits of the *Bt* technology directly. A research team led by Professor Tony Shelton of Cornell University published a paper on *Bt* brinjal in the peer-reviewed journal *Frontiers in Bioengineering and Biotechnology* showed that during the year 2018, 27,012 Bangladeshi farmers harvested benefit due to rapid adaptation of the pest-reducing technology (Shelton et al. 2018). This trend of rapid adaptation proved that farmers are getting benefit from *Bt* technology, and the target pest reduction through built-in toxin mechanism is working successfully in the field level.

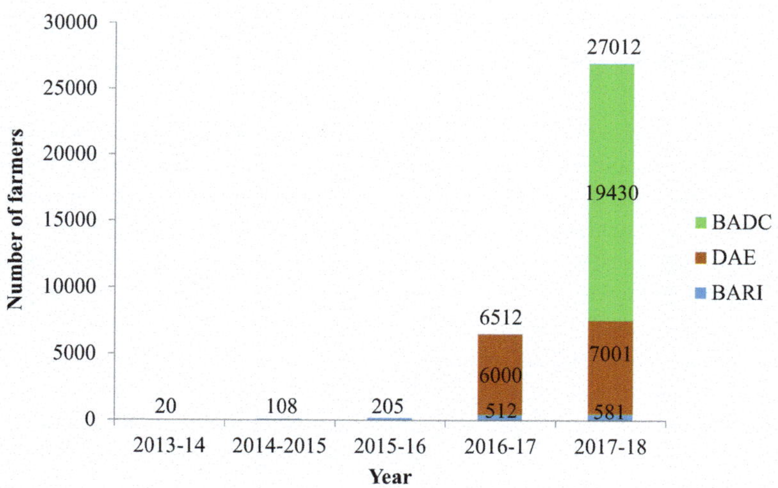

Fig. 23.3 Year-wise adoption rate of *Bt* brinjal in Bangladesh. (Redrawn from Shelton et al. 2018)

23.4.2 Cost of Production

In brinjal cultivation, the main cost is pesticides, but in case of *Bt* brinjal varieties, there was less pesticide required than the regular varieties. Because *Bt* brinjal deters BSFB, so no spraying is required for that pest. Spraying once in a week was enough for *Bt* brinjal to control other paste, whereas non-*Bt* varieties required as often as three times a week. This is financial savings for farmers (Ahmed et al. 2019). The researchers stated features which provided a breakdown of input costs per ha for treatment farmers cultivating *Bt* brinjal (*Bt* begun-4) and control farmers growing non-*Bt* brinjal (ISD-006) (Table 23.2).

The total costs of production for *Bt* brinjal per ha (BDT 72,109) were lower than non-*Bt* brinjal (BDT 81,902). The total gains in terms of input cost (BDT 9793) were mainly due to considerably less spent by *Bt* brinjal farmers on pesticides compared to the non-*Bt* farmers (Ahmed et al. 2019). Similarly, this trend of data was supported by another experimental result which described that pesticides were applied 11 times to *Bt* eggplant, whereas it was 41 times to non-*Bt* eggplant for controlling sucking pests during the whole season. Herein, the *Bt* eggplant farmers selected for this study saved 61% pesticide cost compared to non-*Bt* eggplant farmers. The *Bt* brinjal growers experienced no losses due to BSFB attack and got higher net returns. At the end of this study, Rashid and his co-scientists declared that the *Bt* eggplant technology was good for the farmer. In the future, of course, this technology will potentially improve the socioeconomic status of the *Bt* respondents (Rashid et al. 2018).

23.4.3 Net Return from Brinjal Production

During focus group discussions (FGD) with treatment farmers, a scientist team led by Akhter U. Ahmed of the International Food Policy Research Institute (IFPRI) revealed that although brinjal prices fluctuated with market demand, farmers more often sold *Bt* brinjal to market traders at higher prices than conventional varieties. It

Table 23.2 Input costs per hectare for *Bt* brinjal (*Bt* begun-4) and non-*Bt* brinjal (ISD-006) cultivation (Adapted from Ahmed et al. 2019)

	Cost of production (BDT per ha)	
Inputs	*Bt* brinjal	Non-*Bt* brinjal
Seed/seedlings	5,461	5,539
Fertilizer	30,326	32,026
Irrigation	11,241	11,867
Pesticide	14,852	22,145
Machinery	7,600	8,097
Total hired labor	2,505	2,227
Total cash cost	72,109	81,902

was found that market traders tend to buy *Bt* brinjal at higher prices but also sell at higher prices in the market. In the study period, during IFPRI's monitoring field visits, *Bt* brinjal adopted farmers and monitoring officials from DAE, which frequently reported that farmers received higher prices for *Bt* brinjal compared to traditional brinjal varieties because the physical appearance of *Bt* brinjal was better and had no marks of infestation or holes. Therefore, buyers preferred *Bt* brinjal and paid higher prices (Ahmed et al. 2019). These trends of qualitative results were strongly supported by another study which was conducted by Rashid et al. (2018), and the results were presented in Table 23.3. They revealed that gross returns from *Bt* eggplant cultivation were (Bangladeshi Taka) BDT 394,570/ha as compared to BDT 312,945/ha for non-*Bt* eggplant. In case of net returns, the amount was BDT 179,602/ha for *Bt* eggplant which was five times larger as compared to BDT 29,841/ha for non-*Bt*. The yield difference between the two groups was only 3.02 tons, but the non-*Bt* farmers applied almost three times more pesticides as well as more fertilizer to maintain the yields. The sprayed pesticide costs were BDT 14,215/ha for *Bt* eggplant and BDT 36,057 for non-*Bt* eggplant. As a result, the net returns were higher in case of *Bt* eggplant than non-*Bt* eggplant due to lower production costs and higher yields. Based on total cost, production cost of *Bt* eggplant was BDT 9.26/kg, and for non-*Bt* eggplant, it was BDT 14.20/kg (Rashid et al. 2018).

Previously, another study was conducted in the year 2016–2017 cropping season and compared 505 *Bt* brinjal growing farmers with 350 non-*Bt* brinjal growing farmers. The results indicated a 61% saving in pesticide cost, which translated to a 650% (six-fold) increase in returns, from $2151/ha for *Bt* brinjal as compared to just $357/ha for non-*Bt* brinjal. These cost savings and significant increase in returns are only due to reductions of pesticide costs (Shelton et al. 2018). Other studies suggest that these *Bt* brinjal varieties convey higher yields with lower applications of pesticides. Therefore, *Bt* brinjal farmers received higher revenues than conventional brinjal (ISD-006) (Ahmed et al. 2019). Moreover, an economic analysis proved that all adopted Bt brinjal lines provided higher gross returns compared to their non-*Bt* isolines (Prodhan et al. 2018). Another experimental result showed that all *Bt* and 86% non-*Bt* farmers desired to grow *Bt* eggplant in the next year by receiving higher yield and significant financial returns (Rashid et al. 2018).

Table 23.3 Per hectare return from eggplant production in the study areas (Adopted from Rashid et al. 2018)

	Return (BDT/ha)	
Items	*Bt* eggplant	Non-*Bt* eggplant
Fresh eggplant yield (ton/ha)	23.21	20.19
Gross return	394,570	312,945
Gross margin	248,651	101,590
Net return	179,602	29,841
Benefit cost ratio	1.84	1.11
Production cost (total cost basis) (BDT/kg)	9.26	14.02

23.4.4 Higher Market Price of Bt Brinjal

The major reasons mentioned by the respondents for higher prices of *Bt* brinjal varieties were that they look fresher and most importantly, they have no holes like non-*Bt* eggplants due to pest invasions. This finding contrasts with most of the previous literature which assumed that consumers are willing to pay less for genetically engineered (GE) varieties than non-GE varieties. Here, it was found that when GE products had lucrative observable characteristics, they obtained a positive premium in price for the growers. Surveying the market price *Bt* and non-*Bt* eggplant at both retail and wholesale levels, researchers found that *Bt* prices are higher than non-*Bt* prices at both the retail and wholesale levels (Table 23.4) (Ahsanuzzaman and Zilberman 2018).

A higher wholesale price suggested that farmers were receiving higher prices, while higher retail prices indicated that consumers were willing to pay more for *Bt* brinjal varieties. The retailers also received a higher premium for *Bt* varieties. The results estimated around a 30% price premium at all levels for the *Bt* eggplant varieties (Ahsanuzzaman and Zilberman 2018). Similarly, another result showed that *Bt* brinjal had positive impacts on sales revenue for farmers. Ahmed et al. (2019) revealed that *Bt* varieties gained a 27.3% increase in sales revenue per ha and a 10.9% more sales revenue per kg.

23.5 The Economic Impact of *Bt* Technology Worldwide

Bt crops are plants which are genetically modified (GM) and resistant to specific insect pests. The commercialization of GM crops has occurred at a rapid rate since 1995 when the EPA (Environmental Protection Agency) in the United States gave approval for commercial production and dissemination of the *Bt* products like corn, potato, tobacco, and cotton. These GM crops contain *Cry* toxins which are lethal to certain species of insects belonging to orders such as Coleoptera, Nematoda, Lepidoptera, Hymenoptera, and Diptera. In the year 2016, the total global cultivated area of *Bt* crops reached approximately 185 million ha (Abbas 2018). Additionally, there are several studies estimated the value of using GM crop technology in the agricultural farm level. These studies mainly evaluated impacts on yields, main variable costs of production, gross income, and vital impacts on the production

Table 23.4 Market price information of *Bt* and non-*Bt* eggplant according to Ahsanuzzaman and Zilberman (2018)

	Price (BDT/Kg)		*Bt* premium
Items	*Bt* brinjal	Non-*Bt* brinjal	(%)
Wholesale	15.45	11.7	32
Retail	28.6	22.35	28
Mark up	13.5	10.65	27

system. However, the annual updated analysis regarding *Bt* crops adaptation shows very noteworthy net financial benefits at the farm level since the year 1996 to 2016. Based on these gains, 48% farmers have been benefited in developed countries and 52% of the farmers in developing countries. Among the gains, about 65% have been achieved from yield and production system as well as the rest 35% came from savings of the production cost (Brookes and Barfoot 2018). Furthermore, in Brazil, another study estimated the efficacy of *Bt* maize expressing proteins like *Cry1F* to control *Spodoptera frugiperda* in the field conditions. The result showed that this *Bt* maize technique efficiently controlled seedling cutting injury as well as successfully reduced foliar feeding, and the estimated protection was near 100% from kernel feeding (Moscardini et al. 2020).

23.6 The Maintenance of Environmental Sustainability by the Cultivation of *Bt* Brinjal

Before introduction of new GE plants, environmental risk assessments are conducted on case-by-case basis. In this case, some of the points are taking into account such as plant biology, transgene nature, protein produced by the respective transgene, the phenotypic character of inserted transgene, the ultimate use of the plant, as well as the released environment where it will be cultivated for use (CERA 2011).

23.6.1 Reduction of Pesticide Use Ensures Toxic-Free Environment

Four *Bt* brinjal varieties were developed to resist the BSFB pest. The transgenic *Bt* brinjal successfully repels the BSFB pest, and as a result, farmers can reduce the use of pesticide. Reduction of pesticides application and less exposure of farmers to toxicity reflect positive environmental outcomes (Hautea et al. 2016). Herein, some of the experimental data are graphically presented in Fig. 23.4 based on percentage of farmers using pesticides of varying toxicity levels for BSFB control (Ahmed et al. 2019).

In the Fig. 23.4, data were summarized by constructing a toxicity score, the Pesticide Use Toxicity Score (PUTS), which assigns a score based on the GHS (Globally Harmonized System) oral hazard category of the selected pesticides and the frequency of the use of the respective pesticides. In the GHS Hazard Classification scale, lower numbers (1, 2) correspond to more severe levels of toxicity. For PUTS to be easily interpretable, the GHS scale is inverted so that higher values correspond to higher toxicity levels (Ahmed et al. 2019). Similar trend of results demonstrated that *Bt* eggplant lines containing *Cry1Ac* provide outstanding control against BSFB and can dramatically reduce the need for application of conventional insecticides

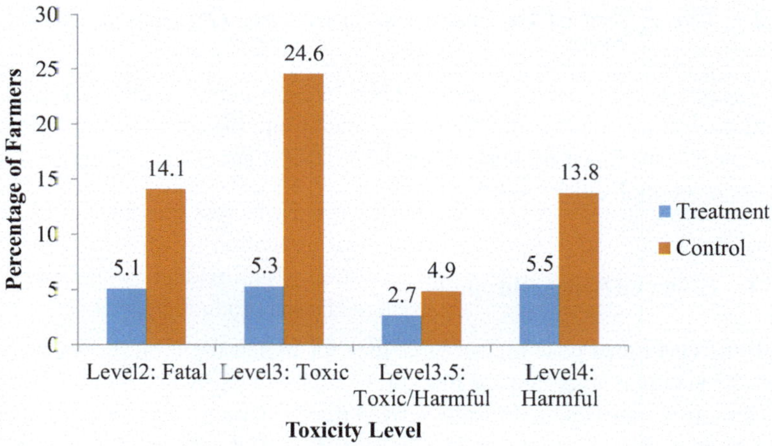

Fig. 23.4 Percentage of farmers using pesticides for brinjal shoot and fruit borer by toxicity level (Redrawn from Ahmed et al. 2019)

(Hautea et al. 2016). Furthermore, adoption of *Bt* eggplant and the reduction of pesticide applications associated with better environmental status (Ahsanuzzaman and Zilberman 2018).

Farmer's knowledge about the harmful effect of conventional pesticides plays an important role on injudicious use of chemicals application to control insect paste in brinjal field. Typically, those who spray the crop are not aware of the hazards of pesticides and do not use any personal protective equipment. These problems have been well documented in Bangladesh and other countries which are associated with our environmental degradation (bteggplant.cornell. Edu/content/facts; bic.searca. org; Del Prado-Lu 2015). These findings were supported by another result which showed about 68.75% brinjal growers have knowledge while 31.25% brinjal growers did not have knowledge about the negative impacts of using higher amounts of insecticides. The brinjal growers are randomly using same insecticides under different brand name (Raza et al. 2018). However, increasing farmer's knowledge about proper use of insecticide depends on many factors; rather it is better to introduce *Bt* brinjal which has built-in mechanism for BSFB control. Finally, according to scientists of the Bangladesh Agricultural Research Institute (BARI) who developed the four varieties, the protein in *Bt* brinjal disrupts the digestive systems of certain pests, causing them to die within 3 days of ingestion and significantly reduce the need to use pesticides (Shelton et al. 2018).

23.6.2 *The Effect of* Bt *Crops on Nontarget Organism (NTO)*

Development of GM crops was started about 25 years before, and various new cultivars of diverse range have already been released into the environment. Transgene cultivars were developed by introducing new desired genes from different sources using advanced molecular toolkit.

23.6.2.1 Effect on Soil Arthropods

Though the GM crops pose higher capacities for increasing crop production, their impact on the nontarget organisms like soil biota is still partly understood. Recently, some laboratory experimental results proved that horizontal gene transfer is occurring between native soil biota and GM crops (Fig. 23.5) (Mandal et al. 2020). A further metagenomics study is needed to precisely understand whether GM technology modulates microbiome in plants and the rhizosphere.

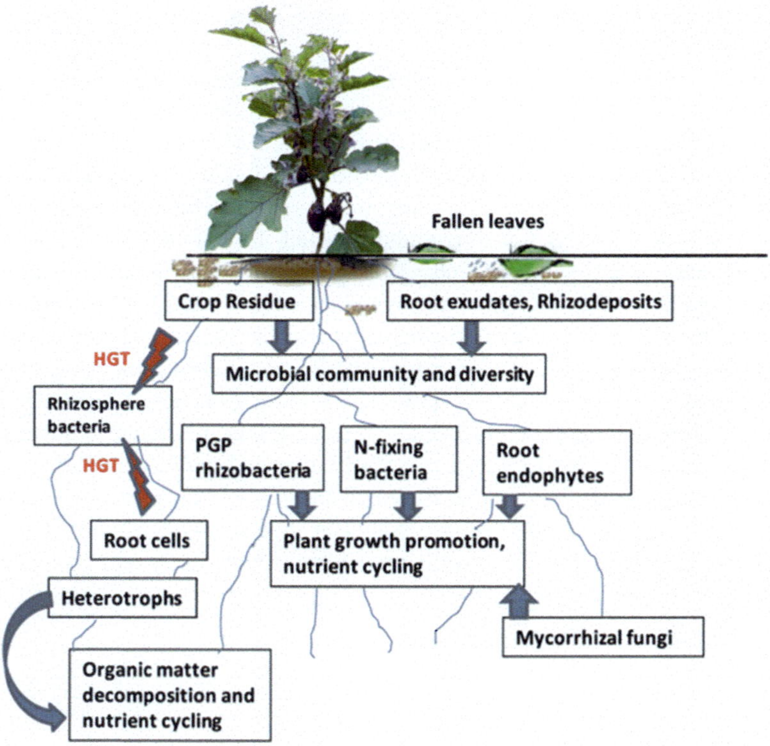

Fig. 23.5 Schematic diagram representing impact of genetically modified crops on soil microbial communities and microbe-mediated processes (PGP: Plant Growth Promoters; HGT: Horizontal Gene Transfer) (Adopted and modified from Mandal et al. 2020)

Transgenic *Bt*-crops have been generated by introducing a *Bacillus thuringiensis* gene, and the respective plant produces a *Cry* toxin to protect crops from insect pests. Mainly, in the rhizospheric zone of the soli, microorganisms are responsible for plant nutrient availability and organic matter decomposition. So, it is relevant to protect these invertebrate taxa from the negative impact of *Cry* toxin. In this connection, a good number of studies have showed comparison between the abundance of populations and biomass of soil microbes in cultivated lands planted with *Bt* crops as well as their traditional counterparts. Herein, some selected studies on Protista, mites, earthworms, nematodes, enchytraeids, and Collembola were reviewed and analyzed critically to strengthen the evidence that whether population richness and biomass of soil invertebrates are altered due to *Bt* crops cultivation in comparison with original cultivars. However, no significant effect of *Cry* protein on soil invertebrates was investigated (Krogh et al. 2020). Previously, the result showed that long-term cultivation of *Bt* cottons had no major effect on the population of soil invertebrates. Impacts on soil invertebrates are considered as a vital aspect of ecological risk assessment, as well as post-release observation of transgene introduced insect-resistant plants (Li et al. 2012). Another study proved that when *Cry1Ac* protein was released into the soil by *Bt* gene, it declined very quickly. Soil particles are absorbed most of the protein and decomposed even within a couple of days. The study also demonstrated that the population sizes, evenness, and diversity of soil microbiota such as fungi, archaea, as well as bacteria were not significantly or consistently affected by *Cry1Ac* protein. These findings suggest that *Bt* protein which resulted through cultivations of *Bt* crops is unable to cause transient or even major persisting changes in soil microbes in the field (Zhaolei et al. 2018). Similarly, transgenic soybean that is producing *Cry1Ac* and *Cry1F* toxins does not adversely affect the nontarget association with soybean (Marques et al. 2018). Therefore, studies on transgenic brinjal in Bangladesh (Prodhan et al. 2018) and the Philippines (Navasero et al. 2016) that are producing *Cry1Ac* protein have also confirmed that it does not destroy nontarget arthropods (Romeis et al. 2019). Despite minor variations that were found in the population size and the distribution pattern across the non-*Bt* and *Bt* brinjal cultivated soil, plant-growth-dependent variability was more prominent in comparison with genetic alteration. However, this study determines that genetic change of brinjal crop has insignificant effect on the fungal population living in the soil (Singh et al. 2013). However, a comprehensive metagenomics study is needed for making a final conclusion on this regard.

23.6.2.2 Effect on Pollinators

As a nontarget organism (NTO), honey bee (*Apis mellifera* L.) plays a very vital role in the pollination of various plants. During their foraging period as pollinator, they are likely to encounter GE crops, especially insect-resistant crops because they produce toxins for particular insect. Thus, it is important to measure probable impacts of these crops on honey bee. Finally, it is concluded that the studied *Cry* proteins do not have adverse effect for the successful survival of honey bees (Ricroch et al. 2018).

23.6.2.3 Effect on Biological Control Agents

Biological control, popularly known as biocontrol, is a method of controlling destructive pest population like weeds, plant disease, insects, and mites using other beneficial organisms. The natural enemies of insect pests can be considered as biological control agents including pathogens, predators, any kind of competitors, and parasitoids. In case of BSFB, natural enemies include predators such as praying mantis, ladybird beetles, earwig, green lacewing, and spiders. In this connection, an experimental result proved that *Bt* brinjal shows excellent pest suppression. But no undesirable nontarget effects on other arthropods were observed in the system, especially those beneficial living organisms that exert important ecosystem services like biocontrol agents (Prodhan et al. 2018). Another study confirmed that the *Cry* proteins of *Bt* crops to control Lepidopteran insect had no detrimental effects to the important natural biocontrol agents of these and other pest species (Shelton et al. 2016).

23.6.3 Establishment and Persistence of Cry1Ac-*Expressing Plants in the Environment*

Environmental risk assessments of GE plants start from its receiving environment where biologically similar non-transformed or host species are existing (OECD 2006). Significant movement of transgenes from GM crops into its wild relatives is evaluated for a new pest-resistant gene in a new crop (Koch et al. 2015). On the other hand, lower fitness was observed due to unintended interruption of endogenous genes. Herein, Indian variety, Swarna, was crossed with Golden Rice and observed a growth disturbance. In this case, precursors for vitamin A-producing gene constructs interfered with own gene of the plants for growth hormone production. The introgression of the transgene interrupted the native plant growth-related *OsAux1* gene function (Bollinedi et al. 2017). However, this effect was not detected in other rice varieties. Actually, such metabolic disruptions do not occur in conventional breeding process. Conventional gene transfer brings gradual changes, whereas genetic rearrangements and abrupt/disruptive additions of additional DNA coding for novel proteins cause drastic metabolic changes, termed as "metabolic meltdown" (Wilson 2018). Moreover, study results concluded that the ecological risk of *Bt* rice regarding the persistence of *Cry* toxins may be much smaller than which were assessed on the basis of ELISA quantification. It was assumed that partially degraded *Cry* proteins retain their immune reactivity but have actually lost their bioactivity (Deng et al. 2019). Another study showed that several plant species were used for genetic modifications, such as alfalfa, maize, creeping bentgrass, oilseed rape, cotton, poplar, eggplant, and rice, which were well known for their persistent nature and gene flow in the regions where they were cultivated (Bauer-Panskus et al. 2013; Devos et al. 2018). Therefore, recent understanding reveals that GM crops impart destructive effects on the environment, such as weediness, modification in crop pervasiveness or invasiveness, the emergence of herbicide and insecticide tolerance, transgene stacking, as well as disturbed biodiversity. To prove these impacts and unveil further facts, a more in-depth view and critical research is required (Tsatsakis et al. 2017). Till today, in case of *Bt* brinjal, there is an insufficient amount of available data regarding establishment and persistence of *Cry1Ac* expressing plants in the environment. But the findings would convey remarkable in-depth information to the scientists, industry people, and policy-makers to realize various facets of struggle and invention of possible assistance for adopting sustainable commercialization of GM crops like brinjal in Bangladesh.

23.7 Conclusions

The BSFB is a destructive insect pest of brinjal which significantly reduces yield and profit of brinjal cultivation. To tackle this notorious pest, farmers intensively use various expensive chemical insecticides that are detrimental to human health, environment, and cost of production. To increase the productivity, reduce the application of insecticides, and increase the farmers' income, Bangladesh released first GM crop *Bt* brinjal by introducing a *Cry1Ac* gene to the genome of four local cultivars of brinjal. In fact, the *Bt* protein is activated within the insect's gut and severely affect the gut membranes of the BSFB larvae. Adoption of this genetically engineered *Bt* eggplant has significantly been increased yield and profitability of brinjal growers. The input costs per hectare of *Bt* brinjal production were lower due to reduced pesticide cost for crop protection. It gives higher yield because no BSFB infestation was found during field level study. Moreover, due to good physical appearance of the products, unit sale price was higher than the non-*Bt*, which gains per unit higher margin and ultimately farmers got more profit. Rapid adoption of *Bt* eggplant by the farmers and consumers also considerably reduces spray volume of pesticides compared to non-*Bt* varieties and thus improves environmental safety. On an average, the non-*Bt* brinjal growers sprayed insecticides 41 times during the whole period of crop cultivation, and the *Bt* adopted farmers applied only 11 times to protect brinjal from other sucking pests. Furthermore, when all the soil invertebrate taxa including collembolans, mites, earthworms, nematodes, and protists were considered together, there was no significant effect of *Bt* technology on these soil invertebrates. Similar results were found in case of other nontargeted insects like pollinator, honey bee. But there were no available data to draw a strong conclusion regarding the durability of this monogenic new technology and any negative effect of *Bt* gene on surrounding environment and biota. It can be concluded that the *Bt* brinjal technology significantly increased yield and farmers' profitability and ensured environmental and human health safety by drastically reducing the requirement for insecticides use for brinjal cultivation.

References

Abbas MST (2018) Genetically engineered (modified) crops (bacillus thuringiensis crops) and the world controversy on their safety. Egypt J Biol Pest Control 28:1–12

Ahmed AU, Hoddinott JF, Islam KMS, Khan ASMMR, Abedin N, Hossain NZ, Ghostlaw J, Parvin A, Quabili W, Rahaman ST, Rahman W (2019) Impact of Bt brinjal (eggplant) technology in Bangladesh. Project Report prepared for the US Agency for International Development (USAID)

Ahsanuzzaman, Zilberman D (2018) Bt eggplant in Bangladesh increases yields and farmers' Incomes, and reduces pesticide use. ARE Update 22:5–8. University of California Giannini Foundation of Agricultural Economics

APAARI (2018) Success story on Bt brinjal in Bangladesh. Asia-Pacific Association of Agricultural Research Institutions, Bangkok, Thailand. xviii+39 p. Available at: https://s.giannini.ucop.edu/uploads/giannini_public/34/0f/340f8c28-cf2d-4246-9f87-84fe661568cb/v22n2_4.pdf

Bangladesh Bureau of Statistics (BBS) (2018) Statistics and information division, Ministry of Planning, Government of the People's Republic of Bangladesh; 29th Series; BBS: Dhaka, Bangladesh, 574p

Bauer-Panskus A, Breckling B, Hamberger S, Then C (2013) Cultivation-independent establishment of genetically engineered plants in natural populations: current evidence and implications for EU regulation. Environ Sci Eur 25:34

Bollinedi H, Prabhu KV, Singh NK, Mishra S, Khurana JP, Singh AK (2017) Molecular and functional characterization of GR2-R1 event based backcross derived lines of Golden Rice in the genetic background of a mega rice variety swarna. PLoS One 12:e0169600

Brookes G, Barfoot P (2018) Farm income and production impacts of using GM crop technology 1996–2016. GM Crops Food 9:59–89

Castagnola AS, Jurat-Fuentes JL (2012) Bt crops: past and future. In: Bacillus thuringiensis biotechnology. Springer, Dordrecht, pp 283–304

Center for Environmental Risk Assessment (CERA), ILSI Research Foundation (2011) A review of the environmental safety of the Cry1Ac protein. Environ Biosafety Res 10:27–49

Crickmore N, Zeigler DR, Feitelson J, Schnepf ESCHERICHIA, Van Rie J, Lereclus D, Baum J, Dean DH (1998) Revision of the nomenclature for the Bacillus thuringiensis pesticidal crystal proteins. Microbiol Mol Biol Rev 62:807–813

de Maagd RA, Bravo A, Crickmore N (2001) How Bacillus thuringiensis has evolved specific toxins to colonize the insect world. Trends Genet 17:193–199

Del Prado-Lu JL (2015) Insecticide residues in soil, water, and eggplant fruits and farmers' health effects due to exposure to pesticides. Environ Health Prev Med 20:53–62

Deng J, Wang Y, Yang F, Liu Y, Liu B (2019) Persistence of insecticidal Cry toxins in Bt rice residues under field conditions estimated by biological and immunological assays. Sci Total Environ 679:45–51

Devos Y, Ortiz-García S, Hokanson KE, Raybould A (2018) Teosinte and maize × teosinte hybrid plants in Europe – environmental risk assessment and management implications for genetically modified maize. Agric Ecosyst Environ 259:19–27

Evdokimov AG, Moshiri F, Sturman EJ, Rydel TJ, Zheng M, Seale JW, Franklin S (2014) Structure of the full-length insecticidal protein Cry1Ac reveals intriguing details of toxin packaging into in vivo formed crystals. Protein Sci 23:1491–1497

Gómez I, Sánchez J, Muñoz-Garay C, Matus V, Gill SS, Soberón M, Bravo A (2014) Bacillus thuringiensis Cry1A toxins are versatile proteins with multiple modes of action: two distinct pre-pores are involved in toxicity. Biochem J 459:383–396

Hautea DM, Taylo LD, Masanga APL, Sison MLJ, Narciso JO, Quilloy RB, Hautea RA, Shotkoski FA, Shelton AM (2016) Field performance of Bt eggplants (Solanum melongena L.) in the Philippines: Cry1Ac expression and control of the eggplant fruit and shoot borer (Leucinodes orbonalis Guenée). PLoS One 11:e0157498

ISAAA (2020) (International service for the acquisition of agri-biotech applications) ISAAA's GM Approval Database. Retrieved from http://www.isaaa.org/gmapprovaldatabase/

James C (2014a) ISAAA briefs brief 49 Global Status of Commercialized Biotech/GM Crops: 2014
James C (2014b) Global status of commercialized biotech/gm crops: 2011. ISAAA brief 43. ISAAA, Ithaca, NY
Javed H, Hussain SS, Javed K, Mukhtar T, Abbasi NA (2017) Comparative infestation of brinjal stem borer (Euzophera perticella) on six aubergine cultivars and correlation with some morphological characters. Pak J Agric Sci 54
Koch MS, Ward JM, Levine SL, Baum JA, Vicini JL, Hammond BG (2015) The food and environmental safety of Bt crops. Front Plant Sci 6:283
Krogh PH, Kostov K, Damgaard CF (2020) The effect of Bt crops on soil invertebrates: a systematic review and quantitative meta-analysis. Transgenic Res:1–12
Li X, Liu B, Wang X, Han Z, Cui J, Luo J (2012) Field trials to evaluate effects of continuously planted transgenic insect-resistant cottons on soil invertebrates. Environ Monit 14:1055–1063
Mandal A, Sarkar B, Owens G, Thakur JK, Manna MC, Niazi NK, Jayaraman S, Patra AK (2020) Impact of genetically modified crops on rhizosphere microorganisms and processes: a review focusing on Bt cotton. Appl Soil Ecol 148:103492
Marques LH, Santos AC, Castro BA, Storer NP, Babcock JM, Lepping MD, Sa V, Moscardini VF, Rule DM, Fernandes OA (2018) Impact of transgenic soybean expressing Cry1Ac and Cry1F proteins on the non-target arthropod community associated with soybean in Brazil. PLoS One 13:e0191567
Moscardini VF, Marques LH, Santos AC, Rossetto J, Silva OA, Rampazzo PE, Castro BA (2020) Efficacy of Bacillus thuringiensis (Bt) maize expressing Cry1F, Cry1A. 105, Cry2Ab2 and Vip3Aa20 proteins to manage the fall armyworm (Lepidoptera: Noctuidae) in Brazil. Crop Protection, 105269
Navasero MV, Candano RN, Hautea DM, Hautea RA, Shotkoski FA, Shelton AM (2016) Assessing potential impact of Bt eggplants on non-target arthropods in the Philippines. PLoS One 11:e0165190
Nuruzzaman MD, Rahman MM, Liu Y, Naidu R (2016) Nanoencapsulation, nano-guard for pesticides: a new window for safe application. J Agric Food Chem 64:1447–1483
OECD (2006) Points to consider for consensus documents on the biology of cultivated plants. Organization for Economic Cooperation and Development (OECD), Paris
Pardo-Lopez L, Soberon M, Bravo A (2013) Bacillus thuringiensis insecticidal three-domain Cry toxins: mode of action, insect resistance and consequences for crop protection. FEMS Microbiol Rev 37:3–22
Prodhan MZH, Hasan MT, Chowdhury MMI, Alam MS, Rahman ML, Azad AK, Hossain MJ, Naranjo SE, Shelton AM (2018) Bt eggplant (Solanum melongena L.) in Bangladesh: fruit production and control of eggplant fruit and shoot borer (Leucinodes orbonalis Guenee), effects on non-target arthropods and economic returns. PLoS One 13:e0205713
Rashid MA, Hasan MK, Matin MA (2018) Socio-economic performance of Bt eggplant cultivation in Bangladesh. Bangladesh J Agric Res 43:187–203
Raza MS, Md AR, Rahaman KMM, Juliana FM, Hossain S, Rahman A, Hossain K, Alam MJ, Asaduzzaman M (2018) Present status of insecticides use for the cultivation of Brinjal in Kushtia region, Bangladesh. Int J Eng Sci 7:44–51
Ricroch A, Akkoyunlu S, Martin-Laffon J, Kuntz M (2018) Assessing the environmental safety of transgenic plants: honey bees as a case study. Academic Press. In Adv Bot Res 86:111–167
Romeis J, Naranjo SE, Meissle M, Shelton AM (2019) Genetically engineered crops help support conservation biological control. Biol Control 130:136–154
Sanahuja G, Banakar R, Twyman RM, Capell T, Christou P (2011) Bacillus thuringiensis: a century of research, development and commercial applications. Plant Biotechnol J 9:283–300
Schnepf E, Crickmore NV, Van Rie J, Lereclus D, Baum J, Feitelson J, Zeigler DR, Dean D (1998) Bacillus thuringiensis and its pesticidal crystal proteins. Microbiol Mol Biol Rev 62:775–806
Shelton AM, Romeis J, Naranjo S, Tian J, Hellmich R (2016) Use of Bt-resistant caterpillars to assess the effect of Cry proteins on beneficial natural enemies. IOBC-WPRS Bulletin 114:51–55

Shelton AM, Hossain MJ, Paranjape V, Azad AK, Rahman ML, Khan ASMMR, Prodhan MZH, Rashid MA, Majumder R, Hossain MA, Hussain SS (2018) Bt eggplant project in Bangladesh: history, present status, and future direction. Front Bioeng Biotechnol 6:106
Shelton AM, Hossain MJ, Paranjape V, Prodhan MZ, Azad AK, Majumder R, Sarwer SH, Hossain MA (2019) Bt brinjal in Bangladesh: the first genetically engineered food crop in a developing country. Cold Spring Harb Perspect Biol 11:a034678
Shirale D, Patil M, Parimi S (2017) Insecticide resistance in field populations of Leucinodes orbonalis (Lepidoptera: Crambidae) in India. Can Entomol 149:399–407
Singh AK, Rai GK, Singh M, Dubey SK (2013) Bacterial community structure in the rhizosphere of a Cry1Ac Bt-Brinjal crop and comparison to its non-transgenic counterpart in the tropical soil. Microb Ecol 66:927–939
Soberón M, Monnerat R, Bravo A (2016) Mode of action of Cry toxins from Bacillus thuringiensis and resistance mechanisms. In: Gopalakrishnakone P, Stiles B, Alape-Girón A, Dubreuil JD, Mandal M (eds) Microbial toxins. Springer Netherlands, Dordrecht, pp 1–13
Song M, Ju J, Luo S, Han Y, Dong Z, Wang Y, Gu Z, Zhang L, Hao R, Jiang L (2017) Controlling liquid splash on superhydrophobic surfaces by a vesicle surfactant. Sci Adv 3:e1602188
Tabashnik BE, Zhang M, Fabrick JA, Wu Y, Gao M, Huang F, Wei J, Zhang J, Yelich A, Unnithan GC, Bravo A (2015) Dual mode of action of Bt proteins: protoxin efficacy against resistant insects. Sci Rep 5:15107
Tsatsakis AM, Nawaz MA, Kouretas D, Balias G, Savolainen K, Tutelyan VA, Golokhvast KS, Lee JD, Yang SH, Chung G (2017) Environmental impacts of genetically modified plants: a review. Environ Res 156:818–833
Wilson A (2018) Goodbye to golden rice? GM trait leads to drastic yield loss and "metabolic meltdown". Bioscience resource project. https://biosciencereso urce.org/goodbye-to-golden-rice-gm-trait-leads–todrastic-yield-loss-and-metabolic-meltdown/. Accessed 8 Dec 2019
Zhao X, Cui H, Wang Y, Sun C, Cui B, Zeng Z (2017) Development strategies and prospects of nano-based smart pesticide formulation. J Agric Food Chem 66:6504–6512
Zhaolei L, Naishun B, Xueping C, Jun C, Manqiu X, Zhiping S, Ming N, Changming F (2018) Soil incubation studies with Cry1Ac protein indicate no adverse effect of Bt crops on soil microbial communities. Ecotoxicol Environ Saf 152:33–41

Chapter 24
Bacillus subtilis: A Multifarious Plant Growth Promoter, Biocontrol Agent, and Bioalleviator of Abiotic Stress

Alka Sagar, S. S. Yadav, R. Z. Sayyed, S. Sharma, and P. W. Ramteke

Abstract Increased agricultural production is the global requirement to nourish the rising people with limited cultivable agricultural land. To achieve this goal, fertilizers and pesticides have been gratuitously used, which has caused environmental impairment. Replacing chemicals with biocontrol agents is an environment-friendly alternative. Microorganisms are useful to plants and carry out the same role as chemical fertilizers and pesticides, performing as a biofertilizer and biopesticide. This chapter discusses the multifarious potential of plant growth-promoting *Bacillus subtilis* in plant growth promotion in different cereals, vegetables, and other plants. Inoculation with *B. subtilis* resulted in healthier seed germination, seedling vigor, and improved growth via both direct and indirect mechanisms. *B. subtilis* forms a thin biofilm on the roots for long-term colonization of the rhizosphere. Since various strains of *B. subtilis* possess several traits that can positively influence plants, they can be utilized in the advancement of innovative, protected, and resourceful seed treatments for the sustainable and environment-friendly strategy to cope with

A. Sagar
Department of Microbiology, Meerut Institute of Engineering and Technology (MIET), Meerut, Uttar Pradesh, India

Sam Higginbottom University of Agriculture Technology and Sciences, Allahabad, Uttar Pradesh, India

S. S. Yadav
Swami Vivekanand Subharti University, Meerut, UP, India

R. Z. Sayyed (✉)
Department of Microbiology, PSGVP Mandal's Arts, Science, and Commerce College, Shahada, Maharashtra, India

S. Sharma
Department of Microbiology, Meerut Institute of Engineering and Technology (MIET), Meerut, Uttar Pradesh, India

P. W. Ramteke
Sam Higginbottom University of Agriculture Technology and Sciences, Allahabad, Uttar Pradesh, India

M. T. Islam et al. (eds.), *Bacilli in Agrobiotechnology*, Bacilli in Climate Resilient Agriculture and Bioprospecting,
https://doi.org/10.1007/978-3-030-85465-2_24

the adverse effects of abiotic and biotic stresses on plants. This is one of the best bacterial champions in implied enzyme manufacturer, and it can be used on a commercialized scale by biotechnology companies.

Keywords *Bacillus subtilis* · Plant growth-promoting (PGP) traits · Biofertilizer · Biopesticide · Stress bioalleviator

24.1 Introduction

Plant growth-promoting bacteria (PGPB) existing in soil (free-living bacteria) and plant roots (rhizosphere bacteria) promote plant development by enhancing pathogen resistance, increasing water uptake, and utilizing nutrients (Yuan et al. 2015), and have been commercialized as PGPB (Goswami et al. 2016). The utilization of PGPR as microbial inoculants is more emphasized nowadays as an environmentally affable method of increasing crop production compared to the use of chemical fertilizers. Their use as bioinoculants for sustainable agricultural manufacturers in drought-prone areas is on the rise (Ojuederie et al. 2019).

B. subtilis is present in the soil and gut of ruminants and humans. It is Gram-positive, rod-shaped, obligate aerobe, and catalase-positive (+) (Hashem et al. 2019). It can form endospores that promote its survival in adverse environments and also secrete metabolites that enhance plant growth and health. Some volatile organic compounds (VOCs) produced by *B. subtilis* strain (GB03) assist plants to recover from stress (Hashem et al. 2019). It stimulates seed germination and supports the general health and vigor of the plant. *B. subtilis* has been regarded as biofertilizers, phytostimulators, and biopesticides (Bhardwaj et al. 2014; Pérez-Montaño et al. 2014). The increasing climatic change and the need to reduce chemical fertilizers lead to the production of abiotic stress-tolerant plants as an international priority, which will require the wide use of PGPB.

24.2 *Bacillus subtilis* as Plant Growth-Promoting Bacterium (PGPB)

The promotion of plant growth and development is achieved by *B. subtilis* through direct and indirect mechanisms. The direct mechanism involves nitrogen fixation, solubilization of complex organic or inorganic nutrients, mobilization of iron via siderophore production, and production of plant growth regulators such as indole acetic acid (IAA), gibberellins, and cytokinin. Based on a different mode of action, direct plant growth-promoting rhizobacteria can be grouped into three categories including biofertilizers, phytostimulators, and rhizoremediators (Wang et al. 2018; Hashem et al. 2019). The indirect mechanism involves the production of various compounds such as hydrogen cyanide (HCN), antibiotics, and volatile compounds that encourage plant development (Radhakrishan et al. 2017) and the induction of acquired systemic resistance. *B. subtilis* improves nutrient uptake, root growth, and the proliferation of plants (Table 24.1 and Fig. 24.1).

Two PGPB initially labeled as PR30 and PR31 isolated from an organic farm, SHUATS, Allahabad, analyzed phylogenetically by 16S rRNA sequences as *B. subtilis* PR30 and *B. subtilis* PR31. These organisms displayed multiple plant growth-promoting (MPGP) traits such as the production of ammonia (NH3), hydrogen cyanide (HCN), siderophore (SD), indole acetic acid (IAA), 1-aminocyclopropane-1-carboxylate deaminase (ACCD), and phosphate solubilization (PS) activity. Also, the organisms were tolerant of high levels of trace elements, a wide range of pH 5-9, and resistant to multiple antibiotics (Sagar 2017).

B. subtilis improves tolerance to biotic stresses. This mechanism of disease resistance begins with the expression of explicit genes and hormones, such as ACCD. ACCD prohibits the ethylene synthesis from its precursor ACC into α-ketobutyrate and ammonia. Ethylene limits root and shoot development and sustains plant homeostasis. The degradation of the ethylene precursor (ACC) by bacterial ACCD helps to relieve plant stress and maintain normal growth under stressful conditions (Glick 2014; Tiwari et al. 2018).

B. subtilis changes the complex form of essential nutrients, such as P and N, to an easily accessible form, such as NH_4^+ and PO_4^{3+}, that is uptaken by plant roots. Some of the *Bacillus* spp. release ammonia from nitrogenous organic matter (Hayat et al. 2010). Ding et al. (2005) reported that some of the *Bacillus* spp. synthesis of the enzyme nitrogenase (EC 1.18.6.1) is controlled by the nifH gene, which can fix atmospheric N2 and supply it to plants to enlarge plant growth and yield by delaying senescence. This property is exploited by scientists to use *B. subtilis* as a potent biofertilizer (Qiao et al. 2014; Garcia-Fraile et al. 2015; Kuan et al. 2016; Chandrasekaran et al. 2019).

Table 24.1 PGP traits of *B. subtilis* and their influence on the physiological functions of crop plants

Characteristics	Physiological functions	References
Phytohormones/IAA	Enhance plant growth, elongation of root and shoot, and cell division	Chowdappa et al. (2013), Xie et al. (2014), Kang et al. (2015b), Shao et al. (2015), Radhakrishnan and Lee (2016), Barnawal et al. (2017)
ACCD	Enhances plant amplification	Xu et al. (2014), Pourbabaee et al. (2016), Glick (2017); Tiwari et al. (2018)
Siderophores	Reduce the movement of toxic ions, assist to solubilize iron as of minerals and organic compounds in the rhizosphere	Patil et al. (2014), Lastochkina et al. (2017), Radhakrishna et al. (2017)
P solubilization	Conversion of inorganic phosphate into soluble phosphates	Sivasakthi et al. (2013), Kang et al. (2014a, 2015a); Kuan et al. (2016)

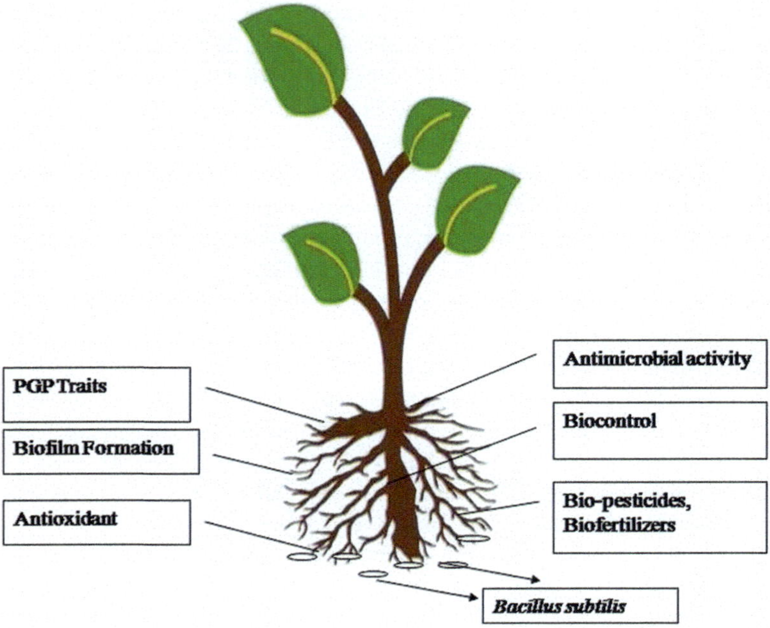

Fig. 24.1 The multifarious potential of *Bacillus subtilis*

24.2.1 Influence of B. subtilis *on Physiological Parameters of Vegetables, Cereals, and Other Plants*

Vegetables and cereals comprise an integral part of human healthy foods. These are important sources of dietary nutrients, such as calcium, magnesium, potassium, iron, beta-carotene, vitamin B complex, vitamin C, vitamin A, vitamin K, and antioxidants, along with providing soluble and insoluble dietary fibers, collectively recognized as non-starch polysaccharides (NSP), for example, cellulose, mucilage, hemicellulose, gums, pectin, etc. Hence, it is recommended to use PGPR in vegetable cultivation to enhance vegetable production without the use of chemicals. Here, an effort is made to enlighten the PGPR role in vegetable production under both regular and decrepit soil (Rai and Nabti 2017). Significant increases in seedling growth of tomato, mung bean, broccoli, and chickpea in response to inoculation with *B. subtilis* have been reported (data not published). Inoculation with PGP *B. subtilis* exhibited a higher percentage of seed germination, seedling vigor, and enhanced growth parameters in vegetables crop seeds such as chickpea (Pusa-256), mung bean (B1Zaid 2012), broccoli (Palam Samridhi), and tomato (NTLI86) compared to non-treated control plants (Qiao et al. 2017; Bahramisharif and Rose 2019), chickpea (Egamberdieva et al. 2017a, b; Abd_Allahet al. 2017; Khan et al. 2019) and mung bean (Hashem et al. 2017; Shen et al. 2017; Akinrinlola et al. 2018; Arif et al. 2019).

A large number of mechanisms are involved in growth promotion and increasing crop yield. The most common is the production of phytohormones, which are IAA and auxin, which enhances plant growth parameters by various mechanisms (Qiao et al. 2017, Bahramisharif and Rose, 2019); phosphate-solubilizing property has an impact on growth parameters in tomato and pepper (Radhakrishnan and Lee 2016; Barnawal et al. 2017). N, P, and K, which are the highest requirement of a plant for its development, are supplied to the plants by these bacteria (Table 24.2).

24.3 Biocontrol Potential of *B. subtilis*

The disease process is an interplay between susceptibility of host or resistance to pathogen environmental conditions and disease-causing potential of a pathogen (Wang et al. 2018). *B. subtilis* profoundly manipulates these interactions and shows its antimicrobial activity through (Ashraf et al. 2019; Lima et al. 2019; Khan et al. 2019) biofilm formation (Gingichashvili et al. 2017; Dragos et al. 2018; Dinh et al. 2019; García et al. 2020; Gingichashvili et al. 2020), which thus act as biocontrol (Bahramisharif and Rose, 2019; Hashem et al. 2019; Zhang et al. 2019) (Table 24.3 and Fig. 24.1).

24.3.1 Mechanisms Involved in Biocontrol

Biocontrol of fungal and bacterial phytopathogens involves the competition for nutrients or colonization sites, cytolytic effects, and antibiotic production (Han et al. 2015; Wang et al. 2015; Mnif et al. 2016). *B. subtilis* produces antimicrobial metabolites such as antibiotics the lipopeptides representing various antifungal and antibacterial antibiotics, which include fengycins, iturins, and surfactins (Mnif and Ghribi 2015); secretes hydrolytic enzymes, which have cell lysis effect; produces endospores; and alters the microenvironment conducive for plant growth, which favors its use as a biocontrol agent.

The success of biocontrol agents depends upon their capability to protect crops against target pathogens and pests; these must be eco-friendly. *B. subtilis* colonizes the root of plants in the rhizosphere; this potential is increased by genetically engineered strains that overexpress one or more characteristics so the strains with several different anti-pathogen traits can act synergistically when used together (Bahramisharif and Rose 2019; Hashem et al. 2019; Wang et al. 2018; Zhang et al. 2019, Dotaniya et al. 2016). *B. subtilis* isolates B4 used as a biofungicide to reduce peanut soil-borne diseases under both greenhouse and field conditions keeping Rizolex-T (fungicide) as standard (Ahmad et al. 2019).

Recovery from oxidative stress-induced damage is possible via antioxidant synthesis, which inhibits ROS formation. *Bacillus*-based bacterial association in plants can reduce the synthesis of ROS in cells via various scavenging enzymes (Ashraf et al. 2019; Lima et al. 2019; Khan et al. 2019) (Table 24.3 and Fig. 24.1). *B. subtilis* checks the growth of pathogens by the following mechanisms: to gain an insight

Table 24.2 Influence of *B. subtilis* inoculation on vegetables and other crops for seed germination and growth parameters

Crop name	Function	References
Tomato	Enhances fruit and grain yield	Xu et al. (2014), Abbasi and Weselowski (2015), Akram et al. (2016), Qiao et al. (2017), and Bahramisharif and Rose (2019)
Pepper plants	Improves photosystem II effectiveness and boost photosynthesis	Samaniego-Gámez et al. (2016)
Chickpea	Enhances chlorophyll, protein, and sugar contents	Patil et al. (2014), Egamberdieva et al. (2017a, b), Abd_Allah et al. (2017) and Khan et al. (2019)
Bean	Augments leaf water content and the parameter of stomata	Li Y et al. (2016a, b), Lima et al. (2019) and de Lima et al. (2019)
Mung bean	Improves growth	Hashem et al. (2017), Shen et al. (2017) and Arif et al. (2019)
Cucumber	Seed germination, plant growth, and yield	Radhakrishnan and Lee (2013, 2014)
Soybean	Increases growth significantly	Kang et al. (2014b) and Akinrinlola et al. (2018)
Pigeon pea	Promotes cardiovascular health and can be build up as a new dietary complement or useful food that avert hypertension	Lee et al. (2015)
Barley	Enhances grain yield, straw, total yield (TY), and plant nutrient element (PNE) content	Baris et al. (2014).
Rice	Enhancing plant growth	Elshakh et al. (2016) and Ahmad et al. (2017)
Maize	Significant increases in root dry weight and increased total biomass	Ahmad et al. (2017), Akinrinlola et al. (2018), Lima et al. (2019), Lima et al. (2019) and Misra and Chauhan (2020)
Wheat	Increases shoot length, increase the crop yield	Pourbabaee et al. (2016), Barnawal et al. (2017), Lastochkina et al. (2017), Reiss (2017), Akinrinlola et al. (2018) and Ashraf et al. (2019)
Arabidopsis thaliana roots	Utilizes carbon-rich root exudates	Allard et al. (2016)
Sesame	Increases the length and biomass of shoot, roots, and leaves	Radhakrishnan and Lee (2016)
Cacao	Stimulates both foliar and root growth	Falcäo et al. (2014)
Timothy plants	Improves the yield	Gagné-Bourque et al. (2016)
Cannabis	Improves marijuana and hemp yield and quality	Lyu et al. (2019)

Table 24.3 Disease control by various strains of *B. subtilis*

Strain no.	Crop	Disease	Causal agent	Mode of action	References
B. subtilis	Tomato	Root knot	Nematodes	Activates ISR	Adam et al. (2014)
B. subtilis	Tomato	*Botrytis* rot	*Botrytis cinerea*	Uses as one of the biocontrol agents for disease suppression	Lee et al. (2014)
4812	Tomato	Bacterial wilt	*R. solanacearum*	Considers as a more effective biocontrol agent	Almoneafy et al. (2013)
IAGS174	Tomato	*Fusarium* wilt	*Fusarium* sp.	ISR	Akram et al. (2014)
OTPB1	Tomato	Early and late blight	*P. infestans*	PGP and ISR	Chowdappa et al. (2013)
QST 713	Tomato	Bacterial spot	*Xanthomonas*	Biofilms formation; surfactin and iturin	Abbasi and Weselowski (2015)
QST 713	Tomato	Bacterial speck	*Pseudomonas syringae*	Reduces disease significantly	Fousia et al. (2016)
B. subtilis	Potato	Black scurf of	*R. solani* (PRS1)	Produces antifungal substances against plant pathogens	Singh et al. (2017)
B. subtilis	Rice	Sheath blight	*F. oxysporum f.* sp. *melonis*, *R. solani*	Effective formulation as an alternative to chemical fungicides	Singh et al. (2017)
NUU4	Chickpea	Root rot	*F. solani*	Reduces the infection	Egamberdieva et al. (2017b)
BS21-1	Chinese cabbage and lettuce	Soft rot	*Pectobacterium carotovorum* SCC1	Uses as biocontrol agents for disease suppression	Lee et al. (2014)
BS21-1	Cucumber	Anthracnose lesions	*Colletotrichum orbiculare*	Uses as biocontrol agents for disease suppression	Lee et al. (2014)
BERA 71	Mung bean	Goid disease	*Macrophomina phaseolina* (Tassi)	Modulate the metabolism of pigments, hormones, antioxidants	Hashem et al. (2017)

(continued)

Table 24.3 (continued)

Strain no.	Crop	Disease	Causal agent	Mode of action	References
LCA1 and M24	Mung bean	Seed sprouts pose	*Salmonella enterica* and *Enterohemorrhagic Escherichia coli* (EHEC)	Production of antimicrobial substance and/or nutrition/space competition	Shen et al. (2017)
PHYS77, PHYS78	Onion	Cape and umbel blights of onion	*Botrytis* sp.	Biocontrol agent	Elyousr et al. (2020)
B. subtilis	*Cannabis* crop	Powdery mildew	*Fusarium*, sooty molds	Growth promotion and disease biocontrol	Lyu et al. (2019)
IIHR BS-2	Carrot	Root-knot nematode and soft rot	*E. carotovorum*	Increases in carrot yield	Rao et al. (2017)
GLB191	Grape	Grape downy mildew	*Plasmopara viticola*	Uses as a new biocontrol product	Yan et al. (2019)
UMAF6639	Melons	Cucurbit powdery mildew	*Podosphaera fusca*	Biological control agent	Garcia-Gutierrez et al. (2013)
Y-1	Apple	Root rot	*Fusarium* sp.	PGP and ISR	Ju et al. (2014)
EA-CB0015	Banana	Black sigatoka	*Mycosphaerella fijiensis*	Reduces the severity of the disease	Gutierrez-Monsalve et al. (2015a, b)
HJ5	Cotton	*Verticillium* wilt	*Verticillium dahlia*	Biofilm formation	Li et al. (2013)
B15	Rice seed	Leaf blight (BLB)	*Xanthomonas oryzae pv. Oryzae*	Enhances activities of defense-related enzymes	Elshakh et al. (2016)
SYX04	Rice	rice blast	*M. oryzae*	Compete for colonization sites; ISR	Mnif et al. (2015)
OH 131.1	Wheat	Head blight	*F. graminearum*	Bioactive metabolites	Dunlap et al. (2015)
QST713	Wheat	Yellow rust	*Puccinia striiformis*	Enhance growth	Reiss and Jørgensen (2017)
Tpb55	Tobacco	Tobacco black shank	*P. nicotianae*	Biofilm formation of antibiotic substances	You et al. (2016) and Han et al. (2015)

into the type of molecules that activate the AHL biosensors, it was decided to test whether these compounds were substrates for the AiiA AHL-lactonase, a quorum quenching enzyme from *Bacillus* sp. The *Bacillus* aiiA gene encoding the AiiA protein was cloned, expressed, and purified to homogeneity as a His6-tagged protein (Degrassi et al. 2007).

Copper hydroxide along with *B. subtilis* QST 713 was tested against *Pseudomonas syringae* under greenhouse conditions. The plant showed protective activity of QST 713, as Pin2 expression was significantly higher in the QST 713-treated plants challenged with Pst compared to the control Pst-inoculated plants (Fousia et al. 2016). Induced systematic tolerance and improve the health of plants (Kang et al. 2014b). *Bacillus* strain 6051produces surfactin, indicating very stable biofilm formation and that it is a good biocontrol agent against pathogenic bacteria (Bais et al. 2004). As previously mentioned, the lipopeptides produced by *B. subtilis* characterize different antifungal and antibacterial antibiotics including fengycins, iturins, and surfactins (Mnif and Ghribi, 2015). *B. subtilis* forms a thin biofilm on the roots for long-term colonization of the rhizosphere. *B. subtilis* locates and colonizes young roots during chemotaxis (Allard et al. 2016). Biofilms consist of a multicellular bacterial community enclosed in a self-secreted matrix. The timing of the formation of a *B. subtilis* biofilm on host roots is also dependent on the promoter of the genes responsible for the production of the matrix when the bacterium initially contacts a root (Beauregard et al. 2013) in the soil; *B. subtilis* forms biofilms on plant roots which help to produce lipopeptides and augment their antimicrobial activity (Davey et al. 2003). *Bacillus* species are used for rhizosphere applications and also function as plant endophytes (McSpadden Gardener and Driks 2004) that also protect plants from pathogens (Romero et al. 2004).

24.3.1.1 Induction of Host Enzymes Such as Peroxidase, Polyphenol Oxidase, Superoxide Dismutase

B. subtilis can secrete antibiotics and hydrolytic enzymes; it can be used as environment modifier in a self-beneficial manner and also produces resistant endospores to sustain itself under adverse conditions (Dotaniya et al. 2016). Antibiotic secretion and release of transcription factors (Malfanova et al. 2012; Beauregard et al. 2013; Allard et al. 2016; Yang et al. 2016a, b). *B. subtilis* serves as a potent insecticide due to the production of lipopeptides (surfactin C14, C15, iturin 15) (S499 strain). The strain BS-Y9 is active against green peach aphid *Myzus persicae*, and the active components are the surfactins isomers of C14leu 7, C14 val7, and C15 leu 7 with leucine moieties showing higher activity. The strain SPB1 lipopeptide showed activity against ectomyelois *Ceratoniae zeller* (Assié et al. 2002; Ghribi et al. 2012; De Faria et al. 2011; Basaid et al. 2020).

24.3.2 B. subtilis *as a Biocontrol Representative in Different Crops*

B. subtilis strains are rhizospheric endophytes and are widely studied as biocontrol agents both in vitro and in greenhouses and fields for sustainable agriculture (Rao et al. 2017; Hashem et al. 2017; Singh et al. 2017; Yan Li et al. 2019; Lyu et al. 2019).

Table 24.3 lists some *B. subtilis* strains as biocontrol agent allied with plant disease control, e.g., *Fusarium* head blight of wheat (Dunlap et al. 2015), wheat take-all (Yang et al. 2015) head blight (Dunlap et al. 2015), yellow rust (Reiss and Jørgensen, 2017), rice blast (Mnif et al. 2015; Yang et al. 2015), leaf blight (BLB) (Elshakh et al. 2016), and sheath blight (Singh et al. 2017). *B. subtilis* is also utilized to manage tomato *Fusarium* wilt (Akram et al. 2016), bacterial spot (Abbasi and Weselowski 2015), and cucumber root rot (Senol et al. 2014). Moreover, *B. subtilis* also controls some post-harvest diseases, e.g., soft rot of fruits (Tang et al. 2014). *B. subtilis* suppresses tobacco black shank (Han et al. 2015; You et al. 2016).

Notably, *Bacillus* species can act as plant endophytes that protect plants against pathogens (Lastochkina et al. 2019) (Elshakh et al. 2016). *B. subtilis* is beneficial for plant productivity (Venkateshwaran et al. 2013) as they promote plant growth by improving plant nutrition, synthesis, and regulation of phytohormones, and suppression of disease-causing organisms.

24.4 Alleviation of Abiotic Stress by *B. subtilis*

Drought and salinity due to industrial pollution engulfing the agricultural land are the most atrocious abiotic factors restricting the yield of crop plants because most crops cannot tolerate these stresses. Microorganisms come in front to save the agricultural land and increase the agriculture produce such as tolerance to saline conditions, synthesis of compatible solutes, production of plant growth-promoting hormones, and biocontrol potential, which interact with crop plants and impart tolerance to stress conditions (Shrivastava and Kumar 2015). Inoculations of plants with these bacteria stimulate root exudation by plants, which in turn promote both bacterial and plant growth (Kakar et al. 2016).

The plants colonized by bacillus absorb more water so it assists plant combat drought and salt stress (Allah et al. 2017; Egamberdieva et al. 2017a, b; Lastochkina et al. 2017; Misra and Chauhan 2020; Woo et al. 2020; Kakar et al. 2016; Kang et al. 2014b) and water stress (Lima et al. 2019; Li et al. 2016a, b). The extenuating property of *Bacillus*-induced physiological alterations in plants is exposed in Table 24.4.

The chemical fungicides and insecticides hamper the number and growth of valuable soil microbes that help to augment plant growth. Thus, the replacement of chemical pesticides with beneficial bacteria, such as *B. subtilis*, can reduce the use of synthetic pesticides and insecticides in modern agriculture. *B. subtilis* is known as protected bacteria. *B. subtilis* constructs endospores, allowing them to germinate under different environmental cues that permit long-term storage, and reduces the intricacy of the formulation procedure (Hashem et al. 2019).

24.4.1 Antioxidant Enzymes in Stress Management

Abiotic stresses impede seed germination, hinder plant growth, and lower crop productivity. When plants are subjected to abiotic stresses such as salinity, drought, temperature extremes, herbicide treatment, and mineral deficiency, excessive salt ions absorbed by the plant break the ion homeostasis of plant cells, which in turn affects the activity of relevant enzymes, resulting in the accumulation of a large number of toxic substances such as reactive oxygen species (ROS)s. A balance between the ROS production and the appeasing action of antioxidants does exist. Any disturbance of the balance results in peroxidation, which causes oxidative damage to macromolecules of cells, viz., lipids, protein, and nucleic acids. Superoxide dismutase (SOD), peroxidase (POD), ascorbate peroxidase (APX), glutathione reductase (GR), and catalase (CAT) are some of the enzymes which act as scavengers of ROS, ensuing enhanced plant resistance to stresses (Hu et al. 2019).

The APX, SOD, and GR degrade H_2O_2 in cell organelles such as chloroplast and mitochondria. Similarly, CAT and POD eliminate H_2O_2 and can counterbalance for ageoxy intermediates and free radicals. SOD curbs relative amounts of O2− and thus diminishes the hazard of −OH radical formation. CAT is the chief scavenger of H_2O_2 in peroxisomes, and it converts this to water and molecular oxygen. CAT and POD also have the same role against oxidative stress by detoxifying H_2O_2 in chloroplast, cytosol, mitochondria, and peroxisome of plant cells. SOD and CAT help GSH in its role as a free radical scavenger physiologically by decreasing levels of hydrogen peroxide, CAT, which is present frequently in glyoxysomes of lipid-storing tissues in plants; this enzyme contains a tetrameric heme which catalyzes the switch of hydrogen peroxide to water and molecular oxygen formed from the β-oxidation of fatty acids. The APX is present in different active isomeric forms in chloroplasts, cytosol, and microsomes of plants, which is one of the vital antioxidant enzymes that scavenge H_2O_2 by using ascorbate for reduction (Abd_Allah et al. 2017; Ashraf et al. 2019; de Lima et al. 2019; Khan et al. 2019).

24.5 Conclusion

The ability of *B. subtilis* to enhance plant growth has been commercialized. *B. subtilis* enhances plant growth through the direct and indirect mechanism. The endospore-forming ability of *Bacillus* allows its application in the harsh environmental conditions such as abiotic and biotic stress. Plant growth-promoting *Bacillus* spp. showed growth of roots, shoots, and leaves as well as enhanced yields. *Bacillus* sp. can be used as biofertilizers, and soil fertility control agents inhibit

Table 24.4 Management of abiotic stress in different crops by *B. subtilis*

Crops	Abiotic stress	Mode of action	References
Wheat	Drought	Increase macronutrients	Barnawal et al. (2017)
Rice	Drought	Induced efficient tolerance and develop the health of plants	Kakar et al. (2016)
Soybean	Salt and drought stress	Enhance plant growth, increase the water, nutrients, antioxidants pigments, and hormones	Kang et al. (2014b)
Chickpea	Salinity	Ameliorate seed germination and plant development, carbohydrates, proteins, and osmolytes	Qurashi and Sabri (2013)
Rice	Salinity	Prompt the expression of the NADP-Me2, EREBP, SOSI, BADH, and SERK1 genes	Nautiyal et al. (2013)
Cucumber	Salinity	Seed germination, plant enlargement, and yield	Radhakrishnan and Lee (2013, 2014)
Soybean	Salinity	Enhance plant growth, increase the water, nutrients, antioxidants, pigments, and hormones	Kang et al. (2014b)
White clover	Salinity	Augmentation of the K^{+}/Na^{+} ratio in plants grown	Han et al. (2014)
Bassia indica	Salt stress	Prompt the synthesis of chlorophylls a and b and carotenoid, increase photosynthesis	Hashem et al. (2015a, b)
Solanum lycopersicum	Salinity	Amend lipid synthesis, specifically that of oleic, linoleic, and linolenic acids as well as phospholipids	Hashem et al. (2015a, b)
Acacia gerrardii Benth	Salt	Induction of osmoregulation	Hashem et al. (2016a, b)
Acacia gerrardii	Salinity	Enhance plant growth	Hashem et al. (2016a, b)
Rice	Salinity	Sledding in lipid synthesis might mitigate lipid peroxidation and oxidative stress, CAT and POD activities are enhanced	Jha and Subramanian (2015)
Chenopodium quinoa	Salinity	Growth and productivity of quinoa could be improved by inoculating	Yang et al. (2016a, b)
Chickpea	Salinity	Renovate plant biomass and the synthesis of photosynthetic pigments	Abd_Allahet al. (2017)
Chickpea	Salinity	Improve plant growth and symbiotic performance and to control root rot disease	Egamberdieva et al. (2017a, b)
Wheat	Salinity	Increase growth	Lastochkina et al. (2017)
Maize	Salinity	Prompt plant response for defense enzymes, chlorophyll, proline, and soluble sugar	Misra and Chauhan (2020)

(continued)

Table 24.4 (continued)

Crops	Abiotic stress	Mode of action	References
Faba bean	Water stress	Efficacy of the closure of stomata and promoted water use efficiency	Li et al. (2016a, b)
Common bean and maize	Water stress	Magnification of leaf water content and the regulation of stomata	Lima et al. (2019)
Mung bean	Heavy metal, lead (Pb)	Improve mung bean growth potential	Arif et al. (2019)

disease-causing pathogenic microbial growth and provoke pest defense systems in plants. Since these PGP *B. subtilis* inoculants exhibited multiple characterizations valuable to the plants, it can be useful for the development of newer, safer, and effective seed treatments as an alternative to chemical fertilizer. *B. subtilis* carves a space in the agriculture market as a potent biofertilizer, biocontrol agent, and mico-remediator of increased drought, salinity, and a whole host of industrial pollutions.

References

Abbasi PA, Weselowski B (2015) Efficacy of Bacillus subtilis QST 713 formulations, copper hydroxide, and their tank mixes on bacterial spot of tomato. Crop Prot 74:70–76

Abd Allah EF, Alqarawia AA, Hashem A, Radhakrishnand R, Al-Huqailb AA, FONA O, Malika JA, Alharbib RI, Egamberdieva D (2017) Endophytic bacterium Bacillus subtilis (BERA 71) improves salt tolerance in chickpea plants by regulating the plant defense mechanisms. J Plant Interact 13(1):37–44. https://doi.org/10.1080/17429145.2017.1414321

Adam M, Heuer H, Hallmann J (2014) Bacterial antagonists of fungal pathogens also control root-knot nematodes by induced systemic resistance of tomato plants. PLoS One 9:e90402. https://doi.org/10.1371/journalpone0090402

Ahmad Z, Wu J, Chen L, Dong W (2017) Isolated Bacillus subtilis strain 330-2 and its antagonistic genes identified by the removing PCR. Sci Rep 7:1777. https://doi.org/10.1038/s41598-017-01940-9

Ahmad AGM, Attia AG, Mohamed MS, Abd Allah HE (2019) Fermentation, formulation and evaluation of PGPR Bacillus subtilis isolate as a bioagent for reducing occurrence of peanut soil-borne diseases. J Integr Agric 18(9):2080–2092

Akinrinlola RJ, Yuen GY, Drijber RA, Adesemoye AO (2018) Evaluation of Bacillus strains for plant growth-promotion and predictability of efficacy by in vitro physiological traits. Intl J Microbiol., 5686874, 11 pages. https://doi.org/10.1155/2018/5686874

Akram W, Anjum T, Ali B (2014) Searching ISR determinants from Bacillus subtilis IAGS174 against Fusarium wilt of tomato. Biocontrol 60:271–280

Akram W, Anjum T, Ali B (2016) Phenylacetic acid is ISR determinant produced by Bacillusfortis IAGS162, which involves extensive re-modulation in metabolomics of tomato to protect against Fusarium wilt. Front Plant Sci 7:498. https://doi.org/10.3389/fpls201600498

AllardMassicotte L, Tessier F, Lecuyer V, Lakshmanan JF, Lucier D, Garneau L, Caudwell H, Vlamakis HP, Bais P, Beauregard (2016) Bacillus subtilis early colonization of Arabidopsis thaliana roots involves multiple chemotaxis receptors. M Bio 7(6)

Almoneafy AA, Ojaghian MR, Seng-fu X, Ibrahim M, Xie GL, Shi Y, Tian WX, Li B (2013) Synergistic effect of acetyl salicylic acid and DL-Beta-aminobutyric acid on biocontrol efficacy of Bacillus strains against tomato bacterial wilt. Trop Plant Pathol 38(2):102–113. https://doi.org/10.1590/s1982-56762013000200003

Arifa MS, Sher TY, Shahzad M, Riaza M, Shahid MR, Muntaha I, AsifaM H (2019) Soli mande-ShafaqatAlia Lead toxicity induced phytotoxic effects on mung bean can be relegated by lead tolerant Bacillus subtilis (PbRB3). Chemosphere 243:70–80

Ashraf A, Bano A, Ali SA (2019) Characterization of plant growth-promoting rhizobacteria from rhizosphere soil of heat-stressed and unstressed wheat and their use as bio-inoculant. Plant Biol 21:762–769

Assie LK, Deleu M, Arnaud L, Paquot M, Thonart P, Gaspar CH, Haubruge E (2002) Insecticide activity of surfactins and iturins from a biopesticide Bacillus subtilis Cohn (S499 strain). Meded Rijksuniv Gent Fak Land bouwkd Toeg ep Biol Wet 67(3):647–655

Bahramisharif A, Rose LE (2019) Efficacy of biological agents and compost on growth and resistance of tomatoes to late blight. Planta 249(3):799–813

Bais HP, Fall R, Vivanco JM (2004) Biocontrol of Bacillus subtilis against infection of Arabidopsis roots by Pseudomonas syringae is facilitated by biofilm formation and surfactin production. Plant Physiol 134(1):307–319. https://doi.org/10.1104/pp.103.028712

Baris O, Sahin F, Turan M, Orhan F, Gulluce M (2014) Use of Plant-growth-promoting rhizobacteria (PGPR) seed inoculation as alternative fertilizer inputs in wheat and barley production. Commun Soil Sci Plan 45:18

Barnawal D, Bharti D, Pandey SS, Pandey A, Chanotiya CS, Kalra A (2017) Plant growth-promoting rhizobacteria enhance wheat salt and drought stress tolerance by altering endogenous phytohormone levels and TaCTR1/TaDREB2 expression. Physiol Plant 161:502–514

Basaid K, Chebli B, Mayad EH, Furze JN, Bouharroud R, Krier F, Paulitz T (2020) Biological activities of essential oils and lipopeptides applied to control plant pests and diseases: a review. Interna J Pest Manag:1–23. https://doi.org/10.1080/09670874.2019.1707327

Beauregard PB, Chai Y, Vlamakis H, Losick R, Kolter R (2013) Bacillus subtilis biofilm induction by plant polysaccharides. Proc Natl Acad Sci U S A 110, E1621–E1630. https://doi.org/10.1073/pnas.1218984110

Bhardwaj D, Ansari MW, Sahoo RK, Tuteja N (2014) Biofertilizers function as key player in sustainable agriculture by improving soil fertility, plant tolerance and crop productivity. Microb Cell Fact 13:66

Chandrasekaran M, Chun SC, Oh JW, Paramasivan M, Saini RK, Sahayarayan JJ (2019) Bacillus subtilis CBR05 for tomato (Solanum lycopersicum) fruits in South Korea as a novel plant probiotic bacterium (PPB): implications from total phenolics, flavonoids, and carotenoids content for fruit quality. Agro 9(12):838. https://doi.org/10.3390/agronomy9120838

Chowdappa P, Kumar SPM, Lakshmi MJ, Upreti KK (2013) Growth stimulation and induction of systemic resistance in tomato against early and late blight by Bacillus subtilis OTPB1 or Trichoderma harzianum OTPB3. Biol Control 65:109–117. https://doi.org/10.1016/jbiocontrol201211009

De Faria AF, Teodoro-Martinez DS, de Oliveira Barbosa GN, Gontijo Vaz B, Serrano Silva Í, Garcia JS, Tótolac MR, Eberlinb MN, Grossmand M, Alves OL, Regina Durrant L (2011) Production and structural characterization of surfactin (C14/Leu7) produced by Bacillus subtilis isolate LSFM-05 grown on raw glycerol from the biodiesel industry. Process Biochem 46(10):1951–1957. https://doi.org/10.1016/j.procbio.2011.07.001

de Lima BC, Moro AL, Santos ACP, Bonifacio A, Araujo ASF, de Araujo FF (2019) Bacillus subtilis ameliorates water stress tolerance in maize and common bean. J Plant Interact 14(1):432–439. https://doi.org/10.1080/17429145.2019.1645896

Davey ME, Caiazza NC, O'Toole GA (2003) Rhamnolipid surfactant production affects biofilm architecture in Pseudomonas aeruginosa PAO1. J Bacteriol 185(3):1027–1036

Degrassi G, Devescovi G, Solis R, Steindler L, Venturi V (2007) Oryza sativa rice plants contain molecules that activate different quorum-sensing N-acyl homoserine lactone biosensors and are sensitive to the specific AiiA lactonase. FEMS Microbiol, 269(2), 213–220. https://doi.org/10.1111/j.1574-6968.2006.00624.x

Ding Y, Wang J, Liu Y, Chen S (2005) Isolation and identification of nitrogen-fixing bacilli from plant rhizospheres in Beijing region. J Appl Microbiol 99:1271–1281

Dinh TL, Akhmetova GR, Martykanova DS, Rudakova NL, Sharipova MR (2019) Influence of divalent metal ions on biofilm formation by Bacillus subtilis. Bio Nano Science 9:521–527

Dotaniya MV, Meena B, Basak RS, Meena (2016) Potassium uptake by crops as well as microorganisms Potassium solubilizing microorganisms for sustainable agriculture. Springer, pp 267–280

Dragos A, Nivedha Lakshmanan N, Marivic Martin M, Horvath B, Gergely Maroti G, Garcıa CF, Lieleg O, Kovacs AT (2018) Evolution of exploitative interactions during diversification in Bacillus subtilis biofilms, FEMS. Microb Ecol 94:fix155

Dunlap CA, Schisler DA, Bowman MJ, Rooney AP (2015) Genomic analysis of Bacillus subtilis OH 1311 and co-culturing with Cryptococcus flavescens for control of Fusarium head blight Plant. Gene 2:1–9

Egamberdieva D, Wirth S, Jabborova D, Räsänen LA, Liao H (2017a) Coordination between Bradyrhizobium and *Pseudomonas* alleviates salt stress in soybean through altering root system architecture. J Plant Inter 12(1):100–107

Egamberdieva D, Wirth S, Shurigin V, Hashem A, Abd Allah EF (2017b) Endophytic bacteria improve plant growth, symbiotic performance of chickpea (*Cicer arietinum* L) and induce suppression of root rot caused by *Fusarium solani* under salt stress. Front Microbiol 8:75. https://doi.org/10.3389/fmicb201701887

Elshakh ASA, Anjum SI, Qiu W, Almoneafy AA, Li W, Yang Z, Cui ZQ, Li B, Chang G, Xie SGL (2016) Controlling and defense-related mechanisms of Bacillus strains against bacterial leaf blight of rice. J Phytopathol 164:534–546. https://doi.org/10.1111/jph12479

Elyousr KAMA, Alamri SAM, Hussein MMA, Hassan MAH, El-Fatah BESA, Hashem M (2020) Molecular disparities among Botrytis species involved in onion umbel blight disease and its management using Bacillus subtilis PHYS7. Egyptian J Biol Pest Cont 30:1

Falcäo LL, Silva-Werneck JO, Vilarinho BR, da Silva JP, Pomella AWV, Marcellino LH (2014) Antimicrobial and plant growth-promoting properties of the cacao endophyte Bacillus subtilis ALB629. J Appl Microbiol 116(6):1584–1592

Fousia S, Paplomatas EJ, Tjamos SE (2016) Bacillus subtilis QST 713 confers protection to tomato plants against Pseudomonas syringae pv tomato and induces plant defense-related genes. J Phytopathol, 164, 264–270 https://doi.org/10.1111/jph12455

Gagné-Bourque F, Bertrand A, Claessens A, Aliferis KA, Jabaji S (2016) Alleviation of drought stress and metabolic changes in timothy (Phleum pratense L) colonized with Bacillus subtilis B26. Front Plant Sci 7:584

García CF, KretschmerM L-ACN, Schönleitner M, DragoŝA KAT, Lieleg O (2020) Metal ions weaken the hydrophobicity and antibiotic resistance of Bacillus subtilis NCIB 3610 biofilms. npj Biofilms and Microbiomes 6:1. https://doi.org/10.1038/s41522-019-0111-8

Garcia-Fraile P, Menendez E, Rivas R (2015) Role of bacterial biofertilizers in agriculture and forestry. AIMS Bioeng 2:183–205. https://doi.org/10.3934/bioeng20153183

García-Gutiérrez L, Zeriouh H, Romero D, Cubero J, Vicente AD, Pérez-García A (2013) The antagonistic strain Bacillus subtilis UMAF6639 also confers protection to melon plants against cucurbit powdery mildew by activation of jasmonate-and salicylic acid-dependent defence responses. Microbiol Biotechnol 6:264–274

Ghribi D, Mesrati LA, Mnif I, Kammoun R, Ayadi I, Saadaoui I, Maktouf S, Ellouze SC (2012) Investigation of antimicrobial activity and statistical optimization of Bacillus subtilis SPB1 biosurfactant production in solid-state fermentation. Biomed Res Int:ID 373682. https://doi.org/10.1155/2012/373682

Gingichashvili S, Duanis-Assaf D, Shemesh M, Featherstone JDB, Feuerstein O, Steinberg D (2017) Bacillus subtilis biofilm development—a computerized study of morphology and kinetics. Front Microbiol, 8: 2072

Gingichashvili S, Duanis-Assaf D, Shemesh M, Featherstone JDB, Feuerstein O, Steinberg D (2020) The adaptive morphology of Bacillus subtilis biofilms: A defense mechanism against bacterial starvation. Microorganisms 8:62. https://doi.org/10.3390/microorganisms8010062
Glick BR (2014) Bacteria with ACC deaminase can promote plant growth and help to feed the world. Microbiol Res 169:30–39. https://doi.org/10.1016/jmicres201309009
Goswami D, Thakker JN, Dhandhukia PC (2016) Portraying mechanics of plant growth-promoting rhizobacteria (PGPR): a review. Cogent Food Agric 2:1–19
Gutierrez-Monsalve JA, Mosquera S, González-Jaramillo LM, Mira JJ, Villegas-Escobar V (2015a) Effective control of black Sigatoka disease using a microbial fungicide based on Bacillus subtilis EA-CB0015 culture. Biol Contr 87:39–46. https://doi.org/10.1016/jbiocontrol201504012
Gutierrez-Monsalve JA, Mosquera S, González-Jaramillo LM, Mira JJ, Villegas-Escobar V (2015b) Effective control of black Sigatoka disease using a microbial fungicide based on Bacillus subtilis EA-CB0015 culture. Biol Cont 87:39–46. https://doi.org/10.1016/j.biocontrol.2015.04.012
Han QQ, Lu XP, Bai JP, Qiao Y, Pare PW, Wang SM, Zhang JL, Wu YN, Pang XP, Xu WB, Wang ZL (2014) Beneficial soil bacterium Bacillus subtilis (GB03) augments salt tolerance of white clover. Front Plant Sci 5:525. https://doi.org/10.3389/fpls201400525
Han Y, Zhang B, Shen Q, You C, Yu Y, Li P, Shang Q (2015) Purification and identification of two antifungal cyclic peptides produced by Bacillus amyloliquefaciens L-H15. Appl Biochem Biotechnol 176:2202–2212. https://doi.org/10.1007/s12010-015-1708-x
Hashem A, Abd Allah EF, Alqarawi AA, AL-Huqail AA, Alshalawi SRM, Wirth S, Egamberdieva D (2015a) Impact of plant growth promoting Bacillus subtilis on growth and physiological parameters of Bassia indica (Indian bassia) grown under salt stress. Pak J Bot 47:1735–1741
Hashem A, Abd Allah EF, Alqarawi AA, Mona S, Whibi Al Alenazi MM, Egamberdieva D, Ahmad P (2015b) Arbuscular mycorrhizal fungi mitigates NaCl adverse effects on Solanum lycopersicum L. Pak J Bot 47(1):327–340
Hashem A, Abd Allah EF, Alqarawi AA, Al-Huqail AA, Wirth S, Egamberdieva D (2016a) The Interaction between arbuscular mycorrhizal fungi and endophytic bacteria enhances plant growth of Acacia gerrardii under salt stress. Front Microbiol 7:1089. https://doi.org/10.3389/fmicb201601089
Hashem A, Abd Allah EF, Alqarawi AA, AL-Huqail AA, Shah MA (2016b) Induction of osmoregulation and modulation of salt stress in Acacia gerrardii Benth by arbuscular mycorrhizal fungi and Bacillus subtilis (BERA 71). Biomed Res 6294098:11. https://doi.org/10.1155/2016/6294098
Hashem A, Abd Allah EF, Alqarawi AA, Radhakrishnan R, Kumar A (2017) Plant defense approach of Bacillus subtilis (BERA 71) against Macrophomina phaseolina (Tassi) Goid in mungbean. Plant-Micro Interact 12(1):390–401
Hashem A, Tabassum B, Abd-Allah EF (2019) Bacillus subtilis: a plant-growth promoting rhizobacterium that also impacts biotic stress. Saudi J Biol Sci 26: 1291–1297
Hayat R, Ali S, Amara U, Khalid R, Ahmed I (2010) Soil beneficial bacteria and their role in plant growth promotion: a review. Ann Microbiol 60:579–598
Hu QP, Liu RX, Liu JX (2019) Effects of Bacillus subtilis QM3 on germination and antioxidant enzymes activities of wheat seeds under salt stress. Open Access Library Journal 6:e5218. https://doi.org/10.4236/oalib.1105218
Jha Y, Subramanian RB (2015) Reduced cell death and improved cell membrane integrity in rice under salinity by root-associated bacteria. Theor Exp Plant Physiol 27:227–235. https://doi.org/10.1007/s40626-015-0047-1
Ju R, Zhao Y, Li J, Jiang H, Liu P, Yang T, BaoZ Z, Zhou B, Liu XZX (2014) Identification and evaluation of a potential biocontrol agent, Bacillus subtilis, against Fusarium sp in apple seedlings. Annals of Microbio 64:377–383
Kakar KU, Ren XL, Nawaz Z, Cui ZQ, Li B, Xie GL, Sun GC (2016) A consortium of rhizobacterial strains and biochemical growth elicitors improve cold and drought stress tolerance in rice (Oryza sativa L). Plant Biol 18:471–483. https://doi.org/10.1111/plb12427
Kang SM, Radhakrishnan R, You YH, Joo GJ, Lee IJ, Lee KE, Kim JH (2014a) Phosphate solubilizing Bacillus megaterium mj1212 regulates endogenous plant carbohydrates and amino

acids contents to promote mustard plant growth. Indian J Microbiol 54:427–433. https://doi.org/10.1007/s12088-014-0476-6

Kang SM, Radhakrishnan R, Khan AL, Kim MJ, Park JM, Kim BR, Lee IJ (2014b) Gibberellin secreting rhizobacterium, Pseudomonas putida H-2-3 modulates the hormonal and stress physiology of soybean to improve the plant growth under saline and drought conditions. Plant Physiol Biochem 84:115–124. https://doi.org/10.1016/jplaphy.2014.09001

Kang SM, Radhakrishnan R, Lee KE, You YH, Ko JH, Kim JH, Lee IJ (2015a) Mechanism of plant growth promotion elicited by Bacillus sp LKE15 in oriental melon. Acta Agric Scand Sect B Soil Plant Sci 65:637–647. https://doi.org/10.1080/09064710.2015.1040830

Kang SM, Radhakrishnan R, Lee I J (2015b) Bacillus amyloliquefaciens sub sp plantarum GR53, a potent biocontrol agent resists Rhizoctonia disease on Chinese cabbage through hormonal and antioxidants regulation. World J Microbiol Biotechnol 31: 1517–1527. https://doi.org/10.1007/s11274-015-1896-0

Khan N, Bano A, Rahman MR, Guo J, Kang Z, Babar M (2019) Comparative physiological, and metabolic analysis reveals a complex mechanism involved in drought tolerance in chickpea (Cicer arietinum L) induced by PGPR, and PGRs. Sci Rep 9:2097. https://doi.org/10.1038/s41598-019-38702-8

Kuan KB, Othman R, Rahim KA, Shamsuddin ZH (2016) Plant growth-promoting rhizobacteria inoculation to enhance vegetative growth, nitrogen fixation and nitrogen remobilization of maize under greenhouse conditions. PLoS One 11:e0152478. https://doi.org/10.1371/journalpone0152478

Lastochkina O, Pusenkova L, Yuldashev R, Babaev M, Garipova S, Blangova D, Khairullin R, Aliniaeifard S (2017) Effects of Bacillus subtilis on some physiological and biochemical parameters of Triticum aestivum L (wheat) under salinity Plant. Physiol Biochem 212:80–88

Lastochkina O, Seifikalhor M, Aliniaeifard S, Baymiev A, Pusenkova L, Garipova S, Kulabuhova D, Maksimov I (2019) Bacillus sp: efficient biotic strategy to control postharvest diseases of fruits and vegetables. Plants, 8(4): 97. https://doi.org/10.3390/plants8040097

Lee SW, Lee SH, Balaraju K, Park KS, Nam KW, Park JW, Park K (2014) Growth promotion and induced disease suppression of four vegetable crops by a selected plant growth-promoting rhizobacteria (PGPR) strain Bacillus subtilis 21-1 under two different soil conditions. Acta Physiol Plant 36:1353–1362. https://doi.org/10.1007/s11738-014-1514-z

Lee BH, Lai YS, Wu SC (2015) Antioxidation, angiotensin converting enzyme inhibition activity, nattokinase, and antihypertension of Bacillus subtilis (natto)-fermented pigeon pea. J Food Drug Anal 23:750–757. https://doi.org/10.1016/jjfda201506008

Li S, Zhang N, Zhang Z, Luo J, Shen B, Zhang R, Shen Q (2013) Antagonist Bacillus subtilis HJ5 controls Verticillium wilt of cotton by root colonization and biofilm formation. Biol Fertil Soils 49:295–303. https://doi.org/10.1007/s00374-012-0718-x

Li C, Sun B, Li Y, Liu C, Wu X, Zhang D, Shi Y, Song Y, Buckler ES, Zhangs Z, Wang T, Li Y (2016a) Numerous genetic loci identified for drought tolerance in the maize nested association mapping populations. BMC Genomics 17:894. https://doi.org/10.1186/s12864-016-3170-8

Li Y, Xu S, Gao J, Pan S, Wang G (2016b) Bacillus subtilis regulation of stomatal movement and instantaneous water use efficiency in Vicia faba. Plant Growth Regul 78:43–55

Lyu D, Backer R, Robinson WG, Smith DL (2019) Plant growth-promoting rhizobacteria for cannabis production: yield, cannabinoid profile and disease resistance. Front Microbiol 10:1761. https://doi.org/10.3389/fmicb201901761

Malfanova N, Franzil L, Lugtenberg B, Chebotar V, Ongena M (2012) Cyclic lipopeptide profile of the plant-beneficial endophytic bacterium Bacillus subtilis HC8. Arch Microbiol 194(11) 893–899. https://doi.org/10.1007/s00203-012-0823-0

McSpadden Gardener BB, Driks A (2004) Overview of the nature and application of biocontrol microbes: Bacillus spp. Phytopathol 94(11):1244

Misra S, Chauhan PS (2020) ACC deaminase-producing rhizosphere competent Bacillus spp. mitigate salt stress and promote Zea mays growth by modulating ethylene metabolism. Biotech 3, 10(3). https://doi.org/10.1007/s13205-020-2104-y

Mnif I, Ghribi D (2015) Potential of bacterial derived biopesticides in pest management. Crop Prot 77:52–64. https://doi.org/10.1016/jcropro201507017

Mnif I, Hammami I, Triki MA, Azabou MC, Ellouze-Chaabouni S, Ghribi D (2015) Antifungal efficiency of a lipopeptide biosurfactant derived from Bacillus subtilis SPB1 versus the phytopathogenic fungus, Fusarium solani Environ. Sci Pollut Res 22:18137–18147

Mnif I, Grau-Campistany A, Coronel-León J, Hammami I, Triki MA, Manresa A, Ghribi D (2016) Purification and identification of Bacillus subtilis SPB1 lipopeptide biosurfactant exhibiting antifungal activity against Rhizoctonia bataticola and Rhizoctonia solani. Environ Sci Pollut Res Int 23(7):6690–6699. https://doi.org/10.1007/s11356-015-5826-3

Nautiyal CS, Srivastava S, Chauhan PS, Seem K, Mishra A, Sopory SK (2013) Plant growth-promoting bacteria Bacillus amyloliquefaciens NBRISN13 modulates gene expression profile of leaf and rhizosphere community in rice during salt stress Plant. Physiol Biochem 66:1–9. https://doi.org/10.1016/jplaphy201301020

Ojuederie OB, Olanrewaju OS, Babalola OO (2019) Plant growth promoting rhizobacterial mitigation of drought stress in crop plants: implications for sustainable agriculture. Agronomy 9:712. https://doi.org/10.3390/agronomy9110712

Patil S, Bheemaraddi CM, Shivannavar TC, Gaddad MS (2014) Biocontrol activity of siderophore producing Bacillus subtilis CTS-G24 against wilt and dry root rot causing fungi in chickpea. IOSR J Agric Vet Sci 7(9):63–68

Pérez-Montaño F, AlíasVillegas C, Bellogín RA, DelCerro P, Espuny MR, Jiménez GI, López-Baena FJ, Ollero FJ, Cubo T (2014) Plant growth promotion in cereal and leguminous agricultural important plants: from microorganism capacities to crop production. Microbiol Res 169:325–336

Pourbabaee AA, Bahmani E, Alikhani HA, Emami S (2016) Promotion of wheat growth under salt stress by halotolerant bacteria containing ACC deaminase. J Agric Sci Technol 18:855–864

Qiao J, Wu H, Huo R, Geo XW, Borriss R (2014) Stimulation of plant growth and biocontrol by Bacillus amyloliquefaciens sub sp. plantarum FZB42 engineered for improved action. Chem Biol Technol Agric 1:12. https://doi.org/10.1186/s40538-014-0012-2

Qiao J, Yu X, Liang X, Liu Y, Borriss R, Liu Y (2017) Addition of plant-growth-promoting Bacillus subtilis PTS-394 on tomato rhizosphere has no durable impact on composition of root microbiome. BMC Microbiol 17:131. https://doi.org/10.1186/s12866-017-1039-x(2017)

Qurashi AW, Sabri AN (2013) Osmolyte accumulation in moderately halophilic bacteria improves salt tolerance of chickpea. Pakt J Bot 45:1011–1016

Radhakrishnan R, Lee IJ (2013) Regulation of salicylic acid, jasmonic acid and fatty acids in cucumber (Cucumis sativus L) by spermidine promotes plant growth against salt-stress. Acta Physiol Plant 35:3315–3322. https://doi.org/10.1007/s11738-013-1364-0

Radhakrishnan R, Lee IJ (2014) Effect of low dose of spermidine on physiological changes in salt stressed cucumber plants. Russ J Plant Physiol 61:90–96. https://doi.org/10.1134/S1021443714010129

Radhakrishnan R, Lee IJ (2016) Gibberellins producing Bacillus methylotrophicus KE2 supports plant growth and enhances nutritional metabolites and food values of lettuce. Plant Physiol Biochem 109:181–189. https://doi.org/10.1016/jplaphy201609018

Rai A, Nabti E (2017) Plant growth-promoting bacteria: importance in vegetable production. In: Zaidi A, Khan MS (eds) Microbial strategies for vegetable production. Springer International Publishing AG 2017 23. https://doi.org/10.1007/978-3-319-54401-4_2

Rao MS, Kamalnath M, Umamaheswari R, Rajinikanth R, Prabu P, Priti K, Grace GN, Chaya MK, Gopalakrishnan C (2017) Bacillus subtilis IIHR BS-2 enriched vermicompost controls root-knot nematode and soft rot disease complex in carrot. Sci Hortic, 218:56–62. https://doi.org/10.1016/j.scienta.2017.01.051

Reiss A, Jørgensen LN (2017) Biological control of yellow rust of wheat (Puccinia striiformis) with Serenader ASO (Bacillus subtilis strain QST713) / Reiss, Antje; Jørgensen, Lise Nistrup I: Crop Protection, Bind 93: 032017, s 1–8

Romero D, Pérez-García A, Rivera M, Cazorla F, De Vicente A (2004) Isolation and evaluation of antagonistic bacteria towards the cucurbit powdery mildew fungus Podosphaera fusca. Appl Microbiol Biotechnol 64(2):263–269

Sagar A (2017) Evaluation of performance of plant growth promoting bacteria (PGPB) containing 1-aminocyclopropane- 1-carboxylate deaminase (ACCD) in improving growth and yield of cereal crops under salinity stress. Sam Higginbottom University of Agriculture, Technology Sciences, Allahabad

Samaniego-Gámez BY, Garruña R, Tun-Suárez JM, Kantun-Can J, Reyes-Ramírez A, Cervantes-Díaz L (2016) Bacillus sp inoculation improves photosystem II efficiency and enhances photosynthesis in pepper plants. Chilean J Agric Res, 76:409–416

Senol M, Nadaroglu H, Dikbas N, Kotan R (2014) Purification of Chitinase enzymes from Bacillus subtilis bacteria TV-125, investigation of kinetic properties and antifungal activity against Fusarium culmorum. Ann Clin Microbiol Antimicrob 13:1–7

Shao JH, Li SQ, Zhang N, Cui XS, Zhou X, Zhang G, Shen Q, Zhang R (2015) Analysis and cloning of the synthetic pathway of the phytohormone indole-3-acetic acid in the plant-beneficial Bacillus amyloliquefaciens SQR9. Microb Cell Fact 14:130. https://doi.org/10.1186/s12934-015-0323-4

Shen Z, Mustaph A, Lin M, Zheng G (2017) Biocontrol of the internalization of Salmonella enterica and Enterohaemorrhagic Escherichia coli in mung bean sprouts with an endophytic Bacillus subtilis. Int J Food Microbiol 250:37–44

Shrivastava P, Kumar R (2015) Soil salinity: a serious environmental issue and plant growth-promoting bacteria as one of the tools for its alleviation. Saudi J Biol Sci 22(2):123–131. https://doi.org/10.1016/jsjbs201412001

Singh N, Raina S, Singh D, Ghosh M, Heflish A (2017) Exploitation of promising native strains of Bacillus subtilis with antagonistic properties against fungal pathogens and their pgpr characteristics. J Plant Pathol 99(1):27–35

Sivasakthi S, Kanchana D, Usharani G, Saranraj P (2013) Production of plant growth-promoting substance by Pseudomonas fluorescens and Bacillus subtilis isolated from paddy rhizosphere soil of Cuddalore district, Tamil Nadu, India. Int J Microbiol Res 4(3):227–233

Tang Q, Bie X, Lu Z, Lv F, Yang T, Qu X (2014) Effects of Fengycin from Bacillus subtilis fmbJ on apoptosis and necrosis in Rhizopus stolonifer. J Microbiol 52:675–680

Tiwari G, Duraivadivel P, Sharma S, Hariprasad P (2018) 1-aminocyclopropane-1-carboxylic acid deaminase producing beneficial rhizobacteria ameliorate the biomass characters of Panicum maximum Jacq. by mitigating drought and salt stress. Sci Rep 8:17513. https://doi.org/10.1038/s41598-018-35565-3

Venkateshwaran M, Volkening JD, Sussman MR, Ané JM (2013) Symbiosis and the social network of higher plants. Curr Opin Plant Biol 16:118–127

Wang T, Liang Y, Wu M, Chen Z, Lin J, Yang L (2015) Natural products from Bacillus subtilis with antimicrobial properties. Chin J Chem Eng 23(4):744–754

Wang X, Zhao D, Shen L, Jing C, Zhang C (2018) Application and mechanisms of Bacillus subtilis in biological control of plant disease. Role of rhizospheric microbes in soil. Springer, pp 225–250

Woo OG, Kim H, Kim JS, Keum HL, Lee K, Sul WJ, Lee JH (2020) Bacillus subtilis strain GOT9 confers enhanced tolerance to drought and salt stresses in Arabidopsis thaliana and Brassica campestris. Plant Physiol Biochem. https://doi.org/10.1016/j.plaphy.2020.01.032

Xie S, Wu HJ, Zang H, Wu L, Zhu Q, Gao X (2014) Plant growth promotion by spermidine-producing Bacillus subtilis OKB105. Mol Plant Microbe Interact 27:655–663. https://doi.org/10.1094/MPMI-01-14-0010-R

Xu M, Sheng J, Chen L, Men Y, Gan L, Guo S, Shen L (2014) Bacterial community compositions of tomato (Lycopersicum esculentum Mill) seeds and plant growth-promoting activity of ACC deaminase producing Bacillus subtilis (HYT-12-1) on tomato seedlings. World J Microbiol Biotechnol 30:835–845. https://doi.org/10.1007/s11274-013-1486-y

Yan L, Héloir M-C, Zhang X, Geissler M, Trouvelot S, Jacquens L, Henkel M, Su X, Fang X, Wang Q, Adrian M (2019) Surfactin and fengycin contribute to the protection of a Bacillus subtilis strain against grape downy mildew by both direct effect and defense stimulation. Mol Plant Pathol 20(8):1037–1050. https://doi.org/10.1111/mpp.12809.1037

Yang LR, Quan X, Xue BG, Goodwin PH, Lu SB, Wang JH et al (2015) Isolation and identification of Bacillus subtilis strain YB-05 and its antifungal substances showing antagonism against Gaeumannomyces graminis var. tritici. Biol Control 85:52–58

Yang A, Akhtar SS, Iqbal S, Amjad M, Naveed M, Zahir ZA, Jacobsen SE (2016a) Enhancing salt tolerance in quinoa by halotolerant bacterial inoculation. Funct Plant Biol 43:632–642. https://doi.org/10.1071/FP15265

Yang SY, Lim DJ, Noh MY, Kim JC, Kim YC, Kim IS (2016b) Characterization of biosurfactants as insecticidal metabolites produced by Bacillus subtilis Y9. Entomol Res 47(1):55–59. https://doi.org/10.1111/1748-5967.12200

You C, Zhang C, Kong F, Feng C, Wang J (2016) Comparison of the effects of biocontrol agent Bacillus subtilis and fungicide metalaxyl–mancozeb on bacterial communities in tobacco rhizospheric. Soil Ecol Eng 91:119–125

Yuan J, Chaparro JM, Manter DK, Zhang R, Vivanco JM, Shen Q (2015) Roots from distinct plant developmental stages are capable of rapidly selecting their own microbiome without the influence of environmental and soil edaphic factors. Soil Biol Biochem 89:206–209

Zhang L-N, Wang D-C, Hu Q, Dai X-Q, Xie Y-S, Li Q, Liu H-M, Guo J-H (2019) Consortium of plant growth-promoting rhizobacteria strains suppresses sweet pepper disease by altering the rhizosphere microbiota. Front Microbiol 10:1668. https://doi.org/10.3389/fmicb.2019.01668

Chapter 25
Bacillus thuringiensis Proteins: Structure, Mechanism and Biological Control of Insect Pests

Swapan Chakrabarty, Panchali Chakraborty, Tofazzal Islam, A. K. M. Aminul Islam, Juel Datta, Tuli Bhattacharjee, Jin Minghui, and Yutao Xiao

Abstract *Bacillus thuringiensis* (Bt) produces a wide variety of insecticidal proteins. It synthesizes δ-endotoxins as parasporal crystalline inclusion bodies during its sporulation and the stationary growth phase. It also produces vegetative insecticidal proteins that are initially given off during the bacterial vegetative growth stage. The insecticidal proteins are widely used as biopesticides either by spraying or by incorporating into transgenic crops. The application of these insecticidal proteins significantly reduced the use of synthetic chemical insecticides. It also controls the major agricultural pests with increased target specificity and environmental safety.

Authors Swapan Chakrabarty, Panchali Chakraborty contributed equally for this chapter.

S. Chakrabarty · J. Minghui · Y. Xiao (✉)
Shenzhen Branch, Guangdong Laboratory for Lingnan Modern Agriculture, Genome Analysis Laboratory of the Ministry of Agriculture, Agricultural Genomics Institute at Shenzhen, Chinese Academy of Agricultural Sciences, Shenzhen, China
e-mail: xiaoyutao@caas.cn

T. Islam
Institute of Biotechnology and Genetic Engineering (IBGE), Bangabandhu Sheikh Mujibur Rahman Agricultural University, Gazipur, Bangladesh

A. K. M. Aminul Islam · P. Chakraborty
Department of Genetics and Plant Breeding, Bangabandhu Sheikh Mujibur Rahman Agricultural University, Gazipur, Bangladesh

J. Datta
China National Rice Research Institute, Chinese Academy of Agricultural Sciences, Hangzhou, China

T. Bhattacharjee
Department of Chemistry, Jahangirnagar University, Savar, Dhaka, Bangladesh

M. T. Islam et al. (eds.), *Bacilli in Agrobiotechnology*, Bacilli in Climate Resilient Agriculture and Bioprospecting,
https://doi.org/10.1007/978-3-030-85465-2_25

This chapter updates the structure and specificity, insecticidal activity, mechanism of action, and application of Bt proteins for biological control of target insect pests in various crops. Furthermore, we discuss the benefits of using Bt biopesticides over chemical insecticides to promote sustainable and eco-friendly agriculture.

Keywords *Bacillus thuringiensis* · insecticidal protein · Bt crops · biopesticide · biological control

25.1 Introduction

Biological control of destructive insect pests is not a new approach. However its importance is increasing day by day due to the emergence and re-emergence of insect resistance to chemical insecticides as well as increasing risk to public health and environmental safety (Usta 2013; Islam et al. 2016, 2019). Insects have a strong capacity of developing genetic resistance which can restrict the prolonged use of insecticidal chemicals or toxins (Jin et al. 2019a; Zhang et al. 2020; Datta et al. 2021). The remarkable progress in functional genomics and genetic engineering makes it possible to know the gene function and to transfer the desired gene from one organism to another organism of a totally different genetic background. An improved understanding of the mechanism of action, target specificity and genetics of the biocontrol agents will provide a basis for their future improvement. Instead of chemical insecticides, different entomopathogenic viruses, bacteria, and fungi can be efficiently applied to manage insect pests. Among the biocontrol agents, Bt is broadly used for controlling insect pests as well as other organisms such as mites, nematodes, and protozoans (Schnepf et al. 1998).

Bt is a spore-forming gram-positive bacterium that can be isolated from soil, water, dead insects, different conifers, and leaves of deciduous trees (Höfte and Whiteley 1989; Palma et al. 2014). It was first isolated and identified in 1901 by a Japanese sericulturist Ishiwata Shigetane in *Bombyx mori* that had flacherie or flaccid disease. Later, it was isolated from infected *Ephestia kuehniella* and rediscovered by German microbiologist Ernst Berliner in 1911 (Berliner 1915; Knowles 1994). Bt is a member of the bacterial group *Bacillus cereus* complex, and they have a strong phylogenetic relationship (Fayad et al. 2019). The entomopathogenic properties of Bt distinguish itself from other members of *B. cereus* group (Helgason et al. 2000). In 1927, Bt was again isolated by Mattes, and it was used for the biological control of *Ostrinia nubilalis* in the next years (Beegle and Yamamoto 1992). Bt produces entomotoxic Crystal (Cry) and Cytolytic (Cyt) proteins or δ-endotoxins as parasporal crystals during its sporulation and the stationary growth phase (Höfte and Whiteley 1989; Schnepf et al. 1998; OECD 2007; Palma et al. 2014). Biopesticides that are produced from insecticidal crystal proteins are widely used to control the major agricultural pests worldwide (Tabashnik et al. 2013; Wu et al. 2019). Since 1996, transgenic crops expressing Cry proteins are comprehensively cultivated, which significantly reduced the use of synthetic insecticides. Bt also synthesizes vegetative insecticidal proteins (Vip) and secreted insecticidal protein

(Sip) during its vegetative growth stage. Novel Vip, first discovered in 1996, can be isolated from the culture media forthrightly (Estruch et al. 1996). Vips differ from Cry toxins in terms of structure and genetic makeup. Their binding sites and mode of action are also different from Cry toxins (Estruch et al. 1996; Yu et al. 1997). Vips have a broad insecticidal range and are compatible with other Cry proteins that facilitate the pyramiding of multiple *Bt* genes encoding different insecticidal proteins with multiplex modes of action. Pyramiding of *Bt* genes greatly increase the susceptibility of major damaging insects and help prevent the rapid emergence of insect resistance to Bt crops (Lee et al. 2006; Sena et al. 2009; Gouffon et al. 2011; Liu et al. 2011). Vip3 proteins exhibit a wide variety of insecticidal activities against pest Lepidoptera (Chakrabarty et al. 2020), and Vip1, Vip2 and Sip are toxic against Coleopteran insects (Donovan et al. 2006; Milne et al. 2008; Hernández-Martínez et al. 2013 de Escudero et al. 2014; Chakroun et al. 2016). However, these variations in function and target specificity among Bt proteins help to overcome the risk of the emergence of insect resistance to Bt toxins. It also provides a worthy alternative to chemical insecticides to control damaging insect pests in an environmentally safe way. This chapter aims to discuss the structure, insecticidal function, and biological control of insect pests by Bt. The benefits of using Bt technology in green farming are also discussed.

25.2 Bt Toxin Diversity

Bt species have been categorized into different subspecies based on phylogenetic and serotyping, phase susceptibility, and plasmid features. Approximately, 100 subspecies of Bt have already been identified to date. Most of these subspecies synthesize more than one Bt toxin that has a specific insect host range. For instance, *Bt* var. *kurstaki*, *Bt* var. *israelensis* and *Bt* var. *morrisoni* show specific insecticidal activity against Lepidopteran, Dipteran, and Coleopteran insects (Schnepf et al. 1998; Sanahuja et al. 2011). Among the δ-endotoxins produced during sporulation, 800 different *Cry* genes under 75 families and 38 different *Cyt* genes under three families have been identified based on their primary amino acid sequence (Jouzani et al. 2017; Sajid et al. 2018) (Fig. 25.1). Cry proteins are active against different insects belonging to the orders Lepidoptera, Coleoptera, Diptera, Homoptera, Hymenoptera, Orthoptera, Mallophaga and some other non-insect invertebrate species such as nematodes, mites, and protozoa (Crickmore et al. 1998; Ye et al. 2012). Bt strains that synthesize Cry5 and Cry6 proteins are found to be active against nematodes (Wei et al. 2003). Cyt toxins are mainly active against Dipteran insects. However, a few Cyt toxins are found to be active against Coleopteran insects (Federici and Bauer 1998; Guerchicoff et al. 2001). The Cry proteins that are not related phylogenetically have been grouped into 3D (three-domain) Cry toxins, Mtx-like (mosquitocidal) Cry toxins and binary (Bin)-like Cry toxins (Crickmore et al. 1998, 2018). Among the Cry toxins, 3D-Cry toxins formed the biggest group that consists of more than 53 different Cry toxin subgroups (Crickmore et al. 2018) (Fig. 25.1). The molecular structure of the members of the 3D-Cry toxin group is globular and

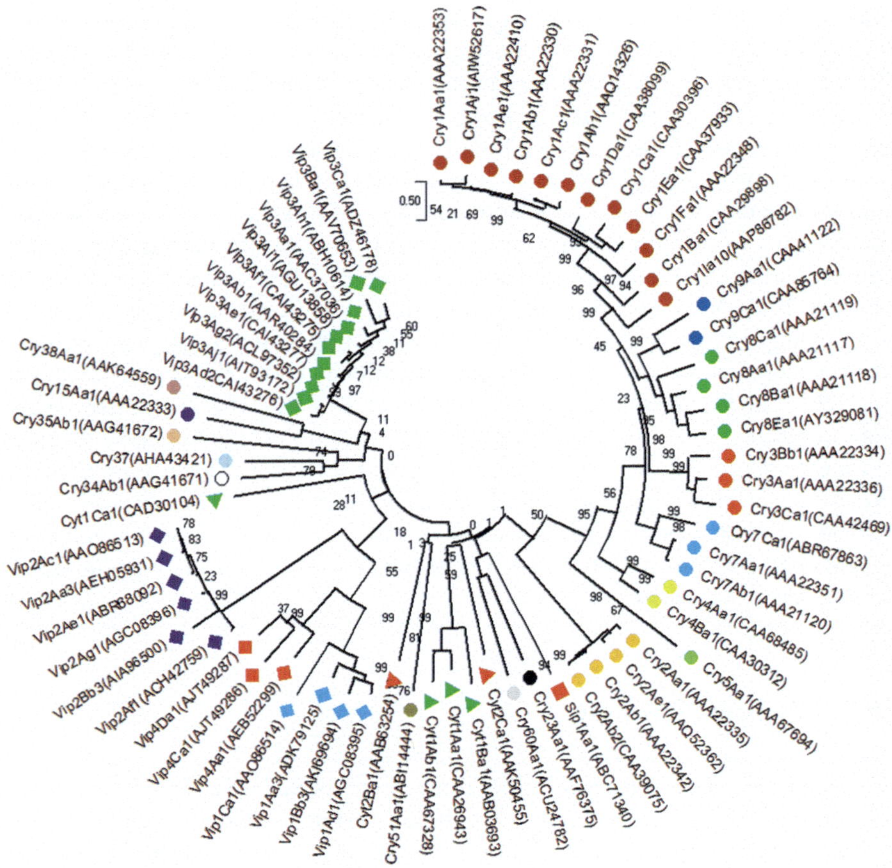

Fig. 25.1 Phylogenetic relationship among Bt toxins. The evolutionary history was inferred by using the Maximum Likelihood method and JTT matrix-based model (Jones et al. 1992). Evolutionary analyses were conducted in MEGA X program (Kumar et al. 2018). The NCBI GenBank accession number for specific protein is presented in the parenthesis

consists of three different domains which are connected by single linkers. The Cry toxins such as Cry34, Cry35 and Cry36 show their insecticidal activity against some Coleopteran larvae. They are named Bin-like Cry toxin as they have some similarities with the binary (Bin) toxins produced by *Lysinibacillus sphaericus*, which are specific to Dipteran pests. These toxins have a pore-forming capacity, and in an acidic environment, they produce ionic channels in the lipid bilayer (Masson et al. 2004). The members of the Mtx-like Cry toxin group have similarities to the Mtx toxin produced by *L. sphaericus* that shows mosquitocidal activity. However, Mtx-like Cry toxins such as Cry15, Cry23, Cry33, Cry37 and Cry38 are active against Coleopteran insects. The toxin Cry60 shows toxic activity against the larvae of Dipteran insects, while Cry51 shows toxicity against the insects belonging to the family Coleoptera and Hemiptera (Crickmore et al. 2018). In contrast to Cry toxins, genetically and structurally different Vip proteins are produced during the vegetative growth stage. Their binding sites and mode of action are also different from Cry toxins (Estruch et al. 1996; Yu et al. 1997). There have been 15 Vip1, 20 Vip2, 110 Vip3, and 1 Vip4 proteins identified so far. The binary toxins that are produced from Vip1 and Vip2 proteins exhibit synergistic entomotoxicity against *Aphis gossypii* and some Coleopteran insects (Sattar and Maiti 2011). The Vip3 proteins are active against a wide range of Lepidopteran insects (Milne et al. 2008; Hernández-Martínez et al. 2013; de Escudero et al. 2014; Chakroun et al. 2016). To date, the insecticidal action of Vip4 protein is unknown (Crickmore et al. 2018; Chakrabarty et al. 2020).

25.3 Structure and Specificity of Bt Proteins

25.3.1 Cry Toxins

The structure of a protein reveals the placements of functional domains that are required for a toxin's specificity and proteolytic stability. Among the Cry proteins, 3D-Cry toxins are mostly used for controlling insect pests and their structures are well studied. The 3D-Cry toxins are produced as ~130 and ~70 kDa protoxins. The 130 kDa protoxin is not responsible for toxicity, but shares a highly conserved C-terminus that is essential for forming intermolecular disulfide bonds at the time of crystal formation (Bietlot et al. 1990; de Maagd et al. 2001). The structure of ~70 kDa protoxin is the same as the N-terminal portion of the large toxin group. In some cases, these small protoxins require the availability of accessory proteins for crystallization (Agaisse and Lereclus 1995; Berry et al. 2002). The C-terminal region is implicated in crystal formation, but in mature toxins, it is cleaved out in the insect midgut. Though there is very little similarity among the amino acid sequences of the Cry proteins, they share an overall similar structural topology. The X-ray crystallography revealed that active 3D-Cry proteins consist of three structurally and functionally distinct domains. Among the Cry proteins, the first 3D structure of Cry3A was elucidated in 1991 (Li et al. 1991), then three-dimensional structures of other members of Cry family such as Cry1Aa, Cry1Ac, Cry2Aa, Cry3Bb1, Cry4Aa, Cry4Ba, Cry5B, Cry7Ca1 and Cry8Ea1 have been revealed (Grochulski et al. 1995; Derbyshire et al. 2001; Galitsky et al. 2001; Morse et al. 2001; Boonserm et al. 2005, 2006; Guo et al. 2009; Hui et al. 2012; Jing et al. 2019). Later, three-dimensional structures of 120 kDa Cry1Ac1 protoxin have been determined, which consist seven different structural domains (DI-DVII) (Evdokimov et al. 2014). Domain I spans at the N-terminal end of the 3D-Cry toxins. It is a bundle of seven α-helix where a central helix α-5 is surrounded by six outer helices. These six helices are amphipathic in nature and are long enough to span the 30 Å thick hydrophobic region of a membrane bilayer. Domain I remains highly conserved, and shares strong structural similarity with the domain of α + PFTs (Pore-Forming Toxins) Colicin A. It is involved in membrane insertion, oligomerization of toxin and pore formation (Berry et al. 2002; Pardo-Lopez et al. 2013; Domínguez-Arrizabalaga et al. 2020; Vílchez 2020). Domain II, a β-prism comprising three anti-parallel β-sheets packed around a hydrophobic core with exposed loop sections, is responsible for receptor recognition in the insect midgut. Domain III is a β-sandwich of two anti-parallel β-sheets that plays a role in receptor binding and formation of pores in the cell membrane (de Maagd et al. 2001; Pardo-Lopez et al. 2013). Domains IV and VI are α-bundles, just as the spectrin domain or the bacterial fibrinogen-binding complement inhibitor. Domains V and VII, on the other hand, are β-rolls like carbohydrate-binding proteins (Evdokimov et al. 2014). Though the functions of domains IV and VII are unknown, they may have important roles in the formation of crystal and stability of toxins in the insect gut. Moreover, recent studies suggest that the domains V and VII could interact with the proteins present in gut membranes, and therefore, are involved in receptor recognition (Zghal et al. 2017).

Though the structure and function of ETX_MTX2 proteins are still not clear, they show similar features with the epsilon toxin of *Clostridium perfringens*. It implies that they may possess an extended β-sheet and the capacity of a forming pore. A single β-stranded domain is found in the crystal structure of Cry23Aa (Fig. 25.2) that is structurally similar to β-pore-forming proaerolysin. The proaerolysin is produced by *Aeromonas hydrophila* and other related bacteria species (Palma et al. 2014; Domínguez-Arrizabalaga et al. 2020). Cry37Aa has a C2 β-sandwich fold, similar to the calcium phospholipid-binding domain identified in human cytosolic phospholipase A2 (Fig. 25.2) (Rydel et al. 2001). Bt toxins belong to the Toxin_10 family include Cry35 and Cry36 toxins. The crystal structure of Cry35Ab1 toxin shows an aerolysin-like fold which has a β-trefoil N-terminal domain containing QxW motifs similar to a carbohydrate-binding domain found in proteins such as ricin and Mtx1 from *L. sphaericus* (Fig. 25.2). The Cry34Ab1 is

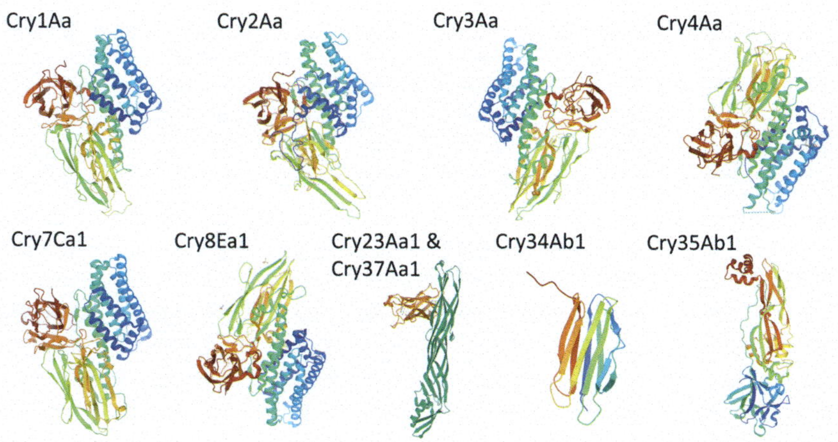

Fig. 25.2 Three-dimensional structures of some Cry toxins produced by Bt. The PDB (Protein Data Bank) accession numbers for Cry1Aa, Cry2Aa, Cry3Aa, Cry4Aa, Cry7Ca1, Cry8Ea1, Cry23Aa1 and Cry37Aa1, Cry34Ab1 and Cry35Ab1 are 1CIY, 1I5P, 1DLC, 2C9K, 5ZI1, 3EB7, 4RHZ, 4JOX and 4JPO, respectively

also a member of aerolysin family that contains a β-sandwich fold and a hydrophobic core (Fig. 25.2) (de Maagd et al. 2003). Cry22 is a non-3D protein, showing structural similarity with the domain III of 3D toxin by containing four cadherin-like proteins and a C-terminal region (Kelker et al. 2014).

25.3.2 Cyt Toxins

The Cyt toxins constitute another group of insecticidal toxins produced by Bt. The size of proteolytically active Cyt protoxins is around 25 kDa (Soberón et al. 2013). Like some Cry proteins, the 3D structure of Cyt proteins has also been known. The 3D structures of Cyt1Aa, Cyt2Aa, and Cyt2Ba revealed that they have a similar structure comprising a single α-β domain with two layers of α-helix hairpins wrapped around a β-sheet (Fig. 25.3) (Li et al. 1996; Cohen et al. 2008, 2011). Analysis of membrane insertion ability with peptides of Cyt1Aa exhibited that the α-helix peptides are major structural components involve in the membrane interaction and also in the toxin oligomerization (Gazit et al. 1997; Promdonkoy et al. 2008), while the β-sheet is involved in membrane insertion by forming an oligomeric pore with a β-barrel structure into the membrane (Li et al. 1996). However, Cyt1Ca is different from other Cyt toxins, as it has a further domain at the C-terminal end with homology to the carbohydrate-binding domain of the ricin (Manasherob et al. 2006).

25.3.3 Vip Toxins

Bt produces novel insecticidal proteins known as Vips during its vegetative growth stage. Their structure, genetics as well as the manner of action are different from Cry proteins (Estruch et al. 1996; Yu et al. 1997). The Vip1 and Vip2 proteins are produced from *B. cereus* strain AB78, and they are transcribed together from a ~4 kb single operon. Vip1 encodes ~100 kDa protein as a protoxin and produce as mature

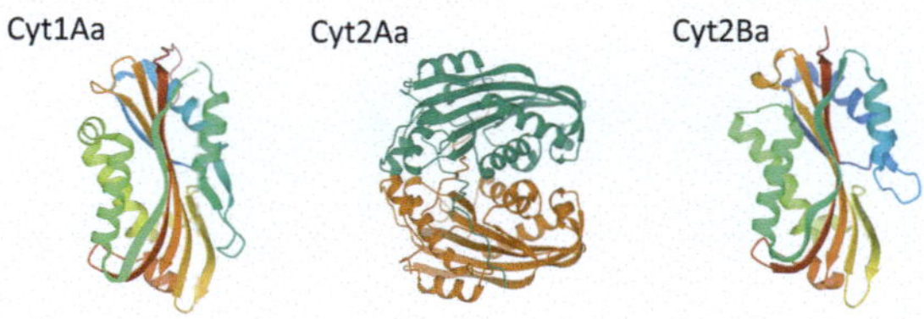

Fig. 25.3 Three-dimensional structures of Cyt toxins produced by Bt. The PDB accession numbers for Cyt1Aa, Cyt2Aa and Cyt2Ba are 3RON, 1CBY, and 2RCI, respectively

toxin of ~80 kDa after secretion. Furthermore, Vip2 generates a ~ 50 kDa trypsin-resistant fragment. The Vip1 and Vip2 together form a binary toxin, and they both are required for toxicity (Warren 1997; Bi et al. 2015). The crystal structure of Vip2 revealed that it is maixed α/β protein and contains two domains (Fig. 25.4). The homology of Vip1 and Vip2 to other binary toxins indicates that they form a classical A + B type binary toxin, where Vip2 is the A domain responsible for cytotoxicity, and Vip1 is the B domain responsible for receptor-binding and translocation of the cytotoxic Vip2 into the susceptible insect cell (de Maagd et al. 2003; Barth et al. 2020). The Vip2 toxin shares structural and sequence similarities with the *Clostridium difficile* toxin CdtA and the *C. perfringens* toxin iota. These toxins possess ADP-ribosyltransferase activity that targets actin protein, and thus the activated form of these toxins could cause cytoskeleton disruption and cell death (Han et al. 1999; Jucovic et al. 2008). The Vip1 binds to the membrane receptor in monomeric form, and their interaction cause a subsequent structural changes followed by the formation of homoheptamers that translocate Vip2 toxin into the cytoplasm through acid endosomes (Barth et al. 2020). After translocation of Vip2 toxin inside the cytoplasm, it destroys filamentous actin and causes cell death by cytoskeleton disruption (Shi et al. 2004, 2007).

The Vip3 proteins are secreted from Bt strain AB88 and AB424 without processing at the N-terminus (Estruch et al. 1996). The number of amino acids in the Vip3A proteins ranges from 657 to 795 with an average of around 787-amino acids, and the average molecular weight is ~89 kDa (Chakroun et al. 2016). The N-terminal region could function as a secretion signal peptide and plays a role in protein activation. It is contingent that the exceptionally conservative N-terminal region of Vip3A protein conserves the protein's structure and control its insecticidal function, whereas the varied C-terminal domain play a role in target precision (Zack et al. 2017). A number of recent studies have reported some important conformational information of Vip3A proteins. The cleavage of Vip3Aa protoxin happened at the primary cleavage site despite the amount of treated trypsin used. However, the resulting peptides of 66 and 19 kDa remained together, indicating the presence of secondary structure with a cluster of β-sheets at the C-terminus (Bel et al. 2017). Alanine scanning revealed five structural domains in the Vip3Af1 protein's 3D structure (Banyuls et al. 2018). In solution, both the protoxin and the functional Vip3 proteins form

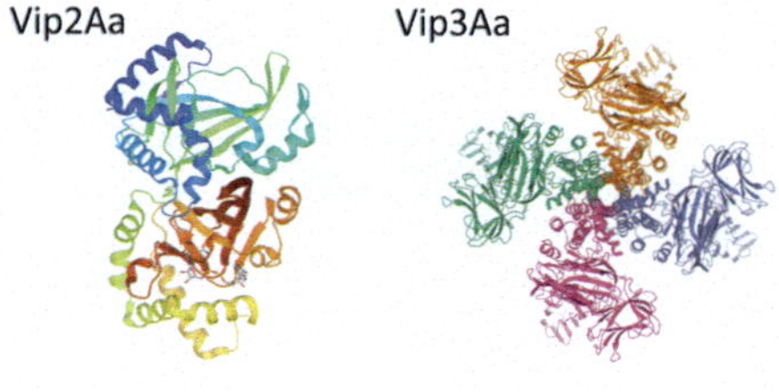

Fig. 25.4 Three-dimensional structures of Vip proteins produced by Bt. The PDB accession numbers for Vip2Aa and Vip3Aa are 1QS2 and 6TFK, respectively

homo-tetramers (Zack et al. 2017). It has also been observed that an interaction between the amino-terminal and the carboxy-terminal is essential for lethal toxicity, as this interaction increases oligomerization and provides proteolytic stability (Zack et al. 2017).

The Vip4 protein is the latterly identified from Bt strain Sbt016. The complete structure and function of Vip4 protein is still unknown (Crickmore et al. 2018). The length of the Vip4Aa1 gene is 2895 bp, and its protein sequence contains 965 amino acid residues with a molecular weight of ~108 kDa. The Vip4Aa protein is phylogenetically close to Vip1 proteins than other Vip proteins. It exhibits 34% sequence similarity with Vip1Aa1 protein. As Vip4 protein resembles the B component of the binary toxin, it is likely that it might interact with an unidentified A component to show its toxicity. Conversely, it might also be possible that the Vip4 is a B component that has lost its A component and forfeit its toxicity (Crickmore et al. 2018; Chakroun et al. 2016). However, deep analysis of the genome of Bt strain Sbt016 may shed some light on the structure and mechanism of this protein.

25.4 Mechanism of Action of Bt Proteins

25.4.1 Mechanism of Action of Cry Proteins

The mechanism of action of Cry toxins has been widely studied, but still there are some controversies. The pore-forming model of Cry toxin activity against various insects is the most commonly accepted mode of action (Schnepf and Whiteley 1981; Zhuang et al. 2002; Bravo et al. 2004; Rodríguez-Almazán et al. 2009) (Fig. 25.5a). The toxin crystals were dissolved at high pH (acid or alkaline, depending on the Cry toxin) and proteolyzed by proteases on the midgut under suitable physicochemical conditions. By traversing through the peritrophic matrix, activated toxins can enter the apical brush border membrane of the insect's midgut. To form pores, Cry toxins need to bind to specific receptors in brush-border membranes (Bravo et al. 2004, 2007, 2005; Jin et al. 2019b). In this case, Domain I of Cry toxins fixes into the cell membrane through its hydrophobic helical hairpin, where highly variable loops from Domain II plays a significant role in the identification of Cry toxin receptors such as aminopeptidase N (APN), alkaline phosphatase (ALP) and ATP-binding cassette (ABC) transporter (Bravo et al. 2004, 2005; Pardo-López et al. 2006; Pigott and Ellar 2007; Adang et al. 2014; Palma et al. 2014; Wu et al. 2019; Jin et al. 2021). Domain III is a crucial structure, involved in receptor binding and pore formation. It interacts to the APN receptor's N-acetylgalactosamine (GalNAc). In *Manduca sexta* and *Bombyx mori*, APN has been identified as a binding receptor for Cry1A toxins (Grochulski et al. 1995; Bravo et al. 2004; Pacheco et al. 2009). Some studies suggested that, for high cytotoxicity, binding of Cry protein to ABC transporters is critical, while the interaction of Cry proteins with the other receptors may increase toxicity (Tanaka et al. 2013; Bretschneider et al. 2016). The detailed mechanism of

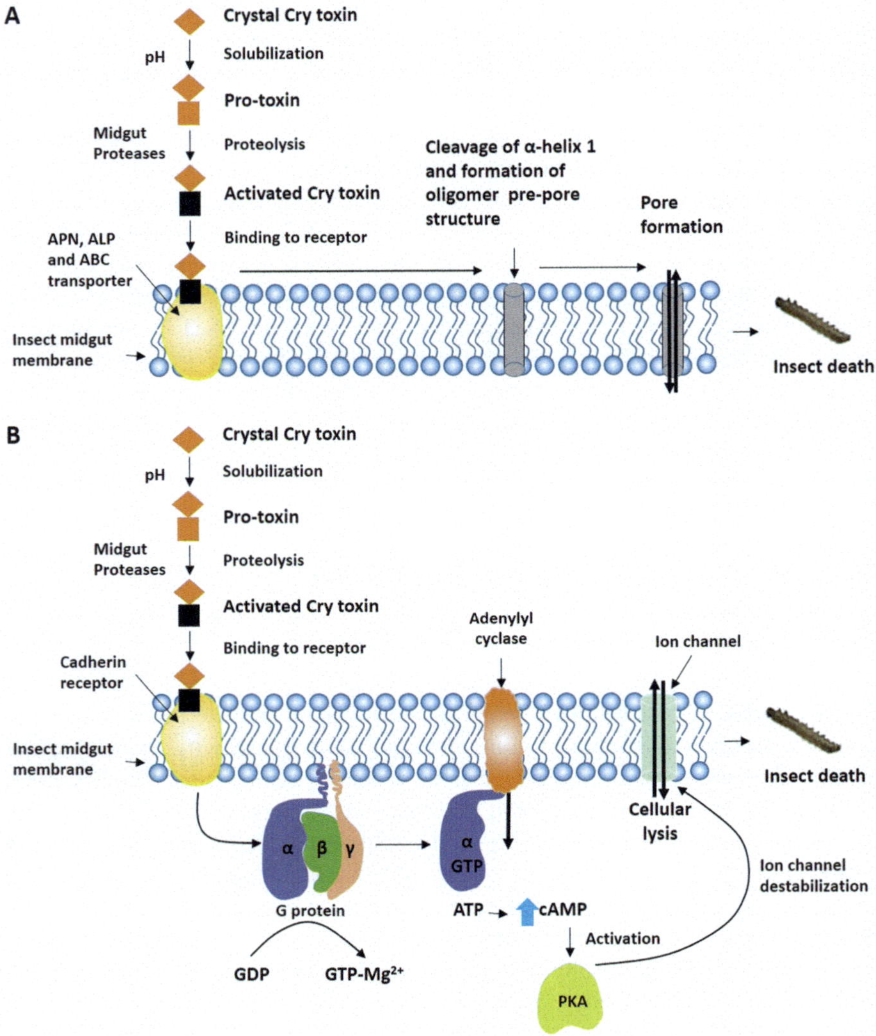

Fig. 25.5 Simplified models of the mechanisms of action of Cry proteins. (**a**) Pore-forming model, (**b**) signaling pathway model

the post-binding events that results in cell death is controversial (Vachon et al. 2012). However, most studies support that insertion of the Cry protein oligomer into the enterocyte membrane is essential to form a cation-selective pore that results in osmotic imbalance and cell death forwarded by reparative water influx through aquaporins (Endo et al. 2017). The midgut epithelial barrier is destroyed by massive enterocyte death that allows invasion of the hemocoel by Bt and other bacteria present in the gut, thus causing the death of the host by septicemia (Raymond et al. 2010; Jurat-Fuentes et al. 2021).

The second proposed model for Cry toxin mode of action is the signaling pathway model. A cell culture toxicity experiment is performed by expressing a cadherin receptor from *M. sexta* in the Sf9 cell line to explore the mechanism of action of Cry toxins in a signaling pathway model (Wickham et al. 1992; Kwa et al. 1998; Castella et al. 2019). The signaling pathway model hypothesizes that cytotoxicity of Cry proteins is reconciled by recognizing and binding to a cadherin receptor, which stimulates an Mg^{2+}-dependent cellular signal cascade pathway that directs cell death (Castella et al. 2019). The binding of Cry toxins to the cadherin receptor stimulates adenylyl cyclase activation that triggers an increase in cAMP and activation of protein kinase A (PKA). These circumstances stimulate a cascade that causes cell death by the formation of an ion channel on the membrane and cytoskeleton destabilization (Fig. 25.5b) (Zhang et al. 2006).

25.4.2 Mechanism of Action of Cyt Proteins

The insecticidal mechanism of Cyt proteins is still unknown, and it is also not clear whether Cyt proteins have a specific receptor for target specificity. Nonetheless, the cytotoxicity mechanism of Cyt toxins follows two main models. In the pore-formation model, similar to the Cry toxins, the monomer of Cyt toxins bind to the specific receptors of the membrane surface. According to this model, Cyt toxins interact straight with saturated membrane lipids such as phosphatidylcholine, phosphatidylethanolamine, and sphingomyelin (Thomas and Ellar 1983; Butko 2003; Mendoza-Almanza et al. 2020). Then, a conformation change occurs in Cyt toxin that supports in recruiting six monomers of Cyt toxin and conjoins the monomers into an open umbrella-like structure. In this structure, the strands span the lipid bilayer crosswise, while the alpha-helices remain on the surface of the membrane. These events result in succeeding membrane permeabilization, pore-formation, and finally insect death (Butko 2003; Mendoza-Almanza et al. 2020).

Another model for the mechanism of action of Cyt proteins is the detergent-effect model. According to this model, the Cyt toxins exhibit their cytotoxicity against target cells through a solubilization effect, where Cyt toxins concentrate on the surface of the cell membrane and dissolve the lipid bilayer in a detergent-like manner (Butko 2003; Mendoza-Almanza et al. 2020). These models are not mutually exclusive, thus, one or both may act against susceptible cells based on the concentration of the toxins (Butko 2003). The detergent effect could be induced only at higher concentration of Cyt toxin, while the oligomerization and pore formation has occurred at low concentrations of Cyt toxin (Butko et al. 1996; Mendoza-Almanza et al. 2020). Thus, at high toxin concentration, the target cell membranes are not able to assemble oligomers; alternately, it forms a toxin-lipid complex in which the integrity of the membrane is entirely lost (Butko 2003; Mendoza-Almanza et al. 2020).

25.4.3 Mechanism of Action of Vip Proteins

The studies on the mechanism of action of Vip proteins against insects are very limited. However, some studies on the mechanism of action of Vip3A proteins have been conducted with the members of the Vip3Aa subfamily. When the larvae fed Vip3A proteins, the midgut cells showed damaging symptoms including cell swelling, lysis of epithelial cells and leakage of cell content (Chakroun and Ferré 2014; Boukedi et al. 2015). Vip3Aa protoxin is hydrolyzed by insect midgut juice into 62–66, 45, 33, and 22 kDa polypeptide fragments in vitro. In susceptible insects, the activated Vip3A can function when it is inserted into the brush border membrane vesicles (BBMV). However, for specificity and insecticidal activity of Vip3A, the hydrolysis and activation by midgut protease is not an essential step. Non-target insects can also use midgut juice to hydrolyze Vip3A protein, resulting in a 65 kDa fragment that has toxic activity against target insects. This variation in the hydrolysis of Vip3A protein by different insects causes a difference in the insecticidal activity of the protein (Chakroun et al. 2012; Hernández-Martínez et al. 2013). According to competitive binding experiments, Vip3A proteins do not have a shared binding site with most of the Cry proteins. Vip3A and Cry toxins bind to the different receptors in the different binding sites on the surface of insect midgut cells (Lee et al. 2003).

The binding proteins of Vip3A have been identified on the surface of Sf21 cells by the yeast two-hybrid system, which revealed that Vip3A could interact with ribosomal S2 protein. In vitro pulldown assay and RNAi were used to validate the interaction of Vip3A and ribosomal S2 protein. Reduced toxicity of Vip3A proteins in Sf21 cells was observed by downregulating the ribosomal S2 protein. Vip3A protein interacts with ribosomal S2 protein that causes Sf21 cells to lyse. It implies that ribosomal S2 protein is Vip3A protein's interacting partner (Singh et al. 2010). Recently, it has been reported that Vip3A could bind to tenascins-like (a family of extracellular glycoproteins) BCW Vip3Aa-Receptor in the midgut tissue of *A. ipsilon* larvae. Vip3A toxin is thought to cause insecticidal toxicity by forming channels or transporting nutrients to the insect intestines (Osman et al. 2019). In Sf9 cells, the fibroblast growth factor receptor-like protein (FGFR) binds Vip3Aa protein. Vip3Aa and Sf-FGFR could co-localize on the surface of Sf9 cells and be internalized into Sf9 cells simultaneously (Jiang et al. 2018a). Vip3Aa binding to Sf-FGFR may also alter an apoptotic signaling pathway and cause apoptosis in Sf9 cell lines and *S. exigua* larvae (Hernández-Martínez et al. 2017; Jiang et al. 2018a; Osman et al. 2019). The Vip3Aa protoxin has been found to bind to the scavenger receptor class C-like protein (Sf-SR-C). The RNAi-mediated knockdown of the expression of SR-C in Sf9 cells and *S. exigua* larvae can reduce the toxicity of Vip3Aa. Furthermore, the Sf-SR-C-mediated endocytosis of the Vip3Aa is linked to its insecticidal action (Jiang et al. 2018b). The existence of pore formation structure of Vip3A has been confirmed in voltage-clamp assays, and the perforation ability of activated Vip3Aa on the BBMV of *Helicoverpa armigera* midgut was confirmed by fluorescence quenching test (Liu et al. 2007). Ion-channel formation is the most widely accepted mechanism of action for activated Vip3Aa toxin.

25.5 Bt Products for Biological Control of Insect Pests

25.5.1 Bt Biopesticides

Bt is one of the most important microbes in agricultural pest control, which has a significant contribution to crop protection as a biopesticide (Lord 2005; Brar et al. 2006; Bravo et al. 2011). It is widely accepted and being used because of its efficacy and safety compared to synthetic chemical pesticides. As Bt pesticides are target-specific, this is not harmful for non-target or other beneficial organisms.

Bt spray has been used to destroy the mosquito larvae (WHO 1999; Bravo et al. 2007). It is extensively used for controlling the pests of organic vegetables such as cauliflower, carrot, lettuce, broccoli, cucumber and tomato. Most of the people eat those vegetables as raw with minimal wash. But no human health hazard has yet been reported (Federici and Siegel 2008). There are a number of Bt commercial pesticides used for controlling insect pests. Bt technologies provide a large and competitive market (Table 25.1). These are effective in controlling destructive Lepidopteran and Coleopteran insect pests (Table 25.1). Most of these products are from wild subspecies *Bt* var. *kurstaki* (Btk). Btk is effective mostly against leaf leaf-feeding Lepidopteran insects. *Bt* var. *aizawai*-derived insecticides are also used for controlling Lepidopterans insects that feed on stored grains. Products that are made from *Bt* var. *san diego* and *Bt* var. *tenebrionis* subspecies are used for the management of Coleopteran insects (Soberon et al. 2009). As soon as Bt subspecies *tenebrionis* was developed, it was formulated as an insecticide to control the Colorado potato beetle. The commercial biopesticides Di-Terra®, M-One®, M-One Plus®, Novodor® and Trident® are manufactured from *Bt* var. *tenebrionis*. It has been reported that Novodor® is effective for the management of *Leptinotarsa decemlineata*. The utility of Novodor® was also examined in laboratory condition on some species of beetles such as *Chrysophtharta bimaculata*, *C. agricola* and *Chrysomela scripta* (Coyle et al. 2000; Beveridge and Elek 2001). *C. scripta* was found to be susceptible to this product in the field condition. The leaf-feeding insects are mostly controlled by Bt-based insecticides, which use natural Bt strains. Bt biopesticides have a limited range of host specificity; thus, advanced studies should be carried out to improve the efficacy of the pesticides through developing new natural strains with more insecticidal properties. Bt subspecies *kurstaki* have been modified, which expresses several *Cry3* genes. These manipulated biopesticides have more host range, including Lepidopteran and Coleopteran pests (Gawron-Burke and Baum 1991).

25.5.2 Bt Crops

Bt biopesticides have many advantages over commercial pesticides but they have some limitations as well. Poor stability and early inactivation under sunlight is major problem for using Bt pesticides. It is also known to be susceptible to the soil

Table 25.1 Bt commercial products used for biological control of insect pests

Commercial Bt biopesticides	Producing company	Bt subspecies	Target insects
Agree®	BioControle	*Bt* var. *kurstaki* and *Bt* var. *aizawai*	Lepidoptera
Bac-control®	Agricontrol	*Bt* var. *kurstaki*	Lepidoptera
BactoSpeine®	Solvay	*Bt* var. *kurstaki*	Lepidoptera
Bactur®	Milenia Agrociências	*Bt* var. *kurstaki*	Lepidoptera
Biobit® and Foray®	Novo-Nordisk	*Bt* var. *kurstaki*	Lepidoptera
Certan®	Novartis	*Bt* var. *kurstaki*	Lepidoptera
Cutlass® and Condor®	Ecogen	*Bt* var. *kurstaki*	Lepidoptera
Delfin®, Javelin® and Thuricide®	Sandoz	*Bt* var. *kurstaki*	Lepidoptera
Dimy Pel®	Dimy	*Bt* var. *kurstaki*	Lepidoptera
DiPel®	Abbott	*Bt* var. *kurstaki*	Lepidoptera
Di-Terra®	Abbott	*Bt* var. *san diego*, *Bt* var. *tenebrionis*	Coleoptera
Ecotech Pro®	Bayer	*Bt* var. *kurstaki*	Lepidoptera
Florbac®	Abbott	*Bt* var. *aizawai*	Lepidoptera
Foil®	Ecogen	*Bt* var. *kurstaki*	Lepidoptera/ Coleoptera
LarvoBt®	Fermone	*Bt* var. *kurstaki*	Lepidoptera
Lepinox®	DuPont	*Bt* var. *kurstaki*	Lepidoptera
M-One® and M-One Plus®	Mycogen	*Bt* var. *san diego*, *Bt* var. *tenebrionis*	Coleoptera
M-Track®	Mycogen	*Bt* var. *tenebrionis*	Coleoptera
MVP®	Mycogen	*Bt* var. *kurstaki*	Lepidoptera
Novodor®	Novo-Nordisk	*Bt* var. *san diego*, *Bt* var. *tenebrionis*	Coleoptera
Nubilacid®	Radonja	*Bt* var. *kurstaki*	Lepidoptera
Spicturin®	Phyllom BioProducts	*Bt* var. *galleriae*	Coleoptera
Steward®	Thermo Trilogy Co	*Bt* var. *kurstaki*	Lepidoptera
Trident®	Sandoz	*Bt* var. *san diego*, *Bt* var. *tenebrionis*	Coleoptera
XenTari®	Abbott	*Bt* var. *aizawai*	Lepidoptera

Source: Adapted from Schünemann et al. (2014), Fernández-Chapa et al. (2019), and Sanahuja et al. (2011)
Bt, *Bacillus thuringiensis*

and environment. As a result, the commercial production of Bt insecticide has become limited. Nowadays, Bt crops are more expedient to use as several *Bt* genes have been identified as well as transgenic technologies have been improved. In Table 25.2, we summarize the crops expressing Bt toxin that has been popularized and commercially produced. Bt transgenic plants are mostly produced in major grain and economic crops such as cotton, maize, potato, rice, soybean, and eggplant. These possess different kinds of *Bt* genes which are mostly used to control Lepidopteran and Coleopteran pests (Table 25.2).

Table 25.2 Commercially available Bt crops expressing Bt genes

Crop	Trade name	Developer	Introduced Bt gene(s)	Bt gene source	Target insects
Cotton	Bollgard™ cotton, Ingard™ and Roundup Ready™ Bollgard™ cotton	Monsanto Company	*Cry1Ac*	*Bt* var. *kurstaki* strain HD73	Lepidoptera
	BXN™ Plus Bollgard™ cotton	Monsanto Company	*Cry1Ac*	*Bt* var. *kurstaki* strain HD73	Lepidoptera
	VIPCOT™	Syngenta	*Vip3A(a)*	*Bt* strain AB88	Lepidoptera
	Bollgard II™ cotton and Roundup Ready™ Flex™ Bollgard II™ cotton	Monsanto Company	*Cry1Ac* and *Cry2Ab2*	*Bt* var. *kurstaki* strain HD73 and *Bt var. kumamotoensis*	Lepidoptera
	Fibermax™ Liberty Link™ Bollgard II™	Bayer CropScience	*Cry1Ac* and *Cry2Ab2*	*Bt* var. *kurstaki* strain HD73 and *Bt* var. *kumamotoensis*	Lepidoptera
	TwinLink™ cotton and Glytol™ × Twinlink™	Bayer CropScience	*Cry1Ab* and *Cry2Ae*	*Bt var. kurstaki* and *Bt* var. *Dakota*	Lepidoptera
	Widestrik™ cotton	Dow AgroSciences LLC	*Cry1Ac* and *Cry1F*	*Bt* var. *kurstaki* strain HD73 and *Bt* var. *aizawai*	Lepidoptera
	Widestrik™ Roundup Ready cotton and Widestrik™ Roundup Ready Flex™ cotton	Monsanto Company and Dow AgroSciences LLC	*Cry1Ac* and *Cry1F*	*Bt* var. *kurstaki* strain HD73 and *Bt* var. *aizawai*	Lepidoptera
	VIPCOT™	Syngenta	*Cry1Ab* and *Vip3A(a)*	*Bt* var. *kurstaki* and *Bt* strain AB88	Lepidoptera
	VIPCOT™ Roundup Ready Flex™ cotton	Syngenta and Monsanto Company	*Cry1Ab* and *Vip3A(a)*	*Bt* var. *kurstaki* and *Bt* strain AB88	Lepidoptera
	Bollgard® III and Bollgard® III × Roundup Ready™ Flex™	Monsanto Company	*Cry1Ac*, *Cry2Ab2* and *Vip3A(a)*	*Bt* var. *kurstaki* strain HD73, *Bt* var. *kumamotoensis* and *Bt* strain AB88	Lepidoptera
	Glytol™ × Twinlink™ × VIPCOT™ cotton	Bayer CropScience	*Cry1Ab*, *Cry2Ae* and *Vip3A(a)*	*Bt* var. *kurstaki, Bt* var. *dakota* and *Bt* strain AB88	Lepidoptera
	Widestrik™ × Roundup Ready Flex™ × VIPCOT™ cotton	Dow AgroSciences LLC	*Cry1Ac*, *Cry1F* and *Vip3A(a)*	*Bt* var. *kurstaki* strain HD73, *Bt* var. *aizawai* and *Bt* strain AB88	Lepidoptera

(continued)

Table 25.2 (continued)

Crop	Trade name	Developer	Introduced Bt gene(s)	Bt gene source	Target insects
Maize	Agrisure™ CB/LL, Agrisure™ GT/CB/LL, NaturGard KnockOut™ and Maximizer™	Syngenta	*Cry1Ab*	*Bt* var. *kurstaki*	Lepidoptera
	YieldGard™, MaizeGard™ and Roundup Ready™ YieldGard™ maize	Monsanto Company	*Cry1Ab*	*Bt* var. *kurstaki*	Lepidoptera
	Mavera™ YieldGard™ Maize	Renessen LLC and Monsanto Company	*Cry1Ab*	*Bt* var. *kurstaki*	Lepidoptera
	Liberty Link™ Yieldgard™ Maize	Monsanto Company and Bayer CropScience	*Cry1Ab*	*Bt* var. *kurstaki*	Lepidoptera
	Bt Xtra™ Maize	Monsanto Company	*Cry1Ac*	*Bt* var. *kurstaki* strain HD73	Lepidoptera
	Herculex™ I, Herculex™ CB and Herculex™ I RR	Dow AgroSciences LLC and DuPont	*Cry1Fa2*	*Bt* var. *aizawai*	Lepidoptera
	YieldGard™ Rootworm RW, MaxGard™ and YieldGard™ VT™ Rootworm™ RR2	Monsanto Company	*Cry3Bb1*	*Bt* var. *kumamotoensis*	Coleoptera
	Starlink™ Maize	Bayer CropScience	*Cry9c*	*Bt* var. *tolworthi* strain BTS02618A	Lepidoptera
	Agrisure™ Viptera	Syngenta	*Vip3Aa20*	*Bt* strain AB88	Lepidoptera
	Agrisure™ RW and Agrisure™ GT/RW	Syngenta	*mCry3A* (modified *Cry3A*)	*Bt* var. *tenebrionis*	Coleoptera
	Agrisure® Duracade™	Syngenta	*ecry3.1Ab* (chimera of *Cry3A* and *Cry1Ab*)	*Bt*	Insects from multiple order
	Optimum™ Intrasect	DuPont	*Cry1Ab and Cry1Fa2*	*Bt* var. *kurstaki* and *Bt* var. *aizawai*	Lepidoptera
	Agrisure™ CB/LL/RW and Agrisure™ 3000GT	Syngenta	*Cry1Ab* and *mCry3A*	*Bt* var. *kurstaki* and *Bt* var. *tenebrionis*	Coleoptera and Lepidoptera
	YieldGard™ VT triple and YieldGard™ plus	Monsanto Company	*Cry1Ab* and *Cry3Bb1*	*Bt* var. *kurstaki* and *Bt* var. *kumamotoensis*	Coleoptera and Lepidoptera

	YieldGard™ VT Pro™ and Genuity® VT Double Pro™	Monsanto Company	*Cry2Ab2* and *Cry1A.105*	*Bt* var. *kumamotoensis*	Lepidoptera
	Optimum™ TRIsect	DuPont	*Cry1Fa2* and *mCry3A*	*Bt* var. *aizawai* and *Bt* var. *tenebrionis*	Coleoptera and Lepidoptera
	Herculex™ RW and Herculex™ RW Roundup Ready™ 2	Dow AgroSciences LLC and DuPont	*Cry34Ab1* and *Cry35Ab1*	*Bt* strain PS149B1	Coleoptera
	Agrisure® Viptera™ 3110	Syngenta	*Cry1Ab* and *Vip3Aa20*	*Bt var. kurstaki* and *Bt* strain *AB88*	Lepidoptera
	Agrisure® Viptera™ 2100	Syngenta	*Cry1Ab* (truncated) and *Vip3Aa20*	*Bt* var. *kumamotoensis* and *Bt* strain *AB88*	Lepidoptera
	Herculex XTRA™	Dow AgroSciences LLC and DuPont	*Cry1F*, *Cry34Ab1* and *Cry35Ab1*	*Bt* var. *aizawai* and *Bt* strain PS149B1	Coleoptera and Lepidoptera
	Genuity® VT Triple Pro™	Monsanto Company	*Cry2Ab2*, *Cry1A.105* and *Cry3Bb1*	*Bt* var. *kumamotoensis*	Coleoptera and Lepidoptera
	Power Core™	Monsanto Company and Dow AgroSciences LLC	*Cry2Ab2*, *Cry1A.105* and *Cry1Fa2*	*Bt* var. *kumamotoensis*, *Bt* var. *aizawai*	Lepidoptera
	Agrisure™ Viptera 3220	Syngenta	*Cry1Ab*, *Cry1Fa2* and *Vip3Aa20*	*Bt* var. *kurstaki*, *Bt* var. *aizawai* and *Bt* strain *AB88*	Lepidoptera
	Power Core™ × MIR162 × Enlist™	Dow AgroSciences LLC	*Cry1Ab*, *Cry1Fa2* and *Vip3Aa20*	*Bt* var. *kurstaki*, *Bt* var. *aizawai* and *Bt* strain *AB88*	Lepidoptera
	Agrisure® Viptera™ 3100, Agrisure® Viptera™ 3111 and Agrisure® Viptera™ 4	Syngenta	*Cry1Ab*, *mCry3A* and *Vip3Aa20*	*Bt* var. *kurstaki, Bt* var. *tenebrionis* and *Bt* strain *AB88*	Coleoptera and Lepidoptera
	Agrisure® Duracade™ 5122	Syngenta	*ecry3.1Ab, mCry3A, Cry1Ab* and *Cry1Fa2*	*Bt*, *Bt* var. *tenebrionis*, *Bt* var. *kurstaki* and *Bt* var. *aizawai*	Insects from multiple order
	Agrisure® Duracade™ 5222	Syngenta	*ecry3.1Ab*, *Cry1Ab*, *Cry1Fa2* and *Vip3Aa20*	*Bt*, *Bt* var. *tenebrionis*, *Bt* var. *kurstaki, Bt* var. *aizawai* and *Bt* strain AB88	Insects from multiple order

(continued)

Table 25.2 (continued)

Crop	Trade name	Developer	Introduced Bt gene(s)	Bt gene source	Target insects
	Agrisure® 3122	Syngenta	*Cry1Ab*, *Cry1Fa2*, *mCry3A*, *Cry34Ab1* and *Cry35Ab1*	*Bt* var. *kurstaki*, *Bt* var. *aizawai*, *Bt* var. *tenebrionis* and *Bt* strain PS149B1	Coleoptera and Lepidoptera
	SmartStax™ Pro × Enlist™	Monsanto Company	*Cry2Ab2*, *Cry1A.105*, *Cry3Bb1*, *Cry1F*, *Cry34Ab1* and *Cry35Ab1*	*Bt* var. *kumamotoensis*, *Bt* var. *aizawai* and *Bt* strain PS149B1	Coleoptera and Lepidoptera
	Genuity® SmartStax™	Monsanto Company and Dow AgroSciences LLC	*Cry2Ab2*, *Cry1A.105*, *Cry3Bb1*, *Cry1Fa2*, *Cry34Ab1* and *Cry35Ab1*	*Bt* var. *kumamotoensis*, *Bt* var. *aizawai* and *Bt* strain PS149B1	Coleoptera and Lepidoptera
Potato	Atlantic NewLeaf™ potato, New Leaf™ Russet Burbank potato, Shepody NewLeaf™ Y potato and Superior NewLeaf™ potato	Monsanto Company	*Cry3A*	*Bt* var. *tenebrionis*	Coleoptera
Rice	Huahui-1	Huazhong Agricultural University	*Cry1Ab* and *Cry1Ac*	*Bt* var. *kurstaki* and *Bt* var. *kurstaki* strain HD73	Lepidoptera
Soybean	Intacta™ Roundup Ready™ 2 Pro	Monsanto Company	*Cry1Ac*	*Bt* var. *kurstaki* strain HD73	Lepidoptera
	Conkesta Enlist E3™ Soybean	Dow AgroSciences LLC	*Cry1Ac* and *Cry1F*	*Bt* var. *kurstaki* strain HD73 and *Bt* var. *aizawai*	Lepidoptera
Eggplant	BARI Bt Begun-1, -2, -3 and -4	Maharashtra Hybrid Seed Company	*Cry1Ac*	*Bt* var. *kurstaki* strain HD73	Lepidoptera

Source: Adapted from International Service for the Acquisition of Agri-biotech Applications (ISAAA) based on *Bt* gene introduced

25.5.2.1 Bt Maize

The *Cry1Ab, Cry1Ac,* and *Cry2A* genes were primarily inserted into the genome of maize to produce transgenic plants (Huang and McGaughey 1999). In the last few years, more advanced research accelerated the development of Bt maize with multiple genes. As an example, the transgenic maize expressing *Cry1Ab* + *Vip3Aa20* or chimeric protein *ecry3.1Ab* + *Cry1Ab* + *Cry1Fa2* + *Vip3Aa20* is resistant against Coleopteran pests (Carriere et al. 2015). In 2017, six genes namely, *Cry2Ab2, Cry1A.105, Cry1F, Cry34Ab1, Cry35Ab1,* and *Cry3Bb1* stacked together and developed another transgenic maize. Nowadays, more transgenic maize expressing multiple *Bt* genes is being produced worldwide (http://www.isaaa.org/gmapprovaldatabase/default.asp). *Chilo partellus* larvae causes more damage to non-Bt maize than the hybrid Bt maize expressing the *Cry1Ab* gene. In laboratory conditions, a 79.4–100% mortality rate was observed in *C. partellus* larvae when feeding with Bt maize (*Cry1Ab*) (Hari et al. 2008). Most Bt maize is effective against Lepidopteran and Coleopteran pests.

25.5.2.2 Bt Cotton

Bt cotton plays an important role in controlling Lepidopteran and Coleopteran pests. First Bt cotton expressing *Cry1Ac* was cultivated in Australia and USA, followed by China. It had great control over a wide range of insects including cotton bollworm (*H. armigera*), pink bollworm (*Pectinophora gossypiella*), and other population of the target pests (Wu and Guo 2005). Cultivation of this single gene transgenic Bt cotton for a long time increases the resistance in targeted insect pests. As a result, at the end of twentieth century, studies were performed with two or more combinations of Bt toxins such as *Cry1Ac* + *Cry2Ae*, *Cry1Ab* + *Vip3A(a)*, *Cry1Ab* + *Cry2Ab2*, *Cry1Ab* + *Cry2Ae*, *Cry1Ac* + *Cry2Ab2*, *Cry1Ac* + *Cry1F* + *Vip3A(a)*, *Cry1Ac* + *Cry2Ab* + *Vip3A(a)*, and *Cry1Ab* + *Cry2Ae* + *Vip3A*. Transgenic cotton expressing *Cry1Ab* + *Cry2Ab2* genes performed better than previously cultivated single toxin expressing cotton (Qiman et al. 1998). Bollgard™ cotton that expresses a combination of Cry1Ac + Cry2Ab2 + Vip3A proteins, shows higher insecticidal activity against *H. virescens, H. zea* and *P. gossypiella* instead of any of the single protein.

25.5.2.3 Bt Soybean

Transgenic Bt expressing *Cry1Ac, Cry1F, Cry1A105,* and *Cry2Ab2* have been developed, which have a great impact in controlling target pests. Monsanto has developed Bt soybean varieties, MON87701 and MON87701RR2Y by introgression of *Cry1Ac* and *EPSPS* genes. These Bt soybeans provide significant protection against *H. armigera* in the whole growing season (Yu et al. 2013).

25.5.2.4 Bt Rice

Bt Huahui-1 is the first Bt rice, which expresses a fusion of Cry1Ab/Cry1Ac protein. It has high insecticidal activity against *Chilo suppressalis*, *Scirpophaga incertulas* and other Lepidopteran pests (Cheng et al. 1998; Shu et al. 2000). A transgenic rice line, Bt-Shanyou 63 was developed expresses a Bt fusion gene derived from *Cry1Ab/Cry1Ac*. Bt-Shanyou 63 exhibited high toxicity against *S. incertulas* and *Cnaphalocrocis medinalis* (Tu et al. 2000). In a recent study, it is suggested that *Cry64Ba* and *Cry64Ca* can be used to produce transgenic rice, which in turn may provide a novel strategy to control Hemipteran pests of rice, *Laodelphax striatellus*, and *Sogatella furcifera* (Liu et al. 2018).

25.5.2.5 Bt Eggplant

Insect-resistant transgenic eggplant can be produced by inserting specific *Bt* gene through genetic engineering. Coleopteran-specific *Cry3B* gene was first successfully incorporated into eggplant, which exhibited resistance to Colorado potato beetle (*L. decemlineata*) (Chen et al. 1995; Arpaia et al. 1997). The main damaging insect of eggplant is eggplant (brinjal) shoot and fruit borer, *Leucinodes orbonalis*. A synthetic *Cry1Ab* gene was transferred to brinjal, which was resistant to *L.orbonalis* larvae (Kumar et al. 1998). Bangladesh is the pioneering country in commercially cultivating Bt eggplant. The commercially available Bt eggplant varieties in Bangladesh are BARI Bt Begun-1, BARI Bt Begun-2, BARI Bt Begun-3 and BARI Bt Begun-4 that express the *Cry1Ac* gene and effective against target Lepidopteran insect, *L. orbonalis* (Shelton et al. 2018). These varieties are getting popular in Bangladesh day by day. Cultivation of Bt brinjal increased an average of 19.6% yield compared to non-Bt brinjal (Shelton et al. 2020). In Chap. 23 of this volume, Paul et al. (2022) reviews the impacts of Bt-brinjal on the economic benefit of the farmers and environmental sustainability in Bangladesh.

Besides, Bt potato expressing *Cry3A* showed entomotoxicity against Coleopteran pests (Table 25.2). As Bt crops are environment-friendly and not harmful to human health, more emphasize should be given to the commercialization of Bt transgenic crop. In addition, the discovery of Bt genes is also needed to prompt the development and commercialization of Bt crops.

25.6 Conclusion

In conclusion, this chapter reviews and updates the structural and functional diversities of Bt toxins and discusses the mechanism of insect control by these natural toxins. Bt-based biopesticides are environment-friendly and effective in controlling various insect pests and nematodes. The application of Bt transgenic technologies in increasing crop production by reducing hazardous synthetic pesticides are also discussed. Stacking of multiple *Bt* genes in various crop plants using recently developed CRISPR-Cas (clustered regularly interspaced short palindromic repeats-CRISPR associated) technology would enhance sustainable agriculture for ensuring food and nutritional security of the ever-increasing population of the world.

References

Adang MJ, Crickmore N, Jurat-Fuentes JL (2014) Diversity of *Bacillus thuringiensis* crystal toxins and mechanism of action. Adv Insect Physiol 47:39–87

Agaisse H, Lereclus D (1995) How does *Bacillus thuringiensis* produce so much insecticidal crystal protein? J Bacteriol 177:6027–6032

Arpaia S, Mennella G, Onofaro V, Perri E, Sunseri F, Rotino GL (1997) Production of transgenic eggplant (*Solanum melongena* L.) resistant to Colorado potato beetle (*Leptinotarsa decemlineata* Say). Theor Appl Genet 95:329–334

Banyuls N, Hernandez-Rodriguez CS, Van Rie J, Ferre J (2018) Critical amino acids for the insecticidal activity of Vip3Af from *Bacillus thuringiensis*: inference on structural aspects. Sci Rep 8:7541

Barth H, Aktories K, Popoff MR, Stiles BG (2020) Binary bacterial toxins: biochemistry, biology, and applications of common *Clostridium* and *Bacillus* proteins. Microbiol Mol Biol Rev 68:373–402

Beegle CC, Yamamoto T (1992) Invitation paper (CP Alexander Fund): history of *Bacillus thuringiensis* Berliner research and development. Can Entomol 124:587–616

Bel Y, Banyuls N, Chakroun M, Escriche B, Ferre J (2017) Insights into the structure of the Vip3Aa insecticidal protein by protease digestion analysis. Toxins 9:131

Berliner E (1915) Über die Schlaffsucht der Mehlmottenraupe (*Ephestia kühniella* Zell.) und ihren Erreger *Bacillus thuringiensis n. sp*. Z Angew Entomol 2:29–56

Berry C, O'Neil S, Ben-Dov E, Jones AF, Murphy L, Quail MA, Holden MT, Harris D, Zaritsky A, Parkhill J (2002) Complete sequence and organization of pBtoxis, the toxin-coding plasmid of *Bacillus thuringiensis* subsp. *israelensis*. Appl Environ Microbiol 68:5082–5095

Beveridge N, Elek JA (2001) Insect and host-tree species influence the effectiveness of a *Bacillus thuringiensis* ssp. *tenebrionis*-based insecticide for controlling chrysomelid leaf beetles. Aust J Entomol 40:386–390

Bi Y, Zhang Y, Shu C, Crickmore N, Wang Q, Du L, Song F, Zhang J (2015) Genomic sequencing identifies novel *Bacillus thuringiensis* Vip1/Vip2 binary and Cry8 toxins that have high toxicity to *Scarabaeoidea* larvae. Appl Microbiol Biotechnol 99:753–760

Bietlot H, Vishnubhatla I, Carey P, Pozsgay M, Kaplan H (1990) Characterization of the cysteine residues and disulphide linkages in the protein crystal of *Bacillus thuringiensis*. Biochem J 267:309–315

Boonserm P, Davis P, Ellar DJ, Li J (2005) Crystal structure of the mosquito-larvicidal toxin Cry4Ba and its biological implications. J Mol Biol 348:363–382

Boonserm P, Mo M, Angsuthanasombat C, Lescar J (2006) Structure of the functional form of the mosquito larvicidal Cry4Aa toxin from *Bacillus thuringiensis* at a 2.8-angstrom resolution. J Bacteriol 188:3391–3401

Boukedi H, Khedher SB, Triki N, Kamoun F, Saadaoui I, Chakroun M, Tounsi S, Abdelkefi-Mesrati L (2015) Overproduction of the *Bacillus thuringiensis* Vip3Aa16 toxin and study of its insecticidal activity against the carob moth *Ectomyelois ceratoniae*. J Invertebr Pathol 127:127–129

Brar SK, Verma M, Tyagi R, Valéro JR (2006) Recent advances in downstream processing and formulations of *Bacillus thuringiensis* based biopesticides. Process Biochem 41:323–342

Bravo A, Gómez I, Conde J, Munoz-Garay C, Sánchez J, Miranda R, Zhuang M, Gill SS, Soberón M (2004) Oligomerization triggers binding of a *Bacillus thuringiensis* Cry1Ab pore-forming toxin to aminopeptidase N receptor leading to insertion into membrane microdomains. Biochim Biophys Acta 1667:38–46

Bravo A, Gill SS, Soberón M (2005) *Bacillus thuringiensis*: mechanisms and use. In: Gilbert IL (ed) Comprehensive molecular insect science. Elsevier, Amsterdam, pp 175–205

Bravo A, Gill SS, Soberón M (2007) Mode of action of *Bacillus thuringiensis* Cry and Cyt toxins and their potential for insect control. Toxicon 49:423–435

Bravo A, Likitvivatanavong S, Gill SS, Soberón M (2011) *Bacillus thuringiensis*: a story of a successful bioinsecticide. Insect Biochem Mol Biol 41:423–431

Bretschneider A, Heckel DG, Pauchet Y (2016) Three toxins, two receptors, one mechanism: mode of action of Cry1A toxins from *Bacillus thuringiensis* in *Heliothis virescens*. Insect Biochem Mol Biol 76:109–117

Butko P (2003) Cytolytic toxin Cyt1A and its mechanism of membrane damage: data and hypotheses. Appl Environ Microbiol 69:2415–2422

Butko P, Huang F, Pusztai-Carey M, Surewicz WK (1996) Membrane permeabilization induced by cytolytic δ-endotoxin CytA from *Bacillus thuringiensis* var. *israelensis*. Biochemistry 35:11355–11360

Carriere Y, Crickmore N, Tabashnik BE (2015) Optimizing pyramided transgenic Bt crops for sustainable pest management. Nat Biotechnol 33:161–168

Castella C, Pauron D, Hilliou F, Zucchini-Pascal N, Gallet A, Barbero P (2019) Transcriptomic analysis of *Spodoptera frugiperda* Sf9 cells resistant to *Bacillus thuringiensis* Cry1Ca toxin reveals that extracellular Ca^{2+}, Mg^{2+} and production of cAMP are involved in toxicity. Biol Open 8:bio037085

Chakrabarty S, Jin M, Wu C, Chakraborty P, Xiao Y (2020) *Bacillus thuringiensis* vegetative insecticidal protein family Vip3A and mode of action against pest lepidoptera. Pest Manag Sci 76:1612–1617

Chakroun M, Ferré J (2014) *In vivo* and *in vitro* binding of Vip3Aa to *Spodoptera frugiperda* midgut and characterization of binding sites by ^{125}I radiolabeling. Appl Environ Microbiol 80:6258–6265

Chakroun M, Bel Y, Caccia S, Abdelkefi-Mesrati L, Escriche B, Ferre J (2012) Susceptibility of *Spodoptera frugiperda* and *S. exigua* to *Bacillus thuringiensis* Vip3Aa insecticidal protein. J Invertebr Pathol 110:334–339

Chakroun M, Banyuls N, Bel Y, Escriche B, Ferre J (2016) Bacterial vegetative insecticidal proteins (Vip) from entomopathogenic bacteria. Microbiol Mol Biol Rev 80:329–350

Chen Q, Jelenkovic G, Chin CK, Billings S, Eherhardt J, Goffreda JC, Day P (1995) Transfer and transcriptional expression of coleopteran *CryIIIB* endotoxin gene of *Bacillus thuringiensis* in eggplant. J Am Soc Hortic Sci 120:921–927

Cheng X, Sardana R, Kaplan H, Altosaar I (1998) *Agrobacterium*-transformed rice plants expressing synthetic *cryIA(b)* and *cryIA(c)* genes are highly toxic to striped stem borer and yellow stem borer. Proc Natl Acad Sci U S A 95:2767–2772

Cohen S, Dym O, Albeck S, Ben-Dov E, Cahan R, Firer M, Zaritsky A (2008) High-resolution crystal structure of activated Cyt2Ba monomer from *Bacillus thuringiensis* subsp. *israelensis*. J Mol Biol 380:820–827

Cohen S, Albeck S, Ben-Dov E, Cahan R, Firer M, Zaritsky A, Dym O (2011) Cyt1Aa toxin: crystal structure reveals implications for its membrane-perforating function. J Mol Biol 413:804–814

Coyle DR, McMillin JD, Krause SC, Hart ER (2000) Laboratory and field evaluations of two *Bacillus thuringiensis* formulations, Novodor and Raven, for control of cottonwood leaf beetle (Coleoptera: Chrysomelidae). J Econ Entomol 93:713–720

Crickmore N, Zeigler D, Feitelson J, Schnepf E, Van Rie J, Lereclus D, Baum J, Dean DH (1998) Revision of the nomenclature for the *Bacillus thuringiensis* pesticidal crystal proteins. Microbiol Mol Biol Rev 62:807–813

Crickmore N, Baum J, Bravo A, Lereclus D, Narva K, Sampson K, Schnepf E, Sun M, Zeigler DR (2018) *Bacillus thuringiensis* toxin nomenclature. http://www.btnomenclature.info/. Accessed on 22 July 2020

Datta J, Monsur MB, Chakraborty P, Chakrabarty S, Sabagh AE, Hossain A, Mondal MF (2021) Obstacle in controlling major rice pests in Asia: insecticide resistance and the mechanisms to confer insecticide resistance. In: Fahad S, Sönmez O, Saud S, Wang D, Wu C, Adnan M, Arif M, Amanullah (eds) Engineering tolerance in crop plants against abiotic stress. CRC Press, Boca Raton

de Escudero IR, Banyuls N, Bel Y, Maeztu M, Escriche B, Munoz D, Caballero P, Ferre J (2014) A screening of five *Bacillus thuringiensis* Vip3A proteins for their activity against lepidopteran pests. J Invertebr Pathol 117:51–55

de Maagd RA, Bravo A, Crickmore N (2001) How *Bacillus thuringiensis* has evolved specific toxins to colonize the insect world. Trends Genet 17:193–199

de Maagd RA, Bravo A, Berry C, Crickmore N, Schnepf HE (2003) Structure, diversity, and evolution of protein toxins from spore-forming entomopathogenic bacteria. Annu Rev Genet 37:409–433

Derbyshire DJ, Ellar DJ, Li J (2001) Crystallization of the *Bacillus thuringiensis* toxin Cry1Ac and its complex with the receptor ligand N-acetyl-D-galactosamine. Acta Crystallogr Sect D 57:1938–1944

Domínguez-Arrizabalaga M, Villanueva M, Escriche B, Ancín-Azpilicueta C, Caballero P (2020) Insecticidal activity of *Bacillus thuringiensis* proteins against coleopteran pests. Toxins 12:430

Donovan WP, Engleman JT, Donovan JC, Baum JA, Bunkers GJ, Chi DJ, Clinton WP, English L, Heck GR, Ilagan OM, Krasomil-Osterfeld KC, Pitkin JW, Roberts JK, Walters MR (2006) Discovery and characterization of Sip1A: a novel secreted protein from *Bacillus thuringiensis* with activity against coleopteran larvae. Appl Microbiol Biotechnol 72:713–719

Endo H, Azuma M, Adegawa S, Kikuta S, Sato R (2017) Water influx via aquaporin directly determines necrotic cell death induced by the *Bacillus thuringiensis* Cry toxin. FEBS Lett 591:56–64

Estruch JJ, Warren GW, Mullins MA, Nye GJ, Craig JA, Koziel MG (1996) Vip3A, a novel *Bacillus thuringiensis* vegetative insecticidal protein with a wide spectrum of activities against lepidopteran insects. Proc Natl Acad Sci U S A 93:5389–5394

Evdokimov AG, Moshiri F, Sturman EJ, Rydel TJ, Zheng M, Seale JW, Franklin S (2014) Structure of the full-length insecticidal protein Cry1Ac reveals intriguing details of toxin packaging into *in vivo* formed crystals. Protein Sci 23:1491–1497

Fayad N, Awad MK, Mahillon J (2019) Diversity of *Bacillus cereus* sensu *lato* mobilome. BMC Genomics 20:436

Federici BA, Bauer LS (1998) Cyt1Aa protein of *Bacillus thuringiensis is* toxic to the cottonwood leaf beetle, *Chrysomela scripta*, and suppresses high levels of resistance to Cry3Aa. Appl Environ Microbiol 64:4368–4371

Federici BA, Siegel JP (2008) Safety assessment of *Bacillus thuringiensis* and Bt crops used in insect control. In: Hammond BG (ed) Food safety of proteins in agricultural biotechnology. CRC Press, New York, pp 45–102

Fernández-Chapa D, Ramírez-Villalobos J, Galán-Wong L (2019) Toxic potential of *Bacillus thuringiensis*: an overview. In: Jia Y (ed) Protecting rice grains in the post-genomic era. IntechOpen, London, pp 183–204

Galitsky N, Cody V, Wojtczak A, Ghosh D, Luft JR, Pangborn W, English L (2001) Structure of the insecticidal bacterial δ-endotoxin Cry3Bb1 of *Bacillus thuringiensis*. Acta Crystallogr Sect D 57:1101–1109

Gawron-Burke C, Baum JA (1991) Genetic manipulation of *Bacillus thuringiensis* insecticidal crystal protein genes in bacteria. In: Setlow JK (ed) Genetic engineering: principles and methods. Springer, Boston, pp 237–263

Gazit E, Burshtein N, Ellar DJ, Sawyer T, Shai Y (1997) *Bacillus thuringiensis* cytolytic toxin associates specifically with its synthetic helices A and C in the membrane bound state. Implications for the assembly of oligomeric transmembrane pores. Biochemistry 36:15546–15554

Gouffon CV, Van Vliet A, Van Rie J, Jansens S, Jurat-Fuentes JL (2011) Binding sites for *Bacillus thuringiensis* Cry2Ae toxin on heliothine brush border membrane vesicles are not shared with Cry1A, Cry1F, or Vip3A toxin. Appl Environ Microbiol 77:3182–3188

Grochulski P, Masson L, Borisova S, Pusztai-Carey M, Schwartz JL, Brousseau R, Cygler M (1995) *Bacillus thuringiensis* CrylA(a) insecticidal toxin: crystal structure and channel formation. J Mol Biol 254:447–464

Guerchicoff A, Delécluse A, Rubinstein CP (2001) The *Bacillus thuringiensis cyt* genes for hemolytic endotoxins constitute a gene family. Appl Environ Microbiol 67:1090–1096

Guo S, Ye S, Liu Y, Wei L, Xue J, Wu H, Song F, Zhang J, Wu X, Huang D, Rao Z (2009) Crystal structure of *Bacillus thuringiensis* Cry8Ea1: an insecticidal toxin toxic to underground pests, the larvae of *Holotrichia parallela*. J Struct Biol 168:259–266

Han S, Craig JA, Putnam CD, Carozzi NB, Tainer JA (1999) Evolution and mechanism from structures of an ADP-ribosylating toxin and NAD complex. Nat Struct Biol 6:932–936

Hari NS, Jindal J, Malhi NS (2008) Resistance of Cry1Ab maize to spotted stemborer *Chilo partellus* (Lepidoptera: Crambidae) in India. Int J Trop Insect Sci 27:223–228

Helgason E, Økstad OA, Caugant DA, Johansen HA, Fouet A, Mock M, Hegna I, Kolstø AB (2000) *Bacillus anthracis*, *Bacillus cereus*, and *Bacillus thuringiensis*-one species on the basis of genetic evidence. Appl Environ Microbiol 66:2627–2630

Hernández-Martínez P, Hernández-Rodríguez CS, Van Rie J, Escriche B, Ferré J (2013) Insecticidal activity of Vip3Aa, Vip3Ad, Vip3Ae, and Vip3Af from *Bacillus thuringiensis* against lepidopteran corn pests. J Invertebr Pathol 113:78–81

Hernández-Martínez P, Gomis-Cebolla J, Ferré J, Escriche B (2017) Changes in gene expression and apoptotic response in *Spodoptera exigua* larvae exposed to sublethal concentrations of Vip3 insecticidal proteins. Sci Rep 7:1–12

Höfte H, Whiteley HR (1989) Insecticidal crystal proteins of *Bacillus thuringiensis*. Microbiol Rev 53:242–255

Huang F, McGaughey WH (1999) Inheritance of resistance to *Bacillus thuringiensis* toxin (Dipel ES) in the European corn borer. Science 284:965–967

Hui F, Scheib U, Hu Y, Sommer RJ, Aroian RV, Ghosh P (2012) Structure and glycolipid binding properties of the nematicidal protein Cry5B. Biochemistry 51:9911–9921

Islam MT, Rahman MM, Pandey P, Jha CK, Aeron A (2016) Bacilli in agrobiotechnology. Springer Nature, Cham

Islam MT, Rahman MM, Pandey P, Boehme MH, Haesaert G (2019) Bacilli in agrobiotechnology: phytostimulation and biocontrol, vol 2. Springer Nature, Cham

Jiang K, Hou X, Han L, Tan T, Cao Z, Cai J (2018a) Fibroblast growth factor receptor, a novel receptor for vegetative insecticidal protein Vip3Aa. Toxins 10:546

Jiang K, Hou XY, Tan TT, Cao ZL, Mei SQ, Yan B, Chang J, Han L, Zhao D, Cai J (2018b) Scavenger receptor-C acts as a receptor for *Bacillus thuringiensis* vegetative insecticidal protein Vip3Aa and mediates the internalization of Vip3Aa via endocytosis. PLoS Pathog 14:e1007347

Jin M, Liao C, Chakrabarty S, Zheng W, Wu K, Xiao Y (2019a) Transcriptional response of ATP-binding cassette (ABC) transporters to insecticides in the cotton bollworm, *Helicoverpa armigera*. Pestic Biochem Physiol 154:46–59

Jin M, Liao C, Chakrabarty S, Wu K, Xiao Y (2019b) Comparative proteomics of peritrophic matrix provides an insight into its role in Cry1Ac resistance of cotton bollworm *Helicoverpa armigera*. Toxins 11:92

Jin M, Yang Y, Shan Y, Chakrabarty S, Cheng Y, Soberón M, Bravo A, Liu K, Wu K, Xiao Y (2021) Two ABC transporters are differentially involved in the toxicity of two *Bacillus thuringiensis* Cry1 toxins to the invasive crop-pest *Spodoptera frugiperda* (JE Smith). Pest Manag Sci 77:1492–1501

Jing X, Yuan Y, Wu Y, Wu D, Gong P, Gao M (2019) Crystal structure of *Bacillus thuringiensis* Cry7Ca1 toxin active against *Locusta migratoria manilensis*. Protein Sci 28:609–619

Jones DT, Taylor WR, Thornton JM (1992) The rapid generation of mutation data matrices from protein sequences. Bioinformatics 8:275–282

Jouzani GS, Valijanian E, Sharafi R (2017) *Bacillus thuringiensis*: a successful insecticide with new environmental features and tidings. Appl Microbiol Biotechnol 101:2691–2711

Jucovic M, Walters FS, Warren GW, Palekar NV, Chen JS (2008) From enzyme to zymogen: engineering Vip2, an ADP-ribosyltransferase from *Bacillus cereus*, for conditional toxicity. Protein Eng Des Sel 21:631–638

Jurat-Fuentes JL, Heckel DG, Ferré J (2021) Mechanisms of resistance to insecticidal proteins from *Bacillus thuringiensis*. Annu Rev Entomol 66:121–140

Kelker MS, Berry C, Evans SL, Pai R, McCaskill DG, Wang NX, Russell JC, Baker MD, Yang C, Pflugrath JW, Wade M, Wess TJ, Narva KE (2014) Structural and biophysical characterization of *Bacillus thuringiensis* insecticidal proteins Cry34Ab1 and Cry35Ab1. PLoS One 9:e112555

Knowles BH (1994) Mechanism of action of *Bacillus thuringiensis* insecticidal δ-endotoxins. Adv Insect Physiol 24:275–308

Kumar PA, Mandaokar A, Sreenivasu K, Chakrabarti SK, Bisaria S, Sharma SR, Kaur S, Sharma RP (1998) Insect-resistant transgenic brinjal plants. Mol Breed 4:33–37

Kumar S, Stecher G, Li M, Knyaz C, Tamura K (2018) MEGA X: molecular evolutionary genetics analysis across computing platforms. Mol Biol Evol 35:1547–1549

Kwa MS, de Maagd RA, Stiekema WJ, Vlak JM, Bosch D (1998) Toxicity and binding properties of the *Bacillus thuringiensis* delta-endotoxin Cry1C to cultured insect cells. J Invertebr Pathol 71:121–127

Lee MK, Walters FS, Hart H, Palekar N, Chen JS (2003) The mode of action of the *Bacillus thuringiensis* vegetative insecticidal protein Vip3A differs from that of Cry1Ab δ-endotoxin. Appl Environ Microbiol 69:4648–4657

Lee MK, Miles P, Chen JS (2006) Brush border membrane binding properties of *Bacillus thuringiensis* Vip3A toxin to *Heliothis virescens* and *Helicoverpa zea* midguts. Biochem Biophys Res Commun 339:1043–1047

Li J, Carroll J, Ellar DJ (1991) Crystal structure of insecticidal δ-endotoxin from *Bacillus thuringiensis* at 2.5 Å resolution. Nature 353:815–821

Li J, Koni PA, Ellar DJ (1996) Structure of the mosquitocidal δ-endotoxin CytB from *Bacillus thuringiensis* sp. *kyushuensis* and implications for membrane pore formation. J Mol Biol 257:129–152

Liu J, Song F, Zhang J, Liu R, He K, Tan J, Huang D (2007) Identification of *vip3A*-type genes from *Bacillus thuringiensis* strains and characterization of a novel *vip3A*-type gene. Lett Appl Microbiol 45:432–438

Liu JG, Yang AZ, Shen XH, Hua BG, Shi GL (2011) Specific binding of activated Vip3Aa10 to *Helicoverpa armigera* brush border membrane vesicles results in pore formation. J Invertebr Pathol 108:92–97

Liu Y, Wang Y, Shu C, Lin K, Song F, Bravo A, Soberón M, Zhang J (2018) Cry64Ba and Cry64Ca, two ETX/MTX2-type *Bacillus thuringiensis* insecticidal proteins active against hemipteran pests. Appl Environ Microbiol 84:e01996–e01917

Lord JC (2005) From Metchnikoff to Monsanto and beyond: the path of microbial control. J Invertebr Pathol 89:19–29

Manasherob R, Itsko M, Sela-Baranes N, Ben-Dov E, Berry C, Cohen S, Zaritsky A (2006) Cyt1Ca from *Bacillus thuringiensis* subsp. *israelensis*: production in *Escherichia coli* and comparison of its biological activities with those of other Cyt-like proteins. Microbiology 152:2651–2659

Masson L, Schwab G, Mazza A, Brousseau R, Potvin L, Schwartz JL (2004) A novel *Bacillus thuringiensis* (PS149B1) containing a Cry34Ab1/Cry35Ab1 binary toxin specific for the western corn rootworm *Diabrotica virgifera virgifera* LeConte forms ion channels in lipid membranes. Biochemistry 43:12349–12357

Mendoza-Almanza G, Esparza-Ibarra EL, Ayala-Luján JL, Mercado-Reyes M, Godina-González S, Hernández-Barrales M, Olmos-Soto J (2020) The cytocidal spectrum of *Bacillus thuringiensis* toxins: from insects to human cancer cells. Toxins 12:301

Milne R, Liu Y, Gauthier D, Van Frankenhuyzen K (2008) Purification of Vip3Aa from *Bacillus thuringiensis* HD-1 and its contribution to toxicity of HD-1 to spruce budworm (*Choristoneura fumiferana*) and gypsy moth (*Lymantria dispar*)(Lepidoptera). J Invertebr Pathol 99:166–172

Morse RJ, Yamamoto T, Stroud RM (2001) Structure of Cry2Aa suggests an unexpected receptor binding epitope. Structure 9:409–417

OECD (Organisation for Economic Co-operation and Development) (2007) Consensus document on safety information on transgenic plants expressing *Bacillus thuringiensis* – derived insect control protein. OECD Pap 7:35–35

Osman GH, Soltane R, Saleh I, Abulreesh HH, Gazi KS, Arif IA, Ramadan AM, Alameldin HF, Osman YA, Idriss M (2019) Isolation, characterization, cloning and bioinformatics analysis of a novel receptor from black cut worm (*Agrotis ipsilon*) of *Bacillus thuringiensis* vip3Aa toxins. Saudi J Biol Sci 26:1078–1083

Pacheco S, Gómez I, Arenas I, Saab-Rincon G, Rodríguez-Almazán C, Gill SS, Bravo A, Soberón M (2009) Domain II loop 3 of *Bacillus thuringiensis* Cry1Ab toxin is involved in a "ping pong" binding mechanism with *Manduca sexta* aminopeptidase-N and cadherin receptors. J Biol Chem 284:32750–32757

Palma L, Muñoz D, Berry C, Murillo J, Caballero P (2014) *Bacillus thuringiensis* toxins: an overview of their biocidal activity. Toxins 6:3296–3325

Pardo-López L, Gómez I, Rausell C, Sánchez J, Soberón M, Bravo A (2006) Structural changes of the Cry1Ac oligomeric pre-pore from *Bacillus thuringiensis* induced by N-acetylgalactosamine facilitates toxin membrane insertion. Biochemistry 45:10329–10336

Pardo-Lopez L, Soberon M, Bravo A (2013) *Bacillus thuringiensis* insecticidal three-domain Cry toxins: mode of action, insect resistance and consequences for crop protection. FEMS Microbiol Rev 37:3–22

Paul SK, Mahmud NU, Islam T (2022) Impacts of Bt brinjal on economic benefit of the farmers and environmental sustainability in Bangladesh. In: Bacilli in agrobiotechnology: plant stress tolerance, bioremediation, and bioprospecting, vol 3. Springer Nature, Cham

Pigott CR, Ellar DJ (2007) Role of receptors in *Bacillus thuringiensis* crystal toxin activity. Microbiol Mol Biol Rev 71:255–281

Promdonkoy B, Rungrod A, Promdonkoy P, Pathaichindachote W, Krittanai C, Panyim S (2008) Amino acid substitutions in αA and αC of Cyt2Aa2 alter hemolytic activity and mosquito-larvicidal specificity. J Biotechnol 133:287–293

Qiman H, Liqun M, Weihong H, Sandui G (1998) Transgenic tobacco plants with a fully synthesized GFM *CryIA* gene provide effective tobacco bollworm (*Heliothis armigera*) control. J Integr Plant Biol 40:228–233

Raymond B, Johnston PR, Nielsen-LeRoux C, Lereclus D, Crickmore N (2010) *Bacillus thuringiensis*: an impotent pathogen? Trends Microbiol 18:189–194

Rodríguez-Almazán C, Zavala LE, Munoz-Garay C, Jimenez-Juarez N, Pacheco S, Masson L, Soberón M, Bravo A (2009) Dominant negative mutants of *Bacillus thuringiensis* Cry1Ab toxin function as anti-toxins: demonstration of the role of oligomerization in toxicity. PLoS One 4:e5545

Rydel T, Sharamitaro J, Brown G, Gouzov V, Seale J, Sturman E, Thoma R, Gruys K, Isaac BB (2001) The crystal structure of a coleopteran insect-active binary Bt protein toxin complex at 2.5 Å resolution. Annual meeting of the American Crystallographic Association. Los Angeles, CA, USA

Sajid M, Geng C, Li M, Wang Y, Liu H, Zheng J, Peng D, Sun M (2018) Whole-genome analysis of *Bacillus thuringiensis* revealing partial genes as a source of novel Cry toxins. Appl Environ Microbiol 8:e00277-18

Sanahuja G, Banakar R, Twyman RM, Capell T, Christou P (2011) *Bacillus thuringiensis*: a century of research, development and commercial applications. Plant Biotechnol J 9:283–300

Sattar S, Maiti MK (2011) Molecular characterization of a novel vegetative insecticidal protein from *Bacillus thuringiensis* effective against sap-sucking insect pest. J Microbiol Biotechnol 21:937–946

Schnepf HE, Whiteley HR (1981) Cloning and expression of the *Bacillus thuringiensis* crystal protein gene in *Escherichia coli*. Proc Natl Acad Sci U S A 78:2893–2897

Schnepf E, Crickmore N, Van Rie J, Lereclus D, Baum J, Feitelson J, Zeigler DR, Dean DH (1998) *Bacillus thuringiensis* and its pesticidal crystal proteins. Microbiol Mol Biol Rev 62:775–806

Schünemann R, Knaak N, Cassal MC, Fiuza LM (2014) Pest management of soybean: sustainable production techniques. In: Sahayaraj K (ed) Basic and applied aspects of biopesticides. Springer, New Delhi, pp 357–373

Sena JAD, Hernández-Rodríguez CS, Ferré J (2009) Interaction of *Bacillus thuringiensis* Cry1 and Vip3A proteins with *Spodoptera frugiperda* midgut binding sites. Appl Environ Microbiol 75:2236–2237

Shelton AM, Hossain MJ, Paranjape V, Azad AK, Rahman ML, Khan ASMMR, Prodhan MZH, Rashid MA, Majumder R, Hossain MA, Hussain SS, Huesing JE, McCandless L (2018) Bt eggplant project in Bangladesh: history, present status, and future direction. Front Bioeng Biotechnol 6:106

Shelton AM, Sarwer SH, Hossain MJ, Brookes G, Paranjape V (2020) Impact of Bt brinjal cultivation in the market value chain in five districts of Bangladesh. Front Bioeng Biotechnol 8:498

Shi Y, Xu W, Yuan M, Tang M, Chen J, Pang Y (2004) Expression of *vip1/vip2* genes in *Escherichia coli* and *Bacillus thuringiensis* and the analysis of their signal peptides. J Appl Microbiol 97:757–765

Shi Y, Ma W, Yuan M, Sun F, Pang Y (2007) Cloning of *vip1/vip2* genes and expression of Vip1Ca/Vip2Ac proteins in *Bacillus thuringiensis*. World J Microbiol Biotechnol 23:501–507

Shu Q, Ye G, Cui H, Cheng X, Xiang Y, Wu D, Gao M, Xia Y, Hu C, Sardana R, Altosaar I (2000) Transgenic rice plants with a synthetic *cry1Ab* gene from *Bacillus thuringiensis* were highly resistant to eight lepidopteran rice pest species. Mol Breed 6:433–439

Singh G, Sachdev B, Sharma N, Seth R, Bhatnagar RK (2010) Interaction of *Bacillus thuringiensis* vegetative insecticidal protein with ribosomal S2 protein triggers larvicidal activity *in Spodoptera frugiperda*. Appl Environ Microbiol 76:7202–7209

Soberon M, Gill SS, Bravo A (2009) Signaling versus punching hole: how do *Bacillus thuringiensis* toxins kill insect midgut cells? Cell Mol Life Sci 66:1337–1349

Soberón M, López-Díaz JA, Bravo A (2013) Cyt toxins produced by *Bacillus thuringiensis*: a protein fold conserved in several pathogenic microorganisms. Peptides 41:87–93

Tabashnik BE, Brévault T, Carrière Y (2013) Insect resistance to Bt crops: lessons from the first billion acres. Nat Biotechnol 31:510–521

Tanaka S, Miyamoto K, Noda H, Jurat-Fuentes JL, Yoshizawa Y, Endo H, Sato R (2013) The ATP-binding cassette transporter subfamily C member 2 in *Bombyx mori* larvae is a functional receptor for cry toxins from *Bacillus thuringiensis*. FEBS J 280:1782–1794

Thomas WE, Ellar DJ (1983) *Bacillus thuringiensis* var *israelensis* crystal delta-endotoxin: effects on insect and mammalian cells *in vitro* and *in vivo*. J Cell Sci 60:181–197

Tu J, Zhang G, Datta K, Xu C, He Y, Zhang Q, Khush GS, Datta SK (2000) Field performance of transgenic elite commercial hybrid rice expressing *Bacillus thuringiensis* δ-endotoxin. Nat Biotechnol 18:1101–1104

Usta C (2013) Microorganisms in biological pest control-a review (bacterial toxin application and effect of environmental factors). Curr Prog Biol Res 13:287–317

Vachon V, Laprade R, Schwartz JL (2012) Current models of the mode of action of *Bacillus thuringiensis* insecticidal crystal proteins: a critical review. J Invertebr Pathol 111:1–12

Vílchez S (2020) Making 3D-Cry toxin mutants: much more than a tool of understanding toxins mechanism of action. Toxins 12:600

Warren GW (1997) Vegetative insecticidal proteins: novel proteins for control of corn pests. In: Carozzi B (ed) Advances in insect control: the role of transgenic plants. CRC Press, London, pp 109–121

Wei JZ, Hale K, Carta L, Platzer E, Wong C, Fang SC, Aroian RV (2003) *Bacillus thuringiensis* crystal proteins that target nematodes. Proc Natl Acad Sci U S A 100:2760–2765

WHO (World Health Organization) (1999) Microbial pest control agent: *Bacillus thuringiensis*. World Health Organization & International Programme on Chemical Safety. https://apps.who.int/iris/handle/10665/42242. Accessed on 20 July 2020

Wickham TJ, Davis T, Granados RR, Shuler ML, Wood HA (1992) Screening of insect cell lines for the production of recombinant proteins and infectious virus in the baculovirus expression system. Biotechnol Prog 8:391–396

Wu KM, Guo YY (2005) The evolution of cotton pest management practices in China. Annu Rev Entomol 50:31–52

Wu C, Chakrabarty S, Jin M, Liu K, Xiao Y (2019) Insect ATP-binding cassette (ABC) transporters: roles in xenobiotic detoxification and Bt insecticidal activity. Int J Mol Sci 20:2829

Ye W, Zhu L, Liu Y, Crickmore N, Peng D, Ruan L, Sun M (2012) Mining new crystal protein genes from *Bacillus thuringiensis* on the basis of mixed plasmid-enriched genome sequencing and a computational pipeline. Appl Environ Microbiol 78:4795–4801

Yu CG, Mullins MA, Warren GW, Koziel MG, Estruch JJ (1997) The *Bacillus thuringiensis* vegetative insecticidal protein Vip3A lyses midgut epithelium cells of susceptible insects. Appl Environ Microbiol 63:532–536

Yu H, Li Y, Li X, Romeis J, Wu K (2013) Expression of Cry1Ac in transgenic Bt soybean lines and their efficiency in controlling lepidopteran pests. Pest Manag Sci 69:1326–1333

Zack MD, Sopko MS, Frey ML, Wang X, Tan SY, Arruda JM, Letherer TT, Narva KE (2017) Functional characterization of Vip3Ab1 and Vip3Bc1: two novel insecticidal proteins with differential activity against lepidopteran pests. Sci Rep 7:1–12

Zghal RZ, Elleuch J, Ali MB, Darriet F, Rebaï A, Chandre F, Jaoua S, Tounsi S (2017) Towards novel Cry toxins with enhanced toxicity/broader: a new chimeric Cry4Ba/Cry1Ac toxin. Appl Microbiol Biotechnol 101:113–122

Zhang X, Candas M, Griko NB, Taussig R, Bulla LA Jr (2006) A mechanism of cell death involving an adenylyl cyclase/PKA signaling pathway is induced by the Cry1Ab toxin of *Bacillus thuringiensis*. Proc Natl Acad Sci U S A 103:9897–9902

Zhang L, Liu B, Zheng W, Liu C, Zhang D, Zhao S, Li Z, Xu P, Wilson K, Withers A, Jones C, Smith J, Chipabika G, Kachigamba D, Nam K, d'Alençon E, Liu B, Liang X, Jin M, Wu C, Chakrabarty S, Yang X, Jiang Y, Liu J, Liu X, Quan W, Wang G, Fan W, Qian W, Wu K, Xiao Y (2020) Genetic structure and insecticide resistance characteristics of fall armyworm populations invading China. Mol Ecol Resour 20:1682–1696

Zhuang M, Oltean DI, Gómez I, Pullikuth AK, Soberón M, Bravo A, Gill SS (2002) *Heliothis virescens* and *Manduca sexta* lipid rafts are involved in Cry1A toxin binding to the midgut epithelium and subsequent pore formation. J Biol Chem 277:13863–13872

Index

M. T. Islam et al. (eds.), *Bacilli in Agrobiotechnology*, Bacilli in Climate Resilient Agriculture and Bioprospecting,
https://doi.org/10.1007/978-3-030-85465-2

C

D

E

The manufacturer's authorised representative in the EU is Springer Nature Customer Service Centre GmbH, Europaplatz 3, 69115 Heidelberg, Germany. If you have any concerns regarding our products, please contact ProductSafety@springernature.com

Printed and bound by CPI Group (UK) Ltd, Croydon, CR0 4YY
30/09/2026
02225233-0016